T0326367

Fish Physiology

Conservation Physiology for the Anthropocene – A Systems Approach

Volume 39A

This is Volume 39A in the

FISH PHYSIOLOGY series

Edited by Anthony P. Farrell, Colin J. Brauner and Erika J. Eliason

Honorary Editors: William S. Hoar and David J. Randall

A complete list of books in this series appears at the end of the volume

Fish Physiology
Conservation Physiology for the Anthropocene – A Systems Approach

Volume 39A

Edited by

Steven J. Cooke
Fish Ecology and Conservation Physiology Laboratory,
Department of Biology, Carleton University, Ottawa, ON, Canada

Nann A. Fangue
Department of Wildlife, Fish and Conservation Biology, University of
California - Davis, Davis, CA, United States

Anthony P. Farrell
Department of Zoology, and Faculty of Land and Food Systems,
The University of British Columbia, Vancouver, British Columbia, Canada

Colin J. Brauner
Department of Zoology, The University of British Columbia,
Vancouver, British Columbia, Canada

Erika J. Eliason
Department of Ecology, Evolution, and Marine Biology,
University of California - Santa Barbara, Santa Barbara, CA, United States

ACADEMIC PRESS

An imprint of Elsevier

Academic Press is an imprint of Elsevier
50 Hampshire Street, 5th Floor, Cambridge, MA 02139, United States
525 B Street, Suite 1650, San Diego, CA 92101, United States
The Boulevard, Langford Lane, Kidlington, Oxford OX5 1GB, United Kingdom
125 London Wall, London, EC2Y 5AS, United Kingdom

First edition 2022

Notices
Knowledge and best practice in this field are constantly changing. As new research and experience broaden our understanding, changes in research methods, professional practices, or medical treatment may become necessary.

Practitioners and researchers must always rely on their own experience and knowledge in evaluating and using any information, methods, compounds, or experiments described herein. In using such information or methods they should be mindful of their own safety and the safety of others, including parties for whom they have a professional responsibility.

To the fullest extent of the law, neither the Publisher nor the authors, contributors, or editors, assume any liability for any injury and/or damage to persons or property as a matter of products liability, negligence or otherwise, or from any use or operation of any methods, products, instructions, or ideas contained in the material herein.

ISBN: 978-0-12-824266-7
ISSN: 1546-5098

For information on all Academic Press publications
visit our website at https://www.elsevier.com/books-and-journals

Publisher: Zoe Kruze
Acquisitions Editor: Sam Mahfoudh
Editorial Project Manager: Jhon Michael Peñano
Production Project Manager:
  Sudharshini Renganathan
Cover Designer: Christian J. Bilbow

Typeset by STRAIVE, India

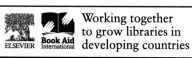

Working together
to grow libraries in
developing countries

www.elsevier.com • www.bookaid.org

Cover images: The American eel (Anguilla rostrata) is considered Endangered by the International Union for the Conservation of Nature. Their complex life cycle and the diverse threats they face necessitate adopting a systems approach to understanding threats and identifying effective conservation strategies. From studies on sensory physiology to inform behavioural guidance at hydropower facilities to studies of swimming performance to inform fish passage facilities or detailed studies of reproductive physiology to enable captive breeding in conservation hatcheries, American eel exemplify the need for and promise of conservation physiology - Dr. Sean J. Landsman.

Contents

Contributors

Sarah L. Alderman (253), Department of Integrative Biology, University of Guelph, Guelph, ON, Canada

Jordanna N. Bergman (1), Fish Ecology and Conservation Physiology Laboratory, Department of Biology, Carleton University, Ottawa, ON, Canada

Nicholas J. Bernier (253), Department of Integrative Biology, University of Guelph, Guelph, ON, Canada

Robin J. Boyd (141), UK Centre for Ecology and Hydrology, MacLean Building, Benson Lane, Crowmarsh Gifford, Wallingford, United Kingdom

Colin J. Brauner (1), Department of Zoology, University of British Columbia, Vancouver, BC, Canada

Richard W. Brill (33), Department of Fisheries Science, Virginia Institute of Marine Science, William & Mary, Gloucester Point, VA, United States

Jacob W. Brownscombe (141), Great Lakes Laboratory for Fisheries and Aquatic Sciences, Fisheries and Oceans Canada, Burlington, ON, Canada

Theodore Castro-Santos (91), U.S. Geological Survey, Eastern Ecological Science Center, S.O. Conte Anadromous Fish Research Center, Turners Falls, MA, United States

Joseph J. Cech Jr. (1), Department of Wildlife, Fish and Conservation Biology, University of California-Davis, Davis, CA, United States

Steven J. Cooke (1, 141), Fish Ecology and Conservation Physiology Laboratory, Department of Biology, Carleton University, Ottawa, ON, Canada

David Deslauriers (141), Institut des sciences de la mer de Rimouski, Université du Québec à Rimouski, Rimouski, QC, Canada

Erika J. Eliason (1, 189), Department of Ecology, Evolution, and Marine Biology, University of California-Santa Barbara, Santa Barbara, CA, United States

Nann A. Fangue (1), Department of Wildlife, Fish and Conservation Biology, University of California-Davis, Davis, CA, United States

Anthony P. Farrell (1), Department of Zoology, University of British Columbia, Vancouver, BC, Canada

Ramon Filgueira (141), Marine Affairs Program, Dalhousie University, Halifax, NS, Canada

Elsa Goerig (91), U.S. Geological Survey, Eastern Ecological Science Center, S.O. Conte Anadromous Fish Research Center, Turners Falls; Museum of Comparative Zoology, Harvard University, Cambridge, MA, United States

Pingguo He (91), School for Marine Science and Technology (SMAST), University of Massachusetts Dartmouth, New Bedford, MA, United States

Timothy M. Healy (435), University of California-San Diego, Scripps Institution of Oceanography, La Jolla, CA, United States

Erika B. Holland (389), Department of Biological Sciences, California State University of Long Beach, Long Beach, CA, United States

Andrij Z. Horodysky (33), Department of Marine and Environmental Science, Hampton University, Hampton, VA, United States

Jennifer D. Jeffrey (389), Department of Biological Sciences, University of Manitoba, Winnipeg, MB, Canada

Ken M. Jeffries (389), Department of Biological Sciences, University of Manitoba, Winnipeg, MB, Canada

George V. Lauder (91), Museum of Comparative Zoology, Harvard University, Cambridge, MA, United States

Michael J. Lawrence (141), Department of Biological Sciences, University of Manitoba, Winnipeg, MB, Canada

Christine L. Madliger (1), Fish Ecology and Conservation Physiology Laboratory, Department of Biology, Carleton University, Ottawa, ON, Canada

Patricia M. Schulte (435), Department of Zoology and Biodiversity Research Centre, The University of British Columbia, Vancouver, BC, Canada

Cara C. Schweitzer (33), Department of Marine and Environmental Science, Hampton University, Hampton, VA, United States

Gail D. Schwieterman (189), Department of Ecology, Evolution, and Marine Biology, University of California-Santa Barbara, Santa Barbara, CA, United States

Jacey C. Van Wert (189), Department of Ecology, Evolution, and Marine Biology, University of California-Santa Barbara, Santa Barbara, CA, United States

Chris M. Wood (321), Department of Zoology, University of British Columbia, Vancouver, BC; Department of Biology, McMaster University, Hamilton, ON, Canada; Rosenstiel School of Marine and Atmospheric Science, University of Miami, Miami, FL, United States

Preface

The world is changing.

Indeed, scholars have now declared that we have entered the Anthropocene epoch—a period distinct from the Holocene. The Anthropocene is defined by the manifold effects of human activities on planet Earth. By any and all measures, humans are changing environmental conditions and contributing to the loss of biodiversity.

Yet, all is not lost. There remains optimism that it is possible to address or mitigate some environmental threats and change our relationship with nature. Doing so could lead to opportunities for restoring biodiversity. In fact, the United Nations just launched (in 2021) the "Decade of Ecosystem Restoration," emphasizing that it is time to act.

Whether it be environmental change or direct interactions between humans (or human infrastructure) and wildlife, organisms can be influenced at various levels of biological organization. Although resource managers and conservation practitioners tend to focus on populations, communities, and ecosystems, it is often processes that play out at the level of the individual and are driven by physiological systems that influence these higher levels of biological organization. For that reason, the discipline of "conservation physiology" has emerged in recognition that physiological tools, knowledge, and concepts can be used to understand and solve conservation problems. Although conservation physiology is relevant to all taxa (including plants and microbes), there has been a particular focus on vertebrates, and more specifically, on fish.

Fish stocks are generally in poor condition around the globe—from the high seas to inland waters. The threats facing fish are many—from overfishing and bycatch, to fragmentation of migration corridors, to existing and emerging environmental pollutants, habitat loss, invasive species, climate change, and so on. In an attempt to stem this trend, a conservation physiology "toolbox" is increasingly being used to help understand the potential mechanisms by which these threats and stressors are influencing fish populations. A firm understanding of the mechanisms at play informs the potential solutions that may mitigate these threats. Thus, to directly address these conservation problems, conservation physiology is being used to develop solutions—from bycatch reduction strategies to restoration plans for fish passage. In other words, conservation physiology extends from problem identification to real-life solutions. To date, however, a synthesis related to conservation physiology specific to wild fish is lacking.

Consequently, the aim of Volume 39 of the *Fish Physiology* series is to generate a synthesis related to the physiology of fish in the Anthropocene. Specifically, in Volume 39A, we consider the ways in which different physiological systems (e.g., sensory physiology, cardiorespiratory) are relevant to conservation physiology. In Volume 39B, we present case studies to explore the ways in which physiology has been or can be used to understand or solve conservation problems (e.g., bycatch, habitat alteration, noise pollution). The first chapter in Volume 39A is written by an established group of scientists and sets the stage for what is meant by a systems approach to conservation. The last chapter of Volume 39B is written by early career researchers and is intended to be a forward-looking perspective on the state and future of conservation physiology of fishes.

Collectively, this volume provides an integrated synthesis that celebrates the successes achieved so far while identifying opportunities to further benefit the management and conservation of wild fish populations. It is our hope that this volume will serve as a resource for learners, fish physiologists, conservation practitioners, and fisheries managers.

We are grateful for the generosity of the many referees who provided peer reviews and thoughtful comments on the various chapters prior to their publication. We are especially appreciative of the authors who generated high-quality chapters during a period of uncertainty and challenges related to the global COVID-19 pandemic. The ideas shared in this book are truly at the frontier of applied fish physiology and conservation. We are fortunate to be part of a caring and supportive community and hope that this volume will serve to further build connections among those working on the conservation physiology of fishes. The team at Elsevier assisted with advancing this project. This book is dedicated to all of the scientists, practitioners, and community members who devote their professional and personal lives to generating and applying evidence to protect and restore fish populations and aquatic ecosystems around the globe.

Steven J. Cooke
Nann A. Fangue
Anthony P. Farrell
Colin J. Brauner
Erika J. Eliason

Abbreviations

11-KT	11-ketotestosterone
17,20βP	17α,20β-dihydroxy-4-pregnen-3-one
20βS	17α,20β,21-trihydroxy-4-pregnen-3-one
3kPZS	3-keto petromyzonol sulphate
αCO$_2$	carbon dioxide solubility coefficient
ABO	air-breathing organ
AC	accessory cell
AhR	aryl hydrocarbon receptor
AS	aerobic scope
ASIC	acid-sensing ion channel
ATP	adenosine triphosphate
BIA	bioelectrical impedance analysis
BLM	Biotic Ligand Model
Ca^{2+}-ATPase	calcium adenosine triphosphatase
CA	carbonic anhydrase
CA	catecholamines
CAc	cytoplasmic carbonic anhydrase
CAIV	membrane-bound carbonic anhydrase
CFTR	cystic fibrosis transmembrane regulator channel
Clc	chloride intracellular ion channel
C$_{max}$	theoretical maximum food consumption
DDT	dichlorodiphenyltrichloroethane
DEB	dynamic energy budget model
DNA	deoxyribonucleic acid
DOC	dissolved organic carbon (DOC)
E2	17β-estradiol
EC50	concentration of a toxicant effective in causing sublethal effects in 50% of the population
ECaC	epithelial calcium channel
ECCC	Environment and Climate Change Canada
ECF	extracellular fluid
EDC	endocrine-disrupting compound/endocrine-disrupting chemical
EE2	17α-ethinylestradiol
EOG	electro-olfactogram

EPOC	post-exercise oxygen consumption
ER	estrogen receptor
FIAM	Free Ion Activity Model
FSH	follicle-stimulating hormone
GFR	glomerular filtration rate
GH	growth hormone
GI	gastrointestinal
GnRH	gonadotropin-releasing hormone
GnRHa	GnRH analogue
GPS	Global Positioning System
GRE	glucocorticoid response element
GSI	gonadal somatic index
GSIM	Gill Surface Interaction Model
HAT	vacuolar-type proton adenosine triphosphatase
hCG	human chorionic gonadotropin
HEA	high environmental ammonia
HPG	hypothalamic-pituitary-gonadal
HPI	hypothalamic-pituitary-interrenal
HPT	hypothalamic-pituitary-thyroid
IBM	individual-based models
ICF	intracellular fluid
IGFBP	IGF binding protein
IGFI	insulin-like growth factor-I
IUCN	International Union for Conservation of Nature
$\textbf{J}_{\textbf{in}}$	influx rate of an ion
$\textbf{J}_{\textbf{max}}$	maximal influx rate of an ion
$\textbf{J}_{\textbf{net}}$	net flux rate of an ion
$\textbf{J}_{\textbf{out}}$	efflux rate of an ion
Km	inverse of affinity = concentration of ion in water at which uptake transport occurs at 50% of the maximal rate
LA50	the gill metal burden associated with 50% mortality on an acute or chronic basis
LC50	concentration of a toxicant lethal to 50% of the population
LH	luteinizing hormone
log K value	conditional equilibrium constant, the negative logarithm (base10) of the concentration at which 50% of the ion is bound to a ligand
$\dot{M}CO_2$	carbon dioxide production rate
$\textbf{Me}^{\textbf{N+}}$	free metal cation
MIH	maturation-inducing hormone
MLR	Multiple Linear Regression Model
MMR	maximum metabolic rate
$\textbf{MO}_2$	rate of oxygen consumption in molar units
$\dot{M}O_2$	oxygen uptake rate

mRNA	messenger ribonucleic acid
MWWE	municipal wastewater effluent
Na$^+$,K$^+$-ATPase	sodium, potassium adenosine triphosphatase
NBC1	sodium bicarbonate co-transporter type one
NCC	sodium, chloride co-transporter
NCX	sodium, calcium exchanger
NHE	sodium proton exchanger
NKA	sodium, potassium adenosine triphosphatase
NKCC	sodium, potassium, two chloride cotransporter
NQ	nitrogen quotient (N-excretion/O_2 consumption)
NSERC	Natural Sciences and Engineering Research Council of Canada
OUC	ornithine urea cycle
PAH	polycyclic aromatic hydrocarbons
PCB	polychlorinated biphenyls
PCDD/F	polychlorinated dibenzodioxins and dibenzofurans
PCO$_2$	partial pressure of carbon dioxide
pCO$_2$	partial pressure of CO_2
PEB	physiological energy budget model
PGF2α	prostaglandin F2α
pK	negative logarithm (base 10) of the equilibrium constant
pK'	negative logarithm (base 10) of the conditional equilibrium constant
PMCA	plasma membrane calcium adenosine triphosphatase
PNH$_3$	partial pressure of ammonia
PO$_2$	partial pressure of oxygen
PPME	pulp and paper mill effluent
Rh	Rhesus glycoprotein
Rhbg	basolateral Rhesus glycoprotein
Rhcg	apical Rhesus glycoprotein
RNA	ribonucleic acid
ROMK	renal outer medullary potassium channel
ROS	reactive oxygen species
SDA	specific dynamic action
SEM	standard error of the mean
SFG	scope for growth
SID	Strong Ion Difference
SMR	standard metabolic rate
T	testosterone
T3	3,5,3'-triiodo-L-thyronine
T4	thyroxine
TCDD	2,3,7,8-tetrachlorodibenzo-para-dioxin
TEP	transepithelial potential
TMAO	trimethylamine oxide

TSD	temperature-dependent sex determination
TSH	thyroid-stimulating hormone
U_{crit}	critical swimming speed
UFR	urine flow rate
US EPA	United States Environmental Protection Agency
UT	facilitated diffusion urea transporter
vH^+-ATPase	vacuolar-type proton adenosine triphosphatase
VRAR	Voluntary Risk Assessment Report
Vtg	vitellogenin
WEB	wisconsin energy budget model
X^{N+}	strong cations
Y^{N-}	strong anions

Chapter 1

Conservation physiology and the management of wild fish populations in the Anthropocene

Steven J. Cooke[a,*], Nann A. Fangue[b], Jordanna N. Bergman[a], Christine L. Madliger[a], Joseph J. Cech Jr.[b], Erika J. Eliason[c], Colin J. Brauner[d], and Anthony P. Farrell[d]

[a]Fish Ecology and Conservation Physiology Laboratory, Department of Biology, Carleton University, Ottawa, ON, Canada
[b]Department of Wildlife, Fish and Conservation Biology, University of California-Davis, Davis, CA, United States
[c]Department of Ecology, Evolution, and Marine Biology, University of California-Santa Barbara, Santa Barbara, CA, United States
[d]Department of Zoology, University of British Columbia, Vancouver, BC, Canada
[*]Corresponding author: e-mail: steven_cooke@carleton.ca

Chapter Outline

Fish Physiology, Vol. 39A. https://doi.org/10.1016/bs.fp.2022.04.001

It is widely regarded that we have entered a new epoch distinct from the Holocene which is defined by the dominance of humans—termed the "Anthropocene." Indeed, for centuries humans have altered aquatic ecosystems by degrading habitats, altering ecosystem structure, and impairing ecosystem function. In the Anthropocene, aquatic ecosystems and their constituent fish populations are exposed to persistent and emerging threats as well as their interactions. Physiological concepts, knowledge and tools have the potential to help understand the threats facing fish and inform the development of science-based management strategies. By understanding the various systems that govern the biology of fish (e.g., cardiorespiratory, endocrine, biomechanics) it is possible to reveal the complex ways in which different environmental conditions and anthropogenic stressors influence organisms and scale up to influence population-level processes. Although the fact that we have entered the Anthropocene is ominous for biodiversity and humanity, conservation physiology offers hope given the growing number of scientists and practitioners focused on understanding and solving conservation problems facing fish and other organisms.

1 The Anthropocene

Humans have dominated the Earth, forever changing it in ways that are hard to fathom (Vitousek et al., 1996). Indeed, many scholars argue that the level of human impact on the planet is so profound that we have now entered a new epoch distinct from the Holocene known as "The Anthropocene" (Crutzen, 2006). While debate continues as to when the Anthropocene began, the date of the Trinity nuclear test on July 16th of 1945 is widely regarded as a symbolic start (Lewis and Maslin, 2015; Smith and Zeder, 2013). No matter when it started, the biological trajectory of the Anthropocene is not going in the "right" direction in terms of the loss of biodiversity and atmospheric pollution, with trends occurring exponentially and coined the "great acceleration" (Steffen et al., 2015). Today human impacts are widespread and devastating. We see plastic pollution throughout the remote Arctic seas, loss of biodiversity associated with clearing of forests and filling of wetlands at a massive scale, the creation of concrete jungles and a pan-global redistribution of organisms. Although the term "Anthropocene" has an inherently negative connotation, there have also been efforts to reframe this as an opportunity to achieve a "good Anthropocene" (Dalby, 2016). In other words, there is still hope.

2 Fish in the Anthropocene

Aquatic ecosystems and their constituent fish populations are not immune to the Anthropocene. In fact, for centuries humans have altered aquatic ecosystems by degrading habitats, altering ecosystem structure, and impairing ecosystem function. Examples abound. The canals and waterways constructed thousands of years ago (e.g., ancient empires in Rome, Egypt and China built canals for transportation, flood control, and water consumption for agriculture; Lin et al., 2020) connected different watersheds and diverted water for human use. Harvesting of stones from essential spawning habitat for fish to create

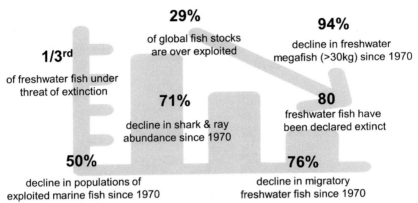

The Status of Fish Populations in the Anthropocene

29%
of global fish stocks
are over exploited

94%
decline in freshwater
megafish (>30kg) since 1970

1/3ʳᵈ
of freshwater fish under
threat of extinction

71%
decline in shark & ray
abundance since 1970

80
freshwater fish have
been declared extinct

50%
decline in populations of
exploited marine fish since 1970

76%
decline in migratory
freshwater fish since 1970

FIG. 1 Key statistics that emphasize the dire state of fish populations in the Anthropocene. Statistics derived from the IUCN Status Reports and WWF Living Planet Index.

human infrastructure is another example (e.g., in Lake Ontario, 43,000t was removed annually in the 1830s which accelerated erosion and altered lake trout spawning habitat; Morrison, 2019). Of course, the outright overharvesting of fish that leads to population collapse is a stand out (e.g., Atlantic cod collapse; Myers et al., 1997 or Blue walleye extinction; Miller et al., 1989). The statistics are dire; wildlife populations continue to decline (Fig. 1). The World Wide Fund for Nature (WWF) reports that 30% of freshwater fish are threatened with extinction, with 80 species already extinct. The situation is similarly grim for marine fishes with the degradation of coastal ecosystems and many fish populations collapsing (Jackson et al., 2001). It is clear that fish—whether freshwater or marine—require drastic action if their populations are to be protected and restored.

3   The threats to fish populations

The specific threats facing fish populations can generally be categorized under five broad headings: habitat loss or alteration, invasive species, overfishing or unsustainable fishing practices, pollution, and climate change (see Fig. 2). Yet, it is also becoming apparent that threats rarely act alone. The intersection of threats (e.g., additive, synergistic, antagonistic; Folt et al., 1999; Jackson et al., 2016) or their cumulative impacts (Crain et al., 2008) are particularly devastating and difficult to address (Schultz, 2010). Climate change is a major threat that overlaps and intersects with all other threats (Hollowed et al., 2013; Myers et al., 2017). There are also emerging stressors ranging from microplastics (Rezania et al., 2018), to noise (Popper and Hawkins, 2019), to light (Davies and Smyth, 2018), and to pathogens (Tompkins et al., 2015) that will further threaten fish populations and the aquatic ecosystems upon which they depend (see Reid et al., 2019). Specific threats vary across fishes, largely based

The Anthropocene from the Perspective of a Fish

FIG. 2 Images depicting various threats and challenges that impact fish and fish habitat in the Anthropocene spanning marine and freshwater systems. Image credits: A—Government of Odisha, CC BY 4.0 via Wikimedia Commons; N—IUCN, Public Domain; T—Downtowngal, CC BY-SA 3.0 via Wikimedia Commons; H—NOAA, Public Domain; R—Tangoev9999, CC BY-SA 3.0 via Wikimedia Commons; O—Erasmus Kamugisha, CC BY-SA 4.0 via Wikimedia Commons; P—US Forest Service, Public Domain; O—Hard Rain Project, Public Domain; C—Randee Daddona, Public Domain; E—European Union, Public Domain; N—US Department of Agriculture, Public domain; E—Great Lakes Fishery Commission, Public Domain.

on aspects like their habitat requirements (especially in freshwater ecosystems versus marine). For example, overharvest in marine systems remains the dominant threat while in freshwater systems it is often a combination of threats (Cooke et al., 2014b; Arthington et al., 2016) and life-history (Reynolds et al., 2005a,b) among other factors (Hutchings and Reynolds, 2004). Identifying and mitigating threats is necessary to ensure that fish populations and the habitats upon which they depend are protected.

4 Physiology connects fish to threats

All of the threats and their pathways of effect influence individual fish at levels ranging from the molecule to the whole organism (Metcalfe et al., 2012; Horodysky et al., 2015; see Fig. 3), with those effects on individuals scaling up to influence population-level processes (Calow and Forbes, 1998). In fact, stressors (direct and indirect) act across, and have the potential to manifest at, different scales (Cooke et al., 2014a). For example, degraded habitats represent impairments in habitat quality that have negative impacts on fish condition and health (Jeffrey et al., 2015). Invasive species alter ecosystem structure and can impact nutrition and space use of endemic species (Lennox et al., 2015). Overfishing or unsustainable fishing practices can certainly lead to mortality via harvest, but there can also be sublethal effects or

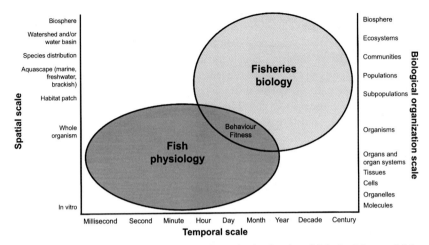

FIG. 3 Complementary temporal, spatial, and organizational scales of fish physiology and fisheries biology. On both the spatial and biological organization scales, the fields of fish physiology and fisheries biology study the organismal level. *Modified from Horodysky, A.Z., Cooke, S.J., Brill, R. W., 2015. Physiology in the service of fisheries science: why thinking mechanistically matters. Rev. Fish Biol. Fish. 25, 425–447.*

delayed mortality induced in bycatch (Davis, 2002) or effects imparted through selective effects of fishing (e.g., selection for certain physiological traits; Hollins et al., 2018). Various types of pollution impair organismal function and can be lethal above thresholds (Heath, 1995). Given that fish are ectotherms, climate change will have diverse effects on fish condition, health, and survival through direct and indirect mechanisms (Pörtner and Farrell, 2008; Whitney et al., 2016). Physiology is the integrator between organism and environment (Fry, 1971), and thus provides a logical suite of systems for exploring how threats influence fish and to identify solutions to conservation problems.

5   Conservation physiology to the rescue?

Conservation physiology is defined as the application of physiological concepts, knowledge, and tools to understand and solve conservation problems (Cooke et al., 2013; Wikelski and Cooke, 2006). It is a nascent, solutions-oriented, and applied discipline increasingly recognized as an important component of conservation science (Cooke et al., 2020b, 2021a; Madliger et al., 2016). This discipline has evolved rapidly over the last decade as evidenced by the launching of a journal (i.e., *Conservation Physiology* by Oxford University Press/Society of Experimental Biology), publishing of a textbook (Madliger et al., 2021a), and an increasing number of individuals who are identifying themselves as conservation physiologists (Madliger et al., 2021b). Although the words "conservation physiology" were not used prior to 2006 in the fish physiology literature, there was certainly applied fish physiology research done in the service of conservation and management for decades which was sometimes termed ecophysiology or under the domain of ecotoxicology. Indeed, Schreck and Scanlon (1977) reviewed the contribution of endocrinology to fisheries management and conservation nearly 30 years prior. Yet, it is only in the last few decades that these ideas became normalized or even institutionalized. Given this recent momentum, here we reflect on the ways in which conservation physiology has been used (or could be used) to address key issues and threats facing fish. We also consider what constitutes a systems approach to conservation physiology and how this may serve as a useful framework for connecting organisms (fish) and issues. We conclude by discussing the merits of achieving a mechanistic approach to conservation and management of fish populations and what is needed to make that happen.

6   Reflections on the evolution of the fish physiology series

The first three volumes in the "Fish Physiology" series were published in 1969. These were very mechanistic in focus and covered a broad range of topics including "Excretion, Ionoregulation, and Metabolism" (Vol. 1), "The Endocrine System" (Vol. 2), and "Reproduction and Growth" (Vol. 3). This was quickly followed by three more volumes in 1970/1971 on "The Nervous System, Circulation, and Respiration" (Vol. 4), "Sensory Systems and Electric Organs" (Vol. 5). and "Environmental Relations and Behavior" (Vol. 6). These six volumes were crucial in setting the stage for the series, defining the state of the field, and providing direction to address the large knowledge gaps in the field of fish physiology. Additionally, the 6th volume laid important foundations for the current field of conservation physiology, in particular, the chapter "The Effect of Environmental Factors on the Physiology of Fish" by F.E.J. Fry. The ideas presented in that chapter, along with many of his other publications, have led to the concept of aerobic scope and its utility in predicting species distributions as a function of the environment and the impact of changes in the environment on individuals and populations, as discussed throughout this volume. For the rest of the series, from 1971 to the present (Vols. 39A and 39B), volumes were progressively more focused on a specific theme, but always with a primary emphasis on mechanistic physiology. However, as the respective fields advanced and mechanisms were clarified, there was a natural interest in how these mechanisms would be impacted by changes in internal and external environmental conditions, with implications to more applied fields such as climate change, toxicology, development of water quality criteria, aquaculture, and conservation. This emphasis is reflected in some of the most recent volumes, such as "Homeostasis and Toxicology of Metals" (Vol. 31), "Organic Chemical Toxicity of Fishes" (Vol. 33), "Aquaculture" (Vol. 38), and "Fish Physiology" (Vols. 39A and 39B).

7   Conservation physiology applications

Here we briefly consider the ways in which fish physiology is relevant to salient issues in conservation. Although there are examples beyond what we describe here (see Cooke et al., 2013), we focus on the threat taxonomy that arises from the CMP IUCN (International Union for Conservation of Nature) Action Taxonomy. We consider how conservation physiology has been used, its future potential, and what needs to be done to further advance these issues.

7.1 Assessing and managing recovery of imperiled species

Most global estimates of fish species imperilment (i.e., endangered, threatened, or vulnerable) hover between 35% and 50% of all species (Helfman et al., 2009). Between 1900 and 2010, 42 documented fish species extinctions have occurred, at an unprecedented rate, 877 times greater than background extinction rates of fishes inferred from the fossil record (Shirey et al., 2018). Today, 87–93 species are identified by the IUCN as extinct, most being freshwater species. Imperilment estimates are heavily biased towards North America and Europe even though 85% of fish species occur in the tropics. Most fish populations, if not formally listed or subject to endangered species laws, are not closely monitored and imperilment may frequently go unnoticed/unassessed (Moyle et al., 2002, 2017). To protect and restore biodiversity, scientists typically argue that management should be proactive, and that it is both time and cost effective to institute management actions before populations become imperiled.

The management of imperiled species is a complicated political process reliant on advancements in fisheries science. Designation and protection of imperiled species at the International scale is coordinated by the IUCN, which also maintains the "Red List" for threatened and endangered species. The Convention on International Trade in Endangered Species of Wild Flora and Fauna (CITES) is an international cooperative designed to protect wildlife from overexploitation through trade. National scale fish conservation is often assigned to federal agencies (e.g., the Department of Fisheries and Oceans (DFO) in Canada). In the U.S., the Endangered Species Act of 1973 (ESA) is implemented by the U.S. Fish and Wildlife Service (USFWS) and/or the National Marine Fisheries Service (NMFS), charged with formally listing a species for protection and developing a species recovery plan. Science-based management has been successful in de-listing four fish species, though this is a fraction of the current 139 species that are listed (https://www.fws.gov/endangered/).

Partnerships between non-governmental organizations, professional societies, and academic institutions are playing an increasingly important role in identifying status and trends in fish populations, including increased reliance on physiological information (Mahoney et al., 2018). For example, the American Fisheries Society (AFS) has imperiled species policies (Shirey et al., 2018) to engage policy makers, and they provide periodic assessments of the conservation status of imperiled North American freshwater and diadromous fishes (see Jelks et al., 2008). It was not until 2017 that all California salmonids were reviewed in their entirety, independent of regulatory agencies, and motivated by concerns of stakeholders and academics. Of the 32 distinct salmonids in California, 15 are recognized by state and federal agencies as threatened or endangered, and one is extinct. In contrast, Moyle et al. (2017), using improved assessment metrics including physiological parameters and their integration with climate change predictions, classified all but two species as imperiled, translating to an extinction rate of 45% in 50 years and 74% in 100 years. Mahoney et al. (2018) directly addresses whether physiological tools are playing a prominent (enough) role in conservation planning, specifically through integration into species recovery plans, concluding a disconnect between available tools and integration of these tools into the process of recovery. Despite classic successes in environmental toxicology or stress detection, or the use of physiology to support determining the how and why of a species decline (reviewed in Wikelski and Cooke, 2006), physiology is often under-utilized or under-acknowledged when utilized. Of the 146 US-ESA recovery plans released from 2005 to 2016, 93% included physiological information, but of those, 56% used only natural history information and did not include physiological information in an actionable form (Mahoney et al., 2018). In Osteichthyes specifically, the incorporation of physiological information into the 23 plans was similar. Barriers proposed include a lack of understanding in the application and value of physiological data, communication difficulties between physiologists and decision-makers, lack of funding, and logistical and ethical constraints when studying endangered species (Madliger et al., 2021b; Mahoney et al., 2018). Several recent reviews and perspective pieces offer suggestions for a more integrative path forward (Cooke et al., 2020a,b, 2021b).

7.2 Invasive species

Invasive species have long been acknowledged as a major cause of global biodiversity declines and animal and plant extinctions (Bellard et al., 2016; Clavero and García-Berthou, 2005; Mollot et al., 2017; Vitousek et al., 1996). Biological invasions negatively impact both freshwater (Hermoso et al., 2011; Light and Marchetti, 2007; Tickner et al., 2020) and marine (Arthington et al., 2016; Green et al., 2012) fishes, resulting in various adverse ecological and evolutionary effects (Pyšek et al., 2020). Numerous options exist for managing invasive species, ranging from total eradication to population control, though most are laborious and can be expensive (Lennox et al., 2015; Rytwinski et al., 2019). The techniques currently used to manage invasive species—the majority of which rely on behavior, ecology, genetics, and/or economic disciplines—however, would benefit from an incorporation of physiology. Because physiological processes underlie an individual's response to its (new) environment, the same characteristics that might have made a species such a strong invader could instead be exploited or manipulated to improve conservation actions (Lennox et al., 2015). Several instances already exist where physiology was used to successfully manage an aquatic invasive species. For example, sensory barriers (e.g., bubble curtains, sound stimuli) have effectively deterred invasive Asian bighead (*Hypophthalmichthys nobilis*) and silver (*Hypophthalmichthys molitrix*) carp by 95% from specific habitats (Pegg and Chick, 2004). In the Laurentian Great Lakes, the Great Lakes Fishery Commission successfully used physiology to dramatically reduce invasive sea lamprey (*Petromyzon marinus*) by developing and applying (1) selective lampricides to kill larvae, (2) lamprey-specific barriers and traps to minimize spawning migrations and to remove adults from the population, respectively, and (3) sterilized males to reduce reproductive potential of spawning populations (Siefkes, 2017). Although here physiology was used to successfully minimize invasive species which have inflicted negative impacts to native fish populations (Phelps et al., 2017; see Siefkes, 2017), it is relatively, and somewhat surprisingly given its success, uncommon for physiology to be incorporated into invasive species management strategies (Lennox et al., 2015; though see Rahel and McLaughlin (2018) for other examples).

The direct and indirect effects of biological invasions are complex and can act synergistically with other threats, in particular climate change (Bellard et al., 2018). Invasive species impacts are predicted to intensify with climate change (Mainka and Howard, 2010), especially from the perspective of invasive aquatic species (see Rahel and Olden, 2008). Conservation physiology tools could be crucial in predicting the potential invasive range of introduced species (Barbet-Massin et al., 2018; Kearney et al., 2009), modifying invasive species distributions, and managing their ongoing impact in nonnative ecosystems (Madliger et al., 2016). Additionally, if we can understand the physiological mechanisms that facilitate biological invasions, and/or identify and exploit physiological vulnerabilities in invasive species (see Cooke et al., 2021b), we could make a major advance in applying conservation physiology to management actions. By identifying traits that could impact whole-organism function (e.g., aerobic scope, thermal tolerances; Chown, 2012), conservation physiology could be used to combat invasions both reactively, by controlling or eradicating present invasive species, or proactively, by preventing invasions from occurring altogether.

7.3 Making fisheries more sustainable

Fishing mortality can arise directly from harvest, with overexploitation being a contributor to biodiversity loss in both freshwater (Allan et al., 2005) and marine (Jackson et al., 2001) systems. In well-managed systems, (over) exploitation is regulated and fisheries can indeed be sustainable (Hilborn, 2007). Yet, some fish that are captured are released because of harvest regulations (e.g., species closures, size limits), because of conservation ethic (mostly in recreational fisheries—termed "catch and release angling"), or because of lack of a market for a given species. Although releasing fish can reduce mortality relative to harvest, a variable portion of released fish will die from the fisheries interactions which represents an important, but sometimes neglected aspect of fishing mortality.

A large body of research has explored the factors that influence the outcome of a fisheries interaction (reviewed in Davis, 2002; Raby et al., 2015). All fisheries interactions induce some level of physiological disturbance which, if not overly severe, is unlikely to lead to mortality. However, if exhaustion occurs as a result of struggling during capture or handling, then fish may die. Davis (2002) synthesized research on fisheries-related stressors and identified the important role of air exposure and water temperature in moderating outcomes. More importantly, this work generates knowledge that can be used to reduce stress and mortality. For example, research has revealed that temporary holding of endangered coho salmon captured in a gillnet in a revival chamber reduces physiological disturbance and expedites recovery while enhancing survival (Farrell et al., 2001). A second area of focus has been on attempting to exploit the sensory physiology of fish in an attempt to increase rates of capture of target species while reducing capture of non-target species (Elmer et al., 2021). The greatest successes in this space have been for sharks on long-line fisheries (Jordan et al., 2013). Sharks are unique relative to teleost fish because of their ampullary electrosensory system which is highly sensitive to low-frequency electrical stimuli. Although they normally use this system for finding prey, certain weak electric signals (e.g., permanent magnets) can also be used to deter sharks which has worked in some fisheries (Jordan et al., 2013).

Additional research is needed to increase the survival of fish that are captured and released as well as to reduce the capture of non-target species. It has become clear that context matters for studies which attempt to reduce stress and mortality; temperature, time, maturation, sex, size, and population can all mediate the extent to which a fish will survive (Raby et al., 2015). As such, it is challenging to develop generalized guidelines that apply across a wide range of species or fisheries. Additional research is needed to understand if and when generalizations are possible as well as efforts that focus on imperiled species. Sensory physiology research has been most useful for reducing capture on non-fish taxa (e.g., birds, marine mammals) given the inherent biological differences between fish and those organisms. However, it is much more challenging to target one fish species yet deter another. For example, Martin and Crawford (2015) revealed that visual warning panels that help organisms detect and avoid gillnets are effective at reducing bycatch of non-fish taxa but less relevant for reducing bycatch on fish. There is more work needed to understand species-specific aspects of vision, olfaction, and other relevant senses to be able to be more tactical with attraction and repulsion. The fisheries of today are more sustainable and responsible because of mechanistic physiological research but many knowledge gaps remain that will be best addressed by working closely with fisheries managers and fishers.

7.4 Identifying pollution thresholds

Chemical, thermal, and particulate pollutants pose threats to fish populations (Heath, 1995), representing an on-going conservation issue in aquatic ecosystems. As with any environmental stressor, pollutants can present challenges to fish that may be compensated by using physiological mechanisms, along with other responses such as behavior. Due to the sensitivity of many physiological systems to environmental perturbations, a variety of physiological metrics have proven useful in documenting the negative consequences of pollution exposure for fishes. For example, in relation to several pollutants (e.g., DDT, atrazine) at sublethal concentrations, carp (*Cyprinus carpio*) show a rise in serum glucose and cortisol, corresponding with signs of exhaustion 72h following exposure (Gluth and Hanke, 1985). Similarly, in white sturgeon (*Acipenser transmontanus*), total DDT, total pesticides, and PCBs are negatively correlated with plasma triglycerides and condition factor, and males further show negative correlations between these tissue burdens and plasma androgens (Feist et al., 2005). Following oil spills, biomarkers of sublethal toxic impacts due to exposure to petroleum hydrocarbons can be established quickly by measuring transcriptomic changes related to inflammation, metabolism, and cardiac function, as well as histology of the heart and liver (Hook et al., 2018). This approach of using complex traits such as performance (e.g., swimming) or tolerance (e.g., hypoxia, warming), to assess the biological and ecological consequences of pollutants are now quite common (McKenzie et al., 2007). There is also continually amassing evidence that the effects of more recently emerging types of pollution, such as light and sound, can be assessed with physiology (Falcón et al., 2020; Mickle and Higgs, 2018). For example, under streetlighting conditions, perch (*Perca fluviatilis*) and roach (*Rutilus rutilus*) show reduced mRNA expression of the gonadotropins luteinizing hormone and follicle stimulating hormone, with the potential for reproductive disruption (Brüning et al., 2018). A meta-analysis by Cox et al. (2018) showed that anthropogenic noise leads to an elevation of cortisol levels and an increase in hearing thresholds. This study further revealed interference with foraging ability, predation risk, and reproductive success, indicating that physiological effects could be a precursor to fitness effects. Further research on low-intensity, localized noise near coral reefs has led to suggestions to regulate the use of outboard motors on boats in sensitive areas based on specific physiological thresholds (e.g., oxygen consumption rates) of physical pollution (Illing and Rummer, 2017).

It is clear that a vast array of tools from the conservation physiology toolbox (Madliger et al., 2018), spanning metabolism and energetics, stress physiology, immune function, and reproductive physiology, are being applied to investigate the sublethal impacts of pollutants, with the goal of applying them as biomarkers. There is potential to further integrate transcriptomics with physiological research to identify functional physiological thresholds that are predictive of organismal responses and sublethal negative effects

(Connon et al., 2018); indeed, transcriptomics has revealed when sublethal thresholds of thermal and osmotic stress lead to adverse impacts on fish species (Connon et al., 2018). With less studied pollution types, such as light pollution, physiological work in other taxa like sea turtles and nesting shorebirds has led to threshold-based management decisions (e.g., Department of the Environment and Energy—Australian Government National Light Pollution Guidelines for Wildlife, 2020), indicating that clear conservation applications could result from undertaking such research.

While there are a growing number of studies that investigate multiple physiological responses simultaneously, it remains uncommon to take a multi-stressor approach. However, fishes are likely to face multiple pollutant types in altered habitats (e.g., light and sound along coastlines) or in combination with other anthropogenic disturbances (e.g., climate change, structural modifications, fishing). By taking a meta-analytic approach, Rodgers et al. (2019) found that contaminants (carbaryl, methylmercury, and selenium), along with other stressors (elevated temperature, salinity, and low food) could impair the growth of threatened green sturgeon (*Acipenser medirostris*), and that contaminants also increased mortality rates. By looking across these stressors individually, the authors were able to suggest management actions related to salt intrusion in nursery habitats and maintaining water temperatures during spawning types, but cited data gaps in relation to pollution exposure such as difficulty linking individual physiology to population declines, lack of data in critical early life periods, and a lack of knowledge of how additional stressors influence uptake, accumulation, toxicity, and excretion (Rodgers et al., 2019). Although we are gaining more understanding of how individual physiological monitoring can scale up to the population level (Bergman et al., 2019), there is still a great deal of work to be done in answering how, when, and where this will be possible. Dynamic Energy Budget theory could provide a fruitful avenue to move from individual to population scales (Pérez-Ruzafa et al., 2018) and, overall, an understanding of transgenerational effects of different pollutants is necessary to translate from shorter-term physiological (and behavioral) responses to population, community, and ecosystem ramifications (Kjelland et al., 2015). Greater integration of behavioral effects and physiological processes will also provide information on how sub-lethal exposure to pollution can alter processes integral to fitness and survival in the wild, and therefore fish populations (Scott and Sloman, 2004).

7.5 Mitigating interactions with water infrastructure

Many aquatic systems currently include water-resource-development structures (e.g., dams, diversion pipes and channels), either for extracting water for agricultural, industrial, or domestic uses, or for hydro-electric power generation. Dams create lentic reservoirs and slow water velocity upstream, smooth out variations in water flow, divert water for irrigation and block or slow fish movements both upstream and downstream, all of which can adversely affect fish populations (Marschall et al., 2011; Raymond, 1979). Physiological and behavioral approaches (e.g., measuring fish-swimming performance, stress-related effects, and responses to water currents) have helped design upstream fish-passage structures around dams to mitigate these barriers' effects (Bunt et al., 2012; Cocherell et al., 2011; Haro et al., 2004). Unfortunately, passage success of downstream-migrating juveniles may be very low, especially if multiple dams and hydro-electric facilities must be navigated (Norrgård et al., 2013). Water diversions (e.g., for irrigation, water-cooling of electrical power plants) also pose significant challenges to resident and migrating fishes, especially juveniles (Moser et al., 2015; Mussen et al., 2014) and physiological studies have been used to examine for potential impacts. Even when water-diversion pipe openings are guarded by fish-exclusion screens, small fishes may suffer stress, injury, or mortality from prolonged screen contacts (Greenwood, 2008; Swanson et al., 2005). For example, relatively weak-swimming delta smelt exposed to multi-vector flows showed increased stress responses (indicated by increased plasma cortisol concentrations), increased screen contact rates, and increased moribund behavior at the highest approach (i.e., flows passing through the fish screen) velocities (Young et al., 2010). Adding detectable vibrations to unseen (e.g., at night) screens could mitigate these effects in some fishes. Mussen and Cech (2019) found that intact, marine surfperch decreased their contacts with an experimental fish screen when it was vibrating, under night-time, darkened conditions. Conclusively, surfperch treated with a streptomycin sulfate solution (to deactivate their lateral line (baroreceptive) system) were not deterred from contacting the screen under these conditions (Mussen and Cech, 2019).

Physiological, behavioral, and modeling methods should be integrated in a systems approach to guide future conservation-related resource-management decisions. For example, whereas upstream passage around dams has been studied quite extensively, many of these studies concentrated on Pacific salmon species (e.g., Keefer et al., 2004), which spawn only once. In contrast, the downstream migration of repeat-spawning adult fishes and the offspring of all anadromous species, including sturgeons, is relatively unstudied (Cooke et al., 2020a). Almost all 25 species of the world's sturgeons desperately need enhanced conservation efforts to prevent (continued) future population declines towards extinctions (Lenhardt et al., 2006). A systems approach, incorporating physiology, behavior (e.g., using biotelemetry), genetics, and population biology should be employed to protect and promote the recovery of these fishes (Cocherell et al., 2011; Cooke et al., 2020a; Heublein et al., 2009; Jager et al., 2016). Besides further fish-passage studies for both upstream and downstream migrants, the removal of dams that are no longer needed must be considered. The decommissioning of a dam on the Sacramento River (California, USA) allowed continued upstream migration of threatened green sturgeon to historic spawning and rearing habitat (Heublein et al., 2009).

Finally, implementation of available results should be incorporated into a systems approach for conservation planning and solutions. For example, coordinating irrigation schedules and velocities so that water diversion intake flows do not entrain (i.e , remove) resident or migrating fishes from natural waterways could have significant conservation effects. Mussen et al. (2014) calculated that by decreasing the intake velocity, but increasing the operation time, of an unscreened water diversion should save many juvenile fishes, while delivering the same volume of irrigation water needed for crops.

7.6 Advancing climate change science

Climate change is causing poleward shifts in fish distribution, changes in fish abundance and diversity, and altered phenology (Roessig et al., 2004). Conservation physiology techniques have enabled researchers to identify the optimal range of environmental conditions for fish performance, to assess the thermal phenotypic plasticity of populations and species, and also make predictions about how they may respond to climate change (i.e., assessing relative vulnerability/resilience). For example, polar and equatorial fish are predicted to be particularly sensitive to warming temperatures. Fish from those regions tend to be adapted to a stable thermal regime and thus be stenotherms (narrow thermal tolerance range). Rummer et al. (2014) evaluated the thermal sensitivity of aerobic scope (maximum—standard metabolic rate) in six species of tropical fishes and found that all were currently living close to their thermal optima and that even small increases in temperature could result in population decline and poleward shifts in distribution. Eliason et al. (2011) compared cardiorespiratory capacity across populations of adult sockeye salmon and found that populations differed in thermal tolerance and current maximum river temperatures exceed the optimal thermal range for every population.

Another goal for conservation physiologists is to identify a biomarker predictive of sublethal effects or mortality. For example, Jeffries et al. (2015) used an integrative approach to compare thermal tolerance in threatened longfin smelt and endangered delta smelt in California. They used RNA sequencing to identify transcriptome-wide profiles indicative of lower thermal tolerance in longfin smelt and suggest that longfin smelt are more vulnerable to warming compared to the delta smelt. A major focus for researchers is to identify the adaptive capacity of fish to climate change. While this is a difficult challenge that requires interdisciplinary research, recent work by Chen and Narum (2021) shows promise. The researchers have used whole genome resequencing to link locally adapted phenotypes (cardiac function and thermal tolerance) with specific genomic regions in redband trout.

Furthermore, we need comprehensive environmental data (e.g., temperature, dissolved oxygen, flow, pH) to characterize the conditions fish currently encounter and to be able to make predictions about the future. Some systems have excellent environmental monitoring in place (e.g., Fraser River, BC, Canada), while it is entirely absent in other systems. Environmental data will enable physiologists to work with modelers to use physiological data to determine if populations are currently persisting at their habitat limits or if they have the capacity to withstand further change. Research needs to be informed by environmental data to ensure the duration, intensity, and scale of the stressor treatments (e.g., temperature exposures) are ecologically relevant. For example, increasingly, researchers are incorporating ecologically relevant fluctuating temperature regimes in their study design, mimicking natural diurnal cycles (e.g. Morash et al., 2018). Field physiology is becoming more prominent, where the lab is brought to the fish and measurements are made in the field on acclimatized individuals (e.g., Barnett et al., 2016; Mochnacz et al., 2017). Technological advances have made biotelemetry and biologgers smaller, longer-lasting, and less expensive which has opened the door to linking physiology with behavior and ecology in free-swimming fish (e.g., Farrell et al., 2008; reviewed in Cooke et al., 2016). Intraspecific variability in environmental tolerance has important implications for climate change mitigation. Individuals within a species vary considerably in environmental tolerance limits, differing across populations, ontogeny, sex, and size (e.g.,Eliason et al., 2011; Hinch et al., 2021). By carefully considering intraspecific variability, we can identify the individuals most resilient and most vulnerable to climate change (Chen et al., 2015; McKenzie et al., 2021; Zillig et al., 2021). To assess the adaptive capacity of fish to climate change, physiologists must continue to work with geneticists to link phenotype with genotype (Chen et al., 2018; Munoz et al., 2015) and assess the potential for transgenerational responses (Donelson et al., 2018).

8   A systems approach

A biological system connects several biologically relevant entities that span scales from molecules to cells to organs to individuals. A collection of individual organisms of the same species comprises a population, while interacting species form communities that are embedded within ecosystems. Conservation decision makers tend to focus on macro-level organization levels like populations and ecosystems (Cooke and O'Connor, 2010). Yet, biological processes that play out at the level of the molecule, gene, cell, and organ ultimately influence organismal fitness which is a driver of population-level processes and demography. Conservation physiology as a discipline is relevant across scales of biological organization (Cooke et al., 2013, 2014a). "A systems approach" implies that there are complex interactions within and external to biological systems which necessitates a holistic approach (Kitano, 2002), something foundational to conservation (Lister, 1998). Given that a systems approach allows one to understand interconnections and cause-effect relationships

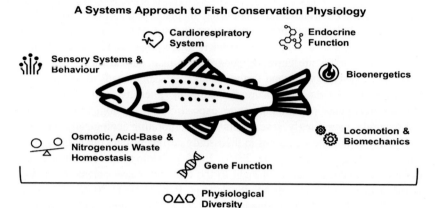

FIG. 4   Schematic illustrating a systems approach to fish conservation physiology.

(Calvert, 2013), it is well suited to addressing complex problems which is a characteristic of most conservation issues (Katagiri, 2003). Here, we briefly consider how different key biological systems (see Fig. 4) are relevant to various applied conservation and management issues.

From the perspective of a fish, a logical starting point is the sensory systems that allow them to sense and respond to changes in their environment. Sensory biology is underpinned by the nervous system and elicits whole organism behaviors (see Chapter 2, Volume 39A: Horodysky et al., 2022). The endocrine system serves to regulate metabolism, growth, tissue function, and reproduction (see Chapter 6, Volume 39A: Bernier and Alderman, 2022). Ensuring the organism maintains some level of homeostasis requires coordination of osmotic, acid-base, and nitrogenous waste systems (see Chapter 7, Volume 39A: Wood, 2022). Fish extract oxygen from the water and distribute it to tissues via the respiratory and circulatory systems (see Chapter 5, Volume 39A: Eliason et al., 2022). "Working" tissues, such as muscles needed for locomotion, are constrained by biomechanics (see Chapter 3, Volume 39A: Castro-Santos et al., 2022). Organisms also operate on simple mass-balance principles whereby bioenergetics determine growth and reproductive potential (see Chapter 4, Volume 39A: Brownscombe et al., 2022). Genes provide instructions for protein synthesis to support various activities including growth (see Chapter 8, Volume 39A: Jeffries et al., 2022). This gross oversimplification emphasizes the inherent interconnections. For example, if an organism is unable to locate prey (requires sensory system) or capture it (requires locomotion) then it lacks the energy needed to allocate resources to tissue repair, growth or reproduction (bioenergetics). The point in characterizing these interactions is to highlight that exploring how individual system components are relevant to conservation can be useful (see Table 1 and subsequent chapters in this volume) but also cannot be done without considering how the system components interact (Fig. 4).

TABLE 1 Examples of how different organismal systems and tools are relevant to identifying and solving conservation problems.

System and tools	Applications—problem Identification	Applications—development of Solutions
Sensory and behavior	– Understanding how climate change will mediate sensory physiology and behavior of fish (Cattano et al., 2018)	– Using knowledge of visual color acuity to identify LEDs for guiding fish away from water intakes (Elvidge et al., 2018) – Improvement of trap catches using pheromones of an invasive species (Johnson et al., 2015) – Development of bycatch deterrence tools for fish (Southwood et al., 2008)
Endocrine	– Quantifying the effects of different anthropogenic stressors on wild fish using glucocorticoids (Baker et al., 2013)	– Identifying thresholds for human disturbance (e.g., boat noise; Wysocki et al., 2006)
Cardiorespiratory	– Predicting how different fish populations will respond to climate change (Nilsson et al., 2009)	– Generating temperature thresholds to inform fisheries management planning (Farrell et al., 2008) – Refining catch-and-release practices for angled fish (Cooke et al., 2001)
Osmotic, acid-base and nitrogenous waste homeostasis	– Identifying the effects of different environmental conditions (e.g., hypoxia) on imperiled fish (Aboagy and Allen 2018)	– Identifying opportunities for restoring habitats for imperiled fish species (Komoroske et al., 2016)
Gene function	– Evaluating how immune function interacts with disease and climate change in wild fish (Miller et al., 2014)	– Identifying toxicant thresholds for imperiled fish species (Jeffries et al., 2015)

Continued

TABLE 1 Examples of how different organismal systems and tools are relevant to identifying and solving conservation problems.—Cont'd

System and tools	Applications—problem Identification	Applications—development of Solutions
Locomotion and biomechanics	– Identifying swimming capacity limits for different sizes and species of fish (Castro-Santos, 2004)	– Incorporation of biomechanics knowledge and principles into fish passage design (Castro-Santos and Haro, 2005) – Identifying optimal flows for migratory fish downstream of dams (Bett et al., 2020) – Refining fishing gear deployment to reduce bycatch using knowledge of swimming capacity (He, 1993)
Bioenergetics	– Development of bioenergetics models to assess the effects of different environmental conditions on fish (Petersen et al., 2008) – Understanding the role of energy depletion in the mortality of diadromous fish (Roscoe et al., 2010)	– Informing predator removal projects for to restore imperiled fish populations (Roby et al., 2003) – Informing stocking decisions for fisheries management (Negus, 1995)

9   On achieving a mechanistic approach to conservation and management

Physiological knowledge, concepts, and tools have long been used to inform fisheries management and conservation but this integration has become codified in a number of ways in the past decade (see Horodysky et al., 2015). In part this is due to the growing number of success stories (see some above; Madliger et al., 2016). Underpinning those successes are creative studies that bring together different knowledge domains (e.g., physiology, behavior, genomics), tools, and endpoints that involve different biological systems (e.g., swimming performance, gene function, osmotic state). Here we have identified some of the biological systems relevant to conservation and management with a focus on the organism, though most fisheries managers and conservation decision makers tend to focus on populations (Cooke and O'Connor, 2010). As such, there remains a disconnect between physiological alterations experienced by the individual organism, their fitness, and how that contributes to population-level processes. There are a growing number of examples that link these domains (see Bergman et al., 2019). For example, Farrell et al. (2008) used a lab and field study of migratory sockeye salmon to demonstrate how lab-derived thermal performance was predictive of the fate of wild salmon tracked with biotelemetry tags. More recently, Moyano et al. (2020) used laboratory-derived thermal tolerance data for juvenile Atlantic herring to explain trends in the population resilience of herring in the wild.

The real strength of physiology is its ability to identify mechanistic causes. Most conservation problems are first noticed based on changes in demography. For example, consider an instance where a fisheries manager observed trends in the decline of a fish population year over year. Without knowing the mechanistic basis for the decline, it is nearly impossible to establish the cause-and-effect needed to take evidence-informed actions to reverse the declines. Physiology, especially when used in an experimental context, can reveal the mechanisms behind declines (Coristine et al., 2014; Madliger et al., 2021a). It also allows one to test different management interventions to see what will work (Cooke et al., 2013). For example, there is a growing body of work focused on captive breeding of imperiled fish species. Successful captive breeding depends on not only keeping the fish alive but in a state where they grow and reproduce. Physiology can be used to assess different feed performance, identify optimal growth conditions, reveal reproductive biology (e.g., what is needed to induce maturation), and inform handling and stocking procedures (see Chapter 1, Volume 39B: Yanagitsuru et al., 2022). Comparative physiological experiments can be used to refine processes and thus increase the likelihood that a captive breeding program will be successful. This is but one example that highlights the value of both a physiological and a systems approach. Organisms depend on fish culturists to provide them with the conditions needed to emulate their life in the wild. That means thinking about their biology from all perspectives—not just from, for example, an endocrine-reproductive perspective.

Ultimately, the most valuable tools for measuring physiological performance and a large number of individuals will be those that can be used in a field setting. This may require some compromise in data quality (i.e., more measurement variability/error) that will need to be acknowledged. Regardless, laboratory studies must increasingly incorporate not only steady-state environmental conditions but also oscillating ones. Computer-controlled environmental chambers will go a long way to injecting environmental realism to the data fish physiologists generate, but high quality physiological studies that are field-based must are also desperately needed. In that way, physiologists, geneticists, and ecologists will be able to work more closely together in a brave new world and with new tools that measure things of importance to fish performance, on a larger number of individuals and using more realistic environmental conditions.

One of the other areas where physiology has been highly relevant to fisheries conservation and management is in terms of understanding their physiological diversity (see Chapter 9, Volume 39A: Schulte and Healy, 2022). Physiological diversity was explored widely from a fundamental comparative physiology perspective going back to the days of Prosser (1955). Yet today, we know that such variation is not just interesting but can be exploited to improve conservation (Spicer and Gaston, 1999). Work on Pacific salmon has revealed population-specific variation in thermal tolerance and migration performance that has been used to refine fisheries planning in years with warm water temperatures (Eliason et al., 2011). Additional work on salmon has revealed that female salmon die at much higher rates than male salmon when faced with various stressors during reproductive migration (Hinch et al., 2021). We also now know that there is a wide range of behavioral and physiological phenotypes within a population perhaps best exemplified by work on the shy-bold continuum (Sneddon, 2003) and pace of life syndromes (Montiglio et al., 2018). In a fisheries context it has been revealed that fisheries select for different phenotypes which can have negative consequences on fish populations (Hollins et al., 2018). Moreover, disease influences individuals and populations differently such that physiological understanding of the immune system may be useful for understanding demography given that disease is becoming a major factor in population dynamics (Hing et al., 2016). With climate change effects already being felt around the globe and our growing recognition of the role of humans as selective forces in fish populations, it can be assumed that there is much opportunity to harness our understanding of fish physiological diversity to benefit fish populations and ensure resilience.

Modelers will need to incorporate intraspecific variability into their models, but to do so fish physiologists will need to greatly increase the number of individuals they test. The evolution of pesticide resistance among insects, for example, informs of the importance of intraspecific variability at a low genetic frequency. Testing a handful of individuals may not be sufficient to reveal the individuals that might be resilient to future climates and conditions. The recent upsurge in measuring oxygen uptake in fishes is a good example of the much-needed trend to increase experimental throughput. Years ago, one fish was tested at a time, but today computer technology (control systems and data analysis) have enabled the multiplexing of 32 simultaneous oxygen uptake tests. Implantable heart rate/ECG loggers can similarly allow a higher throughput with a less intrusive technology, which is a valuable advance given the central importance of increasing heart rate during acute warming of fishes. Ultimately, higher numbers of individuals will reveal the full extent of intraspecific variation in traits such as upper thermal tolerance, so that modelers could for example work with geneticists to tell us how much we must slow the rate of global warming to allow the more tolerant individuals to radiate through the population, a Darwinian experiment in natural selection.

Acknowledgments

Cooke, Brauner, and Farrell are supported by NSERC. Fangue is supported by the California Agricultural Experimental Station of the University of California, Davis (CA-D-WFB-2098-H). We are grateful to Tara Lepine for her assistance with formatting the manuscript and Essie Rodgers and David McKenzie for providing thoughtful reviews of the content.

References

Allan, J.D., Abell, R., Hogan, Z.E.B., Revenga, C., Taylor, B.W., Welcomme, R.L., Winemiller, K., 2005. Overfishing of inland waters. Bioscience 55, 1041–1051.

Arthington, A.H., Dulvy, N.K., Gladstone, W., Winfield, I.J., 2016. Fish conservation in freshwater and marine realms: status, threats and management. Aquat. Conserv. 26, 838–857. https://doi.org/10.1002/aqc.2712.

Baker, M.R., Gobush, K.S., Vynne, C.H., 2013. Review of factors influencing stress hormones in fish and wildlife. J. Nat. Conserv. 21, 309–318.

Barbet-Massin, M., Rome, Q., Villemant, C., Courchamp, F., 2018. Can species distribution models really predict the expansion of invasive species? PLoS One 13, e0193085. https://doi.org/10.1371/journal.pone.0193085.

Barnett, A., Payne, N.L., Semmens, J.M., Fitzpatrick, R., 2016. Ecotourism increases the field metabolic rate of whitetip reef sharks. Biol. Conserv. 199, 132–136.

Bellard, C., Cassey, P., Blackburn, T.M., 2016. Alien species as a driver of recent extinctions. Biol. Lett. 12, 20150623. https://doi.org/10.1098/rsbl.2015.0623.

Bellard, C., Jeschke, J.M., Leroy, B., Mace, G.M., 2018. Insights from modeling studies on how climate change affects invasive alien species geography. Ecol. Evol. 8, 5688–5700. https://doi.org/10.1002/ece3.4098.

Bergman, J.N., Bennett, J.R., Binley, A.D., Cooke, S.J., Fyson, V., Hlina, B.L., Ried, C.H., Vala, M.A., Madliger, C.L., 2019. Scaling from individual physiological measures to population-level demographic change: case studies and future directions for conservation management. Biol. Conserv. 238, 108242. https://doi.org/10.1016/j.biocon.2019.108242.

Bernier, N.J., Alderman, S.L., 2022. Applied aspects of fish endocrinology. Fish Physiol. 39A, 253–320.

Bett, N.N., Hinch, S.G., Bass, A.L., Braun, D.C., Burnett, N.J., Casselman, M.T., Cooke, S.J., Drenner, S.M., Gelchu, A., Harrower, W.L., Ledoux, R., Lotto, A.G., Middleton, C.T., Minke-Martin, W., Patterson, D.A., Zhang, W., Zhu, D.Z., 2020. Using an integrative research approach to improve fish migrations in regulated rivers: a case study on Pacific Salmon in the Seton River, Canada. Hydrobiologia 849, 1–21. https://doi.org/10.1007/s10750-020-04371-2.

Brownscombe, J.W., Lawrence, M.J., Deslauriers, D., Filgueira, R., Boyd, R.J., Cooke, S.J., 2022. Applied fish bioenergetics. Fish Physiol. 39A, 141–188.

Brüning, A., Kloas, W., Preuer, T., Hölker, F., 2018. Influence of artificially induced light pollution on the hormone system of two common fish species, perch and roach, in a rural habitat. Conserv. Physiol. 6, coy016. https://doi.org/10.1093/conphys/coy016.

Bunt, C.M., Castro-Santos, T., Haro, A., 2012. Performance of fish passage structures at upstream barriers to migration. River Res. Appl. 28, 457–478. https://doi.org/10.1002/rra.1565.

Calow, P., Forbes, V.E., 1998. How do physiological responses to stress translate into ecological and evolutionary processes? Comp. Biochem. Physiol. A Mol. Integr. Physiol. 120, 11–16. https://doi.org/10.1016/S1095-6433(98)10003-X.

Calvert, J., 2013. Systems biology, big science and grand challenges. Biosocieties 8, 466–479. https://doi.org/10.1057/biosoc.2013.27.

Castro-Santos, T., 2004. Quantifying the combined effects of attempt rate and swimming capacity on passage through velocity barriers. Can. J. Fish. Aquat. Sci. 61, 1602–1615. https://doi.org/10.1139/f04-094.

Castro-Santos, T., Haro, A., 2005. Biomechanics and fisheries conservation. Fish Physiol. 23, 469–523.

Castro-Santos, T., Goerig, E., He, P., Lauder, G.V., 2022. Applied aspects of locomotion and biomechanics. Fish Physiol. 39A, 91–140.

Cattano, C., Claudet, J., Domenici, P., Milazzo, M., 2018. Living in a high CO2 world: a global meta-analysis shows multiple trait-mediated fish responses to ocean acidification. Ecol. Monogr. 88, 320–335.

Chen, Z., Narum, S.R., 2021. Whole genome resequencing reveals genomic regions associated with thermal adaptation in redband trout. Mol. Ecol. 30 (1), 162–174.

Chen, Z., Snow, M., Lawrence, C., Church, A., Narum, S., Devlin, R., Farrell, A.P., 2015. Selection for upper thermal tolerance in rainbow trout (Oncorhynchus mykiss Walbaum). J. Exp. Biol. 218, 803–812.

Chen, Z., Farrell, A.P., Matala, A., Narum, S.R., 2018. Mechanisms of thermal adaptation and evolutionary potential of conspecific populations to changing environments. Mol. Ecol. 27, 659–674. https://doi.org/10.1111/mec.14475.

Chown, S.L., 2012. Trait-based approaches to conservation physiology: forecasting environmental change risks from the bottom up. Philos. Trans. R. Soc. Lond. B Biol. Sci. 367, 1615–1627. https://doi.org/10.1098/rstb.2011.0422.

Clavero, M., García-Berthou, E., 2005. Invasive species are a leading cause of animal extinctions. Trends Ecol. Evol. 20, 110.

Cocherell, D.E., Kawabata, A., Kratville, D.W., Cocherell, S.A., Kaufman, R.C., Anderson, E.K., Chen, Z.Q., Bandeh, H., Rotondo, M.M., Padilla, R., Churchwell, R., Kavvas, M.L., Cech, J.J., Jr., 2011. Passage performance and physiological stress response of adult white sturgeon ascending a laboratory fishway. J. Appl. Ichthyol. 27, 327–334. https://doi.org/10.1111/j.1439-0426.2010.01650.x.

Connon, R.E., Jeffries, K.M., Komoroske, L.M., Todgham, A.E., Fangue, N.A., 2018. The utility of transcriptomics in fish conservation. J. Exp. Biol. 221, jeb148833.

Cooke, S.J., O'Connor, C.M., 2010. Making conservation physiology relevant to policy makers and conservation practitioners. Conserv. Lett. 3 (3), 159–166.

Cooke, S.J., Philipp, D.P., Dunmall, K.M., Schreer, J.F., 2001. The influence of terminal tackle on injury, handling time, and cardiac disturbance of rock bass. N. Am. J. Fish. Manag. 21, 333–342.

Cooke, S.J., Sack, L., Franklin, C.E., Farrell, A.P., Beardall, J., Wikelski, M., Chown, S.L., 2013. What is conservation physiology? Perspectives on an increasingly integrated and essential science. Conserv. Physiol. 1, cot001.

Cooke, S.J., Arlinghaus, R., Bartley, D.M., Beard, T.D., Cowx, I.G., Essington, T.E., Jensen, O.P., Taylor, W.W., Watson, R., 2014a. Where the waters meet: sharing ideas and experiences between inland and marine realms to promote sustainable fisheries management. Can. J. Fish. Aquat. Sci. 71, 1593–1601.

Cooke, S.J., Killen, S.S., Metcalfe, J.D., McKenzie, D.J., Mouillot, D., Jørgensen, C., Peck, M.A., 2014b. Conservation physiology across scales: insights from the marine realm. Conserv. Physiol. 2, cou024.

Cooke, S.J., Brownscombe, J.W., Raby, G.D., Broell, F., Hinch, S.G., Clark, T.D., Semmens, J.M., 2016. Remote bioenergetics measurements in wild fish: opportunities and challenges. Comp. Biochem. Physiol. A 202, 23–37.

Cooke, S.J., Madliger, C.L., Cramp, R.L., Beardall, J., Burness, G., Chown, S.L., Clark, T.D., Dantzer, B., de la Barrera, E., Fangue, N.A., Franklin, C.E., Fuller, A., Hawkes, L.A., Hultine, K.R., Hunt, K.E., Love, O.P., MacMillan, H.A., Mandelman, J.W., Mark, F.C., Martin, L.B., Mewman, A.E.M., Nicotra, A.B., Robinson, S.A., Ropert-Coudert, Y., Rummer, J.L., Seebacher, F., Todgham, A.E., 2020a. Reframing conservation physiology to be more inclusive, integrative, relevant and forward-looking: reflections and a horizon scan. Conserv. Physiol. 8, coaa016.

Cooke, S.J., Cech, J.J., Glassman, D.M., Simard, J., Louttit, S., Lennox, R.J., Cruz-Font, L., O'Connor, C.M., 2020b. Water resource development and sturgeon (Acipenseridae): state of the science and research gaps related to fish passage, entrainment, impingement and behavioural guidance. Rev. Fish Biol. Fish. 30, 219–244. https://doi.org/10.1007/s11160-020-09596-x.

Cooke, S.J., Madliger, C.L., Bergmann, J.N., Nguyen, V.M., Landsman, S.J., Love, O.P., Rummer, J.L., Franklin, C.E., 2021a. Optimism and opportunities for conservation physiology in the Anthropocene: a synthesis and conclusions. In: Madliger, C.L., Franklin, C.E., Love, O.P., Cooke, S.J. (Eds.), Conservation Physiology: Applications for Wildlife Conservation and Management. Oxford University Press, United Kingdom, pp. 319–329.

Cooke, S.J., Bergman, J.N., Madliger, C.L., Cramp, R.L., Beardall, J., Burness, G., Clark, T.D., Dantzer, B., de la Barrera, E., Fangue, N.A., Franklin, C.E., 2021b. One hundred research questions in conservation physiology for generating actionable evidence to inform conservation policy and practice. Conserv. Physiol. 9, coab009. https://doi.org/10.1093/conphys/coab009.

Coristine, L.E., Robillard, C.M., Kerr, J.T., O'Connor, C.M., Lapointe, D., Cooke, S.J., 2014. A conceptual framework for the emerging discipline of conservation physiology. Conserv. Physiol. 2, cou033. https://doi.org/10.1093/conphys/cou033.

Cox, K., Brennan, L.P., Gerwing, T.G., Dudas, S.E., Juanes, F., 2018. Sound the alarm: a meta-analysis on the effect of aquatic noise on fish behavior and physiology. Glob. Chang. Biol. 24, 3105–3116.

Crain, C.M., Kroeker, K., Halpern, B.S., 2008. Interactive and cumulative effects of multiple human stressors in marine systems. Ecol. Lett. 11, 1304–1315.

Crutzen, P.J., 2006. The "anthropocene". In: Earth System Science in the Anthropocene. Springer, Berlin, Heidelberg, pp. 13–18.

Dalby, S., 2016. Framing the Anthropocene: the good, the bad and the ugly. Anthr. Rev. 3, 33–51.

Davies, T.W., Smyth, T., 2018. Why artificial light at night should be a focus for global change research in the 21st century. Glob. Chang. Biol. 24, 872–882.

Davis, M.W., 2002. Key principles for understanding fish bycatch discard mortality. Can. J. Fish. Aquat. Sci. 59, 1834–1843.

Department of the Environment and Energy, Australian Government, 2020. National Light Pollution Guidelines for Wildlife Including Marine Turtles, Seabirds and Migratory Shorebirds. https://www.environment.gov.au/system/files/resources/2eb379de-931b-4547-8bcc-f96c73065f54/files/national-light-pollution-guidelines-wildlife.pdf. (accessed 23 April 2021).

Donelson, J.M., Salinas, S., Munday, P.L., Shama, L.N., 2018. Transgenerational plasticity and climate change experiments: where do we go from here? Glob. Chang. Biol. 24 (1), 13–34.

Eliason, E.J., Clark, T.D., Hague, M.J., Hanson, L.M., Gallagher, Z.S., Jeffries, K.M., et al., 2011. Differences in thermal tolerance among sockeye salmon populations. Science 332 (6025), 109–112.

Eliason, E.J., Van Wert, J.C., Schwieterman, G.D., 2022. Applied aspects of the cardiorespiratory system. Fish Physiol. 39A, 189–252.

Elmer, L.K., Madliger, C.L., Blumstein, D.T., Elvidge, C.K., Fernández-Juricic, E., Horodysky, A.Z., Johnson, N.S., McGuire, L.P., Swaisgood, R.R., Cooke, S.J., 2021. Exploiting common senses: sensory ecology meets wildlife conservation and management. Conserv. Physiol. 9, coab002.

Elvidge, C.K., Ford, M.I., Pratt, T.C., Smokorowski, K.E., Sills, M., Patrick, P.H., Cooke, S.J., 2018. Behavioural guidance of yellow-stage American eel Anguilla rostrata with a light-emitting diode device. Endanger. Species Res. 35, 159–168.

Falcón, J., Torriglia, A., Attia, D., Viénot, F., Gronfier, C., Behar-Cohen, F., Martinsons, C., Hicks, D., 2020. Exposure to artificial light at night and the consequences for flora, fauna and ecosystems. Front. Neurosci. 14, 1183.

Farrell, A.P., Gallaugher, P.E., Fraser, J., Pike, D., Bowering, P., Hadwin, A.K., Parkhouse, W., Routledge, R., 2001. Successful recovery of the physiological status of coho salmon on board a commercial gillnet vessel by means of a newly designed revival box. Can. J. Fish. Aquat. Sci. 58, 1932–1946.

Farrell, A.P., Hinch, S.G., Cooke, S.J., Patterson, D.A., Crossin, G.T., Lapointe, M., Mathes, M.T., 2008. Pacific salmon in hot water: applying aerobic scope models and biotelemetry to predict the success of spawning migrations. Physiol. Biochem. Zool. 81, 697–708.

Feist, G.W., Webb, M.A., Gundersen, D.T., Foster, E.P., Schreck, C.B., Maule, A.G., Fitzpatrick, M.S., 2005. Evidence of detrimental effects of environmental contaminants on growth and reproductive physiology of white sturgeon in impounded areas of the Columbia River. Environ. Health Perspect. 113, 1675–1682.

Folt, C.L., Chen, C.Y., Moore, M.V., Burnaford, J., 1999. Synergism and antagonism among multiple stressors. Limnol. Oceanogr. 44, 864–877.

Fry, F.E.J., 1971. The effect of environmental factors on the physiology of fish. In: Hoar, W.S., Randall, D.J. (Eds.), Fish Physiology. vol. 6. Academic Press, New York, pp. 1–98.

Gluth, G., Hanke, W., 1985. A comparison of physiological changes in carp, Cyprinus carpio, induced by several pollutants at sublethal concentrations: I. The dependency on exposure time. Ecotoxicol. Environ. Saf. 9, 179–188.

Green, S.J., Akins, J.L., Maljković, A., Côté, I.M., 2012. Invasive lionfish drive Atlantic coral reef fish declines. PLoS One 7, e32596. https://doi.org/10.1371/journal.pone.0032596.

Greenwood, M.F.D., 2008. Fish mortality by impingement on the cooling-water intake screens of Britain's largest direct-cooled power station. Mar. Pollut. Bull. 56, 723–739. https://doi.org/10.1016/j.marpolbul.2007.12.008.

Haro, A., Castro-Santos, T., Noreika, J., Odeh, M., 2004. Swimming performance of upstream migrant fishes in open-channel flow: a new approach to predicting passage through velocity barriers. Can. J. Fish. Aquat. Sci. 61, 1590–1601. https://doi.org/10.1139/f04-093.

He, P., 1993. Swimming speeds of marine fish in relation to fishing gears. ICES Mar. Sci. Symp. 196, 183–189.

Heath, A.G., 1995. Water Pollution and Fish Physiology, second ed. CRC press, Boca Raton.

Helfman, G., Collette, B.B., Facey, D.E., Bowen, B.W., 2009. The Diversity of Fishes: Biology, Evolution and Ecology, second ed. Wiley-Blackwell.

Hermoso, V., Clavero, M., Blanco-Garrido, F., Prenda, J., 2011. Invasive species and habitat degradation in Iberian streams: an analysis of their role in freshwater fish diversity loss. Ecol. Appl. 21, 175–188. https://doi.org/10.1890/09-2011.1.

Heublein, J.C., Kelly, J.T., Crocker, C.E., Klimley, A.P., 2009. Migration of green sturgeon, Acipenser medirostris, in the Sacramento River. Environ. Biol. Fishes 84, 245–258. https://doi.org/10.1007/s10641-008-9432-9.

Hilborn, R., 2007. Moving to sustainability by learning from successful fisheries. Ambio 36, 296–303.

Hinch, S.G., Bett, N.N., Eliason, E.J., Farrell, A.P., Cooke, S.J., Patterson, D.A., 2021. Exceptionally high mortality of adult female salmon: a large-scale pattern and a conservation concern. Can. J. Fish. Aquat. Sci. 78 (6), 639–654.

Hing, S., Narayan, E.J., Thompson, R.A., Godfrey, S.S., 2016. The relationship between physiological stress and wildlife disease: consequences for health and conservation. Wildl. Res. 43 (1), 51–60.

Hollins, J., Thambithurai, D., Koeck, B., Crespel, A., Bailey, D.M., Cooke, S.J., Lindström, J., Parsons, K.J., Killen, S.S., 2018. A physiological perspective on fisheries-induced evolution. Evol. Appl. 11, 561–576.

Hollowed, A.B., Barange, M., Beamish, R.J., Brander, K., Cochrane, K., Drinkwater, K., Foreman, M.G.G., Hare, J.A., Holt, J., Ito, S., Kim, S., King, J.R., Loeng, H., MacKenzie, B.R., Mueter, F.J., Okey, T.A., Peck, M.A., Radchenko, V.I., Rice, J.C., Schirripa, M.J., Yatsu, A., Yamanaka, Y., 2013. Projected impacts of climate change on marine fish and fisheries. ICES J. Mar. Sci. 70, 1023–1037.

Hook, S.E., Mondon, J., Revill, A.T., Greenfield, P.A., Stephenson, S.A., Strzelecki, J., Corbett, P., Armstrong, E., Song, J., Doan, H., Barrett, S., 2018. Monitoring sublethal changes in fish physiology following exposure to a light, unweathered crude oil. Aquat. Toxicol. 204, 27–45.

Horodysky, A.Z., Cooke, S.J., Brill, R.W., 2015. Physiology in the service of fisheries science: why thinking mechanistically matters. Rev. Fish Biol. Fish. 25, 425–447.

Horodysky, A.Z., Schweitzer, C.C., Brill, R.W., 2022. Applied sensory physiology and behavior. Fish Physiol. 39A, 33–90.

Hutchings, J.A., Reynolds, J.D., 2004. Marine fish population collapses: consequences for recovery and extinction risk. Bioscience 54, 297–309.

Illing, B., Rummer, J.L., 2017. Physiology can contribute to better understanding, management, and conservation of coral reef fishes. Conserv. Physiol. 5, cox005.

Jackson, J.B., Kirby, M.X., Berger, W.H., Bjorndal, K.A., Botsford, L.W., Bourque, B.J., Bradbury, R.H., Cooke, R., Erlandson, J., Estes, J.A., Hughes, T.P., Kidwell, S., Lange, C.B.,

Lenihan, H.S., Pandolfi, J.M., Peterson, C.H., Steneck, R.S., Tegner, M.J., Warner, R.R., 2001. Historical overfishing and the recent collapse of coastal ecosystems. Science 293, 629–637.

Jackson, M.C., Loewen, C.J., Vinebrooke, R.D., Chimimba, C.T., 2016. Net effects of multiple stressors in freshwater ecosystems: a meta-analysis. Glob. Chang. Biol. 22, 180–189.

Jager, H.I., Parsley, M.J., Cech, J.J., Jr., McLaughlin, R.L., Forsythe, P.S., Elliott, R.F., Pracheil, B.M., 2016. Reconnecting fragmented sturgeon populations in North American rivers. Fisheries 41, 140–148. https://doi.org/10.1080/03632415.2015.1132705.

Jeffrey, J.D., Hasler, C.T., Chapman, J.M., Cooke, S.J., Suski, C.D., 2015. Linking landscape-scale disturbances to stress and condition of fish: implications for restoration and conservation. Integr. Comp. Biol. 55, 618–630.

Jeffries, K.M., Komoroske, L.M., Truong, J., Werner, I., Hasenbein, M., Hasenbein, S., Fangue, N.A., Connon, R.E., 2015. The transcriptome-wide effects of exposure to a pyrethroid pesticide on the critically endangered delta smelt Hypomesus transpacificus. Endanger. Species Res. 28, 43–60.

Jeffries, K.M., Jeffrey, J.D., Holland, E.B., 2022. Applied aspects of gene function for the conservation of fishes. Fish Physiol. 39A, 389–433.

Jelks, H.L., Walsh, S.J., Burkhead, N.M., Contreras-Balderas, S., Díaz-Pardo, E., Hendrickson, D.A., Lyons, J., Mandrak, N.E., McCormick, F., Nelson, J.S., Platania, S.P., Porter, B.A., Renaud, C.B., Schmitter-Soto, J.J., Taylor, E.B., Warren, M.L., 2008. Conservation status of imperiled North American freshwater and diadromous fishes. Fisheries 33, 372–407.

Johnson, N.S., Tix, J.A., Hlina, B.L., Wagner, C.M., Siefkes, M.J., Wang, H., Li, W., 2015. A sea lamprey (Petromyzon marinus) sex pheromone mixture increases trap catch relative to a single synthesized component in specific environments. J. Chem. Ecol. 41, 311–321.

Jordan, L.K., Mandelman, J.W., McComb, D.M., Fordham, S.V., Carlson, J.K., Werner, T.B., 2013. Linking sensory biology and fisheries bycatch reduction in elasmobranch fishes: a review with new directions for research. Conserv. Physiol. 1, cot002.

Katagiri, F., 2003. Attacking complex problems with the power of systems biology. Plant Physiol. 132, 417–419.

Kearney, M., Porter, W.P., Williams, C., Ritchie, S., Hoffmann, A.A., 2009. Integrating biophysical models and evolutionary theory to predict climatic impacts on species' ranges: the dengue mosquito Aedes aegypti in Australia. Funct. Ecol. 23, 528–538. https://doi.org/10.1111/j.1365-2435.2008.01538.x.

Keefer, M.L., Peery, C.A., Bjornn, T.C., Jepson, M.A., Stuehrenberg, L.C., 2004. Hydrosystem, dam, and reservoir passage rates of adult Chinook salmon and steelhead in the Columbia and Snake rivers. Trans. Am. Fish. Soc. 133, 1413–1439. https://doi.org/10.1577/T03-223.1.

Kitano, H., 2002. Systems biology: a brief overview. Science 295, 1662–1664.

Kjelland, M.E., Woodley, C.M., Swannack, T.M., Smith, D.L., 2015. A review of the potential effects of suspended sediment on fishes: potential dredging-related physiological, behavioral, and transgenerational implications. Environ. Syst. Decis. 35, 334–350.

Komoroske, L.M., Jeffries, K.M., Connon, R.E., Dexter, J., Hasenbein, M., Verhille, C., Fangue, N.A., 2016. Sublethal salinity stress contributes to habitat limitation in an endangered estuarine fish. Evol. Appl. 9, 963–981.

Lenhardt, M., Jaric, I., Kalauzi, A., Cvijanovic, G., 2006. Assessment of extinction risk and reasons for decline in sturgeon. Biodivers. Conserv. 15, 1967–1976. https://doi.org/10.1007/s10531-005-4317-0.

Lennox, R., Choi, K., Harrison, P.M., Paterson, J.E., Peat, T.B., Ward, T.D., Cooke, S.J., 2015. Improving science-based invasive species management with physiological knowledge, concepts, and tools. Biol. Invasions 17, 2213–2227. https://doi.org/10.1007/s10530-015-0884-5.

Lewis, S.L., Maslin, M.A., 2015. Defining the anthropocene. Nature 519, 171–180.

Light, T., Marchetti, M.P., 2007. Distinguishing between invasions and habitat changes as drivers of diversity loss among California's freshwater fishes. Conserv. Biol. 21, 434–446. https://doi.org/10.1111/j.1523-1739.2006.00643.x.

Lin, H.Y., Cooke, S.J., Wolter, C., Young, N., Bennett, J.R., 2020. On the conservation value of historic canals for aquatic ecosystems. Biol. Conserv. 251, 108764.

Lister, N.M.E., 1998. A systems approach to biodiversity conservation planning. Environ. Monit. Assess. 49, 123–155.

Madliger, C.L., Cooke, S.J., Crespi, E.J., Funk, J.L., Hultine, K.R., Hunt, K.E., Rohr, J.R., Sinclair, B.J., Suski, C.D., Willis, C.K., Love, O.P., 2016. Success stories and emerging themes in conservation physiology. Conserv. Physiol. 4, cov057. https://doi.org/10.1093/conphys/cov057.

Madliger, C.L., Love, O.P., Hultine, K.R., Cooke, S.J., 2018. The conservation physiology toolbox: status and opportunities. Conserv. Physiol. 6, coy029.

Madliger, C.L., Love, O.P., Nguyen, V.M., Haddaway, N.R., Cooke, S.J., 2021a. Researcher perspectives on challenges and opportunities in conservation physiology revealed from an online survey. Conserv. Physiol. 9, 1–15.

Madliger, C.L., Franklin, C.E., Love, O.P., Cooke, S.J. (Eds.), 2021b. Conservation Physiology: Applications for Wildlife Conservation and Management. Oxford University Press, USA.

Mahoney, J.L., Klug, P.E., Reed, W.L., 2018. An assessment of the US endangered species act recovery plans: using physiology to support conservation. Conserv. Physiol. 6, coy036.

Mainka, S.A., Howard, G.W., 2010. Climate change and invasive species: double jeopardy. Integr. Zool. 5, 102–111. https://doi.org/10.1111/j.1749-4877.2010.00193.x.

Marschall, E.A., Mather, M.E., Parrish, D.L., Allison, G.W., McMenemy, J.R., 2011. Migration delays caused by anthropogenic barriers: modeling dams, temperature, and success of migrating salmon smolts. Ecol. Appl. 21, 3014–3031. https://doi.org/10.1890/10-0593.1.

Martin, G.R., Crawford, R., 2015. Reducing bycatch in gillnets: a sensory ecology perspective. Glob. Ecol. Conserv. 3, 28–50.

McKenzie, D.J., Garofalo, E., Winter, M.J., Ceradini, S., Verweij, F., Day, N., et al., 2007. Complex physiological traits as biomarkers of the sub-lethal toxicological effects of pollutant exposure in fishes. Philos. Trans. R. Soc. B 362 (1487), 2043–2059.

McKenzie, D.J., Zhang, Y., Eliason, E.J., Schulte, P.M., Nati, J.J.H., Blasco, F.R., Claireaux, G., Farrell, A.P., 2021. Intraspecific variation in tolerance of warming in fishes. J. Fish Biol.

Metcalfe, J.D., Le Quesne, W.J.F., Cheung, W.W.L., Righton, D.A., 2012. Conservation physiology for applied management of marine fish: an overview with perspectives on the role and value of telemetry. Philos. Trans. R. Soc. Lond. B Biol. Sci. 367, 1746–1756.

Mickle, M.F., Higgs, D.M., 2018. Integrating techniques: a review of the effects of anthropogenic noise on freshwater fish. Can. J. Fish. Aquat. Sci. 75, 1534–1541.

Miller, R.R., Williams, J.D., Williams, J.E., 1989. Extinctions of north American fishes during the past century. Fisheries 14, 22–38.

Miller, K.M., Teffer, A., Tucker, S., Li, S., Schulze, A.D., Trudel, M., Juanes, F., Tabata, A., Kaukinen, K.H., Ginther, N.G., Ming, T.J., Cooke, S.J., Hipfner, J.M., Patterson, D.A., Hinch, S.G., 2014. Infectious disease, shifting climates, and opportunistic predators: cumulative factors potentially impacting wild salmon declines. Evol. Appl. 7, 812–855.

Mochnacz, N.J., Kissinger, B.C., Deslauriers, D., Guzzo, M.M., Enders, E.C., Anderson, W.G., et al., 2017. Development and testing of a simple field-based intermittent-flow respirometry system for riverine fishes. Conserv. Physiol. 5 (1).

Mollot, G., Pantel, J.H., Romanuk, T.N., 2017. The effects of invasive species on the decline in species richness: a global meta-analysis. Adv. Ecol. Res. 56, 61–83. https://doi.org/10.1016/bs.aecr.2016.10.002.

Montiglio, P.O., Dammhahn, M., Messier, G.D., Réale, D., 2018. The pace-of-life syndrome revisited: the role of ecological conditions and natural history on the slow-fast continuum. Behav. Ecol. Sociobiol. 72, 1–9.

Morash, A.J., Neufeld, C., MacCormack, T.J., Currie, S., 2018. The importance of incorporating natural thermal variation when evaluating physiological performance in wild species. J. Exp. Biol. 221 (14), jeb164673.

Morrison, B.P., 2019. Chronology of Lake Ontario ecosystem and fisheries. Aquat. Ecosyst. Health Manag. 22, 294–304.

Moser, M.L., Jackson, A.D., Lucas, M.C., Mueller, R.P., 2015. Behavior and potential threats to survival of migrating lamprey ammocoetes and macrophthalmia. Rev. Fish Biol. Fish. 25, 103–116. https://doi.org/10.1007/s11160-014-9372-8.

Moyano, M., Illing, B., Polte, P., Kotterba, P., Zablotski, Y., Gröhsler, T., Hüdepohl, P., Cooke, S.J., Peck, M.A., 2020. Linking individual physiological indicators to the productivity of fish populations: a case study of Atlantic herring. Ecol. Indic. 113, 106146.

Moyle, P.B., van Dyk, C., Tomelleri, J., 2002. Inland Fishes of California. University of California Press.

Moyle, P.B., Lusardi, R., Samuel, P., 2017. State of the Salmonids II: Fish in Hot Water. California Trout, Inc.

Munoz, N.J., Farrell, A.P., Heath, J.W., Neff, B.D., 2015. Adaptive potential of a Pacific salmon challenged by climate change. Nat. Clim. Change 5, 163–166.

Mussen, T.D., Cech Jr., J.J., 2019. Assessing the use of vibrations and strobe lights at fish screens as enhanced deterrents for two estuarine fishes. J. Fish Biol. 95, 238–246. https://doi.org/10.1111/jfb.13776.

Mussen, T.D., Cocherell, D., Poletto, J.B., Reardon, J.S., Hockett, Z., Ercan, A., Bandeh, H., Kavvas, M.L., Cech Jr., J.J., Fangue, N.A., 2014. Unscreened water-diversion pipes pose an entrainment risk to the threatened green sturgeon, *Acipenser medirostris*. PLoS One 9, e86321. https://doi.org/10.1371/journal.pone.0086321.

Myers, R.A., Hutchings, J.A., Barrowman, N.J., 1997. Why do fish stocks collapse? The example of cod in Atlantic Canada. Ecol. Appl. 7, 91–106.

Myers, B.J., Lynch, A.J., Bunnell, D.B., Chu, C., Falke, J.A., Kovach, R.P., Krabbenhoft, T.J., Kwak, T.J., Paukert, C.P., 2017. Global synthesis of the documented and projected effects of climate change on inland fishes. Rev. Fish Biol. Fish. 27, 339–361.

Negus, M.T., 1995. Bioenergetics modeling as a salmonine management tool applied to Minnesota waters of Lake Superior. N. Am. J. Fish. Manag. 15, 60–78.

Nilsson, G.E., Crawley, N., Lunde, I.G., Munday, P.L., 2009. Elevated temperature reduces the respiratory scope of coral reef fishes. Glob. Chang. Biol. 15, 1405–1412.

Norrgård, J.R., Greenberg, L.A., Piccolo, J.J., Schmitz, M., Bergman, E., 2013. Multiplicative loss of landlocked Atlantic salmon Salmo salar L. smolts during downstream migration through multiple dams. River Res. Appl. 29, 1306–1317. https://doi.org/10.1002/rra.2616.

Pegg, M.A., Chick, J.H., 2004. Aquatic Nuisance Species: An Evaluation of Barriers for Preventing the Spread of Bighead and Silver Carp to the Great Lakes. Final Report for the Illinois-Indiana Sea Grant A/SE (ANS)-01–01, Illinois-Indiana Sea Grant, Urbana, IL.

Pérez-Ruzafa, A., Pérez-Marcos, M., Marcos, C., 2018. From fish physiology to ecosystems management: keys for moving through biological levels of organization in detecting environmental changes and anticipate their consequences. Ecol. Indic. 90, 334–345.

Petersen, J.H., DeAngelis, D.L., Paukert, C.P., 2008. An overview of methods for developing bioenergetic and life history models for rare and endangered species. Trans. Am. Fish. Soc. 137, 244–253.

Phelps, Q.E., Tripp, S.J., Bales, K.R., James, D., Hrabik, R.A., Herzog, D.P., 2017. Incorporating basic and applied approaches to evaluate the effects of invasive Asian carp on native fishes: a necessary first step for integrated pest management. PLoS One 12, e0184081. https://doi.org/10.1371/journal.pone.0184081.

Popper, A.N., Hawkins, A.D., 2019. An overview of fish bioacoustics and the impacts of anthropogenic sounds on fishes. J. Fish Biol. 94, 692–713.

Pörtner, H.O., Farrell, A.P., 2008. Physiology and climate change. Science 322, 690–692.

Prosser, C.L., 1955. Physiological variation in animals. Biol. Rev. 30, 229–261.

Pyšek, P., Hulme, P.E., Simberloff, D., Bacher, S., Blackburn, T.M., Carlton, J.T., Dawson, W., Essl, F., Foxcroft, L.C., Genovesi, P., Jeschke, J.M., 2020. Scientists' warning on invasive alien species. Biol. Rev. 95, 1511–1534. https://doi.org/10.1111/brv.12627.

Raby, G.D., Donaldson, M.R., Hinch, S.G., Clark, T.D., Eliason, E.J., Jeffries, K.M., Cook, K.V., Teffer, A., Bass, A.L., Miller, K.M., Patterson, D.A., Farrell, A.P., Cooke, S.J., 2015. Fishing for effective conservation: context and biotic variation are keys to understanding the survival of Pacific salmon after catch-and-release. Integr. Comp. Biol. 55, 554–576.

Rahel, F.J., McLaughlin, R.L., 2018. Selective fragmentation and the management of fish movement across anthropogenic barriers. Ecol. Appl. 28, 2066–2081. https://doi.org/10.1002/eap.1795.

Rahel, F.J., Olden, J.D., 2008. Assessing the effects of climate change on aquatic invasive species. Conserv. Biol. 22, 521–533. https://doi.org/10.1111/j.1523-1739.2008.00950.x.

Raymond, H.L., 1979. Effects of dams and impoundments on migrations of juvenile Chinook salmon and steelhead from the Snake River, 1966 to 1975. Trans. Am. Fish. Soc. 108, 505–529. https://doi.org/10.1577/1548-8659(1979)108<505:EODAIO>2.0.CO;2.

Reid, A.J., Carlson, A.K., Creed, I.F., Eliason, E.J., Gell, P.A., Johnson, P.T., et al., 2019. Emerging threats and persistent conservation challenges for freshwater biodiversity. Biol. Rev. 94 (3), 849–873.

Reynolds, J.D., Dulvy, N.K., Goodwin, N.B., Hutchings, J.A., 2005a. Biology of extinction risk in marine fishes. Proc. R. Soc. B Biol. Sci. 272, 2337–2344.

Reynolds, J.D., Webb, T.J., Hawkins, L.A., 2005b. Life history and ecological correlates of extinction risk in European freshwater fishes. Can. J. Fish. Aquat. Sci. 62, 854–862.

Rezania, S., Park, J., Din, M.F.M., Taib, S.M., Talaiekhozani, A., Yadav, K.K., Kamyab, H., 2018. Microplastics pollution in different aquatic environments and biota: a review of recent studies. Mar. Pollut. Bull. 133, 191–208.

Roby, D.D., Lyons, D.E., Craig, D.P., Collis, K., Visser, G.H., 2003. Quantifying the effect of predators on endangered species using a bioenergetics approach: Caspian terns and juvenile salmonids in the Columbia River estuary. Can. J. Zool. 81, 250–265.

Rodgers, E.M., Poletto, J.B., Gomez Isaza, D.F., Van Eenennaam, J.P., Connon, R.E., Todgham, A.E., Seesholtz, A., Heublein, J.C., Cech Jr., J.J., Kelly, J.T., Fangue, N.A., 2019. Integrating physiological data with the conservation and management of fishes: a meta-analytical review using the threatened green sturgeon (Acipenser medirostris). Conserv. Physiol. 7, coz035.

Roessig, J.M., Woodley, C.M., Cech, J.J., Hansen, L.J., 2004. Effects of global climate change on marine and estuarine fishes and fisheries. Rev. Fish Biol. Fish. 14 (2), 251–275.

Roscoe, D.W., Hinch, S.G., Cooke, S.J., Patterson, D.A., 2010. Behaviour and thermal experience of adult sockeye salmon migrating through stratified lakes near spawning grounds: the roles of reproductive and energetic states. Ecol. Freshw. Fish 19, 51–62.

Rummer, J.L., Couturier, C.S., Stecyk, J.A., Gardiner, N.M., Kinch, J.P., Nilsson, G.E., Munday, P.L., 2014. Life on the edge: thermal optima for aerobic scope of equatorial reef fishes are close to current day temperatures. Glob. Chang. Biol. 20 (4), 1055–1066.

Rytwinski, T., Taylor, J.J., Donaldson, L.A., Britton, J.R., Browne, D.R., Gresswell, R.E., et al., 2019. The effectiveness of non-native fish removal techniques in freshwater ecosystems: a systematic review. Environ. Rev. 27 (1), 71–94.

Schreck, C.B., Scanlon, P.F., 1977. Endocrinology in fisheries and wildlife: biology and management. Fisheries 2, 20–30.

Schulte, P.M., Healy, M., 2022. Physiological diversity and its importance for fish conservation and management in the Anthropocene. Fish Physiol. 39A, 435–477.

Schultz, C., 2010. Challenges in connecting cumulative effects analysis to effective wildlife conservation planning. Bioscience 60, 545–551.

Scott, G.R., Sloman, K.A., 2004. The effects of environmental pollutants on complex fish behaviour: integrating behavioural and physiological indicators of toxicity. Aquat. Toxicol. 68, 369–392.

Shirey, P.D., Roulson, L.H., Bigford, T.E., 2018. Imperiled species policy is a critical issue for AFS. Fisheries 43, 527–532.

Siefkes, M.J., 2017. Use of physiological knowledge to control the invasive sea lamprey (Petromyzon marinus) in the Laurentian Great Lakes. Conserv. Physiol. 5, cox031. https://doi.org/10.1093/conphys/cox031.

Smith, B.D., Zeder, M.A., 2013. The onset of the Anthropocene. Anthropocene 4, 8–13.

Sneddon, L.U., 2003. The bold and the shy: individual differences in rainbow trout. J. Fish Biol. 62, 971–975.

Southwood, A., Fritsches, K., Brill, R., Swimmer, Y., 2008. Sound, chemical, and light detection in sea turtles and pelagic fishes: sensory-based approaches to bycatch reduction in longline fisheries. Endanger. Species Res. 5, 225–238.

Spicer, J., Gaston, K., 1999. Physiological Diversity: Ecological Implications. Wiley-Blackwell.

Steffen, W., Broadgate, W., Deutsch, L., Gaffney, O., Ludwig, C., 2015. The trajectory of the Anthropocene: the great acceleration. Anthr. Rev. 2, 81–98.

Swanson, C., Young, P.S., Cech Jr., J.J., 2005. Close encounters with a fish screen: integrating physiological and behavioral results to protect endangered species in exploited ecosystems. Trans. Am. Fish. Soc. 134, 1111–1123. https://doi.org/10.1577/T04-121.1.

Tickner, D., Opperman, J.J., Abell, R., Acreman, M., Arthington, A.H., Bunn, S.E., Cooke, S.J., Dalton, J., Darwall, W., Edwards, G., Harrison, I., 2020. Bending the curve of global freshwater biodiversity loss: an emergency recovery plan. Bioscience 70, 330–342. https://doi.org/10.1093/biosci/biaa002.

Tompkins, D.M., Carver, S., Jones, M.E., Krkošek, M., Skerratt, L.F., 2015. Emerging infectious diseases of wildlife: a critical perspective. Trends Parasitol. 31, 149–159.

Vitousek, P.M., D'Antonio, C.M., Loope, L.L., Westbrooks, R., 1996. Biological invasions as global environmental change. Am. Sci. 84, 468.

Whitney, J.E., Al-Chokhachy, R., Bunnell, D.B., Caldwell, C.A., Cooke, S.J., Eliason, E.J., Rogers, M.W., Lynch, A.J., Paukert, C.P., 2016. Physiological basis of climate change impacts on North American inland fishes. Fisheries 41, 332–345.

Wikelski, M., Cooke, S.J., 2006. Conservation physiology. Trends Ecol. Evol. 21, 38–46.

Wood, C.M., 2022. Conservation aspects of osmotic, acid-base, and nitrogen homeostasis in fish. Fish Physiol. 39A, 321–388.

Wysocki, L.E., Dittami, J.P., Ladich, F., 2006. Ship noise and cortisol secretion in European freshwater fishes. Biol. Conserv. 128, 501–508.

Yanagitsuru, Y.R., Davis, B.E., Baerwald, M.R., Sommer, T.R., Fangue, N.A., 2022. Using physiology to recover imperiled smelt species. Fish Physiol. 39B (In press).

Young, P.S., Swanson, C., Cech Jr., J.J., 2010. Close encounters with a fish screen III: behavior, performance, physiological stress responses, and recovery of adult delta smelt exposed to two-vector flows near a fish screen. Trans. Am. Fish. Soc. 139, 713–726. https://doi.org/10.1577/T09-029.1.

Zillig, K.W., Lusardi, R.A., Moyle, P.B., Fangue, N.A., 2021. One size does not fit all: variation in thermal eco-physiology among Pacific salmonids. Rev. Fish Biol. Fish. 31 (1), 95–114.

Chapter 2

Applied sensory physiology and behavior

Andrij Z. Horodysky[a,*], Cara C. Schweitzer[a], and Richard W. Brill[b]

aDepartment of Marine and Environmental Science, Hampton University, Hampton, VA, United States
bDepartment of Fisheries Science, Virginia Institute of Marine Science, William & Mary, Gloucester Point, VA, United States
*Corresponding author: e-mail: andrij.horodysky@hamptonu.edu

Chapter Outline

All fish behaviors have their basis in the receipt and processing of sensory information. Finely tuned receptors for and responses to sensory stimuli evolved over millennia, but since the Industrial Revolution, the cuescape of stimuli available to fishes is now changing at a pace faster than the evolution of sensory systems. We therefore posit that sustainable fisheries management requires understanding and predicting the effects of anthropogenic changes on fish populations via: knowing which stimuli are available to fish sensory systems, how the stimuli interact with the morphological structures of relevant neurosensory organs, how physiological performance of a specific neurosensory system transduces the stimuli into actionable information, and most importantly, how all of this is changing in the Anthropocene. Conservation physiologists have successfully applied sensory information and a host of technologies to alternately attract and deter fishes as needed, reduce bycatch, control invasive species, and improve aquaculture. This chapter briefly summarizes the biotic and abiotic stimuli available to

Fish Physiology, Vol. 39A. https://doi.org/10.1016/bs.fp.2022.04.002

fishes, elucidates how fish sensory systems transduce relevant cues from the environment into actionable information, and demonstrates how sensory knowledge has been and can be used to address applied issues in fisheries management. Lastly, this chapter closes with a synthesis of available information to identify a framework for successful applications of sensory-based strategies and to suggest promising new directions for future research that optimize the utility of fish sensory systems as a management tool.

1   Introduction

Aquatic environments produce complex mixtures of physicochemical cues that must be interpreted by fishes, with relevant signals isolated and distinguished from background noise (Kingsford et al., 2002). Individual fishes transduce the abiotic and biotic stimuli into information upon which behavioral decisions are based (Fig. 1). These individual decisions subsequently iterate to become the ecologies of entire populations (Weissburg and Browman, 2005). Global anthropogenic changes are, however, rapidly modifying the sensory landscape by introducing novel cues, increasing background noise, and altering the biochemical cascades that are responsible for transmission and interpretation of critical natural signals (Nagelkerken et al., 2019). Predicting the population level effects of these rapid changes occurring in the Anthropocene's aquatic environments, coupled with increasing harvest of the ocean's biomass and habitat loss, requires mechanistic understanding of how fish populations react to their constantly changing environment (Horodysky et al., 2015; Cooke et al., 2021; see Chapter 1, Volume 39A: Cooke et al., 2022). Understanding the effects of anthropogenic changes on organismal sensory biology is, in turn, critical to effective fisheries management and overall resource conservation as it provides tools necessary to: (1) interpret behavioral responses both at the individual and population levels, (2) suggest approaches to modify behaviors (most relevant to controlling the spread of invasive species, as well as reducing both bycatch and screen impingement), and (3) ultimately predict population level demographic processes associated with natural and anthropogenically-induced environmental changes (Blumstein and Berger-Tal, 2015; Horodysky et al., 2016; Madliger, 2012). Such an approach, however, requires an understanding of neurosensory mechanisms, sensory systems' capabilities, and their interactions with biotic and abiotic stimuli. A mechanistic understanding of the behavioral responses of fishes to sensory stimuli furthermore considers the following obvious tenets: (1) anatomical structures transducing sensory stimuli interact only with the immediate environment and (2) individuals can only react to an environmental variable they can sense (i.e., where there is a direct relationship between afferent nerve activity and the physical variable) (Horodysky et al., 2015, 2016). In addition, understanding and predicting the effects of anthropogenic changes requires knowing which stimuli are available to sensory systems, how the stimuli interact with the morphological structure of relevant neurosensory organs, and how physiological performance of a specific neurosensory system transduces the stimuli into actionable information.

A   Responses to natural cues

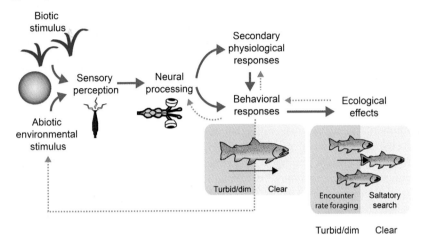

B   Responses to cues of the Anthropocene

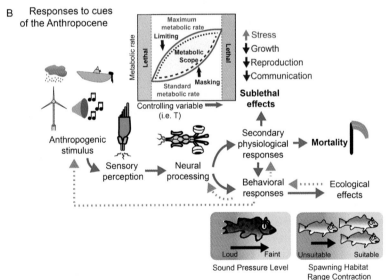

FIG. 1   All fish behaviors have their basis in the receipt and processing of sensory information. A. Physiology provides the mechanistic link between environmental variation and ecological patterns (blue arrows), subject to feedback modulation (gray dashed arrows) (following Horodysky et al., 2015). Natural environmental stimuli (in this example, photons of light) are received by cone photoreceptor cells of the visual system of an individual cutthroat trout (*Oncorhynchus clarkii*); information is passed through and processed by the neural system, resulting in an individual's behavioral response to seek brighter conditions. The movements of multiple light-limited individuals from dim to brightly lit clear water can enable a shift from energetically-costly encounter-rate feeding under dim conditions, such as at night or during turbidity events, to more energetically-favorable visual saltatory foraging under bright conditions (Mazur and Beauchamp, 2006). B. Fish physiological and behavioral responses to the complex

(Continued)

Fishes sense environmental conditions through receptors capable of transducing light, acoustic vibrations, solutes, gasses, temperature, electrical and magnetic fields, bulk water flow, and textures according to species-specific life histories and tasks (e.g., Horodysky et al., 2010, 2013; Kajiura et al., 2010; Kalinoski et al., 2014; Ladich et al., 2006). Chemical and acoustic stimuli are generally detectable the farthest from the stimulus source—followed by visual, mechanoreceptive, and lastly electrical stimuli (Jordan et al., 2013; Lennox et al., 2017) (Fig. 2A). Successful detection of a stimulus requires a signal above a threshold that distinguishes it from background noise (i.e., a minimum signal to noise ratio; Nilsson et al., 2014) (Fig. 2B). But fishes receive constant sensory information via the peripheral nervous system; they are simultaneously hearing, seeing, smelling, tasting, and mechanically interacting with (i.e., "touching" in human terminology) the medium surrounding them (Hara and Zielinski, 2007). These multimodal sensory inputs are integrated within the central nervous system, which prioritizes the sensory input that provides the most direct information. Fishes can thus compensate for the loss or reduced efficacy of individual senses (Goldberg et al., 2018; Valentincic, 2004, 2005), although this ability is obviously limited by their multimodal sensory integration abilities under the current physicochemical conditions. All of this is now changing in the Anthropocene at a pace faster than the evolution of fish sensory systems.

The processing of external stimuli occurs in various levels of the neuraxis to produce multipurpose action patterns. The neuroanatomy of the fish brain has been extensively reviewed (Kotrschal et al., 1998; Yopak, 2012) therefore, we will only briefly overview the fish brain's homologous regions. The fish brain (Fig. 3) can be divided into three sections from anterior to posterior: the forebrain, midbrain, and the hindbrain. The forebrain contains the telencephalon and diencephalon, and specializes in processing stimuli from the cranial nerve I (olfactory) (Schneider, 2014). The midbrain is comprised of the cerebral peduncles, the optic lobes, tectum opticum, torus semicircularis, and the tegmentum (Vernier, 2016). The midbrain receives input from the cranial nerve II (Optic). Furthermore, the midbrain contains several efferent neurons (i.e., signal output) specializing in the integration of visual information and sensory motor integration resulting in goal orientated movements and behaviors (Schneider, 2014). The hindbrain contains the cerebellum, pons, myelencephalon (medulla oblongata), vagal lobe, and the facial

---

FIG. 1—Cont'd   milieu of natural and anthropogenic cues of the Anthropocene can give rise to lethal and sublethal outcomes. In this acoustic example, the presence of anthropogenic aquatic noise (i.e., anthrophony) may mask natural auditory cues, leading to changes in movement patterns (avoidance) and reduced suitable spawning habitat for soniferous fishes. Collectively, the behaviors of an individual in response to its sensory physiology, when iterated over many individuals, can thus become the ecologies of populations (Weissburg and Browman, 2005). *Redrawn and adapted from Horodysky, A.Z., Cooke, S.J., Brill, R.W., 2015. Physiology in the service of fisheries science: Why thinking mechanistically matters. Rev. Fish Biol. Fish. 25, 425–447.*

A **Sensory range (idealized)**

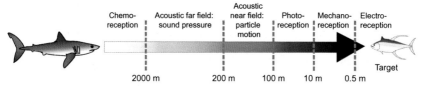

B **Stimulus attenuation**

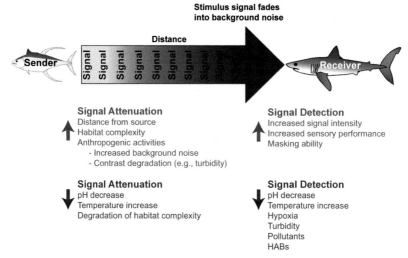

FIG. 2 (A) Idealized relative sensory detection distances (in m) of signals from a target or sender to a receiver demonstrating the ranges of different sensory modalities. Distances are approximate and represent ideals. Distances not to scale. (B) The attenuation of a stimulus signal from sender to receiver. Signal intensity wanes (i.e., attenuates) with distance from source, until the stimulus signal fades into undetectable levels against the background. The ability of fishes to detect signals in the Anthropocene is hampered by increased signal attenuation and background noise due to increased turbidity, marine noise, various anthropogenic activities, and pollutants and HABs. Within phylogentic, physical, and physiological bounds, fishes can improve signal detection by producing signals of higher intensity, increasing sensitivity of sensory receivers, and/or via masked abilities to detect faint target signals amidst a noisier broadband background. *Panel A: redrawn from Jordan, L.K., Mandelman, J.W., McComb, D.M., Fordham, S.V., Carlson, J.K., Werner, T.B., 2013. Linking sensory biology and fisheries by-catch reduction in elasmobranch fishes: a review with new directions for research. Conserv. Physiol. 1, 1–20.*

lobe (Vernier, 2016). The hindbrain receives input from cranial nerves V (somatosensory), cranial VIII (auditory), and cranial nerves VII, IX, and X (gustation) (Soengas et al., 2018).

Our intent is to provide a brief summary of the biotic and abiotic stimuli available to fishes, offer insights into how the sensory systems of fishes transduce relevant cues from the environment into actionable information, and

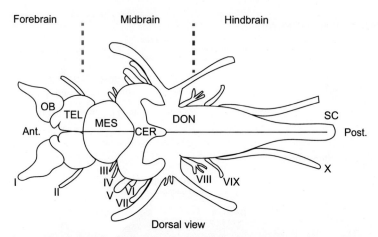

FIG. 3 Schematic diagram of the dorsal view of the brain of American paddlefish (*Polyodon spathula*). Abbreviations are: Ant (anterior), Post. (posterior), OB (olfactory bulb), TEL (telencephalon), MES (mesencephalon), CER (cerebellum), DON (dorsal octavolateral nucleus), SC (spinal chord). Forebrain, midbrain, and hindbrain regions are identified, though not all components of each are visible from the dorsal view. Cranial nerves are I-X are: olfactory (I), optic (II), oculomotor (III), trochlear (IV), trigeminal (V), abducens (VI, ventral to V), facial (VII), auditory (VIII), glossopharyngeal (IX), and vagus (X). *Redrawn and adapted from Pothmann, L., Wilkens, L.A., Schweitzer, C., & Hofmann, M.H., 2009. Two parallel ascending pathways from the dorsal octavolateral nucleus to the midbrain in the paddlefish Polyodon spathula. Brain Res. 1265, 93–102.*

elucidate the novel ways in which sensory knowledge has been and can be used to address applied issues relevant to resource conservation, the spread of invasive species, and effective fisheries management in a now rapidly-changing environment. Our ultimate goal is to synthesize available information to identify successful applications of sensory-based strategies and to suggest promising new directions.

2   Biotic and abiotic stimuli and sensory receptors in fishes

2.1   Mechanosensory

Water is 850 times denser than air, and only 1/10,000 as diffusive; stimuli thus travel differently in air than water. For example, the speed of sound in water is 4.5 times greater than that in air (Au and Suthers, 2014). The high molecular density of water therefore renders it an excellent medium for sound transmission; as a result, sound and vibrations travel faster with less attenuation with distance in water than air, thereby enabling long-distance communication but also impacts at long distances from the noise source (Slabbekoorn et al., 2010). Sound in water is composed of two physically linked components, pressure waves and particle motion, with the relative contribution of each changing as the distance from the sound source increases, and differences in the pathways through which they reach the inner ears of fishes (Fay and Popper, 1975; Higgs and Radford, 2012). Propagation losses occur as sound stimuli spread from their source, and via frequency-dependent absorption by water and the bottom substrates, and scattering by suspended particles and organisms (Nagelkerken et al., 2019). The amplitude of particle motion falls off more rapidly with increasing distance from source than that of a pressure wave. Whereas pressure waves are composed of localized regions of compression and rarefaction and are scalar (having amplitude only), particle motion resulting from a vibrating sound source is a vector (having amplitude and direction, albeit with 180 degrees of ambiguity). The latter enables directional hearing in water (Larsen and Radford, 2018).

The fish mechanosensory system includes: (1) the hearing of sound stimuli through the inner ears; (2) orientation and body motion sensation (the vestibular system), also mediated by the inner ears; and (3) the detection of hydromechanical stimulation in a region near the fish that is mediated by the lateral line (Popper and Schilt, 2008). The fish ear can detect sound stimuli ranging in frequency from below a hundred Hz to upwards of several thousand Hz depending on the species-specific anatomy (Fig. 4), whereas the lateral line

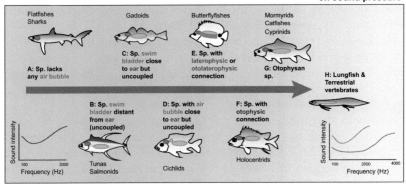

FIG. 4   Gradient of auditory anatomy and performance in fishes, from species lacking anatomical auditory specializations that are sensitive only to particle motion (both graphs, brown line), to fishes with specialized anatomical connections that enable the re-radiation of sound pressure as a particle motion in the otic capsule, which can decrease thresholds and increasing auditory bandwidth (right graph, blue line). Such specialized connections can include multiple pairs of enlarged otoliths and linkages of the auditory system with cephalic bullae, laterophysic or otolaterophysic connections, or otophysic coupling of the swim bladder and ear. The diversity of auditory morphology in some species groups may span multiple categories (i.e. Clupeidae and Sciaenidae). Sound intensity may be either particle acceleration (m s^{-2}, relevant to all species) or sound pressure level (dB re: 1 uPa, relevant only for pressure sensitive species, but historically the unit of choice for auditory comparisons in the literature). *Redrawn from Popper, A. N. Fay, R.R. 2011. Rethinking sound detection by fishes. Hear. Res. 273, 25–36 and Putland, R.L., Montgomery, J.C., Radford, C.A., 2018. Ecology of fish hearing. J. Fish Biol. 1–14.*

primarily detects low frequency water motion and vibrations close to the body. Thus the fish inner ear detects sounds of comparatively higher frequencies and from greater distances than the lateral line (Slabbekoorn et al., 2010).

The auditory system of fishes is well studied and the subject of extensive reviews (Popper, 1983; Popper et al., 2003; Popper and Coombs, 1982; Popper and Fay, 1973, 1993, 2011; Putland et al., 2018), so we present only a rudimentary description to provide context for examples relevant to conservation physiology. The inner ear of fishes is composed of three semicircular canals (horizontal, anterior and posterior) and associated otoliths (in teleost fishes, the saccule, lagena and utricle) or otoconial endorgans (in nonteleost fishes). Additionally, the *macula neglecta* of elasmobranchs may also contribute to vibrational detection (Fay et al., 1974). The semicircular system provides three-dimensional equilibrium orientation, whereas the otoliths and swim bladder may enable detection of sound-source generated pressure waves and particle motion, and the detection of linear acceleration (Popper and Fay, 1997; Putland et al., 2018; Rogers and Cox, 1988). More formally, the otoliths of the inner ear function as biological accelerometers directly detecting inertial differences between otoliths and neurosensory hair cells (Lu and Xu, 2002; Popper and Fay, 1999). Although all fishes can detect particle motion, some also derive information from sound pressure waves (Fig. 4). These can, however, only be detected when converted to mechanical motion through oscillations of the walls of an air-filled space (i.e., the swim bladder or auditory bullae), a process referred to as "pressure-to-displacement transduction" (Popper and Fay, 2011; Radford et al., 2012; Webb, 1998). Furthermore, the resultant mechanical stimuli must reach the hair cells of the ear or lateral line via accessory anatomical structures (i.e., otophysic or laterophysic connections, respectively) (Higgs et al., 2006; Montgomery et al., 1995; Popper and Fay, 1993). Swim bladders or bullae lacking these mechanical couplings are not believed to contribute much to sound pressure sensitivity in most fishes, though this can depend on the thickness of sound-attenuating muscle that separates the swim bladder and the otic capsule (Popper and Fay, 2011; Ramcharitar et al., 2005). Nervous system stimulation generated by sensory hair cells of the inner ear reach the myelencephalon via the acoustic nerve (cranial VIII) (Popper and Fay, 1993; Slabbekoorn et al., 2010).

Interest in the anatomy and function of the lateral line system has a centuries-long history (Coombs and Bleckmann, 2014), has been the subject of extensive reviews (e.g., Coombs et al., 1989a,b, 2014), and has remained an active area of investigation for many decades (e.g., Bleckmann, 2007; Cahn, 1967; Janssen, 2004; Kasumyan, 2003; Mogdans, 2019; Montgomery et al., 1995; van Netten, 2006). The anatomical structures sensing water flow are sensory hair cells (neuromasts) that project a staircase-like flagellar bundle (stereocilia and kinocilia) into a gelatinous cupula. The hair cells are directionally sensitive to mechanical displacement and cupulae are displaced proportionally to fluid flow. Neuromast hair cells are found within the lateral line (canal neuromasts), and both free and in arrays along the epithelium of the head, trunk, and tail (superficial neuromasts). They detect low frequency water displacements (on the scale of mm to cm) and very low sound frequencies (<200 Hz, Weeg and Bass, 2002). The superficial neuromasts have been considered to detect fluid velocity around the animal and canal neuromasts to detect fluid acceleration. The situation is now, however, considered to be more complex; such that the mechanical coupling of water movements relative to the fish, and the combined neural signals from the superficial and canal lateral line neuromasts, sense water acceleration over a bandwidth of less than 1 Hz to up to several hundred Hertz (van Netten, 2006).

The lateral line system has multiple functions including: (1) maintenance of fish spacing during schooling (e.g., Cahn, 1967, 1972), (2) "touch-at-a distance" sense that can guide movements and detect objects (Dijkgraaf, 1963; Windsor et al., 2008, 2010), (3) detection of currents and rheotaxis (Montgomery et al., 1997, 2000), and (4) detection of wakes left by swimming fish (which persist for at least 30 s and may possibly be detectable for up to 3 min). The last function thus aids in prey detection and predator avoidance (e.g., Bleckmann, 2007; Coombs and Montgomery, 2014; Kasumyan, 2003; Montgomery et al., 2014), reproductive and parental behavior (Kasumyan, 2003), and sensing particle displacements (i.e., water bulk flow) produced by a sound source. The last function of the lateral line system operates only in proximity to a sound source (within ~2 m at 100 Hz and ~20 m at 10 Hz), where both particle motion and pressure waves exist (Bleckmann, 2007; Coombs et al., 2014). The rare exceptions are the few fish species, such as Butterflyfishes (Chaetodontidae) and Sweepers (Pempheridae), where parts of the lateral line system are in close association with compressible gas cavities (laterophysic) and/or otoliths (otolaterophysic) and therefore can enhance the detection of pressure waves produced by a sound source (Coombs and Bleckmann, 2014). The lateral line, in conjunction with the olfactory system, also functions in odor plume tracking and prey capture (e.g., Gardiner et al., 2014; Gardiner and Atema, 2007, 2014; New et al., 2001; Pohlmann et al., 2004). Conversely, the lateral line, inner ear, vision, and tactile senses can be simultaneously stimulated in situations where water currents create whole-body accelerations (Montgomery et al., 2000). In brief, the lateral line is involved in multiple behaviors including surface feeding, schooling, obstacle avoidance, and subsurface detection of prey (Coombs et al., 1989a,b).

2.2 Chemosensory

Unlike visual, mechanical, auditory, or electrical signals, chemical stimuli last beyond the moment of production. They thus can disperse great distances from their source depending on water flow which, in turn, is influenced by tidal and other currents, topography, and the movements of organisms.

Chemical stimuli are, however, also subject to dilution, turbulent physical dispersion, and degradation (Hara, 1992a,b; Jordan et al., 2013; Nagelkerken et al., 2019). Chemical stimuli perceived by fish are mostly small molecules with high aqueous solubility (Laberge and Hara, 2001) and are detected though olfaction (smell) and gustation (taste) (Hara, 1986; Marui and Caprio, 1992; Valentincic, 2004, 2005).

The neuroanatomy and functional process of the chemosensory systems in fishes are well delineated (e.g., Hansen and Reutter, 2004; Hara, 1975, 1982, 1992a,b, 2011a,b,c; Laberge and Hara, 2001; Zielinski and Hara, 2001, 2007a), so we again present only a rudimentary description to provide context for examples relevant to conservation physiology. In contrast to the situation in terrestrial vertebrates, where olfaction is considered to be a distance sense and gustation a contact sense, olfaction and gustation in fishes are generally considered distance senses (Hara, 2011a). Although functionally similar, the olfactory and gustatory systems are anatomically different. The olfactory system in fishes originates in an anlage formed by the ectoderm, whereas the taste buds of the gustatory system are primarily endodermic in origin (Hara, 1992a,b). They are also innervated by separate nerves that project to different areas of the brain. Cranial nerve I (olfactory, from the olfactory receptors) projects to the olfactory bulbs in the telencephalon, whereas cranial nerves VII, IX, X (facial, glossopharyngeal, and vagus nerves, respectively from the gustatory receptors) project to the hindbrain (Zielinski and Hara, 2007a,b). Fish with specialized gustatory sensory organs (i.e., barbels) can exhibit enlarged vagal lobes, as observed in cyprinids, or enlarged facial lobes in Siluriformes (Finger, 2009; Kiyohara et al., 2002; Schneider, 2014).

The olfactory system is also unique among the other sensory systems in that it does not include separate receptor cells. Olfactory receptors (OR) are part of nerve endings. In contrast, the gustatory system includes taste buds which are comprised of specialized and innervated taste receptor cells (TRC). TRC have the capability of generating receptor potentials and form synapses with cranial nerves. They can be oral, palatal, laryngeal, branchial, and/or cutaneous. They are also found on specialized appendages such as barbels, or along the entire body (Hara, 1992a,b, 2011f). Generally speaking, there are three different types of gustatory cells that form a fish taste bud: (1) rod-shaped apical protrusion or cilia, (2) ending with microvillar cell or dark cell, and (3) basal cell (Hara, 2011a,b,c,d,e,f). The variability in taste bud size is dependent on thickness of the fish epithelial layer. TRC share functional similarity with ORs in that they respond to amino acids, but they are also sensitive to nucleotides peptides, quaternary ammonium bases, and steroids, with a sensitivity to these compounds equaling or exceeding that of the ORs (Hara, 2011e;Takeda and Takii, 1992). There are, however, also several significant functional differences between the olfactory and gustatory systems. The latter respond to tactile stimuli and are highly sensitive to environmental carbon dioxide (CO_2) levels; implying that the gustatory system may have a role in the control of respiration (e.g., Hara, 2011e; Yamashita et al., 1989). The TRC on the barbels of Ariid catfish species Telostei: Ariidae) are sensitive to pH and CO_2, which may serve as a method of prey detection (Caprio et al., 2014). In brief, the gustatory nerves (i.e., cranial nerves VII, IX, and X) contain chemical- and mechano-sensitive fibers providing fishes with both chemical and tactile information. For the sake of brevity, however, we have excluded description of the solitary chemoreceptor cells (single receptor cells usually imbedded in the epidermis and scattered over the body surface; Whitear, 1992; Hansen and Reutter, 2004) as their function and functional characteristics are less well characterized then TRC. We discuss the magneto-receptor cells present in the *lamina propria* of the olfactory epithelium (Walker et al., 1997, 2007) in a separate section.

Fishes use olfaction to find food, avoid predators, guide migration and orientation, and coordinate spawning (Kotrschal, 2000). Among vertebrates, fishes show the greatest morphological variability in the structure of the olfactory system (Reutter and Kapoor, 2005). Overall, however, symmetrical paired olfactory organs are located in the ethmoidal region of the skull in the dorsal snout. Each consists of an olfactory chamber connecting to the anterior and posterior nares which direct a flow of water through the organ. An olfactory rosette, composed of a series of olfactory lamellae consisting of sensory and non-sensory epithelia, arises from the floor of the olfactory cavity (Laberge and Hara, 2001). The rosette morphology and number of lamellae vary across species. The olfactory receptor neurons (ORNs) are situated within the sensory epithelium, bearing either cilia (ciliated olfactory neurons) or microvilli (microvillar olfactory neurons) (Laberge and Hara, 2001). Also present are the more recently described (e.g., Ferrando et al., 2006; Hansen and Finger, 2000) chemosensory crypt cells that detect both amino acids and sex pheromones (Hamdani and Døving, 2006; Vielma et al., 2008), with the distinction that same receptor appears is expressed by the entire population of cells (i.e., the "one cell type—one receptor" mode of expression; Gaurav Ahuija et al., 2013). Olfactory nerve fibers conduct signals to the olfactory bulb, where they contact bulbar neurons and glomeruli (Hara, 1992a,b). Fish olfactory lamellae also contain polyvalent cation-sensing receptors (CaRs) which provide the ability to detect changes in ambient Ca^{2+}, Mg^+ and Na^+ concentrations (Herrera et al., 2021; Hubbard et al., 2000, 2002; Loretz, 2008; Nearing et al., 2002). It is this mechanism that most likely explains the correlation between estuarine fish distributions and salinity (e.g., Buchheister et al., 2013) and may also be one that salmonids employ for homing to natal streams (Bodznick, 1978).

Olfaction involves the transport of odorants from the external environment to the sensory surface of the olfactory organ and binding of an odorant molecule to an OR which, in turn, induces a transduction cascade and eventually depolarization of the axon sufficient to cause a self-sustaining action potential (i.e., nerve signal to the brain; Hara, 2011c). Odorant molecules are predominately, although not exclusively, single amino acids, which have been well studied, though perhaps as a consequence of their ease of study. Bile acids, gonadal steroids, and prostaglandins are also olfactory stimulants, the latter two acting as pheromones and thus serving for conspecific/heterospecific recognition and to coordinate spawning among conspecifics (Buchinger et al., 2014; Hara, 2011c; Stacey and Sorensen, 2005; Velez et al., 2009). Because the binding affinity of an odorant molecule is specific to a single OR, the specificity of the responsiveness of the olfactory system to the range of odorant molecules is set by the number of different ORs. Specific genes encode single ORs; and fish may have up to 100 OR-encoding genes (Hara, 2011c).

When an item is perceived as potentially edible, the fish will initiate feeding behavior and taste the food item. Once activated by an odorant (olfaction) or tastant (gustation), TRC release activate G-proteins that induce the taste transduction cascade that concludes with the opening of Ca^{2+} channels, the depolarization of the TRC membrane, and ultimately the release of neurotransmitters (Morais, 2017) that have either excitatory or inhibitory effects, resulting in tastants being ingested or rejected (Fig. 5) (Rong et al., 2020; Roper, 2021). Free amino acids evoke the strongest electrophysiological and behavioral responses in fish and can exert either a excitatory response or a

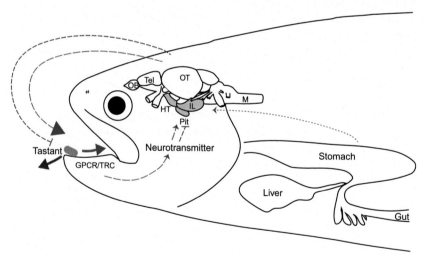

FIG. 5 Feeding behavior is moderated by the hypothalamic regions of the brain (shaded section, HT- hypothalmus, Pit- pituitary gland, IL—inferior lobe), gastrointestinal tract receptors (gray fine-dashed line) and the neuroendocrine system (hunger hormones). A fish tastes a potential food item (tastant) that activates the G-protein coupled receptor (GPCR) in the taste receptor cells (TRC). This activation initiates a transduction cascade that results in the release of neurotransmitters, which can be either stimulatory (green dashed arrow) or inhibitory (red terminal dash). Stimulatory pathway will result in food consumption (solid green arrow) whereas the inhibitory pathway will result in the rejection of the tastant (solid red arrow). Other abbreviations: OB—olfactory bulb, Tel—telencephalon, M—medulla. *Modified and redrawn from Rønnestad, I., Gomes A.S., Murashita, K., Angotzi, R., Jonsson, E., Volkoff, H., 2017. Appetite-controlling endocrine systems in teleosts. Front. Endocrinol. 8:73.*

deterrent effect (Hancz, 2020; Kasumyan and Mouromtsev, 2020; Takeda and Takii, 1992). Fish can finely distinguish differences between chemically similar substances based on taste (Kasumyan and Mouromtsev, 2020). Amino acid preference and palatability are, however, highly species-specific with minimal to no phylogenetic relationship or correlation to a species' ecology or feeding habits (Morais, 2017; Soengas et al., 2018; Takeda and Takii, 1992). Environmental factors can, however, alter foraging and feeding rate as well as food palatability and preference of tastants (Kasumyan, 2019; Rong et al., 2020) by possibly altering the expression of messengers in the taste transduction cascade (Nadermann et al., 2019; Rong et al., 2020).

In brief, fish chemosensory systems have the ability to detect a wide array of compounds: amino and carboxylic acids, nucleotides, steroid hormones, bile salts, prostaglandins, CO_2 and pH, and toxins and other compounds (Hara, 2012; Laberge and Hara, 2001). The broadly tuned fish olfactory system detects a wide array of chemical compounds and enables learning and discrimination of chemical stimuli. By contrast, the gustatory system of fishes detects a relatively small number of chemical stimuli at high sensitivity, and close range, enabling reflexive responses (Lamb and Finger, 1997; Valentincic, 2005). These chemosensory compounds can be enticants, suppressants, stimulants, deterrents, enhancers, or indifferent compounds (Kasumyan and Doving, 2003).

2.3 Photosensory

Daily irradiance in near-surface waters can vary over an intensity range of nine orders of magnitude, but scatter and absorption restrict the spectral bandwidth (color) and intensity (brightness) of downwelling light (Horodysky et al., 2008a; Johnsen, 2012; Lythgoe, 1979; McFarland, 1986; Warrant, 1999). Midday clouds can drop aquatic light intensities by one to two orders of magnitude. During crepuscular periods, however, intensity can change roughly tenfold every 10 min (Cooke et al., 2017). The spectrum of light available to fishes and other aquatic organisms depends on the wavelength-specific optical properties of the water column, as irradiance declines exponentially with depth. Maximal transmission of wavelengths of light visible to humans (400–700 nm) generally occurs at short wavelengths (blue) in oligotrophic freshwaters and clear pelagic seas, and at intermediate (green) wavelengths in coastal waters. In estuarine and most freshwaters, the concentrations of phytoplankton, yellow products of vegetative decay (*Gelbstoffe*), and suspended particulates scatter, absorb, and rapidly attenuate short wavelengths (Lythgoe, 1975, 1988). The spectral distribution in these waters is therefore dimmer and shifted to longer (yellow-red) wavelengths (Jerlov, 1968). The visual systems of fishes sample the properties of light: its luminous intensity, chromatic properties (wavelength), temporal properties (flickering speed), and in some fishes, its direction (polarization) (Horodysky et al., 2008a, 2010). Additionally, fishes possess varying abilities to resolve fine spatial details (acuity) and form images

(Kreysing et al., 2012; Northmore, 2011), depending on environmental attenuation of light stimuli and the structure and function of their visual systems.

Light enters the fish cornea, passes through the lens (where it can be focused and filtered to reduce transmission of certain wavelengths) before reaching the photoreceptors (i.e., rod and cone cells) in the retina. Significant processing of visual signal occurs in the ganglion, amacrine, and horizontal cells of the retina, prior to its passage to the preoptic area or the tectum via the optic nerve (cranial nerve II) (Douglas and Djamgoz, 1990). In some highly light-sensitive fishes, the light passing through the retina may reflect from a guanine-layered *tapetum lucidum* and reradiate through the photoreceptors a second time. Fishes with high luminous sensitivity can have a thicker and better developed tectum compared to fish who are less sensitive (Northmore, 2011). Ocular pigmentation and tapetal reflectance change within a species from individuals inhabiting waters of different optical properties in ways that appear to optimize vision in diverse environments (Best and Nicol, 1967; Litherland et al., 2009).

Many fishes forage visually, using rod photoreceptors during scotopic (dim/dark) conditions to increase sensitivity and form monochromatic images, and anywhere from one to four types of cone photoreceptors under photopic (bright) conditions to form high-resolution, contrasting images (e.g., Horodysky et al., 2008a, 2010). The number, properties and distributions of photoreceptor cells in fish visual systems, their luminous sensitivities, chromatic sensitivities and photopigments, and correlations to the photic properties of habitats have been well characterized (Bowmaker, 1990; Dartnall, 1975; Levine and MacNichol, 1979; McFarland and Munz, 1975; Parkyn and Hawryshyn, 2000), so once again we present only a rudimentary description to provide context for examples relevant to conservation physiology. In brief, visual pigments consist of a large protein (opsin) which binds a retinoic chromophore; this combination largely determines the range of wavelengths to which a visual pigment is most sensitive (Toyama et al., 2008). The visual systems of marine fishes are dominated by rhodopsin (A1, 11-*cis* retinal) photopigments, while those of freshwater fishes are commonly porphyropsin-based (A2, 11-*cis* dehydroretinal); the latter exhibit comparatively more red-shifted absorbance, consistent with the spectral properties of ambient light in freshwaters. Euryhaline and anadromous fishes contain a mix of the two, reflecting the dominant wavelengths of downwelling light in these environments (Toyama et al., 2008; Wald, 1939, 1941). Maximal contrast between an object and the visual background is achieved by a combination of matched and offset cone visual pigments ("Contrast Hypothesis"; Lythgoe, 1968), whereas maximum sensitivity (i.e. maximal photon capture) is conferred by (typically rod) photopigments that match the ambient background's spacelight ("Sensitivity Hypothesis", Bayliss et al., 1936; Clark, 1936). The situation is further complicated by the species-specific density and spacing of receptor cells (i.e., rod and cones) and ganglion cells which determine spatial resolution (i.e., the ability to detect the fine details of an object), as well as the number of receptor cells converging on individual ganglion cells ("spatial summation"). The higher this convergence, the greater the light sensitivity, with the inevitable trade off of decreased spatial resolution and vice versa (e.g., Collin and Pettigrew, 1989; Fritsches et al., 2003; Lisney and Collin, 2008; Litherland and Collin, 2008; Warrant, 1999).

Fishes have radiated across the world's broad range of aquatic habitats with complex photic properties, resulting in a myriad of selective pressures on their visual systems (Collin, 1997; Levine and MacNichol, 1979; Munz, 1977). Unavoidable tradeoffs between visual sensitivity and temporal or spatial resolution render optimal visual performance nearly impossible to maintain over the full range of daily optical conditions (e.g., Horodysky et al., 2013; Warrant, 1999). It is thus not surprising that the characteristics of aquatic light fields are generally reflected in the visual systems of fishes inhabiting them, within phylogenetic and ecological constraints (Guthrie and Muntz, 1993). Deep-dwelling species are characterized by limited monochromatic or dichromatic sensitivity but high luminous sensitivity and slow temporal resolution (Fritsches and Warrant, 2004; Warrant and Locket, 2004). By contrast, the eyes of many coastal and freshwater fishes are faster, less sensitive to dim light, but sensitive to a broader range of wavelengths, with numerous examples of dichromacy, trichromacy, and even tetrachromacy (Horodysky et al., 2010; Levine and MacNichol, 1979). Many fishes can be sensitive to ultraviolet wavelengths, either early in ontogeny to facilitate foraging on translucent prey (many species), or as adults for crepuscular foraging, mating coloration, and/or migration (Losey et al., 1999). Anthropogenic-induced changes of optical properties in aquatic environments are, however, significant and may be occurring at speeds faster than the evolution of fish visual systems (Seehausen et al., 1997; Horodysky et al., 2010).

2.4   Electro- and magneto-sensory

In relation to the lifetime of a fish, the Earth's geomagnetic field is a constant and reliable source of directional information (i.e., a magnetic compass) (Formicki et al., 2019). Local magnetic anomalies (created by significant geological features) provide magnetic cues enabling positional location (i.e., a magnetic map) (Johnsen and Lohmann, 2005; Lohmann, 2010). The field of magnetic orientation in vertebrates in general (e.g., Kirschvink et al., 1985a,b; Wiltschko and Wiltschko, 1995) and fish in particular (e.g., Walker et al., 2002, 2003, 2006, 2007) has been the subject of extensive reviews. Behavioral laboratory studies first evidenced the ability of salmonids (sockeye and Chinook salmon, *Oncorhynchus nerka* and *O. tshawytscha,* respectively) and yellowfin tuna, (*Thunnus albacares*) to sense local changes to the earth's magnetic field (Quinn, 1980; Taylor, 1986, 1987; Walker, 1984). A decade earlier, however, Yuen (1970) demonstrated homing behavior in free-swimming skipjack tuna (*Katsuwonus pelamis*) equivalent to homing in pigeons.

Holland et al. (1990) and Klimley (1993), tracking yellowfin and bigeye (*T. obesus*) tunas and scalloped hammerhead shark (*Sphyrna lewini*), likewise showed that these species were able to make repeated movements away from a fixed point (up to ~10km) and return. These results implied that fish have the ability to sense the earth's magnetic field, including local magnetic anomalies associated with geological features. Subsequent behavioral laboratory studies of rainbow trout (*O. mykiss*), short-tailed stingray (*Dasyatis brevicaudata*), and sandbar shark (*Carcharhinus plumbeus*) demonstrated the presence of a magnetic sense in both teleost and elasmobranch fishes (Anderson et al., 2017; Kirschvink et al., 1985a,b; Mann et al., 1988; Walker et al., 1997; Hudson, 2000). The discovery of crystalline magnetite in the anterior portion of the skull (Kirschvink et al., 1985a,b; Mann et al., 1988; Walker et al., 1984), and subsequently its demonstration in the nasal capsule with associated nervous tissue (Diebel et al., 2000; Walker et al., 1997), has provided mechanistic evidence of magnetite-based magnetic sense. Kalmijn (1982) proposed an alternate electrical-induction hypothesis for elasmobranch fishes where detection of the earth's magnetic field is by electroreceptors responding to currents induced by the fishes' movements through the earth's magnetic field (reviewed by Montgomery and Walker, 2001).

Electroreception in fishes has also been the subject of numerous recent reviews (e.g., Bullock et al., 2005; Collin and Whitehead, 2004; Gardiner et al., 2012; Kajiura et al., 2010; Newton et al., 2019; Tricas and Sisneros, 2004). Electroreceptive systems are present in a phylogenetically broad array of Chondrichthyan and Osteichthyan fishes (Collin and Whitehead, 2004). Sensitivity to electric fields produced by living organs (e.g., prey, predators and conspecifics) is based on the Ampullae of Lorenzini. These are scattered over the head in Chondrichthyan fishes, sturgeons (Acipenseriformes), and catfishes (*Ictalurus* spp.), over the dorsal and ventral surfaces of the pectoral fins in batoid fishes (e.g., Jordan, 2008; Newton et al., 2019; Peters et al., 2001), and in the freshwater paddlefish (*Polyodon spathula*) they extend from the rostrum to the posterior end of the gill cover (Wilkens and Hofmann, 2007). The Ampullae of Lorenzini provide the ability to sense electric fields of 1–$5\,\mu V\,cm^{-1}$, and possibly as low as $5\,nV\,cm^{-1}$ (e.g., Haine et al., 2001; Kajiura et al., 2010; Murray, 1962; Tricas and New, 1998). Ampullae of Lorenzini are derived ontogenetically from dermal placodes that similarly give rise to mechanosensory lateral line system of both elasmobranch and teleost fishes (Gillis et al., 2012; Modrell et al., 2011). There are also numerous weakly electric South American and African fishes (orders Gymnotiformes and Mormyriformes) that employ their electric sense for social communication and to detect field distortions caused by nearby objects (Bullock, 1982; Bullock et al., 2005).

Electrical signals are accompanied with environmental noise (e.g., water flow and animals' own movements) that are filtered through specialized neural mechanisms in the brain (Montgomery and Bodznick, 1994). Research conducted with paddlefish (*P. spathula*), however, suggests that noise may facilitate the detection of weak electrical signals through a phenomenon called stochastic resonance. But when noise levels reach high amplitudes, the target electrical signals are masked and lost (Russell et al., 1999). In both marine and freshwater fishes, information from electroreceptors is transmitted primarily to the dorsal octavolateral nucleus in the medulla oblongata (Newton et al., 2019; Pothmann et al., 2009). The ascending pathways then project to several regions of the brain, including the tectum and telencephalon (Newton et al., 2019).

3 Applied studies of relevant stimuli and senses

It is now well-described how conservation biology can benefit from a mechanistic understanding of sensory physiology and behavior (Blumstein and Berger-Tal, 2015; Domiononi et al., 2020; Elmer et al., 2021; Horodysky et al., 2016; Jordan et al., 2013; Madliger, 2012). More specifically, resource conservation and fisheries management benefit from applied sensory studies that: (1) quantify the range of stimuli that fish perceive, from which potential responses to natural and anthropogenic environmental change can be predicted (Kelley et al., 2018), (2) identify sensory traps (Madliger, 2012), (3) develop fisheries practices and gear configurations that reduce bycatch and discards while maintaining catch rates of targeted species (e.g., Brill et al., 2009; Horodysky et al., 2016), (4) deter entrainment and minimize the environmental disturbance of energy infrastructure technologies and water divergence systems (Noatch and Suski, 2012), (5) predict the range expansion of invasive species (e.g., Hasenei et al., 2020), and (6) guide aquaculture food production as well as stock enhancement and reintroduction of populations of conservation concern (Brill et al., 2019; Pernet and Browman, 2021). In the following sections, we summarize relevant applied studies by sensory modality.

3.1 Mechanosensory

Aquatic soundscapes result from sounds associated with: (1) geological processes (geophony), (2) biological sources (biophony), and (3) human activities (anthrophony). In some situations, anthrophony has come to dominate the Anthropocene soundscape (Duarte et al., 2021), even to sublethal and lethal levels (Popper and Hawkins, 2019). Human activities also alter soundscapes in other ways, both directly and indirectly (Popper and Hawkins, 2019; Putland et al., 2018; Slabbekoorn et al., 2010). Direct effects include overharvesting of biophonous animals (Erisman and Rowell, 2017) and loss of habitat complexity supporting diverse biophonous communities (Duarte et al., 2021). Indirect effects include the reduction of sound absorption and the resulting increases in the acoustic transmission of seawater due to increases in global temperatures and decreases in aquatic pH (Ilyina et al., 2010).

The effect of anthrophony on freshwater and marine fishes has received extensive attention (e.g., Popper and Hawkins, 2019; Putland et al., 2018; Slabbekoorn et al., 2010). Anthropogenic sounds can act as both acute and chronic stressors that induce changes that impact individual physiology and behavior, potentially iterating across individuals to prospective population, stock, and ecosystem effects (Braun, 2015) (Fig. 6). Loud, short duration broadband impulsive sounds from construction (e.g., pile driving) and explosions (e.g., seismic surveys) can cause death, physical injury to the swim bladder and other organs, and impairment of hearing via damage to auditory hair cells. Constant (or long duration) sounds from vessels and structures (e.g., bridges and tunnels, offshore wind energy turbines) can cause physiological stress and changes in behavior, including the displacement of fish from

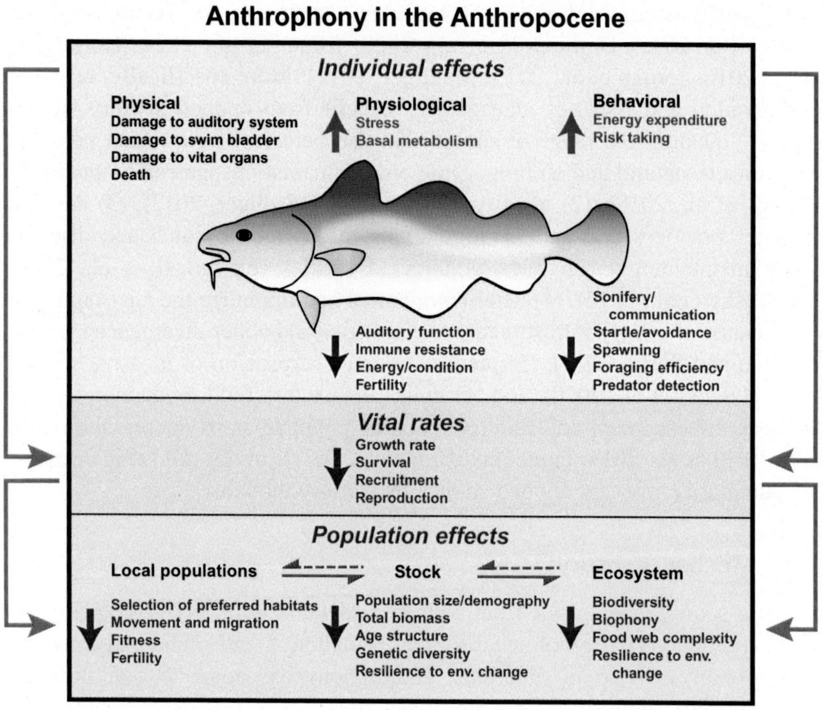

FIG. 6 Effects of rapidly increased anthrophony in the Anthropocene scale from the physical, physiological, and behavioral consequences at the individual level to local populations via changes in vital rates (solid arrows), iterating to changes to stocks and ultimately to ecosystem effects (horizontal arrows), with possible feedbacks (gray arrows). Impacts of anthrophony (green arrows = positive, red arrows = negative) may be most pronounced in species that are soniferous, associated with anthropogenic structures, and/or those that live in urbanized aquatic habitats. This includes major group of fisheries importance, including (but not limited to) gadids, sciaenids, catfishes, and serranids. *Modified and redrawn from Hawkins, A.D., Popper, A.N., 2017. A sound approach to assessing the impact of underwater noise on marine fishes and invertebrates. ICES J. Mar. Sci. 74, 635–651.*

feeding or spawning grounds, changes in fish navigation and migratory abilities, changes in fish sonifery, and masking of important natural sounds (Hawkins and Popper, 2017; Paxton et al., 2017; Popper et al., 2020) (Figs. 6 and 7). Strategies for managing aquatic noise levels include the use of sound-attenuating bubble curtains near sound producing structures (Würsig et al., 2000), regulating engine type and ship speed (McCormick et al., 2018), (ironically) the use of acoustic deterrents to shepherd fishes away from undesired areas (Arimoto, 2013; Arimoto et al., 1993; Götz and Janik, 2013; Merchant, 2019; Zielinski and Sorensen, 2016), and the development of quiet area refuges in biophony hotspots (Williams et al., 2015). The latter consist of temporal suspension of noise-producing activities or spatial closures that coincide with relevant migrations or spawning aggregations of soniferous fishes (Erisman and Rowell, 2017; Horodysky et al., 2008b; Ramcharitar et al., 2011). Multiple authors have recommended that regional

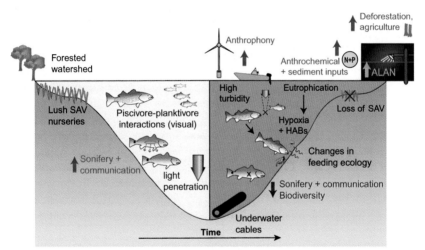

FIG. 7 Complex, rapid, and multifocal changes to the cuescapes of aquatic habitats since the Industrial Revolution (left panel = historical, right panel = present). These include land use changes (deforestation and agriculture), increases in anthrophony and anthrochemical runoff (fertilizers, pesticides, pharmaceuticals, other pollutants), and increased sedimentation. Collectively, these changes have increased turbidity and eutrophication, decreasing visibility and reducing the efficacy of visual predation of piscivores on planktivores, forcing the former to lower-energy benthic food resources (Horodysky et al., 2010). Dramatically increased artificial light at night (ALAN) concentrates predators and prey, when most fish should break up due to light limitation, possibly serving as a sensory trap for planktivores and potentially restructuring food webs (Mazur and Beauchamp, 2006). Similarly, increases in anthrophony can mask auditory signals and impede acoustic communication (Popper and Hawkins, 2019). Increases in pollutants and HABs can affect auditory performance (Lu and Tomchik, 2002), and likely olfactory/gustatory and visual senses as well. Underwater cables can produce electromagnetic frequencies (EMFs) that can be detected by electrosensitive fishes (Kilfoyle et al., 2018). *Redrawn from Horodysky, A.Z., Cooke, S.J., Brill, R.W., 2015. Physiology in the service of fisheries science: Why thinking mechanistically matters. Rev. Fish Biol. Fish. 25, 425–447.*

biophony and migration corridors should be considered when siting locations for offshore wind energy turbines in both their construction and operation phases (Gibson et al., 2017; Kikuchi, 2017; Wahlberg and Westerberg, 2005); we strongly concur. Although offshore wind energy turbines may enhance fish biodiversity and abundance via their artificial reef effect (Langhamer, 2012), the potential disruptive effect of operational sound on successful spawning of soniferous fishes (Gadidae, Sciaenidae, Serranidae, etc.), and possible reductions in growth due to stress, may result in these artificial reefs having a net negative effect. We argue more broadly, as have others (e.g., Popper et al., 2020), that it is incumbent upon fisheries managers operating in the Anthropocene to develop strategies to assess, monitor, regulate, and mitigate the effects of all anthrophony on fisheries resources.

While they obviously contribute to anthrophony (Findlay et al., 2018), there are numerous documented examples of the use of acoustic deterrent devices (ADDs) to resolve fisheries management problems (reviewed in Putland and Mensinger, 2019). ADDs are particularly effective for influencing the behavior of Cyprinids (minnows and carp) because these species have a Weberian apparatus that transduces pressure to particle motion and thus increases their auditory bandwidth (Putland and Mensinger, 2019). In addition to being a method to reduce anthropogenic noise, acoustic bubble curtains can also be a particularly cost-effective ADD. They can deter pressure sensitive fishes (those with morphological adaptations to enhance auditory ability) without affecting the behavior of fishes without such pressure transducing structures (Sager et al., 1987; Zielinski and Sorensen, 2016). In fact, bubble curtains and broadband speakers have both been used to exclude Asian carp species (Grass Carp, *Ctenopharyngodon idella*; silver carp, *Hypophthalmichthys molitrix*; bighead carp, *H. nobilis*; and black carp, *Mylopharyngodon piceus*) from the Laurentian Great Lakes (Murchy et al., 2016; Ruebush et al., 2012; Zielinski and Sorensen, 2016). Some Alosines (Clupeidae:Alosinae) with auditory bullae are able to resolve ultrasonic frequencies (Popper et al., 2004); impingement of alewife (*Alosa pseudoharengus*) and blueback herring (*A. aestivalis*) on power plant screens was decreased more than 80% via the inclusion of ultrasonic transducers emitting pulsed sounds at screen intakes; by contrast, catches of ultrasound-insensitive Atlantic herring (*Clupea harengus*) were reduced with low frequency ADDs (Maes et al., 2004; Ross et al., 1995). Low frequency ADDs have also been used to deter sharks (Chapuis et al., 2019). High frequency ADDs reduce: (1) pinniped and cetacean depredation of fisheries catch (Aneesh et al., 2016; Götz and Janik, 2013; Waples et al., 2013), (2) bycatch in fisheries gear (Barlow and Cameron, 2003), and (3) marine mammal entanglement in shark control nets (Erbe and McPherson, 2012; Götz and Janik, 2014). Anthrophonous noises may, however, attract marine predators to fisheries operations. There is evidence that sperm whales (*Physeter macrocephalus*), which have high frequency hearing, are attracted by the acoustic signature of the cavitation of propeller gear switches during demersal longline haulback (Thode et al., 2007). Pinnipeds can be attracted by acoustic tags applied to fishes and fisheries gear by learning that those sounds are indicators of food (Stansbury et al., 2015). Ultrasonic acoustic tags (i.e., those producing frequency above the range of human hearing) are not, however, effective shark deterrents because they are undetectable by sharks, which are sensitive to low frequencies only (Jordan et al., 2013). By contrast, pulsed intermediate frequency sounds (500–4000 Hz) reduced shark attraction to bait canisters (Chapuis et al., 2019). Lastly, air gun blasts can deter sea turtles (Chelonioidea) from nuclear power plant intakes, but only at extremely loud intensities ($>200\,dB$ re: $1\,\mu Pa$ at $1\,m$; O'Hara and Wilcox, 1990), as turtles are generally most sensitive to sound frequencies from 100 to 1000 Hz (Bartol et al., 1999; Piniak et al., 2012). The frequency hearing range of sea turtles is thus similar to many targeted fisheries species and therefore it is unlikely that acoustic deterrents on pelagic longline gear would prevent sea turtle bycatch without simultaneously reducing catch rates of targeted fish species (Southwood et al., 2008).

Acoustic stimuli can be used to attract fish (Richard, 1968). Some wading birds vibrate their bills in the water for this very purpose (Kushland, 1973; Davis, 2004). Similarly, the traditional Japanese fishing method *"donburi"* uses a conical lead weight cast so it hits the water's surface perpendicularly to maximize the production of sound attractive to fish (Yan et al., 2010). Although it is generally understood that fishing vessel sounds often repel fish and thus bias fisheries surveys (De Robertis and Handegard, 2013; Erickson, 1979), sounds generated by fishing vessels may also be attractive in some cases (Rostad et al., 2006). In fact, some pelagic recreational fishers contend that certain low frequency vessel-radiated noises attract tuna and billfish (Foote, 1980). It is postulated that anchored fish attraction devices (FADs) often have attractive sound signatures due to the oscillation of the floating raft that is coordinated with wave activity, and radiation of these vibrations as they pass down the tether (Dempster and Kingsford, 2003; Yan et al., 2010). Finally, in recreational fisheries, many fishing lures and flies create chugging/popping sounds on the water's surface, and subsurface lures frequently wobble to displace water as they are retrieved. These create low-frequency stimuli that theoretically improve gamefish attraction. Other fishing lures also contain embedded rattles, which recreational fishermen believe attract gamefish (Lennox et al., 2017). The sound frequencies produced by lure rattles are, however, typically well above the auditory frequency ranges of most gamefishes, suggesting that such attraction is mechanistically unlikely (Lennox et al., 2017).

Finally, acoustic conditioning and ranching of fish has been well studied, with some limited successes (Zion and Barki, 2012; Yan et al., 2010). Fish are raised in captivity and conditioned with sound stimuli to feed, or to elicit other desired behaviors. Fish are subsequently released as juveniles into aquaculture pens or into the wild to mature, the same sound stimuli are later used to attract fish for harvest (Lindell et al., 2012; Zion and Barki, 2012).

The applications of such a mechanism are certainly appealing for recalling and capturing escaped aquacultured individuals (Tlusty et al., 2008), or to prosecute commercial fisheries on adults cultured as juveniles and released to the wild. The latter would result in a major cost savings compared to raising fish to marketable size in captivity (Lindell et al., 2012). Acoustic conditioning has been demonstrated to aggregate numerous species of aquacultured fish including rainbow trout (*Oncorhyncus mykiss*; Abbott, 1972), thicklipped mullet (*Creimugil labrosus*; Wright and Eastcott, 1982), common carp (*Cyprinus carpio*; Wright and Eastcott, 1982), red sea bream (*Pagrus major*; Tateda et al., 1985), Atlantic cod (*Gadus morhua*; Oiestad et al., 1987, Ings and Schneider, 1997), tilapia (Cichlidae; Levin and Levin, 1994) and black sea bass (*Centropristis striata*; Lindell et al., 2012). The technique shows some promise over short to moderate temporal intervals of weeks to months (Tlusty et al., 2008), though responses of predators to the behavior of conditioned fish presents conservation concerns (Lindell et al., 2012). Regardless, this is a clever application of sensory biology in fisheries, and warrants further attention.

Humans also alter hydrodynamic conditions and thus the types of hydrodynamic stimuli acting on fishes for recreational, industrial, and domestic purposes (Mogdans, 2019). The lateral line system is well documented to be involved in rheotaxis (e.g., Montgomery et al., 1997, 2000) and to be most likely involved in the ability of fish to hold positions in streams relative to fixed objects, as well as down- and upstream migrations (Mogdans, 2019). The mechanics of flow sensing and movements in response to a current (rheotaxis) are well known, but how this connects to fish behaviors in not. This is the critical missing link—the transfer factor directly connecting an academic interest to an applied solution. But such transfer factors are likely to be both species- and life-stage specific, making the successful application of this idea difficult. Hydrodynamic effects on behavior are of particular interest as they are intimately involved with the effectiveness of manmade structures engineered to enable fish to bypass obstacles (e.g., dams or weirs) or to avoid structures that divert stream flows (e.g., Kasumyan, 2003; Kerr and Kemp, 2019; Meulenbroek et al., 2018). Engineering and construction of effective fish passageways, or structures to create avoidance, depend on knowledge of swimming capacity, the ability to detect created current patterns and thus find the passageway entrance, and hydraulic preferences; all of which are highly specific to both species and life stage (i.e., adults moving upstream or juveniles moving downstream, Schilt, 2007; Williams et al., 2012). Further, vibrational deterrents can cause startle responses and avoidance behaviors in Atlantic (*Salmo salar*) and Chinook salmon (*Oncorhynchus tshawytscha*) and shiner surfperch (*Cymatogaster aggregata*) (Knudsen et al., 1994, 1997; Mussen et al., 2014). Vibrational repulsion could thus reduce impingement of these fishes on water intake screens (Mussen and Cech, 2019). It should be noted, however, that such an approach may not work with all species (Mussen and Cech, 2013), and that after showing initial repulsion fish may later habituate to the constant stimulus (Mussen and Cech, 2019). As vibrational deterrents have not been particularly well studied, we believe that this presents opportunity for future work. Vibrational deterrents, particularly if varied in periodicity and combined with acoustic stimuli, should be investigated further as a deterrent mechanism.

3.2 Chemosensory

Chemical composition of marine habitats and the corresponding chemosensory cues play an important role in structuring populations, communities, and ecosystems, from the migration, foraging, and behavioral ecology of adults to habitat selection and settlement of fish larvae (Atema et al., 2002; Coppock et al., 2020; Hay, 2009; Lecchini et al., 2005; Lecchini and Nakamura, 2013; Paris et al., 2013). Until recently it was largely assumed that chemosensory habitat selection was dominated by olfactory cues. A recent study suggests, however, that gustation may play a more prominent role than previously thought based on hydrodynamics and taste bud ontogeny due to gustation being tied to gill ventilation (Hu et al., 2019).

The olfactory system also functions in predator detection and defensive behaviors. Predators can be detected via the sensing of "alarm kairomone" (substances shed by the skin and mucous of predatory fish); whereas defensive behaviors are elicited by alarm pheromone ("alarm substance" or "*Schreckstoff*") and stress (or fear) pheromone (Døving and Lastein, 2009; Barkhymer et al., 2019). *Schreckstoff* is released when the skin is damaged and stress pheromones are excreted in the appearance of predators (Kasumyan, 2004; Døving et al., 2005; Hara, 2011d). The olfactory system is also intimately involved in reproduction by coordinating spawning and reproductive behaviors, such as nest building (Sorensen, 1992; Stacey and Sorensen, 2005; Hara, 2011d). Although arguably the best documented example of the olfactory system's central role in reproductions is homing of Pacific salmon (*Oncorhynchus* sp.) to natal rivers (e.g., Hasler and Scholz, 1983a,b; Døving and Stabell, 2003; Ueda, 2019). Other sensory cues are, however, involved in the long-distance navigation of salmon back to coastal areas adjacent to their natal rivers (Hinch et al., 2006; Lohmann et al., 2008).

Nagelkerken et al. (2019) recently reviewed the general effects of anthropogenically-induced changes in the sensory landscape, and Lürling (2012) the specific disruption of chemical information conveyance caused by pollutants, so we will once again only provide only brief descriptions. The consequences of exposure to both inorganic (primarily heavy metal ions) and organic pollutants (e.g., pesticides, pharmaceuticals, crude oil, detergents, and oil dispersants) on olfactory and gustatory function have been clearly documented (e.g., Blaxter and Ten Hallers-Tjabbes, 1992; Cave and Kajiura, 2018; Kasumyan, 2019; Kasumyan and Doving, 2003; Klaprat et al., 1992; Sandahl et al., 2004; Tierney et al., 2010). Recent work documents the effects of aquatic acidification on olfactory deficits in fishes

(Porteus et al., 2021). Pollutants can affect the sensitivity of, and gene expression within, ORNs (e.g., Baldwin et al., 2003; Wang et al., 2016); or central nervous system processing of olfactory stimuli (e.g., Schlenker et al., 2019). Pollutants can not only alter taste perception, but can also damage structural integrity, which can lead to disintegration of the taste buds (Kasumyan, 2019), and reduce the ability to distinguish food from nonedible debris (e.g., plastics) (Roch et al., 2020). We contend, however, that for inshore fishes a significant (albeit uninvestigated) problem is the burgeoning (in both temporal and spatial extent) of harmful algal blooms (HABs) resulting from both increasing nutrient inputs from both point and non-point source pollution exacerbated by the effects of directional climate change and ocean acidification (Anderson et al., 2015; Roggatz et al., 2019). The toxins produced by HAB species are highly toxic to fishes (e.g., Sievers, 1969) and may be more so under high temperature and low pH conditions (Roggatz et al., 2019), with at least one of the mechanisms being extensive necrosis and disruption of the gill respiratory epithelium (W. Vogelbein, Virginia Institute of Marine Science, unpublished observations). We suspect equivalent anatomical and functional damage occur in the olfactory and gustatory systems of fishes exposed to HAB toxins, although to the best of our knowledge this has only been documented in auditory systems (Lu and Tomchik, 2002). There are, however, two plausible mitigating processes. First, at least one HAB toxin (saxitoxin) has a potent gustatory stimulus (Yamamori et al., 1988). This, and appropriate behavioral responses, should allow fish avoid areas with concentrations of HAB cells (Konstantine et al., 2017; Rountos et al., 2017). Second, olfactory neurons and gustatory cells are capable of regeneration and regularly turnover within weeks to months (Hansen and Reutter, 2004; Hara, 2011c,f; Zippel, 1993; Zippell et al., 1997). Chemosensory cells damaged by HAB toxins (and other chemosensory toxins) should regenerate, although to the best of our knowledge this also has not been documented. We argue the possible detrimental effects of HAB toxins on chemosensory function (although it may be only temporary) warrants further investigation.

Chemical cues can be used to direct fish behaviors, including attraction and repulsion, presenting opportunities for resource conservation strategies that exploit our understanding of the physiology and functional characteristic of fishes' olfactory and gustatory sensory systems. This can be done in two ways: (1) via methods to direct fish movements either by attraction to specific areas or habitats or repulsion away from areas or structures presenting the danger of injury or death, or (2) by applying methods to reduce interactions of bycatch species with fishing gear. (We will not cover the specific topic of using olfactory cues in methods designed to reduce the spread of the invasive sea lamprey (*Petromyzon marinus*) into the Laurentian Great Lake; Porter et al. (2017), as this will be covered in Volume 39B). Allochthonous chemicals (i.e., those formed elsewhere than *in situ*) are, however, being increasingly released into coastal waters due to anthropogenic regulation of river flows, altered land use, increasing concentrations of pharmaceutical compounds in wastewater, and shifting weather patterns due to climate change (Nagelkerken et al., 2019). All of these serve to complicate the functionality of chemosenses of fish (Fisher et al., 2006), and attempts by fisheries managers to use chemosensory attraction and repulsion techniques.

The primary challenges to using olfactory-gustatory cues to direct fish behaviors involves finding chemical stimuli that are species-specific, economically feasible to deploy, and environmentally safe (Noatch and Suski, 2012). These requirements can be met through the use of pheromones (Sorensen and Stacey, 2004); either anti-predator cures ("alarm kairomone"), alarm pheromone ("alarm substance" or "Schreckstoff"), and stress (or fear) pheromone described previously. Further, species- and response-sensitivity is set by the specificity of the ORs (e.g., Sorensen et al., 1990) and chemical identity of some pheromones is known (Sorensen and Stacey, 2004), making pheromones especially useful in integrated pest management systems (e.g., Noatch and Suski, 2012; Sorensen and Stacey, 2004). Migratory attractant pheromones could be used as potential tools for directing fish movements by guiding desired species around barriers (e.g., dam obstructions) and/or toward desired adult habitat and spawning areas to promoting recovery of depleted populations (e.g., Baker and Hicks, 2003; Lennox et al., 2018; Sorensen and Stacey, 2004), but this is likely cost-prohibitive and has yet to be tested on a larger scale.

The formal search for repellants and replacement baits to: (1) lessen the reliance on declining bait stocks, (2) increase catch rates of targeted species, or (3) reduce bycatch have a long history extending back for over a century (e.g., Bateson, 1890; Januma et al., 1999; Jones, 1992; Lokkeborg et al., 2010). Most recent efforts have, however, generally concentrated on the last objective (e.g., Patanasatienkul et al., 2020; Southwood et al., 2008; Swimmer et al., 2005), and results have been mixed. Two recent extensive analyses (Kumar et al., 2016; Gilman et al., 2020) demonstrated the complexity of the situation. Substitution of natural baits (fish *vs* squid) in pelagic longline fisheries (targeting tunas and swordfish *Xiphias gladius*) to reduce shark, sea turtle and istiophorid billfish bycatch results in tradeoffs; squid bait reduces shark and sea turtle bycatch but increases billfish bycatch, and vice versa. Erickson and Berkeley (2008) showed that artificial bait is equally effective as natural bait in a demersal longline fishery targeting Pacific halibut (*Hippoglossus stenolepis*) and sablefish (*Anoplopoma fimbria*), and eliminated elasmobranch bycatch. Januma et al. (1999) developed an artificial bait for tuna longline fisheries, but catch rates were significantly below those with natural baits, and to the best of our knowledge this has not been subsequently pursued. An effective fisheries chemical deterrent still awaits identification, despite attempts to identify repellants of nontarget fishes, sea turtles, and seabirds using a library of chemicals including alkaloids, pungent and bitter substances, capsaicin, and natural defensive compounds (Gilbert, 1968; Horodysky et al., 2016). Shark repellents composed of rotting shark flesh (releasing ammonium acetate) or copper sulfate appear to be useful although impractical in commercial fishing operations (Hart and Collin, 2015). Shark liver oil slicks dripped during longline sets reduced seabird bycatch, but not fish bycatch (Pierre and Norden, 2006), and is again most likely impractical in commercial fishing operations. Necromones (i.e., semiochemicals derived from putrefied shark flesh) also show promise in repelling sharks (Stroud et al., 2014), but may not be feasible on a commercial scale. There is evidence that applying a secondary plant metabolite (methyl anthranilate) to aquaculture feeds serves as a nonlethal bird repellant which reduces avian pellet-scavenging when feeding juvenile captive-cultured fishes, and the compound does not affect fish growth (Harpaz and Clark, 2006). All aquatic chemical repellents face the same challenges, however: the isolation of compounds that are cost-effective, sustainable, non-toxic, and effective at very low concentrations, so that rapid dispersion by water currents does not negate their repellent effect (Hart and Collin, 2015).

The use of chemosensory stimulants to encourage fish feeding has long been studied to produce better performing artificial baits in recreational fisheries (Carr, 1984, 1996; Jones, 1992; Lokkeborg, 1990) and to increase the palatability of rations used in aquaculture (Mackie and Mitchell, 1985; Takeda and Takii, 1992). For example, Berkley Fishing (berkely-fishing. com) and Fishbites (Carr, 1984; fishbites.com) produce a wide array of synthetic baits, lures, and dips infused with odorants and tastes and they have loyal followings in the recreational fishing community (Cooke et al., 2021). Synthetic baits have several advantages over natural baits in recreational fisheries: (1) they do not require that wild animals are harvested for bait, reducing stress on wild populations and reducing habitat destruction (Miller and Smith, 2012; Sypitkowski et al., 2010), (2) they do not need to be kept alive or refrigerated, (3) they do not contribute to the spread of non-indigenous organisms and pathogens (Cohen et al., 2001). Salmonids and ictalurid catfishes both demonstrate strong reflexive bite responses to the amino acids L-Proline and L-Arginine (Valentincic and Caprio, 1994, 1997) and appear to be the most effective components in the milieu of compounds used in popular synthetic fishing baits and aquaculture feeds (Takeda and Takii, 1992).

Relevant chemosensory stimuli of fish feeds, and their potential effects on feeding behavior, can affect aquaculture productivity in different ways. Chemical attraction leads to faster feeding, reduced waste and costs, and improved water quality, whereas feeding stimulation increases total food consumed and may improve growth and feeding efficiency (Hancz, 2020; Morais, 2017). Aquaculture feed is often derived from fish or other animal protein, and often contain plant-based ingredients with low palatability, none of which are considered highly sustainable (Malcorps et al., 2019). Chemosensory manipulation of aquaculture feed can broaden alternative candidate protein sources that are more sustainable, such as insects selected for their amino acid and fatty acid profiles (Khan et al., 2018). There has been some success substituting animal and plant-based fish feed with insect-based feed, showing minimal adverse effects on food consumption and growth rates of barramundi (*Lates calcarifer*), rainbow trout (*Oncorhynchus mykiss*), tilapia (*Oreochromis niloticus*), and the European sea bass (*Dicentrarchus labrax L.*) (Mastoraki et al., 2020; Shakil Rana et al., 2015; St-Hilaire et al., 2007). Chemosensory manipulation in aquaculture can also prove particularly useful for overcoming specific obstacles such as weaning post-larval stages from live food onto formulated feeds and to mask antimicrobial ingredients (Morais, 2017). We therefore believe that investigation of chemosensory stimulants and repellants still are fruitful paths within the nexus of aquaculture, fisheries science, and fish physiology.

3.3   Photosensory

Light is a central driver of biological processes and systems, affecting physiology and behavior at the individual level and scaling through populations to ecosystem processes (Langbehn and Varpe, 2017). But as mentioned previously, the Anthropocene's visual cuescape is changing at a pace far more rapid than the evolution and adaptation of fish visual systems (Horodysky et al., 2010; Nagelkerken et al., 2019). Aquatic light fields have been altered dramatically over the past few centuries through: (1) the thinning and melting of sea ice in polar regions (Langbehn and Varpe, 2017), (2) land use changes that denude riparian buffers and forests, increasing sedimentation and eutrophication of coastal and freshwaters (Kemp et al., 2005), and (3) point (i.e., dock and bridge lights) and nonpoint (i.e., general nocturnal skyline illumination) sources of artificial light at night (ALAN, often referred to a "light pollution") (Davies and Smyth, 2018; Moore et al., 2006) (Fig. 7). All of these can create a mismatch between species-specific visual abilities and ambient light conditions, and collectively alter the normal circadian rhythms of natural light levels and seasonal changes in daylength. The intrinsic circadian clock exists in nearly all organisms; it synchronizes behaviors (sleep/wake), and physiological processes (e.g., hormone secretion, metabolism, immune system, and DNA repair) to environmental stimuli (Grubisic et al., 2019). Light is one of the strongest environmental timing cues ("*zeitgeber*") that alters, or shifts, the intrinsic clock so that it synchronizes to light/dark cycles. Increased exposure to ALAN can desynchronize and disrupt the circadian rhythm, which can severely impede physiological processes (Brüning et al., 2018). ALAN can also impede prey detection and capture, and can scale through food webs to disrupt important scotoecological (nocturnal) ecosystem services and revert ecosystems back to constant photoecological (daylight) processes with consequences for ecosystem structure and function (Lyytimaki, 2013). Larval reef fish recruits that were exposed to ALAN exhibited endocrine disruption, impeded growth rates, lower fitness, and higher mortality rates

(O'Connor et al., 2019). ALAN-induced circadian disruption can also significantly reduce reproductive rhythms in some fishes (Brüning et al., 2018; Fobert et al., 2019). More specifically, the consequences of anthropogenic changes to aquatic light fields to fishes include documented changes in: (1) circadian rhythm (Brüning et al., 2015), (2) habitat selection/settlement (O'Connor et al., 2019), (3) reproductive communication (Seehausen et al., 1997), (4) recruitment (O'Connor et al., 2019), (5) foraging (Mazur and Beauchamp, 2006), (6) migrations (Vowles and Kemp, 2021), and (7) survival (Schligler et al., 2021). All consequences stem from shifting baselines that cause humans to dramatically alter landscapes to control food production (Kemp et al., 2005), worship of oil-based economies (Arbo and Thuy, 2016), and photophilic human societies that increasingly bathe themselves in light at night to avoid natural darkness (Lyytimaki, 2013). Whereas increased turbidity hampers the ability of visually-foraging piscivores to find their prey, turbidity releases planktivores from predation (Horodysky et al., 2010); except under ALAN, wherein dock and bridge light may serve as sensory traps for forage fishes (Fig. 7). There is little doubt that light pollution (i.e., ALAN), as well as the anthropogenic pollution of light fields (i.e., changes to the spectral quality, quantity, and intensity of light with depth), stand as grand interdisciplinary challenges for conservation physiologists and ecosystem managers.

In contrast, the light-based direction and light-based repulsion of fish offer potent tools for managing lotic, lentic, and marine fisheries. It has long been known that fishes can be attracted to light, either via: (1) overall positive phototaxis, (2) preferences for certain light intensities or light of specific spectral composition, (4) a high abundance of prey attracted to light, (5) visually-mediated schooling for safety, (6) investigatory reflex and curiosity, or (7) due to simple disorientation (Nguyen and Winger, 2019). As a result, fishing fixed and mobile gear with light has become one of the most advanced, efficient, and successful methods for capturing commercially important species on an industrialized scale, accounting for up to 40% of total marine fish production in some regions (Arimoto, 2013; Nguyen and Winger, 2019). Historically, fish were aggregated by light that was produced by shoreline bonfires; this progressed to fishers wading with torches made of bamboo and coconut husk, and in the twentieth century, oil and electrical lamps followed by a wide array of incandescent, fluorescent, halogen, metal halide, and LED lights (Inada and Arimoto, 2007; Nguyen and Winger, 2019). One of the earliest scientifically documented use of underwater lights to attract fish was by Hawaiian fishermen targeting tuna species in the 1920s (Sokimi and Beverly, 2010). Presently one of the largest-scale uses of underwater light during fishing involves disposable phosphorescent light sticks (producing different intensities and wavelengths) placed in proximity to baited hooks in pelagic longline fisheries (Hazin et al., 2005). This practice presents pollution concerns, however, if spent light sticks are discarded at sea (battery-powered LED equivalents are now commercially available; https://www.lindgren-pitman.com/collections/electralume-lights). Low powered LED lights also increase catch-per-unit-effort (CPUE) of target species in demersal pot fisheries (Bryhn et al., 2014). In coastal and freshwater recreational fisheries, anglers often prefer to fish near docklights and bridges at night. Prey schools and/or insect swarms which would otherwise de-aggregate due to nocturnal light limitation instead form aggregations under the bright point source rings of anthropogenic light (Cooke et al., 2017). In the shadowed periphery surrounding the circles of anthropogenic light, associated concentrations of predatory gamefishes (and anglers targeting them) ultimately extend the hours of visual predator-prey and fish-angler arenas (Cooke et al., 2017). Intense non-point-source illumination of the urban night sky can also produce enough aquatic illumination for prey fishes to school and predators to forage visually, potentially changing the structure and function of food webs (Mazur and Beauchamp, 2006). By contrast, the disruption or absence of light caused by shadow-casting overwater structures can delay downstream migrating juvenile salmon (Ono and Simenstad, 2014), via enhanced predation risk perception and increased antipredator behavior (Sabal et al., 2021).

Lure colors are very important to recreational fishers. Fly fishers, for example, rely almost exclusively on visual attraction of gamefish, as olfactory and gustatory attractants are not considered sporting. Well-stocked tackle shops peddle artificial lures, baits, and flies in every imaginable color. Evidence suggests that angler preference of lure colors or brightness changes with optical conditions of (and fishing pressure on) the water body being fished (Lennox et al., 2017). This, in turn, often results in differential success rates for attractor vs. natural colors (Ateşşahin and Cilbiz, 2019; Moraga et al., 2015; Nieman et al., 2020). Numerous devices and schemes for predicting the most effective lure colors have been hawked to anglers (Cooke et al., 2021; Kageyama, 1999), although they generally fail to consider the effects of predator-prey viewing angle relative to the direction of illumination, which may have a dramatic role in contrast discrimination (Cooke et al., 2017). [A more thorough discussion of ideal colouration for crypsis and conspicuousness can be found in Johnsen, 2003 and Johnsen and Sosik, 2003.] Phosphorescent paints are occasionally applied to lures so that they can be charged under bright light and then glow in the dark (Cooke et al., 2021), an attribute that does not resemble many natural prey. Neither does the extremely popular lure color chartreuse—a fluorescent yellow-green whose name originated from the combination of yellow and green French liquer distilled by Carthusian monks in the mid-1700s (Hassid, 1934), and spawned fly fishing icon Bernard "Lefty" Kreh's famed catchphrase about fishing the often-turbid Chesapeake Bay, "if it ain't chartreuse, it ain't no use." In truth, yellow-green wavelengths lie in the middle of the spectral sensitivity curves of many coastal gamefish (Horodysky et al., 2010), as well as the spectral sensitivity of humans. Thus it is not surprising that it is such a popular, if unnatural, lure color. In the United States alone, the production of artificial lures, flies, and dyed natural baits is a multibillion-dollar industry (Hutt et al., 2015) that certainly attracts recreational fishers' spending. Where there is a lot of profit, however, there is a lot of misinformation and pseudoscience. Red monofilament "Cajun fireline" was purported to be "virtually invisible under water" (zebco.com); however red fades quickly to black or dark gray with depth and thus would be exceedingly obvious to almost all gamefish. Similarly, recent promotion of commercial UV-coatings on lures and fly tying materials purport that they can be sensed at greater depth/distance due to UV reflectance and fluorescence (Cooke et al., 2021); unfortunately few gamefish of the sizes targeted by anglers can actually see UV wavelengths. Thus, from the perspective of a gamefish, reflectance of UV light is much less important than fluorescence (i.e., the absorption and readmission of UV light at longer wavelengths, which can intensify lure primary colors). Regardless of the missteps, we believe that there is tremendous commercial and scientific opportunity in open interdisciplinary collaborations between scientists and the recreational fishing industry, particularly for identifying visual stimuli that improve both catch rates and (more importantly for resource conservation and fisheries management) selectivity.

Clearly light can be used to attract fish, but it can also be used to alter the behavior of fish and other marine life in other desirable ways for conservation and management. This requires, however, an understanding of the relationship between light intensity, temporal repetition, and wavelength to species-specific visual function. Flashing lights have been used on their own, and in concert with acoustic bubble curtains, to improve fish passage by either deterring fish from power plant screen intakes or guiding fish away from unfavorable velocity currents (Mussen and Cech, 2013; Vowles and Kemp, 2012). Evaluation of deterrent effects of flashing lights must, however, consider the critical fusion frequency (CFF) at which a flashing light appears as constant illumination which in fishes is species-specific and dependent on light levels and temperature. CFF is 30–60 Hz for many coastal teleost fishes (Horodysky et al., 2013), 5–25 Hz in elasmobranch fishes (McComb et al., 2010; Kalinoski et al., 2014), 10–15 Hz in sea turtles (Horch and Salmon, 2009), and ∼60 Hz in humans.

Using differences in visual function to separate targeted and bycatch species is an obviously appealing approach and there have been some successes to that end. Low-powered flashing LEDs are effective in reducing bycatch of juvenile fish in shrimp, lobster, and groundfish trawls (Grimaldo et al., 2018; Hannah et al., 2015; Melli et al., 2018); salmonids in midwater trawls (Lomeli and Wakefield, 2014); and sea turtles in gillnets (and this technique is now applied nearly worldwide) (Wang et al., 2010, 2013). Trawl fisheries have used light to separate targeted and bycatch species via illuminated "escape rings" in trawl codends and illuminated grids or glow-in-the-dark netting that encourage groundfish species to separate into different codends (Nguyen and Winger, 2019). Seabird bycatch is also reduced when gillnets are illuminated, evidencing that net illumination has strong potential as an effective multi-taxa bycatch mitigation strategy (Mangel et al., 2018).

Unfortunately, attempts to exploit differences in visual function to separate targeted and bycatch species has not been universally successful. Bait color does not consistently and predictably affect bycatch rates of nontarget fishes or sharks (reviewed in Horodysky et al., 2016). Blue-dyed squid baits can reduce seabird bycatch in longlines (Cocking et al., 2008), but does not affect sea turtle catch rates (Swimmer et al., 2005; Yokota et al., 2009). Deploying green chemical light sticks near hooks resulted in the catch rates of target species in Brazilian pelagic longline fisheries, yet were also responsible for the highest incidence of bycatch. In contrast, in the same fishery, use of blue and white chemical light sticks resulted in significantly lower catch rates of both target and bycatch species (Afonso et al., 2021; Hazin et al., 2005). Sea turtle bycatch in coastal gill net fisheries is reduced through the use of Plexiglass shark shapes (Wang et al., 2010). This appears to occur because Plexiglass is opaque to UV light making the shark silhouette visible to sea turtles if their visual systems have sufficient UV-sensitivity under ambient conditions. Unfortunately, this technique also reduces targeted fish catch and as such, would serve as an unpopular resource management tool.

3.4   Electro- and magneto-sensory

Human activities since the Industrial Revolution have also altered electromagnetic cues available to fishes, but this is a relatively new area of research (Klimley et al., 2021). Electrical currents in submarine cables (e.g., telecommunication cables and energy transmission cables) generate detectable low-frequency electromagnetic fields that can penetrate the surrounding water column, albeit over relatively short distances of 1–10 m (Öhman et al., 2007). Direct current (DC) creates only magnetic fields, whereas alternating current (AC) creates both electrical and magnetic fields (Nagelkerken et al., 2019). Near cables carrying AC, induced electromagnetic field anomalies may override geomagnetic field cues for sensitive species (Kavet et al., 2016). Fishes (in particular, elasmobranch fishes), seabirds, sea turtles, and marine mammals can either detect, navigate by, or show behavioral responses to electromagnetic fields (EMFs), with some elasmobranch demonstrating sensitivity down to $0.5–1000 mV m^{-1}$ (Kalmijn, 1982; Kilfoyle et al., 2018). Anthropogenic electromagnetic stimuli can be attractive or repulsive. EMFs of $0.5–100 mV m^{-1}$ attract some species, whereas those over $100 mV m^{-1}$ are avoided (Kalmijn, 1982; Tricas and Gill, 2011). Numerous freshwater fish species are attracted to fyke nets adorned with magnets (Formicki et al., 2004). On the other hand, overhead powerlines across a river have been shown to delay migration in

Atlantic salmon (*S. salar*) and Russian sturgeon (*Acipenser gueldenstaedtii*; Poddubny et al., 1979). Undersea cables are: (1) often bitten by elasmobranch fishes implying that they are somehow triggering a feeding response (Marra, 1989; Taormina et al., 2018), (2) affect the migratory routes and timing of anadromous salmonids (Wyman et al., 2018), and (3) reduce the swimming rates of passing telemetered European eels (*Anguilla anguilla*; Westerberg and Lagenfelt, 2008) presumably due to EMFs.

The situation is a complex one, however, because EMFs from submerged power cables vary as a result of: (1) cable construction and shielding, (2) transmitted power, (3) current (AC vs. DC), (4) frequency and amplitude of AC currents, (4) cable grounding (i.e., seawater ground or a two-wire ground), (5) whether the cable is monopole, bipole, or tri-axial, and (6) the disposition of the cable in the water (i.e., buried, epibenthic, or suspended in the water column) (Kilfoyle et al., 2018). It is clear that this should remain an active area of investigation given the above complexity, the increasing deployment of offshore wind energy technologies, and the ubiquity of underwater and overhead cables in and near freshwater and marine habitats (particularly in urbanized areas and those with highly developed waterfronts) (Basov, 2007; Gill and Kimber, 2005; Gill et al., 2012; Westerberg and Lagenfelt, 2008) (Fig. 7).

The best studied example of the applied use of electromagnetic stimuli in fisheries involves the use of electropositive metals, magnets, and electrogenerating devices in the repulsion of elasmobranch and ancestral fishes from fishing gear. Unfortunately, there have been mixed results. Electropositive metals (e.g., lithium, magnesium, yttrium, the lanthanide metals, or alloys thereof) produce voltage gradients when immersed in seawater that are an order magnitude greater than nanovolt bioelectric fields (McCutcheon and Kajiura, 2013; O'Connell et al., 2014a). Laboratory studies have demonstrated deterrence of spiny dogfish (*Squalus acanthias*) from bait placed near an electropositive metal (but not deterrence of Pacific halibut *Hippoglossus stenolepis*), and longline field trials reduced dogfish catch rates (Stoner and Kaimmer, 2008; Kaimmer and Stoner, 2008). Similar results were achieved with sandbar shark (*Carcharhinus plumbeus*; Brill et al., 2009), juvenile scalloped hammerhead shark (*Sphyrna lewini*; Hutchinson et al., 2012), and Atlantic sturgeon (*Acipenser oxyrhynchus oxyrhynchus*; Bouyoucos et al., 2014); the latter suggesting a possible mechanism to reduce fisheries gear interactions for this endangered species. Paddlefish (*P. spathula*) have been observed avoiding regions that contain metal structures, which can have a negative effect on migratory patterns on rivers with metal bridges and dock structures, or dams with metal locks and associated steel gates (Gurgens et al., 2000). By contrast, other studies have shown no significant deterrent effects of electropositive metals or magnets containing electropositive metals (Godin et al., 2013; Hutchinson et al., 2012; O'Connell et al., 2011, 2014b; Tallack and Mandelman, 2009). A recent study actually found significant increases in blue shark (*Prionace glauca*) catch rates associated with magnets containing electropositive metals (Porsmoguer et al., 2015). Clearly there are interspecific differences in the behavioral effects of electropositive metals which are undoubtedly due to species-specific sensory capabilities and feeding modalities (Hutchinson et al., 2012; Kajiura et al., 2010). It is also likely that environmental conditions play a role, as there is evidence that electropositive metals are effective in deterring batoids and coastal sharks from fishing gear in turbid inshore temperate environments, but ineffective in cold pelagic waters or clear tropical waters (Brill et al., 2009; Hutchinson et al., 2012). There is also the question of feeding facilitation effects when multiple individuals are involved simultaneously in laboratory studies, or in schooling fishes and those that travel in loose aggregations (Brill et al., 2009; Jordan et al., 2011, 2013). It is also unknown (and probably impossible to determine) how often an individual fish in the wild would encounter a hook in proximity to electropositive metal even if this technique were in wide use. As a result, addressing questions about habituation is likely impossible. The use of electropositive metals, and magnets containing electropositive metals, as a deterrent may be impractical in any event. Electropositive metals are neither costly nor exceedingly rare, but they do rapidly corrode in seawater and are potentially flammable (Howard et al., 2018). Magnets containing electropositive metals obviously present a host of additional challenges to gear configuration, deployment, and retrieval (O'Connell et al., 2014a, 2014b). Microprocessor-based battery operated devices programmed to produce electric waveforms of amplitude and frequency know to stimulate elasmobranch fishes' electroreceptors have been shown to reduce bait consumption by sharks in laboratory settings, suggesting their potential use as bycatch reduction devices (Howard et al., 2018); and without the limitations of electropositive metals. Given the successful use of electrical lighting devices in longline fisheries to reduce turtle bycatch and improve target catch on a commercial scale (Hazin et al., 2005; Wang et al., 2013), we believe that studies of microprocessor-based battery-operated devices programmed to produce electric waveforms (which probably could be combined in the same package) could serve multiple purposes. If so, this could be a particularly attractive technology to commercial fishers.

4   Multimodal sensory integration

Fishes obviously receive and process stimuli simultaneously and undoubtedly prioritize responses to the most relevant cues (Gardiner et al., 2012). Thus, while the effects of sensory stimuli may be studied in the laboratory *in singulus*, behavioral decisions are rarely made on information from a single sensory channel. For example, for many visually foraging predatory fishes, visual information may override the efficacy of chemosensory or electroreceptive inputs (or both) during predatory choices in clear waters (Kajiura et al., 2010; Horodysky et al., 2016). This may be one reason why electropositive metals appear to be less effective at deterring open ocean pelagic shark species than their turbid water coastal counterparts, as visual stimulus overrides electrosensory deterrence. Sensory suppression studies reveal differences in fish behavior when one sense is deprived, reduced, or masked because specific sensory modalities are required to direct species specific prey-capture behaviors (Gardiner, 2011; Liao and Chang, 2003; New and Kang, 2000). Specific examples include the integration of multiple cues (e.g., lateral line, visual, and olfactory) when foraging (Gardiner et al., 2014). We argue that it is thus critical to understand how the changing stimuli of the Anthropocene, including both relevant signals and noise, are altering the experiences and responses of fishes and fish populations (de Jong et al., 2018).

4.1 Integrating neurosensory physiology, conservation, and management: A call for fish-centric approaches

Behaviors exhibited by fishes are in reaction to information obtained through their sensory systems. Thus population- and species-specific mechanistic understanding of neurosensory abilities, tolerances, and stressors is fundamentally necessary to: (1) understand catchability, (2) improve stock assessments, (3) describe essential fish habitat, (4) predict rates of post-release mortality, (5) develop effective bycatch reduction strategies, (6) assess anthropogenic impacts, and (7) forecast the population effects of increases in global temperatures and ocean acidification (Horodysky et al., 2015). It is clear that the study of fish neurosensory physiology and behavior will have an increasingly prominent role in the future of sustainable fisheries management (Horodysky et al., 2016; Jordan et al., 2013). We contend that the effectiveness of "sensory-informed" fisheries management and governance requires asking management questions from the fish's perspective. Fish distributions that are governed by the environmental conditions they can measure with their sensory systems are too-often studied by humans applying sampling schemes that are stratified by unrelated spatiotemporal and geopolitical criteria (Horodysky et al., 2015). Dangerous tautologies result from survey-generated inferences that may not have a valid underlying mechanistic basis; it is imperative to know which signals are relevant to fish, and when the "signal" measured by fisheries scientist may instead be a covariate or worse, an unrelated variable from the fish's perspective (Horodysky et al., 2015). To that end, new technologies are inspiring and enabling new hypotheses regarding sensory stimuli and neurosensory performance, as well as new directions of interdisciplinary research (Johnsen, 2007). We propose that the study of sensory stimuli and neurosensory responses will improve our understanding of fish behavior and can be used to address problems in fisheries management. We further suggest that the most significant applied contributions of sensory physiology lie in improving understanding of fish movements in response to natural environmental cues and anthropogenic sensory pollutants, directing fish movements as desired via attractants and dispersants, reducing bycatch, and improving aquaculture best practices.

Sensory intervention tools available to both fishers and fisheries managers can be viewed as levers that can be used to modify behavioral outcomes and thus influence higher-level population processes (Blumstein and Berger-Tal, 2015). After identifying a conservation issue or management problem that may be addressed using sensory mechanisms, research funding enables the study of that question. Completion of relevant studies then provides knowledge of the sensory performance of all species relevant to the issue at hand, and suggests future sensory-based mitigation solutions (Fig. 8). Those prospective solutions must then be evaluated for efficacy in the field, thereafter a comparative evaluation in an adaptive management framework of the practicality and cost effectiveness of the proposed solution determines

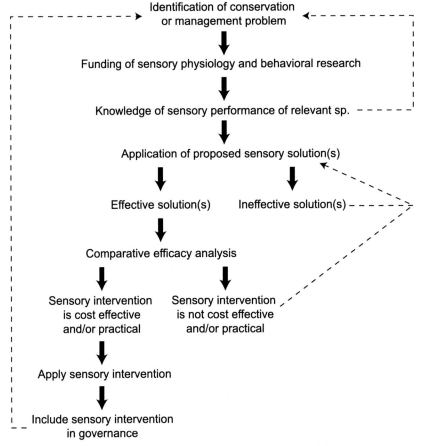

FIG. 8   Schematic framework through which mechanistic knowledge of fish neurosensory performance can be applied to solve fisheries management problems with successful, cost-effective, and enforceable interventions. *Adapted from Blumstein, D.T., Berger-Tal, O., 2015. Understanding sensory mechanisms to develop effective conservation and management tools. Curr. Opin. Behav. Sci. 6, 13–18.*

whether the sensory intervention is adopted by fishers and fisheries managers, and eventually included in governance and resource conservation strategies (Smith et al., 2014.). Ineffective, prohibitively costly, or impractical solutions can be abandoned and other proposed sensory solutions evaluated in their place. Over time, the iterative evaluation of a sensory intervention's success may demonstrate that it solved conservation concerns, or it may highlight other conservation and management issues that need to be addressed. Collectively, such an approach translates many potential mechanistic insights for conservation and management into effective conservation and management interventions (Blumstein and Berger-Tal, 2015).

Yet given the contributions of sensory physiology and behavior to improve fisheries management, few funding mechanisms currently exist to support explicitly applied neurosensory studies. Thus, most insights into fish sensory biology stem from offshoots of academically-driven fundamental studies of basic comparative neurosensory function (naturally often lacking applied focus), or behaviorally-driven agency investigations of deterrents and attractants (often lacking mechanistic insights into the underlying physical and physiological boundaries of neurosensory performance). Given the demonstration of successful sensory deterrents and attractants over the past century, it is beyond the time for government agencies to consider such work eminently fundable and define specific mechanisms for supporting applied sensory ecology. Management agencies may also consider raising revenue via green tax models on industry or resource-users, and rewards for conservation-minded behaviors as economic instruments for conservation (Young, 2005). An example is taxing power plants for thermal pollution and thus creating funding sources for studies to reduce screen impingement. Given the obvious commercial potential of sensory ecology, another appealing approach to underfunding issues lies in the Blue Economy—academic and agency collaborations with industry, co-funded by leveraging public and private sources to develop stakeholder buy-in. Collaborations with recreational and commercial fishers, as well as commercial and recreational fishing gear industries, could lead to the development of more effective lures and baits which increase catch rates of target species while decreasing bycatch. Lastly, NGOs and public crowdsourcing options may be a useful first step for generating seed money for pilot studies and obtaining in-kind stakeholder involvement (Gallo-Cajiao et al., 2018); this could be especially effective for initiating research in prohibitively expensive fisheries applications, such as those prosecuted in the open-ocean or in developing nations (Bower et al., 2017; Horodysky and Graves, 2005).

Acknowledgments

A.Z.H. and C.C.S. were supported by NSF CAREER #1846004; A.Z.H. also received support from NSF #1911928. The authors thank the two reviewers and editorial staff for comments, and M. Henson (Hampton University) for a diligent final review of manuscript references and figures; their combined efforts have collectively strengthened this chapter. This is contribution 4101 of the Virginia Institute of Marine Science, William & Mary.

References

Abbott, R.R., 1972. Induced aggregation of pond-reared rainbow trout (*Salmo gairdneri*) through acoustic conditioning. Trans. Am. Fish. Soc. 101, 35–43.

Afonso, A.S., Mourato, B., Hazin, H., Hazin, F.H.V., 2021. The effect of light attractor color in pelagic longline fisheries. Fish. Res. 235, 105822.

Anderson, C.R., Moore, S.K., Tomlinson, M.C., Silke, J., Cusack, C.K., 2015. Living with harmful algal blooms in a changing world: strategies for modeling and mitigating their effects in coastal marine ecosystems. In: Ellis, J.T., Sherman, D.J. (Eds.), Coastal and Marine Hazards, Risks, and Disasters. Elsevier, Amsterdam, pp. 495–560.

Anderson, J.M., Clegg, T.M., Véras, L.V.M.V.Q., Holland, K.N., 2017. Insight into shark magnetic field perception from empirical observations. Sci. Rep. 7, 11042. https://doi.org/10.1038/s41598-017-11459-8.

Aneesh, K.V., Pravin, K.P., Meenakumari, B., 2016. Bait, bait loss, and depredation in pelagic longline fisheries—a review. Rev. Fish. Sci. Aquac. 24, 295–304.

Arbo, P., Thuy, P.T.T., 2016. Use conflicts in marine ecosystem-based management—the case of oil vs fisheries. Ocean Coast. Manag. 122, 77–86.

Arimoto, T., 2013. Fish behaviour and visual physiology in capture process of light fishing. In: Symposium on Impacts of Fishing on the Environment: ICES-FAO Working Group on Fishing Technology and Fish Behaviour. May 6–10, Bangkok, Thailand (2013).

Arimoto, T., Akiyama, S., Kikuya, K., Kobayashi, H., 1993. Fishherding effect of an air bubble curtain and its application to setnet fisheries. In: Wardle, C., Hollingworth, C.E. (Eds.), Fish Behavior in Relation to Fishing Operations, ICES Marine Science Symposia 196. ICES, Copenhagen, pp. 155–160.

Atema, J., Kingsford, M.J., Gerlach, G., 2002. Larval reef fish could use odour for detection, retention and orientation to reefs. Mar. Ecol. Prog. Ser. 241, 151–160. https://doi.org/10.3354/meps241151.

Ateşşahin, T., Cilbiz, M., 2019. The effect of hook size, spinner colour and fishing season on catching efficiency in angling for rainbow trout, *Oncorhynchus mykiss* (Walbaum, 1792). Pak. J. Zool. 51 (5), 1937–1942.

Au, W.W.L., Suthers, R.A., 2014. Production of Biosonar Signals: Structure and Form. In: Surlykke, A., Nachtigall, P., Fay, R., Popper, A. (Eds.), Biosonar. Springer Handbook of Auditory Research. vol. 51. Springer, New York, NY. https://doi.org/10.1007/978-1-4614-9146-0_3.

Baker, C., Hicks, B.J., 2003. Attraction of migratory inanga (*Galaxias maculatus*) and koaro (*Galaxias brevipinnis*) juveniles to adult galaxiid odours. N. Z. J. Mar. Freshw. Res. 37, 291–299.

Baldwin, D.H., Sandahl, J.F., Labenia, J.S., Scholz, N.L., 2003. Sublethal effects of copper on coho salmon: impacts on nonoverlapping receptor pathways in the peripheral olfactory nervous system. Environ. Toxicol. Chem. 22, 2266–2274.

Barkhymer, A.J., Garrett, S.G., Wisenden, B.D., 2019. Olfactorily-mediated cortisol response to chemical alarm cues in zebrafish *Danio rerio*. J. Fish Biol. 95, 287–292.

Barlow, J., Cameron, G.A., 2003. Field experiments show that acoustic pingers reduce marine mammal bycatch in the California Drift Gill net fishery. Mar. Mamm. Sci. 19, 265–283.

Bartol, S.M., Musick, J.A., Lenhardt, M.L., 1999. Auditory evoked potentials of the loggerhead sea turtle (*Caretta caretta*). Copeia 1999 (3), 836–840.

Basov, B.M., 2007. On electric fields of power lines and on their perception by freshwater fish. J. Ichthyol. 47, 656–661.

Bateson, W., 1890. The sense-organs and perceptions of fishes; with some remarks on the supply of bait. J. Mar. Biol. Assoc. U. K. 1, 225–256.

Bayliss, L.E., Lythgoe, J.N., Tansley, K., 1936. Some forms of visual purple in sea fishes with a note on the visual cells of origin. Proc. R. Soc. Lond. B Biol. Sci. 120, 95–114.

Best, A.C.G., Nicol, J.A.C., 1967. Reflecting cells of the elasmobranch tapetum lucidum. Contrib. Mar. Sci. 12, 172–201.

Blaxter, J.H.S., Ten Hallers-Tjabbes, C.C., 1992. The effect of pollutants on sensory systems and behaviour of aquatic animals. Neth. J. Aquat. Ecol. 26, 43–58.

Bleckmann, H., 2007. The lateral line system of fish. In: Hara, T.J., Zielinski, B.S. (Eds.), "Sensory Systems Neuroscience", Volume 25: Fish Physiology. Elsevier, NY, pp. 411–453.

Blumstein, D.T., Berger-Tal, O., 2015. Understanding sensory mechanisms to develop effective conservation and management tools. Curr. Opin. Behav. Sci. 6, 13–18.

Bodznick, D., 1978. Calcium ion: an odorant for natural water discriminations and the migratory behavior of sockeye salmon. J. Comp. Physiol. 127, 157–166.

Bouyoucos, I., Bushnell, P., Brill, R., 2014. Potential for electropositive metal to reduce interactions of Atlantic sturgeon with fishing gear. Conserv. Biol. 28, 278–282.

Bower, S.D., Danylchuk, A.J., Raghavan, R., et al., 2017. Involving recreational fisheries stakeholders in development of research and conservation priorities for mahseer (*Tor* spp.) of India through collaborative workshops. Fish. Res. 186, 665–671.

Bowmaker, J.K., 1990. Visual pigments of fishes. In: Douglas, R.H., Djamgoz, M.B.A. (Eds.), The Visual System of Fish. Chapman & Hall, London, pp. 82–107.

Braun, C.B., 2015. Signals and noise in the octavolateralis systems: what is the impact of human activities on fish sensory function? Integr. Zool. 10, 4–14.

Brill, R., Bushnell, P., Smith, L., Speaks, C., Sundaram, R., Stroud, E., Wang, J., 2009. The repulsive and feeding-deterrent effects of electropositive metals on juvenile sandbar sharks (*Carcharhinus plumbeus*). Fish. Bull. 107, 298–307.

Brill, R.W., Horodysky, A.Z., Place, A.R., Larkin, M.E.M., Reimschuessel, R., 2019. Effects of dietary taurine level on visual function in European sea bass (*Dicentrarchus labrax*). PLoS One 14 (6), e0214347.

Brüning, A., Holker, F., Franke, S., Preuer, T., Kloas, W., 2015. Spotlight on fish: light pollution affects circadian rhythms of European perch but does not cause stress. Sci. Total Environ. 511, 516–522.

Brüning, A., Kloas, W., Preuer, T., Hölker, F., 2018. Influence of artificially induced light pollution on the hormone system of two common fish species, perch and roach, in a rural habitat. Conserv. Physiol. 6 (1), 1–12.

Bryhn, A.C., Konigson, S.J., Lunneryd, S.G., Bergenius, M.A.J., 2014. Green lamps as visual stimuli affect the catch efficiency of floating cod (*Gadus morhua*) pots in the Baltic Sea. Fish. Res. 157, 187–192.

Buchheister, A., Bonzek, C.F., Gartland, J., Latour, R.J., 2013. Patterns and drivers of the demersal fish community of Chesapeake Bay. Mar. Ecol. Prog. Ser. 481, 161–180.

Buchinger, T.J., Li, W., Johnson, N.S., 2014. Bile salts as semiochemicals in fish. Chem. Senses 39, 647–654.

Bullock, T.H., 1982. Electroreception. Ann. Rev. Neurosci. 5, 121–170.

Bullock, T.H., Hopkins, C.D., Popper, A.N., Fay, R.R., 2005. Electroreception. In: Fay, R.R., Popper, A.N. (Eds.), Springer Handbook of Auditory Research. Springer, NY.

Cahn, P.H., 1967. Lateral Line Detectors. Indiana University Press, Bloomington.

Cahn, P.H., 1972. Sensory factors in the side-to-side spacing and positional orientation of the tuna, *Euthynnus affinis*, during schooling. Fish. Bull. 70, 197–208.

Caprio, J., Shimohara, M., Marui, T., Harada, S., Kiyohara, S., 2014. Marine teleost locates live prey through pH sensing. Science 344 (6188), 1154–1156.

Carr, W.E.S., 1984. Artificial Bait for Aquatic Species. U. S. Patent No. 4,463,018. U. S. Patent Office, Washington, DC.

Carr, W.E.S., 1996. Stimulants of feeding behavior in fish: analyses of tissues of diverse marine organisms. Biol. Bull. 190, 149–160.

Cave, E.J., Kajiura, S.M., 2018. Effect of Deepwater Horizon crude oil water accommodated fraction on olfactory function in the Atlantic stingray, *Hypanus sabinus*. Sci. Rep. 8, 15786.

Chapuis, L., Collin, S.P., Yopak, K.E., McCauley, R.D., Kempster, R.M., Ryan, L.A., Schmidt, C., Kerr, C.C., Gennari, E., Egeberg, C.A., 2019. The effect of underwater sounds on shark behavior. Sci. Rep. 9, 6924.

Clark, R.L., 1936. On the depths at which fishes can see. Ecology 17, 452–456.

Cocking, L., Double, M., Milburn, P., Brando, V., 2008. Seabird bycatch mitigation and blue-dyed bait: a spectral and experimental assessment. Biol. Conserv. 141, 1354–1364.

Cohen, A.N., Weinstein, A., Emmett, M.A., Lau, W., Carlton, J.T., 2001. Investigations into the Introduction of Non-Indigenous Marine Organisms via the Cross-Continental Trade in Marine Baitworms. Report for the U.S. Fish and Wildlife Service, Sacramento CA.

Collin, S.P., 1997. Specialisations of the teleost visual system: adaptive diversity from shallow-water to deep-sea. Acta Physiol. Scand. Suppl. 638, 5–24.

Collin, S.P., Pettigrew, J.D., 1989. Quantitative comparison of the limits on visual spatial resolution set by the ganglion cell layer in twelve species of reef teleosts. Brain Behav. Evol. 34, 184–192.

Collin, S.P., Whitehead, D., 2004. The functional roles of electroreception in nonelectric fishes. Anim. Biol. 54, 1–25.

Cooke, S.J., Lennox, R.J., Bower, S.D., Horodysky, A.Z., Treml, M.K., Stoddard, E., Donaldson, L.A., Danylchuk, A.J., 2017. Fishing in the dark—the science and management of recreational fisheries at night. Bull. Mar. Sci. 93, 519–538.

Cooke, S.J., Venturelli, P., Twardek, W.M., Lennox, R.J., Brownscombe, J.W., Skov, C., Hyder, K., Suski, C.D., Diggles, B.K., Arlinghaus, R., 2021. Technological innovations in the recreational fishing sector: implications for fisheries management and policy. Rev. Fish Biol. Fish. 31, 253–288.

Cooke, S.J., Fangue, N.A., Bergman, J.N., Madliger, C.L., Cech, J.J., Eliason, E.J., Brauner, C.J., Farrell, A.P., 2022. Conservation physiology and the management of wild fish populations in the Anthropocene. Fish Physiol. 39A, 1–31.

Coombs, S., Bleckmann, H., 2014. The gems of the past: a brief history of lateral line research in the context of the hearing sciences. In: Coombs, S., Bleckmann, H., Fay, R.R., Popper, A.N. (Eds.), The Lateral Line System. Springer, NY, pp. 1–16.

Coombs, S., Montgomery, J., 2014. The role of flow and the lateral line in the multisensory guidance of orienting behaviors. In: Bleckmann, H., Mogdans, J., Coombs, S.L. (Eds.), Flow Sensing In Air And Water: Behavioural, Neural, And Engineering Principles Of Operation. Springer, NY, pp. 65–81.

Coombs, S., Görner, P., Münz, H. (Eds.), 1989a. The Mechanosensory Lateral Line: Neurobiology and Evolution. Springer, Berlin, Heidelberg, New York.

Coombs, S., Görner, P., Münz, H., 1989b. The Mechanosensory Lateral Line. Springer-Verlag, New York, p. 724.

Coombs, S., Bleckmann, H., Fay, R.R., Popper, A.N. (Eds.), 2014. The Lateral Line System. Springer, NY.

Coppock, A.G., González-Murcia, S.O., Srinivasan, M., Gardiner, N.M., Jones, G.P., 2020. Different responses of coral and rubble-dwelling coral reef damselfishes (Family:

Pomacentridae) to chemosensory cues from coral reef microhabitats. Mar. Biol. 167 (6), 1–11. https://doi.org/10.1007/s00227-020-03687-w.

Dartnall, H.J.A., 1975. Assessing the fitness of visual pigments for their photic Environments. In: Ali, M.A. (Ed.), Vision in Fishes: New Approaches in Research. Plenum Press, New York, NY, pp. 159–170.

Davies, T.W., Smyth, T., 2018. Why artificial light at night should be a focus for global change research in the 21st century. Glob. Chang. Biol. 24, 872–882.

Davis, W.E., 2004. Black crowned night-heron vibrates bill in water to attract fish. Southeast. Nat. 3, 127–128.

de Jong, K., Amorim, M.C.P., Fonseca, P.J., Heubel, K.U., 2018. Noise affects multimodal communication during courtship in a marine fish. Front. Ecol. Evol. 6, 113.

De Robertis, A., Handegard, N.O., 2013. Fish avoidance of research vessels and the efficacy of noise-reduced vessels: a review. ICES J. Mar. Sci. 70, 34–45.

Dempster, T., Kingsford, M.J., 2003. Homing of pelagic fish to fish aggregation devices (FADs): the role of sensory cues. Mar. Ecol. Prog. Ser. 258, 213–222.

Diebel, C.E., Proksch, R., Greenk, C.R., Neilson, P., Walker, M.M., 2000. Magnetite defines a vertebrate magnetoreceptor. Nature 406, 299–302.

Dijkgraaf, S., 1963. The functioning and significance of the lateral-line organs. Biol. Rev. 38, 51–105.

Domiononi, D.M., Halfwerk, W., Baird, E., Buxton, R.T., Fernández-Juricicm, E., Fristrup, K.M., McKenna, M.F., Mennitt, D.J., Perkin, E.K., Seymoure, B.M., Stoner, D.C., Tennessen, J.B., Toth, C.A., Tyrell, L.P., Wilson, A., Francis, C.D., Carter, N.H., Barber, J.R., 2020. Why conservation biology can benefit from sensory ecology. Nat. Ecol. Evol. 4, 502–511.

Douglas, R.H., Djamgoz, M.B.A., 1990. The Visual System of Fish. Chapman & Hall, New York, p. 526.

Døving, K.B., Lastein, S., 2009. The alarm reaction in fishes—odorants, modulations of responses, neural pathways. Ann. N. Y. Acad. Sci. 1170, 413–423.

Døving, K.B., Stabell, O.B., 2003. Trails in open water: Sensory cues in salmon Migration. In: Collin, S.P., Marshall, N.J. (Eds.), Sensory Processing in Aquatic Environments. Springer-Verlag, New York, pp. 39–52.

Døving, K.B., Hamdan, E.H., Hoglund, E., Kasumyan, A., Tuvikene, A.O., 2005. Review of the chemical and physiological basis of alarm reactions in cyprinids. In: Reutter, K., Kapoor, B.G. (Eds.), Fish Chemosenses. Science Publishers, Enfield, NH, pp. 133–163.

Duarte, C., Chapuis, L., Collin, S., Costa, D., Devassy, R., Eguiluz, V., Erbe, C., Gordon, T., Halpern, B., Harding, H., Havlik, M., Meekan, M., Merchant, N., Miksis-Olds, J., Parsons, M., Predragovic, M., Radford, A., Radford, C., Simpson, S., Slabbekoorn, H., Staaterman, E., Opzeeland, I., Winderen, J., Zhang, X., Juanes, F., 2021. The soundscape of the Anthropocene ocean. Science 371 (6529), eaba4658.

Elmer, L.K., Madliger, C.L., Blumstein, D.T., Elvidge, C.K., Fernández-Juricic, E., Horodysky, A.Z., Johnson, N.S., McGuire, N.P., Swaisgood, R.R., Cooke, S.J., 2021. Exploiting common senses: sensory ecology meets wildlife conservation and management. Conserv. Physiol. 9, coab002. https://doi.org/10.1093/conphys/coab002.

Erbe, C., McPherson, C., 2012. Acoustic characterisation of bycatch mitigation pingers on shark control nets in Queensland, Australia. Endanger. Species Res. 19, 109–121.

Erickson, G.J., 1979. Some frequencies of underwater noise produced by fishing boats affecting albacore catches. J. Acoust. Soc. Am. 66 (1), 296–299.

Erickson, L., Berkeley, S.A., 2008. Methods to reduce bycatch mortality in longline Fisheries. In: Camhi, M.D., Pikitch, E.K., Babcock, E.A. (Eds.), Sharks of the Open Ocean: Biology, Fisheries and Conservation. Blackwell Publishing Ltd., Oxford, UK, pp. 462–471.

Erisman, B.E., Rowell, T.J., 2017. A sound worth saving: acoustic characteristics of a massive fish spawning aggregation. Biol. Lett. 13, 20170656.

Fay, R.R., Popper, A.N., 1975. Modes of stimulation of the teleost ear. J. Exp. Biol. 62, 379–387.

Fay, R.R., Kendall, J.I., Popper, A.N., Tester, A.L., 1974. Vibration detection by the macula neglecta of sharks. Comp. Biochem. Physiol. A Physiol. 47, 1235–1240.

Ferrando, S., Bottaro, M., Gallus, L., Girosi, L., Vacchi, M., Tagliafierro, G., 2006. Observations of crypt neuron-like cells in the olfactory epithelium of a cartilaginous fish. Neurosci. Lett. 403, 280–282.

Findlay, C.R., Ripple, H.D., Coomber, F., Froud, K., Harries, O., van Geel, N.C.F., Calderan, S.V., Benjamins, S., Risch, D., Wilson, B., 2018. Mapping widespread and increasing underwater noise pollution from acoustic deterrent devices. Mar. Pollut. Bull. 135, 1042–1050.

Finger, T.E., 2009. Evolution of gustatory reflex systems in the brainstems of fishes. Integr. Zool. 4 (1), 53–63.

Fisher, H.S., Wong, B.B.M., Rosenthal, G.G., 2006. Alteration of the chemical environment disrupts communication in a freshwater fish. Proc. R. Soc. B Biol. Sci. 273, 1187–1193.

Fobert, E.K., Da Silva, K.B., Swearer, S.E., 2019. Artificial light at night causes reproductive failure in clownfish. Biol. Lett. 15 (7), 20190272.

Foote, K.G., 1980. Some frequencies of underwater noise produced by fishing boats affecting albacore catch. J. Acoust. Soc. Am. 66, 296–299. 1979. J. Acoust Soc. Am. 67, 1064.

Formicki, K., Sadowski, M., Tanski, A., Korzelecka-Orkisz, A., Winnicki, A., 2004. Behaviour of trout (*Salmo trutta* L.) larvae and fry in a constant magnetic field. J. Appl. Ichthyol. 20, 290–294.

Formicki, K., Korzelecka-Orkisz, A., Tański, A., 2019. Magnetoreception in fish. J. Fish Biol. 95, 73–91.

Fritsches, K., Warrant, E., 2004. Do tuna and billfish see in color? Pelagic Fish. Res. Progr. 9 (1), 1–8.

Fritsches, K.A., Marshall, N.J., Warrant, E.J., 2003. Retinal specializations in the blue marlin: eyes designed for sensitivity to low light levels. Mar. Freshw. Res. 54, 333–341.

Gallo-Cajiao, E., Archibald, E.C., Friedman, R., Steven, R., Fuller, R.A., Game, E.T., Morrison, T.H., Ritchie, E.G., 2018. Crowdfunding biodiversity conservation: crowdfunding conservation. Conserv. Biol. 32, 1426–1435.

Gardiner, J., 2011. Multimodal Sensory Integration in Shark Feeding Behaviors. PhD thesis, University of South Florida, Tampa, Fl.

Gardiner, J.M., Atema, J., 2007. Sharks need the lateral line to locate odor sources: rheotaxis and eddy chemotaxis. J. Exp. Biol. 210, 1925–1934.

Gardiner, J.M., Atema, J., 2014. Flow sensing in sharks: lateral line contributions to navigation and prey capture. In: Bleckmann, H., Mogdans, J., Coombs, S.L. (Eds.), Flow Sensing in Air and Water: Behavioural, Neural, and Engineering Principles of Operation. Springer, NY, pp. 127–146.

Gardiner, J., Hueter, R., Maruska, K., Sisneros, J., Casper, B., Mann, D., Dernski, L., 2012. Sensory physiology and behavior of elasmobranchs. In: Carrier, J., Musick, J., Heithaus, M. (Eds.), Biology of Sharks and Their Relatives, second ed. CRC Press, Boca Raton, FL, pp. 349–401.

Gardiner, J.M., Atema, J., Hueter, R.E., Motta, P.J., 2014. Multisensory integration and behavioral plasticity in sharks from different ecological niches. PLoS One 9, e93036.

Gaurav Ahuija, G., Ivandić, L., Saltürk, M., Oka, Y., Nadler, W., Korsching, S.I., 2013. Zebrafish crypt neurons project to a single, identified mediodorsal glomerulus. Zebrafish crypt neurons project to a single, identified mediodorsal glomerulus. Sci. Rep. 3, 2063. https://doi.org/10.1038/srep02063.

Gibson, L., Wilman, E.N., Laurance, W.F., 2017. How green is 'green' energy? Trends Ecol. Evol. 32, 922–935.

Gilbert, P.W., 1968. The shark: barbarian and benefactor. BioScience 18, 946–950.

Gill, A.B., Kimber, J.A., 2005. The potential for cooperative management of elasmobranchs and offshore renewable energy developments in UK waters. J. Mar. Biol. Assoc. UK 85, 1075–1081.

Gill, A.B., Bartlett, M., Thomsen, F., 2012. Potential interactions between diadromous fishes of U.K. conservation importance and the electromagnetic fields and subsea noise rom marine renewable energy developments. J. Fish Biol. 81, 664–695.

Gillis, J.A., Modrell, M.S., Northcutt, R.G., Catania, K.C., Luer, C.A., Baker, C.V.H., 2012. Electrosensory ampullary organs are derived from lateral line placodes in artilaginous fishes. Development 139, 3142–3146.

Gilman, E., Chaloupka, M., Bach, P., Fennell, H., Hall, M., Musyl, M., Piovano, S., Poisson, F., Song, L., 2020. Effect of pelagic longline bait type on species selectivity: a global synthesis of evidence. Rev. Fish Biol. Fish. 30, 535–551.

Godin, A.C., Wimmer, T., Wang, J., Worm, B., 2013. No effect from rare-earth metal deterrent on shark bycatch in a commercial pelagic longline trial. Fish. Res. 143, 131–135.

Goldberg, S.U., Nagelkerken, I., Marangon, E., Bonnet, A., Ferreira, C.M., Connel, S.C., 2018. Ecological complexity buffers the impacts of future climate on marine consumers. Nat. Clim. Chang. 8, 229–233.

Götz, T., Janik, V.M., 2013. Acoustic deterrent devices to prevent pinniped depredation: efficiency, conservation concerns and possible solutions. Mar. Ecol. Prog. Ser. 492, 285–302.

Götz, T., Janik, V.M., 2014. Target-specific acoustic predator deterrence in the marine environment. Anim. Conserv. 18, 102–111.

Grimaldo, E., Sistiaga, M., Herrmann, B., Larsen, R.B., Brinkhof, J., Tatone, I., 2018. Improving release efficiency of cod (*Gadus morhua*) and haddock (*Melanogrammus aeglefinus*) in the Barents Sea demersal trawl fishery by stimulating escape behaviour. Can. J. Fish. Aquat. Sci. 75 (3), 402–416.

Grubisic, M., Haim, A., Bhusal, P., Dominoni, D.M., Gabriel, K.M.A., Jechow, A., Kupprat, F., Lerner, A., Marchant, P., Riley, W., Stebelova, K., van Grunsven, R.H.A., Zeman, M., Zubidat, A.E., Hölker, F., 2019. Light pollution, circadian photoreception, and melatonin in vertebrates. Sustainability (Switzerland) 11 (22), 6400. https://doi.org/10.3390/su11226400.

Gurgens, C., Russell, D.F., Wilkens, L.A., 2000. Electrosensory avoidance of metal obstacles by the paddlefish. J. Fish Biol. 57 (2), 277–290. https://doi.org/10.1006/jfbi.2000.1292.

Guthrie, D.M., Muntz, W.R.A., 1993. Role of vision in fish behavior. In: Pitcher, T.J. (Ed.), Behavior of Teleost Fishes, second ed. Chapman & Hall, London, pp. 89–128.

Haine, O.S., Ridd, P.V., Rowe, R.J., 2001. Range of electrosensory detection of prey by *Carcharhinus melanopterus* and *Himantura granulate*. Mar. Freshw. Res. 52, 291–296.

Hamdani, E.H., Døving, K.B., 2006. Specific projection of the sensory crypt cells in the olfactory system in crucian carp, *Carassius carassius*. Chem. Senses 31, 63–67.

Hancz, C., 2020. Feed efficiency, nutrient sensing and feeding stimulation in aquaculture: a review. Acta Agr. Kapos. 24 (1), 35–54.

Hannah, R.W., Lomeli, M.J.M., Jones, S.A., 2015. Tests of artificial light for bycatch reduction in an ocean shrimp (*Pandalus jordani*) trawl: strong but opposite effects at the footrope and near the bycatch reduction device. Fish. Res. 170, 60–67.

Hansen, A., Finger, T.E., 2000. Phyletic distribution of crypt-type olfactory receptor neurons in fishes. Brain Behav. Evol. 55, 100–110.

Hansen, A., Reutter, K., 2004. Chemosensory systems in fish: structural, functional and ecological aspects. In: von der Emde, G., Mogdans, J., Kapoor, B.G. (Eds.), The Senses of Fish - Adaptations for the reception of natural stimuli. Kluwer Academic Publishers, Boston, pp. 55–89.

Hara, T.J., 1975. Olfaction in fish. Prog. Neurobiol. 5, 271–335.

Hara, T.J. (Ed.), 1982. Chemoreception in Fishes. Elsevier, Amsterdam.

Hara, T.J., 1986. Role of olfaction in fish behaviour. In: Pitcher, T.J. (Ed.), The Behavior of Teleost Fishes. Johns Hopkins University Press, Baltimore, MD, pp. 152–176.

Hara, T.J. (Ed.), 1992a. Fish Chemoreception. Chapman & Hall, London.

Hara, T.J., 1992b. Fish chemoreception. Fish and fisheries series 6. Chapman & Hall, London.

Hara, T.J., 2011a. Chemoreception (smell and taste): An introduction. In: Farrell, A.P. (Ed.), Encyclopedia of Fish Physiology: Volume 1: Sensing the Environment. Academic Press, San Diego, pp. 183–186.

Hara, T.J., 2011b. Morphology of the olfactory (Smell) system in fishes. In: Farrell, A.P. (Ed.), Encyclopedia of Fish Physiology: Volume 1: Sensing the Environment. Academic Press, San Diego, pp. 194–207.

Hara, T.J., 2011c. Neurophysiology of olfaction. In: Farrell, A.P. (Ed.), Encyclopedia of Fish Physiology: Volume 1: Sensing the Environment. Academic Press, San Diego, pp. 208–217.

Hara, T.J., 2011d. Chemosensory behavior. In: Farrell, A.P. (Ed.), Encyclopedia of Fish Physiology: Volume 1: Sensing the Environment. Academic Press, San Diego, pp. 227–235.

Hara, T.J., 2011e. Neurophysiology of gustation. In: Farrell, A.P. (Ed.), Encyclopedia of Fish Physiology: Volume 1: Sensing the Environment. Academic Press, San Diego, pp. 218–226.

Hara, T.J., 2011f. Morphology of the gustatory (taste) system in fishes. In: Farrell, A.P. (Ed.), Encyclopedia of Fish Physiology: Volume 1: Sensing the Environment. Academic Press, San Diego, pp. 187–192.

Hara, T.J., 2012. Fish Chemoreception. Springer, Dordrecht, Netherlands, p. 373.

Hara, T.J., Zielinski, B. (Eds.), 2007. Sensory Systems Neuroscience. Academic Press, San Diego, pp. 97–151. 373 pp.

Harpaz, S., Clark, L., 2006. Effects of addition of a bird repellent to fish diets on their growth and bioaccumulation. Aquacult. Res. 37, 132–138.

Hart, N.S., Collin, S.P., 2015. Shark senses and shark repellants. Integr. Zool. 10, 38–64.

Hasenei, A., Kerstetter, D.W., Horodysky, A.Z., Brill, R.W., 2020. Physiological limits to inshore invasion of Indo-Pacific lionfish (*Pterois* spp.): insights from the functional characteristics of their visual system and hypoxia tolerance. Biol. Invasions 22, 2079–2097.

Hasler, A.D., Scholz, A.T., 1983a. Olfactory Imprinting and Homing in Salmon. Springer, Berlin.

Hasler, A.D., Scholz, A.T., 1983b. Factors influencing smolt transformation: effects of seasonal fluctuations in hormone levels on transitions in morphology, physiology, and behavior. In: Olfactory Imprinting and Homing in Salmon. vol. 14. Springer, Berlin, Heidelberg. Zoophysiology.

Hassid, E., 1934. The History of the Great Chartreuse. Burns, Oates, and Washbourne LTD, London, pp. 1–300.

Hawkins, A.D., Popper, A.N., 2017. A sound approach to assessing the impact of underwater noise on marine fishes and invertebrates. ICES J. Mar. Sci. 74, 635–651.

Hay, M.E., 2009. Marine chemical ecology: chemical signals and cues structure marine populations, communities, and ecosystems. Ann. Rev. Mar. Sci. 1, 193–212.

Hazin, H.G., Hazin, F.H.V., Travassos, P., Erzini, K., 2005. Effect of light-sticks and electralume attractors on surface-longline catches of swordfish (*Xiphias gladius*, Linnaeus, 1959) in the southwest equatorial Atlantic. Fish. Res. 72, 271–277.

Herrera, K.J., Panier, T., Guggiana-Nilo, D., Engert, F., 2021. Larval zebrafish use olfactory detection of sodium and chloride to avoid salt water. Curr. Biol. 31, 782–793.

Higgs, D.M., Radford, C.A., 2012. The contribution of the lateral line to 'hearing' in fish. J. Exp. Biol. 8 (216), 1484–1490.

Higgs, D.M., Lui, Z., Mann, D.A., 2006. Hearing and mechanoreception. In: Evans, D.H. (Ed.), The Physiology of Fishes, third ed. CRC Press, Boca Raton, FL, pp. 391–429.

Hinch, S.G., Cooke, S.J., Healey, M.C., Farrell, A.P., 2006. Behavioural physiology of fish migrations: salmon as a model approach. In: Sloman, K.A., Wilson, R.W., Balshine, S. (Eds.), Behavior and Physiogy of Fish: Volume 24. Elsevier, New York, pp. 239–295.

Holland, K N., Brill, R.W., Chang, R.K.C., 1990. Horizontal and vertical movements of yellowfin and bigeye tuna associated with fish aggregating devices. Fish. Bull. 88, 493–507.

Horch, K., Salmon, M., 2009. Frequency response characteristics of isolated retinas from hatchling leatherback (*Dermochelys coriacea* L.) and loggerhead (*Caretta caretta* L.) sea turtles. J. Neurosci. Methods 178, 276–283.

Horodysky, A.Z., Graves, J.E., 2005. Application of pop-up satellite archival tag technology to estimate postrelease survival of white marlin (*Tetrapturus albidus*) caught on circle and straight-shank ("J") hooks in the western North Atlantic recreational fishery. Fish. Bull. 103, 84–96.

Horodysky, A.Z., Brill, R.W., Warrant, E.J., Musick, J.A., Latour, R.J., 2008a. Comparative visual function in five sciaenid fishes. J. Exp. Biol. 211 (22), 3601–3612.

Horodysky, A.Z., Brill, R.E., Fine, M.L., Musick, J.A., Latour, R.J., 2008b. Acoustic pressure and acceleration thresholds in six sciaenid fishes. J. Exp. Biol. 211 (9), 1504–1511.

Horodysky, A.Z., Brill, R.W., Warrant, E.J., Musick, J.A., Latour, R.J., 2010. Comparative visual function in four piscivorous fishes inhabiting Chesapeake Bay. J. Exp. Biol. 213, 1751–1761.

Horodysky, A.Z., Brill, R.W., Crawford, K.C., Seagroves, E.S., Johnson, A.K., 2013. Comparative visual ecophysiology of mid-Atlantic temperate reef fishes. Biol. Open 2, 1371–1381.

Horodysky, A.Z., Cooke, S.J., Brill, R.W., 2015. Physiology in the service of fisheries science: Why thinking mechanistically matters. Rev. Fish Biol. Fish. 25, 425–447.

Horodysky, A.Z., Cooke, S.J., Graves, J.E., Brill, R.W., 2016. Fisheries conservation on the high seas: linking conservation physiology and fisheries ecology for the management of large pelagic fishes. Conserv. Physiol 4, cou059.

Howard, S., Brill, R., Hepburn, C., Rock, J., 2018. Microprocessor-based prototype bycatch reduction device reduces bait consumption by the spiny dogfish and sandbar shark. ICES J. Mar. Sci. 75, 2235–2244.

Hu, Y., Majoris, J.E., Buston, P.M., Webb, J.F., 2019. Potential roles of smell and taste in the orientation behaviour of coral-reef fish larvae: insights from morphology. J. Fish Biol. 95 (1), 311–323. https://doi.org/10.1111/jfb.13793.

Hubbard, P.C., Barata, E.N., Canário, A.V.M., 2000. Olfactory sensitivity to changes in environmental [Ca2+] in the marine teleost *Sparus aurata*. J. Exp. Biol. 203, 3821–3829.

Hubbard, P.C., Ingleton, P.M., Bendell, L.A., Barata, E.N., Canário, A.V.M., 2002. Olfactory sensitivity to changes in environmental [Ca2+] in the freshwater teleost *Carassius auratus*: an olfactory role for the Ca2+−sensing receptor? J. Exp. Biol. 205, 2755–2764.

Hudson, R.B., 2000. Magnetoreception in the Short-Tailed Stingray, *Dasyatis brevicaudata*. MSc thesis, University of Auckland, New Zealand.

Hutchinson, M., Wang, J.H., Swimmer, Y., Holland, K., Kohin, S., Dewar, H., Wraith, J., et al., 2012. The effects of a lanthanide metal alloy on shark catch rates. Fish. Res. 131–133, 45–51.

Hutt, C., Lovell, S., Steinback, S., 2015. The Economics of Independent Marine Recreational Fishing Bait and Tackle Retail Stores in the United States, 2013. NOAA Technical Memorandum NMFS-F/SPO-151.

Ilyina, T., Zeebe, R.E., Brewer, P.G., 2010. Future ocean increasingly transparent to low frequency sound owing to carbon dioxide emissions. Nat. Geosci. 3, 18–22.

Inada, H., Arimoto, T., 2007. Trends on research and development of fishing light in Japan. J. Illum. Engng. Inst. Jpn. 91, 199–209.

Ings, D.W., Schneider, D.C., 1997. Use of acoustic imprinting in pond based cod aquaculture. In: Canadian Centre for Fisheries Innovation, Project AU, vol. 581, p. 27.

Janssen, J., 2004. Lateral line sensory ecology. In: von der Emde, G., Mogdans, J., Kapoor, B.G. (Eds.), The Senses of Fish - Adaptations for the reception of natural stimuli. Kluwer Academic Publishers, Boston, pp. 231–264.

Januma, S., Kajiwara, Y., Miura, T., Yamamoto, J., Haruyama, M., 1999. Trial use of artificial bait with tuna longline. Bull. Fish. Sci. Hokkaido Univ. 50, 71–76.

Jerlov, N.G., 1968. Optical Oceanography. Elsevier, New York, pp. 4–9.

Johnsen, S., 2003. Lifting the cloak of invisibility: the effects of changing optical conditions on pelagic crypsis. Integr. Comp. Biol. 43 (4), 580–590.

Johnsen, S., 2007. Does new technology inspire new directions? Examples drawn from pelagic visual ecology. Integr. Comp. Biol. 47, 799–807.

Johnsen, S., 2012. The Optics of Life: A Biologist's Guide to Light in Nature. Princeton University Press, Princeton, NJ.

Johnsen, S., Lohmann, K.J., 2005. The physics and neurobiology of magnetoreception. Nat. Rev. Neurosci. 6, 703–712.

Johnsen, S., Sosik, H.M., 2003. Cryptic coloration and mirrored sides as camouflage strategies in near-surface pelagic habitats: implications for foraging and predator avoidance. Limnol. Oceanogr. 48 (3), 1277–1288.

Jones, K.A., 1992. Food search behaviour in fish and the use of chemical lures in commercial and sports fishing. In: Hara, T.J. (Ed.), Fish Chemoreception. Chapman and Hall, London, pp. 288–320.

Jordan, L.K., 2008. Comparative morphology of stingray lateral line canal and electrosensory systems. J. Morphol. 269, 1325–1339.

Jordan, L.K., Mandelman, J.W., Kajiura, S.M., 2011. Behavioral responses to weak electric fields and a lanthanide metal in two shark species. J. Exp. Mar. Biol. Ecol. 409, 345–350.

Jordan, L.K., Mandelman, J.W., McComb, D.M., Fordham, S.V., Carlson, J.K., Werner, T.B., 2013. Linking sensory biology and fisheries by-catch reduction in elasmobranch fishes: a review with new directions for research. Conserv. Physiol. 1, 1–20. https://doi.org/10.1093/conphys/cot002.

Kageyama, C.J., 1999. What Fish See: Understanding Options and Color Shifts for Designing Lures and Flies. Frank Amato Publishers, Portland, OR.

Kaimmer, S., Stoner, A.W., 2008. Field investigation of rare-earth metal as a deterrent to spiny dogfish in the Pacific halibut fishery. Fish. Res. 94, 43–47.

Kajiura, S.M., Cornett, A.D., Yopak, K.E., 2010. Sensory adaptations to the environment: electro-receptors as a case study. In: Carrier, J.C., Musick, J.A., Heithaus, M.R. (Eds.), Sharks and Their Relatives II. Biodiversity, Adaptive Physiology, and Conservation. CRC Press, Boca Raton, pp. 393–433.

Kalinoski, M., Hirons, A., Horodysky, A., Brill, R., 2014. Spectral sensitivity, luminous sensitivity, and temporal resolution of the visual systems in three sympatric coastal shark species. J. Comp. Physiol. A Neuroethol. Sens. Neural Behav. Physiol. 200 (12), 997–1013.

Kalmijn, A.J., 1982. Electric and magnetic field detection in elasmobranch fishes. Science 218, 916–918.

Kasumyan, A.O., 2003. The lateral line in fish: structure, function, and role in behavior. J. Ichthyol. 43, S175–S213.

Kasumyan, A.O., 2004. The olfactory system of fish: structure, function, and role in behavior. J. Ichthyol. 44 (Suppl. 2), S180–S223.

Kasumyan, A.O., 2019. The taste system in fishes and the effects of environmental variables. J. Fish Biol. 95, 155–178. https://doi.org/10.1111/jfb.13940.

Kasumyan, A.O., Doving, K.B., 2003. Taste preferences in fishes. Fish Fish. 4, 289–347.

Kasumyan, A.O., Mouromtsev, G.E., 2020. The teleost fish, blue gourami Trichopodus trichopterus, distinguishes the taste of chemically similar substances. Sci. Rep. 10 (1), 1–10. https://doi.org/10.1038/s41598-020-64556-6.

Kavet, R., Wyman, M.T., Klimley, A.P., 2016. Modeling magnetic fields from a DC power cable buried beneath San Francisco Bay based on empirical measurements. PLoS One 11, e0148543.

Kelley, J.L., Chapuis, L., Davies, W.I.L., Collin, S.P., 2018. Sensory system responses to human-induced environmental change. Front. Ecol. Evol. 6, 95. https://doi.org/10.3389/fevo.2018.00095.

Kemp, W.M., Boynton, W.R., Adolf, J.E., Boesch, D.F., Boicourt, W.C., Brush, G., Cornwell, J.C., Fisher, T.R., Glibert, P.M., Hagy, J.D., et al., 2005. Eutrophication of Chesapeake Bay: historical trends and ecological interactions. Mar. Ecol. Prog. Ser. 303, 1–29.

Kerr, J.R., Kemp, P.S., 2019. Masking a fish's detection of environmental stimuli: application to improving downstream migration at river infrastructure. J. Fish Biol. 95, 228–237.

Khan, S., Khan, R.U., Alam, W., Sultan, A., 2018. Evaluating the nutritive profile of three insect meals and their effects to replace soya bean in broiler diet. J. Anim. Physiol. Anim. Nutr. 102 (2), e662–e668.

Kikuchi, R., 2017. Risk formulation for the sonic effects of offshore wind farms on fish in the EU region. Mar. Pollut. Bull. 60, 172–177.

Kilfoyle, A.K., Jermain, R.F., Dhanak, M.R., Huston, J.P., Spieler, R.E., 2018. Effects of EMF emission from undersea electric cables on coral reef fish. Bioelectromagnetics 39, 35–52.

Kingsford, M.J., Leis, J.M., Shanks, A., Lindeman, K.C., Morgan, S.G., Pineda, J., 2002. Sensory environments, larval abilities and local self-recruitment. Bull. Mar. Sci. 70, 309–340.

Kirschvink, J.L., Jones, D.S., Macfadden, B.J. (Eds.), 1985a. Magnetite Biomineralization and Magnetoreception by Living Organisms: A New Biomagnetism. Plenum, New York.

Kirschvink, J., Walker, K.L., Chang, M.M., Dizon, A.E., Peterson, K.A., 1985b. Chains of single-domain magnetite particles in the chinook salmon, Oncorhynchus tshawytscha. J. Comp. Physiol. A 157, 375–381.

Kiyohara, S., Sakata, Y., Yoshitomi, T., Tsukahara, J., 2002. The "goatee" of goatfish: Innervation of taste buds in the barbels and their representation in the brain. Proc. R. Soc. B Biol. Sci. 269 (1502), 1773–1780. https://doi.org/10.1098/rspb.2002.2086.

Klaprat, D.A., Evans, R.E., Hara, T.J., 1992. Environmental contaminants and chemoreception in fishes. In: Hara, T.J. (Ed.), Fish Chemoreception. Chapman and Hall, London.

Klimley, A.P., 1993. Highly directional swimming by scalloped hammerhead sharks, Sphyrna lewini, and substrate irradiance, temperature, bathymetry and geomagnetic field. Mar. Biol. 117, 1–22.

Klimley, A.P., Putman, N.F., Keller, B.A., Noakes, D., 2021. A call to assess the impacts of electromagnetic fields from subsea cables on the movement ecology of marine migrants. Conserv. Sci Pract. 2021, e436.

Knudsen, F.R., Enger, P.S., Sand, O., 1994. Avoidance responses to low frequency sound in downstream migrating Atlantic salmon smolt, Salmo salar. J. Fish Biol. 45, 227–233.

Knudsen, F.R., Schreck, C.B., Knapp, S.M., Enger, P.S., Sand, O., 1997. Infrasound produces flight and avoidance responses in Pacific juvenile salmonids. J. Fish Biol. 51, 824–829. https://doi.org/10.1111/j.1095-8649.1997.tb02002.x.

Konstantine, J., Rountos, K.J., Gobler, C.J., Pikitch, E.K., 2017. Ontogenetic differences in swimming behavior of fish exposed to the harmful dinoflagellate *Cochlodinium polykrikoides*. Trans. Am. Fish. Soc. 146, 1081–1091.

Kotrschal, K., 2000. Taste(s) and olfaction(s) in fish: a review of specialized sub-systems and central integration. Eur. J. Physiol. 439 (Suppl), R178–R180.

Kotrschal, K., van Staaden, M.J., Huber, R., 1998. Fish brains: evolution and environmental relationships. Rev. Fish Biol. Fish. 8, 373–408.

Kreysing, M., Pusch, R., Haverkate, D., Landsberger, M., Engelmann, J., Ruiter, J., Mora-Ferrer, C., Ulbricht, E., Grosche, J., Franze, K., Streif, S., Schumacher, S., Makarov, F., Kacza, J., Guck, J., Wolburg, H., Bowmaker, J.K., Von Der Emde, G., Schuster, S., Francke, M., 2012. Photonic crystal light collectors in fish retina improve vision in turbid water. Science 336 (6089), 1700–1703. https://doi.org/10.1126/science.1218072.

Kumar, K.V.A., Pravina, P., Meenakumari, B., 2016. Bait, bait loss, and depredation in pelagic longline fisheries—a review. Rev. Fish. Sci. Aquac. 24, 295–304.

Kushland, J.A., 1973. Bill-vibrating: a prey attracting behavior of the snowy egret, Leucophoyx thula. Am. Midl. Nat. 89, 509–512.

Laberge, F., Hara, T.J., 2001. Neurobiology of fish olfaction: a review. Brain Res. Rev. 36 (2001), 46–59.

Ladich, F., Collin, S.P., Moller, P., Kapoor, B.G. (Eds.), 2006. Communication in Fishes. vol. 1. Science Publishers, Enfield, pp. 3–43.

Lamb, C.F., Finger, T.E., 1997. Gustatory control of feeding behavior in goldfish. Physiol. Behav. 57, 483–488.

Langbehn, T.J., Varpe, O., 2017. Sea-ice-loss boosts visual search: fish foraging and changing pelagic interactions in polar oceans. Glob. Chang. Biol. 23, 5318–5330.

Langhamer, O., 2012. Artificial reef effect in relation to offshore renewable energy conversion: state of the art. Scientific World Journal 2012, 386713.

Larsen, O.N., Radford, C., 2018. Acoustic conditions affecting sound communication in air and underwater. In: Slabbekoorn, H., Dooling, R.J., Popper, A.N., Ray, R. (Eds.), Effects of Anthropogenic Noise on Animals. Springer New York, New York, NY, pp. 109–144.

Lecchini, D., Nakamura, Y., 2013. Use of chemical cues by coral reef animal larvae for habitat selection. Aquat. Biol. 19 (3), 231–238.

Lecchini, D., Shima, J., Banaigs, B., Galzin, R., 2005. Larval sensory abilities and mechanisms of habitat selection of a coral reef fish during settlement. Oecologia 143, 326–334.

Lennox, R.J., Alos, J., Cooke, S.J., Horodysky, A.Z., Klefoth, T., Monk, C., Arlinghaus, R., 2017. What makes fish vulnerable to capture by hooks? A conceptual framework and a review of key determinants. Fish Fish. 18, 986–1010.

Lennox, R.J., et al., 2018. One hundred pressing questions on the future of global fish migration science, conservation, and policy. Front. Ecol. Evol. 7, 286.

Levin, L.E., Levin, A.F., 1994. Conditioning as an aid to fish harvest. Aquacult. Eng. 13, 21–210.

Levine, J.S., MacNichol, E.F., 1979. Visual pigments in teleost fishes: effects of habitat, microhabitat, and behavior on visual system evolution. Sens. Processes 3, 95–131.

Liao, I.C., Chang, E.Y., 2003. Role of sensory mechanisms in predatory feeding behavior of juvenile red drum *Sciaenops ocellatus*. Fish. Sci. 69, 317–322.

Lindell, S., Miner, S., Goudey, C., Kite-Powell, H., Page, S., 2012. Acoustic conditioning and ranching of black sea bass *Centropristis striata* in Massachusetts USA. Bull. Fish. Res. Agency 35, 103–110.

Lisney, T.J., Collin, S.P., 2008. Retinal ganglion cell distribution and spatial resolving power in elasmobranchs. Brain Behav. Evol. 72, 59–77.

Litherland, L., Collin, S.P., 2008. Comparative visual function in elasmobranchs: spatial arrangement and ecological correlates of photoreceptor and ganglion cell distributions. Vis. Neurosci. 25, 549–561.

Litherland, L., Collin, S.P., Fritsches, K.A., 2009. Visual optics and ecomorphology of the growing shark eye: a comparison between deep and shallow water species. J. Exp. Biol. 212, 3583–3594.

Lohmann, K.J., 2010. Magnetic-field perception. Nature 464, 1140–1142.

Lohmann, K.J., Lohmann, C.M., Endres, C.S., 2008. The sensory ecology of ocean navigation. J. Exp. Biol. 211, 1719–17287.

Lokkeborg, S., 1990. Reduced catch of under-sized cod (*Gadus morhua*) in longlining by using artificial bait. Can. J. Fish. Aquat. Sci. 47, 1112–1115.

Lokkeborg, S., Fern, A., Humbsorstad, O.-B., 2010. Fish behavior in relation to Longlines. In: He, P. (Ed.), Behavior of Marine Fishes Capture processes and conservation challenges. Wiley-Blackwell, Ames, Iowa, pp. 105–140.

Lomeli, M.J., Wakefield, W.W., 2014. Examining the potential use of artificial illumination to enhance Chinook salmon escapement out a bycatch reduction device in a Pacific hake midwater trawl. In: National Marine Fisheries Service, Northwest Fisheries Science Center Report, Seattle, WA, p. 15.

Loretz, C.A., 2008. Extracellular calcium-sensing receptors in fishes. Comp. Biochem. Physiol. 149A, 225–245.

Losey, G.S., Cronin, T.W., Goldsmith, T.H., Hyde, D., Marshall, N.J., McFarland, W.N., 1999. The UV visual world of fishes: a review. J. Fish Biol. 54, 921–943.

Lu, Z., Tomchik, S., 2002. Effects of a red-tide toxin on fish hearing. J. Comp. Physiol. A 188, 807–813.

Lu, Z., Xu, Z., 2002. Effects of saccular otolith removal on hearing sensitivity of the sleeper goby (*Dormitator latifrons*). J. Comp. Physiol. A 188, 595–602.

Lürling, M., 2012. Infodisruption: pollutants interfering with the natural chemical information conveyance in aquatic systems. In: Brönmark, C., Hansson, L.-A. (Eds.), Chemical Ecology in Aquatic Systems. Oxford University Press, Oxford, pp. 250–271.

Lythgoe, J.N., 1968. Visual pigments and visual range underwater. Vision Res. 8, 997–1012.

Lythgoe, J.N., 1975. Problems of seeing colours under water. In: Ali, M.A. (Ed.), Vision in Fishes: New Approaches in Research. Plenum Press, New York, pp. 619–634.

Lythgoe, J.N., 1979. Ecology of Vision. Clarendon Press, Oxford.

Lythgoe, J.N., 1988. Light and vision in the aquatic environment. In: Atema, J., Fay, R.R., Popper, A.N., Tavolga, W.N. (Eds.), Sensory Biology of Aquatic Animals. Springer-Verlag, New York, pp. 131–149.

Lyytimaki, J., 2013. Nature's nocturnal services: light pollution as a non-recognized challenge for ecosystem services research and management. Ecosyst. Serv. 3, e44–e48.

Mackie, A.M., Mitchell, A.I., 1985. Identification of gustatory feeding stimulants for fish-applications in aquaculture. In: Cowey, C.B., Mackie, A.M., Bell, J.G. (Eds.), Nutrition and Feeding in Fish. Academic Press, London, pp. 177–189.

Madliger, C.L., 2012. Toward improved conservation management: a consideration of sensory ecology. Biodivers. Conserv. 21, 3277–3286.

Maes, J., Turnpenny, A.W.H., Lambert, D.R., Nedwell, J.R., Parmentier, A., Ollevier, F., 2004. Field evaluation of a sound system to reduce estuarine fish intake rates at a power plant cooling water inlet. J. Fish Biol. 64, 938–946.

Malcorps, W., Kok, B., van't Land, M., Fritz, M., van Doren, D., Servin, K., van der Heijden, P., Palmer, R., Auchterlonie, N.A., Rietkerk, M., Santos, M.J., Davies, S.J., 2019. The sustainability conundrum of fishmeal substitution by plant ingredients in Shrimp Feeds. Sustainability 11 (4), 1–19.

Mangel, J.C., Wang, J., Alfaro-Shigueto, J., Pingo, S., Jiminez, A., Carvalho, F., Swimmer, Y., Godley, B.J., 2018. Illuminating gillnets to save seabirds and the potential for multi-taxa bycatch mitigation. R. Soc. Open Sci. 5 (2018), 180254.

Mann, S., Sparks, N.H.C., Walker, M.M., Krishvink, J.L., 1988. Ultrastructure, morphology and organization of biogenic magnetite from sockeye salmon, *Oncorhynchus nerka*: Implications for magnetoreception. J. Exp. Biol. 140, 35–49.

Marra, L.J., 1989. Sharkbite on the SL submarine lightwave cable system: history, causes and resolution. IEEE J. Ocean. Eng. 14, 230–237.

Marui, T., Caprio, J., 1992. Teleost gustation. In: Hara, T.J. (Ed.), Fish Chemoreception. Chapman and Hall, London, pp. 171–198.

Mastoraki, M., Mollá Ferrándiz, P., Vardali, S.C., Kontodimas, D.C., Kotzamanis, Y.P., Gasco, L., Chatzifotis, S., Antonopoulou, E., 2020. A comparative study on the effect of fish meal substitution with three different insect meals on growth, body composition and metabolism of European sea bass (*Dicentrarchus labrax* L.). Aquaculture 528 (May), 735511.

Mazur, M.M., Beauchamp, D.A., 2006. Linking piscivory to spatial-temporal distributions of pelagic prey fishes with a visual foraging model. J. Fish Biol. 69, 151–175.

McComb, D.M., Frank, T.M., Hueter, R.E., Kajiura, S.M., 2010. Temporal resolution and spectral sensitivity of the visual system of three coastal shark species from different light environments. Physiol. Biochem. Zool. 83, 299–307.

McCormick, M.I., Allan, B.J., Harding, H., Simpson, S.D., 2018. Boat noise impacts risk assessment in a coral reef fish but effects depend on engine type. Sci. Rep. 8 (1), 3847.12.

McCutcheon, S.M., Kajiura, S.M., 2013. Electrochemical properties of lanthanide metals in relation to their application as shark repellents. Fish. Res. 147, 47–54.

McFarland, W.N., 1986. Light in the sea: correlations with behaviors of fishes and invertebrates. Am. Zool. 26, 389–401.

McFarland, W.N., Munz, F.W., 1975. Part III: the evolution of photopic visual pigments in fishes. Vision Res. 15, 1071–1080.

Melli, V., Krag, L.A., Herrmann, B., Karlsen, J.D., 2018. Investigating fish behavioural responses to LED lights in trawls and potential applications for bycatch reduction in the *Nephrops*-directed fishery. ICES J. Mar. Sci. 75, 1682–1692.

Merchant, N.D., 2019. Underwater noise abatement: economic factors and policy options. Environ. Sci. Policy 92, 116–123.

Meulenbroek, P., Drexler, S., Nagel, C., Geistler, A., Waidbacher, H., 2018. The importance of a constructed near-nature-like Danube fish by-pass as a lifecycle fish habitat for spawning, nurseries, growing and feeding: a long-term view with remarks on management. Mar. Freshw. Res. 69, 1857–1869.

Miller, R.J., Smith, S.J., 2012. Nova Scotia's bloodworm harvest: assessment, regulation, and governance. Fish. Res. 113, 84–93.

Modrell, M.S., Bemis, W.E., Northcutt, G.R., Davis, M.C., Baker, C.V.H., 2011. Electrosensory ampullary organs are derived from lateral line placodes in bony fishes. Nat. Commun. 2, 496. https://doi.org/10.1038/ncomms1502.

Mogdans, J., 2019. Sensory ecology of the fish lateral-line system: morphological and physiological adaptations for the perception of hydrodynamic stimuli. J. Fish Biol. 95, 53–72.

Montgomery, J.C., Bodznick, D., 1994. An adaptive filter that cancels self-induced noise in the electrosensory and lateral line mechanosensory systems of fish. Neurosci. Lett. 174 (2), 145–148.

Montgomery, J.C., Walker, M.M., 2001. Orientation and navigation in elasmobranchs: which way forward? Environ. Biol. Fishes 60, 109–116.

Montgomery, J., Coombs, S., Halstead, M., 1995. Biology of the mechanosensory lateral line in fishes. Rev. Fish Biol. Fish. 5, 399–416.

Montgomery, J.C., Baker, C.F., Carton, A.G., 1997. The lateral line can mediate rheotaxis in fish. Nature 389, 960–963.

Montgomery, J., Carton, Z.G., Voigt, R., Baker, C., Diebel, C., 2000. Sensory processing of water currents by fishes. Philos. Trans. R. Soc. Lond. B Biol. Sci. 355, 1325–1327.

Montgomery, J.C., Bleckmann, H., Coombs, S., 2014. Sensory ecology and neuroethology of the lateral line. In: Coombs, S., Bleckmann, H., Fay, R.R., Popper, A.N. (Eds.), The Lateral Line System. Springer, New York, NY, pp. 121–150.

Moore, M.V., Kohler, S.J., Cheers, M.S., 2006. Artificial light at night in freshwater habitats and its potential ecological effects. In: Rich, C., Longcore, T. (Eds.), Ecological Consequences of Artificial Night Lighting. Island Press, Washington, pp. 365–384.

Moraga, A.D., Wilson, A.D.M., Cooke, S.J., 2015. Does lure colour influence catch per unit effort, fish capture size and hooking injury in angled largemouth bass? Fish. Res. 172, 1–6.

Morais, S., 2017. The physiology of taste in fish: potential implications for feeding stimulation and gut chemical sensing. Rev. Fish. Sci. Aquac. 25 (2), 133–149.

Munz, F.W., 1977. Evolutionary adaptations of fishes to the photic environment. In: Crescitelli, F. (Ed.), Handbook of Sensory Physiology. Vol. VII/5. Springer-Verlag, Berlin.

Murchy, K.A., Vetterm, B.J., Breym, M.K., Amberg, J.J., Gaikowski, M.P., Mensinger, A.F., 2016. Not all carp are created equal: impacts of broadband sound on common carp swimming behavior. Proc. Meetings Acoust. 27, 010032.

Murray, R.W., 1962. The response of the ampullae of Lorenzini of elasmobranchs to electrical stimulation. J. Exp. Biol. 39, 119–128.

Mussen, T.D., Cech, J.J., 2013. The roles of vision and lateral-line system in Sacramento splittail's fish-screen avoidance behaviors: evaluating vibrating screens as potential fish deterrents. Environ. Biol. Fishes 96, 971–980.

Mussen, T.D., Cech, J.J., 2019. Assessing the use of vibrations and strobe lights at fish screens as enhanced deterrents for two estuarine fishes. J. Fish Biol. 95, 238–246.

Mussen, T.D., Patton, O., Cocherell, D., Ercan, A., Bandeh, H., Kavvas, L., Cech, J.J., Fangue, N.A., 2014. Can behavioral fish guidance devices protect juvenile Chinoook salmon (Oncorhynchus tshawytscha) from entrainment into unscreened water-diversion pipes? Can. J. Fish. Aquat. Sci. 71, 1209–1219.

Nadermann, N., Seward, R.K., Volkoff, H., 2019. Effects of potential climate change -induced environmental modifications on food intake and the expression of appetite regulators in goldfish. Comp. Biochem. Physiol. A Mol. Integr. Physiol. 235 (May), 138–147.

Nagelkerken, I., Doney, S.C., Munday, P.L., 2019. Consequences of anthropogenic changes in the sensory landscape of marine animals. Oceanogr. Mar. Biol. 57, 229–264.

Nearing, J., Betka, M., Quinn, S., Hentschel, H., Elger, M., Baum, M., Bai, M., Chattopadyhay, N., Brown, E.M., Hebert, S.C., Harris, H.W., 2002. Polyvalent cation receptor proteins (CaRs) are salinity sensors in fish. Proc. Natl. Acad. Sci. U. S. A. 99, 9231–9236.

New, J.G., Kang, P.Y., 2000. Multimodal sensory integration in the strike-feeding behavior of predatory fishes. Philos. Trans. R. Soc. Lond. B Biol. Sci. 355, 1321–1324.

New, J.G., Fewkes, L.A., Khan, K.N., 2001. Strike feeding behavior in the muskellunge, Esox masquinongy: contributions of the lateral line and visual sensory systems. J. Exp. Biol. 204, 1207–1221.

Newton, K.C., Gill, A.B., Kajiura, S.M., 2019. Electroreception in marine fishes: chondrichthyans. J. Fish Biol. 95 (1), 135–154.

Nguyen, K.Q., Winger, P.D., 2019. Artificial light in commercial industrialized fishing applications: a review. Rev. Fish. Sci. Aquac. 27, 106–126.

Nieman, C.L., Bruskotter, J.T., Braig, E.C., Gray, S.M., 2020. You can't just use gold: elevated turbidity alters successful lure color for recreational walleye fishing. J. Great Lakes Res. 46, 589–596.

Nilsson, D.E., Warrant, E.J., Johnsen, S., 2014. Computational visual ecology in the pelagic realm. Philos. Trans. R. Soc. B 369, 20130038.

Noatch, M.R., Suski, C.D., 2012. Non-physical barriers to deter fish movements. Environ. Rev. 20, 71–82.

Northmore, D.P.M., 2011. The Optic Tectum. Encyclopedia of Fish Physiology: From Genome to Environment. Elsevier, pp. 131–142.

O'Connell, C.P., Abel, D.C., Stroud, E.M., Rice, P.H., 2011. Analysis of permanent magnets as elasmobranch bycatch reduction devices in hook-and-line and longline trials. Fish. Bull. 109, 394–401.

O'Connell, C.P., Stroud, E.M., He, P., 2014a. The emerging field of electrosensory and semiochemical shark repellents: mechanisms of detection, overview of past studies, and future directions. Ocean Coast. Manag. 97, 2–11.

O'Connell, C.P., He, P., Joyce, J., Stroud, E.M., Rice, P.H., 2014b. Effects of the SMART™ (Selective Magnetic and Repellent-Treated) hook on spiny dogfish catch in a longline experiment in the Gulf of Maine. Ocean Coast. Manag. 97, 38–43.

O'Connor, J.J., Fobert, E.K., Besson, M., Jacob, B., Lecchini, D., 2019. Live fast, die young: behavioural and physiological impacts of light pollution on a marine fish during arval recruitment. Mar. Pollut. Bull. 146, 908–914.

O'Hara, J., Wilcox, J.R., 1990. Avoidance responses of loggerhead turtles, *Caretta caretta*, to low frequency sounds. Copeia 1990, 564–567.

Öhman, M.C., Sigray, P., Westerberg, H., 2007. Offshore windmills and the effects of electromagnetic fields on fish. Ambio 36, 630–633.

Oiestad, V., Pedersen, T., Folkvord, A., Bjordal, A., Kvenseth, P.G., 1987. Automatic feeding and harvesting of juvenile Atlantic cod (*Gadus morhua*) in a pond. Model. Identif. Control. 8, 39–46.

Ono, K., Simenstad, C.A., 2014. Reducing the effect of overwater structures on migrating juvenile salmon: an experiment with light. Ecol. Eng. 71, 180–189.

Paris, C.B., Atema, J., Irisson, J.O., Kingsford, M., Gerlach, G., Guigand, C.M., 2013. Reef odor: a wake up call for navigation in reef fish larvae. PLoS One 8 (8), 1–8. https://doi.org/10.1371/journal.pone.0072808.

Parkyn, D.C., Hawryshyn, C.W., 2000. Spectral and ultraviolet-polarization sensitivity in juvenile salmonids: a comparative analysis using electrophysiology. J. Exp. Biol. 203, 1173–1191.

Patanasatienkul, T., Delphino, K.V.C.M., Thakur, K.K., 2020. Comparing the effectiveness of traditional and alternative baits in Prince Edward Island, Canada lobster fishery. Front. Mar. Sci. 7, 589549. https://doi.org/10.3389/fmars.2020.589549.

Paxton, A.B., Taylor, J.C., Nowacek, D.P., Dale, J., Cole, E., Voss, C.M., Peterson, C.H., 2017. Seismic survey noise disrupted fish use of a temperate reef. Mar. Policy 78, 68–73.

Pernet, F., Browman, H.I., 2021. The Future is now: marine aquaculture in the Anthropocene. ICES J. Mar. Sci. 78, 315–322.

Peters, R.C., Struik, M.L., Bretschneider, F., 2001. Electroreception in freshwater catfish: the biologically adequate stimulus. In: Kapoor, B.G., Hara, T.J. (Eds.), Sensory Biology of Jawed Fishes. Science Publishers Inc., Enfield, NH, pp. 274–296.

Pierre, J.P., Norden, W.S., 2006. Reducing seabird bycatch in longline fisheries using an natural olfactory deterrent. Biol. Conserv. 130, 406–415.

Piniak, W.E.D., Mann, D.A., Eckert, S.A., Harms, C.A., 2012. Amphibious hearing in sea turtles. In: Popper, A.N., Hawkins, A. (Eds.), The Effects of Noise on Aquatic Life. Springer, New York, NY, pp. 83–87.

Poddubny, A.G., Malinin, L.K., Spector, I., 1979. Biotelemetry in Fisheries Research. Moskva. (In Russian), p. 188.

Pohlmann, K., Atema, J., Breithaupt, T., 2004. The importance of the lateral line in nocturnal predation of piscivorous catfish. J. Exp. Biol. 207, 2971–2978.

Popper, A.N., 1983. Organization of the inner ear and processing of acoustic information. In: Northcutt, R.G., Davis, R. (Eds.), Fish Neurobiology. University of Michigan Press, Ann-Arbor, MI, pp. 125–178.

Popper, A.N., Coombs, S., 1982. The morphology and evolution of the ear in Actinopterygian fishes. Am. Zool. 22, 311–328.

Popper, A.N., Fay, R.R., 1973. Sound detection and processing by fish: critical review and major research questions. J. Acoust. Soc. Am. 53, 1515–1529.

Popper, A.N., Fay, R.R., 1993. Sound detection and processing by fish: critical review and major research questions. Brain Behav. Evol. 41, 14–38.

Popper, A.N., Fay, R.R., 1997. Evolution of the ear and hearing: issues and questions. Brain Behav. Evol. 50, 213–221.

Popper, A.N., Fay, R.R., 1999. The auditory periphery in fishes. In: Fay, R.R., Popper, A.N. (Eds.), Comparative Hearing: Fish and Amphibians. Springer, New York, Berlin, Heidelberg, pp. 43–100.

Popper, A.N., Fay, R.R., 2011. Rethinking sound detection by fishes. Hear. Res. 273, 25–36.

Popper, A.N., Hawkins, A.D., 2019. An overview of fish bioacoustics and the impacts of anthropogenic sounds on fishes. J. Fish Biol. 94, 692–713.

Popper, A.N., Schilt, C.R., 2008. Hearing and acoustic behavior (basic and applied). In: Webb, J.F., Fay, R.R., Popper, A.N. (Eds.), Fish Bioacoustics. Springer Science+Business Media, LLC, New York, NY, pp. 17–48.

Popper, A.N., Fay, R.R., Platt, C., Sand, O., 2003. Sound detection mechanisms and capabilities in teleost fishes. In: Collin, S.P., Marshall, N.J. (Eds.), Sensory Processing in Aquatic Environments. Springer-Verlag, NY, pp. 3–38.

Popper, A.N., Plachta, D.T.T., Mann, D.A., Higgs, D., 2004. Response of clupeid fish to ultrasound: a review. ICES J. Mar. Sci. 61, 1057–1061.

Popper, A.N., Hawkins, A.D., Thomsen, F., 2020. Taking the animals' perspective regarding underwater anthropogenic sound. Trends Ecol. Evol. 35, 787–794.

Porsmoguer, S.B., Bănaru, D., Boudouresque, C.F., Dekeyser, I., Almarcha, C., 2015. Hooks equipped with magnets can increase catches of blue shark (*Prionace glauca*) by longline fishery. Fish. Res. 172, 345–351.

Porter, L.L., Hayes, M.C., Jackson, A.D., Burke, B.J., Moser, M.L., Wagner, R.S., 2017. Behavioral responses of Pacific lamprey to alarm cues. J. Fish Wildl. Manag. 8, 101–113.

Porteus, C.S., Roggatz, C.C., Velez, Z., Hardege, J.D., Hubbard, P.C., 2021. Acidification can directly affect olfaction in marine organisms. J. Exp. Biol. 224, jeb237941.

Pothmann, L., Wilkens, L.A., Schweitzer, C., Hofmann, M.H., 2009. Two parallel ascending pathways from the dorsal octavolateral nucleus to the midbrain in the paddlefish *Polyodon spathula*. Brain Res. 1265, 93–102.

Putland, R.L., Mensinger, A.F., 2019. Acoustic deterrents to manage fish populations. Rev. Fish Biol. Fish. 29, 789–807.

Putland, R.L., Montgomery, J.C., Radford, C.A., 2018. Ecology of fish hearing. J. Fish Biol., 1–14. https://doi.org/10.1111/jfb.13867.

Quinn, T.P., 1980. Evidence for celestial and magnetic compass orientation in lake- migrating sockeye salmon fry. J. Comp. Physiol. A 137, 243–248.

Radford, C.A., Montgomery, J.C., Caiger, P., Higgs, D.M., 2012. Pressure and particle motion detection thresholds in fish: a re-examination of salient auditory cues in teleosts. J. Exp. Biol. 215, 3429–3435.

Ramcharitar, J.U., Higgs, D.M., Popper, A.N., 2005. Audition in sciaenid fishes with different swim bladder-inner ear configurations. J. Acoust. Soc. Am. 1 (119), 439–443.

Ramcharitar, J., Gannon, D.P., Popper, A.N., 2011. Bioacoustics of fishes of the family Sciaenidae (Croakers and Drums). Trans. Am. Fish. Soc. 135, 1409–1431.

Reutter, K., Kapoor, B.G., 2005. Fish Chemosenses. Science Publishers Inc, Enfield, NH.

Richard, J.D., 1968. Fish attraction with a pulsed low frequency sound. J. Fish. Res. Board Can. 25, 1441–1452.

Roch, S., Friedrich, C., Brinker, A., 2020. Uptake routes of microplastics in fishes: practical and theoretical approaches to test existing theories. Sci. Rep. 10 (1), 1–12. https://doi.org/10.1038/s41598-020-60630-1.

Rogers, P.H., Cox, M., 1988. Underwater sound as a biological stimulus. In: Atema, J., Fay, R.R., Popper, A.N., Tavolga, W.N. (Eds.), Sensory Biology of Aquatic Animals. Springer-Verlag, New York, NY, pp. 131–149.

Roggatz, C.C., Fletcher, N., Benoit, D.M., Algar, A.C., Doroff, A., Wright, B., Wollenberg Valero, K.C., Hardege, J.D., 2019. Saxitoxin and tetrodotoxin bioavailability increases in future oceans. Nat. Clim. Chang. 9, 840–844.

Rong, J., Tang, Y., Zha, S., Han, Y., Shi, W., Liu, G., 2020. Ocean acidification impedes gustation-mediated feeding behavior by disrupting gustatory signal transduction in the black sea bream, *Acanthopagrus schlegelii*. Mar. Environ. Res. 162, 105182.

Roper, S.D., 2021. Chemical and electrical synaptic interactions among taste bud cells. Curr. Opin. Physio. 20, 118–125.

Ross, Q.E., Dunning, D.J., Menezes, J.K., Kenna, M.J., Tiller, G., 1995. Reducing impingement of alewives with high frequency sound at a power plant intake on Lake Ontario. N. Am. J. Fish. Manag. 15, 378–388.

Rostad, A., Kaarvedt, S., Klejver, T.A., Melle, W., 2006. Fish are attracted to vessels. ICES J. Mar. Sci. 63, 1431–1437.

Rountos, K.J., Gobler, C.J., Pikitch, E.K., 2017. Ontogenetic differences in swimming behavior of fish exposed to the harmful dinoflagellate *Cochlodinium polykrikoides*. Trans. Am. Fish. Soc. 146 (2017), 1081–1091.

Ruebush, B.C., et al., 2012. In-situ tests of sound-bubble-strobe light barrier technologies to prevent range expansions of Asian carp. Aquat. Invasions 7 (1), 37–48.

Russell, D.F., Wilkens, L.A., Moss, F., 1999. Use of behavioural stochastic resonance by paddlefish for feeding. Nature 402, 291–294.

Sabal, M.C., Workman, M.L., Merz, J.E., Palkovacs, E.P., 2021. Shade affects magnitude and tactics of juvenile Chinook salmon antipredator behavior in the migration corridor. Oceologia 197, 89–100.

Sager, D.R., Hocutt, C.H., Stauffer, J.R., 1987. Estuarine fish responses to strobe light, bubble curtains, and strobe light/bubble curtain combinations as influenced by water flow rate and flash frequencies. Fish. Res. 5, 383–399.

Sandahl, J.F., Baldwin, D.H., Jenkins, J.J., Scholz, N.L., 2004. Odor-evoked field potentials as indicators of sublethal neurotoxicity in juvenile coho salmon (*Oncorhynchus kisutch*) exposed to copper, chlorpyrifos, or esfenvalerate. Can. J. Fish. Aquat. Sci. 61, 404–413.

Schilt, C.R., 2007. Developing fish passage and protection at hydropower dams. Appl. Anim. Behav. Sci. 104, 295–325.

Schlenker, L.S., Welch, M.J., Mager, E.M., Stieglitz, J.D., Benetti, D.D., Munday, P.L., Grosell, M., 2019. Exposure to crude oil from the Deepwater Horizon oil spill impairs oil avoidance behavior without affecting olfactory physiology in juvenile mahi-mahi (*Coryphaena hippurus*). Environ. Sci. Technol. 53, 14001–14009.

Schligler, J., Cortese, D., Beldade, R., Swearer, S.E., Mills, S.C., 2021. Long-term exposure to artificial light at night in the wild decreases survival and growth of a coral reef fish. Proc. R. Soc. B 288, 20210454.

Schneider, G.E., 2014. Expansions of the Neuronal Apparatus of Success. In: Brain Structure and its Origins in Development and in Evolution of Behavior and the Mind. The MIT Press, pp. 67–85.

Seehausen, O., Alphen, J.J.M., Witte, F., 1997. Cichlid fish diversity threatened by eutrophication that curbs sexual selection. Science 277, 1808–1811.

Shakil Rana, K.M., Abdus Salam, M., Hashem, S., Salam, M.A., Ariful Islam, M., 2015. Development of Black Soldier Fly Larvae Production Technique as an Alternate Fish Feed 5-Books of colleagues View project Aquaponics Wheatgrass View project Development of Black Soldier Fly Larvae Production Technique as an Alternate Fish Feed. Int. J. Res. Fish. Aquac. 5 (1), 41–47 (https://www.researchgate.net/publication/315574190).

Sievers, A.M., 1969. Comparative study of *Gonyaulax monilata* and *Gymnodinium breve* to annelids, crustaceans, mollusks, and a fish. J. Protozool. 16, 401–404.

Slabbekoorn, H., Bouton, N., van Opzeeland, I., Coersm, A., ten Cate, C., et al., 2010. A noisy spring: the impact of globally rising underwater sound levels on fish. Trends Ecol. Evol. 25, 419–427.

Smith, R.K., Dicks, L.V., Mitchell, R., Sutherland, W.J., 2014. Comparing effectiveness research: the missing link in conservation. Conserv. Evid. 11, 2–6.

Soengas, J.L., Cerdá-Reverter, J.M., Delgado, M.J., 2018. Central regulation of food intake in fish: An evolutionary perspective. J. Mol. Endocrinol. 60 (4), R171–R199. https://doi.org/10.1530/JME-17-0320.

Sokimi, W., Beverly, S., 2010. Small-Scale Fishing Techniques Using Light: A Manual for Fishermen. Secretariat of the Pacific Community Noumea, New Caledonia, p. 54.

Sorensen, P.W., 1992. Hormones, pheromones, and chemoreception. In: Hara, T.J. (Ed.), Fish Chemoreception. Chapman and Hall, London, pp. 199–228.

Sorensen, P.W., Stacey, N.E., 2004. Brief review of fish pheromones and discussion of their possible uses in the control of non-indigenous teleost fishes. N. Z. J. Mar. Freshw. Res. 38, 399–417.

Sorensen, P.W., Hara, T.J., Stacey, N.E., Dulka, J.G., 1990. Extreme olfactory specificity of male goldfish to the preovulatory steroidal pheromone 17α,20β-dihydroxy-4-pregnen-3-one. J. Comp. Physiol. A 166, 373–383.

Southwood, A., Fritsches, K., Brill, R., Swimmer, Y., 2008. Sound, chemical, and light detection in sea turtles and pelagic fishes: sensory-based approaches to bycatch reduction in longline fisheries. Endanger. Species Res. 5, 225–238.

Stacey, N., Sorensen, P., 2005. Reproductive pheromones. In: Sloman, K.A., Wilson, R.W., Balshine, S. (Eds.), Behavior and Physiology of Fish: Volume 24: Fish Physiology. Elsevier, New York, pp. 359–412.

Stansbury, A., Gotz, T., Deecke, V.B., Janik, V.M., 2015. Grey seals use anthropogenic signals from acoustic tags to locate fish: evidence from a simulated foraging task. Proc. R. Soc. Lond. B 282, 20141595.

St-Hilaire, S., Sheppard, C., Tomberlin, J.K., Irving, S., Newton, L., McGuire, M.A., Mosley, E.E., Hardy, R.W., Sealey, W., 2007. Fly prepupae as a feedstuff for rainbow trout, *Oncorhynchus mykiss*. J. World Aquacult. Soc. 38 (1), 59–67. https://doi.org/10.1111/j.1749-7345.2006. 00073.x.

Stoner, A.W., Kaimmer, S.M., 2008. Reducing elasmobranch bycatch: laboratory investigation of rare earth metal and magnetic deterrents with spiny dogfish and Pacific halibut. Fish. Res. 92, 162–168.

Stroud, E.M., O'Connell, C.P., Rice, P.H., et al., 2014. Chemical shark repellent: Myth or fact? The effect of a shark necromone on shark feeding behavior. Ocean Coast. Manag. 97, 50–57.

Swimmer, Y., Arauz, R., Higgins, B., McNaughton, L., McCracken, M., Ballestro, J., Brill, R., 2005. Food color and marine turtle feeding behavior: can blue bait educe turtle bycatch in commercial fisheries? Mar. Ecol. Prog. Ser. 295, 273–278.

Sypitkowski, E., Bohlen, C., Ambrose, W.G., 2010. Estimating the frequency and extent of bloodworm digging in Maine from aerial photography. Fish. Res. 101, 87–93.

Takeda, M., Takii, K., 1992. Gustation and nutrition in fishes: application to aquaculture. In: Hara, T.J. (Ed.), Fish Chemoreception. Chapman and Hall, London, pp. 271–287.

Tallack, S.M.L., Mandelman, J.W., 2009. Do rare earth metals deter spiny dogfish? A feasibility study on the use of mischmetal to reduce the catch of *Squalus acanthias* by hook gear in the Gulf of Maine (USA) ICES. J. Mar. Sci. 66, 315–322.

Taormina, B., Bald, J., Want, A., Thouzeau, G., Lejart, M., Desroy, N., Carlier, A., 2018. A review of potential impacts of submarine power cables on the marine environment: knowledge gaps, recommendations and future directions. Renew. Sustain. Energy Rev. 96 (2018), 380–391.

Tateda, Y., Nakazono, A., Tsukahara, H., 1985. Acoustic conditioning of young red sea bream, *Pagrus major*. Rep. Fish. Res. Lab. Kyushu Univ. 7, 27–36.

Taylor, P.B., 1986. Experimental evidence for geomagnetic orientation in juvenile salmon, *Oncorhynchus tshawytscha* Walbaum. J. Fish Biol. 28, 6070–6623.

Taylor, P.B., 1987. Experimental evidence for juvenile Chinook salmon, *Oncorhynchus tshawytscha* Walbaum orientation at night and in sunlight after a 7E change in latitude. J. Fish Biol. 31, 89–111.

Thode, A., Straley, J., Tiemann, C.O., Folkert, K., O'Connell, V., 2007. Observations of potential acoustic cues that attract sperm whales to longline fishing in the Gulf ofAlaska. J. Acoust. Soc. Am. 122, 1265–1277.

Tierney, K.B., Baldwin, D.H., Hara, T.J., Rosse, P.S., Scholz, N.L., Kennedy, C.J., 2010. Olfactory toxicity in fishes. Aquat. Toxicol. 96, 2–26.

Tlusty, M.F., Andrew, J., Baldwin, K., Bradley, T.M., 2008. Acoustic conditioning for recall/recapture of escaped Atlantic salmon and rainbow trout. Aquaculture 274, 57–64.

Toyama, M., Hironaka, M., Yamahama, Y., Horiguchi, H., Tsukada, O., Uto, N., Ueno, Y., Tokunaga, F., Seno, K., Hariyama, T., 2008. Presence of rhodopsin and porphyropsin in the eyes of 164 fishes, representing marine, diadromous, coastal and freshwater species—a qualitative and comparative study. Photochem. Photobiol. 84, 996–1002.

Tricas, T., Gill, A., 2011. Normandeau, Exponent: Effects of EMFs from undersea power cables on elasmobranchs and other marine species. U.S. Dept. of the Interior, Bureau of Ocean Energy Management, Regulation, and Enforcement, Pacific OCS Region, OCS Study BOEMRE 2011–09, Camarillo, CA.

Tricas, T.C., New, J.G., 1998. Sensitivity and response dynamics of elasmobranch electrosensory primary afferent neurons to near threshold fields. J. Comp. Physiol. A 182, 89–101.

Tricas, T.C., Sisneros, J.A., 2004. Ecological functions and adaptations of the elasmobranch electrosense. In: von der Emde, G., Mogdans, J., Kapoor, B.G. (Eds.), The Senses of Fishes: Adaptations for the Reception of Natural Stimuli. Narosa Publishing House, New Delhi, pp. 308–329.

Ueda, H., 2019. Sensory mechanisms of natal stream imprinting and homing in *Oncorhynchus* spp. J. Fish Biol. 95, 293 303.

Valentincic, T., 2004. Taste and olfactory stimuli and behavior in fishes. In: von der Emde, G., Mogdans, J., Kapoor, B.G. (Eds.), The Senses of Fish - Adaptations for the reception of natural stimuli. Kluwer Academic Publishers, Boston, pp. 90–108.

Valentincic, T., 2005. Olfactory discrimination in fishes. In: Reutter, K., Kapoor, B.G. (Eds.), Fish Chemosenses. Science Publishers, Inc, Enfield, NH, pp. 65–85.

Valentincic, T., Caprio, J., 1994. Chemical and visual control of feeding and escape behaviors in the channel catfish *Ictalurus punctatus*. Physiol. Behav. 55, 45–855.

Valentincic, T., Caprio, J., 1997. Visual and chemical release of feeding behavior in adult rainbow trout. Chem. Senses 22, 375–382.

van Netten, S.M., 2006. Hydrodynamic detection by cupulae in a lateral line canal: functional relations between physics and physiology. Biol. Cybern. 94, 67–85.

Velez, Z., Hubbard, P., Welham, K., Hardege, J., Barata, E., Canário, A.M., 2009. Identification, release and olfactory detection of bile salts in the intestinal fluid of the Senegalese sole (Solea senegalensis). J. Comp. Physiol. A 195, 691–698.

Vernier, P., 2016. The Brains of Teleost Fishes. In: Evolution of Nervous Systems, second ed. vols. 1–4. Issue March https://doi.org/10.1016/B978-0-12-804042-3.00004-X.

Vielma, A., Ardiles, A., Delgado, L., Schmachtenberg, O., 2008. The elusive crypt olfactory receptor neuron: evidence for its stimulation by amino acids and cAMP pathway agonists. J. Exp. Biol. 211, 2417–2422.

Vowles, A.S., Kemp, P.S., 2012. Effects of light on the behavior of brown trout (*Salmo trutta*) encountering accelerating flow: application to donstream fish passage. Ecol. Eng. 47, 247–253.

Vowles, A.S., Kemp, P.S., 2021. Artificial light at night (ALAN) affects the downstream movement behaviour of the critically endangered European eel, *Anguilla anguilla*. Environ. Pollut. 274, 116585.

Wahlberg, M., Westerberg, H., 2005. Hearing in fish and their reaction to sounds from offshore wind farms. Mar. Ecol. Prog. Ser. 288, 295–309.

Wald, G., 1939. On the distribution of vitamins A1 and A2. J. Gen. Physiol. 22, 391–415.

Wald, G., 1941. The visual systems of euryhaline fish. J. Gen. Physiol. 25, 235–245.

Walker, M.M., 1984. Learned magnetic field discrimination in the yellowfin tuna, *Thunnus albacares*. J. Comp. Physiol. A 155, 673–679.

Walker, M.M., Krischvink, J.L., Chang, S.-B.R., Dizon, A.E., 1984. A candidate magnetic sense organ in the yellowfin tuna, *Thunnus albacares*. Science 224, 751–753.

Walker, M.M., Diebel, C.E., Haugh, C.V., Pankhurst, P.M., Montgomery, J.C., Green, C.R., 1997. Structure and function of the vertebrate magnetic sense. Nature 390, 371–376.

Walker, M.M., Dennis, T.E., Kirschvink, J.L., 2002. The magnetic sense and its use in long-distance navigation by animals. Curr. Opin. Neurobiol. 12, 735–744.

Walker, M.M., Diebel, C.E., Kirschvink, J.L., 2003. Detection and use of the earth's magnetic field by aquatic vertebrates. In: Collin, S.P., Marshall, J. (Eds.), Sensory Processing in Aquatic Environments. Springer-Verlag, NY, pp. 53–74.

Walker, M.M., Diebel, C.E., Kirschvink, J.L., 2006. Magnetoreception. Fish Physiol. 25, 337–376. https://doi.org/10.1016/S1546-5098(06)25008-8.

Walker, M.M., Diebel, C.E., Kirschvink, J.L., 2007. Magnetoreception. In: Hara, T.J., Zielinski, B.S. (Eds.), "Sensory Systems Neuroscience", Volume 25: Fish Physiology. Elsevier, NY, pp. 337–376.

Wang, J.H., Fisler, S., Swimmer, Y., 2010. Developing visual deterrents to reduce sea turtle bycatch in gill net fisheries. Mar. Ecol. Prog. Ser. 408, 241–250.

Wang, J., Barkan, J., Fisler, S., Godinez-Reyes, C., Swimmer, Y., 2013. Developing ultraviolet illumination of gillnets as a method to reduce sea turtle bycatch. Biol. Lett. 9, 20130383.

Wang, L., Espinoza, H.M., MacDonald, J.W., Bammler, T.K., Williams, C.R., Yeh, A., Louie, K. W., Marcinek, D.J., Gallagher, E.P., 2016. Olfactory transcriptional analysis of salmon exposed to mixtures of chlorpyrifos and malathion reveal novel molecular pathways of neurobehavioral injury. Toxicol. Sci. 149 (2016), 145–157.

Waples, D.M., Thorne, L.H., Hodge, L.E.W., Burke, E.K., Urian, K.W., et al., 2013. A field test of acoustic deterrent devices used to reduce interactions between bottlenose dolphins and a coastal gillnet fishery. Biol. Conserv. 157, 165–171.

Warrant, E.J., 1999. Seeing better at night: life style, eye design, and the optimum strategy of spatial and temporal summation. Vision Res. 39, 1611–1630.

Warrant, E.J., Locket, N.A., 2004. Vision in the deep sea. Biol. Rev. 79, 671–712.

Webb, J.F., 1998. Laterophysic connection: a unique link between the swimbladder and the lateral line system in Chaetodon (Perciformes: Chaetodontidae). Copeia 1998, 1032–1036.

Weeg, M.S., Bass, A.H., 2002. Frequency response properties of lateral line superficial neuromasts in a vocal fish, with evidence for acoustic sensitivity. J. Neurophysiol. 88, 1252–1262.

Weissburg, M.J., Browman, H.I., 2005. Sensory biology: linking the internal and external ecologies of marine organisms. Mar. Ecol. Prog. Ser. 287, 263–265.

Westerberg, H., Lagenfelt, I., 2008. Sub-sea cables and the migration behavior of the European eel. Fish. Manag. Ecol. 15, 369–375.

Whitear, M., 1992. Solitary chemoreceptor cells. In: Hara, T.J. (Ed.), Fish Chemoreception. Chapman and Hall, London, pp. 103–125.

Wilkens, L.A., Hofmann, M.H., 2007. The paddlefish rostrum as an electrosensory organ: A novel adaptation for plankton feeding. BioScience 57 (5), 399–408.

Williams, J.G., Armstrong, G., Katopodis, C., Lariniere, M., Travade, F., 2012. Thinking like a fish: a key ingredient for development of effective fish passage facilities at river obstructions. River Res. Appl. 28, 407–417.

Williams, R., Erbe, C., Ashe, E., Clark, C.W., 2015. Quiet(er) marine protected areas. Mar. Pollut. Bull. 100 (1), 154–161.

Wiltschko, R., Wiltschko, W., 1995. Magnetic Orientation in Animals. In: Zoophysiology. vol. 33. Springer-Verlag Berlin.

Windsor, S.P., Tan, D., Montgomery, J.C., 2008. Swimming kinematics and hydrodynamic imaging in the blind Mexican cavefish (*Astyanax fasciatus*). J. Exp. Biol. 211, 2950–2959.

Windsor, S.P., Norris, S.E., Cameron, S.M., Mallinson, G.D., Montgomery, J.C., 2010. The flow fields involved in hydrodynamic imaging by blind Mexican cave fish (*Astyanax fasciatus*). Lateral line history part i: open water and heading towards a wall. J. Exp. Biol. 213, 3819–3831.

Wright, D.E., Eastcott, A., 1982. Association of an acoustic signal with operant conditioned feeding responses in thickkipped mullet, *Crenimugil labrosus* (Risso) and common carp, *Cyprinus carpio* (L.). J. Fish Biol. 21, 693–698.

Würsig, B., Green Jr., C.R., Jefferson, T.A., 2000. Development of an air bubble curtain to reduce underwater noise of percussive piling. Mar. Environ. Res. 49, 79–93.

Wyman, M.T., Klimley, A.P., Battleson, R.D., Agosta, T.V., Chapman, E.D., Haverkamp, P.J., Pagel, M.D., Kavet, R., 2018. Behavioral responses by migrating juvenile salmonids to a subsea high-voltage DC power cable. Mar. Biol. 165, 134.

Yamamori, K., Nakamura, M., Matsui, T., Hara, T.J., 1988. Gustatory responses to tetrodotoxin and saxitoxin in fish: a possible mechanism for avoiding marine toxins. Can. J. Fish. Aquat. Sci. 45, 2182–2186.

Yamashita, S., Evans, R.E., Hara, T.J., 1989. Specificity of the gustatory chemoreceptors for CO2 and H+ in rainbow trout (*Oncorhynchus mykiss*). Can. J. Fish. Aquat. Sci. 46, 1730–1734.

Yan, H.Y., Anraku, K., Babaran, R.P., 2010. Hearing in marine fish and its application in fisheries. In: He, P. (Ed.), Behavior of Marine Fishes: Capture Processes and Conservation Challenges. Wiley-Blackwell, Iowa, Ames, pp. 45–64.

Yokota, K., Kiyota, M., Okamura, H., 2009. Effect of bait species and color on sea turtle bycatch and fish catch in a pelagic longline fishery. Fish. Res. 97, 53–58.

Yopak, K.E., 2012. Neuroecology of cartilaginous fishes: the functional implications of brain scaling. J. Fish Biol. 80 (5), 1968–2023. https://doi.org/10.1111/j.1095-8649.2012.03254.x.

Young, C.E.F., 2005. Financial mechanisms for conservation in Brazil. Conserv. Biol. 19, 756–761.

Yuen, H.S.H., 1970. Behavior of skipjack tuna, *Katsuwonus pelamis*, as determined by tracking with ultrasonic devices. J. Fish. Res. Board Can. 27, 2071–2079.

Zielinski, B.S., Hara, T.J., 2001. The neurobiology of fish olfaction. In: Kapoor, B.G., Hara, T.J. (Eds.), Sensory Biology of Jawed Fishes. Science Publishers Inc., Enfield, pp. 347–366.

Zielinski, B.S., Hara, T.J., 2007a. Olfaction. In: Hara, T.J., Zielinski, B.S. (Eds.), Sensory Systems Neuroscience: Volume 25: Fish Physiology. Elsevier, New York, pp. 1–43.

Zielinski, B.S., Hara, T.J., 2007b. Gustation. In: Hara, T.J., Zielinski, B.S. (Eds.), Sensory Systems Neuroscience: Volume 25: Fish Physiology. Elsevier, New York, pp. 45–96.

Zielinski, D.P., Sorensen, P.W., 2016. Bubble curtain deflection screen diverts the movement of both Asian and common carp. N. Am. J. Fish Manag. 36, 267–276.

Zion, B., Barki, A., 2012. Ranching fish using acoustic conditioning: has it reached a dead end? Aquaculture 344–349, 3–11.

Zippel, H.P., 1993. Regeneration in the peripheral and the central olfactory system: a review of morphological, physiological, and behavioral aspects. J. Hirnforsch. 34, 207–229.

Zippell, H.P., Hansen, A., Caprio, J., 1997. Renewing olfactory receptor neurons do not require contact with the olfactory bulb to develop normal responsiveness. J. Comp. Physiol. A 181, 435–447.

Chapter 3

Applied aspects of locomotion and biomechanics

Theodore Castro-Santos[a,*], Elsa Goerig[a,b], Pingguo He[c], and George V. Lauder[b]

[a]*U.S. Geological Survey, Eastern Ecological Science Center, S.O. Conte Anadromous Fish Research Center, Turners Falls, MA, United States*
[b]*Museum of Comparative Zoology, Harvard University, Cambridge, MA, United States*
[c]*School for Marine Science and Technology (SMAST), University of Massachusetts Dartmouth, New Bedford, MA, United States*
[*]*Corresponding author: e-mail: tcastrosantos@usgs.gov*

Chapter Outline

Locomotion is the act and process of moving from place to place, which is fundamental to the life history of all mobile organisms. While the field of biomechanics encompasses the study of the physical constraints of what animals are capable of, ecological contexts require an integrated view that includes ecology and behavior. This chapter provides an overview of some of the areas where locomotion and biomechanics of fish movement interface with the rapidly evolving changes that humans impose on aquatic

Fish Physiology, Vol. 39A. https://doi.org/10.1016/bs.fp.2022.04.003

91

environments. These changes include fundamental alterations to the environment such as altered flows, fragmentation of riverine habitats, and invasive species, but also direct interactions that occur with capture fisheries. We explore each of these areas, considering both challenges and opportunities informed by the study of locomotion and biomechanics, emphasizing how this field can contribute to conservation of fishes in the Anthropocene. We then turn to technology, where important advances are aiding in our understanding of fish movement. In some cases those advances have themselves led to novel technologies, where biomimetic robots and related devices offer novel opportunities, both for conservation and for other pursuits.

1   Introduction

The study of fish locomotion and biomechanics presents humans with challenges and opportunities as we navigate the Anthropocene.[a] These challenges include recognizing and mitigating anthropogenic influences such as habitat quality, quantity, and access, and developing sustainable approaches to harvest fisheries. At the same time, technological advances have enabled humans to better understand the mechanics of aquatic locomotion, providing both challenges and opportunities for conservation and engineering. We begin this chapter with a broad overview of fish swimming and biomechanics, identifying both physiological and behavioral constraints that are particularly relevant to management and conservation. We then describe some challenges confronting fisheries managers and how they relate to locomotion, paying particular attention to topics of habitat fragmentation, invasive species, and harvest fisheries. We conclude with a brief review of recent advances in biomimetic engineering that may offer opportunities for developing sustainable approaches to harvest fisheries and for informing the design of swimming machines and other devices, which may themselves open opportunities for exploration and greater understanding of the world in which we live (Fig. 1).

The way fish move through their environment has captured the imagination of researchers since at least the time of Aristotle (350 BCE) (Aristotle, 1937). To propel themselves through water, fish must exert a force against a medium that deforms continuously. This is in marked contrast to propulsion on land where animals can push off a solid surface to move about. The ability of fishes to move through the dense and viscous aquatic medium with such apparent ease has been influential in stimulating both theory and experiments in the field of fluid mechanics and hydrodynamics of fish locomotion (Alexander, 1983; Lighthill, 1960, 1975; Vogel, 1981).

Fish achieve effective propulsion through a variety of mechanisms, with the unifying characteristic of establishing pressure gradients along their fins and/or

---

[a]Anthropocene is a term used by scientists and nonscientists to highlight the concept that we are living in a time when human activities have significant effects on the global environment. The Anthropocene currently has no formal status in the Divisions of Geologic Time and is not recognized by the USGS. Use of this term is informal.

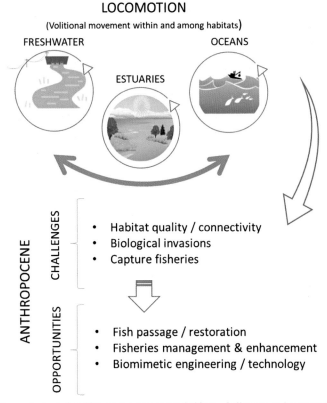

FIG. 1 Conceptual relationships among movement, habitats, challenges, and opportunities associated with the study of locomotion and biomechanics.

bodies, generating force on the surrounding fluid that determines their path of motion. The amount of thrust produced and the efficiency of force transmission is largely driven by body and fin morphology, coupled with skin, skeletal structure, muscular design, and composition (Shadwick and Gemballa, 2006). Most species of fishes move primarily using their body and caudal fins (BCF) for propulsion, while median and paired fins contribute greatly to body stability and maneuvering during locomotion. Body undulation generates thrust through a propulsive wave propagating along the body, which is characterized by wave speed, wavelength, and amplitude of the lateral oscillation (Fig. 2).

The habitats occupied by fishes are hugely diverse, ranging in scale from pan-global marine environments traversed by species like tunas and lamnid sharks (Block, 2005; Block et al., 2001), to the interstitial spaces between the gills of host species and other tiny spaces (Breault, 1991). This diversity of habitats has led to a correspondingly dramatic diversity in both morphology and swimming performance. In some cases, it may be possible to make general predictions about the habitat a fish occupies and its ability to negotiate

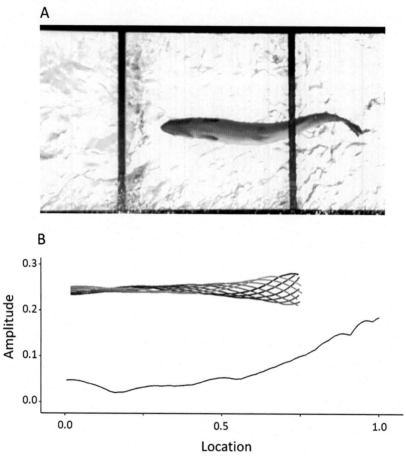

FIG. 2   (A) Atlantic salmon swimming at 10 body lengths per second (BL s^{-1}) in an open flume. (B) Digitized midlines for one tailbeat (overlaid curves, ranging from blue (start of stride) to black (end of stride)), and associated maximum lateral amplitude along the body, presented as proportion of the body length from location 0 (head) to 1 (tail).

certain environments, simply by viewing its external morphology (Webb, 1984, 2006). At the same time, though, diverse adaptations may be constrained by similar physical principles, leading to a surprising level of convergence in kinematics across taxa when fish are performing similar tasks (Di Santo et al., 2021).

There are exceptions to these patterns of diversification and convergence, however. Some fish species have developed morphologies and behaviors that allow them to interact with solid structures within their environments in ways that promote movement or allow them to maintain position, e.g., by attaching to a substrate or a host animal (examples include parasitic lamprey

(*Petromyzon* spp.; Beamish, 1980), remoras (family Echeneidae; Wang et al., 2017), and many benthic species (Webb, 1984, 1989), which use the substrate to escape rapid free-stream flows). Both the general patterns and these exceptions have important implications for how humans and fish will interact during the Anthropocene.

While essential for understanding the mechanics of locomotion, movement itself is not governed solely by external morphology and biomechanics. Other aspects, such as internal structure and physiology, are equally relevant, including vascular circulation (Farrell and Steffensen, 1987), aerobic capacity (Brett, 1964), and the distribution and abundance of aerobic and anaerobic muscle fibers (Jayne and Lauder, 1994; Rome, 1994).

The distribution and abundance of aerobic and anaerobic musculature has important implications for locomotion, and for ecology in general. Fish species that specialize in fast starts and sprinting tend to have proportionally greater amounts of anaerobic muscle, while those that specialize in cruising have more aerobic muscle (McLaughlin and Kramer, 1991). More aerobic muscle requires greater blood flow and overall greater aerobic capacity, which in turn allows those species to sustain greater swim speeds. Moreover, the high metabolic demand of aerobic tissues can necessitate greater activity to ensure sufficient resources are gathered to meet these demands (Brett and Groves, 1979; Ware, 1975, 1980). Provided that sufficient oxygen and nutrients are available to the fish, aerobically fueled swimming can be sustained indefinitely. As we will see below, this has important implications for conservation and management. When fish exceed their maximum sustained swim speed (U_{ms}; Table 1) they begin to recruit anaerobic metabolic pathways and switch to using the generally more abundant white muscle fibers. This is usually done incrementally, allowing fish to prolong their endurance by interspersing bouts of anaerobic swimming with aerobic swimming (Brett, 1964; Peake and Farrell, 2004; Rome et al., 1992a). Thus, "prolonged" swimming describes a narrow band of speeds at which both processes are important in producing thrust. However, as speeds increase further the fish enters "sprint" mode. Sometimes described as "burst" mode, this is powered almost entirely by anaerobic processes and can be sustained only very briefly (the actual limits vary by species, but 20s is a commonly described threshold; Brett, 1964).

The existence of these three modes: "sustained," "prolonged," and "sprint" has been documented for a wide variety of species (Beamish, 1978). Importantly, species that have greater aerobic capacity do not necessarily have greater sprinting ability (Clark et al., 2013). This can be an important factor limiting access to habitats: assumptions that less active species will be less able to traverse barriers may be inaccurate, with potential consequences for management.

Our understanding of swimming modes is influenced by the methods used to measure and describe them. Maximum sustained speeds (U_{ms}) are

TABLE 1 Definitions, metrics, and indices describing fish locomotion—the act and processes associated with moving from place to place, such as swimming, jumping, walking, etc.

Term	Definition, metrics, and indices		Units
Ability	Also Capability or Capacity: the physical and physiological bounds describing the limits of what an organism is capable		
	Endurance	Time to fatigue (T) while executing a given task such as swimming at a constant speed (U_s), described by the relation $\ln(T) = a + bU_s$	s or min | U_s, a, and b
	Speed	• Maximum sustainable swim speed (U_{ms}) • Maximum speed sustained for a fixed interval without fatigue, typically 20–200 min (U_{crit}) • Maximum attainable swim speed (U_{max})	BL s^{-1} or m s^{-1}
	Acceleration	• Maximum acceleration (A_{burst})[a]	BL s^{-2} or m s^{-2}
Performance	The effective execution of a task (e.g., traversing a velocity challenge, avoiding capture, or accessing refuge)		
	Distance/ Height	Maximum distance traveled (D_{max})	BL or m
	Success	Probability of success	Success attempt^{-1} Success h^{-1}
Motivation	The inclination or willingness to perform a task (e.g., move)		
	Attempt rate	Number of attempts per time period or proportion of a population staging attempts per unit time	Attempts h^{-1} P(Attempts) h^{-1}
	Effort	The amount of effort expended to achieve a task	% T | U_s % U_{ms}, U_{crit}, U_{max}, or A_{burst}

[a]Note that U_{burst} is a term that is often used to describe either fast start or sprint speeds (and even in some cases prolonged speeds). We distinguish it here, as sprint and prolonged speeds are defined by endurance relationships and cannot be represented by a single value.

frequently estimated using a process whereby fish are confined within a chamber or tunnel and subjected to flow speeds that increment steadily on a fixed time interval (Brett, 1964). The speed at which fatigue begins to occur within that time interval is hindcast and recorded as the "critical swim speed" (U_{crit}). When the time interval for each velocity increment is sufficiently long (e.g., ~200 min) U_{crit} serves as an estimate of U_{ms}. This technique is excruciatingly slow, however, prompting most researchers to greatly shorten the time intervals. A key consequence of this is that U_{crit} becomes unreliable as an estimate of U_{ms} and may be biased by accumulation of anaerobic byproducts during intervals that preceded the fatigue event (Hammer, 1995; Lee et al., 2003). Prolonged (generally thought of as speeds $> U_{ms}$ that can be sustained for $> \sim 20$ s) and sprint speeds (those that result in fatigue in $< \sim 20$ s) are typically estimated by rapidly increasing flow velocity to a desired test condition and then holding that value constant until the fish fatigues. The relation between flow velocity and endurance characterizes each of these modes.

One consequence of conducting these tests within confined chambers is that fish tend to fatigue at speeds considerably lower than what occurs in more natural settings (Castro-Santos et al., 2013; Tudorache et al., 2007, 2010). Tests performed in larger channels and using volitional swimming against high velocity flows have shown that many species of fish are capable of swimming at speeds nearly double what previous studies had predicted, suggesting that estimates of performance from the literature should be viewed with skepticism (Castro-Santos, 2005; Castro-Santos et al., 2013).

1.1 Temperature and locomotion

Another important distinction between aerobic- and anaerobic-powered locomotion is the importance of temperature. Fish species occupy habitats with temperatures ranging from <0 °C to >44 °C (Bargelloni et al., 1994; Kock, 2005; Lozano-Vilano and De La Maza-Benignos, 2017; Minckley and Minckley, 1986). Most fish are ectotherms, and their metabolic rates vary predictably with temperature. At low temperatures, the scope for activity is reduced and fish tend to be less active. Both basal metabolic rate and scope increase with temperature; however, as temperatures exceed a certain critical value the costs of maintenance increase, and scope declines. Hence, reduced activity is common among fish exposed to temperatures that fall above or below their thermal optima (Brett and Glass, 1973; Brett and Groves, 1979).

The effect of temperature on aerobic swim speeds is much greater than it is on anaerobic speeds. In both cases, chemical reactions occur more rapidly at warmer temperatures. This means that the rates of nutrient delivery, waste removal, and rate of ATP production all increase, raising the potential for power production, as well as the maximum contraction speeds of both red and white muscles (Rome et al., 1992a,b; Wardle, 1975). The benefits of increasing temperature change when thermal optima are exceeded, however:

as temperatures continue to increase, so do maintenance metabolic costs (the actual mechanisms and their relative importance remain a topic of some dispute: Clarke and Fraser, 2004), and these costs add to costs of circulation, etc., which in turn places a greater proportional demand on aerobic scope. This is further aggravated by the fact that at higher temperatures the ability of water to carry oxygen and carbon dioxide diminishes, reducing the efficiency of gas transfer at the tissues and gills. At an upper critical limit, these costs exceed the ability to maintain homeostasis. Thus at both high and low temperatures, the ability to sustain aerobic swimming is compromised, and fish switch to swimming using anaerobic muscle fibers (Pörtner et al., 2017).

The reasons for this transition are not entirely clear, although some explanations appear likely. For one, anaerobic muscles use locally stored glycogen to fuel ATP production and are therefore less dependent on the circulatory system. Furthermore, the number of temperature-dependent reactions required to produce ATP are reduced compared with aerobic metabolism, and with fewer steps in the "supply chain" the overall effect of temperature on muscle kinematics is reduced (Clarke and Fraser, 2004).

These and other factors contribute to a widely observed phenomenon, whereby sustained swimming capacity is strongly influenced by temperature, while sprinting is less so (Haro et al., 2004). An important caveat to this is that recovery between bouts of sprinting does require aerobic processes, and the time required to recover between sprinting bouts is also strongly influenced by temperature (Bayse et al., 2019; Goerig and Castro-Santos, 2017; Kieffer et al., 1994; Wilkie et al., 1997).

This does not mean that anaerobic swimming is unaffected by temperature, however. Fish commonly become sluggish, inactive or "semi-torpid" when the water temperature is near the lower end of their limits (He, 2003; Lagardère and Sureau, 1989; Woodhead, 1964). Temperature may also affect the startle response and reaction to predators, fishing gear, etc. Muscle kinetics are also affected, particularly at very cold temperatures: Özbilgin and Wardle (2002) found that the escape reflex of Haddock (*Melanogrammus aeglefinus*) in the North Sea was much lower at 7 °C compared with that at 12 °C, and attributed this change to reduced contraction speeds of white muscle at colder temperatures. Other studies have indicated that this pattern is widespread (Wardle, 1980; Yanase et al., 2007). Given the ongoing and anticipated changes in temperatures during the Anthropocene further study of the effects of temperature on swimming capacity will be useful for informing management decisions, with important implications for conservation and effectiveness of fishing gears.

1.2 Ability vs performance

Taken together, morphology and physiology determine the limits of "capacity" (Beamish, 1978), or what a given individual fish is capable of doing within its environment (Table 1). And the environmental context matters: metabolic rate and aerobic scope are strongly dependent on temperature; efficiency of force transmission will vary with turbulence; buoyancy is affected by depth, etc.

Importantly, though, ability is not synonymous with performance (Table 1). Performance refers to how well an animal is able to achieve a certain task, and again, each task is context dependent. For example, ability to escape an ambush predator may be governed by acceleration, maneuverability, and maximum speed, while traversing a velocity barrier or escape from pursuit or a fishing trawl may be governed by endurance or maximum sustained speeds. If fish consistently optimize the application of their capabilities to each discrete task with which they are confronted, then capacity and performance would be equivalent, but this is not the case. Instead, an animal in its environment must respond to multiple sensory inputs, its perception of risk, its internal state (e.g., nutritional or reproductive status), and other factors, many of which remain poorly understood.

Each of the issues reviewed above has relevance to various aspects of conservation and management of fishes and fisheries and also to the opportunities afforded to human engineers by improved understanding of principles, constraints, and applications. Throughout, we do not limit our content to the physiology of locomotion. Instead, we cast locomotion and movement in their environmental and applied context, explicitly recognizing that these contexts require that behavior not be decoupled from physiology. We therefore take a holistic approach, integrating behavior and physiology, as occurs in nature.

2   Habitat quality and connectivity

2.1   Syndromes of the anthropocene

Access to adequate habitat is a fundamental metric of habitat quality and is an essential requirement for healthy fish populations. Fish need to access suitable habitats to complete the various stages of their life cycle (Schlosser, 1991). Many factors may partially or completely impede fish movements, as well as the distribution of suitable aquatic habitats. Some of these factors are governed by physical attributes of the environments and locomotor capacity of fishes, while others are linked to fish behavior and ecology.

Although this volume focuses on developments and projections of the period beginning in the mid-20th Century, fundamental changes to aquatic environments from anthropogenic activities long predate this (Hall et al., 2011, 2012). Since 1950, however, fluvial and marine ecosystems have undergone accelerated changes resulting from human activities. Anthropogenic features now pervade many landscapes worldwide, altering aquatic habitat quality and connectivity. Global climate change is also modifying natural processes and physical characteristics of both freshwater and marine environments.

The multiple facets of anthropogenic influences have been synthesized under the term "syndromes" (Meybeck, 2003), which are not specific to a few areas, but instead can be observed in many places around the globe. Each syndrome has its own causes and symptoms, although there can be substantial overlap between them. Below we examine some of these syndromes and how they relate to the ability of fish to move through their environment.

2.1.1 Instability in physical properties of aquatic habitats

Global changes in the Earth's climate due to human activities is currently leading to increased temperatures in most aquatic habitats, along with a reduction in pH and an increase of hypoxic zones, especially in coastal waters where organic matter and nutrients coming from runoff processes are present. Many physiological processes in fish (e.g., growth, metabolic rate, locomotor performance) are temperature- and oxygen-dependent. As a result of species' different physiological tolerances and capacity for adaptation, a modification to the thermal regime will affect overall species distribution. Changes in water temperature may also influence fish behavior, influencing motivation to swim (Bayse et al., 2019; Goerig and Castro-Santos, 2017), triggering migrations (Otero et al., 2014), and altering attack rate during feeding events (Domenici et al., 2019). Changes in the physical properties of aquatic environments are thus likely to affect ecological processes, changing distribution of functional habitats in time and space, and affecting the fish locomotor and sensory/neural systems (Domenici et al., 2019). In addition to the thermal effects on swimming ability and energetics described above, human infrastructures can create novel opportunities for predation, creating barriers to movement and affecting population dynamics (Agostinho et al., 2012; Alcott et al., 2020, 2021).

Changes in the physical properties of aquatic habitats are also often associated with increased instability. In the marine environment, warming temperatures may cause a reduction in ocean ice and a modification in the ocean global circulation (Macdonald and Wunsch, 1996; Maximenko et al., 2013), slowing down some currents and the redistribution of heat around the globe. Resulting thermal barriers may be impenetrable for migratory fish (Otero et al., 2014; Stich et al., 2015), and alteration in the ocean's global circulation may limit distribution of fish larvae that take advantage of ocean currents for transportation (Kettle and Haines, 2006; Smith, 2012).

Altered thermal regimes in rivers coincide with occasional droughts and associated heat waves, where rivers and lakes experience unusually low flow or elevated water temperatures and reduced dissolved oxygen levels. This may create unsuitable areas for fish (too shallow, too warm, too hypoxic, etc.) and cause previously stable habitats to become more variable or intermittent (Meybeck, 2003). This influences bioenergetics and can affect fitness (Friedland et al., 1998, 2005; Lennox et al., 2018). The survival of fish populations in these areas is then dependent on the existence of refuge habitats (deep pools, cold groundwater seepages, colder tributaries) and the ability of fish to access them (Ebersole et al., 2020; Vander Vorste et al., 2020). This ability relates to the species locomotor behavior and ability, but also to the degree of fluvial heterogeneity and connectivity in the watershed (Dugdale et al., 2015; Dzara et al., 2019). Access to these critical refuge habitats (e.g., those necessary for short-term survival) can be as important as access to spawning or breeding habitats for population viability (Schlosser, 1991; Sedell et al., 1990).

2.1.2 Altered flow regimes in rivers

Large or frequent modifications in river flow are a common feature of the Anthropocene epoch; almost two-thirds of world's large rivers are regulated in order to produce energy, control floods, or collect water for human consumption or irrigation purposes, leading to temporal fluctuations and reduction in river flow (Grill et al., 2019). Fluctuations in flow caused by pumped storage facilities and hydropower plants often have a repetitive temporal component, the magnitude and frequency of which both influence aquatic ecosystems. Water collection or diversion for irrigation may affect the flow in a more stable way: a flow reduction of 50% or more leads to "neoarheism," where the river will partially or completely dry up, causing disconnections of the main channel from its delta or tributaries (Meybeck, 2003), making it impossible for fish to move within the river system.

Both short-term fluctuations and reductions in flow affect aquatic ecosystems and organisms living within them. Dams and fluctuations in discharge influence sediment processes (erosion, transport, deposition) and temperature regime, causing instability or chronic alterations to the physical properties of aquatic habitats and their morphology (Capra et al., 2017). As a consequence, specific habitats such as deep pools, sand or gravel bars, oxbow lakes, etc. may become temporarily or permanently inaccessible, leading to a decrease in heterogeneity of the riverscape (Agostinho et al., 2004; Freeman et al., 2001). A reduction in river discharge may also create zones of difficult passage such as shallow rapids and riffles, and impassable falls.

These alterations in flow regimes cause substantial changes to the mosaic of riverine functional habitats (spawning, breeding, refuge, etc.), potentially requiring species to move over longer distances to find adequate habitat. Depending on the frequency and magnitude of the fluctuations, fish may have to quickly modify their habitat selection, adapting to new features or relocating to suboptimal habitats, with associated effects on fitness (Capra et al., 2017). Species and individuals with a higher inclination to move and greater locomotor ability may exhibit higher resilience in hydropeaking rivers, creating traits-based selection in fish populations (Van Looy et al., 2019).

2.1.3 Fragmentation of riverine habitats

Although some river systems are naturally fragmented by features such as waterfalls, rapids, or beaver dams, anthropogenic obstructions are now widely present in many watersheds. Dams and associated reservoirs built for irriga-tion, flood control, or energy production purposes are some of the greatest changes affecting river systems in the Anthropocene. Although dams existed well before the beginning of this epoch (Hall et al., 2011, 2012), the rate of construction of large dams peaked between 1960 and 70 and continues apace today. The International Commission on Large Dams (ICOLD) and the Global Georeferenced Database of Dams (GOODD database; Mulligan et al., 2020) reported 58,000 large dams worldwide in 2020, with a total storage capacity of 7000–8300 km^3, roughly one-sixth of the total annual discharge of all the world's rivers. However, smaller impoundments are even more widespread, numbering nearly 16 million, and with another 8000 km^3 of combined storage capacity (Lehner et al., 2011).

Dams are not the only kind of human-made obstruction in river systems, however. Tide gates and dykes often restrict connectivity between estuarine and freshwater environments (Alcott et al., 2021; Rillahan et al., 2021), and the prevalence of these structures is expected to increase dramatically as nations seek to protect critical infrastructure from rising seas in a context of global climate change (De Vaate et al., 2003; Vincik, 2013; Wright et al., 2016).

Even more prevalent are road-stream crossings, the vast majority of which use culverts to pass flow. The Anthropocene has seen tremendous growth in road construction worldwide: in the United States alone there are more than 6.5 million road-stream crossings (Wieferich, 2022), most of which are believed to pose partial or complete barriers to migration. This pattern is also evident elsewhere, including in developing countries (Makrakis et al., 2012), and is probably the single greatest threat to habitat connectivity in riverine systems worldwide (Grill et al., 2019; Park et al., 2008).

Anthropogenic river features such as dams, tide gates, and culverts have the potential to delay fish movement rates in both the upstream and down-stream direction (Alcott et al., 2021; Goerig et al., 2016; Nyqvist et al., 2016) and create partial or complete barriers to fish movements for various reasons: physical obstructions (e.g., a dam), disconnection in the channel con-tinuity (e.g., a perched culvert), or hydraulic conditions exceeding the fish swimming abilities. This restricts access to complementary and/or critical habitats (e.g., spawning, breeding, or refuge habitats) for fish species and may lead to adverse consequences on individual fitness, and ultimately on population health and survival (Dudgeon, 2011; Fuller et al., 2015). Fragmen-tation of riverine habitats can also confer benefits, for example, by reducing rates of colonization by exotic invasive species (Zielinski et al., 2019).

2.2 Fish passage: Restoring connectivity of riverine systems

To mitigate the effects of dams, culverts, and other barriers, various engineer-ing solutions have been developed. Culverts are usually made of metal, concrete, or plastic pipes, which are often narrower and smoother than the natural stream channel. This creates a constriction in the flow at the inlet and the development of high velocities within the structure, which in turn may cause erosion at the downstream outlet resulting in a "perched" culvert, with the downstream end suspended above the streambed. The disconnection in the channel continuity and excessive water velocities inside the culverts are the two main causes of impeded fish movements. Fish passage through culverts may be improved either by improving design or by retrofitting with structures tailored to the behavior and ability of target species (Duguay et al., 2018b; Park et al., 2008). If resources are available, methods that require less modification of the streambed are favored, such as creation of a stream-simulation culvert (Barnard et al., 2015; Gillespie et al., 2014) or replacement by a bridge. However, the design or retrofitting of culverts remains the most common option due to a lower cost. This may include proper sizing and embedding of the culvert to reduce the slope and the degree of flow constriction, as well as the inclusion of corrugations or baffles in the structure to increase bottom roughness and hence slow down water velocities and increase hydraulic heterogeneity (Duguay et al., 2018b; Feurich et al., 2012; Hotchkiss and Frei, 2007; Wang and Chanson, 2018).

Dams, regardless of their function, generally pose impassable barriers to most riverine fishes. The most drastic way to restore access to fluvial habitats fragmented by dams is to remove the dams themselves. In some contexts, such as aging dams constructed for mill operation or obsolete flow regulation purposes, this is the most logical option. There is currently significant interest in dam removal, frequently requiring extensive design and planning (including the restoration of riverine geomorphic processes (Bellmore et al., 2019; Ryan Bellmore et al., 2017; Wieferich et al., 2021)). Dam removal often has positive effects on fish access to habitat (Hill et al., 2019; Hogg et al., 2015; Magilligan et al., 2016), however, there can also be less desirable consequences, for example, by favoring access to new habitat for invasive species (McLaughlin et al., 2013; Stanley and Doyle, 2003). Sometimes the original structures were placed at falls or similar barriers to movement: in such cases removal may have negligible benefits for providing access; likewise, the surrounding riverscape may have been so altered by the construction and operation of the dam that simple removal fails to restore connectivity.

The dam removal movement coincides with countervailing demands to increase hydropower production to replace fossil fuels. Thus, in many cases dams are likely to persist despite their consequences on fish movements and access to habitat, elevating the need for effective mitigating solutions. Fishways are structures designed to provide a passage route, usually around a dam or similar structure but are sometimes used to provide access past natural barriers as well (Hatry et al., 2013). Although fishways are used to pass fish in both upstream and downstream directions, the challenges and risks associated with the two types differ, leading to distinct designs.

For upstream passage, fishways must dissipate the hydraulic head (difference in water level between the headpond above a dam and the tailrace below it). In the absence of roughness elements, this head translates swiftly to velocities that are impassable to any species. Most upstream fishways dissipate this head using either a series of baffles or weirs and pools. As water flows down the fishway, the roughness elements impart friction and turbulence, converting the potential energy of the initial head into heat and noise (Castro-Santos and Haro, 2006; Clay, 1995).

The design of upstream fishways and technical culverts is predicated largely on the anticipated swimming performance of the fish species it is intended to pass. In practice, however, engineers typically use data on swimming ability for these designs. This is because although data on swimming ability are widely available in the literature, data on actual performance are comparatively rare (Castro-Santos and Haro, 2006; Table 1). Additionally, fishways often have areas with velocities that exceed maximum sustained speeds, interspersed with pools where fish can recover between bouts of anaerobic swimming. It is unclear, however, if these design criteria function as intended for most species; this may be one factor that contributes to a widespread failure of upstream fishways to pass fish effectively (Bunt et al., 2012, 2016; Noonan et al., 2012).

Ideally, fishways and culverts are designed using empirical data on endurance and swimming performance of target species, for example, the maximal distance a species can traverse under a range of flow velocities and environmental conditions (Castro-Santos, 2004; Haro et al., 2004; Weaver, 1963). The design of roughness elements in fishways and culverts may also benefit from considering empirical data on fish kinematics, such as lateral body amplitude during movement (Di Santo et al., 2021; Duguay et al., 2018a).

Another factor that likely contributes to passage success being poorer than expected is that most swimming performance studies are conducted in environments where turbulence is deliberately minimized (Brett, 1964), while most upstream fishways and culverts equipped with baffles or corrugations are specifically designed to dissipate energy through the generation of turbulence (Duguay et al., 2018a,b). While we know of no standard guidance for turbulence structure within fishways and culverts, there is broad recognition that turbulence is important, prompting agencies to size pools and structures such that they remain below specific thresholds of energy dissipation (sometimes called the "energy dissipation factor" or *EDF*), which, depending on structures and species typically ranges from 25 to 240 W m^{-3} (Towler et al., 2015).

While *EDF* provides a measure of energy dissipation derived from first principles, it does not provide details on the characteristics of the associated turbulence. Turbulence is a generic term referring to temporal fluctuations in flow velocity and direction. The characteristics of those fluctuations vary widely, however. At its onset, turbulence comprises vortices (eddies) that vary in intensity, periodicity, orientation, and scale (Lacey et al., 2012;

Tritico and Cotel, 2010). In simple cases, these eddies can be highly structured and reduce the energetic costs of fish locomotion (Liao, 2007; Liao et al., 2003a,b). Although this has been demonstrated in laboratory settings, conditions in the field are rarely as simple. Instead, particularly at high flow velocities, eddies interact and create a highly chaotic structure that reduces swimming efficiency and increases energetic costs (Enders et al., 2003, 2005). Complicating matters further, turbulence does not arise spontaneously but instead is associated with structures, typically located on the channel substrate (e.g., boulders, baffles, corrugations). In some cases, fish can shelter behind these, occupying zones of low-velocity and turbulence (Duguay et al., 2018b; Goerig et al., 2017; Wang and Chanson, 2018). This can enable some individuals, particularly those with smaller body sizes, to traverse otherwise impassable barriers (Goerig et al., 2016). This complexity makes the study of turbulence in the context of swimming ability as challenging as it is important (Duguay et al., 2018b; Enders et al., 2017; Hinch and Rand, 2000), and researchers have called for improved standardization of its measurement in association with fish passage (Castro-Santos et al., 2009; Lacey et al., 2012).

One solution to eliminating hydraulic conditions that exceed swimming performance is to use mechanical means to lift the fish past a barrier. This can take the form of a "fish lift," a sort of elevator that attracts fish into a hopper, then raises them along with a volume of water to the headpond (or sometimes a trucking or similar transport facility), where they are then deposited above the dam (Sprankle, 2005; Travade and Larinier, 2002). Alternatively, navigation locks or specifically-designed fish locks use a similar approach, simply regulating the water level within a chamber to allow passage without requiring fish to ascend a channel volitionally (Júlio Júnior et al., 2009; Travade and Larinier, 2002). While these methods can bypass the limitations of swimming ability, they still require the fish to enter the structures volitionally and remain there while the operation is executed. Although once contained within the structure swimming performance is no longer relevant, conditions below these structures (particularly locks, which must discharge large volumes of water) can be energetically costly. Because of this, passage through locks tends to be poorer than through fishlifts (Finger et al., 2020; Travade and Larinier, 2002). These structures may also create problematic routes of passage for invasive species, as discussed further below.

Downstream fish passage presents engineers with a different suite of challenges. The original work to develop estimates of maximum sustained speeds was intended to determine the maximum flow velocity that downstream migrating salmon smolts could resist at the intakes of hydroelectric dams and other water withdrawals (Brett, 1962, 1964). The thinking was that if velocities could be maintained below U_{ms} salmon would not volitionally pass into the turbine intakes, giving them time to find an alternate and safe route of passage.

Because downstream migrants tend to follow the flow, downstream fishways (often called bypasses) are typically placed close to the turbine intakes. Extensive hydraulic and behavioral studies have shown that flow fields can be manipulated to guide fish to safe passage routes (Bates and Vinsonhaler, 1957; Haefner and Bowen, 2002; Kynard and Horgan, 2001; Shepherd et al., 2007). Then, by regulating the rate of flow acceleration, fish can be enticed to enter these structures. When properly designed, these can entrain fish beyond a point where they are no longer able to sprint out of the structure (Adams et al., 2014; Haro et al., 1998; Kemp et al., 2005) and are conveyed in a jet of water that discharges into a plunge pool of sufficient dimensions to allow for rapid passage without injury (Castro-Santos et al., 2020; USFWS, 2019). Because downstream migrants tend to follow the flow, downstream fishways typically require large discharge. This is inconvenient, because recent work has shown that upstream fishways can be safe conduits for downstream passage, provided that fish can be enticed to enter them (Celestino et al., 2019; Gutfreund et al., 2018), but actual usage is low (Pelicice and Agostinho, 2012).

Despite the extensive efforts and expense applied to the development of fishways and the design of culverts, the performance of these structures is highly variable (Bunt et al., 2012, 2016), and they frequently fail to meet management objectives. While it is common to attribute this to excessive emphasis on a few anadromous species in their initial design, the interplay between fish swimming behavior and ability likewise bears greater scrutiny.

The ability of fish to pass instream barriers in an upstream or downstream direction depends on their locomotor ability, but also on various behavioral aspects such as their motivation to approach, enter and pass a given structure, as well as the strategies employed to navigate challenging hydraulic conditions. The motivation, or willingness to swim (Table 1), can be captured by quantifying the rates at which fish approach and attempt to enter fishways or culverts (Castro-Santos, 2004; Goerig and Castro-Santos, 2017). These rates may depend upon multiple biological (fish species and body size), hydraulic (flow depth and velocity), and environmental factors (temperature, time of day) (Goerig et al., 2020; Goerig and Castro-Santos, 2017; Mensinger et al., 2021). Fish also use behavioral strategies to deal with challenging conditions. They may adopt an exploratory behavior at first then stage repeated attempts to pass a barrier, in which case their probability of success increases with the number of attempts (Castro-Santos, 2004, 2006). They may also optimize their average groundspeed to maximize the distance traversed (Castro-Santos, 2005), or take advantage of turbulence and heterogeneity in hydraulics to select low-velocities areas and move forward without incurring excessive energy costs (Goerig et al., 2017; Wang and Chanson, 2018). It is therefore important to also consider locomotor behavior in fishway and culvert design, and how behavior and ability are related to actual hydrodynamics near and within fish passage structures.

3 Invasive species in river systems

Industrialization, commerce, and transportation systems have provided numerous and novel vectors for conveying species across what were once impermeable barriers of elevation, distance, salinity, temperature, etc. While some of these invasions are the consequence of deliberate introductions gone astray (such as the introduction of Brown Trout (*Salmo trutta*) to North America, and Brook Trout (*Salvelinus fontinalis*) to Europe and western North America (Rahel et al., 2008)), others were purely accidental, either via escapement from aquaculture facilities (e.g., Bighead (*Hypophthalmichthys nobilis*) and Silver Carp (*H. molitrix*) in the Mississippi drainage (Stokstad, 2003)), and in ballast water (Elskus et al., 2017; Treanor et al., 2017), etc.). In some cases, multiple pathways interact, and the success of invasions is mediated both by the vector of transmission and by habitat type and quantity (freshwater habitats are much more vulnerable to invasion (Alcaraz et al., 2005; Beletsky et al., 2017; García-Berthou et al., 2005)).

Navigation has been a major catalyst for these invasions, both in the form of ballast water, and via navigation locks. While ballast water vectors can be mitigated by purging and replacing ballast tanks during transit and chemical treatments (Elskus et al., 2017), this is not true of navigation locks. Locks are pervasive throughout the rivers of the developed world, and have promoted the invasions of Round Goby (*Neogobius melanostomus*) in the Rhine (Roche et al., 2013), invasive carps in the Mississippi and bordering drainages (Stokstad, 2010), and Sea Lamprey (*Petromyzon marinus*) into the Laurentian Great Lakes (Zielinski et al., 2019), to name but a few examples. Novel technologies are in development to limit movements through these pathways, although to date no measure has proven to be effective (Stokstad, 2003, 2010; Zielinski et al., 2019).

Once established, invasive species are exceedingly difficult to control, let alone eradicate. One case in which control can be possible is when invasive species are migratory, using discrete habitats for spawning and growth (Behrens et al., 2017). In this situation, barriers that inhibit movement can be constructed to prevent passage, interrupting key life history stages. In lotic habitats, such efforts sometimes focus on swimming ability and behavior, although these efforts are not always successful: interest in separating Brown Trout and Brook Trout were frustrated by the striking similarity in swimming ability between these two species (Castro-Santos et al., 2013). In the case of Sea Lamprey in the Great Lakes, however, considerable success was achieved by constructing hundreds of barriers on surrounding tributaries.

These barriers were designed to prevent invasive Sea Lamprey from passing, but at sufficiently low heights that highly-valued salmonids were able to pass (Zielinski et al., 2019). This was facilitated by both the ability and propensity of salmonids to leap over obstacles. There was an unintended consequence, however: many non-leaping native species also migrate between the lakes and rivers to spawn and for other activities important to their life history, and these barriers have obstructed these movements (McLaughlin et al., 2013). Current solutions to this dilemma include trap-and-sort facilities, where lamprey are culled from fishways and native species are allowed to proceed upstream (Pratt et al., 2009).

Owing to the overall poor performance of fishways at passing native species, this solution has been deemed inadequate, and significant efforts are underway to develop fishways that selectively block Sea Lamprey while allowing native species to pass (Fig. 3; Zielinski et al., 2019; http:/www. glfc.org/fishpass.php).

Owing to their unique morphology, Sea Lamprey are thought to have reduced swimming ability compared with native species. This suggests that a simple velocity barrier might be a successful approach to selective passage, at least for some strong swimming species (Fig. 4).

However, lamprey have a secret weapon: their sucking disc, used to attach to their prey and for constructing spawning redds, is also an efficient tool for ascending zones of high-velocity flow (Fig. 3). During attachment, the only actual work being performed is the maintenance of sufficient suction to prevent the fish from being dislodged. In this way, they are able to recover from bouts of sprinting and, when ready, to execute another burst of swimming (Castro-Santos et al., 2019). By incrementally sprinting and attaching to a range of substrates they are able to ascend challenges that are impassable by almost any other species (Moser et al., 2011).

This unique ability can be countered by lining channels with a substrate that inhibits attachment. Some studies have shown that this is a promising line of research, although definitive solutions remain elusive (Castro-Santos et al., 2019; McCauley, 1996; Zielinski et al., 2019), and other approaches will be needed for smaller-bodied and weaker-swimming species. Nevertheless, these and similar efforts hold promise for helping to limit access to key habitats and facilitating control. Lessons learned from lamprey may hold clues for the control of other species with related abilities, such as Round Goby (Voegtle et al., 2002; Webb, 1989).

4 Capture fisheries

4.1 The biomechanical foundation of fish capture

Global fisheries production was 178.5 million tons in 2018, with 96.4 million tons from capture fisheries (FAO, 2020). Fish are an important source of nutrition for many countries, providing 3.3 billion people globally with 20% of their animal protein intake in 2018 (FAO, 2020). Fish capture is an essential activity in many nations, especially in coastal and island nations, with a large variety of fishing gears and operational methods, and these uses are likely to continue (He et al., 2021a).

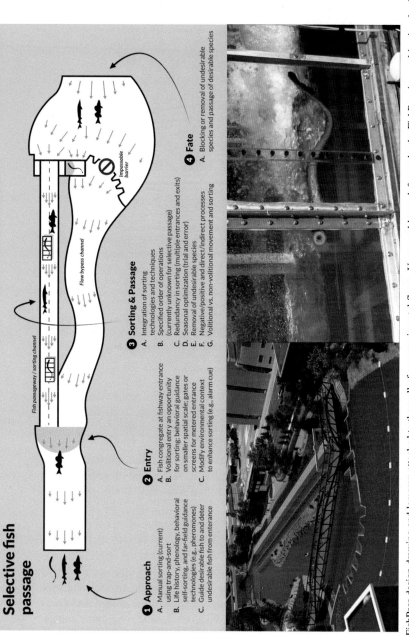

Selective fish passage

Fish passageway / sorting channel

Impossible barrier

Flow bypass channel

① Approach

A. Manual sorting (current) using trap-and-sort
B. Life history, phenology, behavioral for sorting; behavioral guidance technologies (e.g., pheromones)
C. Guide desirable fish to and deter undesirable fish from enterance

② Entry

A. Fish congregate at fishway entrance
B. Volitional entry an opportunity for sorting: behavioral guidance on smaller spatial scale; gates or screens for metered entrance
C. Modify environmental context to enhance sorting (e.g., alarm cue)

③ Sorting & Passage

A. Integration of sorting technologies and techniques
B. Specified order of operations (currently unknown for selective passage)
C. Redundancy in sorting (multiple entrances and exits)
D. Seasonal optimization (trial and error)
E. Removal of undesirable species
F. Negative/positive and direct/indirect processes
G. Volitional vs. non-volitional movement and sorting

④ Fate

A. Blocking or removal of undesirable species and passage of desirable species

FIG. 3 FishPass design drawing and lamprey attached to the wall of an experimental flume. Upper and lower-left panels: the FishPass is a multinational collaboration to develop structures that will selectively pass a range of native fish while obstructing Sea Lamprey. The ultimate intent is to use intrinsic differences in morphology, capacity, and behavior to separate and remove invasive Sea Lamprey while permitting free passage to native species. The design will allow use of free-swimming fish in a field-like setting, while allowing sufficient control to conduct experiments as they pass through the three phases of fish passage: (a) discovery of (or approach to) a potential passage route; (b) entry into the passage structure; and (c) passage through the structure itself. Each of these steps offers an opportunity for failure: passage is only possible by sequentially completing all three phases (Castro-Santos and Haro, 2010; Silva et al., 2018). Lower-right panel: a Sea Lamprey attached to a plexiglass wall in a flume in high-velocity flow (photo by E. Goerig, USGS).

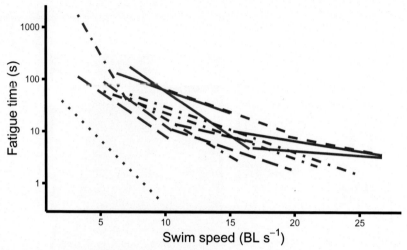

FIG. 4 Comparative sprinting (red) and prolonged (dark blue) endurance curves for a range of species from five orders: Cypriniformes (Iberian barbel (*Luciobarbus bocagei*), straight-mouthed nase (*Pseudochondrostoma duriense*), and white sucker (*Catostomus commersoni*), solid lines); Clupeiformes (American shad (*Alosa sapidissima*), Alewife (*A. phseudoharengus*), and Blueback Herring (*A. aestivalis*), dash-dot lines); Perciformes (Walleye (*Sander vitreum*) and Striped Bass (*Morone saxatilis*), long-dashed lines); Salmoniformes (Brook Trout (*Salvelinus fontinalis*) and Brown Trout (*Salmo trutta*), short-dashed lines); and Petromyzontiformes (Sea Lamprey (*Petromyzon marinus*), dotted line) (Castro-Santos, 2005; Castro-Santos et al., 2013; Sanz-Ronda et al., 2015; Zielinski et al., 2019) . Sea Lamprey have comparatively poor endurance, but their ability to attach to surfaces (Fig. 3) enables them to recover from fatigue without being swept downstream in strong currents.

Given the importance of both commercial and subsistence fishing, it is becoming increasingly important for managers to regulate both the species and sizes captured by fishing gears. In many cases, it is important to selectively target a given size and species in order to optimize future recruitment (Beverton and Holt, 1957). The required selectivity is often not matched by gear performance, however, and vast quantities of fishes and other organisms are unintentionally captured as bycatch and then discarded at sea, often with serious ecological and social consequences (Harrington et al., 2005). Because of this, as well as owing to the associated benefits to the fishers themselves, extensive efforts are underway to improve efficiency and selectivity of fishing gears (Hasselman et al., 2016; He et al., 2021b; Valdemarsen, 2001).

Fishing gears are designed based on the behavior and locomotion ability of target species, and more recently, that of unwanted bycatch species and protected species. Fish locomotion is an important factor in the design and operation of the active gear to catch or release fish, such as trawls, purse seines, and dredges (He, 1993; Wardle, 1986). For passive gears such as gillnets and longlines, swimming activity determines the range of their fishing area, which also directly impacts capture efficiency (He, 2003).

Fishing is the interaction of fish and fishing gear in a physical environment. Environmental conditions affect biology, ecology, and physiology of fish, which governs fish reaction to fishing gears, as well as their ability to escape from them. Environmental conditions also directly impact the type of fishing gear that can be operated. Successful fish capture is managed by using the appropriate fishing gear to catch the intended fish (species and size) at the appropriate time and place (Fig. 5).

4.1.1 Fish capture by trawls: The role of fish locomotion

Understanding fish biomechanics, especially swimming ability and behavior, is crucial in active fishing gears that chase or surround fish for capture, but it is also important for passive fishing gears that rely on fishes' movement into the gear. Here, we take the single boat bottom otter trawl (He et al., 2021a,b; called "otter trawl" or simply "trawl" hereafter), one of the most common active gear types, as an example on how swimming behavior and ability thread the entire capture process from the time the fish is aware of the approaching trawl to the time it is either retained in the codend or has escaped from the trawl. Fig. 6 illustrates the capture process of fish by a trawl. The trawl as well as its influenced area is divided into five zones where fish may behave differently during the capture process.

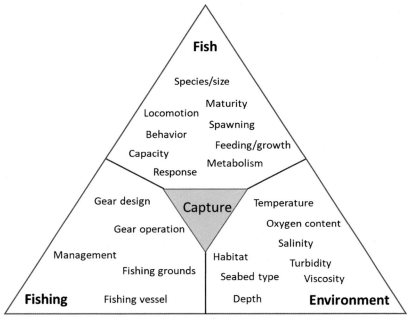

FIG. 5 Three pillars for fish capture. Successful fish capture is about using the right fishing gear to catch the right fish (species and size) at the right time and place.

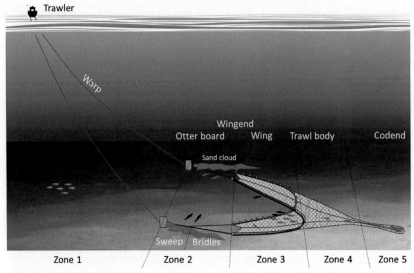

FIG. 6 Typical swimming behavior of fish when approaching and being herded inside an otter trawl. Zone 1 is the pre-trawl zone where fish are alerted of an approaching otter trawl through acoustic and visual stimuli from the trawl. Zone 2 is the herding zone where fish are herded by sweeps and bridles toward the mouth of trawl between the wings. Fish are typically moving away from the towed oblique rope toward the center of the trawl path. The shaded area trialing the otter board are sand clouds. Zone 3 is the concentration zone where fish are concentrated toward the center of the trawl, and turn and swim with the trawl. Zone 4 is the trawl body that gradually narrows and guides fish toward the codend. Zone 5 is the codend, which is often preceded with an "extension piece." Fish are accumulated in the codend where turbulence and motion of the codend stimulate fish to attempt to escape through the meshes.

Zone 1 is the pre-trawl zone, which could extend forward for more than a kilometer from the otter boards. Depending on the type of seabed and the design of the trawl, noises generated by the otter board hitting rocks on the seabed and from the propellor of the trawler can be heard by fish far earlier than the fish can see the trawl. Using stationary hydrophones, Winger (2004) demonstrated that Atlantic Cod (*Gadus morhua*) responded to a survey trawl vessel at 1500 m by slowing their swimming speed to near zero. Acoustic stimuli from the trawler and its trawl may thus raise awareness in the fish, which may look for possible dangers well before the trawl becomes visible (Wardle, 1986).

Zone 2 includes the area from the otter boards to the wingends. For many fish trawls, especially trawls for flatfish, long cables (sweeps and bridles) connect the boards and the wingends to increase the horizontal distance between boards (called door spread). Bridles usually refer to two or more cables that extend from top and lower wingends. They are joined together and connect to the sweep.

Fish often react to moving oblique cables by swimming away from them and are herded toward the path of the trawl. The angle between the cable and the towing direction is called the sweep angle (α in Fig. 7), while the

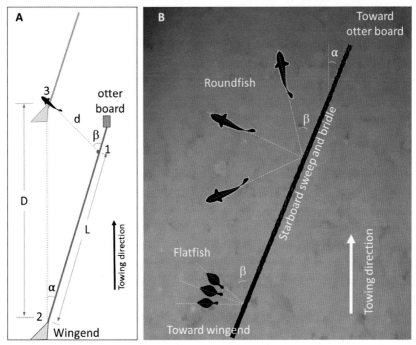

FIG. 7 Fish reaction to sweeps and bridles of a bottom otter trawl. (A) A fish at Position 1 has to be capable of swimming to Position 3 before the wingend moves from Position 2 to 3 for it to be herded into the path of the trawl between the wings. (B) Avoidance angle of fish. Typically, round-fish have a larger range of avoidance angles (β) than flatfish which are usually at around 90 degrees.

angle between the direction of fish avoidance behavior and the cable is the avoidance angle (β in Fig. 7). For fish at Position 1 (ahead of the path of the bridles, but outside of the capture zone defined by the wings, Fig. 7) to be herded into the capture zone, they have to swim the distance (d in Fig. 7) from Position 1 to Position 3 within the time period when the wingend moves forward the distance (D) from Position 2 to Position 3. Swimming speed required (U_s) is related to angles α and β, and to the towing speed of the trawl (U_{Tr}). The endurance time (Table 1) required for a fish to be herded into the trawl path (T_{esc}) is related to α, β, U_{Tr}, as well as to the position of the fish along the length of the cable when fish starts to react (L).

$$U_s = U_{Tr} \frac{\sin(\alpha)}{\sin(\beta)} \qquad (1)$$

$$T_{esc} = \frac{L}{U_{Tr}} \frac{\sin(\beta)}{\sin(\beta - \alpha)} \qquad (2)$$

Towing speed of the trawl and sweep angle directly affect the swimming speed required for herding. Simulations from Eqs. 1 and 2 with a trawl towing speed of 1.5 m/s and reaction distance of 100 m from the wingend are plotted

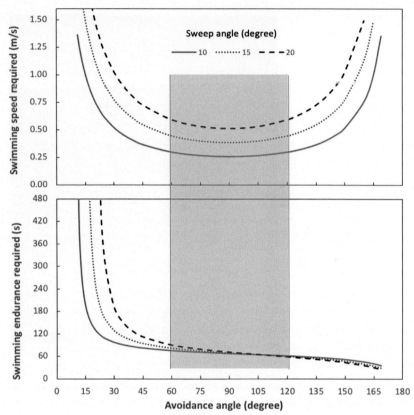

FIG. 8 Swimming speed and endurance required for fish to be herded into the path of trawl as related to sweep angle and avoidance angle for fish that react to a cable towed at 1.5 m/s at 100 m from the wingend.

in Fig. 8. Typically, the lowest swimming speed required is when the fish avoids the trawl by swimming directly away from the cable ($\beta = 90°$), but the changes in the required swimming speed are minimal from 60° to 120° (shaded zone, Fig. 8). Larger and smaller avoidance angles require higher swimming speeds for fish to be herded into the trawl path; if a fish is not capable of doing so, it will be overtaken by the sweep and not herded.

Sweep angles (α) of otter trawls are typically between 10° and 20°. Large sweep angles require fish to swim faster to be herded into the trawl path, which may prove inefficient for catching some fish. Sweep angle >20° was found to reduce catch of Cod and Haddock (Strange, 1984), indicating the importance of matching design and mechanics of the fishing gear with biomechanics of the target fish species.

Fish reacting to longer sweeps are required to swim longer to get away from the cable and into the path of the trawl, which may be challenging for species with lower swimming endurance. As swimming endurance is related to the length of fish (He and Wardle, 1988), among other factors, longer sweeps may result in the catch of more large fish and fewer small fish. Indeed, Engås and Godø (1989) found that longer sweeps resulted in a higher proportion of large Atlantic Cod and Haddock in survey trawls.

Avoidance angles of roundfish (with fusiform or laterally compressed shape) such as Atlantic Cod, Haddock (*Melanogrammus aeglefinus*) and Pollock (*Pollachius virens*) are more variable than those of flatfish, which are more consistent and closer to 90° (Fig. 7). Through underwater telemetry, researchers in Scotland found that Haddock approached and avoided the sweep at larger and more varied angles and at greater distance from the cable (SOAFD, 1998). Flatfish often react at a very short distance from the cable, exhibiting a swim-and-rest behavior, as being periodically "chased" by the approaching cable (Wardle, 1986; Winger et al., 2010).

Zone 3 (Fig. 6) includes the area between the wings and under the square (overhang) of the trawl, just ahead of the groundgear, which contacts the seabed. Fish arriving at Zone 3 typically start to turn and swim with the trawl, with flatfish swimming much closer to the groundgear. The swimming ability of the fish and the towing speed of the trawl strongly influence the duration that fish swim with the trawl in that area.

The variation in swimming ability with body size and species has important implications for selective fishing (He and Wardle, 1988; Wardle, 1977; Wardle and He, 1988). For example, 40 cm-long Saithe (also called Pollock, *Pollachius virens*) can swim for 6 min at 1.5 m/s before being captured by a trawl, while Haddock, Whiting (*Merlangius merlangus*), and Atlantic Cod of similar sizes could only maintain that speed for 2.5, 1 and 0.5 min, respectively. By comparison, endurance of smaller individuals of these species was <1 min at the same towing speed of 1.5 m/s (Main and Sangster, 1981).

Atlantic Mackerel (*Scomber scombrus*, 35–45 cm in length), being a fast swimmer, was observed swimming forward and escaping from the groundfish trawl towed at 1.5 m/s (Main and Sangster, 1981). Midwater trawls targeting Atlantic Mackerel are typically towed much faster, ranging from 1.9 to 2.3 m/s. Even at these high speeds, Mackerel (34 cm) were swimming for 30 to 58 min with a midwater trawl as observed by an underwater camera system (Rosen and Holst, 2013). By comparison, in the laboratory, Atlantic Mackerel of similar size were able to swim at 1.5 m/s for 60 min and at 1.8 m/s for 10 min (He and Wardle, 1988). It must be pointed out that observations at sea of fish swimming speed and duration are opportunistic and often lack precise details of fish size and physiological status (e.g., the level of exhaustion).

Subtle changes in towing speed can make important differences in capture efficiency. Some species will maintain a position just ahead of a trawl and evade capture. To respond to this behavior, a fishing strategy called "power take-off" is sometimes practiced to catch the unexhausted fish by increasing the towing speed at the end of the tow. Conversely, towing speed may be reduced before hauling to avoid catching stronger-swimming fish.

Zone 4 includes the body of the trawl net (Fig. 6), where fish are gradually funneled within the reduced diameter of the net. Fish that have given up swimming in front of groundgear may swim toward the codend, but they may also orient forward again as space becomes narrower. Longer nets can better guide fish toward the codend with less meshing of fish in the net.

Zone 5 is the codend of the trawl where fish are accumulated (Fig. 6). A codend with accumulated catch can block the water flow and create turbulence within the codend. Fish arriving in the codend may start to swim again, at slower swimming speed due to the bucket effect created by the codend with fish. Fish may also attempt to swim out through the open meshes, which may cause scale loss or injuries due to contact with the mesh. He (1993) modeled escape probability of fish through a codend mesh under different towing speeds, mesh sizes, and water temperatures, revealing a higher probability for fish to escape without contacting the netting when the codend has a large mesh size, the towing speed is low, and in areas with warmer waters.

4.2 The role of fish biomechanics in reducing bycatch and discards: A case study

While the definition of bycatch varies among jurisdictions, it generally refers to "the catch of organisms that are not targeted" (Perez Roda et al., 2019). Tropical shrimp trawls in the Gulf of Mexico faced bycatch of sea turtles and juveniles of important commercial fish species such as Red Snapper (*Lutjanus campechamus*). While sea turtle bycatch was significantly reduced through the testing and introduction of various turtle excluder devices (TEDs, also called trawl efficiency devices) in the 1980s and 90s (Jenkins, 2012; Watson et al., 1999), fish bycatch persisted. Researchers in the Gulf of Mexico have conducted a series of studies on topics ranging from behavior of relevant species to gear designs to reduce fish bycatch in shrimp trawls in the last 30 years. The effort led to the development of several types of bycatch reduction devices (BRDs), such as the fisheye, the extended funnel, and the Jones/Davis device (Watson et al., 1999). This case study describes the process and results of the development of the nested cylinder bycatch reduction device (NCBRD), designed to reduce red snapper bycatch through understanding of its biomechanics, especially swimming behavior and ability.

Underwater observations of red snapper found that fish would seek areas of low flow near the bycatch reduction device (BRD), which may provide opportunity for fish to escape from the opening of the device, but very few fish escaped during steady towing of the trawl (Engås et al., 1999). However, as the trawl was slowed during hauling, significant active escape occurred. Due to variabilities in haulback procedures, the rate of bycatch reduction varied among vessels. Understanding the mechanism of fish escape and means to induce fish to escape was thus proposed (Engås et al., 1999).

To further understand swimming behavior of red snapper, Engås and Foster (2002) conducted a laboratory study to examine the effect of "inclined" water flow. When the recirculating swimming tunnel was horizontal, red snapper 9.7–14.4 cm in length maintained a specific position (called bull's eye) while swimming steadily against a flow velocity of 3 body lengths per second ($BL\,s^{-1}$) for 10 min. When the swimming tunnel was tilted (as much as 45°), many fish were not able to maintain the same position and instead showed erratic swimming behavior such as moving up, down and sideways, demonstrating the potential of utilizing inclined water flows to develop more selective devices (Engås and Foster, 2002).

Critical swimming speed (U_{crit}) of red snapper was measured in laboratory at various times of the year, showing a peak in October; importantly, diel period (day and night) did not affect swimming performance (Parsons and Foster, 2007). However, when exposed to a vortex generating bycatch reduction device, significantly larger percentage of fish escaped during the day (higher luminosity) than during the night. Moreover, when light was provided in the test tank (e.g., using a green Cyalume light stick or blue-green LED light), all fish escaped within a short period of time, with the intensity of light negatively correlated to the time they took to escape (Parsons et al., 2012). This indicates that illumination during nighttime trawling can be an effective way to stimulate escape behavior of bycatch species (Parsons and Foster, 2007).

Parsons and Foster (2007) also tested the preference of illuminated and uni-lluminated area in a water tunnel. They found that all fish chose the darker side of the swimming tunnel, indicating negative phototactic response. This provides additional evidence that fish behavior could be partly controlled by manipulating illumination, especially for shrimp trawls in the Gulf of Mexico, where most fishing operations are conducted at night (Parsons and Foster, 2007).

Based on the above and other research, Parsons and his colleagues designed a BRD called nested cylinder bycatch reduction device (NCBRD), as illustrated in Fig. 9 (Parsons and Foster, 2015; Parsons et al., 2012). The NCBRD modifies the flow in the area outside of the inner section and provides illumination at the exit from the mesh sock, allowing Red Snappers to turn and escape. The targeted shrimp species, which have inferior swimming ability than Red Snapper, are pushed to the codend. This design reduced red snapper bycatch by as much as 50% (Parsons and Foster, 2015).

5 Fisheries management and enhancement

5.1 Fisheries surveys

Bottom trawls are the most common fishing gear for fisheries surveys, but pelagic trawls, gillnets, longlines, and pots are also used. Acoustic devices such as echo sounders are used for surveying schooling fish in pelagic environment. In rivers and streams, electrofishing is also used for surveys. More recently, stationary underwater cameras are used to survey coastal and reef fish. While most of these survey methods may be affected by fish behavior,

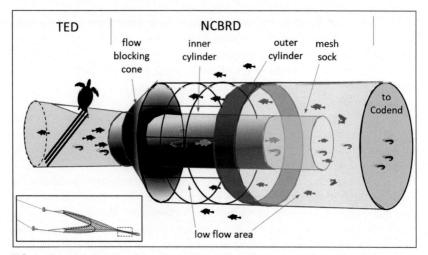

FIG. 9 Schematic drawing of the nested cylinder bycatch reduction device (NCBRD) designed for the Gulf of Mexico and Southeast US shrimp trawl fisheries. The flow-blocking cone reduces flow in the area outside of the inner cylinder. The rear of the exit area is illuminated (yellow). Red snappers turn at the exit area of the mesh sock and seek the low-flow and darkened area, leading to escape in the area between the cone and the out cylinder where there is no netting. Shrimps with poor swimming ability are pushed to the codend. Redrawn based on Parsons et al. (2012) and Parsons and Foster (2015).

especially locomotory habits, we offer the following examples to illustrate how fish biomechanics, especially swimming behavior can affect outcome of fisheries surveys.

5.1.1 Tow duration of bottom trawl surveys

Tow duration of bottom trawl surveys is usually between 15 and 60 min. Evidence had previously been interpreted as indicating there was no effect of tow duration on catch per unit effort (CPUE: Godø et al., 1990; Walsh, 1991; Wieland and Storr-Paulsen, 2006). More recent data challenge this view: Sala (2018) found an effect of tow duration on CPUE as well as the size of fish, comparing 30 and 60 min tows. This likely occurs because of the relation between swimming performance and body size: larger fish may not become exhausted when swimming at the mouth of trawl if their endurance exceeds tow duration. It is thus possible that large fish may be underrepresented when the tow duration of a survey trawl is short. Effect of tow duration on CPUE and size distribution of the catch may be species specific, and more information may help to better-inform design of bottom trawl surveys.

5.1.2 Encountering probability of fish with passive survey gears

Catch by passive gears (also called stationary gear, or fixed gear) such as gill-nets, longlines, and pots rely on the movement of fish which accidentally swim into gillnets, or are attracted to bait in longlines and pots. For gillnets, the faster a fish moves, the more likely the fish will encounter the net, both because of the volume of water encountered (Løkkeborg et al., 2010, 2014; Rudstam et al., 1984), and because of the associated reduction in response time (Engas and Lokkeborg, 1994). Therefore, incorporating probability and speed of fish encountering fixed gear will improve understanding of gear efficiency and hence more accurate stock assessment (Lennox et al., 2017).

5.1.3   Active space in passive survey gears

"Active space" of a passive fishing gear is the potential area within which the fish may be captured by the gear (McQuinn et al., 1988). It is related to the swimming speed of fish and soak time of the gear (the duration the gear is deployed). The effectiveness of these gears varies with season, and it is likely to change as habitats change in response to a warming climate. For example, He (2003) modeled the active space of a gillnet for catching Winter Flounder (*Pseudopleuronectes americanus*) and found that active space would be more than 7 times greater at 4.4 °C than that at −1.2 °C for a gillnet set for 12 h, owing to increased activity of winter flounder at the higher water temperature. Similarly, active gears may become less effective in historically colder regions if the species there become more active and swim with greater endurance at higher temperatures, although such changes may be offset by changes in habitat use. Regardless, temperature-related changes in fish activity and swimming ability are likely to result in changes in vulnerability to both active and passive fishing gears, with associated implications for fish capture, stock assessment surveys and management.

5.2   Stock enhancement

Above we describe various approaches to restoring populations impacted by human activities, whether through fragmentation, habitat alteration, fishing, etc. Often, though, it is necessary to supplement populations in the field and/or to maintain broodstock to safeguard existing genetic diversity. To this end, fish hatcheries have served a vital role for more than a century (Anonymous, 1872, 1877; Clift, 1872). The development of dams and targeted harvest of migratory species, particularly in North America during the 18th and 19th centuries were recognized as important causes of declines in many species. This context, combined with the opening of vast territories of what became public lands and waters prompted significant public investment in the artificial propagation and stocking of both native and non-native species into freshwater lakes and streams (Anonymous, 1877).

The captive rearing of fish has a much deeper history; however, aquaculture has been an important food source for humans for millenia. The earliest records of aquaculture for food production come from China in 3500 BCE; where by the 1980s 2.7 million hectares of freshwater surface area were dedicated to aquaculture for food production (FAO, 1983). The importance of aquaculture continues to grow and will likely remain one of the most important sources of human dietary protein (Teletchea and Fontaine, 2014).

As with many species raised in captivity, freedom of movement and exercise has benefits, both for the health of individual fish and for the quality of the final product, whether it be destined for restoration of wild populations or for food markets. When compared with fish raised under lentic conditions, a range of species subjected to flows on the order of $1\,BL\,s^{-1}$ show increased feed conversion rate, improved growth rates, better fin condition, added mass and changed composition of both red and white muscle, increases in number of dermal layers, and improved shelf life. Not surprisingly, some species show improved survival and fitness, as well as more natural movement patterns when subsequently released to the wild following exercise regimes (Davison, 1997; Jobling et al., 1993; Jørgensen and Jobling, 1993).

6 Biomimetic engineering for fish conservation in the anthropocene

One pervasive problem that has impeded development of effective conservation strategies is the difficulty of observing live fish in their native (or invasive) habitat. Physiology is often studied using *in vitro* methods, or as mentioned above, on whole animals constrained in artificial environments that prevent them from performing behaviors and tasks most relevant to their lived reality.

Advances in telemetry, particularly over the past four decades, have dramatically improved our ability to monitor both large- and small-scale movements of free-ranging animals (Monan, 1985; Monan et al., 1975). Development of large, fixed receiver arrays have promoted an increasing number of studies, offering insights into habitat use, effects of human activities, etc. (Krueger et al., 2018).

Characteristics of local environments, and the sheer vastness of aquatic ecosystems can make such methods impractical, however, and technologies are rapidly evolving that will expand the scope of monitoring. One class of devices that is yielding insights is automated drones. Aerial drones, for example, have allowed surveys of land animals and their movements in locations and over distances that are difficult if not impossible to achieve for human observers (Inman et al., 2019; Schlossberg et al., 2016). Drones are also used in studies of bird migration and behavior (Canal and Negro, 2018; Marinov et al., 2016), and are indispensable for habitat surveys, especially in inaccessible areas. These concepts have recently been expanded to the marine environment. Restrictions on survey crews owing to the Covid-19 pandemic created an opportunity to deploy unmanned drones to survey large tracts of ocean for fisheries surveys that had previously only been performed by human-crewed vessels, and similar efforts are ongoing around the world

(De Robertis et al., 2021). In addition to performing autonomous surveys, such devices also offer opportunities to observe fish in their environment and their response to things like fishing gears, promising to greatly improve our understanding of the mechanical underpinnings of gear efficiency, bycatch, and other challenges.

Nevertheless, technological progress in fish conservation biology is lacking for reasons that must include the many challenges involved in operating mechanical systems underwater, difficulty with vision and acoustic communication in aquatic environments, and the inability of most current robotic systems to function with the speed, independence, and maneuverability needed to operate in often challenging high-velocity or turbulent waters.

These challenges, as well as the general benefits that come from physical and quantitative modeling have prompted many researchers to turn to fish biomechanics and functional morphology for inspiration, with some of the more remarkable devices currently being deployed incorporating design concepts inspired by biological systems (Whitt et al., 2020). The development of fish-inspired robotic systems coupled with currently available technologies such as electronic tags and drones promises to greatly improve our ability to both understand and conserve fish populations in the near future. In addition, unexpected benefits can arise for the analysis of fish biology from mechanical and robotic perspectives. Advances such as the use of fish-skin-like skin coverings to enhance propulsion through the water by both human swimmers and ships, and new vehicles that can navigate complex hydraulic and structural environments at depth are examples of benefits that can result from the study of fish locomotion. In this section, we first summarize the current state of fish-inspired robotic platforms that could be further developed to contribute to fish conservation physiology, and then we describe several specific areas where technological approaches are beginning to make particularly valuable contributions.

6.1 Fish robotics: Current state of the art

Many, if not most, fish-inspired robotic systems are small, low speed platforms that are designed as laboratory-based "scientific demonstrators" constructed to better understand the basic physics of aquatic propulsion (Lauder et al., 2012; Moored et al., 2011). Some of these laboratory systems use simple actuation to drive a flexible surface that has been used to understand the effect of body flexibility, tail shape, and stiffness on swimming speed and efficiency (Fig. 10; Lauder and Tangorra, 2015; Van Buren et al., 2017, 2019). One advantage of such experimental platforms, in addition to their relative simplicity, is their utility for exploring a large parameter space of movement and quantifying the effect of swimming objects to uncover how undulatory wave-like swimming motions affect speed and efficiency (Hertel, 1966; Lauder and Tangorra, 2015; Smits, 2019).

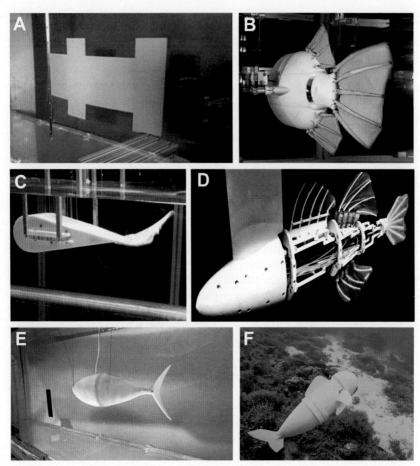

FIG. 10 Current fish robotic systems range from simple flexible plastic models of fish bodies and fins actuated at the leading edge (A) (Matthews and Lauder, 2021) to platforms designed to investigate the function of dorsal, anal, and caudal fins with individually controllable fin rays (B and C: Esposito et al., 2012; Mignano et al., 2019). Whole fish laboratory robotic platforms (D and E) (Wen et al., 2018; Zhu et al., 2019) allow study of the body bending and thrust production during both steady swimming and acceleration behaviors. Katzschmann et al. (2018) designed a free-swimming fish robot (F) for marine reef environments that is controllable remotely by a diver.

Another class of laboratory fish-like robotic systems uses mechanical prototypes of fish fins, modeled explicitly on fish anatomy, to explore how fins generate propulsive forces (Fig. 10). Such studies have included the design and analysis of both pectoral fin and caudal fin robots for propulsion and maneuvering (Esposito et al., 2012; Gottlieb et al., 2010; Tangorra et al., 2011). While progress has been made incorporating these fish fin-like actuators into a whole fish robot, such systems have so far been confined to the laboratory. Using highly simplified fin models based on airfoil shapes attached to a rigid "body" has allowed free-swimming robotic systems to perform well in a swimming pool (Long et al., 2006), but to date such platforms have not been deployed in the field under their own power.

One fish-like robotic system that has been deployed in the field is the SoFi system described by (Katzschmann et al., 2018) who developed a robotic fish using flexible (soft) actuators powered by a hydraulic mechanism. Their explicit goal was to design a robot that could function in marine environments for coral reef exploration (Fig. 10F), and they used acoustic communication from a hand-held controller operated by a nearby diver to control robot trajectory. The SoFi robot is capable of generating tail beat frequencies of up to 1.4 Hz and average swimming speeds of 23.5 cm/s or 0.5 body lengths/s, and the SoFi robot of Katzschmann et al. (2018) represents the most fish-like, controllable, and field-deployable robotic platform developed so far.

One final area of research interest in the design of biomimetic fish-like swimming systems focuses on the skin and specialized surface structures that could enhance swimming performance. Fish exhibit a remarkable diversity of skin surface structures with an array of scale types and textures ranging from shark skin denticles with tooth-like structures that protrude above the epidermis to the scales of ray-finned fishes that possess an enormous variety of ridges, protrusions, and ornamentation (Fig. 11; Reif, 1985; Wainwright and Lauder, 2018; Whitear, 1970). The textured surfaces of fishes most likely have many functions, but the skin surface almost certainly plays a hydrodynamic role in swimming (Lauder et al., 2016). The most common premise of fish skin research is that the surface ornamentation and mucus layer (on bony fishes) reduce drag forces and the cost of transport. By creating a smooth body surface both pressure and friction drag forces could be reduced. For shark skin denticles, experiments on both pieces of skin and biomimetic 3D printed denticle arrays (Fig. 11), moved in a swimming motion by a robotic controller, have shown that the surface texture can both reduce drag and also increase thrust by altering the structure and strength of vortices generated by the tail (Oeffner and Lauder, 2012; Wen et al., 2014, 2015). Studies of airfoils printed with a shark denticle surface have also shown that lift can be enhanced and drag reduced: increasing the lift: drag ratio that is critical to high-performance wing-like function (Domel et al., 2018). To date, biomimetic fish skin has not been applied to freely-swimming robotic systems, but advances in manufacturing technology in the near future may allow application of biomimetic skin to fish robots in the field and contribute to improved locomotor function.

Understanding the physical biology of aquatic propulsion offers multiple benefits and opportunities. The process of developing mechanical systems helps us to understand the forces and dynamics that fish experience in their daily lives, which itself can aid in prioritizing conservation actions. At the same time, though, humans also benefit, potentially developing machines that can traverse challenging environments for a variety of applications such as inspecting infrastructure in hazardous environments.

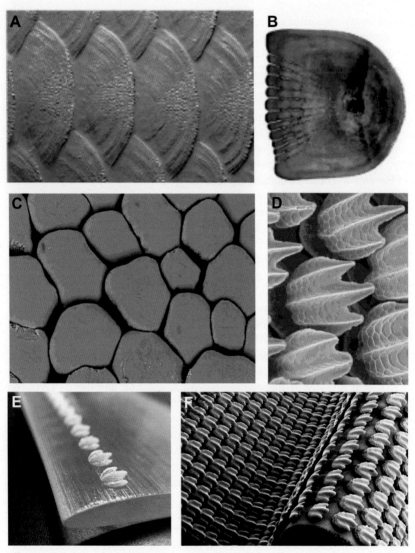

FIG. 11 Fish skin surfaces and biomimetic models. (A) Body surface scales of a bluegill sunfish (*Lepomis macrochrius*) and an isolated ctenoid scale (B; Lauder et al., 2016). (C) Nose denticles from a smooth dogfish shark, (*Mustelus canis*; Ankhelyi et al., 2018), and lateral body denticles (D) covering the skin of a bonnethead shark (*Sphyrna tiburo*; Oeffner and Lauder, 2012). Biomimetic 3D-printed shark denticles attached to the leading edge of an airfoil (E) both improve lift and reduce drag (Domel et al., 2018), while testing a flexible membrane printed with an array of 3D-printed shark denticles (F) improved swimming performance (Wen et al., 2014).

Despite progress over the last 15 years in the development of a variety of fish-like robotic systems, swimming performance still lags far behind that of fishes, and there are several areas in particular where improvements could be achieved. Better implementation of swimming kinematics derived from fishes would likely result in improved swimming speed and maneuverability, as would better streamlining and use of more fish-like body shapes. Robotic systems are often programmed with wave-like movements to generate body undulation without reference to published data on fish kinematics. Many fish species employ very similar body wave motions (Lauder, 2006; Lauder and Tytell, 2006;) which can be implemented into robotic systems to improve swimming performance. Autonomy is a more challenging problem as both power consumption by the robot and communication with a remote controller are difficult problems to solve in underwater locomotion, but improvements in battery technology, artificial intelligence, and machine learning promise to greatly extend the range and autonomy of future fish-like robots.

6.2 Technology for fish conservation biology

Although there are many limitations to current free-swimming fish robotic technology, there are a number of other technological developments that are now being employed to provide insight into fish biology and aid in conservation that are complementary to ongoing robotic research. As mentioned above, drones and airplane tracking are being used to observe and quantify migrating fish populations. In shallow clear waters, aerial observation platforms allow measurement of fish swimming kinematics and speeds, and interindividual spacing and arrangement that avoids the difficulties of attempting to monitor these variables underwater. For example, Kajiura and Tellman (2016) and Porter et al. (2020) have used drones and an airplane to track blacktip shark movements (*Carcharhinus limbatus*) in clear shallows off the Florida coast.

Second, AUV (autonomous underwater vehicle) and ROV (remote underwater vehicle) technology is being used to better understand the physical environment of fishes and to conduct surveys of fish populations (Stoner et al., 2008; Sward et al., 2019). Recent AUV and ROV deployments in the deep sea have led to remarkable discoveries of fish behavior and occurrence: there is no substitute for visual observation and video recordings which are invaluable for documenting novel behaviors (Lundsten et al., 2009; Reisenbichler et al., 2016). Using currently available technology, considerable progress can be made in understanding fish behavior in the field while ongoing development needed to obtain deployable fish-like robots occurs (Bo et al., 2014; Love et al., 2020).

Third, rapid increases in the development of electronic tags have tremendously expanded our knowledge of fish behavior and been largely responsible for the explosion of studies in the field of movement ecology. The challenges of following fish with mechanical devices can be avoided by simply attaching a tracking system to the fish itself. Tags range from relatively simple passive integrated transponder (PIT) tags (Castro-Santos et al., 1996; Mahapatra et al., 2001; Saboret et al., 2021) to increasingly sophisticated tags that report depth, temperature, body position, and motion using inertial measurement units

(IMUs), often with onboard video cameras that provide direct evidence of behavioral interactions and body motion (Block, 2005; Gleiss et al., 2019; Kohler and Turner, 2001).

Finally, future developments in machine vision and image processing will likely enable automated identification and assessment of fish species and movements when sensors are deployed in marine and freshwater habitats. Using pre-programmed three-dimensional data on a diversity of fish shapes and swimming kinematics, it should be possible to develop automatic vision-based screening methods that can discriminate among species at fish counting fences or selective barriers.

In combination with rapid development of the many different technological systems that will enhance our understanding of fish behavior, better understanding of the physical environments in which fish live and the development of new high-performance fish-like robots, when available, promise to greatly expand our understanding of fish biology and knowledge of how fish respond to human-designed systems like trawls, nets, and passage barriers. In the near future, new robotic systems will enable us to "swim with the fishes" and directly observe how fish respond to, associate with, and avoid both natural and artificial components of their environment.

7 Conclusions

The Anthropocene is a time of unprecedented change. These changes have been particularly acute in aquatic environments. Freshwater biodiversity is seriously imperiled (Dudgeon et al., 2006; Reid et al., 2019), and biomass of marine fisheries is at a fraction of what it was just a century ago (Christensen et al., 2014; Pauly et al., 2002). The ability of fish to move will influence their ability to adapt to these changes, and our understanding of both their abilities and limits will profoundly influence the development of effective management solutions. Technological innovations continue to offer improvements in our ability to understand the ecology, physiology, and behavior of fishes, and how these things interact with a changing environment. In the process of developing these advances, we are also learning more about what the natural world has to teach us, and this is particularly true of understanding how fish move through their environment. Some of that information will inform improved management, but in other cases we see advances in technology that are entirely independent of conservation interests. As we navigate this new age, the ability to move through this evolving environment will determine the outcomes for fish populations and humans alike.

Acknowledgments

This chapter is the product of an engaged collaboration, with all coauthors contributing nearly equally to the final product. Important insights, reviews, and comments were provided by Paul Webb, Michael Wagner, Steve Cooke, and two anonymous reviewers. T.C.S. and E.G. were supported in part by a grant from the Great Lakes Fishery

Commission [2017_CAS_54063]; E.G. was supported by a post-doctoral fellowship from the Fonds de Recherche, Nature, et Technologies du Québec; and G.V.L. was supported by grants from the National Science Foundation [IOS-2128033 and EF-2128033], and the Office of Naval Research [N00014-14-1-0533 and N00014-09-1-0352]. Any use of trade, firm, or product names is for descriptive purposes only and does not imply endorsement by the U.S. Government.

References

Adams, N.S., Plumb, J.M., Perry, R.W., Rondorf, D.W., 2014. Performance of a surface bypass structure to enhance juvenile steelhead passage and survival at lower granite dam, Washington. N. Am. J. Fish Manag. 34, 576–594.

Agostinho, A.A., Gomes, L.C., Verissimo, S., Okada, E.K., 2004. Flood regime, dam regulation and fish in the Upper Parana River: effects on assemblage attributes, reproduction and recruitment. Rev. Fish Biol. Fish. 14, 11–19.

Agostinho, A.A., Agostinho, C.S., Pelicice, F.M., Marques, E.E., 2012. Fish ladders: safe fish passage or hotspot for predation? Neotrop. Ichthyol. 10, 687–696.

Alcaraz, C., Vila-Gispert, A., Garcia-Berthou, E., 2005. Profiling invasive fish species: the importance of phylogeny and human use. Divers. Distrib. 11, 289–298.

Alcott, D., Long, M., Castro-Santos, T., 2020. Wait and snap: eastern snapping turtles (*Chelydra serpentina*) prey on migratory fish at road-stream crossing culverts. Biol. Lett. 16, 20200218. https://doi.org/10.1098/rsbl.2020.0218.

Alcott, D., Goerig, E., Rillahan, C., He, P., Castro-Santos, T., 2021. Tide gates form physical and ecological obstacles to river herring (*Alosa* spp.) spawning migrations. Can. J. Fish. Aquat. Sci. 78, 869–880.

Alexander, R.M., 1983. The history of fish mechanics. In: Webb, P.W., Weihs, D. (Eds.), Fish Biomechanics. Praeger, New York, pp. 1–35.

Ankhelyi, M., Wainwright, D.K., Lauder, G.V., 2018. Diversity of dermal denticle structure in sharks: skin surface roughness and three-dimensional morphology. J. Morphol. 279, 1132–1154.

Anonymous, 1872. Proceedings of the American fish culturists' association. Trans. Am. Fish. Soc. 1, 3–21.

Anonymous, 1877. Centennial meeting. Trans. Am. Fish. Soc. 6, 10–45.

Aristotle, 1937. Aristotle: Parts of Animals. Movement of animals. Progression of Animals. Harvard University Press, Cambridge, MA.

Bargelloni, L., Ritchie, P.A., Patarnello, T., Battaglia, B., Lambert, D.M., Meyer, A., 1994. Molecular evolution at subzero temperatures: mitochondrial and nuclear phylogenies of fishes from Antarctica (suborder Notothenioidei), and the evolution of antifreeze glycopeptides. Mol. Biol. Evol. 11, 854–863.

Barnard, R., Yokers, S., Nagygyor, A., Quinn, T., 2015. An evaluation of the stream simulation culvert design method in Washington state. River Res. Appl. 31, 1376–1387.

Bates, D.W., Vinsonhaler, R., 1957. Use of louvers for guiding fish. Trans. Am. Fish. Soc. 86, 38–57.

Bayse, S.M., Mccormick, S.D., Castro-Santos, T., 2019. How lipid content and temperature affect American shad (*Alosa sapidissima*) attempt rate and sprint swimming: implications for overcoming migration barriers. Can. J. Fish. Aquat. Sci. 76, 2235–2244.

Beamish, F.W.H., 1978. Swimming capacity. In: Hoar, W.S., Randall, D.J. (Eds.), Fish Physiology, Vol. VII, Locomotion. Academic Press, London, pp. 101–187.

Beamish, F.W.H., 1980. Biology of the North American anadromous sea lamprey, *Petromyzon marinus*. Can. J. Fish. Aquat. Sci. 37, 1924–1943.

Behrens, J.W., Van Deurs, M., Christensen, E.A., 2017. Evaluating dispersal potential of an invasive fish by the use of aerobic scope and osmoregulation capacity. PLoS One 12, e0176038.

Beletsky, D., Beletsky, R., Rutherford, E.S., Sieracki, J.L., Bossenbroek, J.M., Chadderton, W.L., Wittmann, M.E., Annis, G.M., Lodge, D.M., 2017. Predicting spread of aquatic invasive species by lake currents. J. Great Lakes Res. 43, 14 32.

Bellmore, J.R., Pess, G.R., Duda, J.J., O'connor, J.E., East, A.E., Foley, M.M., Wilcox, A.C., Major, J.J., Shafroth, P.B., Morley, S.A., Magirl, C.S., Anderson, C.W., Evans, J.E., Torgersen, C.E., Craig, L.S., 2019. Conceptualizing ecological responses to dam removal: if you remove it, What's to come? Bioscience 69, 26–39.

Beverton, R.J.H., Holt, S.J., 1957. On the Dynamics of Exploited Fish Populations. The Blackburn Press, Caldwell, NJ.

Block, B.A., 2005. Physiological ecology in the 21st century: advancements in biologging science. Integr. Comp. Biol. 45, 305–320.

Block, B.A., Dewar, H., Blackwell, S.B., Williams, T.D., Prince, E.D., Farwell, C.J., Boustany, A., Teo, S.L.H., Seitz, A., Walli, A., Fudge, D., 2001. Migratory movements, depth preferences, and thermal biology of Atlantic bluefin tuna. Science 293, 1310–1314.

Bo, M., Bava, S., Canese, S., Angiolillo, M., Cattaneo-Vietti, R., Bavestrello, G., 2014. Fishing impact on deep Mediterranean rocky habitats as revealed by ROV investigation. Biol. Conserv. 171, 167–176.

Breault, J.L., 1991. Candirú: Amazonian parasitic catfish. J. Wilderness Med. 2, 304–312.

Brett, J.R., 1962. Some considerations in the study of respiratory metabolism in fish, particularly salmon. J. Fish. Res. Board Can. 19, 1025–1038.

Brett, J.R., 1964. The respiratory metabolism and swimming performance of young sockeye salmon. J. Fish. Res. Board Can. 21, 1183–1226.

Brett, J.R., Glass, N.R., 1973. Metabolic rates and critical swimming speeds of sockeye salmon (*Oncorhynchus nerka*) in relation to size and temperature. J. Fish. Res. Board Can. 30, 379–387.

Brett, J.R., Groves, T.D.D., 1979. Physiological energetics. In: Hoar, W.S., Randall, D.J., Brett, J.-R. (Eds.), Fish Physiology Volume 8: Bioenergetics and Growth. Academic Press, New York, pp. 279–352.

Bunt, C.M., Castro-Santos, T., Haro, A., 2012. Performance of fish passage structures at upstream barriers to migration. River Res. Appl. 28, 457–478.

Bunt, C.M., Castro-Santos, T., Haro, A., 2016. Reinforcement and validation of the analyses and conclusions related to fishway evaluation data from Bunt et al.: "Performance of fish passage structures at upstream barriers to migration". River Res. Appl. 32, 2125–2137.

Canal, D., Negro, J.J., 2018. Use of drones for research and conservation of birds of prey. In: Birds of Prey. Springer.

Capra, H., Plichard, L., Bergé, J., Pella, H., Ovidio, M., Mcneil, E., Lamouroux, N., 2017. Fish habitat selection in a large hydropeaking river: strong individual and temporal variations revealed by telemetry. Sci. Total Environ. 578, 109–120.

Castro-Santos, T., 2004. Quantifying the combined effects of attempt rate and swimming capacity on passage through velocity barriers. Can. J. Fish. Aquat. Sci. 61, 1602–1615.

Castro-Santos, T., 2005. Optimal swim speeds for traversing velocity barriers: an analysis of volitional high-speed swimming behavior of migratory fishes. J. Exp. Biol. 208, 421–432.

Castro-Santos, T., 2006. Modeling the effect of varying swim speeds on fish passage through velocity barriers. Trans. Am. Fish. Soc. 135, 1230–1237.

Castro-Santos, T., Haro, A., 2006. Biomechanics and fisheries conservation. In: Shadwick, R.E., Lauder, G.V. (Eds.), Fish Physiology Volume 23: Fish Biomechanics. Academic Press, New York, pp. 469–523.

Castro-Santos, T., Haro, A., 2010. Fish guidance and passage at barriers. In: Domenici, P., Kapoor, B.G. (Eds.), Fish Locomotion: An Eco-Ethological Perspective. Science Publishers, Enfield, NH, pp. 62–89.

Castro-Santos, T., Haro, A., Walk, S., 1996. A passive integrated transponder (PIT) tag system for monitoring fishways. Fish. Res. 28, 253–261.

Castro-Santos, T., Cotel, A., Webb, P.W., 2009. Fishway evaluations for better bioengineering—an integrative approach. In: Haro, A.J., et al. (Eds.), Challenges for Diadromous Fishes in a Dynamic Global Environment. Symposium 69. American Fisheries Society, Bethesda, MD, pp. 557–575.

Castro-Santos, T., Sanz-Ronda, F.J., Ruiz-Legazpi, J., 2013. Breaking the speed limit—comparative sprinting performance of brook trout (*Salvelinus fontinalis*) and brown trout (*Salmo trutta*). Can. J. Fish. Aquat. Sci. 70, 280–293.

Castro-Santos, T., Goerig, E., Bayse, S.M., 2019. Selective Passage of Sea Lamprey and Native Great Lakes Fish through Hydraulic Barriers. Great Lakes Fishery Commission, 2017_CAS_54063.

Castro-Santos, T., Mulligan, K.B., Kieffer, M., Haro, A.J., 2020. Effects of plunge pool configuration on downstream passage survival of juvenile blueback herring. Aquac. Fish. 6, 135–143.

Celestino, L.F., Sanz-Ronda, F.J., Miranda, L.E., Makrakis, M.C., Dias, J.H.P., Makrakis, S., 2019. Bidirectional connectivity via fish ladders in a large Neotropical river. River Res. Appl. 35, 236–246.

Christensen, V., Coll, M., Piroddi, C., Steenbeek, J., Buszowski, J., Pauly, D., 2014. A century of fish biomass decline in the ocean. Mar. Ecol. Prog. Ser. 512, 155–166.

Clark, T.D., Sandblom, E., Jutfelt, F., 2013. Aerobic scope measurements of fishes in an era of climate change: respirometry, relevance and recommendations. J. Exp. Biol. 216, 2771–2782.

Clarke, A., Fraser, K.P.P., 2004. Why does metabolism scale with temperature? Funct. Ecol. 18, 243–251.

Clay, C.H., 1995. Design of Fishways and Other Fish Facilities. Lewis Publishers, Boca Raton.

Clift, W., 1872. Shad culture. Trans. Am. Fish. Soc. 1, 21–28.

Davison, W., 1997. The effects of exercise training on teleost fish, a review of recent literature. Comp. Biochem. Physiol. A Physiol. 117, 67–75.

De Robertis, A., Levine, M., Lauffenburger, N., Honkalehto, T., Ianelli, J., Monnahan, C.C., Towler, R., Jones, D., Stienessen, S., Mckelvey, D., 2021. Uncrewed surface vehicle (USV) survey of walleye pollock, *Gadus chalcogrammus*, in response to the cancellation of ship-based surveys. ICES J. Mar. Sci. 78, 2797–2808.

De Vaate, A.B., Breukelaar, A.W., Vriese, T., De Laak, G., Dijkers, C., 2003. Sea trout migration in the Rhine delta. J. Fish Biol. 63, 892–908.

Di Santo, V., Goerig, E., Wainwright, D.K., Akanyeti, O., Liao, J.C., Castro-Santos, T., Lauder, G.V., 2021. Convergence of undulatory swimming kinematics across a diversity of fishes. Proc. Natl. Acad. Sci. 118, e2113206118.

Domel, A.G., Saadat, M., Weaver, J., Haj-Hariri, H., Bertoldi, K., Lauder, G.V., 2018. Shark denticle-inspired designs for improved aerodynamics. J. R. Soc. Interface 15, 20170828.

Domenici, P., Allan, B.J.M., Lefrançois, C., Mccormick, M.I., 2019. The effect of climate change on the escape kinematics and performance of fishes: implications for future predator–prey interactions. Conserv. Physiol. 7, coz078. https://doi.org/10.1093/conphys/coz078.

Dudgeon, D., 2011. Asian river fishes in the Anthropocene: threats and conservation challenges in an era of rapid environmental change. J. Fish Biol. 79, 1487–1524.

Dudgeon, D., Arthington, A.H., Gessner, M.O., Kawabata, Z., Knowler, D.J., Lévêque, C., Naiman, R.J., Prieur-Richard, A.H., Soto, D., Stiassny, M.L., Sullivan, C.A., 2006. Freshwater biodiversity: importance, threats, status and conservation challenges. Biol. Rev. Camb. Philos. Soc. 81, 163–182.

Dugdale, S.J., Bergeron, N.E., St-Hilaire, A., 2015. Spatial distribution of thermal refuges analysed in relation to riverscape hydromorphology using airborne thermal infrared imagery. Remote Sens. Environ. 160, 43–55.

Duguay, J., Foster, B., Lacey, J., Castro-Santos, T., 2018a. Sediment infilling benefits rainbow trout passage in a baffled channel. Ecol. Eng. 125, 38–49.

Duguay, J.M., Lacey, R.W.J., Castro-Santos, T., 2018b. Influence of baffles on upstream passage of brook trout and brown trout in an experimental box culvert. Can. J. Fish. Aquat. Sci. 76, 28–41.

Dzara, J.R., Neilson, B.T., Null, S.E., 2019. Quantifying thermal refugia connectivity by combining temperature modeling, distributed temperature sensing, and thermal infrared imaging. Hydrol. Earth Syst. Sci. 23, 2965–2982.

Ebersole, J.L., Quiñones, R.M., Clements, S., Letcher, B.H., 2020. Managing climate refugia for freshwater fishes under an expanding human footprint. Front. Ecol. Environ. 18, 271–280.

Elskus, A.A., Mitchelmore, C.L., Wright, D., Henquinet, J.W., Welschmeyer, N., Flynn, C., Watten, B.J., 2017. Efficacy and residual toxicity of a sodium hydroxide based ballast water treatment system for freshwater bulk freighters. J. Great Lakes Res. 43, 744–754.

Enders, E.C., Boisclair, D., Roy, A.G., 2003. The effect of turbulence on the cost of swimming for juvenile Atlantic salmon (*Salmo salar*). Can. J. Fish. Aquat. Sci. 60, 1149–1160.

Enders, E.C., Boisclair, D., Roy, A.G., 2005. A model of total swimming costs in turbulent flow for juvenile Atlantic salmon (*Salmo salar*). Can. J. Fish. Aquat. Sci. 62, 1079–1089.

Enders, E.C., Castro-Santos, T., Lacey, R.W.J., 2017. The effects of horizontally and vertically oriented baffles on flow structure and ascent performance of upstream-migrating fish. J. Ecohydraul. 2, 38–52.

Engås, A., Foster, D., 2002. The response of red snapper (*Lutjanus campechanus*) and pinfish (*Lagodon rhomboides*) to inclined water flow. Fish. Res. 58, 315–321.

Engås, A., Godø, O.R., 1989. The effect of different sweep lengths on the length composition of bottom-sampling trawl catches. ICES J. Mar. Sci. 45, 263–268.

Engas, A., Lokkeborg, S., 1994. Abundance estimation using bottom gillnet and long-line—the role of fish behaviour. In: Ferno, A., Olsen, S. (Eds.), Marine Fish Behaviour in Capture and Abundance Estimation. Fishing News Books, Oxford, pp. 130–163.

Engås, A., Foster, D., Hataway, B.D., Watson, J.W., Workman, I., 1999. The behavioral response of juvenile red snapper (*Lutjanus campechanus*) to shrimp trawls that utilize water flow modifications to induce escapement. Mar. Technol. Soc. J. 33, 43–50.

Esposito, C., Tangorra, J., Flammang, B.E., Lauder, G.V., 2012. A robotic fish caudal fin: effects of stiffness and motor program on locomotor performance. J. Exp. Biol. 215, 56–67.

FAO, 1983. Freshwater Aquaculture Development in China: Report of the FAO-UNDP Study Tour Organized for French-Speaking African Countries. FAO Fisheries Technical Paper, Food and Agriculture Organization (FAO) Fisheries, Rome.

FAO, 2020. The State of the Worlds Fisheries and Aquaculture, 2020. Sustainability in Action. FAO, Rome.

Farrell, A.P., Steffensen, J.F., 1987. An analysis of the energetic cost of the branchial and cardiac pumps during sustained swimming in trout. Fish Physiol. Biochem. 4, 73–79.

Feurich, R., Boubée, J., Olsen, N.R.B., 2012. Improvement of fish passage in culverts using CFD. Ecol. Eng. 47, 1–8.

Finger, J.S., Riesgraf, A.T., Zielinski, D.P., Sorensen, P.W., 2020. Monitoring upstream fish passage through a Mississippi River lock and dam reveals species differences in lock chamber usage and supports a fish passage model which describes velocity-dependent passage through spillway gates. River Res. Appl. 36, 36–46.

Freeman, M.C., Bowen, Z.H., Bovee, K.D., Irwin, E.R., 2001. Flow and habitat effects on juvenile fish abundance in natural and altered flow regimes. Ecol. Appl. 11, 179–190.

Friedland, K.D., Hansen, L.P., Dunkley, D.A., 1998. Marine temperatures experienced by post-smolts and the survival of Atlantic salmon, *Salmo salar* l., in the North Sea area. Fish. Oceanogr. 7, 22–34.

Friedland, K.D., Chaput, G., Maclean, J.C., 2005. The emerging role of climate in post-smolt growth of Atlantic salmon. ICES J. Mar. Sci. 62, 1338–1349.

Fuller, M.R., Doyle, M.W., Strayer, D.L., 2015. Causes and consequences of habitat fragmentation in river networks. Ann. N. Y. Acad. Sci. 1355, 31–51.

García-Berthou, E., Alcaraz, C., Pou-Rovira, Q., Zamora, L., Coenders, G., Feo, C., 2005. Introduction pathways and establishment rates of invasive aquatic species in Europe. Can. J. Fish. Aquat. Sci. 62, 453–463.

Gillespie, N., Unthank, A., Campbell, L., Anderson, P., Gubernick, R., Weinhold, M., Cenderelli, D., Austin, B., Mckinley, D., Wells, S., 2014. Flood effects on road–stream crossing infrastructure: economic and ecological benefits of stream simulation designs. Fisheries 39, 62–76.

Gleiss, A.C., Schallert, R.J., Dale, J.J., Wilson, S.G., Block, B.A., 2019. Direct measurement of swimming and diving kinematics of giant Atlantic bluefin tuna (*Thunnus thynnus*). R. Soc. Open Sci. 6, 190203.

Godø, O.R., Pennington, M., Vølstad, J.H., 1990. Effect of tow duration on length composition of trawl catches. Fish. Res. 9, 165–179.

Goerig, E., Castro-Santos, T., 2017. Is motivation important to brook trout passage through culverts? Can. J. Fish. Aquat. Sci. 74, 885–893.

Goerig, E., Castro-Santos, T., Bergeron, N.E., 2016. Brook trout passage performance through culverts. Can. J. Fish. Aquat. Sci. 73, 94–104.

Goerig, E., Bergeron, N.E., Castro-Santos, T., 2017. Swimming behaviour and ascent paths of brook trout in a corrugated culvert. River Res. Appl. 33, 1463–1471.

Goerig, E., Wasserman, B.A., Castro-Santos, T., Palkovacs, E.P., 2020. Body shape is related to the attempt rate and passage success of brook trout at in-stream barriers. J. Appl. Ecol. 57, 91–100.

Gottlieb, J., Tangorra, J., Esposito, C., Lauder, G., 2010. A biologically derived pectoral fin for yaw turn maneuvers. Appl. Bionics Biomech. 7, 41–55.

Grill, G., Lehner, B., Thieme, M., Geenen, B., Tickner, D., Antonelli, F., Babu, S., Borrelli, P., Cheng, L., Crochetiere, H., Ehalt Macedo, H., Filgueiras, R., Goichot, M., Higgins, J., Hogan, Z., Lip, B., Mcclain, M.E., Meng, J., Mulligan, M., Nilsson, C., Olden, J.D., Opperman, J.J., Petry, P., Reidy Liermann, C., Sáenz, L., Salinas-Rodríguez, S., Schelle, P., Schmitt, R.J.P., Snider, J., Tan, F., Tockner, K., Valdujo, P.H., Van Soesbergen, A., Zarfl, C., 2019. Mapping the world's free-flowing rivers. Nature 569, 215–221.

Gutfreund, C., Makrakis, S., Castro-Santos, T., Celestino, L.F., Dias, J.H.P., Makrakis, M.C., 2018. Effectiveness of a fish ladder for two Neotropical migratory species in the Parana River. Mar. Freshw. Res. 69, 1848–1856.

Haefner, J.W., Bowen, M.D., 2002. Physical-based model of fish movement in fish extraction facilities. Ecol. Model. 152, 227–245.

Hall, C.J., Jordaan, A., Frisk, M.G., 2011. The historic influence of dams on diadromous fish habitat with a focus on river herring and hydrologic longitudinal connectivity. Landsc. Ecol. 26, 95–107.

Hall, C.J., Jordaan, A., Frisk, M.G., 2012. Centuries of anadromous forage fish loss: consequences for ecosystem connectivity and productivity. Bioscience 62, 723–731.

Hammer, C., 1995. Fatigue and exercise tests with fish. Comp. Biochem. Physiol. 112A, 1–20.

Haro, A., Odeh, M., Noreika, J., Castro-Santos, T., 1998. Effect of water acceleration on downstream migratory behavior and passage of Atlantic salmon smolts and juvenile American shad at surface bypasses. Trans. Am. Fish. Soc. 127, 118–127.

Haro, A., Castro-Santos, T., Noreika, J., Odeh, M., 2004. Swimming performance of upstream migrant fishes in open-channel flow: a new approach to predicting passage through velocity barriers. Can. J. Fish. Aquat. Sci. 61, 1590–1601.

Harrington, J.M., Myers, R.A., Rosenberg, A.A., 2005. Wasted fishery resources: discarded by-catch in the USA. Fish Fish. 6, 350–361.

Hasselman, D.J., Gephard, S.R., Bethoney, N.D., Anderson, E.C., Argo, E.E., Palkovacs, E.P., Willis, T.V., Schultz, T.F., Schondelmeier, B.P., Post, D.M., 2016. Genetic stock composition of marine bycatch reveals disproportional impacts on depleted river herring genetic stocks. Can. J. Fish. Aquat. Sci. 73, 951–963.

Hatry, C., Binder, T.R., Thiem, J.D., Hasler, C.T., Smokorowski, K.E., Clarke, K.D., Katopodis, C., Cooke, S.J., 2013. The status of fishways in Canada: trends identified using the national CanFishPass database. Rev. Fish Biol. Fish. 23, 271–281.

He, P., 1993. Swimming speeds of marine fish in relation to fishing gears. ICES Mar. Sci. Symp. 196, 183–189. 1993. FR 38(4).

He, P., 2003. Swimming behaviour of winter flounder (*Pleuronectes americanus*) on natural fishing grounds as observed by an underwater video camera. Fish. Res. 60, 507–514.

He, P., Wardle, C.S., 1988. Endurance at intermediate swimming speeds of Atlantic mackerel, *Scomber scombrus* L, herring, *Clupea harengus* L, and Saithe, *Pollachius virens* L. J. Fish Biol. 33, 255–266.

He, P., Chopin, F., Suuronen, P., Ferro, R.S.T., Lansley, J., 2021a. Classification and illustrated definition of fishing gears. FAO Fisheries and Aquaculture Technical Paper, No. 672. United Nations Food and Agriculture Organization (FAO), Rome, https://doi.org/10.4060/cb4966en.

He, P., Rillahan, C., Wilsterman, M., 2021b. Measuring swimming capacity of yellowtail and windowpane flounders to provide scientific knowledge for reducing their bycatch in scallop dredges. In: Final report submitted to NOAA Fisheries. University of Massachusetts Dartmouth—SMAST, New Bedford, MA, p. 30. SMAST-CE-REP-2021-100.

Hertel, H., 1966. Structure, Form and Movement. Reinhold, New York, N.Y.

Hill, N.L., Trueman, J.R., Prévost, A.D., Fraser, D.J., Ardren, W.R., Grant, J.W., 2019. Effect of dam removal on habitat use by spawning Atlantic salmon. J. Great Lakes Res. 45, 394–399.

Hinch, S.G., Rand, P.S., 2000. Optimal swimming speeds and forward-assisted propulsion: energy-conserving behaviours of upriver-migrating adult salmon. Can. J. Fish. Aquat. Sci. 57, 2470–2478.

Hogg, R.S., Coghlan, S.M., Zydlewski, J., Gardner, C., 2015. Fish community response to a small-stream dam removal in a Maine coastal river tributary. Trans. Am. Fish. Soc. 144, 467–479.

Hotchkiss, R.H., Frei, C.M., 2007. Design for Fish Passage at Roadway-Stream Crossings: Synthesis Report. United States, Federal Highway Administration.

Inman, V.L., Kingsford, R.T., Chase, M.J., Leggett, K.E., 2019. Drone-based effective counting and ageing of hippopotamus (*Hippopotamus amphibius*) in the Okavango Delta in Botswana. PLoS One 14, e0219652.

Jayne, B.C., Lauder, G.V., 1994. How swiming fish use slow and fast muscle fibers: implications for models of vertebrate muscle recruitment. J. Comp. Physiol. A 175, 123–131.

Jenkins, L.D., 2012. Reducing sea turtle bycatch in trawl nets: a history of NMFS turtle excluder device (TED) research. Mar. Fish. Rev. 74, 26–44.

Jobling, M., Baardvik, B.M., Christiansen, J.S., Jørgensen, E.H., 1993. The effects of prolonged exercise training on growth performance and production parameters in fish. Aquac. Int. 1, 95–111.

Jørgensen, E.H., Jobling, M., 1993. The effects of exercise on growth, food utilisation and osmoregulatory capacity of juvenile Atlantic salmon, *Salmo salar*. Aquaculture 116, 233–246.

Júlio Júnior, H.F., Tós, C.D., Agostinho, A.A., Pavenelli, C.S., 2009. A massive invasion of fish species after eliminating a natural barrier inthe upper rio Paraná basin. Neotrop. Ichthyol. 7, 709–718.

Kajiura, S.M., Tellman, S.L., 2016. Quantification of massive seasonal aggregations of blacktip sharks (*Carcharhinus limbatus*) in Southeast Florida. PLoS One 11, e0150911.

Katzschmann, R.K., Delpreto, J., Maccurdy, R., Rus, D., 2018. Exploration of underwater life with an acoustically controlled soft robotic fish. Sci. Robot. 3, eaar3449.

Kemp, P.S., Gessel, M.H., Williams, J.G., 2005. Fine-scale behavioral responses of Pacific salmonid smolts as they encounter divergence and acceleration of flow. Trans. Am. Fish. Soc. 134, 390–398.

Kettle, A.J., Haines, K., 2006. How does the European eel (*Anguilla anguilla*) retain its population structure during its larval migration across the North Atlantic Ocean? Can. J. Fish. Aquat. Sci. 63, 90–106.

Kieffer, J., Currie, S., Tufts, B., 1994. Effects of environmental temperature on the metabolic and acid-base responses of rainbow trout to exhaustive exercise. J. Exp. Biol. 194, 299–317.

Kock, K.-H., 2005. Antarctic icefishes (Channichthyidae): a unique family of fishes. A review, Part I. Polar Biol. 28, 862–895.

Kohler, N.E., Turner, P.A., 2001. Shark tagging: a review of conventional methods and studies. In: Tricas, T.C., Gruber, S.H. (Eds.), The Behavior and Sensory Biology of Elasmobranch Fishes: An Anthology in Memory of Donald Richard Nelson. Springer, Dordrecht, pp. 191–224.

Krueger, C.C., Holbrook, C.M., Binder, T.R., Vandergoot, C.S., Hayden, T.A., Hondorp, D.W., Nate, N., Paige, K., Riley, S.C., Fisk, A.T., Cooke, S.J., 2018. Acoustic telemetry observation systems: challenges encountered and overcome in the Laurentian Great Lakes. Can. J. Fish. Aquat. Sci. 75, 1755–1763.

Kynard, B., Horgan, M., 2001. Guidance of yearling shortnose and pallid sturgeon using vertical bar rack and louver arrays. N. Am. J. Fish Manag. 21, 561–570.

Lacey, R.W.J., Neary, V.S., Liao, J.C., Enders, E.C., Tritico, H.M., 2012. The IPOS framework: linking fish swimming performance in altered flows from laboratory experiments to rivers. River Res. Appl. 28, 429–443.

Lagardère, J.P., Sureau, D., 1989. Changes in the swimming activity of the sole (*Solea vulgaris* Quensel, 1806) in relation to winter temperatures in a saltmarsh: observations using ultrasonic telemetry. Fish. Res. 7, 233–239.

Lauder, G.V., 2006. Locomotion. In: Evans, D.H., Claiborne, J.B. (Eds.), The Physiology of Fishes, third ed. CRC Press, Boca Raton, pp. 3–46.

Lauder, G.V., Tangorra, J.L., Du, R., Li, Z., Youcef-Toumi, K., Y Alvarado, P.V., 2015. Fish locomotion: biology and robotics of body and fin-based movements. In: Robot Fish—Bio-Inspired Fishlike Underwater Robots. Springer Verlag, Berlin, pp. 25–49.

Lauder, G.V., Tytell, E.D., 2006. Hydrodynamics of undulatory propulsion. In: Shadwick, R.E., Lauder, G.V. (Eds.), Fish Biomechanics. Volume 23 in Fish Physiology. Academic Press, San Diego, pp. 425–468.

Lauder, G.V., Flammang, B.E., Alben, S., 2012. Passive robotic models of propulsion by the bodies and caudal fins of fish. Integr. Comp. Biol. 52, 576–587.

Lauder, G.V., Wainwright, D.K., Domel, A.G., Weaver, J., Wen, L., Bertoldi, K., 2016. Structure, biomimetics, and fluid dynamics of fish skin surfaces. Phys. Rev. Fluids 1, 060502.

Lee, C.G., Farrell, A.P., Lotto, A., Hinch, S.G., Healey, M.C., 2003. Excess post-exercise oxygen consumption in adult sockeye (*Oncorhynchus nerka*) and coho (*O. kisutch*) salmon following critical speed swimming. J. Exp. Biol. 206, 3253–3260.

Lehner, B., Liermann, C.R., Revenga, C., Vörösmarty, C., Fekete, B., Crouzet, P., Döll, P., Endejan, M., Frenken, K., Magome, J., 2011. High-resolution mapping of the world's reservoirs and dams for sustainable river-flow management. Front. Ecol. Environ. 9, 494–502.

Lennox, R.J., Alós, J., Arlinghaus, R., Horodysky, A., Klefoth, T., Monk, C.T., Cooke, S.J., 2017. What makes fish vulnerable to capture by hooks? A conceptual framework and a review of key determinants. Fish Fish. 18, 986–1010.

Lennox, R.J., Eliason, E.J., Havn, T.B., Johansen, M.R., Thorstad, E.B., Cooke, S.J., Diserud, O.-H., Whoriskey, F.G., Farrell, A.P., Uglem, I., 2018. Bioenergetic consequences of warming rivers to adult Atlantic salmon *Salmo salar* during their spawning migration. Freshw. Biol. 63, 1381–1393.

Liao, J.C., 2007. A review of fish swimming mechanics and behaviour in altered flows. Philos. Trans. R. Soc. B Biol. Sci. 362, 1973–1993.

Liao, J.C., Beal, D.N., Lauder, G.V., Triantafyllou, M.S., 2003a. Fish exploiting vortices decrease muscle activity. Science 302, 1566–1569.

Liao, J.C., Beal, D.N., Lauder, G.V., Triantafyllou, M.S., 2003b. The Karman gait: novel body kinematics of rainbow trout swimming in a vortex street. J. Exp. Biol. 206, 1059–1073.

Lighthill, J., 1960. Note on the swimming of slender fish. J. Fluid Mech. 9, 305–317.

Lighthill, J., 1975. Mathematical Biofluiddynamics. Society for Industrial and Applied Mathematics, Philadelphia.

Løkkeborg, S., Fernö, A., Humborstad, O.-B., 2010. Fish behavior in relation to longlines. In: He, P. (Ed.), Behavior of Marine Fishes: Capture Processes and Conservation Challenges. Wiley-Blackwell, Ames, Iowa, pp. 105–158.

Løkkeborg, S., Siikavuopio, S.I., Humborstad, O.-B., Utne-Palm, A.C., Ferter, K., 2014. Towards more efficient longline fisheries: fish feeding behaviour, bait characteristics and development of alternative baits. Rev. Fish Biol. Fish. 24, 985–1003.

Long, J.H., Schumacher, J., Livingston, N., Kemp, M., 2006. Four flippers or two? Tetrapodal swimming with an aquatic robot. Bioinspir. Biomim. 1, 20–29.

Love, M.S., Nishimoto, M.M., Clark, S., Kui, L., Aziz, A., Palandro, D., 2020. A comparison of two remotely operated vehicle (ROV) survey methods used to estimate fish assemblages and densities around a California oil platform. PLoS One 15, e0242017.

Lozano-Vilano, M.L., De La Maza-Benignos, M., 2017. Diversity and status of Mexican killifishes. J. Fish Biol. 90, 3–38.

Lundsten, L., Mcclain, C.R., Barry, J.P., Cailliet, G.M., Clague, D.A., Devogelaere, A.P., 2009. Ichthyofauna on three seamounts off southern and central California, USA. Mar. Ecol. Prog. Ser. 389, 223–232.

Macdonald, A.M., Wunsch, C., 1996. An estimate of global ocean circulation and heat fluxes. Nature 382, 436–439.

Magilligan, F.J., Nislow, K.H., Kynard, B.E., Hackman, A.M., 2016. Immediate changes in stream channel geomorphology, aquatic habitat, and fish assemblages following dam removal in a small upland catchment. Geomorphology 252, 158–170.

Mahapatra, K.D., Gjerde, B., Reddy, P., Sahoo, M., Jana, R., Saha, J., Rye, M., 2001. Tagging: on the use of passive integrated transponder (PIT) tags for the identification of fish. Aquacult. Res. 32, 47–50.

Main, J., Sangster, G.I., 1981. A study of the fish capture process in a bottom trawl by direct observations from a towed underwater vehicle. In: Scotish fisheries research report, 23, pp. 1–23.

Makrakis, S., Castro-Santos, T., Makrakis, M.C., Wagner, R.L., Spagnol-Adames, M., 2012. Culverts in paved roads as suitable passages for Neotropical fish species. Neotrop. Ichthyol. 10, 763–770.

Marinov, M., Pogan, T., Dorosencu, A., Nichersu, L., Alexe, V., Trifanov, C., Bozagievici, R., Tosic, K., Kiss, B., 2016. Monitoring the great White pelican (*Pelecanus onocrotalus* Linnaeus, 1758) breeding population using drones in 2016 the Danube Delta (Romania). Sci. Ann. Danub. Delta Inst. 22, 41–52.

Matthews, D.G., Lauder, G.V., 2021. Fin–fin interactions during locomotion in a simplified biomimetic fish model. Bioinspir. Biomim. 16, 046023.

Maximenko, N., Lumpkin, R., Centurioni, L., 2013. Chapter 12—Ocean surface circulation. In: Siedler, G., Griffies, S.M., Gould, J., Church, J.A. (Eds.), International Geophysics. Academic Press, pp. 283–304.

McCauley, T.C., 1996. Development of an Instream Velocity Barrier to Stop Sea Lamprey (*Petromyzon marinus*) Migrations in Great Lakes Streams. M.S. Thesis, University of Manitoba.

McLaughlin, R.L., Kramer, D.L., 1991. The association between amount of red muscle and mobility in fishes: a statistical evaluation. Environ. Biol. Fishes 30, 369–378.

McLaughlin, R.L., Smyth, E.R.B., Castro-Santos, T., Jones, M.L., Koops, M.A., Pratt, T.C., Vélez-Espino, L.A., 2013. Unintended consequences and trade-offs of fish passage. Fish Fish. 14, 580–604.

McQuinn, I.H., Gendron, L., Himmelman, J.H., 1988. Area of attraction and effective area fished by a whelk (*Buccinum undatum*) trap under variable conditions. Can. J. Fish. Aquat. Sci. 45, 2054–2060.

Mensinger, M.A., Brehm, A.M., Mortelliti, A., Blomberg, E.J., Zydlewski, J.D., 2021. American eel personality and body length influence passage success in an experimental fishway. J. Appl. Ecol. 58, 2760–2769.

Meybeck, M., 2003. Global analysis of river systems: from earth system controls to Anthropocene syndromes. Philos. Trans. R. Soc. Lond. Ser. B Biol. Sci. 358, 1935–1955.

Mignano, A.P., Kadapa, S., Tangorra, J.L., Lauder, G.V., 2019. Passing the wake: using multiple fins to shape forces for swimming. Biomimetics 4, 23. https://doi.org/10.3390/biomimetics4010023.

Minckley, W.L., Minckley, C.O., 1986. *Cyprinodon pachycephalus*, a new species of pupfish (Cyprinodontidae) from the Chihuahuan desert of Northern México. Copeia 1986, 184–192.

Monan, G.E., 1985. Advances in tagging and tracking hatchery salmonids: coded wire tags, multiple-coded and miniature radio tags, and the passive integrated transponder tag. In: Sinderman, C.J. (Ed.), Proceedings of the Eleventh US-Japan Meeting on Aquaculture, Salmon Enhancement, Tokyo, Japan, October 19–20, 1982. Tokyo. U.S. National Marine Fisheries Serrvice, Japan.

Monan, G.E., Johnson, J.H., Esterberg, G.F., 1975. Electronic tags and related tracking techniques aid in study of migrating salmon and steelhead trout in the Columbia River basin. Mar. Fish. Rev. 37, 9–15.

Moored, K.W., Dewey, P.A., Leftwich, M., Bart-Smith, H., Smits, A., 2011. Bioinspired propulsion mechanisms based on manta ray locomotion. Mar. Technol. Soc. J. 45, 110–118.

Moser, M., Keefer, M., Pennington, H., Ogden, D., Simonson, J., 2011. Development of Pacific lamprey fishways at a hydropower dam. Fish. Manag. Ecol. 18, 190–200.

Mulligan, M., Van Soesbergen, A., Sáenz, L., 2020. GOODD, a global dataset of more than 38,000 georeferenced dams. Sci. Data 7, 31.

Noonan, M.J., Grant, J.W., Jackson, C.D., 2012. A quantitative assessment of fish passage efficiency. Fish Fish. 13, 450–464.

Nyqvist, D., Greenberg, L.A., Goerig, E., Calles, O., Bergman, E., Ardren, W.R., Castro-Santos, T., 2016. Migratory delay leads to reduced passage success of Atlantic salmon smolts at a hydroelectric dam. Ecol. Freshw. Fish 26, 707–718. https://doi.org/10.1111/eff.12318.

Oeffner, J., Lauder, G.V., 2012. The hydrodynamic function of shark skin and two biomimetic applications. J. Exp. Biol. 215, 785–795.

Otero, J., Labeelund, J.H., Castro-Santos, T., Leonardsson, K., Storvik, G.O., Jonsson, B., Dempson, J.B., Russell, I.C., Jensen, A.J., Bagliniere, J.L., Dionne, M., Armstrong, J.D., Romakkaniemi, A., Letcher, B.H., Kocik, J.F., Erkinaro, J., Poole, R., Rogan, G., Lundqvist, H., Maclean, J.C., Jokikokko, E., Arnekleiv, J.V., Kennedy, R.J., Niemela, E., Caballero, P., Music, P.A., Antonsson, T., Gudjonsson, S., Veselov, A.E., Lamberg, A., Groom, S., Taylor, B.H., Taberner, M., Dillane, M., Arnason, F., Horton, G., Hvidsten, N.A., Jonsson, I.R., Jonsson, N., Mckelvey, S., Naesje, T.F., Skaala, O., Smith, G.W., Saegrov, H., Stenseth, N.C., Vollestad, L.A., 2014. Basin-scale phenology and effects of climate variability on global timing of initial seaward migration of Atlantic salmon (*Salmo salar*). Glob. Chang. Biol. 20, 61–75.

Özbilgin, H., Wardle, C.S., 2002. Effect of seasonal temperature changes on the escape behaviour of haddock, *Melanogrammus aeglefinus*, from the codend. Fish. Res. 58, 323–331.

Park, D., Sullivan, M., Bayne, E., Scrimgeour, G., 2008. Landscape-level stream fragmentation caused by hanging culverts along roads in Alberta's boreal forest. Can. J. For. Res. 38, 566–575.

Parsons, G.R., Foster, D.G., 2007. Swimming performance and behavior of red snapper: their application to bycatch reduction. Am. Fish. Soc. Symp. 90, 55–75.

Parsons, G.R., Foster, D.G., 2015. Reducing bycatch in the United States Gulf of Mexico shrimp trawl fishery with an emphasis on red snapper bycatch reduction. Fish. Res. 167, 210–215.

Parsons, G.R., Foster, D.G., Osmond, M., 2012. Applying fish behavior to reduce trawl bycatch: evaluation of the nested cylinder bycatch reduction device. Mar. Technol. Soc. J. 46, 26–33.

Pauly, D., Christensen, V., Guenette, S., Pitcher, T.J., Sumaila, U.R., Walters, C.J., Watson, R., Zeller, D., 2002. Towards sustainability in world fisheries. Nature 418, 689–695.

Peake, S.J., Farrell, A.P., 2004. Locomotory behaviour and post-exercise physiology in relation to swimming speed, gait transition and metabolism in free-swimming smallmouth bass (*Micropterus dolomieu*). J. Exp. Biol. 207, 1563–1575.

Pelicice, F.M., Agostinho, C.S., 2012. Deficient downstream passage through fish ladders: the case of Peixe Angical dam, Tocantins River, Brazil. Neotrop. Ichthyol. 10, 705–713.

Perez Roda, M., Gilman, E., Huntingon, T., Kennelly, S.J., Suuronen, P., Chaloupka, M., Medley, P., a. H., 2019. A Third Assessment of Global Marine Fisheries Discards. FAO (Food and Agricultural Organization of the United Nations), Rome.

Porter, M.E., Ruddy, B.T., Kajiura, S.M., 2020. Volitional swimming kinematics of blacktip sharks, *Carcharhinus limbatus*, in the wild. Drones 4, 78.

Pörtner, H.-O., Bock, C., Mark, F.C., 2017. Oxygen- and capacity-limited thermal tolerance: bridging ecology and physiology. J. Exp. Biol. 220, 2685–2696.

Pratt, T., O'connor, L., Hallett, A., McLaughlin, R., Katopodis, C., Hayes, D., Bergstedt, R., 2009. Balancing aquatic habitat fragmentation and control of invasive species: enhancing selective fish passage at sea lamprey control barriers. Trans. Am. Fish. Soc. 138, 652–665.

Rahel, F.J., Bierwagen, B., Taniguchi, Y., 2008. Managing aquatic species of conservation concern in the face of climate change and invasive species. Conserv. Biol. 22, 551–561.

Reid, A.J., Carlson, A.K., Creed, I.F., Eliason, E.J., Gell, P.A., Johnson, P.T.J., Kidd, K.A., Maccormack, T.J., Olden, J.D., Ormerod, S.J., Smol, J.P., Taylor, W.W., Tockner, K., Vermaire, J.C., Dudgeon, D., Cooke, S.J., 2019. Emerging threats and persistent conservation challenges for freshwater biodiversity. Biol. Rev. Camb. Philos. Soc. 94, 849–873.

Reif, W.-E., 1985. Squamation and ecology of sharks. Cour. Forsch. Inst. Senckenberg. 78, 1–255.

Reisenbichler, K.R., Chaffey, M.R., Cazenave, F., Mcewen, R.S., Henthorn, R.G., Sherlock, R.E., Robison, B.H., 2016. Automating MBARI's midwater time-series video surveys: the transition from ROV to AUV. In: OCEANS 2016 MTS/IEEE Monterey. IEEE, pp. 1–9.

Rillahan, C.B., Alcott, D., Castro-Santos, T., He, P., 2021. Activity patterns of anadromous fish below a tide gate: observations from high-resolution imaging sonar. Mar. Coast. Fish. 13, 200–212.

Roche, K.F., Janač, M., Jurajda, P., 2013. A review of Gobiid expansion along the Danube-Rhine corridor—geopolitical change as a driver for invasion. Knowl. Managt. Aquatic Ecosyst. 411, 01–23. https://doi.org/10.1051/kmae/2013066.

Rome, L.C., 1994. The mechanical design of the fish muscular system. In: Maddock, L., Bone, Q., Rayner, J.M.V. (Eds.), Mechanics and Physiology of Animal Swimming. Cambridge University Press, Cambridge, pp. 75–97.

Rome, L.C., Choi, I.H., Lutz, G., Sosnicki, A., 1992a. The influence of temperature on muscle function in the fast swimming scup. I. Shortening velocity and muscle recruitment during swimming. J. Exp. Biol. 163, 259–279.

Rome, L.C., Sosnicki, A., Choi, I.H., 1992b. The influence of temperature on muscle function in the fast swimming scup. II. The mechanics of red muscle. J. Exp. Biol. 163, 281–295.

Rosen, S, Holst, J.C., 2013. DeepVision in-trawl imaging: sampling the water column in four dimensions. Fish. Res. 148, 64–73.

Rudstam, L.G., Magnuson, J.J., Tonn, W.M., 1984. Size selectivity of passive fishing gear: a correction for encounter probability applied to gill nets. Can. J. Fish. Aquat. Sci. 41, 1252–1255.

Ryan Bellmore, J., Duda, J.J., Craig, L.S., Greene, S.L., Torgersen, C.E., Collins, M.J., Vittum, K., 2017. Status and trends of dam removal research in the United States. WIREs Water 4, e1164.

Saboret, G., Dermond, P., Brodersen, J., 2021. Using PIT-tags and portable antennas for quantification of fish movement and survival in streams under different environmental conditions. J. Fish Biol. 99, 581–595.

Sala, A., 2018. Influence of tow duration on catch performance of trawl survey in the Mediterranean Sea. PLoS One 13, e0191662.

Sanz-Ronda, F.J., Ruiz-Legazpi, J., Bravo-Cordoba, F.J., Makrakis, S., Castro-Santos, T., 2015. Sprinting performance of two Iberian fish: *Luciobarbus bocagei* and *Pseudochondrostoma duriense* in an open channel flume. Ecol. Eng. 83, 61–70.

Schlossberg, S., Chase, M.J., Griffin, C.R., 2016. Testing the accuracy of aerial surveys for large mammals: an experiment with African savanna elephants (*Loxodonta africana*). PLoS One 11, e0164904.

Schlosser, I.J., 1991. Stream fish ecology: a landscape perspective. Bioscience 41, 704–712.

Sedell, J.R., Reeves, G.H., Hauer, F.R., Stanford, J.A., Hawkins, C.P., 1990. Role of refugia in recovery from disturbances: modern fragmented and disconnected river systems. Environ. Manag. 14, 711–724.

Shadwick, R.E., Gemballa, S., 2006. Structure, kinematics, and muscle dynamics in undulatory swimming. In: Shadwick, R.E., Lauder, G.V. (Eds.), Fish Biomechanics. Academic Press, San Diego, pp. 241–280.

Shepherd, D., Katopodis, C., Rajaratnam, N., 2007. An experimental study of louvers for fish diversion. Can. J. Civ. Eng. 34, 770 776.

Silva, A.T., Lucas, M.C., Castro-Santos, T., Katopodis, C., Baumgartner, L.J., Thiem, J.D., Aarestrup, K., Pompeu, P.S., O'brien, G.C., Braun, D.C., Burnett, N.J., Zhu, D.Z., Fjeldstad, H.-P., Forseth, T., Rajaratnam, N., Williams, J.G., Cooke, S.J., 2018. The future of fish passage science, engineering, and practice. Fish Fish. 19, 340–362.

Smith, R.J.F., 2012. The Control of Fish Migration. Springer Science & Business Media.

Smits, A.J., 2019. Undulatory and oscillatory swimming. J. Fluid Mech. 874, 1–70. https://doi.org/10.1017/jfm.2019.284.

SOAFD, 1998. Marine Laboratory Aberdeen Biennial Report 1995-1997. Scottish Office of Agriculature and Fisheris Department, pp. 1–64.

Sprankle, K., 2005. Interdam movements and passage attraction of American shad in the lower Merrimack River main stem. N. Am. J. Fish Manag. 25, 1456–1466.

Stanley, E.H., Doyle, M.W., 2003. Trading off: the ecological effects of dam removal. Front. Ecol. Environ. 1, 15–22.

Stich, D.S., Zydlewski, G.B., Kocik, J.F., Zydlewski, J.D., 2015. Linking behavior, physiology, and survival of Atlantic Salmon smolts during estuary migration. Mar. Coast. Fish. 7, 68–86.

Stokstad, E., 2003. Can well-timed jolts keep out unwanted exotic fish? Science 301, 157–159.

Stokstad, E., 2010. Biologists rush to protect Great Lakes from onslaught of carp. Science 327, 932.

Stoner, A.W., Ryer, C.H., Parker, S.J., Auster, P.J., Wakefield, W.W., 2008. Evaluating the role of fish behavior in surveys conducted with underwater vehicles. Can. J. Fish. Aquat. Sci. 65, 1230–1243.

Strange, E.S., 1984. Review of the fishing trials with Granton an Saro deep sea trawl gear 1963–1967. In: Scotish Fisheries Working Apper, no. 8/84.

Sward, D., Monk, J., Barrett, N., 2019. A systematic review of remotely operated vehicle surveys for visually assessing fish assemblages. Front. Mar. Sci. 6, 134.

Tangorra, J., Gericke, T., Lauder, G., 2011. Learning from the fins of ray-finned fishes for the propulsors of unmanned undersea vehicles. Mar. Technol. Soc. J. 45, 65–73.

Teletchea, F., Fontaine, P., 2014. Levels of domestication in fish: implications for the sustainable future of aquaculture. Fish Fish. 15, 181–195.

Towler, B., Mulligan, K., Haro, A., 2015. Derivation and application of the energy dissipation factor in the design of fishways. Ecol. Eng. 83, 208–217.

Travade, F., Larinier, M., 2002. Fish locks and fish lifts. Bull. Fr. Peche Piscic., 102–118.

Treanor, H.B., Ray, A.M., Layhee, M., Watten, B.J., Gross, J.A., Gresswell, R.E., Webb, M.A., 2017. Using carbon dioxide in fisheries and aquatic invasive species management. Fisheries 42, 621–628.

Tritico, H.M., Cotel, A.J., 2010. The effects of turbulent eddies on the stability and critical swimming speed of creek chub (*Semotilus atromaculatus*). J. Exp. Biol. 213, 2284–2293.

Tudorache, C., Viaenen, P., Blust, R., De Boeck, G., 2007. Longer flumes increase critical swimming speeds by increasing burst and glide swimming duration in carp (*Cyprinus carpio*, L.). J. Fish Biol. 71, 1630–1638.

Tudorache, C., O'k Eefe, R.A., Benfey, T.J., 2010. Flume length and post-exercise impingement affect anaerobic metabolism in brook charr Salvelinus fontinalis. J. Fish Biol. 76, 729–733.

USFWS (U.S. Fish and Wildlife Service), 2019. Fish Passage Engineering Design Criteria. USFWS, Northeast Region R5, Hadley, MA.

Valdemarsen, J.W., 2001. Technological trends in capture fisheries. Ocean Coast. Manag. 44, 635–651.

Van Buren, T., Floryan, D., Brunner, D., Senturk, U., Smits, A., 2017. Impact of trailing edge shape on the wake and propulsive performance of pitching panels. Phys. Rev. Fluids 2, 014702.

Van Buren, T., Floryan, D., Bode-Oke, A.T., Han, P., Dong, H., Smits, A., 2019. Foil shapes for efficient fish-like propulsion. In: AIAA Scitech 2019 Forum, p. 1379.

Van Looy, K., Tonkin, J.D., Floury, M., Leigh, C., Soininen, J., Larsen, S., Heino, J., Leroy Poff, N., Delong, M., Jähnig, S.C., Datry, T., Bonada, N., Rosebery, J., Jamoneau, A., Ormerod, S.J., Collier, K.J., Wolter, C., 2019. The three Rs of river ecosystem resilience: resources, recruitment, and refugia. River Res. Appl. 35, 107–120.

Vander Vorste, R., Obedzinski, M., Nossaman Pierce, S., Carlson, S.M., Grantham, T.E., 2020. Refuges and ecological traps: extreme drought threatens persistence of an endangered fish in intermittent streams. Glob. Chang. Biol. 26, 3834–3845.

Vincik, R.F., 2013. Multi-year monitoring to facilitate adult salmon passage through a temperate tidal marsh. Environ. Biol. Fishes 96, 203–214.

Voegtle, B., Larinier, M., Bosc, P., 2002. Experimental study of the climbing capabilities of the goby *Sicyopterus lagocephalus* (Pallas, 1770) for the design of upstream facilities at the Salazie diversion water intakes (Reunion Island). Bull. Fr. Peche Piscic., 109–120.

Vogel, S., 1981. Life in Moving Fluids: The Physical Biology of Flow. Willard Grant Press, Boston.

Wainwright, D.K., Lauder, G.V., 2018. Mucus matters: the slippery and complex surfaces of fish. In: Gorb, E., Gorb, S. (Eds.), Functional Surfaces in Biology III. Springer Verlag, Berlin, pp. 223–246.

Walsh, S.J., 1991. Effect of tow duration on gear selectivity. NAFO Science Council Research Doc 91/84, pp. 1–9.

Wang, H., Chanson, H., 2018. On upstream fish passage in standard box culverts: interactions between fish and turbulence. J. Ecohydraul. 3, 18–29.

Wang, Y., Yang, X., Chen, Y., Wainwright, D.K., Kenaley, C.P., Gong, Z., Liu, Z., Liu, H., Guan, J., Wang, T., Weaver, J.C., Wood, R.J., Wen, L., 2017. A biorobotic adhesive disc for underwater hitchhiking inspired by the remora suckerfish. Sci. Robot. 2, eaan8072.

Wardle, C.S., 1975. Limit of fish swimming speed. Nature 255, 725–727.

Wardle, C.S., 1977. Effects of size on the swimming speeds of fish. In: Pedley, T.J. (Ed.), Scale Effects in Animal Locomotion. Academic Press, New York, pp. 299–313.

Wardle, C.S., 1980. Effect of temperature on the maximum swimming speed of fish. In: Ali, M.A. (Ed.), Environmental Physiology of Fish. Plenum, New York, pp. 519–531.

Wardle, C.S., 1986. Fish behavior and fishing gear. In: Pitcher, T.J. (Ed.), The Behavior of Teleost Fishes. Croom Helm, London and Sydney, pp. 463–495.

Wardle, C.S., He, P., 1988. Burst swimming speeds of mackerel, *Scomber scombrus* L. J. Fish Biol., 471–478.

Ware, D.M., 1975. Growth, metabolism, and optimal swimming speed of a pelagic fish. J. Fish. Res. Board Can. 32, 33–41.

Ware, D.M., 1980. Bioenergetics of stock and recruitment. Can. J. Fish. Aquat. Sci. 37, 1012–1024.

Watson, J., Foster, D., Nichols, S., Shah, A., Scoll-Oenlon, E., Nanc, J., 1999. The development of bycatch reduction technology in the Southeastern United States Shrimp Fishery. Mar. Technol. Soc. J. 33, 51–56.

Weaver, C.R., 1963. Influence of water velocity upon orientation and performance of adult migrating salmonids. Fish. Bull. 63, 97–121.

Webb, P.W., 1984. Form and function in fish swimming. Sci. Am. 251, 72–82.

Webb, P.W., 1989. Station-holding by three species of benthic fishes. J. Exp. Biol. 145, 303–320.

Webb, P W , 2006. Stability and manoeuvrability. In: Fish Biomechanics. Elsevier, London.

Wen, L., Weaver, J.C., Lauder, G.V., 2014. Biomimetic shark skin: design, fabrication, and hydrodynamic function. J. Exp. Biol. 217, 1656–1666.

Wen, L., Weaver, J.C., Thornycroft, P.J.M., Lauder, G.V., 2015. Hydrodynamic function of biomimetic shark skin: effect of denticle pattern and spacing. Bioinspir. Biomim. 10, 1–13.

Wen, L., Ren, Z., Di Santo, V., Kainan, H., Tao, Y., Wang, T., Lauder, G.V., 2018. Understanding fish linear acceleration using an undulatory bio-robotic model with soft fluidic elastomer actuated morphing median fins. Soft Rob. 5, 375–388.

Whitear, M., 1970. The skin surface of bony fishes. J. Zool. 160, 437–454.

Whitt, C., Pearlman, J., Polagye, B., Caimi, F., Muller-Karger, F., Copping, A., Spence, H., Madhusudhana, S., Kirkwood, W., Grosjean, L., Fiaz, B.M., Singh, S., Singh, S., Manalang, D., Gupta, A.S., Maguer, A., Buck, J.J.H., Marouchos, A., Atmanand, M.A., Venkatesan, R., Narayanaswamy, V., Testor, P., Douglas, E., De Halleux, S., Khalsa, S.J., 2020. Future vision for autonomous ocean observations. Front. Mar. Sci. 7, 697. https://doi.org/10.3389/fmars.2020.00697.

Wieferich, D.J., 2022. Database of Stream Crossings in the United States. Data Release, U.S. Geological Survey, https://doi.org/10.5066/P9YX6KTB.

Wieferich, D.J., Duda, J., Wright, J., Uribe, R., Beard, J., 2021. drip.dashboard. 2.1.0. US geological Survey, https://doi.org/10.5066/P9UNIWKF.

Wieland, K., Storr-Paulsen, M., 2006. Effect of tow duration on catch rate and size composition of Northern shrimp (*Pandalus borealis*) and Greenland halibut (*Reinhardtius hippoglossoides*) in the West Greenland bottom trawl survey. Fish. Res. 78, 276–285.

Wilkie, M.P., Brobbel, M.A., Davidson, K., Forsyth, L., Tufts, B.L., 1997. Influences of temperature upon the postexercise physiology of Atlantic salmon (*Salmo salar*). Can. J. Fish. Aquat. Sci. 54, 503–511.

Winger, P.D., 2004. Effect of Environmental Conditions on the Natural Activity Rhythms and Bottom Trawl Catchability of Atlatnic Cod (*Gadus morhua*). PhD Doctoral Dissertation, Memorial University.

Winger, P.D., Eayrs, S., Glass, C.W., 2010. Fish behavior near bottom trawls. In: He, P. (Ed.), Behavior of Marine Fishes: Capture Processes and Conservation Challenges. Wiley-Blackwell, Ames, Iowa, pp. 67–103.

Woodhead, P.M.J., 1964. Changes in the behaviour of the sole,*Solea vulgaris*, during cold winters, and the relation between the winter catch and sea temperatures. Helgoländer Meeresun. 10, 328–342.

Wright, G.V., Wright, R.M., Bendall, B., Kemp, P.S., 2016. Impact of tide gates on the upstream movement of adult brown trout, *Salmo trutta*. Ecol. Eng. 91, 495–505.

Yanase, K., Eayrs, S., Arimoto, T., 2007. Influence of water temperature and fish length on the maximum swimming speed of sand flathead, *Platycephalus bassensis*: implications for trawl selectivity. Fish. Res. 84, 180–188.

Zhu, J., White, C., Wainwright, D.K., Di Santo, V., Lauder, G.V., Bart-Smith, H., 2019. Tuna robotics: a high-frequency experimental platform exploring the performance space of swimming fishes. Sci. Robot. 4, eaax4615.

Zielinski, D.P., McLaughlin, R., Castro-Santos, T., Paudel, B., Hrodey, P., Muir, A., 2019. Alternative Sea lamprey barrier technologies: history as a control tool. Rev. Fish. Sci. Aquac. 27, 438–457.

Chapter 4

Applied fish bioenergetics

Jacob W. Brownscombe[a,*], Michael J. Lawrence[b], David Deslauriers[c], Ramon Filgueira[d], Robin J. Boyd[e], and Steven J. Cooke[f]

[a]*Great Lakes Laboratory for Fisheries and Aquatic Sciences, Fisheries and Oceans Canada, Burlington, ON, Canada*
[b]*Department of Biological Sciences, University of Manitoba, Winnipeg, MB, Canada*
[c]*Institut des sciences de la mer de Rimouski, Université du Québec à Rimouski, Rimouski, QC, Canada*
[d]*Marine Affairs Program, Dalhousie University, Halifax, NS, Canada*
[e]*UK Centre for Ecology and Hydrology, MacLean Building, Benson Lane, Crowmarsh Gifford, Wallingford, United Kingdom*
[f]*Fish Ecology and Conservation Physiology Laboratory, Department of Biology, Carleton University, Ottawa, ON, Canada*
[*]*Corresponding author: e-mail: jakebrownscombe@gmail.com*

Chapter Outline

Fish Physiology, Vol. 39A. https://doi.org/10.1016/bs.fp.2022.04.004

Energy is a fundamental currency of life that can be quantified in organisms to understand how environmental conditions and anthropogenic stressors affect individuals, scaling up to populations and entire ecosystems. Bioenergetics studies have been conducted extensively on fishes, with an historical focus on lab-based experiments relevant to fisheries and aquaculture; however, recent methodological and technological advances are enabling more widespread applications in ecology and conservation including in situ measurement of various aspects of fish bioenergetics in aquatic ecosystems. Much of the utility of bioenergetics is based on a generalized mass-balance equation that describes energy allocation, and there are numerous advanced modeling techniques available to quantify how various environmental and intrinsic biological factors influence fish energetics, as well as relevant endpoints such as growth and reproduction. In this chapter, we discuss (1) key components of fish bioenergetics, (2) available measurement techniques, (3) common modeling techniques, and (4) case studies that highlight some key applications to date. We conclude by discussing current limitations and future research directions in this field. Bioenergetics is an increasingly powerful approach to build mechanistic connections between environmental conditions, stressors, and fish populations that is especially valuable for predicting the responses of fishes to rapidly changing conditions in the Anthropocene.

1 Introduction: History and application

Energy is a fundamental currency of life that can been used to describe ecological systems at diverse levels of organization, from the growth and reproductive rates of individual organisms to population dynamics and the structure of biological communities (Brown et al., 2004; Sibly et al., 2012; Tomlinson et al., 2014). There is long and rich history of bioenergetics studies with fishes. Indeed, of any animal taxa, bioenergetics have perhaps been applied most widely to fish due to tractability of fish as study organisms combined with the applicability of bioenergetics to fisheries and aquaculture (Hansen et al., 1993; Hoar et al., 1979; Jobling, 1995; Tytler and Calow, 1985). Historically, applications of fish bioenergetics focused heavily on developing bioenergetics models to determine optimal conditions for fish growth in aquaculture (Cuenco et al., 1985a,b; Jobling, 1995, 2011), and to estimate fish population dynamics to inform fisheries management (Chipps and Wahl, 2008; Hansen et al., 1993; Hartman and Kitchell, 2008). Energy is a key currency that can be used to optimize aquaculture practices, and to build a mechanistic link between environmental conditions and fish population dynamics for fisheries management (Kitchell et al., 1977). Outside of fisheries and aquaculture, there are also numerous ways in which bioenergetics is relevant to ecology and conservation, from characterizing how environmental factors interact with fish metabolism to influence their behavior, space use, and species distributions (Brownscombe et al., 2017a; Payne et al., 2016;

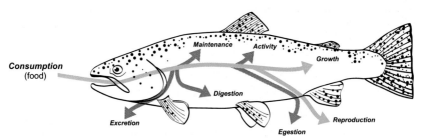

FIG. 1 The general components of bioenergetics mass balance equation illustrated with a fish. Energy is consumed as food, a portion of which becomes wastes (egestion and excretion), and some energy is expended in digestive processes. Assimilated energy is generally allocated first to metabolic maintenance, and a variable amount is allocated to activity (locomotion). Remaining (net) energy is allocated to somatic growth and/or reproduction. Green lines indicate energetic gains, and red energetic costs and wastes. *Graphics were produced by Joanne Dziuba from Environment and Climate Change Canada.*

Pörtner et al., 2017), to understanding the energetic costs of anthropogenic stressors (Watson et al., 2020).

The utility of bioenergetics is based largely upon a generalized energy-balance equation: Energy consumed = metabolism + waste + growth (Brett and Groves, 1979; Winberg, 1956). These components are further distinguished within metabolism (resting + active + digestion) and growth (somatic + reproduction) (Fig. 1; Tytler and Calow, 1985; Jobling, 1995). The allocation of energy to each component can vary widely depending on a combination of environmental (e.g., temperature, diet), physiological (e.g., body size, reproductive status), life history (e.g., age of maturity, reproductive allocation), and behavioral (e.g., activity level) factors. This variation may occur amongst species, populations, or even individuals, and is of major consequence for how fish respond to environmental conditions and anthropogenic stressors. In this chapter, we further discuss the major components of the bioenergetics equation, with a focus on Sections 2.1–2.3, along with how these components can be quantified through Section 3 and integrated through Section 4.

Much of the research to date on fish bioenergetics has occurred in the context of laboratory studies, with a particular focus on aquaculture (Jobling, 1995; Tytler and Calow, 1985). Aquaculture is one potential way to supply growing human populations with food, alleviating pressure from often already overexploited aquatic resources, but can also have an array of negative ecological impacts (Pillay, 2007; Primavera, 2006). Therefore, optimizing aquaculture techniques, including aspects of consumption such as feed quantity, quality, and conditions to achieve efficient, low-environmental-impact farming practices is essential to help align aquaculture with conservation goals.

Fish feeds are generated from environmental resources, and choices of food source (e.g., algae-based vs fish-based proteins) have a major impact on the sustainability of aquaculture (Roques et al., 2020; Turchini et al., 2019). We do not cover fish culture considerations extensively in this chapter, focusing more on considerations for the conservation of wild fish and ecosystems, but this is an important aspect of bioenergetics science and practice for a sustainable Anthropocene.

Accurate estimates of bioenergetics parameters are paramount for effective application in any context, and they are particularly challenging to generate for wild fishes. Historically, most of this information was derived from laboratory studies, and parameter estimates for a given species or population are often unavailable and borrowed from others for bioenergetics models (Chipps and Wahl, 2008), and/or estimated based on known generalized relationships with body size or water temperature (Clarke and Johnston, 1999; Jerde et al., 2019; Killen et al., 2010). Although estimates based on mechanistic relationships improve accuracy, there has been a fairly high level of uncertainty in the accuracy and applicability of bioenergetics models, which was the focus of discourse in the late 20th century (Hansen et al., 1993; Ney, 1993). A primary issue is that bioenergetic estimates often fail to accurately reflect true values for fishes in natural environmental contexts. Although it is exceptionally challenging, there is growing recognition of the need to generate more realistic estimates from wild fish in natural environments, and new technological and experimental developments are increasing our capacity to do so (Cooke et al., 2016; Treberg et al., 2016). Continued advancements in these measurement techniques are paramount for effective application of bioenergetics to conservation—currently available approaches for measuring fish bioenergetics, both in the lab and field, are discussed in this chapter under Section 3, along with their limitations, future directions, and research needs.

Estimates of components of fish bioenergetics can be useful in their own right, but integration into bioenergetics modeling frameworks extends their applicability in many ways. For example, bioenergetics modeling has been used to characterize density-dependent effects on fish characteristics (Taylor et al., 2020), spatial dynamics of energy flow (Hartman and Kitchell, 2008), and predator–prey dynamics to influence fish stocking activity (Jones et al., 1993). Models address the key need to integrate individual measures of fish bioenergetics into a quantified framework to make predictions at whole organismal, population, or ecosystem scales. They also often address shortcomings in information availability, and may be used, for example, to solve for different components such as consumption or growth (Essington et al., 2001; Walters and Essington, 2009). There is a diversity of model types available that differ in assumptions and data requirements for their parameterization and forcing. These differences, as well as the approaches to scale up models from the individual to the population level, are discussed in this chapter under Section 4.

The field of fish bioenergetics is truly at the nexus of fundamental and applied science, based on foundational knowledge of physics and physiology, and widely applied to fisheries and ecosystem management. For these reasons, fish bioenergetics has been the focus of extensive study and synthesis, including a broad array of foundational literature. This chapter aims to build upon those foundations to discuss our current state of knowledge and practice in applied fish bioenergetics with a focus on the challenges presented by the Anthropocene. As a key currency that can link environmental conditions with fish fitness and population dynamics, bioenergetics is well suited for addressing the conservation challenges we face due to rapidly changing conditions and stressors. We discuss the key components of fish bioenergetics and their measurement and modeling approaches, along with a few key examples of how bioenergetics has been applied to pertinent conservation issues (Section 5). Lastly, we conclude with some broader discussion of the field and future research directions for applied fish bioenergetics (Section 6).

2 Bioenergetics components

2.1 Consumption

Consumption (Fig. 1) supplies fish with their nutritional needs, fueling metabolic processes and growth. In the context of fish farming/aquaculture, a large amount of research has gone into understanding relationships between food quantity, nutritional profiles, feeding conditions, and fish growth and health (reviewed in Jobling, 1993, 1995). For wild fish, food is provided by their environment and is therefore influenced by a variety of ecosystem characteristics, primary productivity, and trophic interactions. Individual prey have an energetic value that relates to their energy density and digestive efficiency, which varies amongst species (Hartman and Brandt, 1995; Secor, 2009). Fish must also acquire their food efficiently (i.e., minimum cost with maximum intake) to achieve net energetic gain (Pyke, 1984; Townsend and Winfield, 1985) and therefore, ecological factors such as prey density, distribution, and availability may all impact their bioenergetics. For this reason, fish are generally thought to select for bioenergetically optimal prey, relating to prey energy density and capture costs (Gill and Hart, 1994; Stein et al., 1984). Resource acquisition involves both quantity and quality—fish require specific nutrients (macro- and microscale) for their health and growth, and in some cases must exploit an appropriate balance of food sources to achieve nutrition through "optimal foraging for multiple nutrients" (Houston et al., 2011; Simpson et al., 2004). Factors such as competition and predators may also influence the capacity of fish to exploit certain prey sources, and to do so efficiently (Garvey et al., 2004; Nash et al., 2012). Water temperature may also dictate resource availability, where fish can be excluded from locations due to a temperature-limited metabolic capacity (Guzzo et al., 2017), or may forage in particular locations and times where their metabolism is near-optimal (Brownscombe et al., 2017a). These examples highlight the complex interactions among ecological factors (e.g., presence of predators, ecosystem thermal regime) that can influence fish consumption and more broadly the bioenergetics of wild fish. Indeed, any changes to ecosystem characteristics including factors such as habitat quantity, quality, or the intro-duction of competitors (e.g., invasive species) may affect the capacity of individuals to acquire food effectively and efficiently, in turn influencing their growth and fitness (Giacomini et al., 2013; Mazur et al., 2007). There are many great examples of the applicability of consumption bioenergetics estimates, one of which is highlighted in Section 5.1. Notably, consumption is perhaps the most challenging component to measure remotely in wild fish in terms of what, where, and how much fish are eating (Cooke et al., 2016; but see Section 3 for discussion of a growing number of creative approaches). For this reason, bioenergetics modeling is often used to solve for consumptive rates and needs (Chabot et al., 2016a; Essington et al., 2001; discussed in Section 4).

2.2 Metabolism

Metabolism refers to the chemical reactions involved in maintaining living cells, including the synthesis (anabolism) and breakdown (catabolism) of molecules. Framed within energetics, it is the process by which energy is mobilized for storage and utilization. Metabolic rate represents the rate of energy expenditure, including both baseline energetic needs (i.e., standard metabolic rate; SMR) and active metabolism allocated to exercise, diges-tion, and growth (Fig. 1). Broadly, energy consumed must exceed metabolic needs for surplus available to allocate to growth, a key process for biological fitness (Tytler and Calow, 1985). Indeed, there is evidence that metabolic rate and resource availability combined may be key drivers of fish habitat use (Auer et al., 2020). As poikilotherms (with some exceptions such as tunas), a fish's metabolic rate can be a highly variable component of their energy budget due to environmentally-driven variability in body tem-perature, as well as fish body size and activity level (Clarke and Johnston, 1999; Jerde et al., 2019; Killen et al., 2010). Metabolic rate, and hence ener-getic needs, increase predictably with body size and water temperature; however, activity costs can vary widely—for example, Boisclair and

Leggett (1989) found that activity can comprise 0–40% of a fish's energy budget. There are also significant energetic costs to digestion, which is referred to as specific dynamic action (SDA), encompassing all post-prandial energetic costs (Jobling, 1981). In some cases, such as more sedentary fishes, SDA can comprise a substantial component of their energy budget (Jobling, 1993; Tytler and Calow, 1985). Digestive costs and efficiency can also be highly variable among factors such as water temperature, meal size and quality (Secor, 2009).

As ectotherms, water temperature not only influences their metabolic rate and energetic needs, but also their capacity to mobilize energy for fitness enhancing activities such as digestion, as well as exercise to capture food, to move between key habitats, or to avoid predators (Pörtner et al., 2017). This is because fishes are adapted to operate within a certain temperature window, where various physiological and biochemical processes are optimized—this window can be specific to species, populations, and even individuals. Factors such as water temperature and oxygen availability influence not only the baseline metabolic needs (standard metabolic rate; SMR), but also the maximum rate at which fish can mobilize energy aerobically, their maximum metabolic rate (MMR). Although fish can mobilize energy in the absence of oxygen through anaerobic metabolism, this process is highly inefficient and unsustainable due to the production of potentially harmful waste products, and it can only be used for short periods to accomplish activities such as burst swimming (Blažka, 1958). Therefore, the majority of energetic processes occur through aerobic metabolism, within the window between SMR and MMR, called aerobic scope (AS; Fry, 1947; Brett, 1971; Pörtner et al., 2017). At least some level of AS is therefore required to mobilize energy for the aforementioned fitness enhancing activities. Fish AS generally exhibits two categorically different relationships with water temperature, either peaking at an intermediate temperature and declining at thermal extremes, or increasing continuously with temperature until upper lethal temperatures cause a sharp decline (Clark et al., 2013).

The role of AS in fish ecology is subject of ongoing debate (Jutfelt et al., 2018; Pörtner et al., 2017). There are certain contexts where AS seems to be highly relevant to fish fitness, such as long distance migrations (Eliason et al., 2011), intense exercise demands to pass a dam (Burnett et al., 2014), or when foraging, where there are competing metabolic demands from exercise and digestion (Brownscombe et al., 2017b; Zhang et al., 2018). There is also evidence that temperature and oxygen concentrations may limit the distribution of fishes based on their aerobic performance (Duncan et al., 2020; Payne et al., 2016). Yet, the role of AS is equivocal, for example, some Pacific salmon species and populations appear to have AS thermal performance curves that do not align with environmental thermal regimes

(Clark et al., 2011; MacNutt et al., 2006; Raby et al., 2016). Further, growth rates and fish temperature selection may not correspond to optimal AS (Clark et al., 2013; Gräns et al., 2014; Norin et al., 2014). Indeed, numerous physiological processes, such as digestive enzyme activity, may have varied optima that influence fish bioenergetics and ecology beyond AS (Clark et al., 2013).

Overall, as the process by which energy is expended and gained, metabolism represents an important component of fish bioenergetics that varies widely due to ecological conditions. Hence, metabolic metrics such as metabolic rate and AS represent key metrics with which we can link ecological factors and anthropogenic stressors to fish fitness, to gain insights for effective resource management (Claireaux and Lefrançois, 2007). Although more research is needed on key metrics and mechanisms, metabolic metrics may be particularly useful for predicting, for example, how changes in water temperature and dissolved oxygen may impact species distributions (Duncan et al., 2020; Payne et al., 2016), and access to habitats (Brownscombe et al., 2017b; Guzzo et al., 2017). Metabolic rate and energetic needs play a prominent role in the applied example provided in Section 5.2. As a general pacemaker for biological processes, the relevance of metabolism extends beyond individuals, scaling to describe the characteristics of populations and entire ecosystems in terms of their productivity, diversity, and structure (Brown et al., 2004; Humphries and Mccann, 2014; Sibly et al., 2012). Yet, metabolism is only one component of the bioenergetics equation, which must be considered holistically to maximize the utility of bioenergetics for applied science and resource management (Tomlinson et al., 2014; discussed further in Sections 2.3 and 6).

2.3 Growth

The net energy available to a fish from consumption that is not used by metabolism is allocated to growth, which includes distribution to a variety of somatic tissues and, upon maturation, reproduction. Therefore, combinations of factors that influence consumption and metabolism (e.g., prey availability and energetic value, metabolic rate due to water temperature) ultimately determine the amount of energy available for allocation to growth. As a key focal metric of aquaculture, much of our understanding about fish growth comes from this research context relating it to food quantity and quality, as well as environmental conditions such as water temperature (Jobling, 1995, 2011; Tytler and Calow, 1985). Within these controlled conditions, the above-mentioned factors can often be manipulated to achieve optimal fish growth and health. However, in the wild, fish are presented with diverse and often adverse conditions that result in complex life history trade-offs that influence energy allocation. There is a well-known trade-off between somatic growth and reproductive allocation, ranging from r-selected strategies where species mature relatively quickly and allocate high amounts of resources to reproduction, to k-selected strategies where species tend to grow larger and mature later, allocating relatively higher amounts of energy to somatic growth prior to reproduction (MacArthur and Wilson, 1967; Pianka, 1970; Reznick et al., 2002). Interestingly, these strategies may correlate with metabolic rate (Auer et al., 2018), meaning energetic needs are related to life history processes. The ecological conditions that populations have experienced historically are thought to influence these characteristics to a major extent, especially factors that influence growth and survival such as resource availability and predation rate (Johnson et al., 2011). Sources of mortality may be natural, such as predators, or may be anthropogenic stressors—the most prominent example of this is fisheries-induced evolution, with which high harvest rates (mortality) of large individuals in the population causes a life history shift, coinciding with a suite of genotypic and phenotypic traits of the population (Heino et al., 2015). There is growing recognition that fisheries induced evolution is of major potential consequence for population recovery after overexploitation for a variety of reasons, including growth allocation (Kuparinen and Merilä, 2007; Law, 2007).

Another prominent trade-off exists between fish growth capacity and physical fitness. Perhaps the best example of this is comparing wild and hatchery or farm raised fishes, the latter of which tend to have reduced exercise capacity, high digestive capacity, and high growth rates in captive conditions (Allen et al., 2016; Rosenfeld et al., 2020). This has been identified as likely survival issue for hatchery-reared fish when stocked into the wild, where their glutenous physiological tendencies may make them unprepared for the challenges posed in natural ecosystems. As discussed in Section 2.2, in the wild, aerobic capacity can be important to overcome ecological challenges that demand exercise, such as passing high-velocity water currents or avoiding predators—this capacity may be a key determinant of biological fitness in certain contexts.

Overall, resource allocation to various tissues plays an important role in many applied conservation contexts. Metabolism receives a lot of attention as source of ecological variability that builds links between individuals and broader levels of ecological organization (i.e., populations and ecosystems), but ultimately, this link may also be modulated heavily by growth allocation because it determines how fish will deal with environmental challenges, and hence how energy moves within the ecosystem. For example, species with longer generation times are more vulnerable to anthropogenic stressors such as climate change or overexploitation (Gallagher et al., 2015; Pearson et al., 2014). The contexts in which growth allocation is relevant are highly diverse—from fish stocking to alterations in riverine water flow patterns (e.g., due to dams), to species invasiveness. For example, it is unreasonable to expect that a hatchery reared salmon with high energy stores and limited swimming capacity (Allen et al., 2016; Rosenfeld et al., 2020), would be capable to navigating a high flow fishway to pass a dam and reach spawning habitat (Burnett et al., 2014). At the ecosystem level, the nature of ecological changes will determine the winners and losers, altering energy flow pathways and trophic structure.

3 Measurement

3.1 Consumption and feeding estimates

Consumption estimates in bioenergetic modeling can be made using measurements of the energy content and nutritional composition of food items (i.e., feed or prey) (Fig. 2). Classically, energy content measurements have been made using bomb calorimetry and proximate body composition, but this can also be accomplished with the use of specialized probes. Bomb calorimetry represents a direct measure of a sample's gross energy contents and involves combusting the item while recording the total heat produced (reviewed in Jobling, 1983; Robbins, 2012). Consequently, it has been widely used in dietary-based bioenergetic studies addressing food and prey energy

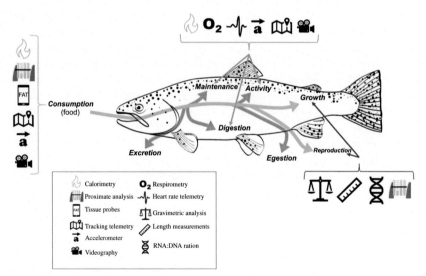

FIG. 2 Overview of how bioenergetic model parameters can be estimated for consumptive (yellow), metabolic (blue), and growth (purple) parameters. Estimation methods include a variety of lethal and non-lethal sampling procedures that attempt to quantify the activity, energy density, or metabolic rate metric of interest.

densities providing some of the foundational characterizations of such (Albo-Puigserver et al., 2017; Brunnschweiler et al., 2018; Hartman and Brandt, 1995). Thus, bomb calorimetry could be thought of as a starting point in conservation-based studies when dietary inputs for bioenergetic models are unknown (Glover et al., 2010; Johnson et al., 2017), especially as energy contents can vary between species, sex, age, and seasonality (Albo-Puigserver et al., 2017; Vollenweider et al., 2011; Wuenschel et al., 2006), which could have implications for model input estimates. Additionally, although it provides an accurate estimation of gross tissue/food energy contents, bomb calorimetry cannot discern individual energetic components (i.e., proteins, carbohydrates, lipids; Battley, 1995), which may be of relevance. The use of radiotracers has been suggested as a method for quantifying consumption rates in fishes (Rowan and Rasmussen, 1996; Sherwood et al., 2000) but it does not appear to have broad use in the literature. Stable isotopes and DNA barcoding may also be implemented to determine dietary preferences and trophic interactions in wild-caught fishes (Amundsen and Sánchez-Hernández, 2019; Nielsen et al., 2018), guiding subsequent energy quantification and modeling.

In contrast to bomb calorimetry, proximate body composition provides an indirect quantification of the total energy density of a tissue or food source (Love, 1970), representing a sum of the total energy provided by individual components in the sample. In doing so, biomolecules, such as carbohydrates, lipids, and proteins, are quantified using standards assays (Bligh and Dyer, 1959; Elliott and Davison, 1975; Paine, 1971), while ash and water contents are simply weighed (Love, 1970). The resulting concentrations of each component are then converted to an energy total using a conversion factor (e.g., 36.4kJ g^{-1} for lipid, 23.6kJ g^{-1} for protein; (Brett, 1995; Brett and Groves, 1979) and summed resulting in an approximation of the sample's energy density (Love, 1970; Paine, 1971). While these estimates can be close to bomb calorimetry measurements (Craig et al., 1978; Henken et al., 1986), it may overestimate tissue energy densities in some instances (Craig, 1977; Craig et al., 1978; Schloesser and Fabrizio, 2015). Regardless, the use of proximate body composition measurements is widespread, being used in much the same manner as bomb calorimetry (Breck, 2008, 2014; Dempson et al., 2004; Lenky et al., 2012) while also providing a finer scale resolution for observing changes in specific energy components (Dempson et al., 2004; Lenky et al., 2012; Niu et al., 2019). Indeed, this has proven useful in determining energy contents of prey species in the natural environment (Ball et al., 2007; Spitz et al., 2010; Vollenweider et al., 2011) which can serve as important model inputs into bioenergetic estimates.

In more recent years, the use of non-invasive techniques in measuring tissue energy densities in fish has been gaining traction. Both bomb calorimetry and proximate body analysis generally require lethal sampling, which could be problematic if the project requires fish to survive post-sampling (i.e., endangered or threatened species, long-term monitoring project, sample storage limitations, etc.). Recently, two mainstream device including bioelectrical impedance analysis (BIA; Cox and Hartman, 2005) and microwave energy meters (aka "fatmeters"; Crossin and Hinch, 2005) have been proposed as non-lethal methods in obtaining tissue energy contents in fishes. Both devices measure tissue water contents as a proxy for tissue energy density (reviewed in Klefoth et al., 2013) wherein the probe is placed in contact with the lateral surface of the animal providing a measurement almost instantaneously. While similar in their approach, both devices have varying degrees of accuracy in relating whole tissue water contents with measures of energy composition. For example, Klefoth et al. (2013) found that while microwave meters generally outperformed BIA meters, both had a weak ability to predict body composition in the study species. Indeed, depending on the context, non-lethal body composition meters have mixed results in predicting tissue energetic status (Cox and Hartman, 2005; Crossin and Hinch, 2005; Klefoth et al., 2013; Pothoven et al., 2008; Whiterod, 2010). It has been recommended that the device be calibrated on the species/setting of interest to ensure that meter readouts are representative of the fish's energetic status (Bayse et al., 2018; Hanson et al., 2010). As technology and techniques continue to improve, non-lethal approaches will likely become a more reliable and widely used method of measuring body energy contents and important tools in conservation and bioenergetic studies. Compared to more traditional approaches, these meters are cheap, easy to use, and highly portable, making them accessible for a variety of field-based studies (Bayse et al., 2018; Crossin et al., 2008; Teffer et al., 2017). They may also enable more widespread data collection, including repeated samples from the same individuals, providing more comprehensive and long-term measures of energy composition and being increasingly used assessments of species of conservation interest (Brosset et al., 2015; Schloesser and Fabrizio, 2016; Whiterod, 2010). Another potential non-lethal indicator of recent consumption and energetic/ nutritional state in fishes is blood plasma sampling for analysis of fatty acids and ketone bodies, with some recent successful applications (e.g., Eldøy et al., 2021; Moorhead et al., 2021).

In addition to the energetic content of consumption, it is also valuable to understand other aspects of fish foraging including strategies, habitat characteristics, energetic costs, and waste production. A variety of biologging/ biotelemetry technologies can aid in remote measurement of fish foraging through estimates of fish space use, as well as integrated sensors that measure acceleration (activity), body temperature, and/or heart rate (Cooke et al., 2016). For example, telemetry-based studies, often done in conjunction with gastric dietary examinations, have been used to infer important foraging habitats of species of conservations concern (Heithaus et al., 2002; Le Pichon et al., 2014; Novak et al., 2017). More recently, acoustic tags have been developed to remotely measure predation (Halfyard et al., 2017; Klinard et al., 2019), which may have increasing applications for understanding fish foraging and consumption. Foraging behaviors can also be measured remotely and directly using sensors such as accelerometers that measure fine scale body movement (Brewster et al., 2018; Brownscombe et al., 2014; Kawabata et al., 2014; Noda et al., 2013). The specific prey types being consumed may also be inferred using accelerometers in some instances (Kawabata et al., 2014). Videography is also highly useful for measuring fish foraging, which can be accomplished with animal-mounted tags (Heithaus et al., 2002; Kudo et al., 2007), or through place-based monitoring at fixed locations (Lawrence et al., 2019b; Mayfield and Cech, 2004; Meynecke et al., 2008), or mobile monitoring through drones (King et al., 2018; Oleksyn et al., 2021). The physiological responses associated with foraging can also be measured using various sensors. For example, with endothermic fishes such as certain shark species and tunas, body temperature loggers can provide remote measures of consumption (Bestley et al., 2008; Clark et al., 2008; Jorgensen et al., 2015). Fish also exhibit a heart rate response to feeding (Eliason et al., 2008), due to the metabolic needs of digestion, which can now be measured readily with heart rate loggers (Shen et al., 2020; Steell et al., 2019). This can generate an estimate of the magnitude of consumption, as well as the energetic costs of digestion (SDA). However, fish heart rate increases for a variety of reasons, most notably from activity, so simultaneous measurement of heart rate and activity is a useful way to untangle digestion vs activity (Clark et al., 2010). Lastly, waste production can be quantified using assessments of renal/branchial/whole body fluxes (e.g., (Bucking and Wood, 2008; Lawrence et al., 2015; Zimmer et al., 2010) or by measuring fecal energy contents (Henken et al., 1986; Page and Andrews, 1973).

3.2 Metabolism estimation

As discussed in Section 2.2, there are a variety of fish metabolic metrics that may be of interest in applied contexts. One of the oldest methods to estimate metabolic rate in animals makes use of direct calorimetry, which involves measuring the formation of heat from metabolic processes in an organism (Kaiyala and Ramsay, 2011; McLean and Tobin, 2007). This technique has been widely used in mammalian and avian studies, but has rarely been applied to fishes due to the poor heat conductance of water and relatively low metabolic rate of most ectothermic fishes (reviewed in van Ginneken and van den Thillart, 2009). However, it can used with small fish in a laboratory setting (Regan et al., 2016; Smith et al., 1978; van Ginneken et al., 1997; Van Waversveld et al., 1989). Challenges with measurement aside, direct calorimetry can measure both aerobic and anaerobic metabolism, which may be useful for studies focused on exercise recovery or stress responses.

Respirometry is the more commonly used method to estimate fish metabolic rate, which involves measurement of the changes in oxygen concentration in the water surrounding the fish in a confined space caused by fish respiration. The decline per unit time in water oxygen concentrations is termed oxygen uptake rate (i.e., MO_2) and serves as a proxy for metabolic rate in fishes (reviewed in Nelson, 2016, including limitations of this approach). This indirect method of measuring metabolism is advantageous as it circumvents the challenges of direct calorimetry and is more tractable (Bolduc et al., 2002; Mochnacz et al., 2017; Svendsen et al., 2016). This enables wider application, for example, in field settings (Bouyoucos et al., 2018; Byrnes et al., 2020; Farrell et al., 2003; Lawrence et al., 2019a) with fish of all sizes ranging from eggs and small zebrafish to large sharks (Fitzgibbon et al., 2008; Ste-Marie et al., 2020; Zimmer et al., 2020), and under a range of experimental treatments (e.g., hypoxia, toxicants, heat shock, swimming performance, taxonomic variation, etc.; Blewett et al., 2016; Clark et al., 2011; Farrell et al., 2003). This is of particular importance for conservation-based projects where sampling is likely to occur in remote field sites and/or is using a non-model species.

Although highly useful, respirometry also has its caveats, perhaps the most major of which is that it only measures aerobic metabolism (Nelson, 2016), limiting its usefulness in scenarios where anaerobic metabolism may be high as in burst/sustained swimming (Lee et al., 2003; Svendsen et al., 2015), exhaustive exercise recovery (Cai et al., 2014; Lawrence et al., 2019a; Zhang et al., 2018), or under low oxygen conditions (Borowiec et al., 2015; Cox et al., 2011; Williams et al., 2019). Anaerobic metabolism can contribute greatly to the cost of transport in a fish and thus, using only $\dot{M}O_2$ in determining this metric can potentially lead to an underestimation of its value (Eliason et al., 2013; Farrell, 2007; Svendsen et al., 2015). This is also relevant to sustained swimming trials (i.e., to measure critical swimming speed; U_{crit}) where $\dot{M}O_2$ can be upwards of 50% higher when anaerobic metabolism is accounted for (Lee et al., 2003). This limitation can be addressed in some cases by accounting for any anaerobic metabolism through measurement of excess post-exercise oxygen consumption (EPOC), which can then be used to adjust $\dot{M}O_2$ (Brett, 1964; Lee et al., 2003; Scarabello et al., 1992; Svendsen et al., 2015). One other caveat of typical respirometry trials is that $\dot{M}O_2$ alone provides no indication of the types of metabolic fuel being used, which influences the relationship between oxygen consumption and energy expenditure. However, it is possible measure $\dot{M}O_2$ and $\dot{M}CO_2$ simultaneously to derive respiratory quotient values for aquatic species (Ferreira et al., 2019; Kieffer et al., 1998; Lauff and Wood, 1996). This is rarely done and requires more extensive experimental methods, which may be worthwhile for improved bioenergetics estimates.

There are a variety of respirometer designs depending on the questions or metrics of interest, fish body size and morphology (Cech, 2011; Clark et al., 2013; McKenzie et al., 2016; Svendsen et al., 2016), with detailed considerations available for estimating SMR (Chabot et al., 2016b), MMR (Little et al., 2020), AS (Clark et al., 2013), and SDA (Chabot et al., 2016a). Interestingly, in some cases MMR estimates can be higher from digestion than encouraged exercise (Steell et al., 2019). These metrics are generated most commonly with static chamber intermittent flow respirometers, but swim tunnel respirometers are also frequently used to generate estimates of fish active metabolism across a range of swim speeds (Bangley et al., 2020; Brett, 1964; Brownscombe et al., 2018; Farrell, 2007; Metcalfe et al., 2016; Nowell et al., 2015; Steffensen et al., 1984). Likely the biggest constraint to swim tunnel studies is with fish participation, which must be able to physically and behaviorally maintain sustained bouts of swimming (reviewed in Norin and Clark, 2016; Lear et al., 2019). Steady-state swimming is also a simplified representation of activity costs for wild fishes, which is influenced by more complex physical maneuvers including costly directional turning and navigation of turbulent water flows (Schakmann et al., 2020).

As discussed in Section 2.2, the aforementioned metabolic metrics can have direct relevance to fish ecology and conservation, especially when measured across a range of relevant environmental conditions such as water temperature. However, these metrics provide even greater insights when estimates are generated in situ with free swimming fish in their ecosystems. Although metabolism is challenging to measure remotely, biologgers with integrated sensors can be calibrated with fish metabolism in the laboratory, and then deployed into wild fish to generate remote estimates. This has been done with electromyogram transmitters (Cooke et al., 2001; Hinch and Rand, 1998), and more recently accelerometer and temperature transmitters using swim tunnel respirometer calibrations (Brodie et al., 2016; Brownscombe et al., 2017b; Thiem et al., 2018; Wilson et al., 2013). For example, Brodie et al. (2016) used this approached to generate real-world estimates to update the metabolic component of bioenergetics models. Brownscombe et al. (2017b) estimated metabolic rate in relation to habitat features and determined that the timing of fish foraging was likely driven by their temperature-specific AS, generating insights into the drivers of fish consumption. Accelerometers are available as transmitters, which provide coarse-scale activity data but enable remote data collection, or as loggers, which can measure fine scale behaviors (including foraging; Brownscombe et al., 2014), but require recapture of the fish to acquire the data, which makes field applications more challenging. In some cases, with larger animals, external pop-off packages may be used for recovery (Whitney et al., 2016), or tag return reward programs (Raby et al., 2018).

As discussed in Section 3.1, heart rate measurement can provide a more comprehensive estimate of metabolic rate beyond active metabolism, but its measurement has been challenging historically, requiring invasive procedures and controlled settings (Farrell and Jones, 1992). More recently, biologgers have been developed and calibrated to measure heart rate in free swimming fish (Davidsen et al., 2019; Hvas et al., 2020; Muller et al., 2020; Prystay et al., 2017; Shen et al., 2020; Svendsen et al., 2021). Effective application relies heavily on proper, secure positioning of the logger in proximity to the pericardial cavity to acquire high quality data. Perhaps the biggest limitation of these loggers is that they only measure heart rate, but stroke volume also modulates cardiac output and metabolism (Farrell et al., 2009; Farrell and Jones, 1992), and therefore this method may be inaccurate in scenarios where stroke volume is heavily modulated. Another limitation is that, unlike accelerometers, which are available with transmitter technology, the only currently available heart rate tags store information in memory, and therefore the fish must be recaptured to acquire the data.

3.3 Characterizing growth in fishes

At its simplest, fish growth can be characterized by changes in weight and/or length by gravimetric measurement at multiple time points (Crane et al., 2016; Hopkins, 1992). These metrics are often used to derive useful measures of growth in relation to certain conditions or environmental characteristics through an estimation of instantaneous growth rates (reviewed in Hopkins, 1992) as well as the von Bertalanffy growth function (reviewed in Essington et al., 2001). Notably, each of these metrics are associated with varying temporal resolutions and assumptions that may make each index more appropriate for a particular context. For example, instantaneous growth rate models exponential growth and is not suitable for life stages where growth has plateaued or is limited (e.g., older/adult fish; Hopkins, 1992). Estimates of age, which can be useful in growth measurements, can be made through determining growth rings in the bony structures of fishes including the otolith of the inner ear, scales, and the cleithrum (Maceina et al., 2007). These direct or modeled measures of fish growth are used extensively in bioenergetic modeling as a primary input (discussed in Section 4).

Another commonly used approach in measuring fish growth patterns is the characterizing tissue RNA:DNA ratios, which are thought to reflect tissue levels of protein biosynthesis (reviewed in Houlihan et al., 1993). Indeed, RNA:DNA ratios have been shown to correspond positively with fish growth rate in empirical trials (Buckley, 1984; Caldarone et al., 2003) and, consequently, have been used in determining fish growth rates across a variety of species and settings (e.g., Chételat et al., 2021; Mercaldo-Allen et al., 2006; Pottinger et al., 2011; Tanaka et al., 2007) including bioenergetics applications (Sewall et al., 2021; Zhou et al., 2001). This method provides an accommodating and simple proxy approach to measuring protein synthesis, and thus growth, where measuring the latter may be challenging (Houlihan et al., 1993). However, in order to ensure more comparable and replicable results among studies, the use of standardization of fluorometric analyses has been suggested (Caldarone et al., 2006). Reproductive investments can be measured using simple weighing of gonadal tissues (i.e., gonadosomatic index; Valdés et al., 2004; Brewer et al., 2008) or by measuring some of the egg properties (e.g., total numbers, energy content; reviewed in Lambert et al., 2003). Further, BIA meters has also been suggested as a tool to track reproductive state throughout the course of the year (Fitzhugh et al., 2010).

4   Modeling approaches

Ecologists have been modeling bioenergetics since at least the 1960s, and there now exists a number of alternative model formulations. These models generally describe the processes outlined in Section 2, but differ in terms of their assumptions, what inputs they require, and what outputs they provide. Most models fit with two approaches, ones that rely on the ecophysiological responses at the organismal level, which are used to estimate the Scope For Growth (SFG) of the individual based on a budget of ingested energy and metabolic costs derived from observed physiological responses (sensu Fry, 1971; Winberg, 1956), and ones that rely on fundamental principles of chemistry, physics, thermodynamics, and evolution, aiming for a mechanistic description of bioenergetics. This section showcases three bioenergetic models that differ in their structures and assumptions: the Wisconsin Energy Budget (hereafter WEB, (Kitchell et al., 1974, 1977), based on the SFG approach; the Dynamic Energy Budget (DEB, (Kooijman, 2010) model, based on a mechanistic approach; and the Physiological Energy Budget (PEB; Sibly et al., 2013) which is also mechanistic but differs from DEB in that its parameters are measurable biological quantities (e.g., Boult et al., 2018; Boyd et al., 2020c). This section also discusses how these models can be scaled up from the individual to the population scale, with particular emphasis on the use of Individual-Based Models (IBM).

4.1   Wisconsin energy budget

The most widely known fish bioenergetics modeling framework is termed the Wisconsin bioenergetics model, first proposed by Dr. James Kitchell at the University of Wisconsin-Madison (Kitchell et al., 1974, 1977). The conceptual model looks to describe specific growth rates by subtracting the energetic costs associated with respiration (standard/active metabolism and SDA), egestion, and excretion from the food consumption rate. Both the respiration and consumption submodels are size- and temperature-dependent, while specific dynamic action, egestion and excretion are typically modeled to be a loss that is a set proportion of the energy associated with ingested food. The model makes use of a P-value, corresponding to the proportion of the theoretical maximum food consumption (C_{max}), to balance the energetic budget. The P-value ranges between 0 and 1 and has often been used as an indicator of habitat quality (Rice et al., 1983), with higher values indicative of greater food abundance. While the works of Kitchell et al. were highly original, the authors relied heavily on fish physiology theory with regards to the allometry of food consumption (Brett, 1971; Elliott, 1976) and respiration (Winberg, 1956). The Kitchell et al. (1977) publication led to the creation of the first fish bioenergetics modeling software (Hewett and Johnson, 1987), which popularized the Wisconsin bioenergetics modeling approach and made fish bioenergetics modeling accessible for 11 species at the time. Since then, three other software versions have been developed (Hanson et al., 1997; Hewett and Johnson, 1992), including the most recent Fish Bioenergetics 4 (Deslauriers et al., 2017), which allows growth and food consumption modeling with 73 species. Over the years, the Wisconsin bioenergetics framework has expanded to allow for contaminant accumulation (PCBs, MMHg; Jackson and Schindler, 1996; Trudel and Rasmussen, 1997; Van Walleghem et al., 2013) and nutrient regeneration (N and P; Kraft, 1992) to be modeled and has the capacity to account for energetic reproductive costs and mortality at the population level. However, WEB does not account for recruitment and cannot be applied to population dynamics studies on its own. Rather, fish are modeled as individuals when using the Wisconsin bioenergetics framework where fish growth or food consumption can be scaled up to the population level using Individual Based Models (see Section 4.4) to infer the impact of an environment at the population level (Rose et al., 1999, 2013). More recently, the use of the Wisconsin energy budget has been applied to elucidate the energetic needs of invasive species (Cerino et al., 2013; Cooke and Hill, 2010) and to evaluate the potential impacts climate change on fish populations (Christianson and Johnson, 2020; Holsman et al., 2019).

4.2 Dynamic energy budget

The Dynamic Energy Budget (DEB) theory is a mechanistic metabolic theory that describes the uptake of energy from the environment and its use for maintenance, development, reproduction, and growth throughout the life cycle of the individual (Kooijman, 2010). Under the assumption that the general processes that govern the organization of metabolism are not species-specific, DEB is built using general principles of chemistry, physics, thermodynamics, and evolution (Jusup et al., 2017; Sousa et al., 2010). This level of generalization and formalism has allowed the broad application of the standard DEB model to more than 2800 species (Add My Pet Portal, 2021). Furthermore, the standard DEB model can be adapted to include species-specific requirements and metabolic processes to simulate specific responses such as the effects of stressors or toxicants on metabolism (Jager, 2020).

The flow of energy in the standard DEB model involves three state variables: reserves, structure, and maturity/reproduction buffer, depending on the life cycle stage of the organism (Fig. 3). In brief, the fraction of the food that is assimilated is incorporated into reserves, a state variable that includes

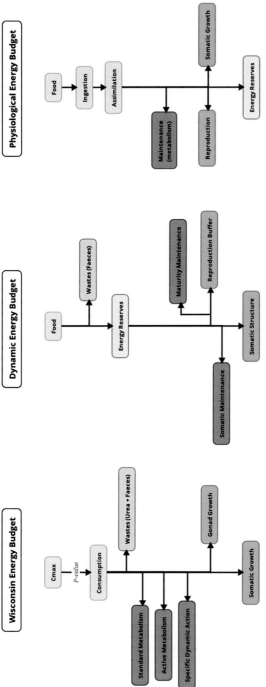

FIG. 3 Conceptual diagrams outlining the structure of three common bioenergetics modeling approaches, including (from left to right) the Wisconsin Energy Budget model, Dynamic Energy Budget model, and Physiological Energy Budget model. Energy intake nodes are marked in green, wastes in gray, energetic costs in red, productive allocation in teal, and energy reserves in yellow.

all compounds that do not require maintenance and are ready to be used as an energy source. The energy in the reserves is mobilized according to the kappa rule: a fraction of this energy (κ) is directed towards somatic maintenance and growth and the remainder (1-κ) toward maturity maintenance and development/reproduction (Fig. 3). All metabolic processes are affected by temperature using an Arrhenius function (Kooijman, 2010). In addition to temperature, the model requires information about food availability, which could be parameterized in different ways, e.g., as a range from 0 to 1, representing starvation and ad libitum feeding, respectively (analogous to the P-value in the Wisconsin model); or as a time series of food availability, depending on available information and specific knowledge about the species. Additional forcing variables could be added to the model depending on species-specific requirements and the purpose of the modeling exercise.

One of the key components of DEB modeling is the parameterization. Although some of the DEB parameters can be estimated using specific experiments (van der Veer et al., 2006), the most current approach is to use mathematical estimation (Filgueira et al., 2020; Lika et al., 2011). The rationale behind mathematical calibration is that a collection of all available data for the species and simultaneous estimation of all DEB parameters would provide a robust parameterization that would minimize the impact of uncertainty that any specific experiment could generate (Lika et al., 2011). Furthermore, this global parameterization is aligned with the general aim of generating a mechanistic model that can be applied to the full range of environmental conditions and used to explore hypothetical scenarios.

Although most DEB research has been carried out on bivalves, it has also been used to simulate metabolic-related responses in fishes, such as growth and other life history traits (Lavaud et al., 2019; Pecquerie et al., 2011), species distribution (Huret et al., 2018; Raab et al., 2013), toxicological effects (Klok et al., 2014; Mounier et al., 2020), and physiological responses (Kooijman and Lika, 2014; van der Veer et al., 2009), the latter of which are crucial to establish a link with other ecosystem-level processes (e.g., Ren et al., 2012; Serpa et al., 2013). Furthermore, DEB theory can be used to reconstruct the feeding history and habitat of an organism (Filgueira et al., 2016; Jusup et al., 2014) or to simulate the effects of hypothetical habitats on bioenergetics, which can be used to project climate change scenarios (Mangano et al., 2019; Teal et al., 2012).

4.3 Physiological energy budget

In 2013, Sibly et al. introduced a dynamic energy budget model—similar but distinct to the DEB model described in the previous section—that has since been labeled the Physiological Energy Budget (PEB). PEB was first developed for use in individual-based models (IBMs) but can be implemented in any simulation framework (e.g., Watson et al., 2020). The model has been applied in a number of contexts and to a wide range of taxa including fish (Boyd et al., 2020c; Mintram et al., 2020; Watson et al., 2020), earthworms (Johnston et al., 2014a,b), porpoises (Nabe-Nielsen et al., 2014), butterflies, and elephants (Boult et al., 2018).

In broad terms, PEB describes the acquisition of energy from food in the environment and its allocation to the processes outlined in Section 2 (as in Fig. 1). If there is sufficient food intake, energy is allocated in the order of maintenance, growth and/or reproduction, and energy storage. If there is a shortfall, growth and/or reproduction rates are reduced accordingly until reserves reach a critical threshold below which all energy is allocated to maintenance. Where energy intake does not cover the costs of maintenance (e.g., when food is short or during periods of fasting), energy is mobilized from the reserves until depleted at which point the animal is considered dead. Rates of energy acquisition and allocation scale with body mass and temperature according to established theoretical relationships.

The key selling points of the PEB approach are its interpretability and consistency with knowledge of animal physiology. The model's parameters are all measurable quantities with biological meaning. This enables simple parameter estimation by fitting its constituent equations to the relevant data individually (i.e., there is no need for "global" calibration as in DEB). PEB also relaxes some of the assumptions made by DEB; in particular, the "kappa" rule whereby a fixed proportion of energy is allocated to reproduction throughout the animal's life. Instead, the modeler is free to decide how surplus energy (i.e., everything left after maintenance is covered) is apportioned between growth and reproduction. For example, Boyd et al. (2020a) assumed that surplus energy is allocated in equal proportion to growth and reproduction, whereas Watson et al. (2020) considered two scenarios in which growth or reproduction were given priority.

4.4 From the individual to the population

In the previous sections, we discussed how bioenergetics models, whatever their specific formulation and assumptions, describe the ways in which individual organisms respond to their environment. For many applications, however, it is useful to understand how these organismal responses manifest at higher levels of organization (e.g., the population or community levels). A popular way to scale up from individual-level responses to higher level phenomena is to use Individual-Based Models (IBMs; sometimes called agent-based models) (DeAngelis and Grimm, 2014; Uchmanski and Grimm, 1996). In IBMs, the system (e.g., a population or community) is represented by its constituent parts (the individuals). The individuals are each characterized by a set of "state variables," such as body size, energy reserves, or geographical location. Detailed models are constructed that describe how the individuals' state variables change in response to their environment and to one another. Bioenergetics models, for example, may be used to describe how body sizes and reproductive outputs change in response to prey availability and temperature. Indeed, DEB (Martin et al., 2012; Pethybridge et al., 2013) and the Wisconsin model (Rose et al., 2013) have been implemented in IBMs, and PEB was originally developed for this purpose (Boyd et al., 2020a,b,c; Sibly et al., 2013). System level properties are obtained by summarizing the state variables of the individuals. For example, spawning stock biomass is the sum of adult body masses, and spatial distribution is a summary of the locations of all individuals.

Bioenergetics models were first merged with IBMs to study fish populations. Early focus was on recruitment: a number of studies looked at the growth rates and survival of young-of-the-year fish, and whether increased mortality could be offset by more food for the survivors (Bartsch and Coombs, 2001; Bartsch et al., 2004; Rose et al., 1999; Scheffer et al., 1995). Since then, bioenergetics IBMs have been developed and applied in a number of contexts. Examples include: predicting population dynamics as a function of food availability and temperature (Politikos et al., 2015b); coupling bioenergetics and optimal foraging submodels to predict the spatial distribution of exploited fish stocks (Boyd et al., 2020c; Politikos et al., 2015a); predicting the combined effects of food availability and chemical exposure on population dynamics (Mintram et al., 2020); estimating the consumption of zooplankton (Utne et al., 2012); assessing potential consequences of future climate and management scenarios for fish stocks and the associated fisheries (Boyd et al., 2020a); and simulating the effects of modified stream flows on the growth and spatial distribution of salmonids (Phang et al., 2016).

4.5 Concluding remarks

In this section, we discussed three bioenergetics modeling approaches that are widely used to address a number of ecological, biological, and physiological questions. Through the use of IBMs it is possible to transpose these questions from the individual to the population level and thus allow for emergent properties to arise (Grimm et al., 2017). This approach is particularly insightful as a hypothesis generator, which can then inform future experimental work and empirical field studies. Importantly, although we presented here three of the most popular models used for bioenergetics modeling of aquatic organisms, others exist and have been used to answer similar questions (e.g., Neill et al., 2004). One observation that emerges from this section is that users are confronted with a choice to make with regards to the best model to use for their specific application. While all three modeling approaches have their own assumptions and data requirements, they all look to answer the same question, that is, how does the energy consumed translate into growth. The described models range in the programming language they use (R, MatLab, STELLA) and the user-friendliness of the model implementation. As it stands, it remains unknown if all models presented here generate similar growth outputs when confronted with the same inputs. As such, an informed decision about the trade-offs associated with each model is necessary before using any given model. The previous section serves to expose key publications that we encourage users new to bioenergetics modeling to explore before deciding on the best approach to use based on the study objectives.

5   Applications

Bioenergetics is a fundamental process underpinning organismal life and fitness and thus has direct relevance to management and conservation. This is particularly salient for fish in which there is a strong relationship between body size and fecundity (a driver of population-level processes) as well as the manifold effects of environmental conditions (especially water temperature) on all facets of organismal and ecosystem function. Here, we briefly review some examples where bioenergetics approaches have been the contexts of invasive species and climate change impacts, as well as to inform fisheries planning and stocking.

5.1   Invasive species impacts: Lionfish in the Caribbean

Modeling the energetics of wild animals is common for native species but can be particularly useful for invasive species in an effort to understand the potential for range expansion by considering the abiotic (e.g., temperature; Brett and Groves, 1979) and biotic (e.g., food availability; Liss et al., 2013) constraints of novel environments (Elith et al., 2010). Moreover, energetics provides insight on the impacts of invasive species on other organisms as well as ecosystems (e.g., via nutrient cycling by invasive species to assess impacts on its new habitat; Lennox et al., 2015). Lionfish are one of the most well-known invasive species given their extensive impacts on coastal marine ecosystems in the Caribbean Sea (Betancur et al., 2011). Bioenergetics approaches have been used to understand and quantify the impacts of lionfish on fish communities and trophic ecology. Côté and Maljković (2010) determined that lionfish consumed native fish at an average rate of 1.44 kills/h emphasizing their voracious predatory behavior. Cerino et al. (2013) created a bioenergetics model for two species of lionfish and then used it to estimate the potential impact of these predators on a reef fish community. Laboratory experiments and the data from Côté and Maljković (2010) were used to parameterize a population-level model where simulations were used to estimate annual consumption. More recently, Steell et al. (2019) investigated the relationship between SDA, scope for activity, metabolic phenotype, temperature and feeding frequency in lionfish. Respirometry revealed that maximum metabolic rate occurred during digestion (not during exhaustive exercise as in more athletic species) and energetic needs were elevated at higher temperatures. This revealed that frequent feeding had a lower and more consistent cost than consuming a single meal but increased the peak of SDA. Notably, lionfish may consume more prey as oceans warm with climate change. It is also possible to model the energetic consequences of different control methods but there are no examples of how such approaches have yet to be applied to lionfish.

5.2   Climate change in the Laurentian Great lakes

Water temperature has manifold effects on fish so it is clear that climate change will have dramatic effects. Given that energetics integrates environmental information and biological processes, it is a logical approach for understanding and predicting climate change impacts in ectotherms such as fish (Matzelle et al., 2015; Tomlinson et al., 2014). In the Laurentian Great Lakes of North America, climate change is predicted to have diverse impacts arising from changes in precipitation and temperature (Mortsch and Quinn, 1996) and will act as a long-term stressor on fish populations (Collingsworth et al., 2017). In fact, some of the earliest work on fish energetics and climate change was conducted in the Great Lakes, where Hill and Magnuson (1990) used bioenergetics simulations to reveal that food web dynamics and the potential for thermoregulation will greatly influence the direction and magnitude of changes in fish growth as the climate warms. Jones et al. (2006) used mechanistic models that couple habitat conditions to population demographics with walleye (*Sander vitreus*). Notably, they revealed that the combined effect of changes in temperature, river hydrology, lake levels, and light penetration can result in very different population outcomes depending on which factors are considered in models emphasizing the diverse and complicated effects of climate change on fish. There is already evidence of changing fish distributions (Cline et al., 2013) as well as impacts on the growth of native fish species (Brandt et al., 2002; Kao et al., 2015). According to bioenergetics modeling, invasive parasitic sea lamprey populations will experience enhanced growth in Lakes Michigan and Huron thus necessitating changes in control efforts (Lennox et al., 2020). Bioenergetics approaches have become a standard tool to predict the effects of climate change on fish and fisheries in the Great Lakes and are increasingly being embraced by the management community to inform decision making related to fisheries planning and restoration in a changing world (Fussell et al., 2016; Lynch et al., 2010).

5.3   Stocking decisions related to freshwater fisheries management

Stocking to supplement wild fish populations or to create and maintain introduced fish populations represent a common tool for the management of freshwater fisheries (Arlinghaus et al., 2016). Stocking is best accomplished when based on science where modeling informs aspects of stocking such as species/ strain selection, the size of stocked fish, and the quantity of stocked fish (Molony et al., 2005). Bioenergetics is relevant to stock enhancement given the inherent links between energetics and the growth and survival of stocked fish as well as their interaction with wild conspecifics or other species (Hansen et al., 1993). One of the earlier applications of bioenergetics to stocking was focused on exploring its utility for salmonine management in Lake Superior. Jones et al. (1993) and Negus (1995) used bioenergetic modeling to reveal that caution should be exercised in future stocking and emphasized the value of the approach for evaluating trade-offs of different management strategies. For example, modeling revealed that there were conflicts between enhancement of native lake trout populations and other (mostly introduced) salmonids based on lack of adequate forage base. Wahl (1999) demonstrated that the source population of muskellunge can affect survival and growth because of thermal variation in bioenergetic variables among muskellunge populations. As such, bioenergetics are often used in hatcheries to make decisions about source populations and feeding in an effort to achieve growth targets (e.g., Csargo et al., 2012). Energetics modeling can also be used to assess the success of stocking. Grausgruber and Weber (2021) used bioenergetics modeling to estimate post-stocking mortality of age-0 walleye and revealed that predators consumed large proportions of stocked walleye. These findings suggested that alternative stocking practices should be considered to achieve recreational fisheries management objectives. Stocking decisions will be even more critical when dealing with conservation stocking focused on the recovery of imperiled species and there is already evidence that bioenergetics is an important approach to inform such decisions. For example, Luecke et al. (2016) used bioenergetics modeling to simulate growth and production of endangered Snake River sockeye salmon to inform stocking decisions and habitat management (e.g., nursery lake fertilization).

6 Conclusions and future directions

Built on the foundations of the fundamental laws of thermodynamics, bioenergetics offers a valuable mechanistic framework linking fish biology with ecological conditions, which has a wide variety of applications for applied resource management and conservation. From our perspective, fish bioenergetics is in a major transitional state from being historically a largely lab-based and aquaculture-focused field, to informing the ecology and conservation of wild fish and their ecosystems. This is not to discount the many insightful studies that have been conducted for decades in this realm, but to say that there is a growing trend toward the study of ecological energetics (Tomlinson et al., 2014). Historically, a major limitation on these efforts has been our capacity to measure bioenergetics in relation to the diverse conditions they experience in the wild, with certain bioenergetic aspects (e.g., consumption) posing greater challenges. Many of these are being overcome with a rapidly growing suite of tools and tactics for quantifying the various components of fish bioenergetics in both the lab and field, discussed here in Section 4. Technologies such as accelerometers and heart rate loggers offer exciting opportunities to remotely measure fish energetics. Further, combined approaches (e.g., energetic measurement through fish tracking and measures of prey consumption through stable isotopes) are likely to provide the greatest insights. These new methods offer opportunities to characterize the complex relationships between environmental factors such as temperature and resource availability to fish bioenergetics (e.g., metabolism, consumption) to understand how these factors influence fish distribution, behavior, growth, reproduction, and population dynamics.

The individual components of bioenergetics, including consumption, metabolism, and growth present opportunities on their own to inform key ecological interactions, with particularly widespread use of metabolism to inform fish ecology. Considered in its entirety, bioenergetics is complex, and modeling approaches address these complexities by integrating the various components in a diversity of creative ways, discussed in Section 3. Despite our growing capacity to measure fish bioenergetics, there will also be challenges and limitations on that front, and modeling addresses these by integrating best known estimates and accounting for uncertainty. It is also difficult to effectively measure a fish's biological fitness (i.e., lifetime reproductive output and success) to link energetics with relevant outcomes. Modeling therefore plays a key role in building this link, as well as scaling these outputs to the population level through tools such as individual-based models. Ultimately, the greatest utility of bioenergetics will likely continue to arise from further advances in modeling approaches, which synthesize the complexities of various aspects of the bioenergetics equation to generate useful outputs.

A variety of applications of fish bioenergetics were referenced throughout this chapter, with more extensive discussion of specific case study applications to lionfish consumption estimates and their relevance to environmental impacts and invasiveness, responses of Great Lakes fishes to climate change, and guiding fish stocking practices. These present just a few of many example applications for fish bioenergetics. As a mechanistic approach linking environmental conditions to fish populations, bioenergetics may be especially useful for predicting the effects of climate change on fish species ranges (Duncan et al., 2020; Lefevre et al., 2021; Payne et al., 2016) and community composition (e.g., warmwater species overtaking cold water species; Hansen et al., 2017; Van Zuiden and Sharma, 2016). Overall, perhaps the greatest documented success in application is informing fisheries management, especially through estimating growth rates to determine harvest regulations (Chipps and Wahl, 2008; Hansen et al., 1993; Hartman and Kitchell, 2008). However, despite the logical applicability of many examples cited in this chapter, there is far less evidence of the direct utility of bioenergetics to other aspects of resource management such as prioritizing conservation areas, habitat protection, or restoration, to which bioenergetics should be highly relevant. This may be, in part, because most study efforts to date are more fundamental in focus, and less commonly aim to generate directly usable model outputs for specific management needs. This is understandable, as there is a need for a foundational fabric prior to useful applications in any scientific field. This is also not unique to energetics; it is a common phenomenon across numerous disciplines, and there is an increasingly recognized need for knowledge cogeneration approaches to produce more tractable applied outcomes (e.g., Nel et al., 2016; Nguyen et al., 2018; Young et al., 2016). It is worth noting that direct applications to resource management often go undocumented and are surely more prevalent than the available literature indicates.

The rapid environmental changes we are eliciting during the Anthropocene are posing increasing conservation challenges, in particular in decision making with uncertainty of future conditions due to factors such as climate change. To address this, there is a need to use mechanistic models that are built upon causal drivers of ecological interactions to produce generalizable outputs. Absent from a unified theory of ecology, bioenergetics presents one of the most promising mechanistic approaches to address these uncertainties and generate educated predictions for decision-making. Continued development and application of fish bioenergetics theory, measurement, and modeling with applied resource management needs in mind will surely make this field integral to achieving a good Anthropocene.

References

Add My Pet Portal, 2021. Available from https://www.bio.vu.nl/thb/deb/deblab/add_my_pet/. (accessed 29 March 2021).

Albo-Puigserver, M., Muñoz, A., Navarro, J., Coll, M., Pethybridge, H., Sánchez, S., Palomera, I., 2017. Ecological energetics of forage fish from the Mediterranean sea: seasonal dynamics and interspecific differences. Deep. Res. Part II Top. Stud. Oceanogr. 140, 74–82. Elsevier Ltd https://doi.org/10.1016/j.dsr2.2017.03.002.

Allen, D., Rosenfeld, J., Richards, J., 2016. Physiological basis of metabolic trade-offs between growth and performance among different strains of rainbow trout. Can. J. Fish. Aquat. Sci. 73 (10), 1493–1506. https://doi.org/10.1139/cjfas-2015-0429.

Amundsen, P.A., Sánchez-Hernández, J., 2019. Feeding Studies Take Guts—Critical Review and Recommendations of Methods for Stomach Contents Analysis in Fish. Blackwell Publishing Ltd, https://doi.org/10.1111/jfb.14151. December 1.

Arlinghaus, R., Lorenzen, K., Johnson, B.M., Cooke, S.J., Cowx, I.G., 2016. Management of freshwater fisheries: addressing habitat, people and fishes. Freshw. Fish. Ecol. 1, 557–579.

Auer, S.K., Dick, C.A., Metcalfe, N.B., Reznick, D.N., 2018. Metabolic rate evolves rapidly and in parallel with the pace of life history. Nat. Commun. 9, 1–6. https://doi.org/10.1038/s41467-017-02514-z.

Auer, S.K., Bassar, R.D., Turek, D., Anderson, G.J., McKelvey, S., Armstrong, J.D., Nislow, K.H., Downie, H.K., Morgan, T.A.J., McLennan, D., Metcalfe, N.B., 2020. Metabolic rate interacts with resource availability to determine individual variation in microhabitat use in the wild. Am. Nat. 196 (2), 132–144. https://doi.org/10.1086/709479.

Ball, J.R., Esler, D., Schmutz, J.A., 2007. Proximate composition, energetic value, and relative abundance of prey fish from the inshore eastern Bering Sea: implications for piscivorous predators. Polar Biol. 30 (6), 699–708. Springer https://doi.org/10.1007/s00300-006-0227-1.

Bangley, C.W., Curtis, T.H., Secor, D.H., Latour, R.J., Ogburn, M.B., 2020. Identifying important juvenile dusky shark habitat in the Northwest Atlantic ocean using acoustic telemetry and spatial modeling. Mar. Coast. Fish. 12 (5), 348–363. John Wiley and Sons Inc https://doi.org/10.1002/mcf2.10120.

Bartsch, J., Coombs, S.H., 2001. An individual-based growth and transport model of the early life-history stages of mackerel (*Scomber scombrus*) in the eastern North Atlantic. Ecol. Model. 138 (1–3), 127–141.

Bartsch, J., Coombs, S.H., Reid, D., 2004. Simulation of mackerel (*Scomber scombrus*) recruitment with an individual-based model and comparison with field data. Fish. Ocean. 13 (6), 380–391.

Battley, E.H., 1995. The advantages and disadvantages of direct and indirect calorimetry. Thermochim. Acta 250 (2), 337–352. Elsevier https://doi.org/10.1016/0040-6031(94)01963-H.

Bayse, S.M., Regish, A.M., McCormick, S.D., 2018. Proximate composition, lipid utilization and validation of a non-lethal method to determine lipid content in migrating American shad *Alosa sapidissima*. J. Fish Biol. 92 (6), 1832–1848. Blackwell Publishing Ltd https://doi.org/10.1111/jfb.13624.

Bestley, S., Patterson, T.A., Hindell, M.A., Gunn, J.S., 2008. Feeding ecology of wild migratory tunas revealed by archival tag records of visceral warming. J. Anim. Ecol. 77 (6), 1223–1233. John Wiley & Sons, Ltd https://doi.org/10.1111/j.1365-2656.2008.01437.x.

Betancur, R.R., Hines, A., Arturo, A.P., Ortí, G., Wilbur, A.E., Freshwater, D.W., 2011. Reconstructing the lionfish invasion: insights into greater Caribbean biogeography. J. Biogeogr. 38 (7), 1281–1293. John Wiley & Sons, Ltd https://doi.org/10.1111/j.1365-2699.2011.02496.x.

Blažka, P., 1958. The anaerobic metabolism of fish. Physiol. Zool. 31 (2), 117–128. https://doi.org/10.1086/physzool.31.2.30155385.

Blewett, T.A., Wood, C.M., Glover, C.N., 2016. Salinity-dependent nickel accumulation and effects on respiration, ion regulation and oxidative stress in the galaxiid fish, *Galaxias maculatus*. Environ. Pollut. 214, 132–141. Elsevier Ltd https://doi.org/10.1016/j.envpol.2016.04.010.

Bligh, E., Dyer, W., 1959. A rapid method of total lipid extraction and purification. Can. J. Biochem. Physiol. 37 (8), 911–917. https://doi.org/10.1139/O59-099.

Boisclair, D., Leggett, W.C., 1989. The importance of activity in bioenergetics models applied to actively foraging fishes. Can. J. Fish. Aquat. Sci. 46 (11), 1859–1867. NRC Research Press Ottawa, Canada https://doi.org/10.1139/f89-234.

Bolduc, M., Lamarre, S., Rioux, P., 2002. A simple and inexpensive apparatus for measuring fish metabolism. Am. J. Physiol. Adv. Physiol. Educ. 26 (1–4), 129–132. American Physiological Society https://doi.org/10.1152/advan.00038.2001.

Borowiec, B.G., Darcy, K.L., Gillette, D.M., Scott, G.R., 2015. Distinct physiological strategies are used to cope with constant hypoxia and intermittent hypoxia in killifish (*Fundulus heteroclitus*). J. Exp. Biol. 218 (8), 1198–1211. Company of Biologists Ltd https://doi.org/10.1242/jeb.114579.

Boult, V.L., Quaife, T., Fishlock, V., Moss, C.J., Lee, P.C., Sibly, R.M., 2018. Individual-based modelling of elephant population dynamics using remote sensing to estimate food availability. Ecol. Modell. 387 (June), 187–195. Elsevier https://doi.org/10.1016/j.ecolmodel. 2018.09.010.

Bouyoucos, I.A., Weideli, O.C., Planes, S., Simpfendorfer, C.A., Rummer, J.L., 2018. Dead tired: evaluating the physiological status and survival of neonatal reef sharks under stress. Conserv. Physiol. 6 (1). https://doi.org/10.1093/conphys/coy053. Oxford University Press.

Boyd, R., Thorpe, R., Hyder, K., Roy, S., Walker, N., 2020a. Potential Consequences of Climate and Management Scenarios for the Northeast Atlantic Mackerel Fishery (August), https://doi. org/10.3389/fmars.2020.00639.

Boyd, R., Walker, N., Hyder, K., Thorpe, R., Roy, S., Sibly, R., 2020b. SEASIM-NEAM: a spatially-explicit agent-based SIMulator of North East Atlantic mackerel population dynamics. MethodsX 7 (2014), 100892. Elsevier B.V https://doi.org/10.1016/j.mex.2020.100892.

Boyd, R.J., Sibly, R., Hyder, K., Walker, N., Thorpe, R., Roy, S., 2020c. Simulating the summer feeding distribution of Northeast Atlantic mackerel with a mechanistic individual-based model. Prog. Oceanogr. 183 (February), 102299. Elsevier https://doi.org/10.1016/j.pocean. 2020.102299.

Brandt, S.B., Mason, D.M., Mccormick, M.J., Lofgren, B., Hunter, T.S., 2002. Climate change: implications for fish growth performance in the Great Lakes. In: American Fisheries Society Symposium. 32, pp. 61–76.

Breck, J.E., 2008. Enhancing bioenergetics models to account for dynamic changes in fish body composition and energy density. Trans. Am. Fish. Soc. 137 (1), 340–356. Wiley https://doi. org/10.1577/t05-240.1.

Breck, J.E., 2014. Body composition in fishes: body size matters. Aquaculture 433, 40–49. Elsevier https://doi.org/10.1016/J.AQUACULTURE.2014.05.049.

Brett, J.R., 1964. The respiratory metabolism and swimming performance of young sockeye salmon. J. Fish. Board Canada 21 (5), 1183–1226. https://doi.org/10.1139/f64-103.

Brett, J.R., 1971. Energetic responses of salmon to temperature. A study of some thermal relations in the physiology and freshwater ecology of sockeye salmon (Oncorhynchus nerkd). Integr. Comp. Biol. 11 (1), 99–113. Oxford Academic https://doi.org/10.1093/icb/11.1.99.

Brett, J., 1995. Ecological energetics. In: Groot, C., Margolis, L., Clark, W. (Eds.), Physiological Ecology of Pacific Salmon. University of British Columbia Press, Vancouver, pp. 1–68.

Brett, J.R., Groves, T.D.D., 1979. Physiological energetics. In: Fish Physiology., https://doi.org/ 10.1016/S1546-5098(08)60029-1.

Brewer, S.K., Rabeni, C.F., Papoulias, D.M., 2008. Comparing histology and gonadosomatic index for determining spawning condition of small-bodied riverine fishes. Ecol. Freshw. Fish 17 (1), 54–58. John Wiley & Sons, Ltd https://doi.org/10.1111/J.1600-0633.2007.00256.X.

Brewster, L.R., Dale, J.J., Guttridge, T.L., Gruber, S.H., Hansell, A.C., Elliott, M., Cowx, I.G., Whitney, N.M., Gleiss, A.C., 2018. Development and application of a machine learning algorithm for classification of elasmobranch behaviour from accelerometry data. Mar. Biol. 165 (4), 62. Springer Verlag https://doi.org/10.1007/s00227-018-3318-y.

Brodie, S., Taylor, M.D., Smith, J.A., Suthers, I.M., Gray, C.A., Payne, N.L., 2016. Improving consumption rate estimates by incorporating wild activity into a bioenergetics model. Ecol. Evol. 6 (8), 2262–2274. https://doi.org/10.1002/ece3.2027.

Brosset, P., Fromentin, J.M., Ménard, F., Pernet, F., Bourdeix, J.H., Bigot, J.L., Van Beveren, E., Pérez Roda, M.A., Choy, S., Saraux, C., 2015. Measurement and analysis of small pelagic fish condition: a suitable method for rapid evaluation in the field. J. Exp. Mar. Bio. Ecol. 462, 90–97. Elsevier https://doi.org/10.1016/j.jembe.2014.10.016.

Brown, J.H., Gillooly, J.F., Allen, A.P., Savage, V.M., West, G.B., 2004. Toward a metabolic theory of ecology. Ecology 85 (7), 1771–1789. https://doi.org/10.1890/03-9000.

Brownscombe, J.W., Gutowsky, L.F.G., Danylchuk, A.J., Cooke, S.J., 2014. Foraging behaviour and activity of a marine benthivorous fish estimated using tri-axial accelerometer biologgers. Mar. Ecol. Prog. Ser. 505, 241–251. https://doi.org/10.3354/mcps10786.

Brownscombe, J.W., Cooke, S.J., Algera, D.A., Hanson, K.C., Eliason, E.J., Burnett, N.J., Danylchuk, A.J., Hinch, S.G., Farrell, A.P., 2017a. Ecology of exercise in wild fish: integrating concepts of individual physiological capacity, behavior, and fitness through diverse case studies. Integr. Comp. Biol. 57, 281–292. https://doi.org/10.1093/icb/icx012.

Brownscombe, J.W., Cooke, S.J., Danylchuk, A.J., 2017b. Spatiotemporal drivers of energy expenditure in a coastal marine fish. Oecologia 183 (3), 689–699. Springer Berlin Heidelberg https://doi.org/10.1007/s00442-016-3800-5.

Brownscombe, J.W., Lennox, R.J., Danylchuk, A.J., Cooke, S.J., 2018. Estimating fish swimming metrics and metabolic rates with accelerometers: the influence of sampling frequency. J. Fish Biol. 93, 207–214. https://doi.org/10.1111/jfb.13652.

Brunnschweiler, J.M., Payne, N.L., Barnett, A., 2018. Hand feeding can periodically fuel a major portion of bull shark energy requirements at a provisioning site in Fiji. Anim. Conserv. 21 (1), 31–35. Blackwell Publishing Ltd https://doi.org/10.1111/acv.12370.

Bucking, C., Wood, C.M., 2008. The alkaline tide and ammonia excretion after voluntary feeding in freshwater rainbow trout. J. Exp. Biol. 211 (15), 2533–2541. The Company of Biologists https://doi.org/10.1242/jeb.015610.

Buckley, L.J., 1984. RNA-DNA ratio: an index of larval fish growth in the sea. Mar. Biol. 80 (3), 291–298. Springer-Verlag https://doi.org/10.1007/BF00392824.

Burnett, N.J., Hinch, S.G., Braun, D.C., Casselman, M.T., Middleton, C.T., Wilson, S.M., Cooke, S.J., 2014. Burst swimming in areas of high flow: delayed consequences of anaerobiosis in wild adult sockeye salmon. Physiol. Biochem. Zool. 87 (5), 587–598. https://doi.org/10.1086/677219.

Byrnes, E.E., Lear, K.O., Morgan, D.L., Gleiss, A.C., 2020. Respirometer in a box: development and use of a portable field respirometer for estimating oxygen consumption of large-bodied fishes. J. Fish Biol. 96 (4), 1045–1050. Blackwell Publishing Ltd https://doi.org/10.1111/jfb.14287.

Cai, L., Liu, G., Taupier, R., Fang, M., Johnson, D., Tu, Z., Huang, Y., 2014. Effect of temperature on swimming performance of juvenile *Schizothorax prenanti*. Fish Physiol. Biochem. 40 (2), 491–498. Kluwer Academic Publishers https://doi.org/10.1007/s10695-013-9860-0.

Caldarone, E.M., St. Onge-Burns, J.M., Buckley, L.J., 2003. Relationship of RNA/DNA ratio and temperature to growth in larvae of Atlantic cod *Gadus morhua*. Mar. Ecol. Prog. Ser. 262, 229–240. Inter-Research https://doi.org/10.3354/meps262229.

Caldarone, E.M., Clemmesen, C.M., Berdalet, E., Miller, T.J., Folkvord, A., Holt, G.J., Olivar, M.P., Suthers, I.M., 2006. Intercalibration of four spectrofluorometric protocols for measuring RNA/DNA ratios in larval and juvenile fish. Limnol. Oceanogr. Methods 4 (5), 153–163. American Society of Limnology and Oceanography Inc https://doi.org/10.4319/lom.2006.4.153.

Cech, J.J., 2011. Techniques in whole animal respiratory physiology introduction further reading apparatus and techniques. In: Encyclopedia of Fish Physiology: From Genome to Environment., https://doi.org/10.1016/B978-0-1237-4553-8.00128-3.

Cerino, D., Overton, A.S., Rice, J.A., Morris, J.A., 2013. Bioenergetics and trophic impacts of the invasive indo-Pacific lionfish. Trans. Am. Fish. Soc. 142 (6), 1522–1534. Taylor & Francis Group https://doi.org/10.1080/00028487.2013.811098.

Chabot, D., Koenker, R., Farrell, A.P., 2016a. The measurement of specific dynamic action in fishes. J. Fish Biol. 88 (1), 152–172. Blackwell Publishing Ltd https://doi.org/10.1111/jfb.12836.

Chabot, D., Steffensen, J.F., Farrell, A.P., 2016b. The determination of standard metabolic rate in fishes. J. Fish Biol. 88 (1), 81–121. https://doi.org/10.1111/jfb.12845.

Chételat, J., Shao, Y., Richardson, M.C., MacMillan, G.A., Amyot, M., Drevnick, P.E., Gill, H., Köck, G., Muir, D.C.G., 2021. Diet influences on growth and mercury concentrations of two salmonid species from lakes in the eastern Canadian Arctic. Environ. Pollut. 268, 115820. Elsevier Ltd https://doi.org/10.1016/j.envpol.2020.115820.

Chipps, S.R., Wahl, D.H., 2008. Bioenergetics modeling in the 21st century: reviewing new insights and revisiting old constraints. Trans. Am. Fish. Soc. 137 (1), 298–313. https://doi.org/10.1577/T05-236.1.

Christianson, K.R., Johnson, B.M., 2020. Combined effects of early snowmelt and climate warming on mountain lake temperatures and fish energetics. Arctic Antarct. Alp. Res. 52 (1), 130–145. Taylor and Francis Ltd https://doi.org/10.1080/15230430.2020.1741199.

Claireaux, G., Lefrançois, C., 2007. Linking environmental variability and fish performance: integration through the concept of scope for activity. Philos. Trans. R. Soc. B Biol. Sci. 362 (1487), 2031–2041. https://doi.org/10.1098/rstb.2007.2099.

Clark, T.D., Taylor, B.D., Seymour, R.S., Ellis, D., Buchanan, J., Fitzgibbon, Q.P., Frappell, P.B., 2008. Moving with the beat: heart rate and visceral temperature of free-swimming and feeding bluefin tuna. Proc. R. Soc. B Biol. Sci. 275 (1653), 2841–2850. Royal Society https://doi.org/10.1098/rspb.2008.0743.

Clark, T.D., Sandblom, E., Hinch, S.G., Patterson, D.A., Frappell, P.B., Farrell, A.P., 2010. Simultaneous biologging of heart rate and acceleration, and their relationships with energy expenditure in free-swimming sockeye salmon (Oncorhynchus nerka). J. Comp. Physiol. B Biochem. Syst. Environ. Physiol. 180 (5), 673–684. https://doi.org/10.1007/s00360-009-0442-5.

Clark, T.D., Jeffries, K.M., Hinch, S.G., Farrell, A.P., 2011. Exceptional aerobic scope and cardiovascular performance of pink salmon (Oncorhynchus gorbuscha) may underlie resilience in a warming climate. J. Exp. Biol. 214 (18), 3074–3081. https://doi.org/10.1242/jeb.060517.

Clark, T.D., Sandblom, E., Jutfelt, F., 2013. Aerobic scope measurements of fishes in an era of climate change: respirometry, relevance and recommendations. J. Exp. Biol. 216 (15), 2771–2782. https://doi.org/10.1242/Jeb.084251.

Clarke, A., Johnston, N.M., 1999. Scaling of metabolic rate with body mass and temperature in teleost fish. J. Anim. Ecol. 68 (5), 893–905. https://doi.org/10.1046/j.1365-2656.1999.00337.x.

Cline, T.J., Bennington, V., Kitchell, J.F., 2013. Climate change expands the spatial extent and duration of preferred thermal habitat for Lake superior fishes. PLoS One 8 (4), 62279. Public Library of Science https://doi.org/10.1371/journal.pone.0062279.

Collingsworth, P.D., Bunnell, D.B., Murray, M.W., Kao, Y.C., Feiner, Z.S., Claramunt, R.M., Lofgren, B.M., Höök, T.O., Ludsin, S.A., 2017. Climate change as a long-term stressor for the fisheries of the Laurentian Great Lakes of North America. Rev. Fish Biol. Fish. 27 (2), 363–391. Springer International Publishing https://doi.org/10.1007/s11160-017-9480-3.

Cooke, S.L., Hill, W.R., 2010. Can filter-feeding Asian carp invade the Laurentian Great Lakes? A bioenergetic modelling exercise. Freshw. Biol. 55 (10), 2138–2152. John Wiley & Sons, Ltd https://doi.org/10.1111/j.1365-2427.2010.02474.x.

Cooke, S.J., Bunt, C.M., Schreer, J.F., Wahl, D.H., 2001. Comparison of several techniques for mobility and activity estimates of smallmouth bass in lentic environments. J. Fish Biol. 58, 573–587. https://doi.org/10.1006/jfbi.2000.1473.

Cooke, S.J., Brownscombe, J.W., Raby, G.D., Broell, F., Hinch, S.G., Clark, T.D., Semmens, J.M., 2016. Remote bioenergetics measurements in wild fish: opportunities and challenges. Comp. Biochem. Physiol. A Mol. Integr. Physiol. 202, 23–37. https://doi.org/10.1016/j.cbpa.2016. 03.022.

Côté, I.M., Maljković, A., 2010. Predation rates of indo-pacific lionfish on bahamian coral reefs. Mar. Ecol. Prog. Ser. 404, 219–225. https://doi.org/10.3354/meps08458.

Cox, M.K., Hartman, K.J., 2005. Nonlethal estimation of proximate composition in fish. Can. J. Fish. Aquat. Sci. 62 (2), 269–275. NRC Research Press Ottawa, Canada https://doi.org/ 10.1139/f04-180.

Cox, G.K., Sandblom, E., Richards, J.G., Farrell, A.P., 2011. Anoxic survival of the Pacific hagfish (*Eptatretus stoutii*). J. Comp. Physiol. B Biochem. Syst. Environ. Physiol. 181 (3), 361–371. Springer https://doi.org/10.1007/s00360-010-0532-4.

Craig, J.F., 1977. The body composition of adult perch, *Perca fluviatilis* in windermere, with reference to seasonal changes and reproduction. J. Anim. Ecol. 46 (2), 617. JSTOR https:// doi.org/10.2307/3834.

Craig, J.F., Kenley, M.J., Talling, J.F., 1978. Comparative estimations of the energy content of fish tissue from bomb calorimetry, wet oxidation and proximate analysis. Freshw. Biol. 8 (6), 585–590. John Wiley & Sons, Ltd https://doi.org/10.1111/j.1365-2427.1978.tb01480.x.

Crane, D.P., Killourhy, C.C., Clapsadl, M.D., 2016. Effects of three frozen storage methods on wet weight of fish. Fish. Res. 175, 142–147. Elsevier https://doi.org/10.1016/j.fishres.2015.11.022.

Crossin, G.T., Hinch, S.G., 2005. A nonlethal, rapid method for assessing the somatic energy content of migrating adult Pacific Salmon. Trans. Am. Fish. Soc. 134 (1), 184–191. Wiley https://doi.org/10.1577/ft04-076.1.

Crossin, G.T., Hinch, S.G., Cooke, S.J., Welch, D.W., Patterson, D.A., Jones, S.R.M., Lotto, A.G., Leggatt, R.A., Mathes, M.T., Shrimpton, J.M., Van Der Kraak, G., Farrell, A.P., 2008. Exposure to high temperature influences the behaviour, physiology, and survival of sockeye salmon during spawning migration. Can. J. Zool. 86 (2), 127–140. https://doi.org/10.1139/Z07-122.

Csargo, I.J., Brown, M.L., Chipps, S.R., 2012. Application of a bioenergetics model for hatchery production: largemouth bass fed commercial feeds. N. Am. J. Aquac. 74 (3), 352–359. Taylor & Francis Group https://doi.org/10.1080/15222055.2012.675998.

Cuenco, M.L., Stickney, R.R., Grant, W.E., 1985a. Fish bioenergetics and growth in aquaculture ponds: I. Individual fish model development. Ecol. Modell. 27 (3–4), 169–190. https://doi. org/10.1016/0304-3800(85)90001-8.

Cuenco, M.L., Stickney, R.R., Grant, W.E., 1985b. Fish bioenergetics and growth in aquaculture ponds: II. Effects of interactions among, size, temperature, dissolved oxygen, unionized ammonia and food on growth of individual fish. Ecol. Modell. 27 (3–4), 191–206. https:// doi.org/10.1016/0304-3800(85)90002-X.

Davidsen, J.G., Dong, H., Linné, M., Andersson, M.H., Piper, A., Prystay, T.S., Hvam, E.B., Thorstad, E.B., Whoriskey, F., Cooke, S.J., Sjursen, A.D., Rønning, L., Netland, T.C., Hawkins, A.D., 2019. Effects of sound exposure from a seismic airgun on heart rate, acceleration and depth use in free-swimming Atlantic cod and saithe. Conserv. Physiol. 7 (1), coz020. https://doi.org/10.1093/conphys/coz020. Oxford University Press.

DeAngelis, D.L., Grimm, V., 2014. Individual-based models in ecology after four decades. F1000 Prime Rep. 6 (June), 39. https://doi.org/10.12703/P6-39.

Dempson, J.B., Schwarz, C.J., Shears, M., Furey, G., 2004. Comparative proximate body composition of Atlantic salmon with emphasis on parr from fluvial and lacustrine habitats. J. Fish Biol. 64 (5), 1257–1271. John Wiley & Sons, Ltd https://doi.org/10.1111/j.0022-1112.2004. 00389.x.

Deslauriers, D., Chipps, S.R., Breck, J.E., Rice, J.A., Madenjian, C.P., 2017. Fish bioenergetics 4.0: an R-based modeling application. Fisheries 42 (11), 586–596. https://doi.org/10.1080/03632415.2017.1377558.

Duncan, M.I., James, N.C., Potts, W.M., Bates, A.E., 2020. Different drivers, common mechanism; the distribution of a reef fish is restricted by local-scale oxygen and temperature constraints on aerobic metabolism. Conserv. Physiol. 8 (1), 1–16. https://doi.org/10.1093/conphys/coaa090.

Eldøy, S., Bordeleau, X., Lawrence, M., Thorstad, E., Finstad, A., Whoriskey, F., Crossin, G., Cooke, S., Aarestrup, K., Rønning, L., Sjursen, A., Davidsen, J., 2021. The effects of nutritional state, sex and body size on the marine migration behaviour of sea trout. Mar. Ecol. Prog. Ser. 665, 185–200. Inter-Research Science Center https://doi.org/10.3354/meps13670.

Eliason, E.J., Higgs, D.A., Farrell, A.P., 2008. Postprandial gastrointestinal blood flow, oxygen consumption and heart rate in rainbow trout (*Oncorhynchus mykiss*). Comp. Biochem. Physiol. A Mol. Integr. Physiol. 149 (4), 380–388. https://doi.org/10.1016/j.cbpa.2008.01.033.

Eliason, E.J., Clark, T.D., Hague, M.J., Hanson, L.M., Gallagher, Z.S., Jeffries, K.M., Gale, M.K., Patterson, D.A., Hinch, S.G., Farrell, A.P., 2011. Differences in thermal tolerance among sockeye salmon population. Science 332, 109–112.

Eliason, E.J., Wilson, S.M., Farrell, A.P., Cooke, S.J., Hinch, S.G., 2013. Low cardiac and aerobic scope in a coastal population of sockeye salmon *Oncorhynchus nerka* with a short upriver migration. J. Fish Biol. https://doi.org/10.1111/jfb.12120.

Elith, J., Kearney, M., Phillips, S., 2010. The art of modelling range-shifting species. Methods Ecol. Evol. 1 (4), 330–342. Wiley https://doi.org/10.1111/j.2041-210x.2010.00036.x.

Elliott, J.M., 1976. Body composition of Brown trout (*Salmo trutta* L.) in relation to temperature and ration size. J. Anim. Ecol. 45 (1), 273. JSTOR https://doi.org/10.2307/3779.

Elliott, J.M., Davison, W., 1975. Energy equivalents of oxygen consumption in animal energetics. Oecologia 19 (3), 195–201. https://doi.org/10.1007/BF00345305.

Essington, T.E., Kitchell, J.F., Walters, C.J., 2001. The von Bertalanffy growth function, bioenergetics, and the consumption rates of fish. Can. J. Fish. Aquat. Sci. 58 (11), 2129–2138. https://doi.org/10.1139/cjfas-58-11-2129.

Farrell, A.P., 2007. Cardiorespiratory Performance During Prolonged Swimming Tests With Salmonids: A Perspective on Temperature Effects and Potential Analytical Pitfalls. Royal Society, https://doi.org/10.1098/rstb.2007.2111.

Farrell, A.P., Jones, D.R., 1992. The heart. In: Hoar, W.S., Randall, D.J., Farrell, A.P. (Eds.), Fish Physiology XII Part A The Cardiovascular System. Academic Press, pp. 1–88.

Farrell, A., Lee, C., Tierney, K., Hodaly, A., Clutterham, S., Healey, M., Hinch, S., Lotto, A., 2003. Field-based measurements of oxygen uptake and swimming performance with adult Pacific salmon using a mobile respirometer swim tunnel. J. Fish Biol. 62, 64–84. https://doi.org/10.1046/j.0022-1112.2003.00010.x.

Farrell, A.P., Eliason, E.J., Sandblom, E., Clark, T.D., 2009. Fish cardiorespiratory physiology in an era of climate change. Can. J. Zool. 87 (10), 835–851. https://doi.org/10.1139/Z09-092.

Ferreira, M.S., Wood, C.M., Harter, T.S., Pont, G.D., Val, A.L., Matthews, P.G.D., 2019. Metabolic fuel use after feeding in the zebrafish (*Danio rerio*): a respirometric analysis. J. Exp. Biol. 222 (4), jeb194217. https://doi.org/10.1242/jeb.194217. Company of Biologists Ltd.

Filgueira, R., Chapman, J.M., Suski, C.D., Cooke, S.J., 2016. The influence of watershed land use cover on stream fish diversity and size-at-age of a generalist fish. Ecol. Indic. 60, 248–257. Elsevier https://doi.org/10.1016/j.ecolind.2015.06.006.

Filgueira, R., Chica, M., Palacios, J.J., Strohmeier, T., Lavaud, R., Agüera, A., Damas, S., Strand, Ø., 2020. Embracing multimodal optimization to enhance dynamic energy budget parameterization. Ecol. Modell. 431, 109139. Elsevier B.V https://doi.org/10.1016/j.ecolmodel.2020.109139.

Fitzgibbon, Q.P., Baudinette, R.V., Musgrove, R.J., Seymour, R.S., 2008. Routine metabolic rate of southern bluefin tuna (*Thunnus maccoyii*). Comp. Biochem. Physiol. A Mol. Integr. Physiol. 150 (2), 231–238. Elsevier Inc https://doi.org/10.1016/j.cbpa.2006.08.046.

Fitzhugh, G.R., Wuenschel, M.J., McBride, R.S., 2010. Evaluation of bioelectrical impedance analysis (BIA) to measure condition and energy allocated to reproduction in marine fishes. J. Phys. Conf. Ser. 224 (1), 012137. IOP Publishing https://doi.org/10.1088/1742-6596/224/1/012137.

Fry, F., 1947. Effects of the environment on animal activity. Publ. Ontario Fish. Res. Lab. 55, 1–62.

Fry, F., 1971. The effect of environmental factors on the physiology of fish. Fish Physiol. 6, 1–98.

Fussell, K.M.D., Smith, R.E.H., Fraker, M.E., Boegman, L., Frank, K.T., Miller, T.J., Tyson, J.T., Arend, K.K., Boisclair, D., Guildford, S.J., Hecky, R.E., Höök, T.O., Jensen, O.P., Llopiz, J.K., May, C.J., Najjar, R.G., Rudstam, L.G., Taggart, C.T., Rao, Y.R., Ludsin, S.A., 2016. A Perspective on Needed Research, Modeling, and Management Approaches That Can Enhance Great Lakes Fisheries Management Under Changing Ecosystem Conditions. International Association of Great Lakes Research, https://doi.org/10.1016/j.jglr.2016.04.007.

Gallagher, A.J., Hammerschlag, N., Cooke, S.J., Costa, D.P., Irschick, D.J., 2015. Evolutionary Theory as a Tool for Predicting Extinction Risk., https://doi.org/10.1016/j.tree.2014.12.001.

Garvey, J.E., Ostrand, K.G., Wahl, D.H., 2004. Energetics, predation, and ration affect size-dependent growth and mortality of fish during winter. Ecology 85 (10), 2860–2871. https://doi.org/10.1890/03-0329.

Giacomini, H.C., Shuter, B.J., Lester, N.P., 2013. Predator bioenergetics and the prey size spectrum: do foraging costs determine fish production? J. Theor. Biol. 332, 249–260. https://doi.org/10.1016/j.jtbi.2013.05.004.

Gill, A.B., Hart, P.J.B., 1994. Feeding behaviour and prey choice of the threespine stickleback: the interacting effects of prey size, fish size and stomach fullness. Anim. Behav. 47 (4), 921–932. https://doi.org/10.1006/anbe.1994.1124.

Glover, D.C., DeVries, D.R., Wright, R.A., Davis, D.A., 2010. Sample preparation techniques for determination of fish energy density via bomb calorimetry: an evaluation using largemouth bass. Trans. Am. Fish. Soc. 139 (3), 671–675. Wiley https://doi.org/10.1577/t09-110.1.

Gräns, A., Jutfelt, F., Sandblom, E., Jönsson, E., Wiklander, K., Seth, H., Olsson, C., Dupont, S., Ortega-Martinez, O., Einarsdottir, I., Björnsson, B.T., Sundell, K., Axelsson, M., 2014. Aerobic scope fails to explain the detrimental effects on growth resulting from warming and elevated CO2 in Atlantic halibut. J. Exp. Biol. 217, 711–717. https://doi.org/10.1242/jeb.096743.

Grausgruber, E.E., Weber, M.J., 2021. Using bioenergetics to estimate consumption of stocked age-0 walleye by a suite of piscivores. North Am. J. Fish. Manag. 41 (2), 383–398. John Wiley and Sons Inc https://doi.org/10.1002/nafm.10523.

Grimm, V., Ayllón, D., Railsback, S.F., 2017. Next-generation individual-based models integrate biodiversity and ecosystems: yes we can, and yes we must. Ecosystems 20 (2), 229–236.

Guzzo, M.M., Blanchfield, P.J., Rennie, M.D., 2017. Behavioral responses to annual temperature variation alter the dominant energy pathway, growth, and condition of a cold-water predator. Proc. Natl. Acad. Sci. U. S. A. 114 (37), 9912–9917. https://doi.org/10.1073/pnas.1702584114.

Halfyard, E.A., Webber, D., Del Papa, J., Leadley, T., Kessel, S.T., Colborne, S.F., Fisk, A.T., 2017. Evaluation of an acoustic telemetry transmitter designed to identify predation events. Methods Ecol. Evol. 8 (9), 1063–1071. https://doi.org/10.1111/2041-210X.12726.

Hansen, M.J., Boisclair, D., Brandt, S.B., Hewett, S.W., Kitchell, J.F., Lucas, M.C., Ney, J.J., 1993. Applications of bioenergetics models to fish ecology and management: where do we go from here? Trans. Am. Fish. Soc. 122 (5), 1019–1030. https://doi.org/10.1577/1548-8659(1993)122<1019:aobmtf>2.3.co;2.

Hansen, G.J.A., Read, J.S., Hansen, J.F., Winslow, L.A., 2017. Projected shifts in fish species dominance in Wisconsin lakes under climate change. Glob. Chang. Biol. 23 (4), 1463–1476. John Wiley & Sons, Ltd https://doi.org/10.1111/GCB.13462.

Hanson, P.C., Johnson, T.B., Schindler, D.E., Kitchell, J.F., 1997. Fish Bioenergetics 3.0 Software for Windows. Univ. Wisconsin Cent. Limnol. Sea Grant Institute, Madison, Wisconsin. Tech. Rep. WISCU-T-97-001.

Hanson, K.C., Ostrand, K.G., Gannam, A.L., Ostrand, S.L., 2010. Comparison and validation of nonlethal techniques for estimating condition in juvenile salmonids. Trans. Am. Fish. Soc. 139 (6), 1733–1741. Wiley https://doi.org/10.1577/t10-014.1.

Hartman, K.J., Brandt, S.B., 1995. Estimating energy density of fish. Trans. Am. Fish. Soc. 124 (3), 347–355. https://doi.org/10.1577/1548-8659(1995)124<0347:eedof>2.3.co;2.

Hartman, K.J., Kitchell, J.F., 2008. Bioenergetics modeling: progress since the 1992 symposium. Trans. Am. Fish. Soc. 137 (1), 216–223. https://doi.org/10.1577/T07-040.1.

Heino, M., Díaz Pauli, B., Dieckmann, U., 2015. Fisheries-induced evolution. Annu. Rev. Ecol. Evol. Syst. 46, 461–480. https://doi.org/10.1146/annurev-ecolsys-112414-054339.

Heithaus, M.R., Dill, L.M., Marshall, G.J., Buhleier, B., 2002. Habitat use and foraging behavior of tiger sharks (*Galeocerdo cuvier*) in a seagrass ecosystem. Mar. Biol. 140 (2), 237–248. Springer https://doi.org/10.1007/s00227-001-0711-7.

Henken, A.M., Lucas, H., Tijssen, P.A.T., Machiels, M.A.M., 1986. A comparison between methods used to determine the energy content of feed, fish and faeces samples. Aquaculture 58 (3–4), 195–201. Elsevier https://doi.org/10.1016/0044-8486(86)90085-2.

Hewett, S.W., Johnson, B.L., 1987. A Generalized Bioenergetics Model of Fish Growth for Microcomputers. Univ. Wisconsin, Sea Grant Institute, Ann Arbor, Michigan.

Hewett, S.W., Johnson, B.L., 1992. A Generalized Bioenergetics Model of Fish Growth for Microcomputers. Univ. Wisconsin Sea Grant Institute, Madison, Wisconsin. UW Sea Grant Tech. Rep.WIS-SG-92-250. 79 pp.

Hill, D.K., Magnuson, J.J., 1990. Potential effects of global climate warming on the growth and prey consumption of Great Lakes fish. Trans. Am. Fish. Soc. 119 (2), 265–275.

Hinch, S.G., Rand, P.S., 1998. Swim speeds and energy use of upriver-migrating sockeye salmon (*Oncorhynchus nerka*): role of local environment and fish characteristics. Can. J. Fish. Aquat. Sci. 55 (8), 1821–1831. https://doi.org/10.1139/cjfas-55-8-1821.

Hoar, W.S., Randall, D.J., Brett, J.R. (Eds.), 1979. Bioenergetics and growth. In: Fish Physiogy. Vol. 8. Academic Press, San Diego, US.

Holsman, K.K., Aydin, K., Sullivan, J., Hurst, T., Kruse, G.H., 2019. Climate effects and bottom-up controls on growth and size-at-age of Pacific halibut (*Hippoglossus stenolepis*) in Alaska (USA). Fish. Oceanogr. 28 (3), 345–358. Blackwell Publishing Ltd https://doi.org/10.1111/fog.12416.

Hopkins, K.D., 1992. Reporting fish growth: a review of the basics. J. World Aquac. Soc. 23 (3), 173–179. John Wiley & Sons, Ltd https://doi.org/10.1111/j.1749-7345.1992.tb00766.x.

Houlihan, D.F., Mathers, E.M., Foster, A., 1993. Biochemical correlates of growth rate in fish. In: Fish Ecophysiology. Springer Netherlands, pp. 45–71, https://doi.org/10.1007/978-94-011-2304-4_2.

Houston, A.I., Higginson, A.D., McNamara, J.M., 2011. Optimal foraging for multiple nutrients in an unpredictable environment. Ecol. Lett. 14 (11), 1101–1107. https://doi.org/10.1111/j.1461-0248.2011.01678.x.

Humphries, M.M., Mccann, K.S., 2014. Metabolic ecology. J. Anim. Ecol. 83, 7–19. https://doi.org/10.1111/1365-2656.12124

Huret, M., Tsiaras, K., Daewel, U., Skogen, M.D., Gatti, P., Petitgas, P., Somarakis, S., 2018. Variation in life-history traits of European anchovy along a latitudinal gradient: a bioenergetics modelling approach. Mar. Ecol. Prog. Ser. 2018, 95–112. Inter-Research https://doi.org/10.3354/meps12574.

Hvas, M., Folkedal, O., Oppedal, F., 2020. Heart rate bio-loggers as welfare indicators in Atlantic salmon (*Salmo salar*) aquaculture. Aquaculture 529 (April), 735630. Elsevier https://doi.org/10.1016/j.aquaculture.2020.735630.

Jackson, L.J., Schindler, D.E., 1996. Field estimates of net trophic transfer of PCBs from prey fishes to Lake Michigan salmonids. Environ. Sci. Technol. 30 (6), 1861–1865. American Chemical Society https://doi.org/10.1021/es950464a.

Jager, T., 2020. Revisiting simplified DEBtox models for analysing ecotoxicity data. Ecol. Modell. 416, 108904. Elsevier B.V https://doi.org/10.1016/j.ecolmodel.2019.108904.

Jerde, C.L., Kraskura, K., Eliason, E.J., Csik, S.R., Stier, A.C., Taper, M.L., 2019. Strong evidence for an intraspecific metabolic scaling coefficient near 0.89 in fish. Front. Physiol. 10, 1166. https://doi.org/10.3389/fphys.2019.01166.

Jobling, M., 1981. The influences of feeding on the metabolic rate of fishes: a short review. J. Fish Biol. 18 (4), 385–400. https://doi.org/10.1111/j.1095-8649.1981.tb03780.x.

Jobling, M., 1983. A short review and critique of methodology used in fish growth and nutrition studies. J. Fish Biol. 23 (6), 685–703. John Wiley & Sons, Ltd https://doi.org/10.1111/j.1095-8649.1983.tb02946.x.

Jobling, M., 1993. Bioenergetics: feed intake and energy partitioning. In: Fish Ecophysiology., https://doi.org/10.1007/978-94-011-2304-4_1.

Jobling, M., 1995. Fish bioenergetics. Oceanogr. Lit. Rev. 9 (42), 785. Pergamon Press. Available from https://www.infona.pl/resource/bwmeta1.element.elsevier-2effe3d3-5432-3e26-9f74-239ed62512b1. (accessed 20 March 2017).

Jobling, M., 2011. Energetic models | bioenergetics in aquaculture settings. In: Encyclopedia of Fish Physiology., https://doi.org/10.1016/B978-0-12-374553-8.00152-0.

Johnson, J.B., Bagley, J.C., Evans, J.P., Pilastro, A., Schlupp, I., 2011. Ecological drivers of life-history divergence. In: Ecology and Evolution of Poeciliid Fishes. University of Chicago Press, Chicago.

Johnson, B.M., Pate, W.M., Hansen, A.G., 2017. Energy density and dry matter content in fish: new observations and an evaluation of some empirical models. Trans. Am. Fish. Soc. 146 (6), 1262–1278. John Wiley and Sons Inc https://doi.org/10.1080/00028487.2017.1360392.

Johnston, A.S.A., Holmstrup, M., Hodson, M.E., Thorbek, P., Alvarez, T., Sibly, R.M., 2014a. Earthworm distribution and abundance predicted by a process-based model. Appl. Soil Ecol. 84, 112–123. Elsevier B.V https://doi.org/10.1016/j.apsoil.2014.06.001.

Johnston, A.S., Hodson, M.E., Thorbek, P., Alvarez, T., Sibly, R.M., 2014b. An energy budget agent-based model of earthworm populations and its application to study the effects of pesticides. Ecol. Modell. 280, 5–17. Elsevier B.V https://doi.org/10.1016/j.ecolmodel.2013.09.012.

Jones, M.L., Koonce, J.F., O'Gorman, R., 1993. Sustainability of hatchery-dependent salmonine fisheries in lake Ontario: the conflict between predator demand and prey supply. Trans. Am. Fish. Soc. 122 (5), 1002–1018. https://doi.org/10.1577/1548-8659(1993)122<1002:sohdsf>2.3.co;2.

Jones, M.L., Shuter, B.J., Zhao, Y., Stockwell, J.D., 2006. Forecasting Effects of Climate Change on Great Lakes Fisheries: Models That Link Habitat Supply to Population Dynamics Can Help. NRC Research Press, Ottawa, Canada, https://doi.org/10.1139/f05-239.

Jorgensen, S.J., Gleiss, A.C., Kanive, P.E., Chapple, T.K., Anderson, S.D., Ezcurra, J.M., Brandt, W.T., Block, B.A., 2015. In the belly of the beast: resolving stomach tag data to link temperature, acceleration and feeding in white sharks (*Carcharodon carcharias*). Anim. Biotelemetry 3 (1), 1–10. BioMed Central Ltd https://doi.org/10.1186/s40317-015-0071-6.

Jusup, M., Klanjšček, T., Matsuda, H., 2014. Simple measurements reveal the feeding history, the onset of reproduction, and energy conversion efficiencies in captive bluefin tuna. J. Sea Res. 94, 144–155. Elsevier B.V https://doi.org/10.1016/j.seares.2014.09.002.

Jusup, M., Sousa, T., Domingos, T., Labinac, V., Marn, N., Wang, Z., Klanjšček, T., 2017. Physics of Metabolic Organization. Elsevier B.V, https://doi.org/10.1016/j.plrev.2016.09.001.

Jutfelt, F., Norin, T., Ern, R., Overgaard, J., Wang, T., McKenzie, D.J., Lefevre, S., Nilsson, G.E., Metcalfe, N.B., Hickey, A.J.R., Brijs, J., Speers-Roesch, B., Roche, D.G., Gamperl, A.K., Raby, G.D., Morgan, R., Esbaugh, A.J., Gräns, A., Axelsson, M., Ekström, A., Sandblom, E., Binning, S.A., Hicks, J.W., Seebacher, F., Jørgensen, C., Killen, S.S., Schulte, P.M., Clark, T.D., 2018. Oxygen- and capacity-limited thermal tolerance: blurring ecology and physiology. J. Exp. Biol. 221, jeb169615. https://doi.org/10.1242/jeb.169615.

Kaiyala, K.J., Ramsay, D.S., 2011, March 1. Direct Animal Calorimetry, The Underused Gold Standard for Quantifying the Fire of Life. Elsevier Inc, https://doi.org/10.1016/j.cbpa.2010.04.013.

Kao, Y.C., Madenjian, C.P., Bunnell, D.B., Lofgren, B.M., Perroud, M., 2015. Temperature effects induced by climate change on the growth and consumption by salmonines in lakes Michigan and Huron. Environ. Biol. Fishes 98, 1089–1104. https://doi.org/10.1007/s10641-014-0352-6.

Kawabata, Y., Noda, T., Nakashima, Y., Nanami, A., Sato, T., Takebe, T., Mitamura, H., Arai, N., Yamaguchi, T., Soyano, K., 2014. Use of a gyroscope/accelerometer data logger to identify alternative feeding behaviours in fish. J. Exp. Biol. 217 (Pt. 18), 3204–3208. https://doi.org/10.1242/jeb.108001.

Kieffer, J.D., Alsop, D., Wood, C.M., 1998. A respirometric analysis of fuel use during aerobic swimming at different temperatures in rainbow trout (*Oncorhynchus mykiss*). J. Exp. Biol. 201 (22), 3123–3133. The Company of Biologists https://doi.org/10.1242/jeb.201.22.3123.

Killen, S.S., Atkinson, D., Glazier, D.S., 2010. The intraspecific scaling of metabolic rate with body mass in fishes depends on lifestyle and temperature. Ecol. Lett. 13 (2), 184–193. https://doi.org/10.1111/j.1461-0248.2009.01415.x.

King, A.J., George, A., Buckle, D.J., Novak, P.A., Fulton, C.J., 2018. Efficacy of remote underwater video cameras for monitoring tropical wetland fishes. Hydrobiologia 807 (1), 145–164. Springer International Publishing https://doi.org/10.1007/s10750-017-3390-1.

Kitchell, J.F., Koonce, J.F., Magnuson, J.J., O'Neill, R.V., JR, H.H.S., Booth, R.S., 1974. Model of fish biomass dynamics. Trans. Am. Fish. Soc. 103 (4), 786–798.

Kitchell, J.F., Stewart, D.J., Weininger, D., 1977. Applications of a bioenergetics model to yellow perch (*Perca flavescens*) and walleye (*Stizostedion vitreum vitreum*). J. Fish. Res. Board Can. 34, 1922–1935.

Klefoth, T., Skov, C., Aarestrup, K., Arlinghaus, R., 2013. Reliability of non-lethal assessment methods of body composition and energetic status exemplified by applications to eel (*Anguilla anguilla*) and carp (*Cyprinus carpio*). Fish. Res. 146, 18–26. Elsevier https://doi.org/10.1016/j.fishres.2013.03.010.

Klinard, N.V., Matley, J.K., Fisk, A.T., Johnson, T.B., 2019. Long-Term Retention of Acoustic Telemetry Transmitters in Temperate Predators Revealed by Predation Tags Implanted in Wild Prey Fish (August). pp. 2015–2019, https://doi.org/10.1111/jfb.14156.

Klok, C., Nordtug, T., Tamis, J.E., 2014. Estimating the impact of petroleum substances on survival in early life stages of cod (*Gadus morhua*) using the dynamic energy budget theory. Mar. Environ. Res. 101 (1), 60–68. Elsevier Ltd https://doi.org/10.1016/j.marenvres.2014.09.002.

Kooijman, S.A.L.M., 2010. Dynamic Energy Budget Theory for Metabolic Organisation. Cambridge University Press, https://doi.org/10.1098/rstb.2010.0167.

Kooijman, S.A.L.M., Lika, K., 2014. Comparative energetics of the 5 fish classes on the basis of dynamic energy budgets. J. Sea Res. 94, 19–28. Elsevier B.V https://doi.org/10.1016/j.seares.2014.01.015.

Kraft, C.E., 1992. Estimates of phosphorus and nitrogen cycling by fish using a bioenergetics approach. Can. J. Fish. Aquat. Sci. 49 (12), 2596–2604. NRC Research Press Ottawa, Canada https://doi.org/10.1139/f92-287.

Kudo, T., Tanaka, H., Watanabe, Y., Naito, Y., Otomo, T., Miyazaki, N., 2007. Use of fish-borne camera to study chum salmon homing behavior in response to coastal features. Aquat. Biol. 1 (1), 85–90.

Kuparinen, A., Merilä, J., 2007. Detecting and Managing Fisheries-Induced Evolution., https://doi.org/10.1016/j.tree.2007.08.011.

Lambert, Y., Yaragina, N.A., Kraus, G., Marteinsdottir, G., Wright, P.J., 2003. Using environmental and biological indices as proxies for egg and larval production of marine fish. J. Northw. Atl. Fish. Sci 33, 115–159. Available from http://journal.nafo.int. (accessed 5 July 2021).

Lauff, R.F., Wood, C.H., 1996. Respiratory gas exchange, nitrogenous waste excretion, and fuel usage during aerobic swimming in juvenile rainbow trout. J. Comp. Physiol. B Biochem. Syst. Environ. Physiol. 166 (8), 501–509. Springer Verlag https://doi.org/10.1007/BF02338293.

Lavaud, R., Thomas, Y., Pecquerie, L., Benoît, H.P., Guyondet, T., Flye-Sainte-Marie, J., Chabot, D., 2019. Modeling the impact of hypoxia on the energy budget of Atlantic cod in two populations of the Gulf of Saint-Lawrence, Canada. J. Sea Res. 143, 243–253. Elsevier B.V https://doi.org/10.1016/j.seares.2018.07.001.

Law, R., 2007. Fisheries-induced evolution: present status and future directions. Mar. Ecol. Prog. Ser. 335, 271–277. https://doi.org/10.3354/meps335271.

Lawrence, M.J., Wright, P.A., Wood, C.M., 2015. Physiological and molecular responses of the goldfish (*Carassius auratus*) kidney to metabolic acidosis, and potential mechanisms of renal ammonia transport. J. Exp. Biol. 218 (13), 2124–2135. Company of Biologists Ltd https://doi.org/10.1242/jeb.117689.

Lawrence, M.J., Eliason, E.J., Zolderdo, A.J., Lapointe, D., Best, C., Gilmour, K.M., Cooke, S.J., 2019a. Cortisol modulates metabolism and energy mobilization in wild-caught pumpkinseed (*Lepomis gibbosus*). Fish Physiol. Biochem. 45 (6), 1813–1828. Springer Netherlands https://doi.org/10.1007/s10695-019-00680-z.

Lawrence, M.J., Zolderdo, A.J., Godin, J.G.J., Mandelman, J.W., Gilmour, K.M., Cooke, S.J., 2019b. Cortisol does not increase risk of mortality to predation in juvenile bluegill sunfish: a manipulative experimental field study. J. Exp. Zool. A Ecol. Integr. Physiol. 331 (4), 253–261. John Wiley and Sons Inc https://doi.org/10.1002/jez.2257.

Le Pichon, C., Trancart, T., Lambert, P., Daverat, F., Rochard, E., 2014. Summer habitat use and movements of late juvenile European flounder (*Platichthys flesus*) in tidal freshwaters: results from an acoustic telemetry study. J. Exp. Mar. Bio. Ecol. 461, 441–448. Elsevier https://doi.org/10.1016/j.jembe.2014.09.015.

Lear, K.O., Whitney, N.M., Brewster, L.R., Gleiss, A.C., 2019. Treading water: respirometer choice may hamper comparative studies of energetics in fishes. Mar. Freshw. Res. 70 (3), 437–448. CSIRO https://doi.org/10.1071/MF18182.

Lee, C.G., Farrell, A.P., Lotto, A., MacNutt, M.J., Hinch, S.G., Healey, M.C., 2003. The effect of temperature on swimming performance and oxygen consumption in adult sockeye (*Oncorhynchus nerka*) and coho (*O. kisutch*) salmon stocks. J. Exp. Biol. 206 (18), 3239–3251. https://doi.org/10.1242/jeb.00547.

Lefevre, S., Wang, T., McKenzie, D.J., 2021. The role of mechanistic physiology in investigating impacts of global warming on fishes. J. Exp. Biol. 224, jeb238840. https://doi.org/10.1242/jeb.238840.

Lenky, C., Eisert, R., Oftedal, O.T., Metcalf, V., 2012. Proximate composition and energy density of nototheniid and myctophid fish in McMurdo Sound and the Ross Sea, Antarctica. Polar Biol. 35 (5), 717–724. Springer https://doi.org/10.1007/s00300-011-1116-9.

Lennox, R., Choi, K., Harrison, P.M., Paterson, J.E., Peat, T.B., Ward, T.D., Cooke, S.J., 2015. Improving science-based invasive species management with physiological knowledge, concepts, and tools. Biol. Invasions 17 (8), 2213–2227. Kluwer Academic Publishers https://doi.org/10.1007/s10530-015-0884-5.

Lennox, R.J., Bravener, G.A., Lin, H.Y., Madenjian, C.P., Muir, A.M., Remucal, C.K., Robinson, K.F., Rous, A.M., Siefkes, M.J., Wilkie, M.P., Zielinski, D.P., Cooke, S.J., 2020. Potential Changes to the Biology and Challenges to the Management of Invasive Sea Lamprey Petromyzon Marinus in the Laurentian Great Lakes Due to Climate Change. Blackwell Publishing Ltd, https://doi.org/10.1111/gcb.14957.

Lika, K., Kearney, M.R., Freitas, V., van der Veer, H.W., van der Meer, J., Wijsman, J.W.M., Pecquerie, L., Kooijman, S.A.L.M., 2011. The "covariation method" for estimating the parameters of the standard dynamic energy budget model I: philosophy and approach. J. Sea Res. 66 (4), 270–277. Elsevier https://doi.org/10.1016/j.seares.2011.07.010.

Liss, S.A., Sass, G.G., Suski, C.D., 2013. Spatial and temporal influences on the physiological condition of invasive silver carp. Conserv. Physiol. 1 (1), cot017. https://doi.org/10.1093/conphys/cot017. Oxford University Press.

Little, A.G., Dressler, T., Kraskura, K., Hardison, E., Hendriks, B., Prystay, T., Farrell, A.P., Cooke, S.J., Patterson, D.A., Hinch, S.G., Eliason, E.J., 2020. Maxed out: optimizing accuracy, precision, and power for field measures of maximum metabolic rate in fishes. Physiol. Biochem. Zool. 93 (3), 243–254. https://doi.org/10.1086/708673.

Love, R.M., 1970. The chemical biology of fishes. With a key to the chemical literature. In: Chem. Biol. Fishes. With a key to Chem. Lit. Academic Press Inc, London: New York.

Luecke, C., Wurtsbaugh, W.A., Budy, P., Gross, H.P., Steinhart, G., 2016. Simulated growth and production of endangered Snake River sockeye salmon: assessing management strategies for the nursery lakes. Fisheries 21 (6), 18–25. https://doi.org/10.1577/1548-8446(1996)021<0018:SGAPOE>2.0.CO;2.

Lynch, A.J., Taylor, W.W., Smith, K.D., 2010. The influence of changing climate on the ecology and management of selected Laurentian Great Lakes fisheries. J. Fish Biol. 77 (8), 1764–1782. Blackwell Publishing Ltd https://doi.org/10.1111/j.1095-8649.2010.02759.x.

MacArthur, R.H., Wilson, E.O., 1967. The Theory of Island Biogeography. Princeton Univ. Press, Princeton, N.J.

Maceina, M.J., Boxrucker, J., Buckmeier, D.L., Gangl, R.S., Lucchesi, D.O., Isermann, D.A., Martinez, P.J., 2007. Current status and review of freshwater fish aging procedures used by state and provincial fisheries agencies with recommendations for future directions. Fisheries 32 (7), 329–340.

MacNutt, M.J., Hinch, S.G., Lee, C.G., Phibbs, J.R., Lotto, A.G., Healey, M.C., Farrell, A.P., 2006. Temperature effects on swimming performance, energetics, and aerobic capacities of mature adult pink salmon (*Oncorhynchus gorbuscha*) compared with those of sockeye salmon (*Oncorhynchus nerka*). Can. J. Zool. 84, 88–97. https://doi.org/10.1139/z05-181.

Mangano, M.C., Giacoletti, A,, Sarà, G , 2019. Dynamic energy budget provides mechanistic derived quantities to implement the ecosystem based management approach. J. Sea Res. 143, 272–279. Elsevier B.V https://doi.org/10.1016/j.seares.2018.05.009.

Martin, B.T., Zimmer, E.I., Grimm, V., Jager, T., 2012. Dynamic energy budget theory meets individual-based modelling: a generic and accessible implementation. Methods Ecol. Evol. 3 (2), 445–449. https://doi.org/10.1111/j.2041-210X.2011.00168.x.

Matzelle, A.J., Sarà, G., Montalto, V., Zippay, M., Trussell, G.C., Helmuth, B., 2015. A bioenergetics framework for integrating the effects of multiple stressors: opening a "black box" in climate change research. In: American Malacological Bulletin. American Malacological Society, pp. 150–160, https://doi.org/10.4003/006.033.0107.

Mayfield, R.B., Cech, J.J., 2004. Temperature effects on green sturgeon bioenergetics. Trans. Am. Fish. Soc. 133 (4), 961–970. Wiley https://doi.org/10.1577/t02-144.1.

Mazur, M.M., Wilson, M.T., Dougherty, A.B., Buchheister, A., Beauchamp, D.A., 2007. Temperature and prey quality effects on growth of juvenile walleye pollock *Theragra chalcogramma* (Pallas): a spatially explicit bioenergetics approach. J. Fish Biol. 70 (3), 816–836. https://doi.org/10.1111/j.1095-8649.2007.01344.x.

McKenzie, D.J., Axelsson, M., Chabot, D., Claireaux, G., Cooke, S.J., Corner, R.A., De Boeck, G., Domenici, P., Guerreiro, P.M., Hamer, B., 2016. Conservation physiology of marine fishes: state of the art and prospects for policy. Conserv. Physiol. 4, 1–20. https://doi.org/10.1093/conphys/cow046Introduction.

McLean, J.A., Tobin, G., 2007. Animal and Human Calorimetry. Cambridge University Press.

Mercaldo-Allen, R., Kuropat, C., Caldarone, E.M., 2006. A model to estimate growth in young-of-the-year tautog, *Tautoga onitis*, based on RNA/DNA ratio and seawater temperature. J. Exp. Mar. Bio. Ecol. 329 (2), 187–195. Elsevier https://doi.org/10.1016/j.jembe.2005.08.015.

Metcalfe, J.D., Wright, S., Tudorache, C., Wilson, R.P., 2016. Recent advances in telemetry for estimating the energy metabolism of wild fishes. J. Fish Biol. 88 (1), 284–297. https://doi.org/10.1111/jfb.12804.

Meynecke, J.O., Poole, G.C., Werry, J., Lee, S.Y., 2008. Use of PIT tag and underwater video recording in assessing estuarine fish movement in a high intertidal mangrove and salt marsh creek. Estuar. Coast. Shelf Sci. 79 (1), 168–178. Academic Press https://doi.org/10.1016/j.ecss.2008.03.019.

Mintram, K.S., Maynard, S.K., Brown, A.R., Boyd, R., Johnston, A.S.A., Sibly, R.M., Thorbek, P., Tyler, C.R., 2020. Applying a mechanistic model to predict interacting effects of chemical exposure and food availability on fish populations. Aquat. Toxicol. 224 (November 2019), 105483. Elsevier https://doi.org/10.1016/j.aquatox.2020.105483.

Mochnacz, N.J., Kissinger, B.C., Deslauriers, D., Guzzo, M.M., Enders, E.C., Gary Anderson, W., Docker, M.F., Isaak, D.J., Durhack, T.C., Treberg, J.R., 2017. Development and testing of a simple field-based intermittent-flow respirometry system for riverine fishes. Conserv. Physiol. 5 (1). https://doi.org/10.1093/conphys/cox048. Oxford University Press.

Molony, B.W., Lenanton, R., Jackson, G., Norriss, J., 2005. Stock enhancement as a fisheries management tool. Rev. Fish Biol. Fish. 13 (4), 409–432. Springer Science and Business Media LLC https://doi.org/10.1007/s11160-005-1886-7.

Moorhead, S.G., Gallagher, A.J., Merly, L., Hammerschlag, N., 2021. Variation of body condition and plasma energy substrates with life stage, sex, and season in wild-sampled nurse sharks *Ginglymostoma cirratum*. J. Fish Biol. 98 (3), 680–693. https://doi.org/10.1111/jfb.14612.

Mortsch, L.D., Quinn, F.H., 1996. Climate change scenarios for Great Lakes Basin ecosystem studies. Limnol. Oceanogr. 41 (5), 903–911. American Society of Limnology and Oceanography Inc https://doi.org/10.4319/lo.1996.41.5.0903.

Mounier, F., Pecquerie, L., Lobry, J., Sardi, A.E., Labadie, P., Budzinski, H., Loizeau, V., 2020. Dietary bioaccumulation of persistent organic pollutants in the common sole *Solea solea* in the context of global change. Part 1: revisiting parameterisation and calibration of a DEB model to consider inter-individual variability in experimental and natural conditions. Ecol. Modell. 433, 109224. Elsevier B.V https://doi.org/10.1016/j.ecolmodel.2020.109224.

Muller, C., Childs, A.R., Duncan, M.I., Skeeles, M.R., James, N.C., Van Der Walt, K.A., Winkler, A.C., Potts, W.M., 2020. Implantation, orientation and validation of a commercially produced heart-rate logger for use in a perciform teleost fish. Conserv. Physiol. 8 (1). https://doi.org/10.1093/conphys/coaa035. Oxford University Press.

Nabe-Nielsen, J., Sibly, R.M., Tougaard, J., Teilmann, J., Sveegaard, S., 2014. Effects of noise and by-catch on a Danish harbour porpoise population. Ecol. Modell. 272, 242–251. Elsevier B.V https://doi.org/10.1016/j.ecolmodel.2013.09.025.

Nash, K.L., Graham, N.A.J., Januchowski-Hartley, F.A., Bellwood, D.R., 2012. Influence of habitat condition and competition on foraging behaviour of parrotfishes. Mar. Ecol. Prog. Ser. 457, 113–124. https://doi.org/10.3354/meps09742.

Negus, M.T., 1995. Bioenergetics modeling as a salmonine management tool applied to Minnesota waters of lake superior. North Am. J. Fish. Manag. 15 (1), 60–78. https://doi.org/10.1577/1548-8675(1995)015<0060:bmaasm>2.3.co;2.

Neill, W.H., Brandes, T.S., Burke, B.J., Craig, S.R., Dimichele, L.V., Duchon, K., et al., 2004. Ecophys.Fish: a simulation model of fish growth in time-varying environmental regimes. Rev. Fish. Sci. 12 (4), 233–288.

Nel, J.L., Roux, D.J., Driver, A., Hill, L., Maherry, A.C., Snaddon, K., Petersen, C.R., Smith-Adao, L.B., Van Deventer, H., Reyers, B., 2016. Knowledge co-production and boundary work to promote implementation of conservation plans. Conserv. Biol. 30 (1), 176–188. https://doi.org/10.1111/cobi.12560.

Nelson, J.A., 2016. Oxygen consumption rate v. rate of energy utilization of fishes: a comparison and brief history of the two measurements. J. Fish Biol. 88 (1), 10–25. https://doi.org/10.1111/jfb.12824.

Ney, J.J., 1993. Bioenergetics modeling today: growing pains on the cutting edge. Trans. Am. Fish. Soc. 122 (5), 736–748. https://doi.org/10.1577/1548-8659(1993)122<0736:bmtgpo>2.3.co;2.

Nguyen, V.M., Young, N., Cooke, S., 2018. Applying a knowledge-action framework for navigating barriers to incorporating telemetry science into fisheries management and conservation: a qualitative study. Can. J. Fish. Aquat. Sci. 75 (10), 1733–1743. Available from https://doi.org/10.1139/cjfas-2017-0303.

Nielsen, J.M., Clare, E.L., Hayden, B., Brett, M.T., Kratina, P., 2018. Diet Tracing in Ecology: Method Comparison and Selection. British Ecological Society, https://doi.org/10.1111/2041-210X.12869.

Niu, K.M., Khosravi, S., Kothari, D., Lee, W.D., Lim, J.M., Lee, B.J., Kim, K.W., Lim, S.G., Lee, S.M., Kim, S.K., 2019. Effects of dietary multi-strain probiotics supplementation in a low fishmeal diet on growth performance, nutrient utilization, proximate composition, immune parameters, and gut microbiota of juvenile olive flounder (*Paralichthys olivaceus*). Fish Shellfish Immunol. 93, 258–268. Academic Press https://doi.org/10.1016/j.fsi.2019. 07.056.

Noda, T., Kawabata, Y., Arai, N., Mitamura, H., Watanabe, S., 2013. Monitoring escape and feeding behaviours of cruiser fish by inertial and magnetic sensors. PLoS One 8 (11), 1–13. https://doi.org/10.1371/journal.pone.0079392.

Norin, T., Clark, T.D., 2016. Measurement and relevance of maximum metabolic rate in fishes. J. Fish Biol. 88 (1), 122–151. https://doi.org/10.1111/jfb.12796.

Norin, T., Malte, H., Clark, T.D., 2014. Aerobic scope does not predict the performance of a tropical eurythermal fish at elevated temperatures. J. Exp. Biol. 217 (2), 244–251. https://doi.org/10.1242/jeb.089755.

Novak, A.J., Carlson, A.E., Wheeler, C.R., Wippelhauser, G.S., Sulikowski, J.A., 2017. Critical foraging habitat of Atlantic sturgeon based on feeding habits, prey distribution, and movement patterns in the Saco river estuary, Maine. Trans. Am. Fish. Soc. 146 (2), 308–317. Taylor and Francis Inc https://doi.org/10.1080/00028487.2016.1264472.

Nowell, L.B., Brownscombe, J.W., Gutowsky, L.F.G., Murchie, K.J., Suski, C.D., Danylchuk, A.J., Shultz, A., Cooke, S.J., 2015. Swimming energetics and thermal ecology of adult bonefish (*Albula vulpes*): a combined laboratory and field study in Eleuthera, the Bahamas. Environ. Biol. Fishes 98 (11). https://doi.org/10.1007/s10641-015-0420-6.

Oleksyn, S., Tosetto, L., Raoult, V., Joyce, K.E., Williamson, J.E., 2021. Going Batty: The Challenges and Opportunities of Using Drones to Monitor the Behaviour and Habitat Use of Rays. MDPI AG, https://doi.org/10.3390/drones5010012.

Page, J.W., Andrews, J.W., 1973. Interactions of dietary levels of protein and energy on channel catfish (*Ictalurus punctatus*). J. Nutr. 103 (9), 1339–1346. Oxford Academic https://doi.org/10.1093/jn/103.9.1339.

Paine, R.T., 1971. The measurement and application of the calorie to ecological problems. Annu. Rev. Ecol. Syst. 2 (1), 145–164. Annual Reviews https://doi.org/10.1146/annurev.es.02.110171.001045.

Payne, N.L., Smith, J.A., van der Meulen, D.E., Taylor, M.D., Watanabe, Y.Y., Takahashi, A., Marzullo, T.A., Gray, C.A., Cadiou, G., Suthers, I.M., 2016. Temperature dependence of fish performance in the wild: links with species biogeography and physiological thermal tolerance. Funct. Ecol. 30 (6), 903–912. https://doi.org/10.1111/1365-2435.12618.

Pearson, R.G., Stanton, J.C., Shoemaker, K.T., Aiello-Lammens, M.E., Ersts, P.J., Horning, N., Fordham, D.A., Raxworthy, C.J., Ryu, H.Y., Mcnees, J., Akçakaya, H.R., 2014. Life history and spatial traits predict extinction risk due to climate change. Nat. Clim. Chang. 4 (3), 217–221. https://doi.org/10.1038/nclimate2113.

Pecquerie, L., Johnson, L.R., Kooijman, S.A.L.M., Nisbet, R.M., 2011. Analyzing variations in life-history traits of Pacific salmon in the context of dynamic energy budget (DEB) theory. J. Sea Res. 66 (4), 424–433. Elsevier https://doi.org/10.1016/j.seares.2011.07.005.

Pethybridge, H., Roos, D., Loizeau, V., Pecquerie, L., Bacher, C., 2013. Responses of European anchovy vital rates and population growth to environmental fluctuations: an individual-based modeling approach. Ecol. Model. 250, 370–383. https://doi.org/10.1016/j.ecolmodel.2012.11.017.

Phang, S.C., Stillman, R.A., Cucherousset, J., Britton, J.R., Roberts, D., Beaumont, W.R.C., Gozlan, R.E., 2016. FishMORPH—an agent-based model to predict salmonid growth and distribution responses under natural and low flows. Sci. Rep. 6 (July), 1–13. Nature Publishing Group https://doi.org/10.1038/srep29414.

Pianka, E.R., 1970. On r-and K-selection. Am. Nat. 104 (940), 592–597.

Pillay, T.V.R., 2007. Aquaculture and the environment: second edition. In: Aquaculture and the Environment, second ed., https://doi.org/10.1002/9780470995730.

Politikos, D., Huret, M., Petitgas, P., 2015a. A coupled movement and bioenergetics model to explore the spawning migration of anchovy in the Bay of Biscay. Ecol. Model. 313 (October), 212–222. https://doi.org/10.1016/j.ecolmodel.2015.06.036.

Politikos, D., Somarakis, S., Tsiaras, K., Giannoulaki, M., Petihakis, G., Machias, A., Triantafyllou, G., 2015b. Simulating anchovy's full life cycle in the northern Aegean Sea (eastern Mediterranean): a coupled hydro-biogeochemical-IBM model. Prog. Oceanogr. 138, 399–416. Elsevier Ltd https://doi.org/10.1016/j.pocean.2014.09.002.

Pörtner, H.O., Bock, C., Mark, F.C., 2017. Oxygen- & capacity-limited thermal tolerance: bridging ecology & physiology. J. Exp. Biol. 220 (15), 2685–2696. https://doi.org/10.1242/jeb.134585.

Pothoven, S.A., Ludsin, S.A., Höök, T.O., Fanslow, D.L., Mason, D.M., Collingsworth, P.D., Van Tassell, J.J., 2008. Reliability of bioelectrical impedance analysis for estimating whole-fish energy density and percent lipids. Trans. Am. Fish. Soc. 137 (5), 1519–1529. Wiley https://doi.org/10.1577/t07-185.1.

Pottinger, T.G., Cook, A., Jürgens, M.D., Rhodes, G., Katsiadaki, I., Balaam, J.L., Smith, A.J., Matthiessen, P., 2011. Effects of sewage effluent remediation on body size, somatic RNA: DNA ratio, and markers of chemical exposure in three-spined sticklebacks. Environ. Int. 37 (1), 158–169. Elsevier Ltd https://doi.org/10.1016/j.envint.2010.08.012.

Primavera, J.H., 2006. Overcoming the impacts of aquaculture on the coastal zone. Ocean Coast. Manag. 49 (9–10), 531–545. https://doi.org/10.1016/j.ocecoaman.2006.06.018.

Prystay, T.S., Eliason, E.J., Lawrence, M.J., Dick, M., Brownscombe, J.W., Patterson, D.A., Crossin, G.T., Hinch, S.G., Cooke, S.J., 2017. The influence of water temperature on sockeye salmon heart rate recovery following simulated fisheries interactions. Conserv. Physiol. 5 (1), cox050. https://doi.org/10.1093/conphys/cox050. Oxford University Press.

Pyke, G.H., 1984. Optimal foraging theory: a critical review. Annu. Rev. Ecol. Syst. 15, 523–575.

Raab, K., Llope, M., Nagelkerke, L.A.J., Rijnsdorp, A.D., Teal, L.R., Licandro, P., Ruardij, P., Dickey-Collas, M., 2013. Influence of temperature and food availability on juvenile European anchovy *Engraulis encrasicolus* at its northern boundary. Mar. Ecol. Prog. Ser. 488, 233–245. https://doi.org/10.3354/meps10408.

Raby, G.D., Casselman, M.T., Cooke, S.J., Hinch, S.G., Farrell, A.P., Clark, T.D., 2016. Aerobic scope increases throughout an ecologically relevant temperature range in coho salmon. J. Exp. Biol. 219, 1922–1931. https://doi.org/10.1242/jeb.137166.

Raby, G.D., Vandergoot, C.S., Hayden, T.A., Faust, M.D., Kraus, R.T., Dettmers, J.M., Cooke, S.J., Zhao, Y., Fisk, A.T., Krueger, C.C., 2018. Does behavioural thermoregulation underlie seasonal movements in Lake Erie walleye? Can. J. Fish. Aquat. Sci. 75 (3), 488–496. https://doi.org/10.1139/cjfas-2017-0145.

Regan, M.D., Gill, I.S., Richards, J.G., 2016. Calorespirometry reveals that goldfish prioritize aerobic metabolism over metabolic rate depression in all but near-anoxic environments. J. Exp. Biol. 220, 564–572.

Ren, J.S., Stenton-Dozey, J., Plew, D.R., Fang, J., Gall, M., 2012. An ecosystem model for optimising production in integrated multitrophic aquaculture systems. Ecol. Modell. 246 (C), 34–46. Elsevier https://doi.org/10.1016/j.ecolmodel.2012.07.020.

Reznick, D., Bryant, M.J., Bashey, F., 2002. R- and K-Selection Revisited: The Role of Population Regulation in Life-History Evolution., https://doi.org/10.1890/0012-9658(2002)083 [1509:raksrt]2.0.co;2.

Rice, J.A., Breck, J.E., Bartell, S.M., Kitchell, J.F., 1983. Evaluating the constraints of temperature, activity and consumption on growth of largemouth bass. Environ. Biol. Fishes 9 (3–4), 263–275. Kluwer Academic Publishers https://doi.org/10.1007/BF00692375.

Robbins, C., 2012. Wildlife Feeding and Nutrition. Elsevier.

Roques, S., Deborde, C., Richard, N., Skiba-Cassy, S., Moing, A., Fauconneau, B., 2020. Metabolomics and Fish Nutrition: A Review in the Context of Sustainable Feed Development., https://doi.org/10.1111/raq.12316.

Rose, K.A., Cowan, J.H., Clark, M.E., Houde, E.D., Wang, S.B., 1999. An individual-based model of bay anchovy population dynamics in the mesohaline region of Chesapeake Bay. Mar. Ecol. Prog. Ser. 185, 113–132. https://doi.org/10.3354/meps185113.

Rose, K.A., Kimmerer, W.J., Edwards, K.P., Bennett, W.A., 2013. Individual-based modeling of delta smelt population dynamics in the upper San Francisco estuary: I. model description and baseline results. Trans. Am. Fish. Soc. 142 (5), 1238–1259. https://doi.org/10.1080/00028487.2013.799518.

Rosenfeld, J., Richards, J., Allen, D., Van Leeuwen, T., Monnet, G., 2020. Adaptive trade-offs in fish energetics and physiology: insights from adaptive differentiation among juvenile salmonids. Can. J. Fish. Aquat. Sci. 1255 (March), 1–13. https://doi.org/10.1139/cjfas-2019-0350.

Rowan, D.J., Rasmussen, J.B., 1996. Measuring the bioenergetic cost of fish activity in situ using a globally dispersed radiotracer (137Cs). Can. J. Fish. Aquat. Sci. 53 (4), 734–745. https://doi.org/10.1139/f95-046.

Scarabello, M., Heigenhauser, G.J., Wood, C.M., 1992. Gas exchange, metabolite status and excess post-exercise oxygen consumption after repetitive bouts of exhaustive exercise in juvenile rainbow trout. J. Exp. Biol. 167 (1), 155–169. The Company of Biologists https://doi.org/10.1242/jeb.167.1.155.

Schakmann, M., Steffensen, J.F., Bushnell, P.G., Korsmeyer, K.E., 2020. Swimming in unsteady water flows: is turning in a changing flow an energetically expensive endeavor for fish? J. Exp. Biol. 223 (6), jeb212795. https://doi.org/10.1242/JEB.212795/VIDEO-2. Company of Biologists Ltd.

Scheffer, M., Baveco, J.M., Deangelis, D.L., Rose, K.A., Vannes, E.H., 1995. Super-individuals A simple solution for modeling large populations on an individual basis. Ecol. Model. 80 (1995), 161–170. https://doi.org/10.1016/0304-3800(94)00055-M.

Schloesser, R.W., Fabrizio, M.C., 2015. Relationships among proximate components and energy density of juvenile Atlantic estuarine fishes. Trans. Am. Fish. Soc. 144 (5), 942–955. Taylor and Francis Inc https://doi.org/10.1080/00028487.2015.1052557.

Schloesser, R.W., Fabrizio, M.C., 2016. Temporal dynamics of condition for estuarine fishes in their nursery habitats. Mar. Ecol. Prog. Ser. 557, 207–219. Inter-Research https://doi.org/10.3354/meps11858.

Secor, S.M., 2009. Specific dynamic action: a review of the postprandial metabolic response. J. Comp. Physiol. B Biochem. Syst. Environ. Physiol. 179, 1–56. https://doi.org/10.1007/s00360-008-0283-7.

Serpa, D., Ferreira, P.P., Ferreira, H., da Fonseca, L.C., Dinis, M.T., Duarte, P., 2013. Modelling the growth of white seabream (*Diplodus sargus*) and gilthead seabream (*Sparus aurata*) in semi-intensive earth production ponds using the dynamic energy budget approach. J. Sea Res. 76, 135–145. Elsevier https://doi.org/10.1016/j.seares.2012.08.003.

Sewall, F., Norcross, B.L., Heintz, R.A., 2021. Growth, condition, and swimming performance of juvenile Pacific herring with winter feeding rations. Can. J. Fish. Aquat. Sci. 78 (7), 881–893. https://doi.org/10.1139/cjfas-2020-0293. Canadian Science Publishing.

Shen, Y., De Hoog, F., Clark, T.D., Severati, A., 2020. Estimating heart rate and detecting feeding events of fish using an implantable biologger. 19th ACM/IEEE International Conference on Information Processing in Sensor Networks (IPSN), pp. 37–48.

Sherwood, G.D., Rasmussen, J.B., Rowan, D.J., Brodeur, J., Hontela, A., 2000. Bioenergetic costs of heavy metal exposure in yellow perch (*Perca flavescens*): in situ estimates with a radio-tracer (137Cs) technique. Can. J. Fish. Aquat. Sci. 57 (2), 441–450. National Research Council of Canada https://doi.org/10.1139/f99-268.

Sibly, R.M., Brown, J.H., Kodric-Brown, A., 2012. Metabolic ecology: a scaling approach. In: Metabolic Ecology: A Scaling Approach., https://doi.org/10.1002/9781119968535.

Sibly, R.M., Grimm, V., Martin, B.T., Johnston, A.S.A., Kulakowska, K., Topping, C.J., Calow, P., Nabe-Nielsen, J., Thorbek, P., Deangelis, D.L., 2013. Representing the acquisition and use of energy by individuals in agent-based models of animal populations. Methods Ecol. Evol. 4 (2), 151–161. https://doi.org/10.1111/2041-210x.12002.

Simpson, S.J., Sibly, R.M., Lee, K.P., Behmer, S.T., Raubenheimer, D., 2004. Optimal foraging when regulating intake of multiple nutrients. Anim. Behav. 68 (6), 1299–1311. https://doi.org/10.1016/j.anbehav.2004.03.003.

Smith, R.R., Rumsey, G.L., Scott, M.L., 1978. Net energy maintenance requirements of salmonids as measured by direct calorimetry: effect of body size and environmental temperature. J. Nutr. 108 (6), 1017–1024. Oxford Academic https://doi.org/10.1093/jn/108.6.1017.

Sousa, T., Domingos, T., Poggiale, J.C., Kooijman, S.A.L.M., 2010. Dynamic Energy Budget Theory Restores Coherence in Biology. Royal Society, https://doi.org/10.1098/rstb.2010.0166.

Spitz, J., Mourocq, E., Schoen, V., Ridoux, V., 2010. Proximate composition and energy content of forage species from the Bay of Biscay: high- or low-quality food? ICES J. Mar. Sci. 67 (5), 909–915. Oxford Academic https://doi.org/10.1093/icesjms/fsq008.

Steell, S.C., Van Leeuwen, T.E., Brownscombe, J.W., Cooke, S.J., Eliason, E.J., 2019. An appetite for invasion: digestive physiology, thermal performance and food intake in lionfish (Pterois spp.). J. Exp. Biol. 222 (19), jeb209437. https://doi.org/10.1242/jeb.209437.

Steffensen, J.F., Johansen, K., Bushnell, P.G., 1984. An automated swimming respirometer. Comp. Biochem. Physiol. A Physiol. 79 (3), 437–440.

Stein, R.A., Goodman, C.G., Marschall, E.A., 1984. Using time and energetic measures of cost in estimating prey value for fish predators. Ecology 65 (3), 702–715. https://doi.org/10.2307/1938042.

Ste-Marie, E., Watanabe, Y.Y., Semmens, J.M., Marcoux, M., Hussey, N.E., 2020. A first look at the metabolic rate of Greenland sharks (*Somniosus microcephalus*) in the Canadian Arctic. Sci. Rep. 10 (1), 1–8. Nature Research https://doi.org/10.1038/s41598-020-76371-0.

Svendsen, J.C., Tirsgaard, B., Cordero, G.A., Steffensen, J.F., 2015. Intraspecific variation in aerobic and anaerobic locomotion: Gilthead sea bream (*Sparus aurata*) and Trinidadian guppy (*Poecilia reticulata*) do not exhibit a trade-off between maximum sustained swimming speed and minimum cost of transport. Front. Physiol. 6 (FEB), 43. Frontiers Media S.A https://doi.org/10.3389/fphys.2015.00043.

Svendsen, M.B.S., Bushnell, P.G., Steffensen, J.F., 2016. Design and setup of intermittent-flow respirometry system for aquatic organisms. J. Fish Biol. 88 (1), 26–50. https://doi.org/10.1111/jfb.12797.

Svendsen, E., Føre, M., Økland, F., Gräns, A., Hedger, R.D., Alfredsen, J.A., Uglem, I., Rosten, C.M., Frank, K., Erikson, U., Finstad, B., 2021. Heart rate and swimming activity as stress indicators for Atlantic salmon (*Salmo salar*). Aquaculture 531, 735804. Elsevier B.V https://doi.org/10.1016/j.aquaculture.2020.735804.

Tanaka, Y., Gwak, W.S., Tanaka, M., Sawada, Y., Okada, T., Miyashita, S., Kumai, H., 2007. Ontogenetic changes in RNA, DNA and protein contents of laboratory-reared Pacific bluefin tuna *Thunnus orientalis*. Fish. Sci. 73 (2), 378–384. Springer https://doi.org/10.1111/j.1444-2906.2007.01345.x.

Taylor, T.N., Cross, B.K., Moore, B.C., 2020. Modeling brook trout carrying capacity in whi Lake, Washington, using bioenergetics. North Am. J. Fish. Manag. 40 (1), 84–104. https://doi.org/10.1002/nafm.10378.

Teal, L.R., van Hal, R., van Kooten, T., Ruardij, P., Rijnsdorp, A.D., 2012. Bio-energetics under-pins the spatial response of North Sea plaice (*Pleuronectes platessa* L.) and sole (*Solea solea* L.) to climate change. Glob. Chang. Biol. 18 (11), 3291–3305. John Wiley & Sons, Ltd https://doi.org/10.1111/j.1365-2486.2012.02795.x.

Teffer, A.K., Hinch, S.G., Miller, K.M., Patterson, D.A., Farrell, A.P., Cooke, S.J., Bass, A.L., Szekeres, P., Juanes, F., 2017. Capture severity, infectious disease processes and sex influ-ence post-release mortality of sockeye salmon bycatch. Conserv. Physiol. 5 (1), cox017. https://doi.org/10.1093/conphys/cox017. Oxford University Press.

Thiem, J.D., Wooden, I.J., Baumgartner, L.J., Butler, G.L., Forbes, J., Taylor, M.D., Watts, R.J., 2018. Abiotic drivers of activity in a large, free-ranging, freshwater teleost, Murray cod (*Maccullochella peelii*). PLoS One 13 (6), e0198972.

Tomlinson, S., Arnall, S.G., Munn, A., Bradshaw, S.D., Maloney, S.K., Dixon, K.W., Didham, R.K., 2014. Applications and implications of ecological energetics. Trends Ecol. Evol. 29 (5), 280–290. Elsevier Ltd https://doi.org/10.1016/j.tree.2014.03.003.

Townsend, C.R., Winfield, I.J., 1985. The application of optimal foraging theory to feeding beha-viour in fish. In: Fish Energetics., https://doi.org/10.1007/978-94-011-7918-8_3.

Treberg, J.R., Killen, S.S., MacCormack, T.J., Lamarre, S.G., Enders, E.C., 2016. Estimates of metabolic rate and major constituents of metabolic demand in fishes under field conditions: methods, proxies, and new perspectives. Comp. Biochem. Physiol. A Mol. Integr. Physiol. 202, 10–22. Elsevier Inc https://doi.org/10.1016/j.cbpa.2016.04.022.

Trudel, M., Rasmussen, J.B., 1997. Modeling the elimination of mercury by fish. Environ. Sci. Technol. 31 (6), 1716–1722. American Chemical Society https://doi.org/10.1021/es960609t.

Turchini, G.M., Trushenski, J.T., Glencross, B.D., 2019. Thoughts for the future of aquaculture nutrition: realigning perspectives to reflect contemporary issues related to judicious use of marine resources in aquafeeds. N. Am. J. Aquac. 81 (1), 13–39. https://doi.org/10.1002/naaq.10067.

Tytler, P., Calow, P., 1985. Fish Energetics: New Perspectives. Croom Helm Ltd, Sydney, Australia.

Uchmanski, J., Grimm, V., 1996. Individual-Based Modelling in Ecology: What Makes the Difference?: 437–440.

Utne, K.R., Hjøllo, S.S., Huse, G., Skogen, M., 2012. Estimating the consumption of *Calanus finmarchicus* by planktivorous fish in the Norwegian Sea using a fully coupled 3D model system. Mar. Biol. Res. 8 (August), 527–547. https://doi.org/10.1080/17451000.2011.642804.

Valdés, P., García-Alcázar, A., Abdel, I., Arizcun, M., Suárez, C., Abellán, E., 2004. Seasonal changes on gonadosomatic index and maturation stages in common pandora *Pagellus erythri-nus* (L.). Aquac. Int. 12 (4), 333–343. 2004 124. Springer https://doi.org/10.1023/B:AQUI.0000042136.91952.9E.

van der Veer, H.W., Cardoso, J.F.M.F., van der Meer, J., 2006. The estimation of DEB parameters for various Northeast Atlantic bivalve species. J. Sea Res. 56 (2), 107–124. Elsevier https://doi.org/10.1016/j.seares.2006.03.005.

van der Veer, H.W., Cardoso, J.F.M.F., Peck, M.A., Kooijman, S.A.L.M., 2009. Physiological performance of plaice *Pleuronectes platessa* (L.): a comparison of static and dynamic energy budgets. J. Sea Res. 62 (2–3), 83–92. Elsevier https://doi.org/10.1016/j.seares.2009.02.001.

van Ginneken, V., van den Thillart, G., 2009. Metabolic Depression in Fish Measured by Direct Calorimetry: A Review. Elsevier, https://doi.org/10.1016/j.tca.2008.09.027.

van Ginneken, V.J.T., Addink, A.D.F., Van Den Thillart, G.E.E.J.M., Körner, F., Noldus, L., Buma, M., 1997. Metabolic rate and level of activity determined in tilapia (*Oreochromis mossambicus* Peters) by direct and indirect calorimetry and videomonitoring. Thermochim. Acta 291 (1–2), 1–13. Elsevier https://doi.org/10.1016/s0040-6031(96)03106-1.

Van Walleghem, J.L.A., Blanchfield, P.J., Hrenchuk, L.E., Hintelmann, H., 2013. Mercury elimination by a top predator, *Esox lucius*. Environ. Sci. Technol. 47 (9), 4147–4154. American Chemical Society https://doi.org/10.1021/es304332v.

Van Waversveld, J., Addink, A.D.F., Van den Thillart, G., 1989. Simultaneous direct and indirect calorimetry on normoxic and anoxic goldfish. J. Exp. Biol. 142 (1), 325–335. The Company of Biologists https://doi.org/10.1242/jeb.142.1.325.

Van Zuiden, T.M., Sharma, S., 2016. Examining the effects of climate change and species invasions on Ontario walleye populations: can walleye beat the heat? Divers. Distrib. 22 (10), 1069–1079. John Wiley & Sons, Ltd https://doi.org/10.1111/DDI.12468.

Vollenweider, J.J., Heintz, R.A., Schaufler, L., Bradshaw, R., 2011. Seasonal cycles in whole-body proximate composition and energy content of forage fish vary with water depth. Mar. Biol. 158 (2), 413–427. Springer https://doi.org/10.1007/s00227-010-1569-3.

Wahl, D.H., 1999. An ecological context for evaluating the factors influencing muskellunge stocking success. North Am. J. Fish. Manag. 19 (1), 238–248. https://doi.org/10.1577/1548-8675(1999)019<0238:aecfet>2.0.co;2.

Walters, C., Essington, T., 2009. Recovery of bioenergetics parameters from information on growth: over-view of an approach based on statistical analysis of tagging and size-at-age data. Science 3 (1), 1–17. https://doi.org/10.2174/1874401X01003020052.

Watson, J., Hyder, K., Cooke, S.J., Roy, S., Sily, R.M., 2020. Assessing the sublethal impacts of anthropogenic stressors on fish: an energy-budget approach. Fish Fish. 21 (5), 1034–1045. https://doi.org/10.1111/faf.12487.

Whiterod, N.R., 2010. The bioenergetic implications for Murray cod (*Maccullochella peelii*) below a large hypolimnetic-releasing impoundment. Doctoral Dissertation. Charles Sturt University.

Whitney, N.M., White, C.F., Gleiss, A.C., Schwieterman, G.D., Anderson, P., Hueter, R.E., Skomal, G.B., 2016. A novel method for determining post-release mortality, behavior, and recovery period using acceleration data loggers. Fish. Res. 183, 210–221. Elsevier B.V https://doi.org/10.1016/j.fishres.2016.06.003.

Williams, K.J., Cassidy, A.A., Verhille, C.E., Lamarre, S.G., MacCormack, T.J., 2019. Diel cycling hypoxia enhances hypoxia tolerance in rainbow trout (Oncorhynchus mykiss): evidence of physiological and metabolic plasticity. J. Exp. Biol. 222 (14), jeb206045. https://doi.org/10.1242/jeb.206045. Company of Biologists Ltd.

Wilson, S.M., Hinch, S.G., Eliason, E.J., Farrell, A.P., Cooke, S.J., 2013. Calibrating acoustic acceleration transmitters for estimating energy use by wild adult Pacific salmon. Comp. Biochem. Physiol. A Mol. Integr. Physiol. 164 (3), 491–498. Elsevier Inc https://doi.org/10.1016/j.cbpa.2012.12.002.

Winberg, G.G., 1956. Rate of metabolism and food requirements of fishes. Fish. Res. Board Canada Transl. Ser. 194, 202.

Wuenschel, M.J., Jugovich, A.R., Hare, J.A., 2006. Estimating the energy density of fish: the importance of ontogeny. Trans. Am. Fish. Soc. 135 (2), 379–385. Wiley https://doi.org/10.1577/t04-233.1.

Young, N., Nguyen, V.M., Corriveau, M., Cooke, S.J., Hinch, S.G., 2016. Knowledge users' perspectives and advice on how to improve knowledge exchange and mobilization in the case of a co-managed fishery. Environ. Sci. Policy 66, 170–178. https://doi.org/10.1016/j.envsci.2016.09.002.

Zhang, Y., Claireaux, G., Takle, H., Jørgensen, S.M., Farrell, A.P., 2018. A three-phase excess post-exercise oxygen consumption in Atlantic salmon *Salmo salar* and its response to exercise training. J. Fish Biol. 92 (5), 1385–1403. Blackwell Publishing Ltd https://doi.org/10.1111/jfb.13593.

Zhou, B.S., Wu, R.S.S., Randall, D.J., Lam, P.K.S., 2001. Bioenergetics and RNA/DNA ratios in the common carp (*Cyprinus carpio*) under hypoxia. J. Comp. Physiol. B Biochem. Syst. Environ. Physiol. 171 (1), 49–57. Springer https://doi.org/10.1007/s003600000149.

Zimmer, A.M., Nawata, C.M., Wood, C.M., 2010. Physiological and molecular analysis of the interactive effects of feeding and high environmental ammonia on branchial ammonia excretion and Na+ uptake in freshwater rainbow trout. J. Comp. Physiol. B Biochem. Syst. Environ. Physiol. 180 (8), 1191–1204. Springer https://doi.org/10.1007/s00360-010-0488-4.

Zimmer, A.M., Mandic, M., Rourke, K.M., Perry, S.F., 2020. Breathing with fins: do the pectoral fins of larval fishes play a respiratory role? Am. J. Physiol. Regul. Integr. Comp. Physiol. 318 (1), R89–R97. American Physiological Society https://doi.org/10.1152/AJPREGU.00265.2019.

Chapter 5

Applied aspects of the cardiorespiratory system

Erika J. Eliason*, Jacey C. Van Wert, and Gail D. Schwieterman

Department of Ecology, Evolution, and Marine Biology, University of California-Santa Barbara, Santa Barbara, CA, United States
Corresponding author: e-mail: erika.eliason@lifesci.ucsb.edu

Chapter Outline

The cardiorespiratory system distributes oxygen and other factors (nutrients, wastes, hormones etc.) around the body and thus plays a central role in mediating many physiological processes such as digestion, locomotion, and reproduction. Building from a rich body of foundational research, cardiorespiratory physiology techniques are now being used for applied, conservation purposes. These techniques can evaluate performance at the whole animal, organ, or cellular levels. At the whole animal level, metabolism and energetic capacities are routinely measured. Aerobic scope (the maximum energetic capacity of a fish beyond maintenance; absolute aerobic scope = maximum metabolic rate − standard metabolic rate) is used to identify optimal conditions for fish. At the level of the heart, heart rate biologging and biotechnology techniques enable researchers to monitor performance in free-ranging fish, while Arrhenius breakpoint tests can measure the optimal and upper thermal limits for the heart. At the cellular level, blood sampling is an easy and effective way to gain a wealth of information about the physiological status of the fish. Several case studies are presented to illustrate how

Fish Physiology, Vol. 39A. https://doi.org/10.1016/bs.fp.2022.04.005

cardiorespiratory physiology techniques have been effectively put into practice (Pacific salmon conservation, shark fisheries-induced mortality, pelagic fishes responding to oil). We conclude with some recommendations to continue moving the field forward and advance fish conservation.

1 Introduction

All fish require energy to be able to grow, move, and reproduce. The nutrients and oxygen that power these processes must be obtained from the environment and efficiently delivered to the necessary tissues for a fish to thrive. Oxygen flows from the environment to the mitochondria through a series of diffusion and convection steps known as the "oxygen transport cascade." In brief, oxygen diffuses from the environment across the respiratory surface (gills, lung, air bladder, etc.) into the blood and (usually) binds to hemoglobin: a means of transportation that allows blood to deliver oxygen to tissues where it can diffuse into the cell and subsequently, the mitochondria. Within the cell, energy in the form of ATP is produced from nutrients (proteins, lipids, carbohydrates) either anaerobically (without oxygen) or aerobically (with oxygen). This process is executed by the "cardiorespiratory system," which encapsulates both the cardiovascular and respiratory systems and involves the integrated function of gas exchange surfaces, blood, heart(s), vasculature, and a control system. Beyond oxygen, the cardiorespiratory system also serves to transport other respiratory gases (CO_2, NH_3), hormones, metabolites, nutrients, immune factors, and wastes among active tissues. Thus, the cardiorespiratory system plays a central role in supporting a wide array of physiological processes such as digestion, locomotion, osmoregulation, reproduction, and immune function.

A vast body of literature has considered the importance of understanding cardiorespiratory function in fishes. F.E.J. Fry introduced the concept that fish have a "scope for activity" between their basic metabolic needs (standard metabolic rate) and maximum capacity (maximum metabolic rate), which is the energetic capacity for activities such as locomotion, growth, fighting (Fry, 1947). He devised a classification system for environmental factors as having lethal, controlling, limiting, masking, and directing influence on fish performance and their scope for activity (Fry, 1971). Building on this foundational framework, many fish researchers have considered how standard and maximum metabolism and aerobic scope change in response to environmental stressors (e.g., dissolved oxygen, salinity, temperature, pH, pollutants; Bouyoucos et al., 2019; Ern et al., 2014; Lefevre, 2016; Schulte, 2015). Extensive literature has examined how the cardiovascular system responds to external and internal perturbations such as temperature (Eliason and Anttila, 2017), oxygen (Stecyk, 2017), pollution (Incardona and Scholz, 2017), and disease (Powell and Yousaf, 2017). Similarly, many studies have focused on the cellular responses to stressors such as salinity (Evans and Kültz, 2020), temperature (Iwama et al., 1999), and oxygen (Richards, 2009).

Developing the ideas rooted in this fundamental research has generated significant interest in using cardiorespiratory research techniques to address applied conservation problems, which is the focus of this chapter.

In an applied context, the cardiorespiratory system has been used to evaluate the invasive potential of species (Marras et al., 2015; Steell et al., 2019), imperiled species vulnerability (Jeffries et al., 2016), thermal tolerance and climate change resilience (Eliason et al., 2011; Gilbert et al., 2020; Nilsson et al., 2009; Rummer et al., 2014), habitat alteration impacts (Cortese et al., 2021; Downie et al., 2021), pollution thresholds (Hook et al., 2018; Illing and Rummer, 2017), fishing practices (Marshall et al., 2012; Raby et al., 2015), aquaculture practices (Brijs et al., 2019; Yasuda et al., 2012; Zupa et al., 2021), and cumulative stressor impacts (Somo et al., 2020). Here, we detail the primary methods and techniques currently used in an applied setting before describing several case studies where cardiorespiratory research can be used to conserve or manage fish populations. We conclude by identifying emerging research interests and discussing the direction that must be taken in order to successfully integrate cardiorespiratory physiology into effective management and conservation.

2 Methods

While the field of cardiorespiratory research is vast, several experimental approaches are well-suited to investigate physiological responses to anthropogenic and environmental stressors. Here, we summarize some of the most common techniques used in applied investigations. These can be broadly broken down into three levels of biological organization: whole animal, organ, and cellular.

2.1 Whole animal

Every individual fish is the product of innumerable organ, cellular, and subcellular physiological processes and reactions. By measuring metrics of whole-animal performance, researchers can capture the ultimate product of these interwoven processes.

2.1.1 Respirometry

Metabolism is the sum of all chemical reactions in a fish's body, including aerobic and anaerobic reactions, that convert food to energy and building blocks for growth and body maintenance (Norin and Speers-Roesch, 2020). Metabolism is measured as metabolic rate, or energy turnover, and can be assessed in two primary ways: (1) calorimetry, which measures heat released from the body and provides the most accurate estimation of metabolic rate, or (2) respirometry, which estimates whole-animal rates of oxygen (O_2) consumption or carbon dioxide (CO_2) excretion to indirectly estimate metabolic rate (Ferrannini, 1988; Nelson, 2016). Although calorimetry provides a more accurate estimation of metabolic rate, it is rarely used in studies with fish because of the technical challenges it presents (Nelson, 2016, but see Regan et al., 2013, Regan et al., 2017). Instead, respirometry is the typical method of choice (Ferrannini, 1988). The major caveats of using O_2 consumption are that it does not allow for instantaneous measurements of aerobic metabolism, nor does it determine the true cost of anaerobic metabolism (instead, it relies on back-calculations—e.g., excess post exercise oxygen consumption [EPOC], discussed below). Nevertheless, respirometry has been used for nearly a century and is becoming increasingly prevalent because of technological advancements and increased accessibility (Kieffer, 2010; Steffensen et al., 1984; Svendsen et al., 2016b). Respirometry played a pivotal role in early experimental physiology and has since been instrumental to conservation physiology (e.g., Section 3.1; Clark et al., 2012, 2013; Donelson et al., 2011; Eliason et al., 2011; Raby et al., 2015; Rummer et al., 2013; Steell et al., 2019).

During respirometry, a fish is placed in a respirometer (sealed chamber), and oxygen uptake rates are measured over time (oxygen consumption; $\dot{M}O_2$). Importantly, it is the oxygen uptake and not oxygen being used by tissues that is being measured (Chabot et al., 2016). The apparatus is defined primarily by its method of replenishing oxygenated water in the sealed chamber, including closed (static respirometer, swimming respirometer), open (flow-through respirometer), or intermittent (intermittent-flow respirometer) (Blazka et al., 1960; Brett, 1964; Clark et al., 2013; Fry, 1971; Steffensen, 1989; Svendsen et al., 2016b). Of the respirometers, the swimming respirometer, or "swim flume" has been the most foundational to fish physiology and conservation (Brett, 1964; Fry, 1971). This swim flume consists of a closed loop of pipe with a pump powered by a variable speed motor that pushes water in laminar flow past the fish, creating a current that allows for oxygen measurements during swimming (Brett, 1964; Clark et al., 2013; Fry, 1971). However, respirometer type and design considerations (e.g., chamber size, flush rate and time, chamber mixing) are highly species-specific and described further in Svendsen et al. (2016a,b). Fish size, life history, and swimming style are important considerations for protocol design (Clark et al., 2013; Svendsen et al., 2016a,b). For example, swimming respirometers are insufficient for unsteady swimmers or sedentary species, and static respirometers are unsuitable for obligate ram ventilating fish because they are unable to ventilate their gills while motionless (Clark et al., 2013; Killen et al., 2017; Peake and Farrell, 2006).

Respirometry can be used to infer several key physiological values that can be applied more broadly to conservation. Standard metabolic rate (SMR) is the minimum measurable aerobic metabolism and represents the $\dot{M}O_2$ required for subsistence (e.g., maintaining ion gradients, cardiac pumping, ventilation, and protein turnover) (Brett and Groves, 1979; Fry, 1971; Schurmann and Steffensen, 1997). Accurate measurement of SMR requires that fish remain undisturbed in the respirometer chamber for 24–48 h, and are quiescent, post-absorptive, and non-reproductive (Chabot et al., 2016; Clark et al., 2013; Fry and Hart, 1948; Jobling, 1981; Secor, 2009). Logistical and experimental constraints can make it challenging to measure true SMR (see review by Chabot et al., 2016). Some fish are in a biological state (e.g., sexually mature spawning salmon) that prevents the measurement of SMR, and thus measured baseline metabolism is often termed Resting metabolic rate (Eliason et al., 2011; Lee et al., 2003; Steinhausen et al., 2008). Routine metabolic rate (RMR) may be reported, representing the cost to "act normal" including spontaneous movements, though typically measurements are made on quiescent fish protected from disturbances (Norin and Speers-Roesch, 2020). Note that the term RMR is used in the literature to refer to both resting metabolic rate and routine metabolic rate, so readers should exercise caution when comparing "RMR" across studies.

The maximum metabolic rate (MMR) is the upper metabolic rate threshold a fish can reach. This value represents the maximum capacity of the cardiorespiratory system to supply O_2 to tissues (Fry, 1971; Norin and Clark, 2016). A fish typically reaches MMR during maximal swimming, but might reach MMR in other ways, including during recovery (e.g., barramundi [*Lates calcarifer*]), or during digestion (e.g., lionfish [*Pterois* spp.]) (Norin and Clark, 2016; Steell et al., 2019). The two primary methods for estimating MMR are (1) the critical swim speed protocol and (2) the exhaustive chase protocol: (1) The critical swim speed protocol (U_{crit}) requires a swimming respirometer, where the fish first swims in steady-state (aerobic) and then transitions to unsteady (anaerobic) exhaustive exercise. Water velocity is incrementally increased and paired with concurrent $\dot{M}O_2$ measurements until the fish is exhausted and stops swimming, coined "U_{crit}." Because O_2 consumption is measured during exercise, the U_{crit} swim test is thought to provide the most accurate, highest MMR measurement (Farrell and Steffensen, 1987; Raby et al., 2020; Roche et al., 2013; Rummer et al., 2016; Shultz et al., 2011). Several studies, however, suggest the exhaustive chase protocol can provide comparable MMR measurements, though this equivalence might be species or life-stage specific (Killen et al., 2017; Little et al., 2020; Zhang et al., 2020). (2) The exhaustive chase protocol involves a manual chase in a circular tank of sufficient size, allowing the fish to burst and glide, a form of anaerobic swimming, for a set time (3–5 min) or until exhaustion (Clark et al., 2013; Donaldson et al., 2010b; Killen et al., 2017; Little et al., 2020; Roche et al., 2013; Rummer et al., 2016). For some dorso-ventrally flattened fishes, chasing must involve repeatedly turning the fish over to induce a "righting" response rather than burst and glide (Di Santo, 2016; Schwieterman et al., 2019). Some protocols include an additional minute of air exposure to ensure complete exhaustion and mimic fisheries interactions (e.g., hook removal, photography), before transferring the fish to a static respirometer and immediately measuring $\dot{M}O_2$ (Clark et al., 2012; Cooke et al., 2014; Donaldson et al., 2010b; Little et al., 2020; Norin et al., 2014; Roche et al., 2013). MMR measured after the exhaustive chase assumes that MMR occurs during the aerobic recovery phase (undergoes EPOC to repay the O_2 recovery costs from anaerobic swimming) (Killen et al., 2017). For best accuracy, the lifestyle and life stage of a fish must be considered when measuring and interpreting MMR (Norin and Clark, 2016; Raby et al., 2020; Reidy et al., 1995; Roche et al., 2013; Svendsen et al., 2016a,b).

The difference between these two limits (MMR–SMR) yields the absolute aerobic scope (AAS), which is the capacity for a fish to increase aerobic metabolism beyond basic needs/SMR (Figs. 1A and 2B) (Bushnell et al., 1994; Clark et al., 2011; Fry, 1947). This scope encompasses the energy to perform fitness enhancing activities, such as swimming, defending territory, competing for mates, foraging for food, digesting a meal, etc. AAS is influenced by intrinsic and extrinsic parameters and has been measured extensively

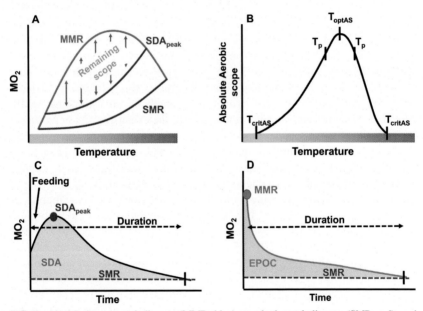

FIG. 1 (A) Maximum metabolic rate (MMR, blue), standard metabolic rate (SMR, red), and peak oxygen consumption rate (MO_2) during specific dynamic action (SDA) (teal) across temperatures. A constraint on remaining scope develops as temperature warms, at very warm temperatures fish do not have sufficient scope to digest a meal and perform other activities. (B) Absolute aerobic scope (AAS = MMR − SMR) thermal performance curve (TPC) with upper and lower critical thermal temperature (T_{critAS}), upper and lower pejus temperature (T_p) (here defined as 90% thermal optima [T_{optAS}]), and optimal temperature at peak AS (T_{optAS}). (C) Digestion response post-feeding. SDA is integrated beneath the MO_2 curve to SMR (or often resting metabolic rate, RMR). Peak MO_2 during SDA is highlighted (SDA_{peak}), generally occurring within several hours post-feeding. (D) Recovery response after exercise. Excess post oxygen consumption (EPOC) is integrated beneath the MO_2 curve to SMR, starting immediately after the fish ceases swimming and ending once the fish returns to SMR (or often RMR).

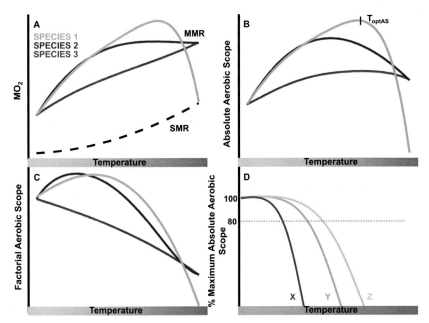

FIG. 2 Conceptual figure depicting how differences in maximum metabolic rate (MMR) curves result in different absolute aerobic scope (AAS = MMR − SMR) and factorial aerobic scope (FAS = MMR/SMR) thermal performance curves and influence overall performance. This example depicts three species (Species 1 [green], 2[blue], 3[purple]), but differences in MMR can also vary within a species. (A) Standard metabolic rate (SMR) increases exponentially with temperature, MMR has empirically been shown to increase (purple), plateau (blue), or collapse (green) at high temperatures. (B) Absolute aerobic scope with T_{optAS} noted for Species 1 and (C) Factorial aerobic scope based on the same exponential SMR but respective MMR curves for each species in panel A. Species 1 displays a collapse in both absolute and factorial aerobic scope, while species 3 displays a broad and consistent AAS across temperatures. (D) Percentage of maximum absolute aerobic scope (AAS) across temperatures, with curves (green spectrum) demonstrating how performance can still vary *within* Species 1 (e.g., across populations, sex, individuals). In this example, all groups (green X, green Y, green Z) have the same T_{opt} for aerobic scope, but Green X has the lowest upper thermal tolerance and Green Z has the highest upper thermal tolerance. Note the change in the x-axis temperature bar, as panel D represents the upper thermal tolerance starting from T_{optAS} in panel (B). The threshold value (in this example, 80%), is a proposed percent of AAS needed for some spawning adult salmon, but this threshold is expected to vary across species, life stages, etc.

to assess local adaptation, behavior, thermal tolerance, changes in biogeographic patterning, phenotypic plasticity, and vulnerability to external disturbances (Clark et al., 2013; Eliason et al., 2011; Farrell, 2016; Hvas et al., 2017; Killen et al., 2014; Metcalfe et al., 2016a; Pörtner, 2010; Schulte, 2015; Zhang et al., 2017, 2018a,b).

External factors that may affect AAS include disease, fishing, pollutants, carbon dioxide and temperature, in addition to other factors (Fitzgibbon et al., 2007; Gomez Isaza et al., 2020; Grans et al., 2014; Hvas et al., 2017;

Molina et al., 2020; Schulte, 2015). The environmental factor that has received the most research attention is temperature. The AAS curve is based on the dependence of SMR and MMR with temperature (Fig. 1B) and has been a critical constituent of conservation physiology because it can predict the theoretical capacity of a fish to perform aerobic activities across temperatures (Brett, 1971; Claireaux and Lefrançois, 2007; Farrell, 2016; Farrell et al., 2008; Fry, 1947). The AAS curve is dynamic and can change in shape and thermal range within a lifespan, across populations and species, and with thermal acclimation (Farrell, 2016). The critical thermal temperatures (T_{critAS}) are reached when AAS is zero at the cold and hot extremes, where it is considered a temporary state that is sustained by anaerobic metabolism (Fig. 1B). T_{critAS} has been emphasized as an important parameter to fisheries management because it represents the thermal extreme that induces mortality (Farrell, 2016; Pörtner, 2001; Pörtner and Farrell, 2008). However, a more useful threshold might be one that is sublethal experimentally, but environmentally would likely lead to a decline in fish performance. The pejus temperature (T_p) is a rate transition-temperature where AAS first starts to decline and should have significant conservation implications because it defines the point at which performance starts to become impaired (Fig. 1B) (Farrell, 2016). T_p is also important because the area between the upper and lower T_p defines the optimal range of temperatures for maximum AAS, where peak aerobic activity is predicted to occur (Brett, 1971). The AAS curve can also be visualized with the percent of maximum AAS (T_{optAS}) across temperatures (Fig. 2D). Determining the percentage of AAS necessary to complete some activity can be a useful metric for predicting thermal performance. For example, adult sockeye salmon (*Oncorhynchus nerka*) require 80–90% of AAS to complete the spawning migration, which has been directly applied to the management of fisheries (Fig. 5; see Section 4.6) (Eliason et al., 2011, 2013a,b; Farrell et al., 2008). However, this threshold can vary within the life cycle, and across populations and species and requires further development (see in Section 4.4) (Farrell, 2016). The mechanistic underpinnings of the collapse in aerobic scope at temperature extremes has received considerable research attention and vigorous debate and has been extensively reviewed elsewhere (see Jutfelt et al., 2018; Pörtner and Farrell, 2008; Pörtner and Knust, 2007; Portner et al., 2017; Schulte, 2015, for discussion on the Oxygen- and Capacity-Limited Thermal Tolerance Hypothesis, OCLTT).

Studies have found discrepancies in the assumption that optimal AAS (T_{optAS}) reflects optimal existence (Clark et al., 2011; Grans et al., 2014; Halsey et al., 2018; Hvas et al., 2017; Norin et al., 2014). There are several important considerations. First, the timescales of temperature exposure are essential to consider (Schulte et al., 2011) yet are inconsistent across studies, which makes comparisons challenging (Lefevre et al., 2021). Further, MMR is often thought to plateau or collapse with temperature and result in a decrease in AAS (Chabot et al., 2016; Eliason et al., 2011; Rummer et al., 2014), but there are also cases where MMR continues to increase (Fig. 2A) (Poletto et al., 2017), resulting in different AAS and factorial aerobic scope (FAS) response curves (Fig. 2B and C). Furthermore, optimal temperatures can vary substantially across traits (e.g., growth, swimming, behavior) and performance requirements, and thus the ecological relevance of AAS must be taken in context of a fish's environment (Clark et al., 2013; Cooke et al., 2021; Farrell, 2016; Norin et al., 2014). For example, Clark et al. (2011) found the T_{optAS} of pink salmon (*Oncorhynchus gorbuscha*) to be 21 °C, but this is not an optimal temperature for all traits and reproduction would likely fail at this temperature. Hvas et al. (2017) showed that adult Atlantic salmon (*Salmo salar*) swim maximally at 18 °C while the AAS remained high at 23 °C. Work by Grans et al. (2014) on Atlantic halibut (*Hippoglossus hippoglossus*) highlighted a mismatch between growth and optimal AAS. This lack of congruence between important traits and optimal AAS indicates the need for reassessment and reframing of performance and AAS.

Not all fish need their entire AAS to thrive, so AAS might not be the most relevant proxy for determining the capacity to thrive in the wild (Halsey et al., 2018). FAS is a measure of the proportional amount of scope available (MMR/SMR) to perform aerobic functions that scale with SMR, such as digestion and swimming (Clark et al., 2013; Farrell, 2016). FAS reveals how a biological factor might constrain maximum capacity in relation to SMR. A high FAS indicates low maintenance costs (therefore greater capacity for other activities) and a low FAS indicates high maintenance costs (reduced capacity for other activities) (Farrell, 2016). See Section 4.4, for more information on FAS.

An important but often underappreciated metric in ecophysiology is specific dynamic action (SDA), which is the rise in metabolism during digestion (Jobling, 1981; McCue, 2006; Secor, 2009). The response represents the increase in aerobic metabolism following ingestion that is required to digest and assimilate nutrients (McCue, 2006). It is measured by feeding an individual some quantity of food and immediately measuring the postprandial $\dot{M}O_2$ response for hours to days, depending on meal size, food quality, species, and temperature (Eliason et al., 2007; McCue, 2006; Pang et al., 2011; see review by Chabot et al., 2016; McKenzie et al., 2013). SDA generally has a peak $\dot{M}O_2$ value several hours after feeding (SDA_{peak}) and is complete when the fish has returned to baseline $\dot{M}O_2$, and thus SDA is calculated as the area under $\dot{M}O_2$ trace bounded by baseline metabolism (or other baseline level, e.g., routine MR) (Fig. 1C) (Jobling, 1994). Other responses measured with SDA are time to peak SDA, overall duration, and the SDA coefficient (SDA/meal energy) (Secor, 2009). The SDA response can occupy a major portion of AAS in fishes (Fig. 1A), and for some individuals, SDA_{peak} can exceed measured activity costs following exhaustive exercise and represent the true MMR (e.g., lionfish; Steell et al., 2019). One tradeoff that exists is the growth efficiency and meal size; a larger meal can improve growth efficiency but may constrain AAS (Norin and Clark, 2017). This may have implications on how fish prioritize aerobic capacity under suboptimal conditions. The SDA response has been used to predict the performance of fishes in different temperature scenarios in aquaculture settings (Frisk et al., 2013; McCue, 2006), with only a few studies that have paired SDA with AAS in a broader ocean warming context (Flikac et al., 2020; Sandblom et al., 2014; Steell et al., 2019; Tirsgaard et al., 2015a) or ocean acidification scenarios (Tirsgaard et al., 2015b). Nevertheless, because of its significant role in a fish's daily energy expenditure (Fig. 1A), SDA has major ecological implications that may be a useful performance metric in future thermal tolerances studies (Lefevre et al., 2021; Sandblom et al., 2014; Steell et al., 2019).

The recovery profile reveals a fish's capability to recover, thrive, and survive after an exhaustive exercise event. The primary metric, EPOC, measures the oxygen required to re-establish homeostasis following exhaustive anaerobic exercise (Lee et al., 2003). This increase in oxygen uptake is necessary to restore cardiorespiratory, hormone, metabolite, and osmoregulatory balance that was disrupted during anaerobic metabolism (Gaesser and Brooks, 1984; Kieffer, 2000; Milligan, 1996). Recovery requires time and energy and has been measured by magnitude and duration. Total EPOC encompasses the overall magnitude of recovery and is the integration between the $\dot{M}O_2$ trace and baseline metabolism following exhaustive exercise until the fish returns to baseline metabolism (Fig. 1D). The duration can be measured at several timescales with different phases of recovery and are discussed in Zhang et al. (2018a,b). A more recent metric that shows promise in ecological relevance is the time to MMR_{50}, or time to recover to 50% of MMR (Kraskura et al., 2021). This metric can occur within minutes to hours depending on the level of prior stress and indicates when an individual may be able to resume normal activity (Kraskura et al., 2021). Recovery has broad ecological and conservation implications. The time to restore muscle energy stores and clear metabolites will limit performance and determine the potential frequency to maximally perform (Milligan, 1996). A prolonged recovery means missed opportunities (feeding, mating, migrating) or increased susceptibility to predators, influencing long-term success within a lifetime and across generations (Eliason and Farrell, 2016). Metrics related to metabolic recovery are relatively easy to obtain and hold great potential in the broad context of anthropogenic interactions. With changing environments and unpredictable stressors, the ability to rapidly recover from disturbances means life or death and ultimate success.

Experiments are now bringing respirometry to the sides of streams and more remote locations because of recent advancements in respirometry equipment and data acquisition software (Barnett et al., 2016; Bouyoucos et al., 2018; Chung et al., 2020; Farrell et al., 2003; Gilbert et al., 2020; Little et al., 2020; Mochnacz et al., 2017; Svendsen et al., 2016b; Verhille et al., 2016).

Field-based assessments can be more reliable in measuring relevant oxygen uptake rates and thermal performance because fish are acclimatized and are less affected by transport and maintenance in laboratory conditions (Treberg et al., 2016). Additionally, in situ metabolic measurements are possible by calibrating some metric (e.g., electromyogram [EMG], telemetry, heart rate, ventilation, dynamic body acceleration [DBA]) to $\dot{M}O_2$ consumption in the lab and observing fish in the field to estimate a field metabolic rate (FMR) in fishes (Byrnes et al., 2021; Clark et al., 2010; Hinch and Rand, 1998; Lucas et al., 1993; Metcalfe et al., 2016b; Prystay et al., 2020; Treberg et al., 2016). These methods present their own challenges but have the potential to grow in application (Homyack, 2010; Treberg et al., 2016). Most importantly, as respirometry gains traction across disciplines and is applied to management, it must be measured reliably and reported with repeatability and transparency (Clark et al., 2013).

2.2 Organ—Heart

The heart is essential for pumping blood to distribute oxygen, nutrients, immune factors, hormones, and wastes throughout the body. This section briefly describes two techniques used to measure heart rate via recording the electrocardiogram (ECG) waveform because these are broadly available and applicable for applied research. However, many other valuable techniques used to measure cardiac structure and function are not detailed here, including microscopy, histology, and *in vivo* surgical intervention (e.g., ventral aorta flow probes).

2.2.1 Heart rate biologging and biotelemetry

Broadly speaking, biotelemetry and biologging technology allows us to measure behavior and physiology in free-swimming individual fish in their natural (or semi-natural), complex environment (Cooke et al., 2012a,b; Treberg et al., 2016). Heart rate biologgers have been used to examine energy use (Clark et al., 2010), survival/fitness consequences (Prystay et al., 2019, 2020; Twardek et al., 2021), stress (Lovén Wallerius et al., 2019), recovery from fisheries interactions (Donaldson et al., 2010b; Prystay et al., 2017; Raby et al., 2015), and fish welfare/aquaculture practices (Brijs et al., 2018, 2019, 2021). In biotelemetry, a signal is transmitted from the fish to a receiver whereas in biologging, the information is stored locally on a device in the fish. Both systems have their pros and cons. When using biotelemetry, fish must pass close to a receiver to transmit the data, which can be challenging in open environments that require numerous receivers. Biologging requires that the loggers be successfully recovered to download the data, limiting experimental design options. In addition, biologging technology tends to be quite expensive, thus, to date, most studies are still performed in a relatively controlled setting where tags are ensured to be recovered. However, the advantages of these technologies are numerous, allowing fish researchers to measure heart rate, acceleration, temperature, depth, salinity (and more) in fishes living a wide variety of natural conditions. Accordingly, there is great potential for heart rate biologgers to be used for conservation purposes (e.g., determine stressor thresholds, assess fishing practices, improve fish welfare).

Currently, heart rate is measured more frequently using biologgers (Brijs et al., 2018, 2019; Donaldson et al., 2010b; Føre et al., 2021; Prystay et al., 2017, 2019, 2020; Raby et al., 2015; Zrini and Gamperl, 2021) rather than via biotelemetry (Anderson et al., 1998). Devices are commercially available for purchase, including heart rate loggers from Star-Oddi (Denmark), and biotelemetry tags from Thelma BioTel (Norway) and Transonic (USA); however, some research groups develop their own systems in-house. Most of these devices transmit or record the electrocardiogram (ECG) waveform (e.g., at 100 Hz) and/or the processed mean heart rate (e.g., in beats per minute [BPM]) at a set timing interval (e.g., every 10 min). The frequency of the timing interval will depend on the species in question, battery/memory of the device and the duration of the experiment.

Heart rate tags are implanted into the peritoneal cavity via surgery. Aseptic surgical techniques should be followed, including regular equipment sterilization, even in a field setting (see Cooke et al., 2012a,b, for details and recommendations for tag implantation). One major assumption of tagging studies is that the tags do not adversely affect the behavior and physiology of the fish (Cooke et al., 2012a,b). As such, recovery from surgery and anesthesia are important to consider. Studies with rainbow trout and Atlantic salmon using chemical anesthesia report from 4 days and up to 2 weeks recovery time (Brijs et al., 2018, 2019; Føre et al., 2021; Zrini and Gamperl, 2021). In contrast, other studies have electro-sedated wild sockeye salmon (Prystay et al., 2020), smallmouth bass (*Micropterus dolomieu*) (Prystay et al., 2019), and Chinook salmon (*Oncorhynchus tshawytscha*) (Twardek et al., 2021) using fish handling gloves (Smith-Root Inc., Washington, DC) directly in the field. These studies tended to observe more rapid recovery times of ~10 h (Twardek et al., 2021), though often the loggers did not record the initial recovery period to conserve battery (e.g., 9–24 h [Prystay et al., 2019; Prystay et al., 2020]). Recovery time is likely to be influenced by numerous factors, including the type of sedation, invasiveness and duration of surgery, water temperature, fish species and size (Føre et al., 2021).

2.2.2 Arrhenius breakpoint methods

As ectotherms, temperature has a profound effect on the physiological processes of fishes. Understanding the upper thermal limits of individuals, populations, and species is, therefore, essential to effective management in the face of climate change and associated ocean and freshwater warming (Horodysky et al., 2015; McKenzie et al., 2016). Determining thermal ranges, optimum temperatures (T_{opt}), and upper thermal limits (T_{crit}) has often relied upon measures of aerobic metabolism (see Section 2.1.1), although this does not provide an underlying mechanistic explanation for differing thermal responses. Numerous studies have found correlations between cardiac function and upper thermal limit (e.g., Eliason et al., 2013b; Ekström et al., 2021), lending confidence to the assertion that cardiac function (and thus oxygen delivery) may be a driver of upper thermal limits in fishes (Eliason and Anttila, 2017; Pörtner, 2010).

One method used to determine the upper thermal limits of the heart is the Arrhenius Breakpoint Temperature (ABT) test on maximum heart rate (f_{Hmax}). To conduct this test, a fish is anesthetized and injected with atropine (to block vagal tone, removing the "brake" on the heart) and isoproterenol (a beta-adrenergic antagonist, pressing the "accelerator"), thus achieving f_{Hmax}. Electrodes are placed beneath the skin to continuously record ECGs, and the fish is gradually warmed until the heart becomes arrhythmic. From the resulting ECG, multiple metrics of thermal tolerance can be calculated: Maximum f_{Hmax}, temperature at maximum heart rate (T_{peak}), temperature at the onset of arrythmias (T_{arr}), and the temperature as which the rate of increase in f_{Hmax} slows, as determined through Arrhenius breakpoint analysis (T_{ABT}) (Casselman et al., 2012; Gilbert and Farrell, 2021). The actual ABT analysis is performed on the rising phase of the heart rate only and does not describe the decline in heart rate observed following T_{peak}. Arrhenius plots (natural log of maximum heart rate vs. the inverse of test temperature in Kelvin) are made for each individual (Anttila et al., 2013).

This relatively simple, rapid test can provide a wealth of information that is relevant in applied contexts. For example, T_{ABT} was found to be associated with a population's preferred temperature (T_{pref}) (Casselman et al., 2012), though more work needs to be done to evaluate if this holds true across species (e.g., see Ferreira et al., 2014; Hansen et al., 2016). T_{arr} is typically 2–6.7 °C lower than CT_{max} and can be used to determine life stage-, species-, and population-differences in thermal sensitivity (Chen et al., 2013; Drost et al., 2016; Sidhu et al., 2014). Because increases in maximum heart rate indicate changes to cardiorespiratory function and thus metabolic capacity, the ABT test can measure the acclimation potential of different species to climate change scenarios (Anttila et al., 2014; Chen et al., 2018a; Gilbert and Farrell, 2021). It has also been applied to questions regarding diet and transgenic mutations in salmon aquaculture, with consequences for transgenic fishes who are released into the wild population (Chen et al., 2015), as well as to assess the impact of crude oil spills on juvenile rainbow trout and European sea bass (*Dicentrarchus labrax*) (Anttila et al., 2017). In these cases, changes in T_{ABT}, T_{peak}, and T_{arr} can indicate reduced thermal tolerance of the heart, which is assumed to translate into reductions in whole animal thermal tolerance. When paired with other indicators, such as metabolic rate or protein kinase activity, this method can help determine mechanisms driving thermal limitations (Anttila et al., 2013).

The ABT test is relatively simple to conduct, provides a suite of information in a mere few hours of experimentation, and has been adapted for field settings (Chen et al., 2018a). It therefore has the potential to easily provide essential information regarding the thermal tolerances that can be used to understand species- and population-specific vulnerabilities to global warming. In contrast to respirometry, these experiments can be conducted more quickly, thus enabling higher sample sizes and increased statistical confidence. The ABT test is not suitable for all situations, however. Some have found ECG recordings in saltwater to be unstable, but this can be overcome with implantable heart rate loggers (Skeeles et al., 2020) or with subcutaneous electrodes. This test does have a terminal endpoint if measuring T_{arr}, and thus may not be appropriate for endangered or protected species. If T_{peak} is a suitable response variable, however, the fish may be able to recover and be released following the test (Chen et al., 2018a).

2.3 Cellular

To understand the mechanisms driving changes in organ or whole animal physiological performance, we often need to examine specific processes at the cellular level.

2.3.1 Blood

Blood analysis has broad applications, including determination of reproductive status (see Chapter 6, Volume 39A: Bernier and Alderman, 2022), digestion, immune function, osmoregulation (see Chapter 7, Volume 39A: Wood, 2022), as well as acute and chronic stress responses (Roche and Bogé, 1996). Here, we will briefly describe some of the most common parameters that can be measured from a blood draw and their physiological significance for management and conservation of fishes.

Blood can be obtained from an individual in multiple ways, including through cannulation of the dorsal aorta (Soivio and Okari, 1976), cardiac puncture (Blaxhall and Daisley, 1973), caudal venipuncture (Lawrence et al., 2020), pectoral venipuncture (in elasmobranch fishes; Phillips et al., 2016), severance of the caudal peduncle (Hesser, 1960), decapitation to access the dorsal aorta (Allen, 1994), and dorsal gill incision (Watson et al., 1989). The method of blood collection will depend on the size of the fish, as well as the parameters of interest. For example, it may be difficult to obtain blood via caudal venipuncture in fish that are only a few centimeters in length. Likewise, accurate measures of the partial pressure of oxygen (PO_2) rely on obtaining isolated venous or arterial blood that has not been air exposed, rendering techniques such as cannulation of the dorsal aorta preferable to caudal peduncle severance.

Once a blood sample has been obtained, there are a suite of different variables that can provide information on an individual's physiological state. For example, stress frequently threatens an individual's acid-base balance (see Chapter 7, Volume 39A: Wood, 2022). A generalized acidosis, measured as a drop in plasma pH (pH_e), results from air exposure and increases in metabolic expenditure. Acidosis decreases both blood oxygen affinity (Bohr Effect) and maximum oxygen carrying capacity (Root Effect) (Jensen, 2004; Root, 1931), hindering oxygen binding at the gills and subsequent oxygen delivery at a time when oxygen demand at the tissues is high. Measuring pH_e, PO_2, the partial pressure of carbon dioxide (PCO_2), and bicarbonate concentrations ($[HCO_3^-]$) can help determine the extent of disruptions to acid-base balance (see Brill et al., 1992; Cameron and Davis, 1970; Tucker, 1967, for techniques). Decreases in venous PO_2 in particular can signify increased oxygen consumption by the tissue, which could lead to insufficient oxygen for the heart (Farrell and Clutterham, 2003). While calculations of the full Henderson-Hasselbalch equation have typically only been measured in controlled laboratory settings, technological advances in high resolution, handheld pH meters are enabling pH_e measurements in field settings, thus enhancing the ability to detect disruptions to acid-base balance under true environmental or anthropogenic stressors (Gallagher et al., 2010; Harter et al., 2015; Talwar et al., 2017).

Hematocrit (Hct) and hemoglobin concentration ([Hb]) are both metrics that can provide further information about the secondary stress response (i.e., the changes in metabolism, respiration, acid-base status, and cellular responses in response to the primary neuroendocrine stress response). Hct, the proportion of the blood composed of red blood cells (RBCs), increases following stress due to (1) an increase in the number of circulating RBCs via splenic contraction, and (2) an increase in RBC volume (e.g., RBC swelling) due to beta-adrenergically stimulated sodium-hydrogen exchangers and the subsequent influx of osmotically active Na^+ (Jensen, 1987; Nikinmaa, 2012). Both processes increase blood-oxygen transport capacity, with splenic contraction increasing the density of Hb-O_2 binding sites per unit volume, and the latter restoring intracellular pH (pH_i) to improve Hb-O_2 affinity and maximum carrying capacity (Jensen, 2004). Thus, an increase in both Hct and [Hb] indicates that multiple physiological responses have occurred to meet high oxygen demand. Hct is measured by centrifuging a small blood sample in a microcapillary tube and then calculating the percentage of blood volume comprising red blood cells (as opposed to white blood cells and plasma). Hb can be measured via a handheld point-of-care device, the Hemocue, with the caveat that correction equations are necessary to account for differences between human and fish erythrocytes (Andrewartha et al., 2016; Clark et al., 2008a; Schwieterman et al., 2019). It can also be measured spectrophotometrically using the traditional cyanomethemoglobin method (Dacie and Lewis, 1975).

Metabolites are also useful in determining the level of metabolic expenditure. Lactate, a metabolic byproduct, can be used to infer high levels of anaerobic metabolism, and can be immediately measured using a point-of-care device (i-Stat; Gallagher et al., 2010; Harter et al., 2015; Lactate Plus; Stoot et al., 2014), using enzyme assays (Marbach and Weil, 1967), or from $-80\,°C$ frozen plasma using a benchtop analyzer (e.g., Critical Care Xpress; Marshall et al., 2012; YSI 2300; Farrell et al., 2001a). Similarly, the catecholamine-induced mobilization of glucose stores from the liver, evidenced by increased glucose concentrations in the blood (hyperglycemia), can indicate a high energy demand (Skomal and Bernal, 2010; Wells and Pankhurst, 1999), although this metric can be largely impacted by feeding and diet (Barton et al., 1988).

Changes in electrolyte concentrations indicate cellular damage and may have downstream effects. For example, the leakage of intracellular K^+ from heavily working skeletal muscle and its accumulation in the blood (hyperkalemia) can negatively impact cardiac function and may be associated with delayed mortality (Hanson et al., 2006; Schwieterman et al., 2021). Increases in plasma $[Ca^{2+}]$, $[Mg^{2+}]$, $[Cl^-]$, and $[Na^+]$ may also be the result of muscle cell damage or intracellular acidosis, driving these electrolytes into the blood. Changes in $[Cl^-]$ and $[Na^+]$ could also demonstrate a loss of osmoregulatory ability (Cliff and Thurman, 1984; Moyes et al., 2006; Wells et al., 1986), which can be fatal if prolonged. In field studies, electrolyte concentrations are typically measured using an i-stat point-of-care device (Gallagher et al., 2010; Harter et al., 2015), or the plasma is separated and frozen for later analysis using benchtop analyzers (e.g., Critical Care Xpress; Schlenker et al., 2016; model 510 flame photometer; 4,425,000 Haake Buchler digital chloridometer; Farrell et al., 2001a,b).

Heat shock proteins (Hsp) are inducible proteins that protect the cell from stressors that cause protein malfolding. Hsp concentrations increase with the stress response and help identify stressors that cause intracellular changes (Currie et al., 1999; Iwama et al., 1999). Changes in Hsp concentrations can be measured from blood samples, as well as from the liver, muscle, brain, spleen, kidney, and fin tissue (see Mohanty et al., 2018, for a review).

Hormones (e.g., cortisol, estradiol, testosterone) are commonly measured to assess the stress response or reproductive status of an individual (Wendelaar-Bonga, 1997). Hormones can be highly sensitive to stressors, with plasma cortisol levels being widely accepted as an indicator of acute (but not chronic) stress in teleost fish (see Sadoul and Geffroy, 2019 for techniques). Cortisol levels vary widely with season, life stage, and fasted state, but have been employed to understand a species' response to marine heat waves, hypoxia, pollution, and other anthropogenic stressors (Alfonso et al., 2020; Hontela et al., 1992; O'Connor et al., 2011), though appropriate interpretation is required (Romero and Beattie, 2022). Reproductive hormones isolated from blood plasma samples can help to reveal sex, age at maturation, and reproductive status, all of which can help identify important spawning and parturition habitats (McKinley et al., 1998; Sulikowski et al., 2016).

The benefit of blood analysis is its broad applicability and the ease of sample collection, facilitating rapid assessments of an individuals' stress response. These metrics can be used to inform regulations regarding fishing gear, handling time, and seasonal closures which aim to reduce fishing mortality (Koldkjaer et al., 2004; Martins et al., 2018). One major drawback of this method is the relative difficulty in obtaining baseline or control values for a specific species or life stage, as obtaining a blood sample in the field usually involves the inherently stressful processes of capture and restraint. Particularly for teleost fishes, handling time and air exposure should be minimized and avoided, respectively, to ensure results are as close to unstressed conditions as possible. Increases in medical technology are enabling more field-based measures of blood parameters, which may alleviate some of these concerns. The medical devices used for blood analysis were, however, originally designed and optimized for use with human blood. There are fundamental differences between human and fish blood (e.g., fish have nucleated red blood cells, temperature is typically much cooler). As such, devices must be calibrated for use with fish blood and even then, some variables yield inaccurate results (Harter et al., 2015; Schwieterman et al., 2019; Stoot et al., 2014).

2.3.2 Cellular metabolites

A myriad of different cellular metabolites could be assessed and considered here. We focus on those that relate to the oxygen transport cascade and have been used in an applied context. Several tissue metabolites are of applied interest because they can reflect activity and recovery dynamics of a fish. ATP is the energy currency of the cell; it is produced aerobically via oxidative phosphorylation and anaerobically via glycolysis. Both ATP and Phosphocreatine (PCr) are important energy sources; their stores are rapidly depleted during exercise and then restored during recovery. Tissue (e.g., ventricle, muscle, liver) glycogen indicates the stored energy available to the fish. Glycolysis (anaerobic process that uses glucose to produce two ATP, two NADH, and two pyruvate) produces ATP rapidly (e.g., to fuel rapid burst swimming) and in the absence of oxygen (e.g., during hypoxia). Pyruvate is converted to lactate by lactate dehydrogenase, and thus lactate is indicative of anaerobic metabolism. Lactate is also an important source of energy and is converted back to pyruvate by lactate dehydrogenase during recovery.

Some tissue samples can be biopsied non-lethally from fish. For example, a small gill clip may be collected from the end of the filaments or a small muscle sample (~3 mm) may be biopsied from the lateral body wall. However, for very small fishes or certain tissues (e.g., heart, liver), some samples can only be obtained from euthanized individuals. Tissue samples must be rapidly flash frozen in liquid nitrogen and stored at $-80\,°C$ until analysis because metabolites can change quickly. Detailed assay techniques can be found in the iconic book on analytical biochemistry: *"Methods of Enzymatic Analysis"* edited by Bergmeyer (1983).

3 Applied case studies

There are a few clear examples where cardiorespiratory research has been effectively implemented into management or conservation action with fishes. Here, we report two clear case studies from Pacific salmon and another from pelagic fishes and oil spills. We also report on promising avenues emerging to assist with shark catch and release mortality estimates.

3.1 Pacific salmon

Pacific salmon (*Onchorynchus* spp.) have a longstanding cultural significance in indigenous communities, an integral role in fisheries and economies across the North Pacific, and a perpetual, nearly priceless input to the ecosystems of the Pacific Northwest (Cooke et al., 2012a,b; Janetski et al., 2009). Their value is shared across people, borders, and systems and has made them a primary focus in conservation efforts. The anadromous lifecycle of salmon makes them particularly vulnerable to stressors. Pacific salmon are semelparous which means they have one chance at spawning before they die. During their final migration (from the ocean to their natal rivers), salmon stop eating and have a finite amount of energy to develop gonads and swim upstream, often traversing hydraulic challenges across long distances (up to 3200 km in the Yukon River chinook salmon) (Eiler et al., 2015; Eliason et al., 2011). Pacific salmon have varying levels of homing, spawning site fidelity and migration timing (Dittman and Quinn, 1996). Since the glaciation retreat has opened new spawning areas, each species has evolved genetically and phenotypically distinct populations displaying local adaptation to their specific environmental conditions (Crossin et al., 2004; Eliason et al., 2011; Lee et al., 2003, Ramstad et al., 2004, see review by Zillig et al., 2021).

What started off as scientific inquiry into fish diversity and physiology with salmonids as a model system soon became highly relevant in the conservation of Pacific salmon (Brett et al., 1958; Fry, 1948). Populations are in severe decline across their wide latitudinal range from California to Alaska in the Eastern Pacific (Grant et al., 2019; Moyle et al., 2017). Anthropogenic effects (e.g., overfishing, climate change, habitat modification) have exacerbated salmon mortalities to the point of extirpation or near extinction for many populations (Bowerman et al., 2021; Moyle et al., 2017; Quiñones et al., 2015). Multi-year droughts in California have increased the demand for freshwater, an essential habitat to migrating salmon, compounding the tension and unresolved issues of water rights and best management practices (Lennox et al., 2019). In California, nearly 90% of extant salmonids are in danger of extinction within 50–100 years if trends continue (Moyle et al., 2017). In the center of their range in southern British Columbia, Canada, nearly all species are declining in abundance (Grant et al., 2019). Recent and ongoing efforts toward understanding fisheries interactions and thermal physiology using cardiorespiratory physiology have been successfully applied to the conservation and management of this iconic group of fish.

3.1.1 Effects of fisheries on capture, release, and recovery

A migrating adult salmon can interact with fisheries in both marine and freshwater environments. Pacific salmon can interact and be released from fisheries for a few reasons: they escape, they are released by the fisherman for conservation (catch and release- it is mandated by law), or they are not the targeted species and released as bycatch. Fishing methods vary (angling, gill net, purse seine, beach seine, etc.) and cause different levels of physiological disturbance due to exhaustive exercise, air exposure, injury, and handling stress (Cooke and Suski, 2005; Davis, 2002). During this interaction, fish are likely to undergo the primary stress response, where stress hormones including adrenaline, noradrenaline, and cortisol are released into bloodstream and induce physiological responses (e.g., glucose release to fuel aerobic tissues, elevated cardiac output to increase O_2 delivery) (Barton, 2002). Additionally, the interaction involves anaerobic metabolism during moments of hypoxia and burst exercise, resulting in low plasma pH, metabolite buildup, and muscle energy depletion (Wood, 1991). These physiological changes can lead to acidosis and sometimes failure to recover, leading to immediate, short-term, or delayed mortality and are characterized as fishing-related incidental mortality (FRIM) (Wood, 1991; see review by Patterson et al., 2017).

The methods employed before or during capture have significant physiological effects on Pacific salmon. For example, the longer gill nets are deployed (soak time), the greater the physiological disturbance (via plasma and muscle for physiological indices) and mortality (Farrell et al., 2000; Buchanan et al., 2002). Donaldson et al. (2010b) used heart rate loggers to show that coho salmon corralled in confined spaces for 30 min had elevated heart rates for more than 11 h, whereas fish corralled for 10 min returned to resting heart rates 5 h sooner. This work revealed that corralling time makes a difference in recovery rates and that heart rate recovery is a sensitive metric that can be monitored continuously for a more precise recovery timepoint compared to plasma indices (Cooke et al., 2012a,b; Donaldson et al., 2010b). Later, Raby et al. (2015) used heart rate loggers in coho salmon during beach seine capture simulations and showed that, paired with respirometry, blood sampling, and tissue sampling, beach seine capture time and handling time should be minimized, especially at warmer water temperatures. Combining telemetry with physiological stress response (plasma cortisol, glucose, ions, and osmolality), Donelson et al. (2011) compared the effects of different fishery capture gear (beach seine or angling) on mortality and showed that longer holding periods in net pens resulted in only 3% of fish completing their migration, compared with 52% that were released immediately after beach seining and 36% after immediate release from angling. Managers could then use these mortality estimates to promote improved capture and release methods and minimized handling (Cooke et al., 2012a,b; Donelson et al., 2011).

Cardiorespiratory physiology has also made significant contributions toward improving the recovery environment of Pacific salmon. For example, past work used metabolic measurements as recovery indicators to show that the recovery environment is important in determining recovery fate (Donaldson et al., 2013; Farrell et al., 2001a,b; Milligan, 1996; Milligan et al., 2000; Patterson et al., 2017). During recovery, fish increase their O_2 uptake to restore homeostasis (EPOC; as discussed in Section 2.1.1) and regain swim performance, which is limited during this time (Milligan, 1996). Milligan et al. (2000) took blood samples at different recovery timepoints in exercised rainbow trout under two different recovery scenarios (static vs. swimming) using dorsal aortic cannulations and terminal muscle samples for cellular analyses to show that sustained swimming following exercise enhances recovery compared to when the fish remains static. Water flow past the gills likely facilitated O_2 uptake and mediated recovery. Later, Farrell et al. (2001a,b) created the Fraser recovery box, a recovery tank for bycatch coho salmon (*Oncorhynchus kisutch*) that forced ventilation in fish by jetting water toward the fish's mouth. After a gillnet capture simulation, they measured plasma and muscle for ions, metabolites, and energy and showed that coho salmon bycatch were under severe metabolic exhaustion (e.g., high lactate, low PCr, glycogen and ATP) and stress (high cortisol) but that their recovery was alleviated in the Fraser recovery box (Farrell et al., 2001a,b). The Fraser recovery box was then incorporated in regulations for commercial fishing vessels for coho bycatch (Farrell et al., 2001a,b). Another version of a recovery environment was later established in recreational fisheries, which was a portable recovery bag designed to facilitate recovery of Pacific salmon after capture stress by providing water flow to fish (Donaldson et al., 2013). Authors tested the stress and metabolic responses of fish (e.g., cortisol, ions, lactate) by sampling blood at variable timepoints in recovery under different velocities of water flow and showed that any velocity was effective in reducing metabolic exhaustion and stress (Donaldson et al., 2013).

A promising avenue for predicting recovery is by using a bioindicator associated with mortality. Plasma lactate has been measured following many fisheries simulations, with levels varying significantly across methods (e.g., tangle net induced a ~15 mmol/L response, angling ~4 mmol/L) (Cooke et al., 2012a,b) (Fig. 3 and Table 1). Lactate levels have different immediate responses to varying capture stressors, as well as different rates of recovery based on capture type (e.g., net purse seine resulted in higher values during the first few hours of recovery, while salmon recovered to low lactate levels after troller capture within several hours) (Fig. 3). Importantly, Pacific salmon

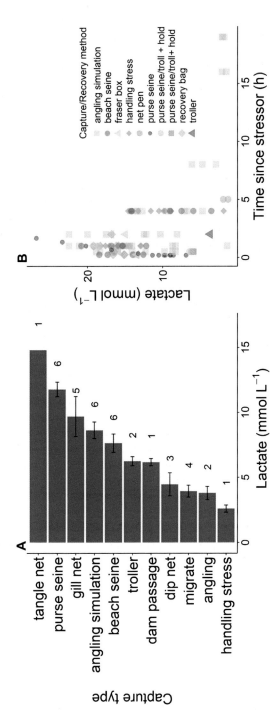

FIG. 3 The effect of different capture stressors on plasma lactate in Pacific salmon. (A) Plasma lactate values (expressed as mean ± SEM) across capture treatments measured immediately after a stressor. Values are pooled and extracted from Pacific salmon capture studies (both field and lab). Number of studies related to each capture type is shown beside each bar (Total N = 37). (B) Recovery plasma lactate levels in Pacific salmon following capture and holding methods. Data are pooled values extracted from 17 Pacific salmon studies representing 2021 salmon.

TABLE 1 Summary table of studies (N = 49) that sampled plasma lactate concentrations on migrating adult British Columbia Pacific salmon, including 5 species: Chinook (N = 1), chum (N = 2), coho (N = 10), pink (N = 5), and sockeye salmon (N = 34).

Species	Habitat	Treatment types	Lactate (mmol L^{-1})	Study
Chinook	Freshwater	Migration simulation, thermal	1.4–8.2	Clark et al. (2008a,b)
Chum	Freshwater	Fisheries capture	2.6–8.1	Raby et al. (2013)
Chum	Marine	Fisheries capture	6.3–15.8	Cook et al. (2018)
Coho	Freshwater	Fisheries capture	1.4–22.5	Clark et al. (2012), Kraskura et al. (2021), Raby et al. (2015), Donaldson et al. (2010a,b), Davis and Schreck (1997)
Coho	Freshwater	Holding	1.7–3.9	Little et al. (2020)
Coho	Marine	Fisheries capture	0.9–26.8	Cook et al. (2019), Raby et al. (2015), Farrell et al. (2001a,b)
Pink	Freshwater	Fisheries capture	0.8–19.2	Donaldson et al. (2013, 2014), Raby et al. (2013)
Pink	Freshwater	Holding	1.7–9.3	Jeffries et al. (2012a)
Pink	Freshwater	Migration	2.8–6.4	Pon et al. (2012)
Sockeye	Freshwater	Fisheries capture	0.4–18.6	Cooke et al. (2012a,b), Donaldson (unpublished), Donaldson et al. (2014), Gale et al. (2011, 2014), Robinson et al. (2013), Raby et al. (2015), Nguyen et al. (2014), Mathes et al. (2010), Eliason et al. (2020), Pon et al. (2009), Young et al. (2006), Donaldson et al. (2012), Teffer et al. (2017) Dick et al. (2018), Roscoe et al. (2011), Crossin et al. (2008)
Sockeye	Freshwater	Holding and/or thermal	0.6–23.7	Jeffries et al. (2012a,b), Gale et al. (2011, 2014)

Sockeye	Freshwater	Migration	2.3–2.9	Pon et al. (2009), Minke-Martin et al. (2018) (data unavailable), Middleton et al. (2018) (data unavailable)
Sockeye	Freshwater	Migration simulation	0.8–15.5	Jain et al. (1998), Eliason et al. (2013a,b), Wagner et al. (2005, 2006), Farrell et al. (1998)
Sockeye	Marine	Fisheries capture	5.9–10.6	Cooke et al. (2012a,b), Donaldson (unpublished), Crossin et al. (2007), Drenner et al. (2018)
Sockeye	Marine	Fisheries capture, migration	6.9–11.4	Cooke et al. (2006a,b, 2008)
Sockeye	Marine	Migration	7.4–10.8	Crossin et al. (2009), Drenner et al. (2015) (data unavailable)
Sockeye	Marine	Migration simulation	9.6–9.9	Wagner et al. (2006)

Studies represent salmon sampled from freshwater (N=38) and marine habitat (N=13), in both field (N=28) and lab (N=22) experimental settings. Note sample sizes do not add up to the total number of studies represented (N=49) because some studies sampled multiple species or sampled from both habitats. Treatment types include migration simulations (laboratory U_{crit}/swim flume protocol) (N=7), thermal simulations (laboratory holding and exposure to altered temperature) (N=5), fisheries capture (capture and/or recovery event [e.g., purse seine, gill net, laboratory chase protocol, Fraser box, holding pen, etc.] (N=33), holding (laboratory setting) (N=3), and migration (intercept migration and sample for blood, tag, track success) (N=9).

that surpass lactate levels of 10–13 mmol/L cannot repeat swim performance, and plasma lactate levels must return beneath these values before fish can resume swimming or migration (Farrell et al., 1998; Jain and Farrell, 2003; Stevens and Black, 1966). Lactate is one physiological response variable that shows consistent trends with survivorship and timely arrival to spawning grounds: salmon that survive en route to spawning grounds consistently had lower lactate levels than individuals that did not arrive or were delayed, and survivor levels were recorded below 10 mmol/L in 7 primary tagging or moribund studies (Fig. 4). Thus, plasma lactate values of 10–13 in wild migrating salmon may be predictive of mortality (Crossin et al., 2009). Lactate levels might also help to explain mortality differences in male and female Pacific salmon, where following a beach seine capture event, males recovered below the 10–13 mmol/L threshold levels within 4 h, while female lactate levels remained elevated (Eliason et al., 2020). Another potential indicator for mortality is the expression of transcription factors using non-lethally sampled gill tissue, where certain genes involved with apoptosis or homeostasis have been correlated with temperature-induced mortality in Pacific salmon holding studies (Jeffries et al., 2012b, 2014). Though this technique been applied in experimental settings, it has the potential to be expanded into the field as a

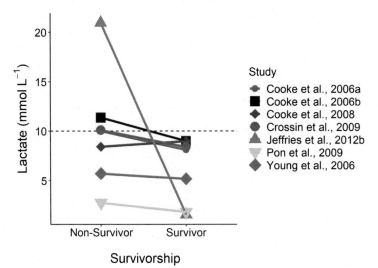

FIG. 4 Plasma lactate values for Pacific salmon captured in the field and sampled for blood, then tagged and tracked en route to spawning grounds (Cooke et al., 2006a,b, 2008; Crossin et al., 2009; Pon et al., 2009; Young et al., 2006), or tracked for survival in a laboratory setting (Jeffries et al., 2012b). Salmon are categorized as survivors if they passed upriver locations specific to the study or survived lab treatments and non-survivors if they did not reach or delayed arrival to these locations or were categorized as moribund in the laboratory setting. The horizontal dashed line at 10 mmol L^{-1} denotes the lactate threshold value required to repeat swim performance (Farrell et al., 1998; Jain and Farrell, 2003; Stevens and Black, 1966).

stress indicator (Akbarzadeh et al., 2018), already having shown utility in measuring rapid immune response and screening for pathogens in Pacific salmon (Connon et al., 2018; Miller et al., 2014). Other biomarkers (gross somatic energy, reproductive status) measured through blood sampling and an energy meter have also been associated with migration success (Young et al., 2006). Using a variety of cardiorespiratory metrics, these studies showed that higher magnitude capture stressors result in greater physiological impairment and often mortality (Donaldson et al., 2013).

3.1.2 Thermal performance of Pacific salmon in warming rivers

Temperature is a major factor threatening spawning migration success (Farrell et al., 2008; Keefer et al., 2008; Martins et al., 2011; Patterson et al., 2016). Across the Pacific salmon range, large proportions of returning runs have died migrating upstream during high river temperatures (Bowerman et al., 2021; Macdonald et al., 2010; Martins et al., 2011; Westley, 2020). Though the exact mechanism of temperature-induced mortality is unknown, one hypothesis suggests that warmer waters increase the aerobic demand, and fish may no longer have the capacity to sufficiently transport O_2 to the tissues that are necessary to complete migration (Eliason and Farrell, 2016). This increased metabolic demand, in combination with fasting conditions typical for migrating adult Pacific salmon, may cause greater energy store depletion, resulting in failed migrations and premature mortalities (Fenkes et al., 2016).

Intrinsic traits across multiple levels (i.e., species, population, life stage, sex, body size) make some individuals more susceptible than others to thermal-induced mortalities. There is growing evidence that variability in thermal tolerance exists interspecifically (Lee et al., 2003; Richter and Kolmes, 2005; Verhille et al., 2016) and intraspecifically (Abe et al., 2019; Chen et al., 2013; Eliason et al., 2011; Lee et al., 2003). For example, pink salmon have a larger aerobic scope and T_{optAS} at higher temperatures compared to sockeye, chum (*Oncorhynchus keta*), and coho adult salmon (Abe et al., 2019; Clark et al., 2011; Eliason et al., 2011; Lee et al., 2003). Species also are locally adapted to thermal history and migration conditions. Populations of sockeye salmon that experience more difficult migrations are associated with greater ventricle size, coronary supply, sarco/endoplasmic reticulum Ca^{2+}ATPase (SERCA) activity, higher AAS, and T_{optAS} (Anttila et al., 2019; Eliason et al., 2011). Populations of chum salmon in Japan and coho and sockeye salmon in British Columbia also differed in optimal thermal range of AAS, corresponding with environmental temperatures historically experienced (Abe et al., 2019; Eliason et al., 2011; Lee et al., 2003). Currently the Pacific Salmon Commission (PSC) pairs AAS thermal performance curves of different Fraser River, BC populations with river temperatures and flows to better predict mortalities and manage harvest rates accordingly (Hague et al., 2011; Lee et al., 2003; Macdonald et al., 2010) (Fig. 5). Values such

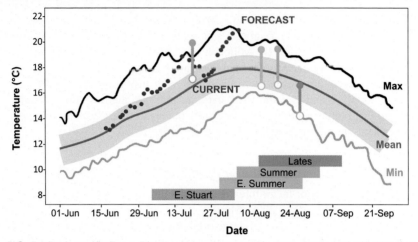

FIG. 5 Stock-specific Fraser River sockeye salmon aerobic scope data paired with historical and projected Fraser River temperatures. Modified from the Pacific Salmon Commission Panel newsletter released on July 29, 2014. The blue dots indicate the actual river temperatures on those dates and the red dots show the forecasted temperatures. The optimal range of temperatures for swimming performance for each stock group (determined from Eliason et al., 2011) is indicated by the orange, green, light blue and dark blue vertical lines (upper filled value is upper T_p (i.e., 90% of T_{opt}), lower open value is T_{optAS}). Above this temperature range, we expect high population-specific migration failure. The colored horizontal bars at the bottom of the figure show when each stock group is migrating through the river. Mean (gray), min (light gray), and max (black) temperatures are from 1971 to 2000; shaded gray area representing the standard deviation. Managers meet twice a week during the salmon migration season to discuss escapement numbers and determine catch allotment. The managers use this data to help with management decisions to balance harvest with conservation.

as T_p can be used as thresholds, that when river temperatures surpass this, harvest numbers may be limited (Cooke et al., 2012a,b; Eliason et al., 2011).

Tolerance also varies across lifecycle. For example, Chilko sockeye salmon have lowest thermal tolerance as embryos (Whitney et al., 2013) but highest thermal tolerance as adults (Eliason et al., 2011). Sex-specific thermal effects on migration success are also present. Migrating female adult salmon have higher en route mortalities than males, especially when exposed to secondary stressors (Bass et al., 2018; Crossin et al., 2008; Gale et al., 2014; Robinson et al., 2013; Teffer et al., 2017, see review by Hinch et al., 2021). Differences might be attributed to prolonged recovery, as shown in female sockeye salmon that had elevated metabolic and stress indicators (e.g., lactate, K^+, glucose, cortisol) compared to males during recovery from capture and tagging stress (Eliason et al., 2020). Other causes of female-biased mortality might come from increased energy exhaustion and reliance on anaerobic effort (Burnett et al., 2014), reduced cardiac capacity and AAS (Clark et al., 2011; Sandblom et al., 2009), elevated cortisol levels (Cook et al., 2011; Gale et al., 2011; Jeffries et al., 2012b; Little et al., 2020), and a compromised immune response (Bass et al., 2017; Teffer et al., 2017) exacerbated by increased temperature. Size also plays a role in thermal response. Clark et al. (2012) used respirometry, heart rate loggers, and blood parameters to measure size-specific responses to acute thermal exposure and fisheries capture simulation. They found that smaller coho salmon have faster recoveries (reduced EPOC, faster recovery of plasma testosterone to baseline levels) than larger individuals during this multi-stressor experiment. In another study, Clark et al. (2008a,b) combined respirometry, blood flow probes, and arterial/venous cannulations to measure metabolic rate, heart metrics (cardiac output, heart rate, stroke volume), blood parameters (e.g., lactate, glucose, cortisol, ions) and partial pressure of O_2 during water heating. The authors identified an interesting trend with a limited sample size where the larger Chinook salmon experienced cardiac arrythmias and based on blood oxygen partial pressure and oxygen content, likely switched to anaerobic metabolism sooner at high temperatures. This trend might play a role in shifting sizes of salmon (Oke et al., 2020) and have significant implications in the future of salmon, warranting follow up work. These studies using cardiorespiratory physiology identified differences in performance across biological levels. Certain groups (e.g., Fraser River females, larger individuals) are more vulnerable to warming rivers and management can use this information to modify harvest numbers to protect these iconic species.

3.2 Shark fisheries-induced mortality

Chondrichthyan fishes (including elasmobranch fishes and chimaeras) are differentiated from their teleost counterparts by characteristics including cartilaginous endoskeletons, internal fertilization, paired spiracles, and separate internal and external gill openings (Carrier et al., 2012; Compagno, 1990; Ebert et al., 2013). Their 400-million year-long evolutionary history has likely contributed to the success of elasmobranch fishes at colonizing almost all aquatic environments from freshwater rivers to the abyssal planes of the deep sea, though the majority of species are fully marine (Compagno et al., 2005; Klimley, 2013). Although elasmobranch fishes occupy a broad scope of ecological niches, ranging from benthic foragers to apex predators, they are overrepresented at high trophic levels and thus play an important role in maintaining ecosystem health through top-down effects (Heithaus et al., 2010). Life spans of elasmobranch fishes vary dramatically, with some coastal sharks (e.g., blacknose shark, *Carcharhinus acronotus*) living only 20 years, while the Greenland shark (*Somniosus microcephalus*) lives to 400 years (Driggers et al., 2004; Nielsen et al., 2016). Elasmobranch fishes display all forms of reproduction (i.e., viviparity, oviparity, and ovoviviparity), but generally have low fecundity with individual litter sizes ranging from 2 to 100 depending on the species (Carrier et al., 2004). Worldwide, one in four elasmobranch species are under threat of extinction largely due to a combination of high fishing mortality and habitat degradation (Dulvy et al., 2014).

Because of their long lifespans and low fecundity, elasmobranch populations are particularly vulnerable to exploitation. (Dulvy et al., 2014; Hoenig and Gruber, 1990; Pratt Jr. and Casey, 1990). Commercial fishing operations only target a few species of elasmobranch fishes (e.g., spiny dogfish; *Squalus acanthias,* in the northeast USA), but many are caught recreationally (e.g., common thresher shark, *Alopias vulpinus,* in Southern California), and even more interact with commercial fishing gear as non-target bycatch (e.g., blue sharks, *Prionace glauca,* in the pelagic longline fishery) (Campana et al., 2009; Haugen et al., 2017; Sepulveda et al., 2015). For the latter two situations, unknown percentages of individuals die post-release from the physiological stress of capture, rendering these individuals incapable of reproducing and contributing to the population. Quantifying the total mortality associated with fishing activities, an essential component of sustainable management practices, therefore relies upon accurate measures of mortality both from landed and released catch. A primary goal for management bodies is to identify at-vessel metrics that can predict post-release mortality in these released individuals (NOAA, 2018). Here, we will focus on attempts to employ cardiorespiratory physiology to predict and measure post-release mortality.

Blood is an obvious choice for investigating the physiological disturbance that accompanies capture events. Capture typically involves exhaustive exercise, air exposure, and tissue trauma, all of which are associated with measurable changes in blood gases, acid-base balance, and metabolite and electrolyte concentrations (see Section 2.3.1). Although elasmobranch fishes do not produce cortisol like teleost fishes, changes in electrolyte concentrations are known to be caused by muscle degradation and are indicative of physiological stress (Frick et al., 2010c; Schwieterman et al., 2021).

Captive and at-vessel studies have played an important role in establishing protocols and determining metrics relevant to management and conservation. For example, Marshall et al. (2012) found correlations between both blood pH and K^+, and moribund status of 11 different elasmobranch species captured on longline, suggesting these were good bioindicators for post release mortality. In another study, air exposure caused significant disruptions to blood acid–base properties (e.g., decline in pH; elevation in PCO_2), metabolites (e.g., elevated whole-blood lactate) and electrolytes (e.g., elevated plasma K^+) in captive little skates (*Leucoraja erinacea*) (Cicia et al., 2012). Acidosis was corroborated as an important metric in predicting post-release mortality by captive studies done on gummy sharks (*Mustelus antarcticus*) and Port Jackson sharks (*Heterodontus portusjacksoni*) (Frick et al., 2010b), as well as field studies on five species of carcharhinid elasmobranchs (Mandelman and Skomal, 2009).

At-vessel studies on blacktip sharks (*Carcarhinus limbatus*) and mako sharks (*Isurus oxyrinchus*) have also suggested lactate may be a good

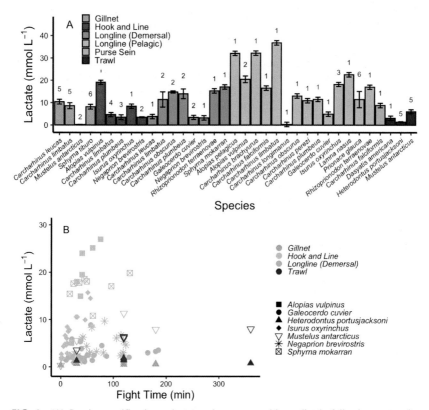

FIG. 6 (A) Species specific plasma lactate values measured immediately following capture by multiple gear types. The number of studies is denoted above each bar, with associated standard errors when applicable. (B) Individual plasma lactate values measured following capture in when fight time was measured.

indicator of post-release mortality (Weber et al., 2020; Wells and Davie, 1985). Lactate at the time of capture varies widely across different species and gear types (Fig. 6), likely due to differences in species ventilation mode (buccal pumping vs. obligate ram ventilating), the amount of restraint the gear imposes, and the nature of the species (SEDAR 65 RD06). Within a specific gear type lactate does show trends of increasing with fight time, supporting the idea that it can be used as an indicator of physiological stress. Plasma lactate does, however, peak several hours after exhaustive exercise as demonstrated by serial sampling on captive sand tiger sharks (*Carcharias taurus*), Port Jackson sharks, Australian swellsharks (*Cephaloscyllium laticeps*), spiny dogfish, and dusky shark (*Carcarhinus obscurus*) (Cliff and Thurman, 1984; Frick et al., 2010a; Kneebone et al., 2013; Mandelman and Farrington, 2007). For gear types where fight time is uncertain (e.g., longline), at-vessel lactate values may be measured before or after the peak of physiological distress.

When paired with tagging data, researchers have definitively linked at-vessel blood parameters with post-release mortality events. For example, pop-off satellite archival tags (PSATs) were used to identify mortality events in blacktip sharks in the Gulf of Mexico recreational shark fishery. Short term post-release mortalities (i.e., within an hour of capture) were correlated with high blood lactate and low Hct (Mohan et al., 2020). In a similar study employing PSATs in the recreational fishery in California, Heberer et al. (2010) found increased lactate and high Hct were indicative of mortality in common thresher sharks (*A. vulpinus*), showing the species-specific nature of stress responses and physiological thresholds. Surviving blue sharks caught on longline were differentiated from their moribund counterparts with significant differences in lactate, Hsp, $[K^+]$, $[Mg^{2+}]$, and $[Ca^{2+}]$ (Moyes et al., 2006). Shortfin mako sharks captured in the Australian recreational fishery showed correlations between increased lactate and glucose with higher sea surface temperatures and longer fight times in a study using Survivorship Pop-up Archival Transmitting (sPAT) (French et al., 2015). Acoustic tags have been used to study post-release mortality in juvenile sand tiger sharks, where blood metrics were not found to be indicative of mortality events (Kneebone et al., 2013). Another kind of tag, acceleration data loggers (ADLs), have also been used to quantify mortality and recovery times, enabling more nuanced interpretations of the level of stress experienced by blacktip sharks caught on rod and reel (Whitney et al., 2017). For short-term post-release mortality events, manual tracking may also prove informative. Danylchuk et al. (2014) studied juvenile lemons sharks (*Negaprion brevirostris*) and found that blood ions disruption correlated with fight times, and may be useful in predicting post release mortality, although water temperature was a better indicator. More studies pairing these physiological metrics with technological measures of post-release mortality will aid in meta-analyses to find trends across species and gear types.

Currently, management measures incorporating post-release mortality estimates rely largely on tagging studies, although the most recent South East Data Assessment and Review (SEDAR) on sharks did include physiological blood stress studies in their literature review (Courtney and Mathers, 2019). High levels of blood lactate upon capture were also anecdotally used in discussion of changes to the Florida shore shark fishing regulations, which ultimately mandated the cutting of the line for prohibited shark species (Shiffman, 2020). In Australia, the shark assessment explicitly states blood stress indices can be used to predict post-release mortality (Woodhams and Harte, 2018), although these studies should be validated with tagging or penning studies.

3.3 Pelagic fishes and oil

Offshore oil drilling has expanded globally, meaning there are greater risks of large-scale oil spills. In the past few decades, incidents including the *Castillo de Bellver* oil spill (1983), the Incidents at the Nowruz Oil Field (1983), the Exxon-Valdez oil spill (1989), and the Deepwater Horizon oil spill (2010) each released millions of gallons of crude oil into the ocean, reaching hundreds of thousands of square kilometers of ocean surface and untold volumes of water. Large scale oil spills such as these are environmental disasters with long term ecological consequences impacting both nearshore fisheries, as well as pelagic fishes that depend on coastal habitats for spawning and nursery habitats (Pasparakis et al., 2019; Peterson et al., 2003). Oil spills are particularly difficult environmental disasters to study, as their widespread and spontaneous occurrence challenge the ability to obtain accurate and detailed baseline data. Although each oil spill releases different proportions of toxic compounds, they all affect a suite of ecological processes. A general understanding of the impacts of crude oil contaminants on fish populations is essential for quantifying the ecological impacts of these human disasters.

Crude oil is particularly dangerous because of polycyclic aromatic hydrocarbons (PAHs) which are highly lipophilic and persist in the environment (Cram et al., 2006). Once released into the environment, these PAHs can be modified through weathering (Esbaugh et al., 2016) and UV exposure (Roberts et al., 2017), changing their toxicity. Additionally, the use of oil dispersant increases the PAH uptake rate in larval fishes, thus reaching lethal levels more quickly (Couillard et al., 2005; Ramachandran et al., 2004). Measuring the population and community impacts of oil spills over time is logistically complex, and as such inferences from physiological studies at organismal and cellular levels of biological organization are crucial (Grosell and Pasparakis, 2021). While oil is known to affect a suite of physiological processes (see Grosell and Pasparakis, 2021 for a review), here, we will focus solely on cardiac impairments that are relevant for conservation and mitigation efforts.

Early life stage (ELS) cardiac impairment, or cardiotoxicity, is one of the most well-studied endpoints associated with crude oil exposure in fishes. Following oil exposure, ELSs of marine fishes frequently present with decreased heart rates and fluid buildup anterior to the yolk sack (i.e., pericardial edema; Fig. 7) (Incardona et al., 2014; Xu et al., 2016). Pericardial edema is correlated with high mortality during ELSs, as well as reduced cardiac output for surviving individuals (Esbaugh et al., 2016; Khursigara et al., 2017). Four ring PAHs, such as pyrene and chrysene, are also known to produce toxic metabolites that impair cardiac development (Incardona et al., 2006). These lethal and sub-lethal effects underscore the short and long-term ecological significance of oil spills (Barron et al., 2020). This type of research can identify the exposure thresholds used to inform best practices regarding dispersant usage (Ramachandran et al., 2004) and help predict when populations may be deleteriously affected by years of poor recruitment (Szedlmayer and Mudrak, 2014). This work has been used by management to support the National Resource Damage Assessment in determining the ecological impacts of anthropogenic activities.

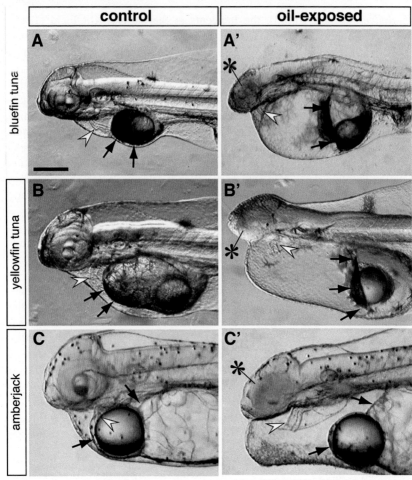

FIG. 7 The impact of oil on (A) bluefin tuna, (B) yellowfin tuna, and (C) amberjack embryos under exposure to oil and corresponding PAHs (8.5, 3.4, and 13.8 µg/L $\sum$PAH, respectively). Control fishes (A–C) show normal development and location of the yolk sac. Oil-exposed fishes (A′–C′) show circulatory failure and corresponding fluid accumulation (edema). Black arrows indicate the anterior edges of the yolk mass, which are pushed posteriorly in oil-exposed fishes. White arrows indicate the heart. Reduction of the eyes is shown by asterisks. (Scale bar, 0.2 mm). *Figure originally published by Incardona et al. (2014).*

In adult fishes, where the cardiorespiratory systems are fully developed at the time of oil exposure, disruptions to cardiac function are the primary concern. At a cellular level, 3-ringed (tricyclic) PAH exposure inhibits K+ movement into the cell, thus lengthening cardiac action potentials and decreasing heart rate (Brette et al., 2014, 2017; Esbaugh et al., 2016; Incardona et al., 2014). Studies on isolated bluefin and yellowfin tuna (*Thunnus orientalis* and *T. albacares*, respectively) cardiomyocytes found oil exposure decreased calcium currents through L-type calcium channels and impaired calcium cycling (Brette et al., 2014, 2017), which explains decreases in contractile force and stroke volume. Decreases in both heart rate and stroke volume will reduce overall cardiac output, limiting oxygen delivery and perhaps contributing to the decrease in maximum metabolic rate and aerobic scope observed in coho salmon (*O. kisutch*), red drum (*Sciaenops ocellatus*), and mahi mahi (*Coryphaena hippurus*) (Johansen and Esbaugh, 2017; Mager et al., 2014; Stieglitz et al., 2016; Thomas and Rice, 1987). These cardiac impairments may also provide a mechanistic explanation for reduced maximum sustained swim speed (U_{crit}) observed across taxa, although damage to the gill epithelium (Katsumiti et al., 2009) or reduction in mitochondrial efficiency (Kirby et al., 2019) may also be driving this result. These cardiac impairments may lead to reductions in predator avoidance, post-release survivorship, or competitive foraging success, all of which may reduce standing stock biomass and spawning success rates (Johansen and Esbaugh, 2017). As managers set catch quotas for areas affected by oil spill disasters, the precautionary principle should be employed (SEDAR, 2013).

Understanding ecological areas of vulnerability may help first responders decide what actions to take to minimize the environmental impact (Grosell and Pasparakis, 2021). More research is needed, however. There is a high level of species specificity exhibited in the physiological responses to oil exposure; indeed Pasparakis et al. (2019) found differences of nearly three orders of magnitude in the growth and mortality responses of sensitive pelagic species compared to those of more tolerant near-shore species. As such, there is a need for research on a greater diversity of species (e.g., mahi mahi are grossly overrepresented in the literature; Pasparakis et al., 2019). Further, the interactions between oil exposure and other environmental stressors such as warm temperatures and hypoxia are still under investigation (Khursigara et al., 2019). Expanding this research to more species and life stages will be informative for predicting recruitment failure in a diversity of important stocks, and thus promote sustainable management.

4 Moving the field forward

4.1 Context matters

Intraspecific variability is critical to consider when developing climate change mitigation, pollutant thresholds, fishing recommendations, and habitat rehabilitation. This is because stressor sensitivity differs across populations, ontogeny, sex, and with body size (see Chapter 9, Volume 39A: Schulte and Healy, 2022). Populations specifically differ in thermal tolerance, aerobic and cardiovascular capacity (Eliason et al., 2011; McKenzie et al., 2021; Zillig et al., 2021) and fisheries capture stress response and survival (Donaldson et al., 2010a,b). Across the lifecycle, responses to thermal stress

(Dahlke et al., 2020; Fowler et al., 2009; Pörtner and Farrell, 2008) and capture stress (Raby et al., 2015) vary considerably. Sex-related diversity is often overlooked but can have important management implications (Hanson et al., 2008). For example, female Pacific salmon have much higher mortality rates compared to males (see review by Hinch et al., 2021). Body size can play a profound role in ecology (e.g., predator-prey dynamics), physiology (e.g., metabolic scaling), evolution (e.g., fitness consequences), and economics (e.g., harvest value). Yet, a decrease in body size is proposed as the third universal response to climate warming (Daufresne et al., 2009; Gardner et al., 2011). Global reports of decreasing body size in fishes (e.g., Oke et al., 2020) are concerning and there is much interest in understanding the causes and consequences of these trends for conservation and management (i.e., to improve prediction, prevention, and mitigation).

There is also considerable inter-individual variability in cardiorespiratory performance (Metcalfe et al., 2016a), for example, aerobic scope varies with boldness (Binder et al., 2016), dominance (Killen et al., 2014), maternal effects (Banet et al., 2019), and parasite load (Hvas et al., 2017). Moving forward, researchers and managers must not overlook the importance of context when planning and interpreting research. By carefully considering intraspecific variability, conservation physiologists can identify the individuals, life stages, and populations most resilient and most vulnerable to environmental stressors.

4.2 Environmental realism

The vulnerability of a fish to an environmental stressor is related to its sensitivity (e.g., physiology) and exposure (e.g., duration and intensity) (Schulte, 2014). Accordingly, it is essential to have a comprehensive understanding of the current environmental conditions that fish encounter across their life cycle (e.g., temperature, flow, pH, dissolved oxygen, salinity). While some systems have widespread environmental monitoring in place, others are lacking. Data loggers (e.g., temperature, dissolved oxygen, pH) can be inexpensive and easy for an individual researcher to deploy to obtain habitat-specific information on their study system. Environmental data can be used to inform experimental design to ensure treatments are ecologically realistic and mimicking nature. Accordingly, the research output will be more applicable to conservation efforts. Using environmental data, modelers can collaborate with conservation physiologists to evaluate current environmental conditions, determine if fish populations currently persist at their habitat limits, and make predictions about whether they have the capacity to withstand further change (Sections 4.4 and 4.5).

Aquatic environments are highly variable (e.g., temperature, pH, oxygen, light, salinity can fluctuate over many temporal and spatial scales: diurnal, tidal, seasonal, decadal, and episodic with changing weather patterns and upwelling events/marine heat waves). While researchers have traditionally acclimated fish to simple, stable, average treatment regimens (e.g., static temperatures, static oxygen levels), increasingly, researchers are appreciating that fish physiological performance is influenced by environmental variability (Morash et al., 2018). For example, researchers are using fluctuating temperature regimens to mimic natural diurnal cycles for respirometry experiments (Cooper et al., 2021; Morash et al., 2018) and multistressor treatments to examine interactive effects (Targett et al., 2019). Similarly, field respirometry is also becoming more prominent, where measurements are performed directly in the field on acclimatized individuals (Gilbert et al., 2020; Verhille et al., 2016) and can incorporate fluctuating, natural environmental variability. Biotelemetry and biologging go a step further to monitor free-swimming fish in multi-stressor wild environments (Prystay et al., 2019, 2020), though admittedly high tag costs have meant that most heart rate logging studies to date have still been performed in controlled environments to ensure tag retrieval. These types of field physiology studies are extremely promising to drive further advances in the discipline.

4.3 Coupling techniques

A common refrain is that despite rapid growth in the field of conservation physiology, there is a dearth of concrete examples where research has been put into practice. However, some of the best examples come from integrative and transdisciplinary approaches, coupling research methods from multiple levels of biological organization and across fields of study, which enables researchers to effectively produce information useful for management and policy makers (e.g., Farrell et al., 2008). As technology becomes smaller, longer lasting, and inexpensive, ever more researchers will be able to embrace the use of biologgers and biotelemetry methods to remotely monitor free-swimming fish in the wild. These tags can be calibrated with respirometry to determine FMR (Treberg et al., 2016). An important step forward will be coupling tagging techniques with tissue sampling (e.g., blood, gill, muscle) to identify reliable biomarkers indicative of stressor thresholds. As described above (Section 2.2.1), blood can yield a wealth of information (Lawrence et al., 2020), and plasma lactate is a promising bioindicator of survival. Transcriptomics have great potential to provide a cellular level understanding of fish responses to environmental stressors (see Chapter 8, Volume 39A: Jeffries et al., 2022; Connon et al., 2018) and to detect biomarkers for thermal stress and survival (von Biela et al., 2020). Another critical interdisciplinary question is to identify whether fish have the adaptive capacity to keep pace with climate change, by coupling conservation genomics, physiology, environmental data, and modeling (e.g., Chen and Narum, 2021; Chen et al., 2018a,b; Crozier et al., 2008).

4.4 How much aerobic scope does a fish need to thrive?

As discussed above (Section 2.1.1), absolute aerobic scope (AAS = MMR − SMR) is the aerobic capacity of a fish to perform activities beyond maintenance. An important distinction is that aerobic scope is a *capacity*, not an energy *budget*: fish do not necessarily utilize their full aerobic capacity. A major open question remains how much aerobic scope does a fish need to thrive? To answer this question, we need to couple fieldwork and tagging studies (e.g., accelerometry) with lab studies measuring aerobic metabolism to measure FMR (Section 2.1.1). These types of experiments can reveal how much scope a fish uses to perform day to day activities. This has been estimated to be ~80–90% of maximum aerobic scope for an adult migrating salmon when they are maximally swimming upstream to return to their spawning grounds (e.g., Eliason et al., 2011; Farrell et al., 2008), and this information is useful for managers (Fig. 5). However, adult migrating salmon are not "typical" fish. Also, the adult salmon upriver migration represents only ~2% of the life cycle. It is unknown how much aerobic scope is necessary for the remainder of the salmonid life cycle. For example, it is unknown what percent of maximum aerobic scope is necessary for a juvenile coho salmon fry to thrive in a stream, or as a growing adult in the marine environment. Beyond salmon, many fishes may rarely utilize the full extent of their aerobic capacity (Farrell, 2016). This question should be explored across species, populations, and life stages.

In order to thrive, a fish must have sufficient energy available to support maintenance costs and also be able to support other activities: locomotion costs to forage for food, digestion costs to assimilate nutrients, vigilance costs to assess for predators and competitors, recovery costs to reestablish homeostasis after exhaustively chasing after a mate/competitor or escaping from a predator, etc. Many of these activities can co-occur (e.g., a fish can simultaneously swim and digest a meal). The energy distribution will vary for different fish, depending on their life history and priorities. However, the energetic costs of the various activities are largely unknown. The capacity of the heart is insufficient to maximally perfuse all capillary beds simultaneously. Wild fish must partition blood flow among the metabolically demanding tissues (e.g., swimming muscles, stomach and intestines, gonads) to accomplish different tasks simultaneously (e.g., swim, digest, sexually mature). Advances in this area have been relatively stagnant in fish given that many regional blood flow distribution methods are unreliable (e.g., colored and radiolabeled microspheres; Farrell et al., 2001a,b) or require fish to be immobilized (nuclear magnetic resonance [NMR], e.g., Pörtner and Knust, 2007; epi-illumination microscopy, e.g., Nilsson et al., 1994). Only a subset of organ systems in a small number of fishes have been directly assessed using flow probes (e.g., coronary blood flow to the ventricle Ekström et al., 2017, gut blood flow to the stomach and intestines; Farrell et al., 2001a,b). An essential step forward is to determine which physiological activities are most important for a given species at a given life stage and the cost of these activities.

Aerobic scope is influenced by environmental stressors (e.g., temperature, dissolved oxygen, salinity, pollutants). As environmental stressors limit the overall energetic capacity, fish performance will become compromised. For example, fish may consume less food as temperatures warm, possibly because they have insufficient scope to support SDA (Jutfelt et al., 2021, see Fig. 1C). Thus, sublethal increases in temperature could lead to decreases in growth via a reduction in feed intake, possibly to protect sufficient capacity for locomotion (Jutfelt et al., 2021; Norin and Clark, 2017; but see a notable exception in invasive lionfish, Steell et al., 2019). Warming temperatures also prolonged recovery from exhaustive exercise (e.g., measured via heart rate loggers and $\dot{M}O_2$/EPOC as increased recovery costs and duration; Kraskura et al., 2020; Prystay et al., 2017), which could delay migration, increase susceptibility to predation, defer feeding and mating opportunities, and relinquish territory. Similarly, gut blood flow and metabolism decreased during hypoxia in rainbow trout (Eliason and Farrell, 2014), suggesting that digestion and growth would be impaired under hypoxic conditions. Thus, sublethal hypoxia levels could lead to similar reductions in growth via a similar mechanism (Pichavant et al., 2000). Other stressors (salinity, pollutants etc.) can similarly constrain metabolism and performance. Accordingly, a suite of cardiorespiratory traits relating to fitness including aerobic scope (AAS and FAS), SDA, and EPOC may be excellent bioindicators of sublethal physiological impairment for conservation physiologists.

One promising way forward may be to consider which is most important and ecologically relevant for a particular species of fish at a particular life stage: the metabolic ceiling (i.e. MMR) or the metabolic floor (i.e. SMR). If maximum metabolic capacity is critical to thrive (e.g., actively swimming adult sockeye salmon to reach distant spawning grounds), then AAS may be the critical factor to consider when setting a threshold to thrive (e.g., 90% of AAS; Eliason et al., 2011). However, if fish rarely require maximum capacity to thrive and maintenance metabolism is more of a concern (e.g., juvenile salmonids), then FAS (MMR/SMR) may be a better choice for setting a threshold. FAS allows the evaluation of aerobic scope as a proportion of maintenance costs, thus it can reveal when a metabolic constraint may develop (see Fig. 8). An FAS of 2 means that the fish can double their metabolism above maintenance levels. This is only enough energy to digest an average size meal (e.g., 2% of body mass) for a rainbow trout (Eliason et al., 2008) with no remaining scope for other activities during digestion. This low FAS is indicative of a high maintenance metabolism, and thus a high feed intake would be necessary just to support baseline metabolic requirements with limited or no scope for growth. More time spent foraging and feeding to support higher SMR could also mean lost opportunity costs for the fish (e.g., territory, mates) and increased susceptibility to predation.

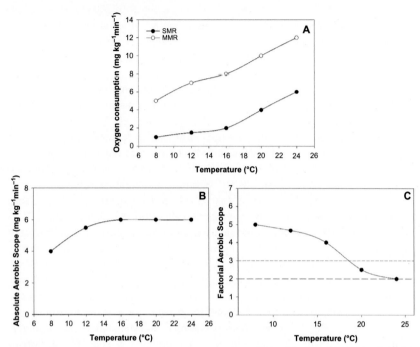

FIG. 8 Hypothetical data for one population of juvenile salmonids tested across a range of temperatures. (A) standard metabolic rate (SMR), maximum metabolic rate (MMR), (B) absolute aerobic scope (AAS) and (C) factorial aerobic scope (FAS). Empirically, researchers often observe that both SMR and MMR increase with increasing temperatures (see A), and thus AAS will display a plateau across temperatures (see B) (e.g., Poletto et al., 2017). However, FAS decreases with warming (see C) as the maintenance costs (i.e., SMR) increase with increasing temperature. Thus, even though the fish have the same *absolute* aerobic scope between 16 and 24 °C, fish must allocate a greater proportion of their energy and nutrient intake simply to support maintenance metabolism at the higher temperatures (as indicated by a lower FAS). For well-studied salmonid species, we predict that immature juvenile fish require a minimum FAS of 2 to survive (aerobically meet maintenance energy requirements with some minimal scope for digestion or swimming) and an FAS of 3 to thrive (grow, digest, swim, behave normally). Note that to effectively generate these thermal performance curves, fish should ideally be tested at several temperatures (i.e., at least 4 or more).

An FAS of 3 means that a fish can triple their metabolism above maintenance levels. This would presumably allow a rainbow trout sufficient energy to digest a meal with leftover scope to support other activities during digestion (e.g., vigilance, locomotion etc.). An FAS of 4 or more would allow for even more activities. For an immature rainbow trout and other similar salmonids, an FAS threshold of 2 may be necessary to *survive* (i.e., aerobically meet basic maintenance requirements over a period of days/weeks), while an FAS of 3 may be necessary to *thrive* (i.e., grow, recover quickly, behave normally; see Fig. 8). The FAS threshold to thrive will vary across fishes and life stages, given that different species have different energetic capacities and life history strategies (e.g., invasive lionfish consume large meals and SDA exhausts their full aerobic capacity, Steell et al., 2019). Clearly, more empirical work is needed before broadly applying these ideas across diverse fishes.

4.5 Thermal safety margins (TSM) and functional warming tolerance (FWT)

Ultimately, management agencies and practitioners want to know the habitat requirements and climate change vulnerability of fishes. Several studies have defined Thermal Safety Margins (TSM) as the difference between the upper critical thermal tolerance (CT_{max}) and the current maximum environmental temperature (e.g., Pinsky et al., 2019; Sunday et al., 2014; see Fig. 9). However, this overestimates the true environmental tolerance for a fish because the CT_{max} test is an acute thermal test that warms the fish at unnatural rates

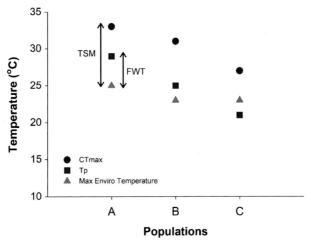

FIG. 9 Thermal safety margin (TSM) is calculated as the difference between the critical thermal maximum (CT_{max}) and the maximum environmental temperature. TSM overestimates the thermal safety for a fish population because a fish cannot thrive (swim, grow, defend territory) at the incipient lethal temperatures indicated by a CT_{max} protocol. Functional Warming Tolerance (FWT) is calculated as the difference between the temperature of physiological impairment (T_p; i.e., when growth, digestion, recovery and/or aerobic scope becomes limiting—for example, could be 80–90% of AAS for adult salmon or FAS of 2 or 3 for juvenile salmonids) and the maximum environmental temperature. FWT provides a more useful estimate of thermal vulnerability for managers. In this example, Populations A and B have the same TSM (8 °C) but different FWT (A = 4 °C; B = 2 °C). Populations B and C experience the same maximum environmental temperature (23 °C) but have different TSM (B = 8 °C; C = 4 °C) and FWT (B = 2 °C; C = −2 °C). Thus, Population A is the most resilient to future warming, while Population C is already experiencing temperatures exceeding their optimal threshold to thrive and they are highly vulnerable to continued warming.

(usually by 0.3 C/min) until the fish loses equilibrium, and it therefore has no functional meaning for wild populations. This test is indicative of an incipient lethal threshold and does not determine functional thermal tolerance (Lefevre et al., 2021). Alternatively, Functional Warming Tolerance (FWT) can be calculated by measuring the difference between the temperature of physiological impairment (when growth, digestion, reproduction, behavior, recovery, and/or aerobic scope becomes limiting [e.g., $T_p = 80\%$ or 90% of AAS for adult migration salmon or potentially FAS = 3 for juvenile salmonids]) and the current maximum environmental temperature (Fig. 9). This could be a useful metric for managers to define habitat requirements and functional limits for fishes. Other environmental stressors can similarly be assessed (e.g., dissolved oxygen levels, pollutants, multistressors).

4.6 Future outlook

The recent IPCC projections (IPCC, 2021) and perilous state of freshwater fish (aka the world's forgotten fishes; WWF, 2021) are sounding alarm bells for fish physiologists and conservation biologists around the globe. Our aquatic systems are in trouble (see Chapter 1, Volume 39A: Cooke et al., 2022), but cardiorespiratory physiology techniques can help identify underlying mechanisms, tolerance thresholds, and population vulnerability to further environmental degradation. Evidence suggests that conservation science is most useful for decision-making when it is co-produced by researchers and practitioners (Baker et al., 2020; Cooke et al., 2021). Through collaboration with knowledge holders, rightsholders, and stakeholders, cardiorespiratory physiology research can be incorporated into action to benefit fishes.

References

Abe, T.K., Kitagawa, T., Makiguchi, Y., Sato, K., 2019. Chum salmon migrating upriver adjust to environmental temperatures through metabolic compensation. J. Exp. Biol. 222, jeb186189.

Akbarzadeh, A., Günther, O.P., Houde, A.L., Li, S., Ming, T.J., Jeffries, K.M., Hinch, S.G., Miller, K.M., 2018. Developing specific molecular biomarkers for thermal stress in salmonids. BMC Genomics 19, 749.

Alfonso, S., Gesto, M., Sadoul, B., 2020. Temperature increase and its effects on fish stress physiology in the context of global warming. J. Fish Biol. 98 (6), 1496–1508. https://doi.org/10.1111/jfb.14599.

Allen, P., 1994. Evaluation of a technique for sampling blood from small fishes. Comp. Biochem. Physiol. A-Mol. Integr. Physiol. 107 (2), 413–418.

Anderson, W., Booth, R., Beddow, T., McKinley, R., Finstad, B., Økland, F., Scruton, D., 1998. Remote monitoring of heart rate as a measure of recovery in angled Atlantic salmon, Salmo salar (L.). Hydrobiologia 371, 233–240.

Andrewartha, S.J., Munns, S.L., Edwards, A., 2016. Calibration of the HemoCue point-of-care analyser for determining haemoglobin concentration in a lizard and a fish. Conserv. Physiol. 4 (1), cow006. https://doi.org/10.1093/conphys/cow006.

Anttila, K., Casselman, M.T., Schulte, P.M., Farrell, A.P., 2013. Optimum temperature in juvenile salmonids: connecting subcellular indicators to tissue function and whole-organism thermal optimum. Physiol. Biochem. Zool. 86 (2), 245–256. https://doi.org/10.1086/669265.

Anttila, K., Couturier, C.S., Overli, O., Johnsen, A., Marthinsen, G., Nilsson, G.E., Farrell, A.P., 2014. Atlantic salmon show capability for cardiac acclimation to warm temperatures. Nat. Commun. 5, 4252. https://doi.org/10.1038/ncomms5252.

Anttila, K., Mauduit, F., Le Floch, S., Claireaux, G., Nikinmaa, M., 2017. Influence of crude oil exposure on cardiac function and thermal tolerance of juvenile rainbow trout and European sea bass. Environ. Sci. Pollut. Res. Int. 24 (24), 19624–19634. https://doi.org/10.1007/s11356-017-9609-x.

Anttila, K., Farrell, A.P., Patterson, D.A., Hinch, S.G., Eliason, E.J., 2019. Cardiac SERCA activity in sockeye salmon populations: an adaptive response to migration conditions. Can. J. Fish. Aquat. Sci. 76, 1–5.

Baker, Z., Ekstrom, J.A., Meagher, K.D., Preston, B.L., Bedsworth, L., 2020. The social structure of climate change research and practitioner engagement: evidence from California. Glob. Environ. Chang. 63, 102074.

Banet, A.I., Healy, S.J., Eliason, E.J., Roualdes, E.A., Patterson, D.A., Hinch, S.G., 2019. Simulated maternal stress reduces offspring aerobic swimming performance in Pacific salmon. Conserv. Physiol. 7 (1). https://doi.org/10.1093/conphys/coz095.

Barnett, A., Payne, N.L., Semmens, J.M., Fitzpatrick, R., 2016. Ecotourism increases the field metabolic rate of whitetip reef sharks. Biol. Conserv. 199, 132–136.

Barron, M.G., Vivian, D.N., Heintz, R.A., Yim, U.H., 2020. Long-term ecological impacts from oil spills: comparison of Exxon Valdez, Hebei Spirit, and Deepwater horizon. Environ. Sci. Technol. 54 (11), 6456–6467. https://doi.org/10.1021/acs.est.9b05020.

Barton, B.A., 2002. Stress in fishes: a diversity of responses with particular reference to changes in circulating corticosteroids. Integr. Comp. Biol. 42, 517–525.

Barton, B.A., Schreck, C.B., Fowler, L.G., 1988. Fasting and diet content affect stress-induced changes in plasma glucose and cortisol in juvenile Chinook salmon. Prog. Fish Cult. 50 (1), 16–22.

Bass, A.L., Hinch, S.G., Patterson, D.A., Cooke, S.J., Farrell, A.P., 2018. Location-specific consequences of beach seine and gillnet capture on upriver-migrating sockeye salmon migration behavior and fate1. Can. J. Fish. Aquat. Sci. 75, 2011–2023. https://doi.org/10.1139/cjfas-2017-0474.

Bass, A.L., Hinch, S.G., Teffer, A.K., Patterson, D.A., Miller, K.M., 2017. A survey of microparasites present in adult migrating Chinook salmon (Oncorhynchus tshawytscha) in south-western British Columbia determined by high-throughput quantitative polymerase chain reaction. J. Fish. Dis. 40, 453–477. https://doi.org/10.1111/jfd.12607.

Bergmeyer, H.U., 1983. Methods of Enzymatic Analysis. Academic Press, New York.

Bernier, N.J., Alderman, S.L., 2022. Applied aspects of fish endocrinology. Fish Physiol. 39A, 253–320.

Binder, T.R., Wilson, A.D.M., Wilson, S.M., Suski, C.D., Godin, J.-G.J., Cooke, S.J., 2016. Is there a pace-of-life syndrome linking boldness and metabolic capacity for locomotion in bluegill sunfish? Anim. Behav. 121, 175–183. https://doi.org/10.1016/j.anbehav.2016.09.006.

Blazka, P., Volf, M., Cepela, M., 1960. A new type of respirometer for the determination of the metabolism of fish in an active state. Physiol. Bohemoslov. 9(6), 553 - 558.

Blaxhall, P.C., Daisley, K.W., 1973. Routin heamatological methods for use with fish blood. J. Fish Biol. 5, 771–781.

Bouyoucos, I.A., Weideli, O.C., Planes, S., Simpfendorfer, C.A., Rummer, J.L., 2018. Dead tired: evaluating the physiological status and survival of neonatal reef sharks under stress. Conserv Physiol 6, coy053.

Bouyoucos, I.A., Simpfendorfer, C.A., Rummer, J.L., 2019. Estimating oxygen uptake rates to understand stress in sharks and rays. Rev. Fish Biol. Fish. 29 (2), 297–311.

Bowerman, T.E., Keefer, M.L., Caudill, C.C., 2021. Elevated stream temperature, origin, and individual size influence Chinook salmon prespawn mortality across the Columbia River basin. Fish. Res. 237, 105874.

Brett, J.R., 1964. The respiratory metabolism and swimming performance of young sockeye salmon. J. Fish. Res. Board Can. 21 (5), 1183–1226.

Brett, J.R., 1971. Energetic responses of salmon to temperature. A study of some thermal relations in the physiology and freshwater ecology of sockeye salmon (*Oncorhynchus nerka*). Am. Zool. 11, 99–113. https://doi.org/10.1093/icb/11.1.99.

Brett, J.R., Groves, T., 1979. Physiological energetics. Fish Physiol. 8, 280–352.

Brett, J.R., Hollands, M., Alderdice, D.F., 1958. The effect of temperature on the cruising speed of Young Sockeye and Coho Salmon. J. Fish. Res. Board Can. 15, 587–605.

Brette, F., Machado, B., Cros, C., Incardona, J.P., Scholz, N.L., Block, B.A., 2014. Crude oil impairs cardiac excitation-contraction coupling in fish. Science 343 (6172), 772–776.

Brette, F., Shiels, H.A., Galli, G.L., Cros, C., Incardona, J.P., Scholz, N.L., Block, B.A., 2017. A novel cardiotoxic mechanism for a pervasive global pollutant. Sci. Rep. 7 (1), 1–9.

Brijs, J., Sandblom, E., Axelsson, M., Sundell, K., Sundh, H., Huyben, D., Broström, R., Kiessling, A., Berg, C., Gräns, A., 2018. The final countdown: continuous physiological welfare evaluation of farmed fish during common aquaculture practices before and during harvest. Aquaculture 495, 903–911.

Brijs, J., Sandblom, E., Rosengren, M., Sundell, K., Berg, C., Axelsson, M., Gräns, A., 2019. Prospects and pitfalls of using heart rate bio-loggers to assess the welfare of rainbow trout (Oncorhynchus mykiss) in aquaculture. Aquaculture 509, 188–197.

Brijs, J., Føre, M., Gräns, A., Clark, T., Axelsson, M., Johansen, J., 2021. Bio-sensing technologies in aquaculture: how remote monitoring can bring us closer to our farm animals. Philos. Trans. R. Soc. B 376 (1830), 20200218.

Brill, R.W., Bushnell, P.G., Jones, D.R., Shimizu, M., 1992. Effects of acute temperature change, *in vivo* and *in vitro*, on the acid-base status of blood from yellowfin tuna (*Thunnus albacares*). Can. J. Zool. 70, 654–662.

Buchanan, S., Farrell, A.P., Fraser, J., Gallaugher, P., Routledge, R., 2002. Reducing gill-net mortality of incidentally caught coho salmon. N. Am. J. Fish. Manag. 22, 1270–1275. https://doi.org/10.1577/1548-8675(2002)022<1270:RGNMOI>2.0.CO;2.

Burnett, N.J., Hinch, S.G., Braun, D.C., Casselman, M.T., Middleton, C.T., Wilson, S.M., Cooke, S.J., 2014. Burst swimming in areas of high flow: delayed consequences of anaerobiosis in wild adult sockeye salmon. Physiol. Biochem. Zool. 87, 587–598. https://doi.org/10.1086/677219.

Bushnell, P.G., Steffensen, J.F., Schurmann, H., Jones, D.R., 1994. Exercise metabolism in two species of cod in arctic waters. Polar Biol. 14, 43–48. https://doi.org/10.1007/BF00240271.

Byrnes, E.E., Lear, K.O., Brewster, L.R., Whitney, N.M., Smukall, M.J., Armstrong, N.J., Gleiss, A.C., 2021. Accounting for body mass effects in the estimation of field metabolic rates from body acceleration. J. Exp. Biol. 224 (7), jeb233544.

Cameron, J.N., Davis, J.C., 1970. Gas exchange in rainbow trout (*Salmo gairdneri*) with varying blood oxygen capacity. J. Fish. Res. Board Can. 27 (6).

Campana, S.E., Joyce, W., Manning, M.J., 2009. Bycatch and discard mortality in commercially caught blue sharks *Prionace glauca* assessed using archival satellite pop-up tags. Mar. Ecol. Prog. Ser. 387, 241–253.

Carrier, J.C., Pratt Jr., H.L., Castro, J.I., 2004. Reproductive biology of elasmobranchs. In: Carrier, J.C., Musick, J.A., Heithaus, M.R. (Eds.), Biology of Sharks and their Relatives. Marine Biology, vol. 10. CRC Press, Boca Raton, pp. 269–286.

Carrier, J.C., Musick, J.A., Heithaus, M.R., 2012. Biology of Sharks and their Relatives. CRC Press.

Casselman, M.T., Anttila, K., Farrell, A.P., 2012. Using maximum heart rate as a rapid screening tool to determine optimum temperature for aerobic scope in Pacific salmon Oncorhynchus spp. J. Fish Biol. 80 (2), 358–377. https://doi.org/10.1111/j.1095-8649.2011.03182.x.

Chabot, D., Koenker, R., Farrell, A.P., 2016. The measurement of specific dynamic action in fishes: measuring sda in fishes. J. Fish Biol. 88, 152–172.

Chen, Z., Narum, S.R., 2021. Whole genome resequencing reveals genomic regions associated with thermal adaptation in redband trout. Mol. Ecol. 30 (1), 162–174.

Chen, Z., Anttila, K., Wu, J., Whitney, C.K., Hinch, S.G., Farrell, A.P., 2013. Optimum and maximum temperatures of sockeye salmon (Oncorhynchus nerka) populations hatched at different temperatures. Can. J. Zool. 91 (5), 265–274. https://doi.org/10.1139/cjz-2012-0300.

Chen, Z., Devlin, R.H., Farrell, A.P., 2015. Upper thermal tolerance of wild-type, domesticated and growth hormone-transgenic coho salmon Oncorhynchus kisutch. J. Fish Biol. 87 (3), 763–773. https://doi.org/10.1111/jfb.12736.

Chen, Z., Farrell, A.P., Matala, A., Hoffman, N., Narum, S.R., 2018a. Physiological and genomic signatures of evolutionary thermal adaptation in redband trout from extreme climates. Evol. Appl. 11 (9), 1686–1699. https://doi.org/10.1111/eva.12672.

Chen, Z., Farrell, A.P., Matala, A., Narum, S.R., 2018b. Mechanisms of thermal adaptation and evolutionary potential of conspecific populations to changing environments. Mol. Ecol. 27 (3), 659–674.

Chung, M., Jørgensen, K.M., Trueman, C.N., Knutsen, H., Jorde, P.E., Grønkjær, P., 2020. First measurements of field metabolic rate in wild juvenile fishes show strong thermal sensitivity but variations between sympatric ecotypes. Oikos 130, 287–299. oik.07647.

Cicia, A.M., Schlenker, L.S., Sulikowski, J.A., Mandelman, J.W., 2012. Seasonal variations in the physiological stress response to discrete bouts of aerial exposure in the little skate, *Leucoraja erinacea*. Comp. Biochem. Physiol. A Mol. Integr. Physiol. 162 (2), 130–138.

Claireaux, G., Lefrançois, C., 2007. Linking environmental variability and fish performance: integration through the concept of scope for activity. Philos. Trans. R. Soc. B 362, 2031–2041.

Clark, T.D., Eliason, E.J., Sandblom, E., Hinch, S.G., Farrell, A.P., 2008a. Calibration of a hand-held haemoglobin analyser for use on fish blood. J. Fish Biol. 73 (10), 2587–2595. https://doi.org/10.1111/j.1095-8649.2008.02109.x.

Clark, T.D., Sandblom, E., Cox, G.K., Hinch, S.G., Farrell, A.P., 2008b. Circulatory limits to oxygen supply during an acute temperature increase in the Chinook salmon (*Oncorhynchus tshawytscha*). Am. J. Phys. Regul. Integr. Comp. Phys. 295, R1631–R1639.

Clark, T.D., Sandblom, E., Hinch, S.G., Patterson, D.A., Frappell, P.B., Farrell, A.P., 2010. Simultaneous biologging of heart rate and acceleration, and their relationships with energy expenditure in free-swimming sockeye salmon (*Oncorhynchus nerka*). J. Comp. Physiol. B. 180, 673–684.

Clark, T.D., Jeffries, K.M., Hinch, S.G., Farrell, A.P., 2011. Exceptional aerobic scope and cardiovascular performance of pink salmon (Oncorhynchus gorbuscha) may underlie resilience in a warming climate. J. Exp. Biol. 214, 3074–3081.

Clark, T.D., Donaldson, M.R., Pieperhoff, S., Drenner, S.M., Lotto, A., Cooke, S.J., Hinch, S.G., Patterson, D.A., Farrell, A.P., 2012. Physiological benefits of being small in a changing world: responses of Coho Salmon (Oncorhynchus kisutch) to an acute thermal challenge and a simulated capture event. PLoS One 7, e39079.

Clark, T.D., Sandblom, E., Jutfelt, F., 2013. Aerobic scope measurements of fishes in an era of climate change: respirometry, relevance and recommendations. J. Exp. Biol. 216, 2771–2782.

Cliff, G., Thurman, G.D., 1984. Pathological and physiological effects of stress during capture and transport in the juvenile dusky shark, *Carcharhinus obscurus*. Comp. Biochem. Physiol. A Mol. Integr. Physiol. 78 (1), 167–173.

Compagno, L.J., 1990. Alternative life-history styles of cartilaginous fishes in time and space. Environ. Biol. Fish 28 (1–4), 33–75.

Compagno, L., Dando, M., Fowler, S., 2005. Sharks of the World. Princeton Field Guides. Princeton University Press, Princeton.

Connon, R.E., Jeffries, K.M., Komoroske, L.M., Todgham, A.E., Fangue, N.A., 2018. The utility of transcriptomics in fish conservation. J. Exp. Biol. 221 (2), jeb148833.

Cooke, S.J., Suski, C.D., 2005. Do we need species-specific guidelines for catch-and-release recreational angling to effectively conserve diverse fishery resources? Biodivers. Conserv. 14, 1195–1209.

Cooke, S.J., Hinch, S.G., Crossin, G.T., Patterson, D.A., English, K.K., Healey, M.C., Shrimpton, J.M., Van Der Kraak, G., Farrell, A.P., 2006a. Mechanistic basis of individual mortality in Pacific salmon during spawning migrations. Ecology 87, 1575–1586. https://doi.org/10.1890/0012-9658(2006)87[1575:MBOIMI]2.0.CO;2.

Cook, K.V., Hinch, S.G., Drenner, S.M., Raby, G.D., Patterson, D.A., Cooke, S.J., 2019. Dermal injuries caused by purse seine capture result in lasting physiological disturbances in coho salmon. Comp. Biochem. Physiol. A Mol. Integr. Physiol. 227, 75–83. https://doi.org/10.1016/j.cbpa.2018.09.026.

Cook, K.V., McConnachie, S.H., Gilmour, K.M., Hinch, S.G., Cooke, S.J., 2011. Fitness and behavioral correlates of pre-stress and stress-induced plasma cortisol titers in pink salmon (Oncorhynchus gorbuscha) upon arrival at spawning grounds. Horm. Behav. 60, 489–497. https://doi.org/10.1016/j.yhbeh.2011.07.017.

Cook, K.V., Hinch, S.G., Watson, M.S., Patterson, D.A., Reid, A.J., Cooke, S.J., 2018. Experimental capture and handling of chum salmon reveal thresholds in injury, impairment, and physiology: best practices to improve bycatch survival in a purse seine fishery. Fish. Res. 206, 96–108.

Cooke, S.J., Hinch, S.G., Crossin, G.T., Patterson, D.A., English, K.K., Healey, M.C., Macdonald, J.S., Shrimpton, J.M., Young, J.L., Lister, A., Van Der Kraak, G., Farrell, A.P., 2008. Physiological correlates of coastal arrival and river entry timing in late summer Fraser River sockeye salmon (Oncorhynchus nerka). Behav. Ecol. 19, 747–758. https://doi.org/10.1093/beheco/arn006.

Cooke, S.J., Hinch, S.G., Crossin, G.T., Patterson, D.A., English, K.K., Shrimpton, J.M., 2006b. Physiology of individual late-run Fraser River sockeye salmon (Oncorhynchus nerka) sampled in the ocean correlates with fate during spawning migration. Can. J. Fish. Aquat. Sci. 63 (7), 1469–1480.

Cooke, S.J., Hinch, S.G., Donaldson, M.R., Clark, T.D., Eliason, E.J., Crossin, G.T., Raby, G.D., Jeffries, K.M., Lapointe, M., Miller, K., et al., 2012a. Conservation physiology in practice: how physiological knowledge has improved our ability to sustainably manage Pacific salmon during up-river migration. Philos. Trans. R. Soc. B 367, 1757–1769.

Cooke, S.J., Hinch, S., LuCaS, M.C., Lutcavage, M., 2012b. Biotelemetry and Biologging. Fisheries Techniques, third ed. American Fisheries Society, Bethesda, Maryland, pp. 819–860.

Cooke, S.J., Messmer, V., Tobin, A.J., Pratchett, M.S., Clark, T.D., 2014. Refuge-seeking impairments Mirror metabolic recovery following fisheries-related stressors in the Spanish flag snapper (*Lutjanus carponotatus*) on the great barrier reef. Physiol. Biochem. Zool. 87, 136–147.

Cooke, S.J., Nguyen, V.M., Chapman, J.M., Reid, A.J., Landsman, S.J., Young, N., Hinch, S.G., Schott, S., Mandrak, N.E., Semeniuk, C.A., 2021. Knowledge co-production: a pathway to effective fisheries management, conservation, and governance. Fisheries 46, 89–97.

Cooke, S.J., Fangue, N.A., Bergman, J.N., Madliger, C.L., Cech Jr., J.J., Eliason, E.J., Brauner, C.J., Farrell, A.P., 2022. Conservation physiology and the management of wild fish populations in the Anthropocene. Fish Physiol. 39A, 1–31.

Cooper, C.J., Kristan, W.B., Eme, J., 2021. Thermal tolerance and routine oxygen consumption of convict cichlid, Archocentrus nigrofasciatus, acclimated to constant temperatures (20° C and 30° C) and a daily temperature cycle (20° C → 30° C). J. Comp. Physiol. B. 191 (3), 479–491.

Cortese, D., Norin, T., Beldade, R., Crespel, A., Killen, S.S., Mills, S.C., 2021. Physiological and behavioural effects of anemone bleaching on symbiont anemonefish in the wild. Funct. Ecol. 35 (3), 663–674.

Couillard, C.M., Lee, K., Légaré, B., King, T.L., 2005. Effect of dispersant on the composition of the water-accommodated fraction of crude oil and its toxicity to larval marine fish. Environ. Toxicol. Chem.: Int. J. 24 (6), 1496–1504.

Courtney, D., Mathers, A., 2019. An Updated Literature Review of Post-Release Live-Discard Mortality Rate Estimates in Sharks for Use in SEDAR 65. SEDAR65-DW0. North Charleston, SC.

Cram, S., Ponce De Leon, C.A., Fernandez, P., Sommer, I., Rivas, H., Morales, L.M., 2006. Assesment of trace elements and organic pollutants from a marine oil complex into the coral reef system of Cayo Arcas, Mexico. Environ. Monit. Assess. 121 (1-3), 127–149. https://doi.org/10.1007/s10661-005-9111-7.

Crossin, G.T., Hinch, S.G., Farrell, A.P., Higgs, D.A., Lotto, A.G., Oakes, J.D., Healey, M.C., 2004. Energetics and morphology of sockeye salmon: effects of upriver migratory distance and elevation. J. Fish Biol. 65, 788–810.

Crossin, G.T., Hinch, S.G., Cooke, S.J., Welch, D.W., Batten, S.D., Patterson, D.A., Van Der Kraak, G., Shrimpton, J.M., Farrell, A.P., 2007. Behaviour and physiology of sockeye salmon homing through coastal waters to a natal river. Mar. Biol. 152, 905–918. https://doi.org/10.1007/s00227-007-0741-x.

Crossin, G.T., Hinch, S.G., Cooke, S.J., Welch, D.W., Patterson, D.A., Jones, S.R.M., Lotto, A.G., Leggatt, R.A., Mathes, M.T., Shrimpton, J.M., et al., 2008. Exposure to high temperature influences the behaviour, physiology, and survival of sockeye salmon during spawning migration. Can. J. Zool. 86, 127–140.

Crossin, G.T., Hinch, S.G., Cooke, S.J., Cooperman, M.S., Patterson, D.A., Welch, D.W., Hanson, K.C., Olsson, I., English, K.K., Farrell, A.P., 2009. Mechanisms influencing the timing and success of reproductive migration in a capital breeding Semelparous fish species, the Sockeye Salmon. Physiol. Biochem. Zool. 82, 635–652.

Crozier, L.G., Hendry, A.P., Lawson, P.W., Quinn, T.P., Mantua, N.J., Battin, J., Shaw, R.G., Huey, R.B., 2008. PERSPECTIVE: potential responses to climate change in organisms with complex life histories: evolution and plasticity in Pacific salmon. Evol. Appl. 1 (2), 252.

Currie, S., Tufts, B.L., Moyes, C.D., 1999. Influence of bioenergetic stress on heat shock protein gene expression in nucleated red blood cells of fish. Am. J. Phys. Regul. Integr. Comp. Phys. 276 (4), R990–R996.

Dacie, J., Lewis, S., 1975. Practical Haematology, fifth ed. The English language book society and Churchill livingstone.

Dahlke, F.T., Wohlrab, S., Butzin, M., Pörtner, H.-O., 2020. Thermal bottlenecks in the life cycle define climate vulnerability of fish. Science 369 (6499), 65–70.

Danylchuk, A.J., Suski, C.D., Mandelman, J.W., Murchie, K.J., Haak, C.R., Brooks, A.M., Cooke, S.J., 2014. Hooking injury, physiological status and short-term mortality of juvenile lemon sharks (*Negaprion bevirostris*) following catch-and-release recreational angling. Conserv Physiol 2 (1), cot036. https://doi.org/10.1093/conphys/cot036.

Daufresne, M., Lengfellner, K., Sommer, U., 2009. Global warming benefits the small in aquatic ecosystems. Proc. Natl. Acad. Sci. U. S. A. 106 (31), 12788–12793.

Davis, M.W., 2002. Key principles for understanding fish bycatch discard mortality. Can. J. Fish. Aquat. Sci. 59, 1834–1843.

Davis, L.E., Schreck, C.B., 1997. The energetic response to handling stress in juvenile coho salmon. Trans. Am. Fish. Soc. 126, 248–258. https://doi.org/10.1577/1548-8659(1997) 126<0248:TERTHS>2.3.CO;2.

Di Santo, V., 2016. Intraspecific variation in physiological performance of a benthic elasmobranch challenged by ocean acidification and warming. J. Exp. Biol. 219 (11), 1725–1733.

Dick, M., Eliason, E.J., Patterson, D.A., Robinson, K.A., Hinch, S.G., Cooke, S.J., 2018. Short-term physiological response profiles of tagged migrating adult sockeye salmon: a comparison of gastric insertion and external tagging methods. Trans. Am. Fish. Soc. 147, 300–315. https://doi.org/10.1002/tafs.10027.

Dittman, A.H., Quinn, T.P., 1996. Homing in Pacific Salmon: mechanisms and ecological basis. J. Exp. Biol. 199, 83–91.

Donaldson, M.R., Hinch, S.G., Patterson, D.A., Farrell, A.P., Shrimpton, J.M., Miller-Saunders, K.M., Robichaud, D., Hills, J., Hruska, K.A., Hanson, K.C., et al., 2010a. Physiological condition differentially affects the behavior and survival of two populations of sockeye Salmon during their freshwater spawning migration. Physiol. Biochem. Zool. 83, 446–458.

Donaldson, M.R., Clark, T.D., Hinch, S.G., Cooke, S.J., Patterson, D.A., Gale, M.K., Frappell, P.B., Farrell, A.P., 2010b. Physiological responses of free-swimming adult coho salmon to simulated predator and fisheries encounters. Physiol. Biochem. Zool. 83 (6), 973–983.

Donaldson, M.R., Hinch, S.G., Raby, G.D., Patterson, D.A., Farrell, A.P., Cooke, S.J., 2012. Population-specific consequences of fisheries-related stressors on adult sockeye Salmon. Physiol. Biochem. Zool. 85, 729–739.

Donaldson, M.R., Hinch, S.G., Jeffries, K.M., Patterson, D.A., Cooke, S.J., Farrell, A.P., Miller, K.M., 2014. Species- and sex-specific responses and recovery of wild, mature pacific salmon to an exhaustive exercise and air exposure stressor. Comp. Biochem. Physiol. A Mol. Integr. Physiol. 173, 7–16.

Donaldson, M.R., Raby, G.D., Nguyen, V.N., Hinch, S.G., Patterson, D.A., Farrell, A.P., Rudd, M.A., Thompson, L.A., O'Connor, C.M., Colotelo, A.H., McConnachie, S.H., Cook, K.V., Robichaud, D., English, K.K., Cooke, S.J., 2013. Evaluation of a simple technique for recovering fish from capture stress: integrating physiology, biotelemetry, and social science to solve a conservation problem. Can. J. Fish. Aquat. Sci. 70, 90–100. https://doi.org/10.1139/cjfas-2012-0218.

Donelson, J.M., Munday, P.L., McCormick, M.I., Nilsson, G.E., 2011. Acclimation to predicted ocean warming through developmental plasticity in a tropical reef fish. Glob. Chang. Biol. 17, 1712–1719.

Downie, A.T., Phelps, C.M., Jones, R., Rummer, J.L., Chivers, D.P., Ferrari, M.C., McCormick, M.I., 2021. Exposure to degraded coral habitat depresses oxygen uptake rate during exercise of a juvenile reef fish. Coral Reefs 40, 1361–1367.

Drenner, S.M., Hinch, S.G., Furey, N.B., Clark, T.D., Li, S., Ming, T., Jeffries, K.M., Patterson, D.A., Cooke, S.J., Robichaud, D., Welch, D.W., Farrell, A.P., Miller, K.M., 2018. Transcriptome patterns and blood physiology associated with homing success of sockeye salmon during their final stage of marine migration. Can. J. Fish. Aquat. Sci. 75, 1511–1524. https://doi.org/10.1139/cjfas-2017-0391.

Drenner, S.M., Hinch, S.G., Martins, E.G., Furey, N.B., Clark, T.D., Cooke, S.J., Patterson, D.A., Robichaud, D., Welch, D.W., Farrell, A.P., Thomson, R.E., 2015. Environmental conditions and physiological state influence estuarine movements of homing sockeye salmon. Fish. Oceanogr. 24, 307–324. https://doi.org/10.1111/fog.12110.

Driggers, W.I., Carlson, J.K., Cullum, B., Dean, J.M., Oakley, D., Ulrich, G., 2004. Age and growth of the blacknose shark, *Carcharhinus acronotus*, in the western North Atlantic Ocean with comments on reagional variation in growth rates. Environ. Biol. Fish 71, 171–178.

Drost, H.E., Fisher, J., Randall, F., Kent, D., Carmack, E.C., Farrell, A.P., 2016. Upper thermal limits of the hearts of Arctic cod *Boreogadus saida*: adults compared with larvae. J. Fish Biol. 88 (2), 718–726. https://doi.org/10.1111/jfb.12807.

Dulvy, N.K., Fowler, S.L., Musick, J.A., Cavanagh, R.D., Kyne, P.M., Harrison, L.R., Carlson, J.K., Davidson, L.N.K., Fordham, S.V., Malcolm, P.F., Pollock, C.M., Simpfendorfer, C.A., Burgess, G.H., Carpenter, K.E., Compagno, L.J.V., Ebert, D.A., Gibson, C., Heupel, M.R., Livingstone, S.R., Sanciangco, J.C., Stevens, J.D., Valenti, S., White, W.T., 2014. Extinction risk and conservation of the world's sharks and rays. elife 2014 (3), e00590.

Ebert, D.A., Fowler, S.L., Compagno, L.J., 2013. Sharks of the World: A Fully Illustrated Guide. Wild Nature Press.

Eiler, J.H., Evans, A.N., Schreck, C.B., 2015. Migratory patterns of wild Chinook Salmon Oncorhynchus tshawytscha returning to a large, free-flowing river basin. PLoS One 10, e0123127.

Ekström, A., Axelsson, M., Gräns, A., Brijs, J., Sandblom, E., 2017. Influence of the coronary circulation on thermal tolerance and cardiac performance during warming in rainbow trout. Am. J. Phys. Regul. Integr. Comp. Phys. 312 (4), R549–R558. https://doi.org/10.1152/ajpregu.00536.2016.

Ekström, A., Sundell, E., Morgenroth, D., Sandblom, E., 2021. Adrenergic tone benefits cardiac performance and warming tolerance in two teleost fishes that lack a coronary circulation. J. Comp. Physiol. B. 191, 701–709. https://doi.org/10.1007/s00360-021-01359-9.

Eliason, E.J., Anttila, K., 2017. Temperature and the cardiovascular system. In: Fish Physiology. vol 36. Elsevier, pp. 235–297.

Eliason, E.J., Farrell, A.P., 2014. Effect of hypoxia on specific dynamic action and postprandial cardiovascular physiology in rainbow trout (*Oncorhynchus mykiss*). Comp. Biochem. Physiol. A Mol. Integr. Physiol. 171, 44–50. https://doi.org/10.1016/j.cbpa.2014.01.021.

Eliason, E.J., Farrell, A.P., 2016. Oxygen uptake in Pacific salmon *Oncorhynchus* SPP.: when ecology and physiology meet: OXYGEN UPTAKE IN ONCORHYNCHUS SPP. J. Fish Biol. 88, 359–388.

Eliason, E.J., Higgs, D.A., Farrell, A.P., 2007. Effect of isoenergetic diets with different protein and lipid content on the growth performance and heat increment of rainbow trout. Aquaculture 272, 723–736.

Eliason, E.J., Higgs, D.A., Farrell, A.P., 2008. Postprandial gastrointestinal blood flow, oxygen consumption and heart rate in rainbow trout (Oncorhynchus mykiss). Comp. Biochem. Physiol. A Mol. Integr. Physiol. 149, 380–388.

Eliason, E.J., Clark, T.D., Hague, M.J., Hanson, L.M., Gallagher, Z.S., Jeffries, K.M., Gale, M.K., Patterson, D.A., Hinch, S.G., Farrell, A.P., 2011. Differences in thermal tolerance among sockeye Salmon populations. Science 332, 109–112.

Eliason, E.J., Wilson, S.M., Farrell, A.P., Cooke, S.J., Hinch, S.G., 2013a. Low cardiac and aerobic scope in a coastal population of sockeye salmon *Oncorhynchus nerka* with a short upriver migration: thermal tolerance in *Oncorhynchus nerka*. J. Fish Biol. 82, 2104–2112.

Eliason, E.J., Clark, T.D., Hinch, S.G., Farrell, A.P., 2013b. Cardiorespiratory collapse at high temperature in swimming adult sockeye salmon. Conserv Physiol 1 (1), cot008. https://doi.org/10.1093/conphys/cot008.

Eliason, E.J., Dick, M., Patterson, D.A., Robinson, K.A., Lotto, J., Hinch, S.G., Cooke, S.J., 2020. Sex-specific differences in physiological recovery and short-term behaviour following fisheries capture in adult sockeye salmon (Oncorhynchus nerka). Can. J. Fish. Aquat. Sci. 77 (11), 1749–1757.

Ern, R., Huong, D.T.T., Cong, N.V., Bayley, M., Wang, T., 2014. Effect of salinity on oxygen consumption in fishes: a review. J. Fish Biol. 84 (4), 1210–1220. https://doi.org/10.1111/jfb.12330.

Esbaugh, A.J., Mager, E.M., Stieglitz, J.D., Hoenig, R., Brown, T.L., French, B.L., Linbo, T.L., Lay, C., Forth, H., Scholz, N.L., Incardona, J.P., Morris, J.M., Benetti, D.D., Grosell, M., 2016. The effects of weathering and chemical dispersion on Deepwater horizon crude oil toxicity to mahi-mahi (Coryphaena hippurus) early life stages. Sci. Total Environ. 543 (Pt A), 644–651. https://doi.org/10.1016/j.scitotenv.2015.11.068.

Evans, T.G., Kültz, D., 2020. The cellular stress response in fish exposed to salinity fluctuations. J. Exp. Zool. A Ecol. Integr. Physiol. 333 (6), 421–435.

Farrell, A.P., 2016. Pragmatic perspective on aerobic scope: peaking, plummeting, pejus and apportioning. J. Fish Biol. 88, 322–343.

Farrell, A.P., Clutterham, S.M., 2003. On-line venous oxygen tensions in rainbow trout during graded exercise at two acclimation temperatures. J. Exp. Biol. 206 (Pt 3), 487–496. https://doi.org/10.1242/jeb.00100.

Farrell, A.P., Lee, C.G., Tierney, K., Hodaly, A., Clutterham, S., Healey, M., Hinch, S., Lotto, A., 2003. Field-based measurements of oxygen uptake and swimming performance with adult Pacific salmon using a mobile respirometer swim tunnel. J. Fish Biol. 62, 64–84. https://doi.org/10.1046/j.1095-8649.2003.00010.x.

Farrell, A.P., Steffensen, J.F., 1987. An analysis of the energetic cost of the branchial and cardiac pumps during sustained swimming in trout. Fish Physiol. Biochem. 4, 73–79.

Farrell, A.P., Gamperl, A.K., Birtwell, I.K., 1998. Prolonged swimming, recovery and repeat swimming performance of mature sockeye salmon Oncorhynchus nerka exposed to moderate hypoxia and pentachlorophenol. J. Exp. Biol. 201, 2183–2193.

Farrell, A.P., Gallaugher, P., Clarke, C., DeLury, N., Kreiberg, H., Parkhouse, W., Routledge, R., 2000. Physiological status of coho salmon (Oncorhynchus kisutch) captured in commercial nonretention fisheries. Can. J. Fish. Aquat. Sci. 57, 1668–1678.

Farrell, A.P., Gallaugher, P.E., Fraser, J., Pike, D., Bowering, P., Hadwin, A.K.M., Parkhouse, W., Routledge, R., 2001a. Successful recovery of the physiological status of coho salmon on board a commercial gillnet vessel by means of a newly designed revival box. Can. J. Fish. Aquat. Sci. 58 (10), 1932–1946. https://doi.org/10.1139/cjfas-58-10-1932.

Farrell, A.P., Thorarensen, H., Axelsson, M., Crocker, C.E., Gamperl, A.K., Cech, J.J., 2001b. Gut blood flow in fish during exercise and severe hypercapnia. Comp. Biochem. Physiol. A Mol. Integr. Physiol. 128 (3), 551–563.

Farrell, A.P., Hinch, S.G., Cooke, S.J., Patterson, D.A., Crossin, G.T., Lapointe, M., Mathes, M.T., 2008. Pacific salmon in hot water: applying aerobic scope models and biotelemetry to predict the success of spawning migrations. Physiol. Biochem. Zool. 81 (6), 697–709. https://doi.org/ 10.1086/592057.

Fenkes, M., Shiels, H.A., Fitzpatrick, J.L., Nudds, R.L., 2016. The potential impacts of migratory difficulty, including warmer waters and altered flow conditions, on the reproductive success of salmonid fishes. Comp. Biochem. Physiol. A Mol. Integr. Physiol. 193, 11–21.

Ferrannini, E., 1988. The theoretical bases of indirect calorimetry: a review. Metabolism 37, 287–301.

Ferreira, E.O., Anttila, K., Farrell, A.P., 2014. Thermal optima and tolerance in the eurythermic goldfish (Carassius auratus): relationships between whole-animal aerobic capacity and maximum heart rate. Physiol. Biochem. Zool. 87 (5), 599–611. https://doi.org/10.1086/677317.

Fitzgibbon, Q.P., Strawbridge, A., Seymour, R.S., 2007. Metabolic scope, swimming performance and the effects of hypoxia in the mulloway, Argyrosomus japonicus (Pisces: Sciaenidae). Aquaculture 270 (1–4), 358–368.

Flikac, T., Cook, D.G., Davison, W., 2020. The effect of temperature and meal size on the aerobic scope and specific dynamic action of two temperate New Zealand finfish Chrysophrys auratus and Aldrichetta forsteri. J. Comp. Physiol. B. 190, 169–183.

Føre, M., Svendsen, E., Økland, F., Gräns, A., Alfredsen, J.A., Finstad, B., Hedger, R.D., Uglem, I., 2021. Heart rate and swimming activity as indicators of post-surgical recovery time of Atlantic salmon (Salmo salar). Anim. Biotelemetry 9 (1), 1–13.

Fowler, S.L., Hamilton, D., Currie, S., 2009. A comparison of the heat shock response in juvenile and adult rainbow trout (Oncorhynchus mykiss)—implications for increased thermal sensitivity with age. Can. J. Fish. Aquat. Sci. 66 (1), 91–100.

French, R.P., Lyle, J., Tracey, S., Currie, S., Semmens, J.M., 2015. High survivorship after catch-and-release fishing suggests physiological resilience in the endothermic shortfin mako shark (Isurus oxyrinchus). Conserv Physiol 3 (1), cov044. https://doi.org/10.1093/conphys/cov044.

Frick, L.H., Reina, R.D., Walker, T.I., 2010a. The phyiological response of port Jackson sharks and Australian swellsharks to sedation, gill-net capture, and repeated sampling in captivity. N. Am. J. Fish Manag. 29, 127–139.

Frick, L.H., Reina, R.D., Walker, T.I., 2010b. Stress related physiological changes and post-release survival of Port Jackson sharks (Heterodontus portusjacksoni) and gummy sharks (Mustelus antarcticus) following gill-net and longline capture in captivity. J. Exp. Mar. Biol. Ecol. 385 (1–2), 29–37.

Frick, L.H., Walker, T.I., Reina, R.D., 2010c. Trawl capture of Port Jackson sharks, Heterodontus portusjacksoni, and gummy sharks, Mustelus antarcticus, in a controlled setting: effects of tow duration, air exposure and crowding. Fish. Res. 106 (3), 344–350. https://doi.org/ 10.1016/j.fishres.2010.08.016.

Frisk, M., Steffensen, J.F., Skov, P.V., 2013. The effects of temperature on specific dynamic action and ammonia excretion in pikeperch (Sander lucioperca). Aquaculture 404–405, 65–70.

Fry, F.E.J., 1947. Effects of the Environment on Animal Activity. vol. 68 Publications of the Ontario Fisheries Research Laboratory, pp. 1–63.

Fry, F.E.J., 1948, January. Temperature relations of salmonids. Proc. Can. Comm. Freshw. Fish Res. 1st Meeting, App. D 1 (6), 1–6.

Fry, F.E.J., 1971. The effect of environmental factors on the physiology of fish. In: Hoar, W.S., Randall, D.J. (Eds.), Fish Physiology. vol. 6. Academic Press, New York, pp. 1–98.

Fry, F., Hart, J.S., 1948. The relation of temperature to oxygen consumption in the goldfish. Biol. Bull. 94 (1), 66–77.

Gaesser, G.A., Brooks, G.A., 1984. Metabolic bases of excess post-exercise oxygen consumption: a review. Med. Sci. Sports Exerc. 16, 29–43.

Gale, M.K., Hinch, S.G., Cooke, S.J., Donaldson, M.R., Eliason, E.J., Jeffries, K.M., Martins, E.G., Patterson, D.A., 2014. Observable impairments predict mortality of captured and released sockeye salmon at various temperatures. Conserv. Physiol. 2, cou029. https://doi.org/10.1093/conphys/cou029.

Gale, M.K., Hinch, S.G., Eliason, E.J., Cooke, S.J., Patterson, D.A., 2011. Physiological impairment of adult sockeye salmon in fresh water after simulated capture-and-release across a range of temperatures. Fish. Res. 112, 85–95.

Gallagher, A.J., Frick, L.H., Bushnell, P.G., Brill, R.W., Mandelman, J.W., 2010. Blood gas, oxygen saturation, pH, and lactate values in elasmobranch blood measured with a commercially available portable clinical analyzer and standard laboratory instruments. J. Aquat. Anim. Health 22 (4), 229–234. https://doi.org/10.1577/H10-012.1.

Gardner, J.L., Peters, A., Kearney, M.R., Joseph, L., Heinsohn, R., 2011. Declining body size: a third universal response to warming? Trends Ecol. Evol. 26 (6), 285–291.

Gilbert, M.J.H., Farrell, A.P., 2021. The thermal acclimation potential of maximum heart rate and cardiac heat tolerance in Arctic char (Salvelinus alpinus), a northern cold-water specialist. J. Therm. Biol. 95, 102816. https://doi.org/10.1016/j.jtherbio.2020.102816.

Gilbert, M.J., Harris, L.N., Malley, B.K., Schimnowski, A., Moore, J.-S., Farrell, A.P., 2020. The thermal limits of cardiorespiratory performance in anadromous Arctic char (Salvelinus alpinus): a field-based investigation using a remote mobile laboratory. Conserv. Physiol. 8 (1), coaa036.

Gomez Isaza, D.F., Cramp, R.L., Franklin, C.E., 2020. Thermal acclimation offsets the negative effects of nitrate on aerobic scope and performance. J. Exp. Biol. 223 (16), jeb224444.

Grans, A., Jutfelt, F., Sandblom, E., Jonsson, E., Wiklander, K., Seth, H., Olsson, C., Dupont, S., Ortega-Martinez, O., Einarsdottir, I., et al., 2014. Aerobic scope fails to explain the detrimental effects on growth resulting from warming and elevated CO2 in Atlantic halibut. J. Exp. Biol. 217, 711–717.

Grant, S.C., MacDonald, B.L., Winston, M.L., 2019. State of the Canadian Pacific Salmon: Response to Changing Climate and Habitats. Department of Fisheries and Oceans.

Grosell, M., Pasparakis, C., 2021. Physiological responses of fish to oil spills. Annu. Rev. Mar. Sci. 13, 137–160. https://doi.org/10.1146/annurev-marine-040120-094802.

Hague, M.J., Ferrari, M.R., Miller, J.R., Patterson, D.A., Russell, G.L., Farrell, A.P., Hinch, S.G., 2011. Modelling the future hydroclimatology of the lower Fraser River and its impacts on the spawning migration survival of sockeye salmon. Glob. Chang. Biol. 17, 87–98.

Halsey, L.G., Killen, S.S., Clark, T.D., Norin, T., 2018. Exploring key issues of aerobic scope interpretation in ectotherms: absolute versus factorial. Rev. Fish Biol. Fish. 28, 405–415.

Hansen, A.K., Byriel, D.B., Jensen, M.R., Steffensen, J.F., Svendsen, M.B.S., 2016. Optimum temperature of a northern population of Arctic charr (Salvelinus alpinus) using heart rate Arrhenius breakpoint analysis. Polar Biol. 40 (5), 1063–1070. https://doi.org/10.1007/s00300-016-2033-8.

Hanson, L.M., Obradovich, S., Mouniargi, J., Farrell, A.P., 2006. The role of adrenergic stimulation in maintaining maximum cardiac performance in rainbow trout (Oncorhynchus mykiss) during hypoxia, hyperkalemia and acidosis at 10° C. J. Exp. Biol. 209 (Pt 13), 2442–2451. https://doi.org/10.1242/jeb.02237.

Hanson, K.C., Gravel, M.A., Graham, A., Shoji, A., Cooke, S.J., 2008. Sexual variation in fisheries research and management: when does sex matter? Rev. Fish. Sci. 16 (4), 421–436. https://doi.org/10.1080/10641260802013866.

Harter, T.S., Morrison, P.R., Mandelman, J.W., Rummer, J.L., Farrell, A.P., Brill, R.W., Brauner, C.J., 2015. Validation of the i-STAT system for the analysis of blood gases and acid-base status in juvenile sandbar shark (Carcharhinus plumbeus). Conserv Physiol 3 (1), cov002. https://doi.org/10.1093/conphys/cov002.

Haugen, J.B., Curtis, T.H., Fernandes, P.G., Sosebee, K.A., Rago, P.J., 2017. Sexual segregation of spiny dogfish (Squalus acanthias) off the northeastern United States: implications for a male-directed fishery. Fish. Res. 193, 121–128.

Heberer, C., Aalbers, S.A., Bernal, D., Kohin, S., DiFiore, B., Sepulveda, C.A., 2010. Insights into catch-and-release survivorship and stress-induced blood biochemistry of common thresher sharks (Alopias vulpinus) captured in the southern California regreationl fishery. Fish. Res. 106 (3), 495–500.

Heithaus, M.R., Frid, A., Vaudo, J.J., Worm, B., Wirsing, A.J., 2010. Unraveling the ecological importance of elasmobranchs. In: Sharks and their Relatives II. CRC Press, pp. 627–654.

Hesser, E.F., 1960. Methods for routine fish hematology. Prog. Fish Cult. 22 (4), 164–171.

Hinch, S.G., Bett, N.N., Eliason, E.J., Farrell, A.P., Cooke, S.J., Patterson, D.A., 2021. Exceptionally high mortality of adult female salmon: a large-scale pattern and a conservation concern. Can. J. Fish. Aquat. Sci. 99 (999), 1–16. cjfas-2020-0385.

Hinch, S.G., Rand, P.S., 1998. Swim speeds and energy use of upriver-migrating sockeye salmon (Oncorhynchus nerka): role of local environment and fish characteristics. Can. J. Fish. Aquat. Sci. 55 (8), 1821–1831. https://doi.org/10.1139/f98-067.

Hoenig, J.M., Gruber, S.H., 1990. Life-history patterns in the elasmobranchs: implications for fisheries management. In: Pratt Jr., H.L., Gruber, S.H., Taniuchi, T. (Eds.), Elasmobranchs as Living Resources. NOAA Technical Report NMFS. 90, pp. 1–16.

Homyack, J.A., 2010. Evaluating habitat quality of vertebrates using conservation physiology tools. Wildl. Res. 37, 332.

Hontela, A., Rasmussen, J.B., Audet, C., Chevalier, G., 1992. Impaired cortisol stress response in fish from environments polluted by PAHs, PCBs, and mercury. Arch. Environ. Contam. Toxicol. 22 (3), 278–283.

Hook, S.E., Mondon, J., Revill, A.T., Greenfield, P.A., Stephenson, S.A., Strzelecki, J., Corbett, P., Armstrong, E., Song, J., Doan, H., 2018. Monitoring sublethal changes in fish physiology following exposure to a light, unweathered crude oil. Aquat. Toxicol. 204, 27–45.

Horodysky, A.Z., Cooke, S.J., Brill, R.W., 2015. Physiology in the service of fisheries science: why thinking mechanistically matters. Rev. Fish Biol. Fish. 25 (3), 425–447.

Hvas, M., Karlsbakk, E., Mæhle, S., Wright, D.W., Oppedal, F., 2017. The gill parasite Paramoeba perurans compromises aerobic scope, swimming capacity and ion balance in Atlantic salmon. Conserv. Phys. 5 (1), 1–12. https://doi.org/10.1093/conphys/cox066.

Illing, B., Rummer, J.L., 2017. Physiology can contribute to better understanding, management, and conservation of coral reef fishes. Conserv. Physiol. 5 (1), 1–11.

Incardona, J.P., Scholz, N.L., 2017. 6—environmental pollution and the fish heart. In: Gamperl, A.K., Gillis, T.E., Farrell, A.P., Brauner, C.J. (Eds.), Fish Physiology. vol 36. Academic Press, pp. 373–433. https://doi.org/10.1016/bs.fp.2017.09.006.

Incardona, J.P., Day, H.L., Collier, T.K., Scholz, N.L., 2006. Developmental toxicity of 4-ring polycyclic aromatic hydrocarbons in zebrafish is differentially dependent on AH receptor isoforms and hepatic cytochrome P4501A metabolism. Toxicol. Appl. Pharmacol. 217 (3), 308–321. https://doi.org/10.1016/j.taap.2006.09.018.

Incardona, J.P., Gardner, L.D., Linbo, T.L., Brown, T.L., Esbaugh, A.J., Mager, E.M., Stieglitz, J.D., French, B.L., Labenia, J.S., Laetz, C.A., Tagal, M., Sloan, C.A., Elizur, A., Benetti, D.D., Grosell, M., Block, B.A., Scholz, N.L., 2014. Deepwater horizon crude oil impacts the developing hearts of large predatory pelagic fish. Proc. Natl. Acad. Sci. U. S. A. 111 (15), E1510–E1518. https://doi.org/10.1073/pnas.1320950111.

Iwama, G.K., Vijayan, M.M., Forsyth, R.B., Ackerman, P.A., 1999. Heat shock proteins and physiological stress in fish. Am. Zool. 39 (6), 901–909.

Jain, K.E., Farrell, A.P., 2003. Influence of seasonal temperature on the repeat swimming performance of rainbow trout Oncorhynchus mykiss. J. Exp. Biol. 206, 3569–3579.

Jain, K.E., Birtwell, I.K., Farrell, A.P., 1998. Repeat swimming performance of mature sockeye salmon following a brief recovery period: a proposed measure of fish health and water quality. Can. J. Zool. 76, 14.

Janetski, D.J., Chaloner, D.T., Tiegs, S.D., Lamberti, G.A., 2009. Pacific salmon effects on stream ecosystems: a quantitative synthesis. Oecologia 159, 583–595.

Jeffries, K.M., Hinch, S.G., Martins, E.G., Clark, T.D., Lotto, A.G., Patterson, D.A., Cooke, S.J., Farrell, A.P., Miller, K.M., 2012a. Sex and proximity to reproductive maturity influence the survival, final maturation, and blood physiology of Pacific Salmon when exposed to high temperature during a simulated migration. Physiol. Biochem. Zool. 85, 62–73. https://doi.org/10.1086/663770.

Jeffries, K.M., Hinch, S.G., Sierocinski, T., Clark, T.D., Eliason, E.J., Donaldson, M.R., Li, S., Pavlidis, P., Miller, K.M., 2012b. Consequences of high temperatures and premature mortality on the transcriptome and blood physiology of wild adult sockeye salmon (Oncorhynchus nerka): temperature stress and mortality in Sockeye Salmon. Ecol. Evol. 2, 1747–1764.

Jeffries, K.M., Hinch, S.G., Sierocinski, T., Pavlidis, P., Miller, K.M., 2014. Transcriptomic responses to high water temperature in two species of Pacific salmon. Evol. Appl. 7, 286–300.

Jeffries, K.M., Connon, R.E., Davis, B.E., Komoroske, L.M., Britton, M.T., Sommer, T., Todgham, A.E., Fangue, N.A., 2016. Effects of high temperatures on threatened estuarine fishes during periods of extreme drought. J. Exp. Biol. 219 (11), 1705–1716.

Jeffries, K.M., Jeffrey, J.D., Holland, E.B., 2022. Applied aspects of gene function for the conservation of fishes. Fish Physiol. 39A, 389–433.

Jensen, F.B., 1987. Influences of exercise-stress and adrenaline upon intra-and extracellular acid-base status, electrolyte composition and respiratory properties of blood in tench (Tinca tinca) at different seasons. J. Comp. Physiol. B. 157 (1), 51–60.

Jensen, F.B., 2004. Red blood cell pH, the Bohr effect, and other oxygenation-linked phenomena in blood O_2 and CO_2 transport. Acta Physiol. Scand. 182, 215–227.

Jobling, M., 1981. The influences of feeding on the metabolic rate of fishes: a short review. J. Fish Biol. 18, 385–400.

Jobling, M., 1994. Fish Bioenergetics. Springer Verlag.

Johansen, J.L., Esbaugh, A.J., 2017. Sustained impairment of respiratory function and swim performance following acute oil exposure in a coastal marine fish. Aquat. Toxicol. 187, 82–89. https://doi.org/10.1016/j.aquatox.2017.04.002.

Jutfelt, F., Norin, T., Åsheim, E., Rowsey, L., Andreassen, A., Morgan, R., Clark, T., Speers-Roesch, B., 2021. Aerobic scope protection reduces ectotherm growth under warming. Funct. Ecol. 35 (7), 1397–1407. 1365-2435.13811.

Jutfelt, F., Norin, T., Ern, R., Overgaard, J., Wang, T., McKenzie, D.J., Lefevre, S., Nilsson, G.E., Metcalfe, N.B., Hickey, A.J.R., Brijs, J., Speers-Roesch, B., Roche, D.G., Gamperl, A.K., Raby, G.D., Morgan, R., Esbaugh, A.J., Gräns, A., Axelsson, M., Ekström, A., Sandblom, E., Binning, S.A., Hicks, J.W., Seebacher, F., Jørgensen, C., Killen, S.S.,

Schulte, P.M., Clark, T.D., 2018. Oxygen- and capacity-limited thermal tolerance: blurring ecology and physiology. J. Exp. Biol. 221, jeb169615. https://doi.org/10.1242/jeb.169615.

Katsumiti, A., Domingos, F.X., Azevedo, M., da Silva, M.D., Damian, R.C., Almeida, M.I., de Assis, H.C., Cestari, M.M., Randi, M.A., Ribeiro, C.A., Freire, C.A., 2009. An assessment of acute biomarker responses in the demersal catfish Cathorops spixii after the vicuna oil spill in a harbour estuarine area in southern Brazil. Environ. Monit. Assess. 152 (1–4), 209–222. https://doi.org/10.1007/s10661-008-0309-3.

Keefer, M.L., Peery, C.A., Heinrich, M.J., 2008. Temperature-mediated en route migration mortality and travel rates of endangered Snake River sockeye salmon. Ecol. Freshw. Fish 17, 136–145. https://doi.org/10.1111/j.1600-0633.2007.00267.x.

Khursigara, A.J., Perrichon, P., Martinez Bautista, N., Burggren, W.W., Esbaugh, A.J., 2017. Cardiac function and survival are affected by crude oil in larval red drum, Sciaenops ocellatus. Sci. Total Environ. 579, 797–804. https://doi.org/10.1016/j.scitotenv.2016.11.026.

Khursigara, A.J., Ackerly, K.L., Esbaugh, A.J., 2019. Oil toxicity and implications for environmental tolerance in fish. Comp Biochem. Physiol. C Toxicol. Pharmacol. 220, 52–61. https://doi.org/10.1016/j.cbpc.2019.03.003.

Kieffer, J.D., 2000. Limits to exhaustive exercise in fish. Comp. Biochem. Physiol. A Mol. Integr. Physiol. 126, 161–179.

Kieffer, J.D., 2010. Perspective—exercise in fish: 50+years and going strong. Comp. Biochem. Physiol. A Mol. Integr. Physiol. 156, 163–168.

Killen, S.S., Mitchell, M.D., Rummer, J.L., Chivers, D.P., Ferrari, M.C., Meekan, M.G., McCormick, M.I., 2014. Aerobic scope predicts dominance during early life in a tropical damselfish. Funct. Ecol. 28 (6), 1367–1376.

Killen, S.S., Norin, T., Halsey, L.G., 2017. Do method and species lifestyle affect measures of maximum metabolic rate in fishes?: maximum metabolic rate in fishes. J. Fish Biol. 90, 1037–1046.

Kirby, A.R., Cox, G.K., Nelson, D., Heuer, R.M., Stieglitz, J.D., Benetti, D.D., Grosell, M., Crossley 2nd, D.A., 2019. Acute crude oil exposure alters mitochondrial function and ADP affinity in cardiac muscle fibers of young adult Mahi-mahi (Coryphaena hippurus). Comp Biochem. Physiol. C Toxicol. Pharmacol. 218, 88–95. https://doi.org/10.1016/j.cbpc.2019.01.004.

Klimley, A.P., 2013. The Biology of Sharks and Rays. University of Chicago Press.

Kneebone, J., Chisholm, J., Bernal, D., Skomal, G., 2013. The physiological effects of capture stress, recovery, and post-release survivorship of juvenile sand tigers (Carcharinas taurus) caught on rod and reel. Fish. Res. 147, 103–114.

Koldkjaer, P., Pottinger, T.G., Perry, S.F., Cossins, A.R., 2004. Seasonality of the red blood cell stress response in rainbow trout (Oncorhynchus mykiss). J. Exp. Biol. 207 (2), 357–367.

Kraskura, K., Hardison, E., Little, A., Dressler, T., Prystay, T., Hendriks, B., Farrell, A., Cooke, S., Patterson, D., Hinch, S., 2020. Sex-specific differences in swimming, aerobic metabolism and recovery from exercise in adult coho salmon (Oncorhynchus kisutch) across ecologically relevant temperatures. Conservation. Phys. Ther. 9 (1), coab016.

Kraskura, K., Hardison, E.A., Little, A.G., Dressler, T., Prystay, T.S., Hendriks, B., Farrell, A.P., Cooke, S.J., Patterson, D.A., Hinch, S.G., Eliason, E.J., 2021. Sex-specific differences in swimming, aerobic metabolism and recovery from exercise in adult coho salmon (Oncorhynchus kisutch) across ecologically relevant temperatures. Conserv. Physiol. 9, coab016. https://doi.org/10.1093/conphys/coab016.

Lawrence, M.J., Raby, G.D., Teffer, A.K., Jeffries, K.M., Danylchuk, A.J., Eliason, E.J., Hasler, C.T., Clark, T.D., Cooke, S.J., 2020. Best practices for non-lethal blood sampling of fish via the caudal vasculature. J. Fish Biol. 97 (1), 4–15. https://doi.org/10.1111/jfb.14339.

Lee, C.G., Devlin, R.H., Farrell, A.P., 2003. Swimming performance, oxygen consumption and excess post-exercise oxygen consumption in adult transgenic and ocean-ranched coho salmon. J. Fish Biol. 62, 753–766.

Lefevre, S., 2016. Are global warming and ocean acidification conspiring against marine ectotherms? A meta-analysis of the respiratory effects of elevated temperature, high CO_2 and their interaction. Conserv. Physiol. 4 (1), 1–31. https://doi.org/10.1093/conphys/cow009.

Lefevre, S., Wang, T., McKenzie, D.J., 2021. The role of mechanistic physiology in investigating impacts of global warming on fishes. J. Exp. Biol. 224 (Suppl_1), jeb238840.

Lennox, R.J., Paukert, C.P., Aarestrup, K., Auger-Méthé, M., Baumgartner, L., Birnie-Gauvin, K., Bøe, K., Brink, K., Brownscombe, J.W., Chen, Y., et al., 2019. One hundred pressing questions on the future of global fish migration science, conservation, and policy. Front. Ecol. Evol. 7, 286.

Little, A., Dressler, T., Kraskura, K., Hardison, E., Hendriks, B., Prystay, T., Farrell, A.P., Cooke, S.J., Patterson, D.A., Hinch, S.G., et al., 2020. Maxed out: optimizing accuracy, precision and power for field measures of maximum metabolic rate in fishes. Physiol. Biochem. Zool. 3, 243–254.

Lovén Wallerius, M., Gräns, A., Koeck, B., Berger, D., Sandblom, E., Ekström, A., Arlinghaus, R., Johnsson, J.I., 2019. Socially induced stress and behavioural inhibition in response to angling exposure in rainbow trout. Fish. Manag. Ecol. 26 (6), 611–620. https://doi.org/10.1111/fme.12373.

Lucas, M.C., Johnstone, A.D., Priede, I.G., 1993. Use of physiological telemetry as a method of estimating metabolism of fish in the natural environment. Trans. Am. Fish. Soc. 122 (5), 822–833.

Macdonald, J.S., Patterson, D.A., Hague, M.J., Guthrie, I.C., 2010. Modeling the influence of environmental factors on spawning migration mortality for sockeye Salmon fisheries Management in the Fraser River, British Columbia. Trans. Am. Fish. Soc. 139, 768–782.

Mager, E.M., Esbaugh, A.J., Stieglitz, J.D., Hoenig, R., Bodinier, C., Incardona, J.P., Scholz, N.L., Benetti, D.D., Grosell, M., 2014. Acute embryonic or juvenile exposure to Deepwater horizon crude oil impairs the swimming performance of mahi-mahi (Coryphaena hippurus). Environ. Sci. Technol. 48 (12), 7053–7061. https://doi.org/10.1021/es501628k.

Mandelman, J.W., Farrington, M.A., 2007. The physiological status and mortality associated with otter-trawl capture, transport, and captivity of an exploited elasmobranch, *Squalus acanthias*. ICES J. Mar. Sci. 64 (1), 122–130.

Mandelman, J.W., Skomal, G.B., 2009. Differential sensitivity to capture stress assessed by blood acid–base status in five carcharhinid sharks. J. Comp. Physiol. B. 179 (3), 267–277.

Marbach, E.P., Weil, M.H., 1967. Rapid enzymatic measurement of blood lactate and pyruvate: use and significance of metaphosphoric acid as a common precipitant. Clin. Chem. 13 (4), 314–325.

Marras, S., Cucco, A., Antognarelli, F., Azzurro, E., Milazzo, M., Bariche, M., Butenschön, M., Kay, S., Di Bitetto, M., Quattrocchi, G., 2015. Predicting future thermal habitat suitability of competing native and invasive fish species: from metabolic scope to oceanographic modelling. Conserv. Physiol. 3 (1), cou059.

Marshall, H., Field, L., Afiadata, A., Sepulveda, C., Skomal, G., Bernal, D., 2012. Hematological indicators of stress in longline-captured sharks. Comp. Biochem. Physiol. A Mol. Integr. Physiol. 162, 121–129.

Martins, E.G., Hinch, S.G., Patterson, D.A., Hague, M.J., Cooke, S.J., Miller, K.M., Lapointe, M.-F., English, K.K., Farrell, A.P., 2011. Effects of river temperature and climate warming on stock-specific survival of adult migrating Fraser River sockeye salmon (Oncorhynchus nerka): TEMPERATURE EFFECTS ON ADULT SALMON SURVIVAL. Glob. Chang. Biol. 17, 99–114.

Martins, C.L., Walker, T.I., Reina, R.D., 2018. Stress-related physiological changes and post-release survival of elephant fish (Callorhinchus milii) after longlining, gillnetting, angling and handling in a controlled setting. Fish. Res. 204, 116–124.

Mathes, M.T., Hinch, S.G., Cooke, S.J., Crossin, G.T., Patterson, D.A., Lotto, A.G., Farrell, A.P., 2010. Effect of water temperature, timing, physiological condition, and lake thermal refugia on migrating adult Weaver Creek sockeye salmon (*Oncorhynchus nerka*). Can. J. Fish. Aquat. Sci. 67, 70–84. https://doi.org/10.1139/F09-158.

McCue, M.D., 2006. Specific dynamic action: a century of investigation. Comp. Biochem. Physiol. A Mol. Integr. Physiol. 144, 381–394.

McKenzie, D.J., Estivales, G., Svendsen, J.C., Steffensen, J.F., Agnèse, J.-F., 2013. Local adaptation to altitude underlies divergent thermal physiology in tropical killifishes of the genus Aphyosemion. PLoS One 8, e54345.

McKenzie, D.J., Axelsson, M., Chabot, D., Claireaux, G., Cooke, S.J., Corner, R.A., De Boeck, G., Domenici, P., Guerreiro, P.M., Hamer, B., Jorgensen, C., Killen, S.S., Lefevre, S., Marras, S., Michaelidis, B., Nilsson, G.E., Peck, M.A., Perez-Ruzafa, A., Rijnsdorp, A.D., Shiels, H.A., Steffensen, J.F., Svendsen, J.C., Svendsen, M.B., Teal, L.R., van der Meer, J., Wang, T., Wilson, J.M., Wilson, R.W., Metcalfe, J.D., 2016. Conservation physiology of marine fishes: state of the art and prospects for policy. Conserv Physiol 4 (1), cow046. https://doi.org/10.1093/conphys/cow046.

McKenzie, D.J., Zhang, Y., Eliason, E.J., Schulte, P.M., Claireaux, G., Blasco, F.R., Nati, J.J., Farrell, A.P., 2021. Intraspecific variation in tolerance of warming in fishes. J. Fish Biol. 98, 1536–1555.

McKinley, S., Van Der Kraak, G., Power, G., 1998. Seasonal migrations and reproductive patterns in the lake sturgeon, *Acipenser fulvescens*, in the vincinity of hydroelectric stations in northern Ontario. Environ. Biol. Fish 5 (1), 245–256.

Metcalfe, N.B., Van Leeuwen, T.E., Killen, S.S., 2016a. Does individual variation in metabolic phenotype predict fish behaviour and performance? J. Fish Biol. 88 (1), 298–321. https://doi.org/10.1111/jfb.12699.

Metcalfe, J.D., Wright, S., Tudorache, C., Wilson, R.P., 2016b. Recent advances in telemetry for estimating the energy metabolism of wild fishes: telemetry for estimating energy metabolism. J. Fish Biol. 88, 284–297. https://doi.org/10.1111/jfb.12804.

Middleton, C.T., Hinch, S.G., Martins, E.G., Braun, D.C., Patterson, D.A., Burnett, N.J., Minke-Martin, V., Casselman, M.T., 2018. Effects of natal water concentration and temperature on the behaviour of up-river migrating sockeye salmon. Can. J. Fish. Aquat. Sci. 75, 2375–2389. https://doi.org/10.1139/cjfas-2017-0490.

Miller, K.M., Teffer, A., Tucker, S., Li, S., Schulze, A.D., Trudel, M., Juanes, F., Tabata, A., Kaukinen, K.H., Ginther, N.G., et al., 2014. Infectious disease, shifting climates, and opportunistic predators: cumulative factors potentially impacting wild salmon declines. Evol. Appl. 7, 812–855.

Milligan, C.L., 1996. Metabolic recovery from exhaustive exercise in rainbow trout. Comp. Biochem. Physiol. A Physiol. 113, 51–60.

Milligan, C.L., Hooke, G.B., Johnson, C., 2000. Sustained swimming at low velocity following a bout of exhaustive exercise enhances metabolic recovery in rainbow trout. J. Exp. Biol. 203, 921–926.

Minke-Martin, V., Hinch, S.G., Braun, D.C., Burnett, N.J., Casselman, M.T., Eliason, E.J., Middleton, C.T., 2018. Physiological condition and migratory experience affect fitness-related outcomes in adult female sockeye salmon. Ecol. Freshw. Fish 27, 296–309. https://doi. org/10.1111/eff.12347.

Mochnacz, N.J., Kissinger, B.C., Deslauriers, D., Guzzo, M.M., Enders, E.C., Anderson, W.G., Docker, M.F., Isaak, D.J., Durhack, T.C., Treberg, J.R., 2017. Development and testing of a simple field-based intermittent flow respirometry system for riverine fishes. Conserv. Physiol. 5, 1–13. https://doi.org/10.1093/conphys/cox048.

Mohan, J.A., Jones, E.R., Hendon, J.M., Falterman, B., Boswell, K.M., Hoffmayer, E.R., Wells, R.J.D., 2020. Capture stress and post-release mortality of blacktip sharks in recreational charter fisheries of the Gulf of Mexico. Conserv Physiol 8 (1), coaa041. https://doi.org/10.1093/conphys/coaa041.

Mohanty, B.P., Mahanty, A., Mitra, T., Parija, S.C., Mohanty, S., 2018. Heat shock proteins in stress in teleosts. In: Regulation of Heat Shock Protein Responses. Springer, pp. 71–94.

Molina, J.M., Finotto, L., Walker, T.I., Reina, R.D., 2020. The effect of gillnet capture on the metabolic rate of two shark species with contrasting lifestyles. J. Exp. Mar. Biol. Ecol. 526, 151354.

Morash, A.J., Neufeld, C., MacCormack, T.J., Currie, S., 2018. The importance of incorporating natural thermal variation when evaluating physiological performance in wild species. J. Exp. Biol. 221 (14), jeb164673.

Moyes, C.D., Fragoso, N., Musyl, M.K., Brill, R.W., 2006. Predicting postrelease survival in large pelagic fish. Trans. Am. Fish. Soc. 135, 1389–1397.

Moyle, P., Lusardi, R., Samuel, P., Katz, J., 2017. State of the Salmonids: Status of California's Emblematic Fishes 2017.

Nelson, J.A., 2016. Oxygen consumption rate *v.* rate of energy utilization of fishes: a comparison and brief history of the two measurements: oxygen consumption and metabolic rate. J. Fish Biol. 88, 10–25.

Nguyen, V.M., Martins, E.G., Robichaud, D., Raby, G.D., Donaldson, M.R., Lotto, A.G., Willmore, W.G., Patterson, D.A., Farrell, A.P., Hinch, S.G., Cooke, S.J., 2014. Disentangling the roles of air exposure, gill net injury, and facilitated recovery on the postcapture and release mortality and behavior of adult migratory sockeye salmon (*Oncorhynchus nerka*) in freshwater. Physiol. Biochem. Zool. 87, 125–135. https://doi.org/10.1086/669530.

Nielsen, J., Hedeholm, R.B., Heinemeier, J., Bushnell, P.G., Christiansen, J.S., Olsen, J., Ramsey, C.B., Brill, R.W., Simon, M., Steffensen, K.F., 2016. Eye lens radiocarbon reveals centuries of longevity in the Greenland shark (*Somniosus microcephalus*). Science 353 (6300), 702–704.

Nikinmaa, M., 2012. Vertebrate Red Blood Cells: Adaptations of Function to Respiratory Requirements. vol. 28 Springer Science & Business Media.

Nilsson, G.E., Crawley, N., Lunde, I.G., Munday, P.L., 2009. Elevated temperature reduces the respiratory scope of coral reef fishes. Glob. Chang. Biol. 15 (6), 1405–1412.

Nilsson, G.E., Hylland, P., Lofman, C.O., 1994. Anoxia and adenosine induce increased cerebral blood flow rate in crucian carp. Am. J. Physiol. Regul. Integr. Comp. Physiol. 267 (2), R590–R595.

NOAA F, 2018. Bycatch Reduction Engineering Program FY 2017 & 2018 Report to Congress.

Norin, T., Speers-Roesch, B., 2020. Chapter 10: Metabolism. The Physiology of Fishes, fifth ed. CRC Press.

Norin, T., Clark, T.D., 2016. Measurement and relevance of maximum metabolic rate in fishes: maximum metabolic rate in fishes. J. Fish Biol. 88, 122–151.

Norin, T., Clark, T.D., 2017. Fish face a trade-off between 'eating big' for growth efficiency and 'eating small' to retain aerobic capacity. Biol. Lett. 13, 20170298.

Norin, T., Malte, H., Clark, T.D., 2014. Aerobic scope does not predict the performance of a tropical eurythermal fish at elevated temperatures. J. Exp. Biol. 217, 244–251.

O'Connor, E.A., Pottinger, T.G., Sneddon, L.U., 2011. The effects of acute and chronic hypoxia on cortisol, glucose and lactate concentrations in different populations of three-spined stickleback. Fish Physiol. Biochem. 37 (3), 461–469. https://doi.org/10.1007/s10695-010-9447-y.

Oke, K.B., Cunningham, C.J., Westley, P.A.H., Baskett, M.L., Carlson, S.M., Clark, J., Hendry, A.P., Karatayev, V.A., Kendall, N.W., Kibele, J., Kindsvater, H.K., Kobayashi, K.M., Lewis, B., Munch, S., Reynolds, J.D., Vick, G.K., Palkovacs, E.P., 2020. Recent declines in salmon body size impact ecosystems and fisheries. Nat. Commun. 11 (1), 4155. https://doi.org/10.1038/s41467-020-17726-z.

Pang, X., Cao, Z.-D., Fu, S.-J., 2011. The effects of temperature on metabolic interaction between digestion and locomotion in juveniles of three cyprinid fish (Carassius auratus, Cyprinus carpio and Spinibarbus sinensis). Comp. Biochem. Physiol. A Mol. Integr. Physiol. 159, 253–260.

Pasparakis, C., Esbaugh, A.J., Burggren, W., Grosell, M., 2019. Impacts of Deepwater horizon oil on fish. Comp Biochem. Physiol. C Toxicol. Pharmacol. 224, 108558. https://doi.org/10.1016/j.cbpc.2019.06.002.

Patterson, D.A., Cooke, S.J., Hinch, S.G., Robinson, K.A., Young, N., Farrell, A.P., Miller, K.M., 2016. A perspective on physiological studies supporting the provision of scientific advice for the management of Fraser River sockeye salmon (*Oncorhynchus nerka*). Conserv. Physiol. 4, cow026. https://doi.org/10.1093/conphys/cow026.

Patterson, D.A., Robinson, K.A., Lennox, R.J., Nettles, T.L., Eliason, E.J., Raby, G.D., Chapman, J.M., Cook, K.V., Donaldson, M.R., Bass, A.L., et al., 2017. Review and evaluation of fishing-related incidental mortality for Pacific salmon. DFO Can. Sci. Advis. Sec. Res. Doc., 164.

Peake, S.J., Farrell, A.P., 2006. Fatigue is a behavioural response in respirometer-confined smallmouth bass. J. Fish Biol. 68, 1742–1755.

Peterson, C.H., Rice, S.D., Short, J.W., Esler, D., Bodkin, J.L., Ballachey, B.E., Irons, D.B., 2003. Long-term ecosystem response to the Exxon Valdez oil spill. Science 302 (5653), 2082–2086.

Phillips, B.E., Christiansen, E.F., Stoskopf, M.K., Broadhurst, H., George, R., Harms, C.A., 2016. Comparison of hematology, plasma biochemistry, and blood gas variables between 2 venipuncture sites in southern stingrays (Dasyatis americana). Vet. Clin. Pathol. 45 (4), 627–633. https://doi.org/10.1111/vcp.12424.

Pichavant, K., Person-Le-Ruyet, J., Le Bayon, N., Severe, A., Le Roux, A., Quemener, L., Maxime, V., Nonnotte, G., Boeuf, G., 2000. Effects of hypoxia on growth and metabolism of juvenile turbot. Aquaculture 188 (1-2), 103.

Pinsky, M.L., Eikeset, A.M., McCauley, D.J., Payne, J.L., Sunday, J.M., 2019. Greater vulnerability to warming of marine versus terrestrial ectotherms. Nature 569 (7754), 108–111.

Poletto, J.B., Cocherell, D.E., Baird, S.E., Nguyen, T.X., Cabrera-Stagno, V., Farrell, A.P., Fangue, N.A., 2017. Unusual aerobic performance at high temperatures in juvenile Chinook salmon, *Oncorhynchus tshawytscha*. Conserv. Physiol. 5. https://doi.org/10.1093/conphys/cow067.

Pon, L.B., Hinch, S.G., Cooke, S.J., Patterson, D.A., Farrell, A.P., 2009. Physiological, energetic and behavioural correlates of successful fishway passage of adult sockeye salmon Oncorhynchus nerka in the Seton River, British Columbia. J. Fish Biol. 74, 1323–1336. https://doi.org/10.1111/j.1095-8649.2009.02213.x.

Pon, L.B., Hinch, S.G., Suski, C.D., Patterson, D.A., Cooke, S.J., 2012. The effectiveness of tissue biopsy as a means of assessing the physiological consequences of fishway passage. River Res. Appl. 28 (8), 1266–1274.

Pörtner, H., 2001. Climate change and temperature-dependent biogeography: oxygen limitation of thermal tolerance in animals. Naturwissenschaften 88, 137–146.

Pörtner, H.O., 2010. Oxygen- and capacity-limitation of thermal tolerance: a matrix for integrating climate-related stressor effects in marine ecosystems. J. Exp. Biol. 213 (6), 881–893. https://doi.org/10.1242/jeb.037523.

Pörtner, H.-O., Bock, C., Mark, F.C., 2017. Oxygen- and capacity-limited thermal tolerance: bridging ecology and physiology. J. Exp. Biol. 220, 2685–2696. https://doi.org/10.1242/jeb.134585.

Pörtner, H.O., Farrell, A.P., 2008. Ecology: physiology and climate change. Science 322 (5902), 690–692. https://doi.org/10.1126/science.1163156.

Pörtner, H.O., Knust, R., 2007. Climate change affects marine fishes through the oxygen limitation of thermal tolerance. Science 315 (5808), 95–97. https://doi.org/10.1126/science.1135471.

Powell, M.D., Yousaf, M.N., 2017. Cardiovascular effects of disease: parasites and pathogens. In: Fish Physiology. vol. 36. Elsevier, pp. 435–470.

Pratt Jr., H.L., Casey, J.G., 1990. Shark reproductive strategies as a limiting factor in directed fisheries, with a review of Holden's method of estimating growth parameters. In: Pratt Jr., H.L., Gruber, S.H., Taniuchi, T. (Eds.), Elasmobranchs as Living Resources. vol. Technical Report 90. NOAA, pp. 97–110.

Prystay, T.S., Eliason, E.J., Lawrence, M.J., Dick, M., Brownscombe, J.W., Patterson, D.A., Crossin, G.T., Hinch, S.G., Cooke, S.J., 2017. The influence of water temperature on sockeye salmon heart rate recovery following simulated fisheries interactions. Conserv. Physiol. 5 (1), 1–12.

Prystay, T.S., Lawrence, M.J., Zolderdo, A.J., Brownscombe, J.W., de Bruijn, R., Eliason, E.J., Cooke, S.J., 2019. Exploring relationships between cardiovascular activity and parental care behavior in nesting smallmouth bass: a field study using heart rate biologgers. Comp. Biochem. Physiol. A Mol. Integr. Physiol. 234, 18–27.

Prystay, T.S., de Bruijn, R., Peiman, K.S., Hinch, S.G., Patterson, D.A., Farrell, A.P., Eliason, E.J., Cooke, S.J., 2020. Cardiac performance of free-swimming wild sockeye salmon during the reproductive period. Integr. Organismal Biol. 2 (1), obz031.

Quiñones, R.M., Grantham, T.E., Harvey, B.N., Kiernan, J.D., Klasson, M., Wintzer, A.P., Moyle, P.B., 2015. Dam removal and anadromous salmonid (Oncorhynchus spp.) conservation in California. Rev. Fish Biol. Fish. 25, 195–215.

Raby, G.D., Cooke, S.J., Cook, K.V., McConnachie, S.H., Donaldson, M.R., Hinch, S.G., Whitney, C.K., Drenner, S.M., Patterson, D.A., Clark, T.D., et al., 2013. Resilience of pink Salmon and Chum Salmon to simulated fisheries capture stress incurred upon arrival at spawning grounds. Trans. Am. Fish. Soc. 142, 524–539.

Raby, G.D., Clark, T.D., Farrell, A.P., Patterson, D.A., Bett, N.N., Wilson, S.M., Willmore, W.G., Suski, C.D., Hinch, S.G., Cooke, S.J., 2015. Facing the river gauntlet: understanding the effects of fisheries capture and water temperature on the physiology of Coho Salmon. PLoS One 10 (4), e0124023. https://doi.org/10.1371/journal.pone.0124023.

Raby, G.D., Doherty, C.L.J., Mokdad, A., Pitcher, T.E., Fisk, A.T., 2020. Post-exercise respirometry underestimates maximum metabolic rate in juvenile salmon. Conserv. Physiol. 8, coaa063. https://doi.org/10.1093/conphys/coaa063.

Raby, G.D., Hinch, S.G., Patterson, D.A., Hills, J.A., Thompson, L.A., Cooke, S.J., 2015. Mechanisms to explain purse seine bycatch mortality of coho salmon. Ecol. Appl. 25, 1757–1775. https://doi.org/10.1890/14-0798.1.

Ramachandran, S.D., Hodson, P.V., Khan, C.W., Lee, K., 2004. Oil dispersant increases PAH uptake by fish exposed to crude oil. Ecotoxicol. Environ. Saf. 59 (3), 300–308. https://doi.org/10.1016/j.ecoenv.2003.08.018.

Ramstad, K.M., Woody, C.A., Sage, G.K., Allendorf, F.W., 2004. Founding events influence genetic population structure of sockeye salmon (Oncorhynchus nerka) in Lake Clark, Alaska. Mol. Ecol. 13, 277–290.

Regan, M.D., Gosline, J.M., Richards, J.G., 2013. A simple and affordable calorespirometer for assessing the metabolic rates of fishes. J. Exp. Biol. 216 (24), 4507–4513.

Regan, M.D., Gill, I.S., Richards, J.G., 2017. Calorespirometry reveals that goldfish prioritize aerobic metabolism over metabolic rate depression in all but near-anoxic environments. J. Exp. Biol. 220 (4), 564–572.

Reidy, S.P., Nelson, J.A., Tang, Y., Kerr, S.R., 1995. Post-exercise metabolic rate in Atlantic cod and its dependence upon the method of exhaustion. J. Fish Biol. 47, 377–386.

Richards, J.G., 2009. Metabolic and molecular responses of fish to hypoxia. In: Fish Physiology. vol. 27. Elsevier, pp. 443–485.

Richter, A., Kolmes, S.A., 2005. Maximum temperature limits for Chinook, coho, and chum salmon, and steelhead trout in the Pacific Northwest. Rev. Fish. Sci. 13, 23–49. https://doi.org/10.1080/10641260590885861.

Roberts, A.P., Alloy, M.M., Oris, J.T., 2017. Review of the photo-induced toxicity of environmental contaminants. Comp Biochem. Physiol. C Toxicol. Pharmacol. 191, 160–167. https://doi.org/10.1016/j.cbpc.2016.10.005.

Robinson, K.A., Hinch, S.G., Gale, M.K., Clark, T.D., Wilson, S.M., Donaldson, M.R., Farrell, A.-P., Cooke, S.J., Patterson, D.A., 2013. Effects of post-capture ventilation assistance and elevated water temperature on sockeye salmon in a simulated capture-and-release experiment. *Conservation*. Phys. Ther. 1, cot015.

Roche, H., Bogé, G., 1996. Fish blood parameters as a potential tool for identification of stress caused by environmental factors and chemical intoxication. Mar. Environ. Res. 41 (1), 27–43.

Roche, D.G., Binning, S.A., Bosiger, Y., Johansen, J.L., Rummer, J.L., 2013. Finding the best estimates of metabolic rates in a coral reef fish. J. Exp. Biol. 216, 2103–2110.

Romero, L. Michael, Beattie, Ursula K., 2022. Common mythos of glucocorticoid function in ecology and conservation. J. Exp. Zool. A Ecol. Integr. Physiol. 337 (1), 7–14. https://doi.org/10.1002/jez2459.

Root, R., 1931. The respiratory function of the blood of marine fishes. Biol. Bull. 61 (3), 427–456.

Rummer, J.L., Stecyk, J.A.W., Couturier, C.S., Watson, S.-A., Nilsson, G.E., Munday, P.L., 2013. Elevated CO2 enhances aerobic scope of a coral reef fish. Conserv. Physiol. 1, cot023.

Rummer, J.L., Couturier, C.S., Stecyk, J.A.W., Gardiner, N.M., Kinch, J.P., Nilsson, G.E., Munday, P.L., 2014. Life on the edge: thermal optima for aerobic scope of equatorial reef fishes are close to current day temperatures. Glob. Chang. Biol. 20 (4), 1055–1066. https://doi.org/10.1111/gcb.12455.

Roscoe, D.W., Hinch, S.G., Cooke, S.J., Patterson, D.A., 2011. Fishway passage and post-passage mortality of up-river migrating sockeye salmon in the Seton River, British Columbia. River Res. Appl. 27, 693–705. https://doi.org/10.1002/rra.1384.

Rummer, J.L., Binning, S.A., Roche, D.G., Johansen, J.L., 2016. Methods matter: considering locomotory mode and respirometry technique when estimating metabolic rates of fishes. Conserv Physiol 4, 1–13. https://doi.org/10.1093/conphys/cow008.

Sadoul, B., Geffroy, B., 2019. Measuring cortisol, the major stress hormone in fishes. J. Fish Biol. 94 (4), 540–555. https://doi.org/10.1111/jfb.13904.

Sandblom, E., Cox, G.K., Perry, S.F., Farrell, A.P., 2009. The role of venous capacitance, circulating catecholamines, and heart rate in the hemodynamic response to increased temperature and hypoxia in the dogfish. Am. J. Physiol. Regul. Integr. Comp. Physiol. 296 (5), R1547–R1556.

Sandblom, E., Grans, A., Axelsson, M., Seth, H., 2014. Temperature acclimation rate of aerobic scope and feeding metabolism in fishes: implications in a thermally extreme future. Proc. R. Soc. B Biol. Sci. 281, 20141490.

Schlenker, L.S., Latour, R.J., Brill, R.W., Graves, J.E., 2016. Physiological stress and post-release mortality of white marlin (*Kajikia albida*) caught in the United States recreational fishery. Conserv Physiol 4 (1), cov066. https://doi.org/10.1093/conphys/cov066.

Schulte, P.M., 2014. What is environmental stress? Insights from fish living in a variable environment. J. Exp. Biol. 217 (1), 23–34. https://doi.org/10.1242/jeb.089722.

Schulte, P.M., 2015. The effects of temperature on aerobic metabolism: towards a mechanistic understanding of the responses of ectotherms to a changing environment. J. Exp. Biol. 218 (12), 1856–1866. https://doi.org/10.1242/jeb.118851.

Schulte, P.M., Healy, M., 2022. Physiological diversity and its importance for fish conservation and management in the Anthropocene. Fish Physiol. 39A, 435–477.

Schulte, P.M., Healy, T.M., Fangue, N.A., 2011. Thermal performance curves, phenotypic plasticity, and the time scales of temperature exposure. Integr. Comp. Biol. 51, 691–702. https://doi.org/10.1093/icb/icr097.

Schurmann, H., Steffensen, J.F., 1997. Effects of temperature, hypoxia and activity on the metabolism of juvenile Atlantic cod. J. Fish Biol. 50, 1166–1180.

Schwieterman, G.D., Bouyoucos, I.A., Potgieter, K., Simpfendorfer, C.A., Brill, R.W., Rummer, J.L., 2019. Analysing tropical elasmobranch blood samples in the field: blood stability during storage and validation of the HemoCue(R) haemoglobin analyser. Conserv Physiol 7 (1), coz081. https://doi.org/10.1093/conphys/coz081.

Schwieterman, G.D., Winchester, M.M., Shiels, H.A., Bushnell, P.G., Bernal, D., Marshall, H.M., Brill, R.W., 2021. The effects of elevated potassium, acidosis, reduced oxygen levels, and temperature on the functional properties of isolated myocardium from three elasmobranch fishes: clearnose skate (*Rostroraja eglanteria*), smooth dogfish (*Mustelus canis*), and sandbar shark (*Carcharhinus plumbeus*). J. Comp. Physiol. B. 191 (1), 127–141. https://doi.org/10.1007/s00360-020-01328-8.

Secor, S.M., 2009. Specific dynamic action: a review of the postprandial metabolic response. J. Comp. Physiol. B. 179, 1–56.

SEDAR, 2013. SEDAR 31—Gulf of Mexico red snapper stock assessment report. North Charleston SC.

Sepulveda, C.A., Heberer, C., Aalbers, S.A., Spear, N., Kinney, M., Bernal, D., Kohin, S., 2015. Post-release survivorship studies on common thresher sharks (*Alopias vulpinus*) captured in the southern California recreational fishery. Fish. Res. 161, 102–108.

Shiffman, D.S., 2020. Recreational shark fishing in Florida: how research and strategic science communication helped to change policy. Conserv. Sci. Practice 2 (4), e174. https://doi.org/10.1111/csp2.174.

Shultz, A.D., Murchie, K.J., Griffith, C., Cooke, S.J., Danylchuk, A.J., Goldberg, T.L., Suski, C.D., 2011. Impacts of dissolved oxygen on the behavior and physiology of bonefish: implications for live-release angling tournaments. J. Exp. Mar. Biol. Ecol. 402, 19–26.

Sidhu, R., Anttila, K., Farrell, A.P., 2014. Upper thermal tolerance of closely related danio species. J. Fish Biol. 84 (4), 982–995. https://doi.org/10.1111/jfb.12339.

Skeeles, M.R., Winkler, A.C., Duncan, M.I., James, N.C., van der Walt, K.A., Potts, W.M., 2020. The use of internal heart rate loggers in determining cardiac breakpoints of fish. J. Therm. Biol. 89, 102524. https://doi.org/10.1016/j.jtherbio.2020.102524.

Skomal, G., Bernal, D., 2010. Physiological responses to stress in sharks. In: Carrier, J.C., Heithaus, L.I., Musick, J.A. (Eds.), Sharks and Their Relatives II. CRC Press, Boca Raton, FL, p. 713.

Soivio, A., Okari, A., 1976. Haematological effects of stress on a teleost, *Esoc lucius* L. J. Fish Biol. 8, 397–411.

Somo, D.A., Onukwufor, J.O., Wood, C.M., Richards, J.G., 2020. Interactive effects of temperature and hypoxia on diffusive water flux and oxygen uptake rate in the tidepool sculpin, Oligocottus maculosus. Comp. Biochem. Physiol. A Mol. Integr. Physiol. 250, 110781.

Stecyk, J.A., 2017. Cardiovascular responses to limiting oxygen levels. In: Fish Physiology. vol 36. Elsevier, pp. 299–371.

Steell, S.C., Van Leeuwen, T.E., Brownscombe, J.W., Cooke, S.J., Eliason, E.J., 2019. An appetite for invasion: digestive physiology, thermal performance and food intake in lionfish (Pterois spp.). J. Exp. Biol. 222 (19), jeb209437.

Steffensen, J.F., 1989. Some errors in respirometry of aquatic breathers: how to avoid and correct for them. Fish Physiol. Biochem. 6, 49–59.

Steffensen, J.F., Johansen, K., Bushnell, P.G., 1984. An automated swimming respirometer. J. Comp. Physiol. B. 79, 437–440.

Steinhausen, M.F., Sandblom, E., Eliason, E.J., Verhille, C., Farrell, A.P., 2008. The effect of acute temperature increases on the cardiorespiratory performance of resting and swimming sockeye salmon (Oncorhynchus nerka). J. Exp. Biol. 211, 3915–3926.

Stevens, E.D., Black, E.C., 1966. The effect of intermittent exercise on carbohydrate metabolism in rainbow trout, *Salmo gairdneri*. J. Fish. Res. Board Can. 23, 471–485. https://doi.org/10.1139/f66-039.

Stieglitz, J.D., Mager, E.M., Hoenig, R.H., Benetti, D.D., Grosell, M., 2016. Impacts of Deepwater horizon crude oil exposure on adult mahi-mahi (Coryphaena hippurus) swim performance. Environ. Toxicol. Chem. 35 (10), 2613–2622. https://doi.org/10.1002/etc.3436.

Stoot, L.J., Cairns, N.A., Cull, F., Taylor, J.J., Jeffrey, J.D., Morin, F., Mandelman, J.W., Clark, T.D., Cooke, S.J., 2014. Use of portable blood physiology point-of-care devices for basic and applied research on vertebrates: a review. Conserv. Physiol. 2 (1), cou011. https://doi.org/10.1093/conphys/cou011.

Sulikowski, J.A., Wheeler, C.R., Gallagher, A.J., Prohaska, B.K., Langan, J.A., Hammerschlag, N., 2016. Seasonal and life-stage variation in the reproductive ecology of a marine apex predator, the tiger shark Galeocerdo cuvier, at a protected female-dominated site. Aquat. Biol. 24 (3), 175–184. https://doi.org/10.3354/ab00648.

Sunday, J.M., Bates, A.E., Kearney, M.R., Colwell, R.K., Dulvy, N.K., Longino, J.T., Huey, R.B., 2014. Thermal-safety margins and the necessity of thermoregulatory behavior across latitude and elevation. Proc. Natl. Acad. Sci. U. S. A. 111 (15), 5610–5615.

Svendsen, M.B.S., Bushnell, P.G., Christensen, E.A.F., Steffensen, J.F., 2016a. Sources of variation in oxygen consumption of aquatic animals demonstrated by simulated constant oxygen consumption and respirometers of different sizes. J. Fish Biol. 88, 51–64.

Svendsen, M.B.S., Bushnell, P.G., Steffensen, J.F., 2016b. Design and setup of intermittent-flow respirometry system for aquatic organisms. J. Fish Biol. 88, 26–50.

Szedlmayer, S.T., Mudrak, P.A., 2014. Influence of Age-1 conspecifics, sediment type, dissolved oxygen, and the Deepwater horizon oil spill on recruitment of Age-0 red snapper in the

Northeast Gulf of Mexico during 2010 and 2011. N. Am. J. Fish Manag. 34 (2), 443–452. https://doi.org/10.1080/02755947.2014.882457.

Talwar, B., Bouyoucos, I.A., Shipley, O., Rummer, J.L., Mandelman, J.W., Brooks, E.J., Grubbs, R.D., 2017. Validation of a portable, waterproof blood pH analyser for elasmobranchs. Conserv Physiol 5 (1), cox012. https://doi.org/10.1093/conphys/cox012.

Targett, T.E., Grecay, P.A., Dixon, R.L., 2019. Growth of the estuarine fish Fundulus heteroclitus in response to diel-cycling hypoxia and acidification: interaction with temperature. Can. J. Fish. Aquat. Sci. 76 (8), 1295–1304.

Teffer, A.K., Hinch, S.G., Miller, K.M., Patterson, D.A., Farrell, A.P., Cooke, S.J., Bass, A.L., Szekeres, P., Juanes, F., 2017. Capture severity, infectious disease processes and sex influence post-release mortality of sockeye salmon bycatch. Conserv. Physiol. 5, 1–33. https://doi.org/10.1093/conphys/cox017.

Thomas, R.E., Rice, S.D., 1987. Effect of water-soluble fraction of Cook inlet crude oil on swimming performance and plasma cortisol in juvenile coho salmon (*Oncorhynchus kisutch*). Comp. Biochem. Physiol. C: Comp. Pharmacol. 87 (1), 177–180.

Tirsgaard, B., Behrens, J.W., Steffensen, J.F., 2015a. The effect of temperature and body size on metabolic scope of activity in juvenile Atlantic cod Gadus morhua L. Comp. Biochem. Physiol. A Mol. Integr. Physiol. 179, 89–94.

Tirsgaard, B., Moran, D., Steffensen, J.F., 2015b. Prolonged SDA and reduced digestive efficiency under elevated CO2 may explain reduced growth in Atlantic cod (Gadus morhua). Aquat. Toxicol. 158, 171–180.

Treberg, J.R., Killen, S.S., MacCormack, T.J., Lamarre, S.G., Enders, E.C., 2016. Estimates of metabolic rate and major constituents of metabolic demand in fishes under field conditions: methods, proxies, and new perspectives. Comp. Biochem. Physiol. A Mol. Integr. Physiol. 202, 10–22.

Tucker, V.A., 1967. Method for oxygen content and dissociation curves on microliter blood samples. J. Appl. Physiol. 23 (3), 410–414.

Twardek, W., Ekström, A., Eliason, E., Lennox, R., Tuononen, E., Abrams, A., Jeanson, A., Cooke, S., 2021. Field assessments of heart rate dynamics during spawning migration of wild and hatchery-reared Chinook salmon. Philos. Trans. R. Soc. B 376 (1830), 20200214.

Verhille, C.E., English, K.K., Cocherell, D.E., Farrell, A.P., Fangue, N.A., 2016. High thermal tolerance of a rainbow trout population near its southern range limit suggests local thermal adjustment. Conserv. Physiol. 4 (1), cow057.

von Biela, V.R., Bowen, L., McCormick, S.D., Carey, M.P., Donnelly, D.S., Waters, S., Regish, A.M., Laske, S.M., Brown, R.J., Larson, S., 2020. Evidence of prevalent heat stress in Yukon River Chinook salmon. Can. J. Fish. Aquat. Sci. 77 (12), 1878–1892.

Wagner, G.N., Hinch, S.G., Kuchel, L.J., Lotto, A., Jones, S.R., Patterson, D.A., Macdonald, J.S., Kraak, G.V.D., Shrimpton, M., English, K.K., Larsson, S., Cooke, S.J., Healey, M.C., Farrell, A.P., 2005. Metabolic rates and swimming performance of adult Fraser River sockeye salmon (*Oncorhynchus nerka*) after a controlled infection with *Parvicapsula minibicornis*. Can. J. Fish. Aquat. Sci. 62, 2124–2133. https://doi.org/10.1139/f05-126.

Wagner, G.N., Kuchel, L.J., Lotto, A., Patterson, D.A., Shrimpton, J.M., Hinch, S.G., Farrell, A.P., 2006. Routine and active metabolic rates of migrating adult wild sockeye salmon (*Oncorhynchus nerka* Walbaum) in seawater and freshwater. Physiol. Biochem. Zool. 79, 100–108. https://doi.org/10.1086/498186.

Watson, C.F., Baer, K.N., Benson, W.H., 1989. Dorsal gill incision: a simple method for obtaining blood samples in small fish. Environ. Toxicol. Chem.: Int. J. 8 (5), 457–461.

Weber, D.N., Janech, M.G., Burnett, L.E., Sancho, G., Frazier, B.S., 2020. Insights into the origin and magnitude of capture and handling-related stress in a coastal elasmobranch Carcharhinus limbatus. ICES J. Mar. Sci. 78 (3), 910–921.

Wells, R., Davie, P., 1985. Oxygen binding by the blood and hematological effects of capture stress in two big game-fish: mako shark and striped marlin. Comp. Biochem. Physiol. A Mol. Integr. Physiol. 81 (3), 643–646.

Wells, R.M., Pankhurst, N.W., 1999. Evaluation of simple instruments for the measurement of blood glucose and lactate, and plasma protein as stress indicators in fish. J. World Aquacult. Soc. 30 (2), 276–284.

Wells, R.M.G., McIntyre, R.H., Morgan, A.K., Davie, P.S., 1986. Physiological stress responses in big gamefish after capture observations on plasma chemistry and blood factors. Comp. Biochem. Physiol. A Mol. Integr. Physiol. 84A (3), 565–571.

Wendelaar-Bonga, S.E., 1997. The stress response in fish. Physiol. Rev. 77 (3), 591–625.

Westley, P.A.H., 2020. Documentation of en route mortality of summer chum salmon in the Koyukuk River, Alaska and its potential linkage to the heatwave of 2019. Ecol. Evol. 10, 10296–10304.

Whitney, C.K., Hinch, S.G., Patterson, D.A., 2013. Provenance matters: thermal reaction norms for embryo survival among sockeye salmon Oncorhynchus nerka populations. J. Fish Biol. 82, 1159–1176.

Whitney, N.M., White, C.F., Anderson, P.A., Hueter, R.E., Skomal, G.B., 2017. The physiological stress response, postrelease behavior, and mortality of blacktip sharks (*Carcharhinus limbatus*) caught on circle and J-hooks in the Florida recreational fishery. Fish. Bull. 115 (4), 532–544.

Wood, C.M., 1991. Acid-base and ion balance, metabolism, and their interactions, after exhaustive exercise in fish. J. Exp. Biol. 160 (1), 285–308.

Wood, C.M., 2022. Conservation aspects of osmotic, acid-base, and nitrogen homeostasis in fish. Fish Physiol. 39A, 321–388.

Woodhams, J., Harte, C., 2018. Shark assessment report 2018. Canberra., https://doi.org/10.25814/5beb798826ad7.

Xu, E.G., Mager, E.M., Grosell, M., Pasparakis, C., Schlenker, L.S., Stieglitz, J.D., Benetti, D., Hazard, E.S., Courtney, S.M., Diamante, G., Freitas, J., Hardiman, G., Schlenk, D., 2016. Time- and oil-dependent transcriptomic and physiological responses to deepwater horizon oil in Mahi-Mahi (Coryphaena hippurus) embryos and larvae. Environ. Sci. Technol. 50 (14), 7842–7851. https://doi.org/10.1021/acs.est.6b02205.

Yasuda, T., Komeyama, K., Kato, K., Mitsunaga, Y., 2012. Use of acceleration loggers in aquaculture to determine net-cage use and field metabolic rates in red sea bream Pagrus major. Fish. Sci. 78 (2), 229–235.

Young, J.L., Hinch, S.G., Cooke, S.J., Crossin, G.T., Patterson, D.A., Farrell, A.P., Lister, A., Healey, M.C., English, K.K., 2006. Physiological and energetic correlates of en route mortality for abnormally early migrating adult sockeye salmon (Oncorhynchus nerka) in the Thompson River, British Columbia. Can. J. Fish. Aquat. Sci. 63, 11.

Zhang, Y., Mauduit, F., Farrell, A.P., Chabot, D., Ollivier, H., Rio-Cabello, A., Claireaux, G., 2017. Exposure of European sea bass (Dicentrarchus labrax) to chemically dispersed oil has a chronic residual effect on hypoxia tolerance but not aerobic scope. Aquat. Toxicol. 191, 95–104.

Zhang, Y., Claireaux, G., Takle, H., Jørgensen, S.M., Farrell, A.P., 2018a. A three-phase excess post-exercise oxygen consumption in Atlantic salmon *Salmo salar* and its response to exercise training: THREE-PHASE EPOC IN *S. SALAR*. J. Fish Biol. 92, 1385–1403.

Zhang, Y., Gilbert, M.J.H., Farrell, A.P., 2020. Measuring maximum oxygen uptake with an incremental swimming test and by chasing rainbow trout to exhaustion inside a respirometry chamber yield the same results. J. Fish. Biol. 97 (1), 28–38. https://doi.org/10.1111/jfb.14311.

Zhang, Y., Healy, T.M., Vandersteen, W., Schulte, P.M., Farrell, A.P., 2018b. A rainbow trout Oncorhynchus mykiss strain with higher aerobic scope in normoxia also has superior tolerance of hypoxia. J. Fish Biol. 92 (2), 487–503.

Zillig, K.W., Lusardi, R.A., Moyle, P.B., Fangue, N.A., 2021. One size does not fit all: variation in thermal eco-physiology among Pacific salmonids. Rev. Fish Biol. Fish. 31 (1), 95–114. https://doi.org/10.1007/s11160-020-09632-w.

Zrini, Z.A., Gamperl, A.K., 2021. Validating star-Oddi heart rate and acceleration data storage tags for use in Atlantic salmon (Salmo salar). Anim. Biotelemetry 9 (1), 1–15.

Zupa, W., Alfonso, S., Gai, F., Gasco, L., Spedicato, M.T., Lembo, G., Carbonara, P., 2021. Calibrating accelerometer tags with oxygen consumption rate of rainbow trout (Oncorhynchus mykiss) and their use in aquaculture facility: a case study. Animals 11 (6), 1496.

Chapter 6

Applied aspects of fish endocrinology

Nicholas J. Bernier* and Sarah L. Alderman
Department of Integrative Biology, University of Guelph, Guelph, ON, Canada
*Corresponding author: e-mail: nbernier@uoguelph.ca

Chapter Outline

Endocrine systems are regulators of physiological responses to environmental conditions, acting as key transmitters of external and internal cues, and can therefore provide valuable insights to help address pressing issues in fish conservation biology. In this review, after a brief overview of the endocrine systems involved in regulating stress, growth, and reproduction, we examine how fish endocrinologists are developing and applying new tools to monitor, conserve, and assist threatened and endangered wild fish populations. Specifically, we provide examples of how endocrine signals are used to guide the development of conservation hatcheries, to reveal how exposure to environmental stressors can affect development and growth, to enable assisted reproduction, to mitigate the impacts of climate change and endocrine-disrupting chemicals on fish reproduction, and to facilitate the management of invasive species. We also examine how non-invasive sampling techniques, profiling of steroid hormones, and the

Fish Physiology, Vol. 39A. https://doi.org/10.1016/bs.fp.2022.04.006

integration of endocrinology with emerging fields such as ecotoxicogenomics and host-microbiome interactions will have impacts on future conservation efforts. Finally, we identify limitations for the broader application of endocrinology in fish conservation and opportunities for fish endocrinologists to make meaningful contributions to the most urgent conservation challenges of our time.

1 Introduction

The ongoing and accelerating loss of fish biodiversity in freshwater and marine ecosystems has brought about an urgent need to develop and apply new tools to monitor, conserve, and assist wild fish populations that are threatened and endangered. As a messenger system involved in the regulation and coordination of all biological processes, the endocrine system can provide key insights into the threats posed by anthropogenic and environmental stressors, as well as the mechanisms by which animals cope. Our understanding of fish endocrine systems continues to grow from its early promise in fisheries management (Schreck and Scanlon, 1977) as novel technologies, model species, and applications are studied, including the identification of key endocrine-related endpoints that can be used to guide the management and conservation of species at risk. In this review, after a brief overview of the endocrine systems involved in regulating stress, growth, and reproduction, we examine how endocrine signals can be used in the development of conservation hatcheries, to monitor the development and growth of wild fish, to control and assist the reproduction of threatened species, to assess the impact of climate change and endocrine-disrupting chemicals on fish reproduction, and to manage invasive species. Finally, looking into the future, we discuss how non-invasive sampling techniques, hormonal profiling, and the integration of endocrine systems in multidisciplinary approaches can benefit and advance the field of fish conservation physiology.

2 Overview of endocrine systems with applications to conservation physiology

The following section is a primer in fish endocrinology focused specifically on the stress, reproductive, and growth-regulating hormones, i.e., the endocrine signals that are more commonly measured and used in fish conservation. As such, for reviews on the hormonal systems involved in ionic regulation, drinking, food intake, digestion, metabolism, and cardiovascular control, we refer the reader to recent volumes of the *Fish Physiology* series (Bernier et al., 2009a; Gamperl et al., 2017; Grosell et al., 2010; McCormick et al., 2012).

2.1 Hormonal control of stress

Fish, like all vertebrates, respond to stressors by initiating the primary stress response which culminates in elevated levels of catecholamines and corticosteroids in the blood. Together, these hormones coordinate a multisystem physiological response that helps the fish meet and overcome a challenge to homeostasis (Barton, 2002; Barton and Iwama, 1991; Gorissen and Flik, 2016). Because this integrated response includes energetic and behavioral changes, as well as the potential for immune (Khansari et al., 2018) and reproductive inhibition (Pankhurst, 2016), assessing and monitoring the stress status of wild animals, including fish, is a recognized and valued tool in conservation physiology (McCormick and Romero, 2017; Madliger and Love, 2014). Moreover, the conserved and generalized nature of the primary stress response dictates a prescribed physiological response that is largely independent of stressor type, but that varies predictably with stressor magnitude and duration; thus, quantifying stress-related endpoints in wild fish populations can be used in broad contexts, some of which are described in this chapter.

The catecholamine (CA) hormones, epinephrine and norepinephrine, are synthesized from tyrosine in head kidney chromaffin cells of teleost fish and then released from secretory vesicles into the circulation following stimulation by the sympathetic nervous system (Fabbri and Moon, 2016; Reid et al., 1998); although it should be noted that the localization of peripheral catecholaminergic cells and their regulation by neurotransmitters and various blood borne factors differs across fish groups (Fabbri et al., 1998; Nilsson, 1983; Perry and Bernier, 1999). Once in the circulation, CAs initiate a suite of physiological changes, colloquially called the "fight or flight" response, that increase the availability and transport of metabolic fuels (Fig. 1). In the liver, for example, CAs bind to G-protein-coupled α- and β-adrenergic receptors and stimulate glycogenolysis through activation of glycogen phosphorylase (Fabbri et al., 1998). At the same time, increased plasma CAs can have a profound effect on hemoglobin oxygen affinity in some fish groups (Harter and Brauner, 2017). The appearance of CAs in the blood after a stressor is, necessarily, incredibly fast (seconds to minutes), which challenges attempts to collect meaningful baseline values, and therefore limits their usefulness as bioindicators of stress in wild populations (Sopinka et al., 2016). Nevertheless, the mechanisms of action and downstream effects of elevated CAs are relevant to conservation physiology due to the increasing opportunity for this system to be perturbed by pharmaceuticals in the aquatic environment. Drugs that target the highly conserved vertebrate adrenergic receptors (i.e., the β-blockers propranolol, atenolol, metoprolol, and sotalol that are widely prescribed in humans as a treatment for hypertension) continually enter the aquatic environment in bioactive forms that reach concentrations capable of eliciting biological responses in fish (e.g., Ings et al., 2012).

In contrast to CAs, a stress-induced increase in plasma corticosteroids occurs with some time lag from stressor onset (minutes to hours). This important difference means that accurate species-specific baseline values can be acquired, and as a result corticosteroids are a standard and widely used bioindicator of stress in both laboratory and field studies (Sopinka et al., 2016). The time lag is due to the hierarchical hormone cascade, the hypothalamic-pituitary-interrenal (HPI) axis, that regulates on-demand synthesis from

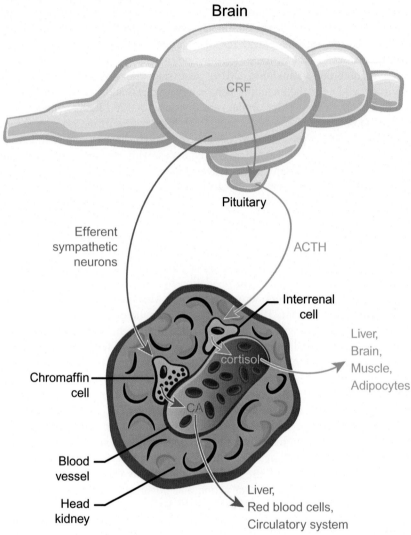

FIG. 1 Schematic representation of the neuroendocrine control of stress in fish. The brain-sympathetic-chromaffin cell pathway (blue) culminates in a release of the catecholamines (CA), epinephrine and norepinephrine, from the chromaffin cells of the head kidney into the circulation. The hypothalamic-pituitary-interrenal (HPI) axis (green) culminates in the release of glucocorticoids (e.g., cortisol) from the head kidney into the circulation. Together, CA and glucocorticoids initiate energy mobilization and other homeostatic responses by interacting with specific receptors that are widely expressed throughout the body, including in the indicated target tissues. Not shown is the cortisol-mediated negative feedback that occurs across the HPI axis to inhibit corticotropin-releasing factor (CRF), adrenocorticotropic hormone (ACTH), and cortisol release. Artwork generated by Ian Smith (University of Guelph).

cholesterol (Fig. 1). Corticosteroids are then distributed via the circulation to target tissues, where their primarily genomic actions are mediated by intracellular receptors. The activation, regulation, and outcome of HPI axis activation and corticosteroid signaling involves the integrated actions of numerous players (neuropeptides, receptors, enzymes, accessory proteins; see Bernier et al., 2009b; Faught et al., 2016 for reviews). For the purpose of this chapter, what is important to note is that this inherent complexity contributes to within and between individual differences in the endocrine stress response (Romero and Beattie, 2022), and that the involvement of so many protein components underscores the heritability of the endocrine stress response (Heath et al., 1993; Øverli et al., 2005; Pottinger and Carrick, 1999) and species-specific differences in maximal stress-induced corticosteroid levels (Barton, 2002; Barton and Iwama, 1991).

Cortisol is the dominant corticosteroid hormone produced in fishes, with the notable exceptions of 1α-hydroxycorticosterone in elasmobranchs and 11-deoxycortisol in cyclostomes. Importantly, fish lack aldosterone synthase, which in tetrapods converts cortisol to aldosterone, and therefore the corticosteroids synthesized by the head kidney interrenal cells have dual function as both glucocorticoids (energy homeostasis) and mineralocorticoids (osmotic homeostasis). As glucocorticoids, a primary function of corticosteroids in fish is to stimulate gluconeogenesis in the liver via upregulation of the major enzymes in this pathway, and glucocorticoid signaling is further implicated in the regulation of glycogen, lipid, and protein stores (Mommsen et al., 1999). Moreover, sustained elevations in circulating corticosteroid levels facilitates the redirection of energy utilization toward essential life-sustaining functions by inhibiting the digestive (Barton et al., 1987; Pfalzgraff et al., 2021), immune (Fabbri and Moon, 2016; Philip and Vijayan, 2015; Pickering and, Pottinger, 1989), and reproductive systems (Pankhurst, 2016; Pankhurst and Van Der Kraak, 1997). As mineralocorticoids, corticosteroids contribute to the maintenance of ion and water homeostasis by regulating ion transporter expression in ionocytes such as the chloride cells of the gills (McCormick et al., 2008). This is especially relevant in euryhaline fish species (Young et al., 1989), and variation in baseline cortisol predicts the timing and success of migrations (Birnie-Gauvin et al., 2019). An appreciation for the breadth of physiological functions under the regulatory control of corticosteroids emphasizes why measuring other known correlates of stress (e.g., growth, immune response, reproduction) will provide more comprehensive information on the stress status of fish (Baker et al., 2013; MacDougall-Shackleton et al., 2019; Romero and Beattie, 2022). This is relevant to management and conservation initiatives including best practices in conservation hatcheries (see Section 3.1) and population monitoring via non-invasive hormone measurements (see Section 4.1).

Given the strong integration of the HPI axis with other physiological systems, repeated or chronic stress exposure may pose a multi-pronged fitness challenge to wild fish populations by reducing growth and body condition (O'Connor et al., 2011; Sadoul and Vijayan, 2016; Vargas-Chacoff et al., 2021), increasing susceptibility to disease (Maule et al., 1987), and limiting reproductive success (Algera et al., 2017; McConnachie et al., 2012; O'Connor et al., 2009).

2.2 Hormonal control of reproduction

Neuroendocrine regulation of fish reproduction is governed by the hypothalamic-pituitary-gonadal (HPG) axis and is a critical component of conservation initiatives given its direct link to individual fitness and population dynamics. For example, manipulation of the HPG axis can be used to support fish culture and restocking programs of threatened species (Kim et al., 2020; Peñaranda et al., 2018; Zadmajid et al., 2018; see Section 3.3). Moreover, disruption of normal HPG axis activity through environmental contamination can be a driving factor in population decline (Abdel-Moneim et al., 2015; Kidd et al., 2007; see Section 3.5). The fact that the maturation and function of the HPG axis is influenced by growth, metabolism, stress, and other physiological attributes (Fuzzen et al., 2011; Zohar et al., 2010) further underscores the need for comprehensive understanding of HPG axis regulation in fish conservation.

The HPG axis in fish, as in all vertebrates, involves a cascade of hormones that govern the development, maturation, and function of the reproductive system (Fig. 2). Stimulatory and inhibitory signals from hypothalamic neurons control the release of the gonadotropic hormones, follicle-stimulating hormone (FSH) and luteinizing hormone (LH), from pituitary gonadotropes. Among the many neuropeptides and neurotransmitters involved in this process, stimulation by gonadotropin-releasing hormones (GnRH) and inhibition by dopamine are considered the primary hypothalamic regulators of gonadotropes (Zohar et al., 2010). Dopamine also inhibits GnRH secretion (Bryant et al., 2016) and is therefore a potent inhibitory signal of the HPG axis. In turn, the activities of gonadotropic and dopaminergic neurons are influenced by other neuromodulators (e.g., kisspeptin, neuropeptide Y, gonadotropin-inhibitory hormone) and hormones (e.g., cortisol, sex steroids) to integrate the reproductive system with other physiological systems (Dufour et al., 2020; Fuzzen et al., 2011; Zohar et al., 2010). Once in circulation gonadotropic hormones induce changes in gonadal tissue to support gamete development and sexual maturation. While FSH stimulates the early stages of gamete development, LH specifically, stimulates ovulation in females and spermiation in males. Both gonadotropic hormones also stimulate the production of sex hormones. In males, the predominant steroids produced are the androgens testosterone (T) and its more biologically active metabolite 11-ketotestosterone (11-KT). In females, testosterone is also produced but it

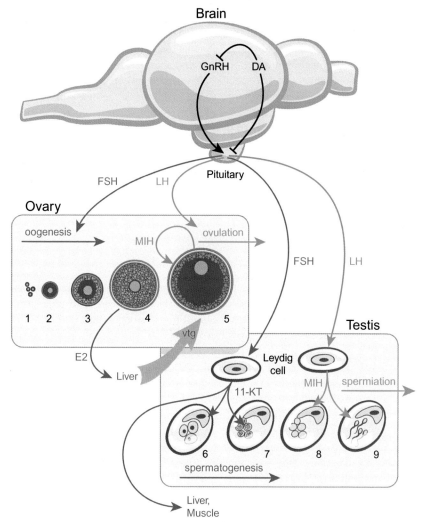

FIG. 2 Schematic representation of the hypothalamic-pituitary-gonadal (HPG) axis that regulates multiple aspects of sexual maturation and reproductive success in fish. The secretion of follicle-stimulating hormone (FSH) and luteinizing hormone (LH) from the pituitary is regulated by multiple stimulatory (arrow) and inhibitory (blunt-end) hypothalamic signals, including gonadotropin-releasing hormone (GnRH) and dopamine (DA). Inhibition of GnRH neurons by DA occurs in most fishes. FSH and LH stimulate gametogenesis and the production of estrogens (e.g., estradiol, E2), androgens (e.g., 11-ketotestosterone, 11-KT), and progestins (e.g., maturation-inducing hormone, MIH) from target cells in the gonads. In the ovary, oogenesis defines the successive maturation of an oogonium (1) to a primary oocyte (2), pre-vitellogenic oocyte (3), vitellogenic oocyte (4), and mature oocyte (5) ready for LH-mediated ovulation. The increased yolk deposition in later growth stages is driven by E2-mediated production of vitellogenin (Vtg) in the liver, one of many E2 targets. In the testis, spermatogenesis defines the successive maturation of spermatogonia (6) to spermato-cytes (7), spermatids (8), and spermatozoa (9), which occurs within the seminiferous tubules and is supported by the Sertoli cells (yellow cell alongside developing sperm). The rise in circulating 11-KT also drives the development of a male-specific phenotype via interactions with widely expressed androgen receptors, including in the liver and muscle. Gonad schematics inspired by Alix et al. (2020). Artwork generated by Ian Smith (University of Guelph).

is converted to 17β-estradiol (E2) by the enzyme aromatase (Tokarz et al., 2015). Increased circulating levels of E2 and 11-KT are critical for sexual maturation, including proliferation of oogonia and spermatogonia, respectively, as well as the development of sex-specific phenotypes and behaviors. In females, for example, E2 promotes in hepatocytes the synthesis of the egg yolk protein, vitellogenin (Vtg). In males, increased muscle growth, aggression, and secondary sex characteristics are mediated by 11-KT. Unlike other vertebrates, teleosts also produce maturation-inducing hormones (MIH) in response to gonadal stimulation by LH, either 17α,20β-dihydroxy-4-pregnen-3-one (17,20βP) or 17α,20β,21-trihydroxy-4-pregnen-3-one (20βS). These progesterone-like steroids induce final oocyte maturation and spermiation, enhance sperm motility, and act as male pheromones in some species (Scott et al., 2010; Tokarz et al., 2015).

2.3 Hormonal control of growth and metabolism

As an integrated response of external environmental conditions and internal physiological status, growth in fish can be an important measure of population health and habitat quality. Among the multiple hormones that contribute to regulating the indeterminate growth of fish, the growth hormone (GH)-insulin-like growth factor-I (IGFI) axis and thyroid hormones play prominent roles across all life stages (Fuentes et al., 2013; Power et al., 2001) (Fig. 3). Previous laboratory studies established that components of the GH-IGFI axis can be used as indices of growth and increasingly these biomarkers are being used to monitor the growth of wild fish (Beckman, 2011; Duguid et al., 2018; Picha et al., 2008; see Section 3.2). In contrast, the thyroid axis in fish is particularly sensitive to environmental and anthropogenic stressors and components of this axis may be suitable biomarkers of endocrine disruption (Carr and Patiño, 2011; Deal and Volkoff, 2020; Jarque and Piña, 2014; Nugegoda and Kibria, 2017).

The GH-IGFI axis plays a key role in regulating somatic growth across fishes (Picha et al., 2008; Reinecke et al., 2005). Stored in the pituitary somatotropes, GH is secreted via a multifactorial control system of hypothalamic, pituitary, and peripheral origins which integrates information related to energy metabolism, feeding, and food availability (Chang and Wong, 2009). In fed fish, the growth-stimulating effects of GH are primarily via the stimulation of IGFI from the liver and other tissues, and to a lesser degree from the direct muscle protein synthesizing actions of GH (Nordgarden et al., 2006). In contrast, in fasted fish, plasma GH levels rise and promote lipolysis (Bergan et al., 2015). These nutrient status-dependent effects of GH are explained by the differential expression of GH receptor subtypes linked to signaling cascades that either stimulate the production and release of IGFI or lipolysis (Bergan-Roller and Sheridan, 2018). At the tissue level, activation of IGF receptors by circulating or locally produced IGFI can

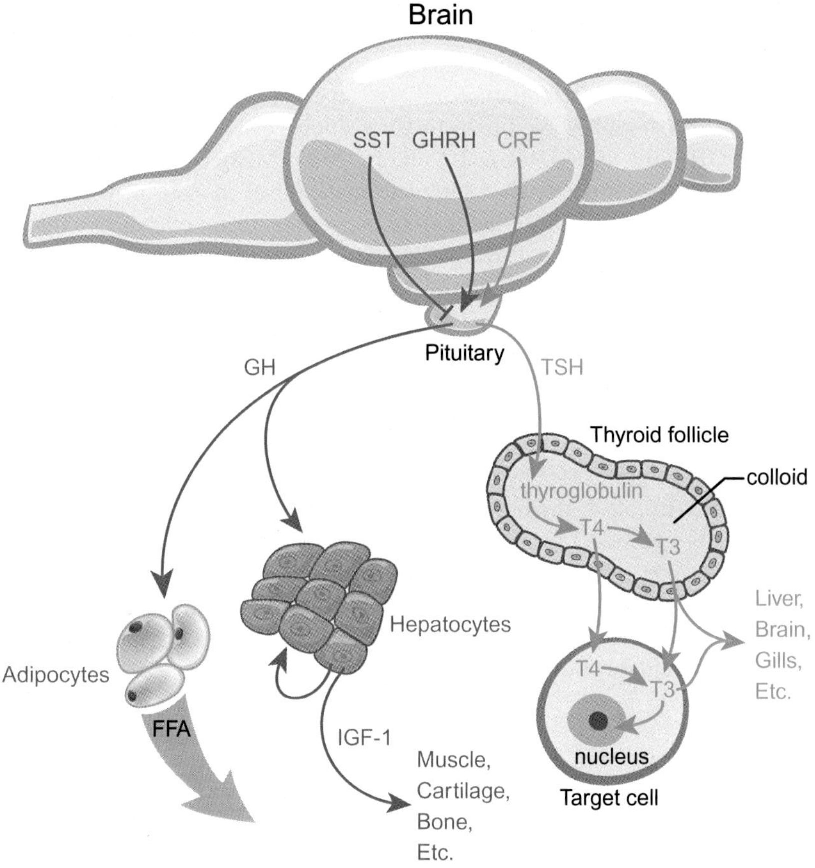

FIG. 3 Schematic representation of the endocrine control of growth and metabolism in fish. The growth hormone (GH)—insulin-like growth factor-1 (IGF-1) axis (left) and the thyroid axis (right) work independently and synergistically (not shown) throughout the lifespan of a fish. Both axes are regulated by complex stimulatory (arrows; e.g., GH-releasing hormone, GHRH; corticotropin-releasing factor, CRF) and inhibitory (blunt-end; e.g., somatostatin, SST) signals from the hypothalamus. Depending on nutritional status, pituitary secretion of GH can signal either catabolic (e.g., lipolysis to release free fatty acids, FFA) or anabolic (e.g., hyperplasia) effects directly on target tissues, or indirectly via IGF-1. Receptors for both GH and IGF-1 are broadly distributed and therefore similar effects occur in multiple cell types beyond those shown here. The release of pituitary thyroid-stimulating hormone (TSH) signals the production and release of thyroid hormones (thyroxine, T4; triiodothyronine, T3) from thyroid follicles, the distribution of which is species dependent. The biologically active T3 regulates gene transcription via thyroid receptors that are widespread in fish tissues. Artwork generated by Ian Smith (University of Guelph).

stimulate protein synthesis as well as cellular proliferation, differentiation, and survival (Reindl and Sheridan, 2012). While multiple IGF binding proteins (IGFBPs) regulate the availability of free IGFI in fish, and IGFBPs can potentiate and/or inhibit IGFI actions (Allard and Duan, 2018), under various physiological conditions plasma IGFI levels are positively correlated with specific growth rate (Beckman, 2011; Picha et al., 2008).

Thyroid hormones directly contribute to the regulation of growth in fish via their actions on differentiation and organogenesis during embryonic development and post-embryonic life history transitions, and indirectly by stimulating the GH-IGFI axis (Deal and Volkoff, 2020; McMenamin and Parichy, 2013). Thyroid hormones also stimulate metabolism in fish in response to cold, an essential function of the thyroid axis to maintain performance during cold acclimation (Little and Seebacher, 2014). As in other vertebrates, the synthesis and metabolism of thyroid hormones in fishes is regulated by the hypothalamic-pituitary-thyroid (HPT) axis. Under the control of multiple hypothalamic factors, the pituitary thyrotropes release thyroid-stimulating hormone (TSH) into the circulation (Bernier et al., 2009b). In return, TSH stimulates the synthesis and release of pro-hormone thyroxine (T4) from subpharyngeal and renal thyroid follicles (Geven et al., 2007). Circulating T4 is converted in target tissues by type 1 and type 2 deiodinase enzymes into bioactive 3,5,3′-triiodo-L-thyronine (T3) (Jarque and Piña, 2014). T3 binds to two broadly expressed nuclear receptors that mediate the diverse actions of the thyroid axis by promoting or repressing gene transcription (Nelson and Habibi, 2009). Finally, both T3 and T4 have negative feedback effects on the hypothalamic and pituitary levels of the HPT axis and are inactivated by a type 3 deiodinase (Bernier et al., 2009b; Jarque and Piña, 2014).

3 Applied aspects of endocrine systems

3.1 Fish culture

Fish culture can play an important role in the management and protection of endangered species. According to the 2021 version of the International Union for Conservation of Nature's red list, 3210 fish species are threatened, or ~9% of the estimated number of fish species. Among those endangered, hundreds now depend on hatcheries to provide juveniles produced for species protection or restoration (Flagg and Nash, 1999; Froehlich et al., 2017; Taylor et al., 2017). While the use of hatcheries to recover wild populations can contribute to conservation efforts, it may also have unintended consequences. The juvenile fish produced by hatcheries differ from their wild counterparts and can therefore have negative effects on wild fish populations (Araki et al., 2008; Brown and Day, 2002; Flagg et al., 2000; Rand et al., 2012). Still, research in the field of fish endocrinology has played an important role in the development of conversation hatcheries and in their on-going efforts at optimizing the fitness of released fish. For example, fundamental and translational research in fish reproductive endocrinology has been key to the success of conservation hatcheries in reliably producing juveniles from wild fish (Zohar, 2021) (see Section 3.3 for more details). Similarly, research on stress and growth-regulating hormones is playing an important role in identifying the physiological and behavioral differences between hatchery and wild fish and is guiding the development of novel conservation hatchery strategies. Numerous studies have compared the endocrine stress response between hatchery and wild fish. In general, cultured fish have a reduced cortisol response to an acute stressor relative to their wild counterparts. For example, hatchery rainbow trout (*Oncorhynchus mykiss*; Woodward and Strange, 1987), chinook salmon (*Oncorhynchus tshawytscha*; Mazur and Iwama, 1993; Salonius and Iwama, 1993), ayu (*Plecoglossus altivelis*; Awata et al., 2011), rainbowfish (*Melanoteania duboulayi*; Zuberi et al., 2011), Eurasian perch (*Perca fluviatilis*; Douxfils et al., 2011) and fighting fish (*Betta splendens*; Verbeek et al., 2008) have lower cortisol levels following acute stress relative to wild fish. Since the cortisol response to stressors in fish has a heritable component (Heath et al., 1993; Øverli et al., 2005; Pottinger and Carrick, 1999), artificial selection for fish with a higher tolerance to the frequent episodes of stress encountered in hatcheries (e.g., netting, air exposure, chasing, crowding, grading) may result in the production of fish with muted stress responses. However, environmental conditions can also affect the responsiveness and sensitivity of fish to stressors. For example, among three groups of coho salmon (*Oncorhynchus kisutch*) sharing a similar genetic background and acclimated to common garden hatchery conditions for 7 months, the wild and colonized groups (i.e., hatchery fish previously transported into a natural environment as fry and re-captured as smolts) had higher cortisol levels following acute air exposure relative to hatchery fish (Salonius and Iwama, 1993). In the same coho salmon stock, wild and colonized smolts consistently had a larger springtime increase in plasma cortisol and a greater saltwater tolerance than hatchery fish (Shrimpton et al., 1994a,b). These results suggest that natural rearing environments favor the survival of fish with a heightened response to stressors and environmental cues, and that colonization of hatchery fish into natural environments may be needed to improve the success of conservation hatcheries.

Several studies have also evaluated alternative rearing practices to reduce the effects of stressors in conservation hatcheries. Relative to a barren environment, hatchery tank enrichments (e.g., the addition of substrate, plants, or shelter) have been shown to reduce resting plasma cortisol levels (Cogliati et al., 2019; Näslund et al., 2013; Rosengren et al., 2017) and the endocrine stress response to simulated hatchery disturbances (Barcellos et al., 2009; Braithwaite and Salvanes, 2005; Cogliati et al., 2019; Marcon et al., 2018; Rosengren et al., 2017). In contrast, enriched tank environment either had no effect (Näslund et al., 2013; Pounder et al., 2016; Wilkes et al., 2012) or were associated with larger post-stress plasma cortisol levels in other studies (Batzina et al., 2014; Boerrigter et al., 2016; Zubair et al., 2012). Among the factors that are known to contribute to the varying effects of tank enrichment (Näslund and Johnsson, 2016), the capacity of environmental complexity to reduce the endocrine stress response may require lower rearing densities (Cogliati et al., 2019; Rosengren et al., 2017). Interestingly, whether environmental enrichment is provided in hatchery tanks or in a more natural setting also appears to have a differential effect on the endocrine stress response of fish. In juvenile chinook salmon, relative to fish reared in barren hatchery tanks, early rearing in seminatural channels resulted in fish with higher plasma cortisol levels in response diverse stressors (Garner et al., 2011; Madison et al., 2015a), suggesting that the stress responsiveness of channel-reared fish is more akin to those of wild fish. While the above studies suggest that environmental enrichment can affect HPI axis regulation of hatchery fish, it remains to be determined whether these alternative rearing practices translate into improved survival in the wild. In general, while fish reared in semi-natural environments usually perform and survive better in the wild than tank-reared fish, the benefits of hatchery tank enrichment to post-release performance and survival have been mixed (for review, see Näslund and Johnsson, 2016). Future work is needed to assess the relationships between rearing practices, stress responsiveness, and the capacity of hatchery fish to survive in natural environments.

Identifying the neuroendocrine basis of the behavioral differences between hatchery and wild fish may also help conservation hatcheries increase post-release survival rates and minimize the impact of stock supplementation on wild fish populations. The behavioral effects of hatchery selection are well known (for reviews, see Huntingford, 2004; Milla et al., 2021; Olla et al., 1998). Domestication in fish leads to a reduction in predator avoidance (Berejikian, 1995), to decreased foraging abilities (Brown et al., 2003), and promotes aggression and boldness (Rhodes and Quinn, 1998; Riley et al., 2005; Sundström et al., 2003). These behavioral effects of domestication may be at least partially mediated by alterations in brain monoaminergic activity (i.e., dopamine, norepinephrine, and serotonin) involved in the control of behavioral and endocrine stress response in fish (Winberg et al., 2016). For example, in response to standardized stressors, hatchery brown trout (*Salmo trutta*) have lower brain serotonergic and dopaminergic activity than their wild counterparts (Lepage et al., 2000). Similarly, the unpredictable chronic stress of hatchery environments leads to lower hypothalamic catecholaminergic and brain stem serotonergic responses to an acute stressor in Atlantic salmon (*Salmo salar*; Vindas et al., 2016). Although still poorly understood, comparison of hatchery- and wild-reared Atlantic salmon also suggests that the stimulus-deprived hatchery environment can lead to a reduction in neuroplasticity (Mes et al., 2018). In contrast, environmental enrichments can enhance neural plasticity, promote survival-related behaviors, and alter both basal and stress-induced levels of brain monoamines (Arechavala-Lopez et al., 2020; Batzina et al., 2014; Höglund et al., 2005; Mes et al., 2019; Salvanes et al., 2013; Ullah et al., 2020). As such, we suggest that conservation programs may benefit from the study of neurohormones involved in the regulation of behavioral flexibility, and from research efforts directed toward a greater understanding of the role of environmental complexity in shaping these neuronal circuits.

Hormonal bioindicators of early sexual maturation and growth are also contributing to the development of novel conservation hatchery strategies. Historically, based on the observation that fish size and growth rate are positively correlated with higher post-release smolt survival, juvenile salmonid conservation hatcheries produced fish that were larger and with higher body condition than their wild counterparts (Beckman et al., 1999; Tipping, 1997). However, higher post-release smolt survival rates do not necessarily translate into higher smolt-to-adult return rates (Beamish et al., 2008; Morita et al., 2006). Factors predicted to contribute to this lower overall return of hatchery fish include early sexual maturation of males and reduced propensity to migrate. The faster growth rates and higher energy reserves of hatchery salmonids are associated with increased rates of precocious male maturation (Larsen et al., 2006; Shearer et al., 2006; Vainikka et al., 2012). Recent studies have shown that non-lethal measurement of springtime plasma levels of 11-KT can be used to predict the proportion of chinook salmon and steelhead trout males maturing at the time of smolt release (Medeiros et al., 2018; Middleton et al., 2019), a measure that may prove beneficial for reducing the impact of hatchery programs on precocious sexual maturation. The faster growth rates of hatchery-reared juvenile salmonids are also associated with freshwater residualism, i.e., the failure to out-migrate as smolts, relative to wild fish populations (Chittenden et al., 2010; Davidsen et al., 2014; Vainikka et al., 2012). In chinook salmon, the seasonal growth pattern of wild juveniles is characterized by a marked anabolic to catabolic shift in the autumn that results in a depletion of body lipids and a cessation of growth through the winter, followed by a spring anabolic phase that promotes smolt development and the deposition of energy reserves prior to smolting (Beckman et al., 2000). This dynamic seasonal shift in growth is paralleled by changes in plasma IGFI, with levels dropping throughout the fall, reaching a low in the winter, and increasing again in the spring (Beckman et al., 2000). Importantly, salmon hatcheries that best mimic the above wild fish seasonal growth pattern have higher smolt-to-adult return rates (Beckman et al., 2017; Harbicht et al., 2020; Harstad et al., 2018). In chinook salmon hatcheries, fish with the highest summer and autumn growth rates have higher plasma IGFI and adiposity levels and produce fish with the highest rates of precocious male maturation. In contrast, time-matched wild fish are significantly smaller, leaner, have much lower plasma IGFI levels, and greatly reduced rates of precocious maturation (Larsen et al., 2006). Therefore, further evaluation of plasma IGFI levels may be useful in the assessment of smolt quality and fitness, and for the management of salmonid conservation hatcheries.

The above examples highlight how the integration of endocrinology with physiology and behavior can help guide the refinement of conservation hatcheries practices to produce fish with more natural hormonal cycles prior to release and to improve post-release survival. Similarly, comparing the endocrine profile of hatchery fish to their wild counterpart post-release may provide a benchmark of successful fish culture conditions and contribute to reintroduction and restoration efforts.

3.2 Development and growth monitoring

Hormones are key regulators of development and life history transitions in fishes. Through their pleiotropic actions, hormones integrate internal and external cues to coordinate the complex physiological and behavioral changes that characterize early development and the larval to juvenile transition (Holzer and Laudet, 2015; McMenamin and Parichy, 2013). Key among these signals for proper embryonic development are maternal thyroid hormones and cortisol (Deal and Volkoff, 2020; Nesan and Vijayan, 2013). For example, in zebrafish embryos, while maternal thyroid hormones play essential roles in neuron differentiation and survival in the brain and spinal cord (Campinho et al., 2014), the knockdown of maternal glucocorticoid receptors prevents mesoderm formation (Nesan et al., 2012) and broadly disrupts organogenesis (Pikulkaew et al., 2011). As such, environmental conditions or endocrine disrupting compounds that interfere with the maternal transfer or actions of thyroid hormones and glucocorticoids during embryonic development may induce developmental defects and compromise larval viability.

Thyroid hormones also play crucial roles in orchestrating the morphological changes associated with the larval to juvenile transition in teleosts, and disruption to the thyroid axis during this sensitive life history transition can have marked effects on larval recruitment (McMenamin and Parichy, 2013). In coral reef fishes, metamorphosis from a pelagic plankton-eating dispersal larval stage to a grazing reef-associated juvenile fish involves a surge in thyroid axis activity, and dramatic changes in pigmentation, digestive tract morphology, and behavior (McCormick et al., 2002). In the convict surgeonfish (*Acanthurus triostegus*), consistent with the notion that thyroid hormones control metamorphosis, algal grazing, and remodeling of the digestive tract for herbivory are stimulated by T3 injections and repressed by thyroid receptor blockade (Holzer et al., 2017). Similarly, manipulation of thyroid hormone signaling in *A. triostegus* demonstrated that thyroid hormones control the development of sensory structures and modulate the behavioral responses and vulnerability to predation (Besson et al., 2020). The fact that exposure of *A. triostegus* larvae to the pesticide chlorpyrifos, a common reef pollutant, reduces T3 levels, impairs metamorphosis, and leads to a T3-reversible increase in predation, highlights the sensitivity of coral reef fish larval recruitment to thyroid disruption (Besson et al., 2020; Holzer et al., 2017). Moreover, given the importance of algal grazing by coral reef fishes to maintain coral health (Hughes et al., 2007), thyroid disruption during larval recruitment may have larger consequences for the conservation of coral reef ecosystems (Holzer et al., 2017).

Smoltification in salmon, i.e., the transformation of a freshwater-acclimated parr into a seawater tolerant smolt, provides another example of a hormone-dependent life history transition that is critical for recruitment and population sustainability (Björnsson et al., 2011). Driven by seasonal changes in photoperiod and temperature, increases in cortisol, thyroid hormones, GH, and IGFI, coordinate the changes in morphology, physiology, and behavior that pre-adapt the smolt for entry into seawater (McCormick, 2013). Acting primarily on the gills, intestine, integument, and brain, the individual and synergistic actions of the GH-IGFI, HPI, and HPT axes promote lipid mobilization, growth stimulation, salinity tolerance, imprinting, silvering, downstream migration, and schooling behaviors (McCormick, 2013). While the multi-hormonal regulation of smoltification may provide increased flexibility to respond to variable environmental conditions, it also makes the process more vulnerable to a wide range of endocrine-disrupting compounds (EDCs). For example, exposure to environmentally relevant concentrations of estrogenic compounds and xenobiotics (e.g., flame-retardants, pesticides, persistent organic pollutants, or acid and aluminum) can affect the circulating levels of one or more of the above hormones, and impair the growth, downstream migration, seawater tolerance, and olfactory function of juvenile salmonids (Arkoosh et al., 2017; Bangsgaard et al., 2006; Breves et al., 2018; Duffy et al., 2014; Fairchild et al., 1999; Lerner et al., 2007, 2012; Lower and Moore, 2007; Monette et al., 2008). As a result, the disruptive effects of environmental stressors on the endocrine signals that regulate smolt development have the potential to reduce early marine survival and contribute to variation in adult returns between years and among rivers (McCormick et al., 2009).

Rapid growth during the larval and juvenile stages is a major determinant of recruitment to adult fish populations (Pepin, 2016). Both laboratory and field studies show that larger and faster growing juvenile fish gain a survival advantage over smaller conspecifics (Sogard, 1997). Characterized by better energy reserves, the larger members of a cohort may have an enhanced resistance to starvation, a reduced vulnerability to predators, and a greater tolerance of environmental stressors (Shuter and Post, 1990; Sogard, 1997). Therefore, growth monitoring is an important management tool for the conservation of fish populations, and increasingly endocrine indices of growth are being used to assess the growth of field-captured fish.

Among the components of the GH-IGFI axis, IGFI is the most well-established and validated endocrine biomarker of growth in fish. Plasma IGFI levels are higher in fed than in fasted fish, and positively correlated with growth rate in multiple fish species (Andrews et al., 2011; Beckman, 2011; Beckman et al., 2004a; Hack et al., 2019; Picha et al., 2008; Shimizu et al., 2009). In salmonids, laboratory studies have demonstrated that IGFI provides a robust index of recent relative growth rates (4 days to 2 weeks) (Caldarone et al., 2016; Duguid et al., 2018), and plasma IGFI levels are now used to assess the growth of field-captured fish and to investigate the factors that influence growth during the early marine residence of juvenile fish (Chamberlin et al., 2017; Ferriss et al., 2014; Kaneko et al., 2015). For example, in wild subyearling Chinook salmon (*O. tschwytscha*), higher plasma IGFI levels are closely associated with faster scale-derived growth rates, fuller stomachs, and stomach contents which demonstrate the early adoption of piscivory, key factors known to contribute to early marine survival (Davis et al., 2020). Similarly, plasma IGFI levels used in combination with otolith microchemistry to identify recent habitat use revealed that estuaries provide heterogeneous but overall greater growth opportunities than lake habitats for diadromous dolly varden (*Salvelinus malma*) (Bond et al., 2014). Interestingly, plasma IGFI concentrations have also been used to identify poor growth environments in the coastal waters of British Columbia across five species of juvenile Pacific salmon (Journey et al., 2018). Overall, while there is good evidence that IGFI is a useful bioindicator of growth in juvenile salmonids, it remains to be determined whether circulating levels of IGFI can also be used to monitor the growth of other fish species in the wild.

Emerging evidence suggests that the plasma levels of specific IGFI binding proteins may also serve as indices of growth in fish. In general, the mRNA and protein levels of IGFBPs with larger molecular weights (e.g., IGFBP-2b) decrease in abundance in response to fasting and are positively related to growth rate (Beckman et al., 2004b; Kelley et al., 2002; Peterson and Waldbieser, 2009; Shimizu and Dickhoff, 2017). In contrast, the expression and circulating levels of smaller molecular weight IGFBPs (e.g., IGFBP-1a and -1b) increase in response to fasting and stress, and are inversely related to growth rate (Hack et al., 2019; Madison et al., 2015b; Peterson and Waldbieser, 2009; Shimizu and Dickhoff, 2017; Shimizu et al., 2009). In wild fish, the utility of IGFBP-1a as an inverse index of growth has recently been demonstrated in out-migrating chum (*Oncorhynchus keta*) (Kaneko et al., 2019a) and coho (*O. kisutch*) (Kaneko et al., 2019b) post-smolts. For example, in coastal British Columbia, plasma IGFBP-1b levels were the highest in juvenile coho salmon from regions with poor ocean conditions and were associated with reduced IGFI levels and stomach contents (Kaneko et al., 2019b). Although more research is needed to understand how environmental conditions and nutritional states differentially regulate the various IGFBPs of fish, results from the above studies suggests that circulating IGFBP levels have the potential to contribute to fish stock assessment and conservation.

In contrast, despite the clear growth stimulatory effects of GH treatment in fish and GH overexpression in GH-transgenic fish (Devlin et al., 2001), plasma GH levels may have limited utility as a bioindicator of instantaneous growth in fish. The discordant relationship between the circulating levels of GH and growth in fish stems in part from the dual role of GH in the regulation of anabolic and catabolic metabolic pathways, and from the context-dependent actions of GH (Bergan-Roller and Sheridan, 2018; see Section 2.3 for more details).

3.3 Reproductive control

Successful reproduction is essential for population viability, yet wild fish face a variety of threats to reproduction. For example, overfishing can reduce the abundance of breeding adults, greatly reduce recruitment, and ultimately cause populations to collapse (Myers et al., 1994). Similarly, spawning habitat loss, barriers to migration and spawning grounds, or competition with invasive species have led to the need for reproductive assistance to reduce the loss of genetic diversity and the risk of extinction (Reid and Hall, 2003; Swanson et al., 2008). Climate change (see Section 3.4) and EDCs (see Section 3.5) can also affect all levels of the reproductive axis in fish and the threats to reproduction from these environmental stressors are well documented (Carnevali et al., 2018; Overturf et al., 2015; Servili et al., 2020). Driven primarily by the need to overcome reproduction-related barriers and close the life cycle of economically desirable species for aquaculture, a variety of techniques have been developed in the field of reproduction biology to enable the predictable reproduction of captive fish (Zohar, 2021). While the breeding programs of aquaculture operations and conservation hatcheries have opposite goals, i.e., select for specific genotypes vs maintain genetic diversity, the techniques developed to facilitate fish reproduction in commercial hatcheries have found a variety of applications for fish conservation. In this section, we provide a few select examples of applications of research in fish reproductive control for the conservation of fish.

Various fish species cannot reproduce in captivity as they lack the required environmental and biological cues of spawning habitats that are needed for the final maturation of gametes. Working on the premise that this inability to reproduce in captivity was caused by a hormonal failure, fish endocrinologists in the 1970s set out to discover the hormones of the HPG axis in fish and to identify ways to manipulate them (Zohar, 2021). Early on, it was shown that human chorionic gonadotropin (hCG) could be used to induce final oocyte maturation, ovulation, and spawning (Donaldson and Hunter, 1983), suggesting that the failure of captive fish to undergo final oocyte maturation may be due to a lack of pituitary gonadotropin (LH) release. Once confirmed, this discovery prompted the identification, characterization, and synthesis of synthetic GnRHs to induce LH release from the pituitary and thereby stimulate spawning in captive fish (Mylonas et al., 2017; Zohar and Mylonas, 2001). Combining these GnRH analogues (GnRHa) with a dopamine receptor antagonist led to the formulation of Ovaprim, a potent ovulating/spermiating agent used to promote and facilitate reproduction in a variety of teleost species (Dufour et al., 2010; Peter et al., 1988; Yanong et al., 2009). Further technical advancements, such as the use of homologous recombinant proteins and gene therapy for *in vivo* gonadotropin delivery, should permit a more targeted use of hormonal treatments to solve species-specific reproductive problems (Molés et al., 2020). In general, hormonal manipulations are now used in aquaculture and captive broodstock programs to advance and synchronize maturation in both sexes; to enhance fecundity, fertility, and embryo survival; to promote the development of secondary sex characteristics; and to counter the reproductive behavioral deficiencies exhibited by captively-reared fish that are released into natural environments (Berejikian et al., 2003; Mylonas et al., 2010; Zohar and Mylonas, 2001).

The conservation and restoration of various fish species and stocks with rapidly declining populations may also benefit from hormonal treatments. With a single opportunity to reproduce and complex life cycles, semelparous and diadromous fish species, such as Pacific salmon and anguillid eels, are particularly sensitive to reproductive barriers. For example, several freshwater eel populations are now listed as critically endangered, so there is an urgent need to further develop and improve protocols for their artificial propagation (Burgerhout et al., 2019; Jacoby et al., 2015). In captivity, however, anguillid eels fail to spontaneously reach sexual maturity, remaining in the pre-pubertal, or silver, life stage. Injections of hCG can induce testicular development and spermatogenesis in male silver *Anguilla* species (Herranz-Jusdado et al., 2019; Lokman et al., 2016). Similarly, a combination of pituitary extracts and a MIH (i.e., 17,20βP) injected into female silver eels helps induce vitellogenesis, follicular maturation and ovulation (Kottmann et al., 2020). A better understanding of the natural triggers involved in eel gametogenesis and the physiological roles played by hormonal transmitters of external cues, such as melatonin, may further improve the success of these maturation protocols (Burgerhout et al., 2019). Such interventions are likely to help in conservation efforts of other at-risk fish species that do not reproduce well in captive environments. For example, pituitary extracts and GnRHa preparations are widely used to induce spermiation in several species of sturgeon and gar (Alavi et al., 2012; Mendoza Alfaro et al., 2008). Although the specific mechanisms of oocyte maturation have yet to be resolved (Hasegawa et al., 2022), ovulation and spawning can also be induced by treatment with GnRHa in these species (Mendoza Alfaro et al., 2008; Mohammadzadeh et al., 2021). Finally, assisted reproductive techniques are also needed for the conservation of many elasmobranch species that are experiencing rapid declines (Ferretti et al., 2010). With a small litter size, a long gestation period and slow sexual maturation, most shark and skate populations are being depleted by fisheries faster than they can reproduce (Frisk et al., 2001; McPhie and Campana, 2009). Consistent with its effects in numerous teleosts species, Kim et al. (2020) demonstrated for the first time in elasmobranchs that Ovaprim can effectively induce semen release and follicular maturation in mature banded houndshark (*Triakis scyllium*) and white-tip shark (*Carcharhinus longimanus*). Although this study did not assess the quality of the germ cells produced, nor whether the semen can be used for successful internal fertilization of the females, the results are a promising step toward the development of a hormone-induced artificial insemination protocol for endangered elasmobranchs.

3.4 Climate change

Climate change-driven alterations in the abiotic characteristics of aquatic ecosystems can have broad and complex effects on the endocrine systems of fishes. While the effects vary significantly with habitat, latitude and depth, climate change is generally slowly raising seawater temperature and increasing ocean acidification; increasing the frequency of combined low flow, elevated temperature, and hypoxic events in riverine habitats; and warming the surface temperature of lakes which increases thermal stratification and hypolimnetic oxygen depletion (Adrian et al., 2009; Bopp et al., 2013; Butcher et al., 2015). Nutrient enrichment is also a major driver of large scale, seasonal, and expanding hypoxic zones in lakes and coastal marine ecosystems around the world (Diaz and Rosenberg, 2008; Howarth et al., 2011; Jenny et al., 2016). While previous studies in fish have shown that elevated temperatures, hypoxia, and acidification can affect stress (Bernier and Craig, 2005; Chadwick et al., 2015; Goikoetxea et al., 2021; Petochi et al., 2011; Vargas-Chacoff et al., 2018) and growth-regulating (Deal and Volkoff, 2020; Kajimura and Duan, 2007; Kamei, 2020) hormones, far more have characterized the effects of these environmental stressors on the hormones of the HPG axis. In general, elevated temperatures and hypoxia can have serious impacts on several critical reproductive processes (Alix et al., 2020; Pankhurst and Munday, 2011; Servili et al., 2020; Pankhurst and Van Der Kraak, 1997). Therefore, given the relevance of successful recruitment to conservation, here we review the known effects of climate-driven warming, deoxygenation and acidification on reproductive hormones and their consequences for reproduction and sex determination in fish.

Although fish generally benefit reproductively from increases in temperature within their physiological tolerance range, temperatures above species-specific thermal maxima can inhibit reproduction via effects at all levels of the HPG axis. While the magnitude and temporal nature of the temperature change are expected to influence species-specific responses, consistent effects on components of the HPG axis in fish exposed to high or prolonged elevations in temperature have been observed. For example, at the hypothalamic level, high temperatures can inhibit the gene expression of GnRH and kisspeptins (Elisio et al., 2012; Okuzawa and Gen, 2013; Shahjahan et al., 2013, 2017), and stimulate gonadotropin-inhibiting hormone (Bock et al., 2021). Similarly, in the pituitary, exposure to high temperatures can reduce transcript abundance of the GnRH receptor and the gonadotropin subunits FSHβ and LHβ (Elisio et al., 2012; Okuzawa and Gen, 2013; Pérez et al., 2011; Shahjahan et al., 2017; Soria et al., 2008; Taranger et al., 2015). In the ovaries, high temperature inhibits FSH and LH receptor mRNA levels (Anderson et al., 2019; Bock et al., 2021; Elisio et al., 2012; Soria et al., 2008) and suppress the expression of several key genes involved in sex steroid synthesis (Anderson et al., 2012, 2019; Bock et al., 2021; Elisio et al., 2012; Mazzeo et al., 2014). Moreover, multiple studies have shown that fish reared at high temperatures can have lower levels of plasma E2, T, 11-KT, and MIH (Bock et al., 2021; Elisio et al., 2012; García-López et al., 2006; Mazzeo et al., 2014; Okuzawa and Gen, 2013; Pankhurst et al., 2011; Pérez et al., 2011; Vikingstad et al., 2016). In females, the reduced production of E2 is likely due to a down-regulation of gonadal aromatase expression since exposure to high temperatures increases methylation of the aromatase gene promoter and reduces its expression (Navarro-Martín et al., 2011). In fact, the observation that high temperatures inhibit E2 synthesis in isolated ovarian follicles suggests that the effects of temperature on the HPG axis is primarily via direct effects on gonadal steroidogenesis and indirect effects on the hypothalamic and pituitary levels that are mediated by sex steroid feedback loops (Watts et al., 2004). Reduced circulating levels of E2 also likely explain why temperature-stressed female fish typically have lower circulating levels of Vtg and liver Vtg expression (Bock et al., 2021; Clark et al., 2005; King et al., 2007; Mahanty et al., 2019; Pankhurst et al., 2011; Pérez et al., 2011). Therefore, in thermally challenged female fish, while reduced E2 synthesis and Vtg sequestration may inhibit the progression of oocyte development and egg quality, MIH synthesis inhibition may delay or inhibit oocyte maturation, ovulation, and spawning. Similarly, the reduced circulating levels of androgens and MIH in males that experience high temperatures likely contribute to impaired spermatogenesis and sperm maturation, respectively. Importantly, beyond species differences, factors such as sex, thermal insult severity and duration, sexual maturity, age, nutritional status, and others, are known to modulate the effects of high temperatures on the HPG axis (Alix et al., 2020; Pankhurst and Munday, 2011; Servili et al., 2020). Lastly, while it is known that stressors can inhibit reproduction in fish, the extent to which the HPI axis and cortisol contribute to high temperature-mediated reproductive suppression is uncertain (Fuzzen et al., 2011; Pankhurst, 2016).

Hypoxia is now a common occurrence in coastal waters and lakes worldwide (Breitburg et al., 2018; Tellier et al., 2022) and these abiotic conditions can impair reproduction in fish, especially in species that reproduce during the summer (Pankhurst, 2016; Servili et al., 2020; Wu et al., 2003). Although fish avoid lethal levels of dissolved oxygen, sublethal hypoxia is known to suppress reproduction in wild fish populations (Cheek et al., 2009; Friesen et al., 2012a; Thomas et al., 2007, 2015). Importantly, hypoxic conditions that cause reproductive impairment can also induce DNA methylome modifications that cause transgenerational epigenetic impairment of male and female reproduction (Lai et al., 2018, 2019). However, since hypoxia tolerance varies widely among fish species (Rogers et al., 2016), the hypoxic conditions (severity and duration) required to inhibit reproduction in one species may be tolerated by another. Therefore, care should be exercised in extrapolating how specific dissolved oxygen concentrations may affect reproduction across species. As observed for the effects of high temperature on reproduction, chronic hypoxia appears to have broad effects at all levels of the HPG axis in fish. For example, chronic hypoxia exposure can reduce hypothalamic GnRH gene expression (Lu et al., 2014; Thomas et al., 2007), suppress pituitary FSHβ, LHβ, and estrogen receptor transcript abundance (Lu et al., 2014), and is associated with reduced plasma LH levels (Thomas et al., 2007; Wang et al., 2008). In the gonads, hypoxia exposure has sex-specific effects on the mRNA levels of the receptors for FSH, LH, and estrogen, and reduces the expression of several key genes involved in steroidogenesis (Lu et al., 2014)—effects which may be regulated through the actions of several miR-NAs (Lai et al., 2016). Across several species, studies have consistently observed that fish exposed to chronic hypoxia have lower levels of plasma sex steroids, including E2, T, 11-KT, and MIH (Dabrowski et al., 2003; Landry et al., 2007; Lu et al., 2014; Thomas and Rahman, 2009; Thomas et al., 2007; Wu et al., 2003). In Atlantic croaker (*Micropogonias undulatus*), in addition to reducing circulating levels of 20βS, chronic hypoxia reduces the protein levels of the membrane progestin receptor that mediates the non-genomic actions of this MIH in both oocytes and sperm (Thomas and Rahman, 2009). Interestingly, hypoxia impaired signaling from progestin and E2 membrane receptors, while enhancing signaling from membrane androgen receptors and increasing the proportion of atretic and apoptotic ovarian follicles in this species (Ondricek and Thomas, 2018). Finally, hypoxia can downregulate the gene expression of specific liver estrogen receptors and *vtg*, and reduce plasma Vtg levels (Lu et al., 2014; Thomas et al., 2007). Consequently, as observed in temperature-stressed fish, the widespread endocrine disruption associated with chronic hypoxia exposure in fish inhibits oogenesis and spermatogenesis, impairs gamete maturation and egg fertilization, delays spawning, and reduces hatching success and larval survival.

In contrast to the detrimental effects of high temperatures and hypoxic conditions, ocean acidification may only have a limited direct impact on fish reproduction. To date, experiments using partial pressures of CO_2 (pCO_2) up to the levels predicted to occur in 100 years ($\sim$1000 ppm CO_2) do not support wide-spread impacts on reproductive performance. Overall, the effects of elevated pCO_2 on reproductive output in fishes are equivocal. For example, while exposure to elevated CO_2 reduced the rate of paired spawning in ocellated wrasse (*Symphodus ocellatus*; Milazzo et al., 2016) and the number of egg clutches in spiny damselfish (*Acanthochromis polyacanthus*; Welch and Munday, 2016), it increased reproductive output in orange clownfish (*Amphiprion percula*; Welch and Munday, 2016) and three-spined stickleback (*Gasterosteus aculeatus*; Schade et al., 2014). In cinnamon anemonefish (*Amphiprion melanopus*) high pCO_2 increased breeding activity in one study (Miller et al., 2013) but had no effect in another (Miller et al., 2015). Similarly, high pCO_2 conditions stimulated reproductive output in two-spotted goby (*Gobiusculus flavescens*) in one study (Faria et al., 2018) and had no effect in another (Forsgren et al., 2013). The effects of elevated pCO_2 on gamete quality and survival also appears to be species- and condition-specific. In general, elevated

CO_2 levels do not affect sperm motility in most species of fish examined (Frommel et al., 2010; Inaba et al., 2003). Although high CO_2 can increase egg loss and embryonic abnormalities in some species (Forsgren et al., 2013; Miller et al., 2015) it has no effect in others (Munday et al., 2009, 2011). So far, very few studies have characterized the effects of elevated CO_2 on the HPG axis of fish. While short-term exposure (3–9 days) to elevated CO_2 produced elevated E2 levels in muscle tissue and increased liver E2 receptor gene expression in juvenile Atlantic cod (*Gadus morhua*; Preus-Olsen et al., 2014), long-term exposure (10 months) had no effect on plasma E2 levels in cinnamon anemonefish (Miller et al., 2015). Whether elevated CO_2 levels affect other components of the reproductive endocrine axis in fish remains to be ascertained.

In addition to their effects on reproduction, both elevated temperatures and hypoxic conditions can have marked effects on sex determination in fish. In dozens of species across various families, and in both domesticated and wild fish populations, elevated temperatures during the developmental period of primary sex determination results in the masculinization of genotypic females (for reviews, see Geffroy and Wedekind, 2020; Ospina-Álvarez and Piferrer, 2008). For example, in wild populations of juvenile southern flounder (*Paralichthys lethostigma*) living in North Carolina (USA), where a thermal range of $>5\,°C$ naturally occurs across nursery habitats, the cooler northern waters consistently produced a lower proportion of males (37–67%) relative to southern habitats with warmer temperatures (86–94%). Moreover, flounders reared in the laboratory under temperature conditions that mimicked those of natural habitats recapitulated sex ratio differences observed across the wild populations (Honeycutt et al., 2019). Both epigenetic and stress-related mechanisms have been shown to contribute to temperature-dependent sex determination (TSD) in fish. In several species, warm temperature-induced methylation of the *aromatase* gene promoter decreases E2 production and contributes to TSD (Ortega-Recalde et al., 2020). High temperature-induced increases in cortisol production have also been implicated as an important factor influencing TSD in fish (Geffroy and Wedekind, 2020). In brief, while activation of the glucocorticoid response element (GRE) by cortisol and its receptor in the promoter of *aromatase* and the *fsh receptor* results in a down-regulation of these feminizing genes (Hayashi et al., 2010; Yamaguchi et al., 2010), activation of the GRE in the promoter of the *dmrt1a* gene, a transcription factor involved in sex determination, promotes masculinization (Adolfi et al., 2019; Castañeda Cortés et al., 2019). Both field and laboratory studies have also shown that hypoxia exposure can lead to male-biased sex ratios in fish (Cheung et al., 2014; Robertson et al., 2014; Shang et al., 2006; Thomas and Rahman, 2012). Atlantic croakers collected from numerous hypoxic sites in the northern Gulf of Mexico not only had a male-biased sex ratio ($>60\%$), but 19% of the ovaries collected in the hypoxic region contained male germ cells (Thomas and Rahman, 2012). Mechanistically, hypoxia-induced masculinization is due to decreases in brain and ovarian aromatase activity and an ensuing disruption in the balance of sex steroids (Ivy et al., 2017; Shang et al., 2006; Thomas and Rahman, 2012). In the long-term, male-biased sex ratios may diminish effective population size, thus depleting genetic diversity and reducing the potential of fish populations to adapt to changing environments (Geffroy and Wedekind, 2020). Thus, understanding how climate change disrupts sex steroid production will be important for the future conservation of wild fish populations.

3.5 Endocrine-disrupting chemicals

Endocrine disrupting chemicals (EDCs) pose a pervasive and ubiquitous challenge to aquatic ecosystems. An EDC is defined as any exogenous compound that interferes with the normal activities of hormone systems, including hormone synthesis, cellular action, and bodily elimination (La Merrill et al., 2020). As such, EDCs comprise a large and growing list of chemicals that interfere with endocrine systems through direct (e.g., receptor agonists) or indirect (e.g., cell toxicity) mechanisms, and drive the specific or nonspecific dysfunction of that system (Fig. 4). Examples of EDCs include legacy pollutants (e.g., polychlorinated biphenyls, PCB; dichlorodiphenyltrichloroethane, DDT)

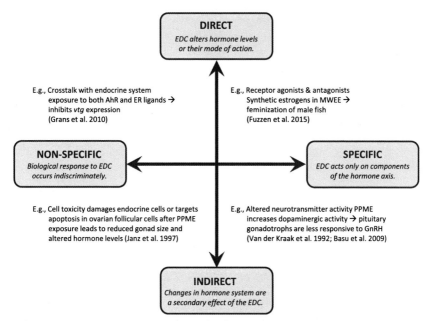

FIG. 4 Endocrine-disrupting chemicals (EDC) affect hormone systems in many ways and may not be mutually exclusive. The vertical axis describes the capacity of EDCs to interact primarily or secondarily with components of endocrine systems. The horizontal axis depicts the degree of specificity by which EDC alters the functionality of endocrine systems. Examples for each quadrant that are discussed in this chapter are provided.

as well as chemicals that are deposited daily into water systems from industrial (e.g., flame retardants, plasticizers), agricultural (e.g., pesticides, herbicides), and other human activities (e.g., hydrocarbons, pharmaceuticals, personal care products). Decades of concerted global efforts to research, assess, monitor, and regulate EDCs have led to established standardized testing protocols for environmental effects monitoring that help to identify and mitigate the risks of EDCs on humans and wildlife (Hecker and Hollert, 2011; Parrott et al., 2006). In fish, for example, the estrogenic potential of aquatic pollutants is defined using *in vitro* or *in vivo* Vtg assays, gonad histopathological assessment, and multigeneration reproductive screening (Hecker and Hollert, 2011).

Conservation concerns are inherent for fish populations exposed to EDCs given the central role of hormone systems in development, growth, life-history transitions, and reproduction. It is well beyond the scope of this chapter to offer a comprehensive review of all categories of EDCs and their various impacts and implications for fish conservation, and many insightful reviews are available for interested readers (Delbes et al., 2022; Goksøyr, 2006; Le Page et al., 2011; Mennigen et al., 2011; Segner, 2011; Söffker and Tyler, 2012; Waye and Trudeau, 2011). Therefore, given that estrogenic effects dominate the literature on EDCs in fish, we highlight three case studies of global relevance that demonstrate some similarities, differences, and challenges posed by EDCs in fish conservation.

3.5.1 Case study 1: Municipal wastewater effluent

Municipal wastewater effluent (MWWE) is a complex mixture of anthropogenic wastes and is the largest contributor of pharmaceuticals and personal care products into the aquatic environment (Aus der Beek et al., 2016). Many of these compounds are found in surface waters at concentrations capable of exerting biological effects, including endocrine disruption, in non-target organisms. Pharmaceuticals, for example, are designed to modify biological systems that are highly conserved across animal taxa, such as E2 receptors. Well-known examples are the synthetic E2 used in oral contraceptives (17α-ethinylestradiol, EE2; Servos et al., 2005), and the E2 receptor antagonist used for hormone therapy in breast cancer patients (tamoxifen; Orias et al., 2015). These and other sex steroid agonists and antagonists are held accountable for the occurrence of mixed gonadal tissue (i.e., both gamete types found in testes or ovaries) in fish populations living downstream of MWWE outfalls around the globe (Abdel-Moneim et al., 2015). Transcriptional responses that promote or inhibit sex-specific phenotypes underpin the development of the intersex condition. Male fathead minnow (*Pimephales promelas*) exposed to EE2, for example, were characterized by decreased gonadal expression of genes involved in testes development (e.g., *anti-Mullerian hormone*) and androgen biosynthesis (e.g., *cytochrome P450 17, cyp17*), and decreased hepatic expression of growth-promoting genes (e.g., *igf1*). At the same time, these same fish showed increased expression of genes involved in estrogen biosynthesis (e.g., *cyp19a*) and signaling (e.g., *er*, *vtg*) in the testes and liver, respectively (Filby et al., 2007). Over time, these transcriptional responses to environmental EE2 exposure may lead to a feminized phenotype that can include, in severe cases, the presence of full-grown follicles within testicular tissue and reduced fertilization success (Fuzzen et al., 2015; Harris et al., 2011; Jobling et al., 2002a,b).

Despite strong laboratory and field evidence supporting MWWE-induced feminization of male fish, population-level impacts remain challenging to confirm (Matthiessen et al., 2018; Mills and Chichester, 2005). A 3-year dosing study in the Environmental Lakes Area of Ontario, Canada confirmed that chronic exposure to low, environmentally relevant concentrations of EE2 caused feminization of male fish (Vtg production, intersex) and reproductive impairment (adult-skewed age distribution) in a wild population of fathead minnows (Kidd et al., 2007). The population remained collapsed for 2 years after the EE2 dosing ceased in the lake, with only a handful of fathead minnows collected during sampling efforts (Kidd et al., 2007); however, individual and population effects in the lake were species-specific, suggesting that fish with longer generation times, like white sucker (*Catostomus commersonii*) and lake trout (*Salvelinus namaycush*), may be more buffered against chronic EE2 exposure (Kidd et al., 2014; Palace et al., 2009). Moreover, feminization of male fathead minnows was no longer observed for 3 years once the EE2 dosing ceased, and age-class distribution of the population recovered by year 4 (Blanchfield et al., 2015).

3.5.2 Case study 2: Pulp and paper mill effluents

The manufacturing of paper products from raw wood has received international attention for the environmental impacts of mill effluents, including endocrine disruption on fish reproduction. Fish living in the aquatic receiving environments of pulp and paper mills are often characterized by reduced circulating sex hormones (McMaster et al., 1991; Van Der Kraak et al., 1992) and smaller gonads (McMaster et al., 1991; Munkittrick et al., 1991; Sandstrom, 1994), but effects can be species-specific (Chiang et al., 2011; Gibbons et al., 1998; Karels et al., 2001). Importantly, these biological effects are coincident with changes in population sex ratios (Larsson et al., 2000), lower fecundity (McMaster et al., 1991; Munkittrick et al., 1991) and population declines (Sandstrom, 1994). Multiple mechanisms of endocrine disruption are likely to underlie these effects. For example, white sucker exposed to pulp and paper mill effluents (PPME) from a Canadian mill had lower circulating LH levels and a blunted response to GnRH injection (Van Der Kraak et al., 1992). In turn, this impaired pituitary function may be linked to the capacity of some PPME to disrupt neurotransmitter signaling, including dopamine, as shown using *in vitro* competitive binding assays in goldfish (*Carassius auratus*) brain homogenates (Basu et al., 2009). Consistent with the pleiotropic effects of sex steroids and the complex regulation of the HPG axis, changes in the expression of reproduction-related genes (Costigan et al., 2012), alterations in steroid synthesis and metabolism (Gibbons et al., 1998; Leusch and MacLatchy, 2003; McMaster et al., 1995; Van Der Kraak et al., 1992) and histopathological changes in gonad tissues (Castro et al., 2018; Janz et al., 1997) have all been observed in PPME-exposed fish.

PPME are complex mixtures of chemicals containing many bioactive substances such as wood-derived compounds (e.g., phytosterols, lignin) and organochlorines produced by the bleaching process (e.g., polychlorinated dibenzodioxins and dibenzofurans, PCDD/F) (Hewitt et al., 2007; Lindholm-Lehto et al., 2015; Singh and Chandra, 2019). Given the complexity and mill-to-mill variability in effluent composition (e.g., depending on wood source, pulping process, bleaching agent, effluent treatment infrastructure), efforts to define the causative agents of PPME toxicity and establish regulatory guidelines around them have been challenging (see reviews by Hewitt et al., 2007; McMaster et al., 2006). Moreover, the presence of multiple and as-yet unidentified EDCs in PPME are often reported (Larsson et al., 2006; Martel et al., 2017; Orrego et al., 2017). Nevertheless, altered reproductive endpoints in fish exposed to PPME around the globe highlight the universality of the problem and emphasize the utility of using endocrine metrics to quantify, monitor, and regulate the pulp and paper industry (Barra et al., 2021; Martel et al., 2017; McMaster et al., 2006; Parrott et al., 2006; Ussery et al., 2021).

3.5.3 Case study 3: The aryl hydrocarbon receptor (AhR) and endocrine disruption

Dioxin (2,3,7,8-tetrachlorodibenzo-para-dioxin, TCDD) and dioxin-like chemicals (e.g., PCB; PCDD/F; polycyclic aromatic hydrocarbons, PAH) are well-known and ubiquitous environmental pollutants. These contaminants are found at especially high concentrations in the soils and sediments of industrialized areas and pose a constant challenge to human and animal health, including aquatic organisms, due to continuous inputs (e.g., industrial discharge, atmospheric deposition, oil spills), bioaccumulation and trophic biomagnification, and the staggeringly slow degradation of certain legacy pollutants (Weber et al., 2008). Congeners of dioxin-like chemicals with planar conformations are agonists of the aryl hydrocarbon receptor (AhR), a ligand-activated transcription factor that regulates the expression of thousands of gene targets (Denison and Nagy, 2003). Dysregulation of gene expression mediated by AhR can compromise normal cell function and manifest in a range of toxic responses, including endocrine disruption (Anulacion et al., 2013). For example, studies using fish cell cultures and model species have shown that individual PAH or mixtures containing PAH (e.g., crude oil) alter sex hormone synthesis (Arukwe et al., 2008; Lee et al., 2017; Liu et al., 2018; Monteiro et al., 2000), decrease circulating hormone levels (Arukwe et al., 2008; Booc et al., 2014; Kennedy and Smyth, 2015; Tintos et al., 2007), and reduce fertilization and hatching success (Booc et al., 2014). Consistent with these indicators of reproductive impairment, *in vitro* studies using fish hepatocytes exposed in combination to ER and AhR agonists (e.g., Gräns et al., 2010; Mortensen and Arukwe, 2007; Yan et al., 2012; but see Mortensen and Arukwe, 2008) support the conclusion of inhibitory crosstalk between the E2 receptor and AhR signaling pathways that disrupts target gene expression, as described in mammalian models (Matthews and Gustafsson, 2006; Fig. 5).

The killifish (*Fundulus heteroclitus*) has been well studied as a model for physiological and multigenerational impacts of exposure to dioxin-like chemicals. Killifish are a small estuarine fish found abundantly along the Atlantic coast of North America, including in some of the most contaminated estuaries on the continent. Plasma Vtg in female killifish and gonadal somatic index (GSI) in male and female killifish were negatively correlated with sediment PAH concentrations throughout Chesapeake Bay, a large estuary and designated Superfund site in the District of Columbia, USA (Pait and Nelson, 2009). Similarly, male and female killifish from the heavily industrialized Newark Bay in New Jersey showed reduced gonadal maturation and GSI relative to fish sampled at a clean reference site, and these results were associated with increased AhR activation in the liver and elevated PAH metabolites in the bile (Bugel et al., 2010). In the New Bedford harbor of Massachusetts, one of the highest ranked PCB-contaminated sites in the

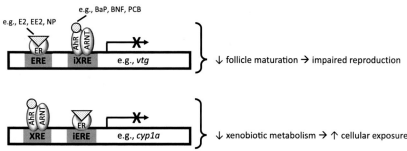

FIG. 5 A proposed mechanism for reciprocal inhibition of estrogen receptor (ER) and aryl hydrocarbon receptor (AhR) gene targets in fish. Both ER and AhR are ligand-activated transcription factors. Activation of ER by endogenous (e.g., 17β-estradiol, E2) or exogenous (e.g., 17α-ethinylestradiol, EE2; nonylphenol, NP) ligands initiates transcription of genes containing estrogen response elements (ERE) in the gene promoter region, such as *vitellogenin (vtg)*. Activation of AhR by exogenous ligands (e.g., benzo[a]pyrene, BaP; β-naphthoflavone, BNF; dioxin-like pentachlorobiphenyls, PCB) promotes the formation of a protein complex with the AhR-nuclear translocator (ARNT) and subsequent transcription of genes containing xenobiotic response elements (XRE), such as *cytochrome P450 type 1a (cyp1a)*, to facilitate xenobiotic metabolism. Simultaneous exposure to ER- and AhR-agonists inhibits the downstream activation of target genes (see text for references), suggesting the involvement of inhibitory response elements (iERE, iXRE). The outcome of such reciprocal inhibition is a decrease in the typical response to either receptor alone, leading to biological consequences (e.g., impaired reproduction; increased bio-load and cellular exposure to toxicants).

USA, resident female killifish had lower liver *vtg* mRNA levels, decreased circulating sex hormone and lower GSI during the reproductive season compared to females from an uncontaminated site (Greytak et al., 2005). Remarkably, despite clear evidence of endocrine disruption and the assumption of reproductive impairment, killifish populations continue to persist in these polluted areas. Their success has been attributed to genomic adaptations within local populations that have resulted in increased tolerance to industrial pollutants acting via the AhR (Greytak et al., 2005; Whitehead et al., 2012). In contrast to the prolific killifish, and of relevance to conservation, the potential for evolutionary adaptation to dioxin-like chemicals or other pollutants is less likely in fish species with slower generation times, smaller population sizes, and limited genetic variation (Whitehead et al., 2017).

3.6 Management of invasive species

Pheromones may have a variety of applications for the management of invasive fish species. Defined as chemicals that mediate behavioral interactions between individuals of the same species, pheromones are broadly used by fish as anti-predator cues, social cues, and reproductive cues (Sorensen and Stacey, 2004; Stacey and Sorensen, 2011). Pheromones in fish are produced and released into the environment by various cell types and glands. In return, they are detected by highly sensitive receptors in the olfactory epithelium of the nares that can activate specific behavioral and physiological responses (Hara, 2011; Laberge and Hara, 2001). For example, derivatives of bile acids, sex steroids and prostaglandins in fish have been implicated in predator avoidance, kin and gender recognition, migration to and aggregation on spawning sites, and the coordination of reproductive processes (Kamio et al., 2022; Stacey and Sorensen, 2002). In general, since they are naturally occurring in the environment, do not persist and can be used at low concentrations because of their high potency and specificity, pheromones are regarded as a relatively safe technique for the control of invasive species (Fredricks et al., 2021). As a result, pheromones have been proposed to facilitate trapping, to disrupt and reduce reproductive success, to promote the success of sterilized fish, to disrupt migration and to repel non-indigenous species from ideal spawning sites or from entering specific waterways (Corkum and Belanger, 2007; Kamio et al., 2022; Sorensen and Johnson, 2016; Sorensen and Stacey, 2004).

Sea lamprey (*Petromyzon marinus*) rely on several complex pheromone mixtures to complete their life cycle, and the use of pheromones may therefore be a useful tool to mitigate the effects of this invasive parasitic species on the native fish of the Laurentian Great Lakes. In brief, larvae residing in streambeds release at least four different migratory pheromones that attract adults upstream during the spawning migration (Fine and Sorensen, 2008; Fissette et al., 2021; Li et al., 2018). Spermiating males in return release at least 12 different sex pheromones that stimulate sexual maturation in both sexes and elicit mate search and spawning behavior in mature females (Fissette et al., 2021; Li et al., 2017). Although still a work in progress, significant advances have been made toward the incorporation of pheromones into the management of invasive sea lamprey which currently relies primarily on lampricides to kill larvae and barriers to block adults from spawning beds (Fissette et al., 2021; Siefkes, 2017). For example, both synthetic and mixtures of migratory pheromones can elicit electro-olfactogram (EOG) responses and partially mediate stream-finding behaviors in lakes near the mouths of streams (Burns et al., 2011; Meckley et al., 2014). Using a combination of migratory and sex pheromones, both males and females were captured using pheromone-baited traps (Wagner et al., 2006). Specifically, under management scenarios, Johnson et al. (2013) demonstrated that barrier-integrated traps baited with the male sex pheromone, 3-keto petromyzonol sulfate (3kPZS), increased trapping efficiency by 10%. Traps located en route to spawning grounds and baited with a mixture containing all the pheromones from sexually mature males increased trapping efficiency by an additional 10% (Johnson et al., 2015). While ongoing studies are identifying optimal application rates and the specific biological and environmental factors that maximize trapping efficiency with 3kPZS, future research is needed to develop more effective and economical pheromone mixtures (Fissette et al., 2021; Johnson et al., 2020).

Pheromones may also be an effective tool to control the populations of several invasive teleost species such as the round goby (*Neogobius melanostomous*; Corkum et al., 2004). Mature male and female round goby release various pheromones that are detected by conspecifics (Gammon et al., 2005). Screening over 100 steroids and prostaglandins using EOG, Murphy et al. (2001) identified 19 steroids that elicited responses in mature fish. Several of these steroids are synthesized by the testes of sexually mature males (Arbuckle et al., 2005) and have life-stage specific effects on female attraction (Farwell et al., 2017). Moreover, although these pheromones elicit EOG responses in round goby, they do not evoke olfactory sensory responses in various non-target species that share the same habitat, suggesting their specificity for pheromone trapping (Ochs et al., 2013). Reproductive pheromones may also serve for the management of various invasive cyprinid species. Among teleosts, the physiological and behavioral roles of reproductive pheromones is perhaps best understood in goldfish (*C. auratus*) and common carp (*Cyprinus carpio*; Stacey and Sorensen, 2002). For instance, ovulated common carp are known to release a mixture of prostaglandin F2α (PGF2α) and metabolites that attract males and mediate spawning interactions (Lim and Sorensen, 2011). Measurement of PGF2α in natural waters has also been used, together with eDNA, to confirm the presence, infer gender, and assess the relative abundance of common carp induced to aggregate at specific locations using food (Ghosal et al., 2018). In the field, female common carp implanted with PGF2α capsules release a biologically relevant mixture of pheromones that attract mature males (Lim and Sorensen, 2012). A similar strategy has now been devised to specifically attract silver carp (*Hypophthalmichthys molitrix*; Sorensen et al., 2019), a highly invasive species in the Mississippi River. Finally, pheromone trapping may be a useful technique for controlling populations of invasive largemouth bass (*Micropterus salmoides*). While the specific sex pheromone involved remains to be identified, trapping experiments have shown that the bile from sexually mature largemouth bass males can specifically attract and increase the capture rate of mature females (Fujimoto et al., 2020).

4 Future applications of endocrine systems in conservation physiology

Despite a recognition that hormones play a key role in maintaining homeostasis and regulating reproductive functions, the intrusive and sometimes invasive nature of the traditional methods used to monitor basal circulating levels of hormones in captive and wild animals limit their applicability to conservation physiology. As such, for several decades now, there has been considerable interest in developing non-invasive techniques to monitor hormones (e.g., Monfort, 2002; Scott and Ellis, 2007). The publication of methods to measure corticosteroids in human hair (Bévalot et al., 2000) further spurred the development of innovative approaches to measure steroids in a variety of inert matrices (Aerts et al., 2015; Ellis et al., 2013; Meyer and Novak, 2012; Romero and Fairhurst, 2016). Although the application of non-invasive methods to study fish endocrinology is in its infancy, the validation of these techniques and the development of standard protocols will facilitate the study of fish reproduction, welfare, and conservation. Similarly, since the regulation of most physiological processes involves multiple hormones, the parallel development of techniques to perform hormonal profiling in traditional and novel matrices will provide greater mechanistic insight into the functions of endocrine systems and how these are disrupted by environmental and anthropogenic stressors. Finally, advances in the availability of genomic information for non-model fish species and in genomic technologies are giving researchers new tools to integrate the application of endocrine systems in conservation physiology. Here we highlight the multiomics approach used in ecotoxicogenomics to predict the adverse effects of EDCs on wild fish, and how the use of omics and hormonal measurements can be integrated to provide insight into the emerging field of endocrine-microbiome interactions and to promote its application to conservation measures.

4.1 Non-Invasive monitoring of steroids

Traditionally, studies that measure hormones in fish to gain insight into physiological processes or behaviors quantify hormones in the plasma component of blood samples. Among the different approaches that can be used to sample blood in fish, the most common technique is caudal puncture (Houston, 1990; Jeffrey et al., 2022). While the invasive nature of blood sampling through the caudal vasculature can be minimized when best practices are followed (Lawrence et al., 2020), several considerations may constrain the use of blood samples for the study of endocrine systems in the field and its application to conservation physiology. For example, since capture and handling stressors are known to affect circulating hormone levels, blood samples must be withdrawn within a few minutes of disturbing the animal (Clark et al., 2011; Gamperl et al., 1994; Pankhurst, 2011; Wendelaar Bonga, 1997)—a challenging prerequisite for a variety of field settings. Blood samples also provide a snapshot of circulating hormone levels at the time of sampling and may therefore have limited value as an integrative long-term measure of stress and welfare. Moreover, blood sampling is impractical for small-bodied fish species and ill-advised for at-risk fish populations. Therefore, taking advantage of the fact that steroids are generally stable chemical substances, lipid-soluble, and excreted via the kidney and gut, endocrinologists have developed several non-invasive techniques to quantify glucocorticoids and sex steroids in alternative matrices. For instance, excreted steroid hormones (i.e., saliva, urine, feces) as well as those deposited in keratinized tissues (i.e., hair, feathers) are now routinely used to quantify steroids and gain insight into the stress and reproductive physiology of various amphibian, avian, and mammalian species (Behringer and Deschner, 2017; Bortolotti et al., 2008; Gormally and Romero, 2020; Heimbürge et al., 2019; Palme, 2019; Romero and Fairhurst, 2016). Similarly, although their development and application lags for fish, several techniques are now available for the quantification of steroids in novel, less invasive matrices. In general, these techniques vary in terms of invasiveness, sample collection practicality and field application. The quantification of steroids in novel matrices may also permit hormone monitoring on different time scales (Fig. 6), thereby broadening the context for integrating this information into conservation initiatives.

The least invasive technique for measuring steroids in fish involves the monitoring of tank water. In fish, free steroids passively diffuse from the bloodstream into the water across the gills (Ellis et al., 2005; Vermeirssen and Scott, 1996). As a result, the rate of steroid release into the surrounding holding water is generally proportional to their concentration in the plasma across multiple steroid classes. For example, the release rate of E2, T, 11-KT, androstenedione, 17,20βP, and cortisol have been shown to be proportional to plasma or whole-body measures (Ellis et al., 2004; Friesen et al., 2012b; Graham et al., 2018; Sebire et al., 2007; Stacey et al., 1989; White et al., 2017). Although the quantification of steroids in fish-holding water requires more planning and validation than typical plasma sampling, this technique has the distinct advantages of eliminating the need to handle and anesthetize fish to collect samples, allowing the simultaneous measurement of

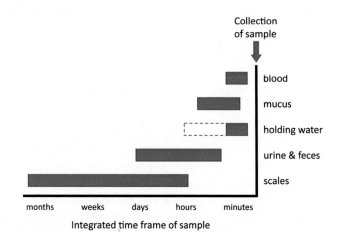

FIG. 6 Alternative matrices may permit cortisol monitoring on different time scales in fish. While blood samples provide a snapshot of circulating cortisol levels at the time of sampling, the general delay in the appearance of peak cortisol in skin mucus relative to peak plasma levels suggest that this matrix may integrate cortisol levels on a slightly longer timeframe than blood. The timeframe integrated by quantifying cortisol in holding water depends on the technique used. Since the rate of cortisol release into the surrounding holding water is generally proportional to its plasma concentration, holding water cortisol levels will integrate over a similar timeframe as blood when using a flow-through holding tank. In contrast, holding water cortisol levels can integrate over a longer time frame when using a static holding tank. Although the factors that affect the lag-time between a stressor and subsequent elevations in free and conjugated urine and fecal corticosteroids in fish are poorly understood, urine and feces may integrate corticosteroid levels on a timeframe of several hours to a few days. Finally, though scales may not show short-lived fluctuations in plasma cortisol, scales may permit cortisol monitoring on a timeframe of hours to months and therefore could serve as an indicator of chronic stress in fish.

multiple steroid hormones from a single sample, and permitting temporal profiling of hormonal responses (Blackwell and Ankley, 2021; Graham et al., 2018; Kidd et al., 2010; Scott and Ellis, 2007; Scott et al., 2008). To date, the quantification of steroids in fish-holding water has been used, among other applications, to study stress and welfare (Ellis et al., 2004, 2007; Fanouraki et al., 2008; Friesen et al., 2012a,b; Leong et al., 2009; White et al., 2017), identify EDCs (Blackwell and Ankley, 2021), assess reproductive stage (Friesen et al., 2012a,b; Graham et al., 2018; Sebire et al., 2007), and detect pheromone communication (Scott and Sorensen, 1994; Stacey et al., 1989). Overall, while the quantification of steroids in holding water is primarily limited to controlled laboratory settings (Ellis et al., 2013), its broad application for the non-invasive study of steroids in various physiological systems and behavioral settings highlight its utility to address important issues in conservation biology.

Whereas free steroids in holding water are primarily derived from the gills, the main excretory routes of conjugated sex steroids, i.e., sulfated and glucoronidated steroids, are the feces and urine (Vermeirssen and Scott, 1996).

Similarly, the main excretory route for clearance of cortisol metabolites and conjugates is the hepato-biliary-fecal route (Mommsen et al., 1999; Scott et al., 2014). Therefore, the quantification of fecal corticosteroid levels may offer an additional method to monitor stress in fish (Cao et al., 2017; Lupica and Turner, 2009; Simontacchi et al., 2008; Turner et al., 2003). Since evacuation of the gastrointestinal (GI) track in fish is protracted, especially in cold-water species (e.g., GI tract evacuation in Atlantic salmon takes up to 48 h at ~13 °C; Aas et al., 2017), fecal corticosteroid levels are unlikely to be influenced by sampling procedures. However, while minimally invasive, the collection of feces from fish tanks by siphon, by manually stripping feces from the abdomen, or by following individuals in the field using scuba may be intrusive and limit the utility of this technique for the collection of repeated samples across time (Ellis et al., 2013). Moreover, to date, it remains unclear how factors known to affect GI evacuation such as water temperature, meal size, feed type, satiety, activity, or metabolic rate affect the lag-time between a stressor and the subsequent elevation in free and conjugated fecal corticosteroids and their metabolites. Similarly, although direct sampling of urine by stripping to quantify sex steroids and their conjugates (Oliveira et al., 1996, 2001) is a minimally invasive technique, the intrusive nature of the procedure may limit its applicability for repeated sampling, and much work remains to determine how such measurements relate to circulating levels (Ellis et al., 2013). Yet, the quantification of conjugated sex steroids of urinary provenance in fish-holding water can provide unique insights into the physiological and behavioral strategies used by fish to communicate social and reproductive cues (Almeida et al., 2005; Hirschenhauser et al., 2008; Keller-Costa et al., 2014; Scott and Sorensen, 1994; Sorensen et al., 2005; Stacey et al., 1989). Therefore, the development of novel sensitive methods for the simultaneous non-invasive quantification of multiple free and conjugated steroids in fish-holding water (e.g., Blackwell and Ankley, 2021) should provide insight into the complex signals used by fish for chemical communication.

Epidermal mucus can also be used to monitor steroids in fish. Mucus collected from the skin contains both sex steroids and cortisol, and the levels are correlated with circulated plasma concentrations (Bertotto et al., 2010; Carbajal et al., 2019a; Schultz et al., 2005; Simontacchi et al., 2008). Although the collection of skin mucus samples requires fish handling and sedation, the technique is minimally invasive, and can readily be used in both laboratory and field settings (Carbajal et al., 2019b; Ellis et al., 2013). As such, the quantification of mucus 11-KT levels has served to determine fish sex, to assess annual reproductive cycles, and to identify whether environmental or hormonal manipulations can stimulate the HPG axis (Barkowski and Haukenes, 2014; Schultz et al., 2007). Mucus cortisol levels have also been used as a physiological indicator of stress in response to various acute disturbances (Bertotto et al., 2010; De Mercado et al., 2018; Guardiola et al., 2016; Khansari et al., 2018; Kroska et al., 2021; Simontacchi et al., 2008). Whether mucus and plasma cortisol levels provide an opportunity to examine the effects of acute stressors at different time scales remains equivocal. While stressors elicited a similar time course between plasma and mucus cortisol levels in some studies (Carbajal et al., 2019a; Khansari et al., 2018), the appearance of peak cortisol in skin mucous was significantly delayed relative to peak plasma cortisol levels in others (Guardiola et al., 2016; Kroska et al., 2021). In general, several questions remain regarding the potential effects of stressors on mucus production as well as the route and factors that affect the movement of steroids from blood to mucus (Ellis et al., 2013; Pérez-Sánchez et al., 2017; Shephard, 1994).

Recent studies have suggested that fish scales may provide an integrative measure of cortisol production. While acute stressors do not appear to affect scale cortisol content, high and sustained circulating cortisol levels do, and the rates of cortisol accumulation and clearance are much slower in scales than in plasma (Aerts et al., 2015; Carbajal et al., 2019a; Hanke et al., 2019; Laberge et al., 2019). To date, scale cortisol content has been used to assess the effects of unpredictable chronic stress (Laberge et al., 2019), chronic thermal stress (Hanke et al., 2019), sub-optimal aquaculture conditions and impaired welfare (Weirup et al., 2021), and the long-term cortisol dynamics between social group members in a field setting (Culbert et al., 2021). Importantly, despite its utility as an indicator of chronic stress, fish scale cortisol content is unlikely to allow a retrospective historical assessment of stress status as far back in time as feather or hair corticosteroid content (Romero and Fairhurst, 2016; Russell et al., 2012). While specific studies are needed to determine the turnover of steroids in fish scales, the dynamic exchange of cortisol between the circulation and the scale matrix suggest that scale cortisol only allows assessment of the most recent integrated cortisol status of fish (Laberge et al., 2019). The quantity of scale material needed to provide a single determination may also impose limitations on the use of this approach as a non-invasive technique for measuring steroids in fish. Although only a few scales may be needed to determine scale cortisol content in large fish, in smaller fish, most, and in some cases, all scales may be needed to reach current assay thresholds (e.g., Aerts et al., 2015; Culbert et al., 2021; Laberge et al., 2019). Finally, beyond the need to handle and sedate fish to remove scales, sampling procedures should also consider the spatial heterogeneity of scale cortisol content (Laberge et al., 2019).

4.2 Hormonal profiling

Physiological processes are regulated by the actions of multiple endocrine signals, therefore quantifying these hormonal networks (i.e., hormonal profiling) may offer valuable mechanistic insight into animal physiology. While used for the analysis of select hormones since the 1960s, recent technical advances in mass spectrometry have facilitated the simultaneous quantification of multiple classes of steroids and their metabolites from a single experimental sample (Boggs et al., 2016; Kaabia et al., 2018; Olesti et al., 2021; Son et al., 2020). This technique has been successfully applied to investigations of steroid-related endocrine diseases (Schiffer et al., 2019), and recent initiatives emphasize the suitability of steroid profiling in the field of conservation physiology for reproductive and stress assessment of endangered mammalian species (e.g., Azevedo et al., 2020; Galligan et al., 2018, 2020; Graham et al., 2021; Kumar and Govindhaswamy, 2019; Legacki et al., 2020). In fish, steroid profiling has been used to identify EDCs and their mode of action (Flores-Valverde et al., 2010; Labadie and Budzinski, 2006a,b), to identify novel biomarkers of aging (Dabrowski et al., 2020) and female maturation status (Lokman et al., 1998; Zhou et al., 2017), and to understand the mechanism of action of novel drugs (Kim et al., 2021); however, it has yet to be applied to the conservation of threatened species. Mass spectrometric techniques have also been developed to perform an integrated analysis of peptide hormones and neuropeptides, i.e., for peptidomic analysis (Baggerman et al., 2004; Romanova and Sweedler, 2015). To date, however, these techniques have largely been used to characterize the complement of neuropeptides and peptide hormones of fish brains (Hu et al., 2016; Van Camp et al., 2017) and have yet to find applications in the field of conservation endocrinology. While the simultaneous measurement of multiple classes of hormones is not without significant technical challenges, the study of hormonal profile variation over time in a variety of biological matrices will advance the study of hormone function, accelerate the discovery of novel stress and reproductive biomarkers, and facilitate the incorporation of endocrine measures in conservation management.

4.3 Multisystem integration of endocrinology in conservation physiology

4.3.1 Ecotoxicogenomics

Fish are established sentinels of aquatic pollution and are routinely used to screen for endocrine disrupting properties of new chemicals. The advent of genomic technologies has facilitated a deep and comprehensive understanding of tissue-specific biological responses to chemical exposure in model fish species exposed under controlled laboratory conditions, resulting in detailed knowledge on modes of action and adverse outcomes (Caballero-Gallardo et al., 2016; Hook, 2010; Martyniuk et al., 2020; Tyler et al., 2008). In the emerging field of ecotoxicogenomics this mechanistic knowledge is placed into an ecological framework to facilitate interpretation of complex gene-environment interactions for the purposes of risk assessment, site remediation, and ecosystem management. For example, 'omics tools can be used to screen for and predict the impact of contaminant exposure on local fish populations, and to understand species distributions in and around contaminated sites based on the genetic underpinnings of pollutant tolerance (Reid and Whitehead, 2016).

The rapid adoption of ecotoxicogenomics as an interdisciplinary approach to tackle environmental issues is fueled by the increased availability of whole-sequenced genomes for non-model species, technological advances in mass-spectrometry-based proteomics and metabolomics, and the development of bioinformatic tools to facilitate analysis and interpretation of massive datasets. The overall aim is to capitalize on insights from laboratory studies to help define the relationships between water quality and physiological status of resident fish. For example, molecular signatures consistent with HPG axis disruption were detected in the ovarian and hepatic transcriptomes of fathead minnow caged downstream of three WWTP in Minnesota, USA (Berninger et al., 2014; Martinović-Weigelt et al., 2014). Functional analyses of the enriched gene sets indicated dysregulation of E2 signaling pathways, cholesterol biosynthesis and steroid metabolism in the liver (Martinović-Weigelt et al., 2014), and the regulation of oocyte meiosis and gonadotropin signaling pathways in the ovary (Berninger et al., 2014). Importantly, these transcriptional responses were evident after only 4 days of exposure, and varied with site-specific differences in watershed dynamics, land use patterns, and urban population densities that generated unique contaminant profiles and exposure scenarios at each site (Berninger et al., 2014; Martinović-Weigelt et al., 2014). Thus, a measured genomic response can be used as a predictive early indicator of exposure to EDCs where known adverse outcomes of reproductive impairment are expected to impact population recruitment.

Ecotoxicogenomics has a bright future in assessing endocrine disruption for conservation efforts. The holistic nature of 'omics technologies means that multiple endocrine systems can be monitored simultaneously by identifying enriched gene sets or functional pathways in target tissues that are hormonally regulated. This is especially advantageous in the context of complex natural habitats, where wild fish are likely to experience multiple stressors and chemical mixtures of known and unknown endocrine disrupting properties (Hook, 2010). In contrast, targeted approaches that quantify biomarkers of specific endocrine disruption (e.g., circulating vitellogenin or hepatic *vtg* expression in male fish) requires a priori knowledge of causative agents, and a narrow focus on predicted effects may bias interpretation of water quality or even miss underlying issues that are yet to be identified (Hook et al., 2014). An additional strength of ecotoxicogenomics is that it enables early detection of endocrine disruption in advance of adverse organismal outcomes, which may expedite habitat remediation and initiate conservation strategies before critical impacts to populations occur.

4.3.2 Endocrine-microbiome interactions

All animals carry with them a diverse and plentiful community of microorganisms, particularly in the gastrointestinal tract. The gut microbiome has long been known to participate in chemical digestion and thus facilitate nutrient acquisition, but increasingly it is being linked to a seemingly infinite array of host physiological and behavioral processes. A role for the gut microbiome in health and disease is now widely accepted and is mediated in part by cross-talk with host endocrine systems (Cusick et al., 2021; de Weerth, 2017; Neuman et al., 2015). For example, gut microbes can synthesize bioactive catecholamines *de novo* (Asano et al., 2012), and microbial enzymes (i.e., glucoronidases and sulfatases) can reactivate conjugated hormone metabolites entering the gut lumen from the bile to promote reabsorption over excretion (Ervin et al., 2019, 2020). Comparative physiologists are only beginning to scratch the surface on the mechanisms and outcomes of host-microbiome interactions, but it is clear that many current ecosystem threats can influence the composition of the gut microbiome (Sullam et al., 2012), including increased temperature (Kohl and Yahn, 2016), habitat degradation (San Juan et al., 2020; Watson et al., 2019), and pollutants (Adamovsky et al., 2021; Chen et al., 2018), as well as stress and glucocorticoids in general (Noguera et al., 2018; Stothart et al., 2016).

In fish, studies of gut microbiomes are largely centered on aquaculture applications, but lessons learned here may be transferable to captive breeding and conservation programs. Antibiotic treatments on fish farms are used to prevent and control disease outbreaks and are linked to reduced microbial diversity in the gut and increased representation of opportunistic pathogenic strains (Navarrete et al., 2008). As an alternative approach, supplementation of diets with prebiotics alone, or in combination with probiotics, can improve fish growth and reduce disease outbreaks by fortifying the abundance and diversity of beneficial gut microbes (Clements et al., 2014; Egerton et al., 2018). There is some evidence to suggest an endocrine link to these benefits. For example, great sturgeon (*Huso huso*) fed prebiotic supplements had elevated plasma T4 and TSH along with greater feed conversion efficiency (Adel et al., 2016), suggesting a link between endocrine-controlled growth and the gut microbiome. Complex interactions between stress, immune responses, and the gut microbiome are also likely to contribute to the health of aquaculture species and the success of captive breeding programs. For example, a study in Atlantic salmon (*S. salar*) exposed to a mild, repeated handling stressor showed a strong association between fecal cortisol and the diversity and structure of the gut microbiome, including reduced representation of beneficial strains and increased levels of pathogenic strains in stressed fish relative to controls (Uren Webster et al., 2020). Building fundamental understanding of microbiome-host interactions in fish health, particularly with respect to the endocrine systems, could have considerable impact on conservation efforts. Importantly, the gut microbiome can be qualitatively and quantitatively evaluated via non-lethal fecal sampling and is therefore amenable to long-term monitoring and intervention programs.

5 Conclusions

Fish endocrinologists are making important contributions to the field of conservation physiology, bringing a wealth of laboratory and fundamental research to applied and emerging challenges. In this chapter, we offer many examples that demonstrate how quantification and manipulation of hormone systems can provide insight into the well-being, growth, and reproductive status of fish, ultimately advancing conservation goals by:

1. Guiding the development of novel conservation hatchery strategies that will minimize the impact of cultured fish on wild populations and help maintain the genetic diversity of wild fish;
2. Revealing how exposure to EDCs and environmental stressors can delay or prevent normal development, result in a mismatch between phenotype and environment, and affect the survival and recruitment of individuals in degrading ecosystems;
3. Identifying habitats and environmental factors that influence growth of wild fish to support life history transitions;
4. Enabling the assisted reproduction of threatened species;
5. Mitigating the impacts of climate change and EDCs on reproduction, sex determination, and population viabilities by establishing mechanistic links between endocrine bioindicators, fish reproduction, and larval recruitment;
6. Facilitating the quantification and removal of invasive species while simultaneously minimizing impacts of such procedures on native species.

Looking ahead, the use of alternative matrices (e.g., holding water, feces, mucus, scales) to simultaneously quantify multiple free and conjugated steroids holds great promise as a non-invasive approach for monitoring the stress and reproductive physiology of captive and wild fish on different time scales. Indeed, the conserved structure of steroid hormones across fish species, their tendency to accumulate in inert matrices, and the commercial availability of tools for steroid quantification favor an increased effort to integrate steroid profiling into conservation practices. Given the pervasive effects of abiotic factors and EDCs on the synthesis of steroid hormones, and the broad physiological consequences of impaired steroidogenesis on fish, we suggest that such efforts could offer valuable insight into fish recruitment mechanisms and help to monitor the impact of anthropogenic and environmental stressors on fish. However, broadscale adoption of such initiatives will first require the development of more sensitive, high-throughput, and cost-effective hormonal analyses, followed by the mobilization of these tools out of researcher's hands and into those of resource managers.

Fish endocrinology has already helped contribute to policy change by providing standardized quantitative endpoints to set water quality standards for certain EDCs around the globe; however, focused attention on a few key limitations and knowledge gaps are needed for fish endocrinology to become ingrained in conservation and restoration initiatives, and to bring about sustained impact on fish conservation. First, fish are the most speciose group of vertebrates, yet most study is carried out in single species or defined groups of fish (e.g., salmonids). Even within the strong evolutionary framework of vertebrate endocrine systems, important species differences emerge that could skew management decisions without efforts to build a more taxonomically diverse foundation in fish endocrinology. Second, hormone systems function to integrate within and across physiological systems. As such, linking endocrine signals with other biological endpoints, including molecular, biochemical, and physiological biomarkers, will help build a better understanding of chronic sublethal detrimental health effects. Here, approaches using functional genomics and bioinformatics will be essential. Finally, researchers must look beyond biomarker identification to the validation and deployment of feasible tools for specific conservation applications. This will help resource managers use endocrine tools for evidence-based decisions within a regulatory framework.

The greatest environmental challenges of the 21st century include biodiversity loss, the climate crisis, and pollution. Conservation endocrinology is gaining momentum as a dedicated sub-discipline and stands to make meaningful contributions to the conservation problems inherent in these environmental challenges by providing resource managers with tools to assess the impact of natural and anthropogenic factors on the stress, growth, and reproductive status of fish, as well as the means to develop breeding programs for vulnerable populations.

References

Aas, T.S., Sixten, H.J., Hillestad, M., Sveier, H., Ytrestøyl, T., Hatlen, B., Åsgård, T., 2017. Measurement of gastrointestinal passage rate in Atlantic salmon (*Salmo salar*) fed dry or soaked feed. Aquac. Rep. 8, 49–57.

Abdel-Moneim, A., Coulter, D.P., Mahapatra, C.T., Sepúlveda, M.S., 2015. Intersex in fishes and amphibians: population implications, prevalence, mechanisms and molecular biomarkers. J. Appl. Toxicol. 35, 1228–1240.

Adamovsky, O., Bisesi, J.H., Martyniuk, C.J., 2021. Plastics in our water: fish microbiomes at risk? Comp. Biochem. Physiol. Part D Genomics Proteomics 39, 100834.

Adel, M., Nayak, S., Lazado, C.C., Yeganeh, S., 2016. Effects of dietary prebiotic GroBiotic®-A on growth performance, plasma thyroid hormones and mucosal immunity of great sturgeon, *Huso huso* (Linnaeus, 1758). J. Appl. Ichthyol. 32, 825–831.

Adolfi, M.C., Fischer, P., Herpin, A., Regensburger, M., Kikuchi, M., Tanaka, M., Schartl, M., 2019. Increase of cortisol levels after temperature stress activates dmrt1a causing female-to-male sex reversal and reduced germ cell number in medaka. Mol. Reprod. Dev. 86, 1405–1417.

Adrian, R., O'Reilly, C.M., Zagarese, H., Baines, S.B., Hessen, D.O., Keller, W., Livingstone, D.M., Sommaruga, R., Straile, D., Van Donk, E., Weyhenmeyer, G.A., Winder, M., 2009. Lakes as sentinels of climate change. Limnol. Oceanogr. 54, 2283–2297.

Aerts, J., Metz, J.R., Ampe, B., Decostere, A., Flik, G., De Saeger, S., 2015. Scales tell a story on the stress history of fish. PLoS One 10, e0123411.

Alavi, S.M.H., Hatef, A., Pšenička, M., Kašpar, V., Boryshpolets, S., Dzyuba, B., Cosson, J., Bondarenko, V., Rodina, M., Gela, D., Linhart, O., 2012. Sperm biology and control of reproduction in sturgeon: (II) sperm morphology, acrosome reaction, motility and cryopreservation. Rev. Fish Biol. Fish. 22, 861–886.

Algera, D.A., Brownscombe, J.W., Gilmour, K.M., Lawrence, M.J., Zolderdo, A.J., Cooke, S.J., 2017. Cortisol treatment affects locomotor activity and swimming behaviour of male smallmouth bass engaged in paternal care: a field study using acceleration biologgers. Physiol. Behav. 181, 59–68.

Alix, M., Kjesbu, O.S., Anderson, K.C., 2020. From gametogenesis to spawning: how climate-driven warming affects teleost reproductive biology. J. Fish Biol. 97, 607–632.

Allard, J.B., Duan, C., 2018. IGF-binding proteins: why do they exist and why are there so many? Front. Endocrinol. (Lausanne) 9, 1–12.

Almeida, O.G., Miranda, A., Frade, P., Hubbard, P.C., Barata, E.N., Canário, A.V.M., 2005. Urine as a social signal in the Mozambique tilapia (*Oreochromis mossambicus*). Chem. Senses 30, i309–i310.

Anderson, K., Swanson, P., Pankhurst, N., King, H., Elizur, A., 2012. Effect of thermal challenge on plasma gonadotropin levels and ovarian steroidogenesis in female maiden and repeat spawning Tasmanian Atlantic salmon (*Salmo salar*). Aquaculture 334, 205–212.

Anderson, K., Luckenbach, J.A., Yamamoto, Y., Elizur, A., 2019. Impacts of Fsh, Igf1, and high temperature on the expression of genes involved in steroidogenesis, cell communication, and apoptosis in isolated coho salmon previtellogenic ovarian follicles. Aquaculture 506, 60–69.

Andrews, K.S., Beckman, B.R., Beaudreau, A.H., Larsen, D.A., Williams, G.D., Levin, P.S., 2011. Suitability of insulin-like growth factor 1 (IGF1) as a measure of relative growth rates in lingcod. Mar. Coast. Fish. 3, 250–260.

Araki, H., Berejikian, B.A., Ford, M.J., Blouin, M.S., 2008. Fitness of hatchery-reared salmonids in the wild. Evol. Appl. 1, 342–355.

Arbuckle, W.J., Bélanger, A.J., Corkum, L.D., Zielinski, B.S., Li, W., Yun, S.S., Bachynski, S., Scott, A.P., 2005. In vitro biosynthesis of novel 5β-reduced steroids by the testis of the round goby, Neogobius melanostomus. Gen. Comp. Endocrinol. 140, 1–13.

Arechavala-Lopez, P., Caballero-Froilán, J.C., Jiménez-García, M., Capó, X., Tejada, S., Saraiva, J.L., Sureda, A., Moranta, D., 2020. Enriched environments enhance cognition, exploratory behaviour and brain physiological functions of *Sparus aurata*. Sci. Rep. 10, 11252.

Arkoosh, M.R., Van Gaest, A.L., Strickland, S.A., Hutchinson, G.P., Krupkin, A.B., Dietrich, J.P., 2017. Alteration of thyroid hormone concentrations in juvenile Chinook salmon (*Oncorhynchus tshawytscha*) exposed to polybrominated diphenyl ethers, BDE-47 and BDE-99. Chemosphere 171, 1–8.

Arukwe, A., Nordtug, T., Kortner, T.M., Mortensen, A.S., Brakstad, O.G., 2008. Modulation of steroidogenesis and xenobiotic biotransformation responses in zebrafish (*Danio rerio*) exposed to water-soluble fraction of crude oil. Environ. Res. 107, 362–370.

Asano, Y., Hiramoto, T., Nishino, R., Aiba, Y., Kimura, T., Yoshihara, K., Koga, Y., Sudo, N., 2012. Critical role of gut microbiota in the production of biologically active, free catecholamines in the gut lumen of mice. Am. J. Physiol. Gastrointest. Liver Physiol. 303, 1288–1295.

Aus der Beek, T., Weber, F.-A., Bergmann, A., Hickmann, S., Ebert, I., Hein, A., Küster, A., 2016. Pharmaceuticals in the environment—global occurrences and perspectives. Environ. Toxicol. Chem. 35, 823–835.

Awata, S., Tsuruta, T., Yada, T., Iguchi, K., 2011. Effects of suspended sediment on cortisol levels in wild and cultured strains of ayu *Plecoglossus altivelis*. Aquaculture 314, 115–121.

Azevedo, A., Wauters, J., Kirschbaum, C., Serra, R., Rivas, A., Jewgenow, K., 2020. Sex steroids and glucocorticoid ratios in Iberian lynx hair. Conserv. Physiol. 8, coaa075.

Baggerman, G., Verleyen, P., Clynen, E., Huybrechts, J., De Loof, A., Schoofs, L., 2004. Peptidomics. J. Chromatogr. B Analyt. Technol. Biomed. Life Sci. 803, 3–16.

Baker, M.R., Gobush, K.S., Vynne, C.H., 2013. Review of factors influencing stress hormones in fish and wildlife. J. Nat. Conserv. 21, 309–318.

Bangsgaard, K., Madsen, S.S., Korsgaard, B., 2006. Effect of waterborne exposure to 4-tert-octylphenol and 17β-estradiol on smoltification and downstream migration in Atlantic salmon, *Salmo salar*. Aquat. Toxicol. 80, 23–32.

Barcellos, L.J.G., Kreutz, L.C., Quevedo, R.M., da Rosa, J.G.S., Koakoski, G., Centenaro, L., Pottker, E., 2009. Influence of color background and shelter availability on jundiá (*Rhamdia quelen*) stress response. Aquaculture 288, 51–56.

Barkowski, N.A., Haukenes, A.H., 2014. Investigating the utility of measuring 11α-ketotestosterone and vitellogenin in surface mucus as an alternative to plasma samples in assessments of the reproductive axis of white bass. N. Am. J. Aquac. 76, 112–118.

Barra, R.O., Chiang, G., Saavedra, M.F., Orrego, R., Servos, M.R., Hewitt, L.M., McMaster, M.E., Bahamonde, P., Tucca, F., Munkittrick, K.R., 2021. Endocrine disruptor impacts on fish from Chile: the influence of wastewaters. Front. Endocrinol. (Lausanne) 12, 208.

Barton, B.A., 2002. Stress in fishes: a diversity of responses with particular reference to changes in circulating corticosteroids. Integr. Comp. Biol. 42, 517–525.

Barton, B.A., Iwama, G.K., 1991. Physiological changes in fish from stress in aquaculture with emphasis on the response and effects of corticosteroids. Annu. Rev. Fish Dis. 1, 3–26.

Barton, B., Schreck, C., Barton, L., 1987. Effects of chronic cortisol administration and daily acute stress on growth, physiological conditions, and stress responses in juvenile rainbow trout. Dis. Aquat. Organ. 2, 173–185.

Basu, N., Ta, C.A., Waye, A., Mao, J., Hewitt, M., Arnason, J.T., Trudeau, V.L., 2009. Pulp and paper mill effluents contain neuroactive substances that potentially disrupt neuroendocrine control of fish reproduction. Environ. Sci. Technol. 43, 1635–1641.

Batzina, A., Dalla, C., Tsopelakos, A., Papadopoulou-Daifoti, Z., Karakatsouli, N., 2014. Environmental enrichment induces changes in brain monoamine levels in gilthead seabream *Sparus aurata*. Physiol. Behav. 130, 85–90.

Beamish, R.J., Sweeting, R.M., Lange, K.L., Neville, C.M., 2008. Changes in the population ecology of hatchery and wild coho salmon in the Strait of Georgia. Trans. Am. Fish. Soc. 137, 503–520.

Beckman, B.R., 2011. Perspectives on concordant and discordant relations between insulin-like growth factor 1 (IGF1) and growth in fishes. Gen. Comp. Endocrinol. 170, 233–252.

Beckman, B.R., Dickhoff, W.W., Zaugg, W.S., Sharpe, C., Hirtzel, S., Schrock, R., Larsen, D.A., Ewing, R.D., Palmisano, A., Schreck, C.B., Mahnken, C.V.W., 1999. Growth, smoltification, and smolt-to-adult return of spring chinook salmon from hatcheries on the Deschutes River, Oregon. Trans. Am. Fish. Soc. 128, 1125–1150.

Beckman, B.R., Larsen, D.A., Sharpe, C., Lee-Pawlak, B., Schreck, C.B., Dickhoff, W.W., 2000. Physiological status of naturally reared juvenile spring chinook salmon in the Yakima River: seasonal dynamics and changes associated with smolting. Trans. Am. Fish. Soc. 129, 727–753.

Beckman, B.R., Fairgrieve, W., Cooper, K.A., Mahnken, C.V.W., Beamish, R.J., 2004a. Evaluation of endocrine indices of growth in individual postsmolt coho salmon. Trans. Am. Fish. Soc. 133, 1057–1067.

Beckman, B.R., Shimizu, M., Gadberry, B.A., Cooper, K.A., 2004b. Response of the somatotropic axis of juvenile coho salmon to alterations in plane of nutrition with an analysis of the relationships among growth rate and circulating IGF-I and 41 kDa IGFBP. Gen. Comp. Endocrinol. 135, 334–344.

Beckman, B.R., Harstad, D.L., Spangenberg, D.K., Gerstenberger, R.S., Brun, C.V., Larsen, D.A., 2017. The impact of different hatchery rearing environments on smolt-to-adult survival of spring chinook salmon. Trans. Am. Fish. Soc. 146, 539–555.

Behringer, V., Deschner, T., 2017. Non-invasive monitoring of physiological markers in primates. Horm. Behav. 91, 3–18.

Berejikian, B.A., 1995. The effects of hatchery and wild ancestry and experience on the relative ability of steelhead trout fry (*Oncorhynchus mykiss*) to avoid a benthic predator. Can. J. Fish. Aquat. Sci. 52, 2476–2482.

Berejikian, B.A., Fairgrieve, W.T., Swanson, P., Tezak, E.P., 2003. Current velocity and injection of GnRHa affect reproductive behavior and body composition of captively reared offspring of wild chinook salmon (*Oncorhynchus tshawytscha*). Can. J. Fish. Aquat. Sci. 60, 690–699.

Bergan, H.E., Kittilson, J.D., Sheridan, M.A., 2015. Nutritional state modulates growth hormone-stimulated lipolysis. Gen. Comp. Endocrinol. 217–218, 1–9.

Bergan-Roller, H.E., Sheridan, M.A., 2018. The growth hormone signaling system: insights into coordinating the anabolic and catabolic actions of growth hormone. Gen. Comp. Endocrinol. 258, 119–133.

Bernier, N.J., Craig, P.M., 2005. CRF-related peptides contribute to stress response and regulation of appetite in hypoxic rainbow trout. Am. J. Physiol. Regul. Integr. Comp. Physiol. 289, 982–990.

Bernier, N.J., Van Der Kraak, G., Farrell, A.P., Brauner, C.J., 2009a. Fish Physiology: Fish Neuroendocrinology. Academic Press, London. vol. 28.

Bernier, N.J., Flik, G., Klaren, P.H.M., 2009b. Regulation and contribution of corticotropic, melanotropic and thyrotropic axes to the stress response in fishes. In: Bernier, N.J., Van Der Kraak, G., Farrell, A.P., Brauner, C.J. (Eds.), Fish Neuroendocrinology, Fish Physiology. vol. 28. Academic Press, London, pp. 235–311.

Berninger, J.P., Martinović-Weigelt, D., Garcia-Reyero, N., Escalon, L., Perkins, E.J., Ankley, G.T., Villeneuve, D.L., 2014. Using transcriptomic tools to evaluate biological effects across effluent gradients at a diverse set of study sites in Minnesota, USA. Environ. Sci. Technol. 48, 2404–2412.

Bertotto, D., Poltronieri, C., Negrato, E., Majolini, D., Radaelli, G., Simontacchi, C., 2010. Alternative matrices for cortisol measurement in fish. Aquacult. Res. 41, 1261–1267.

Besson, M., Feeney, W.E., Moniz, I., François, L., Brooker, R.M., Holzer, G., Metian, M., Roux, N., Laudet, V., Lecchini, D., 2020. Anthropogenic stressors impact fish sensory development and survival via thyroid disruption. Nat. Commun. 11, 3614.

Bévalot, F., Gaillard, Y., Lhermitte, M.A., Pépin, G., 2000. Analysis of corticosteroids in hair by liquid chromatography-electrospray ionization mass spectrometry. J. Chromatogr. B Biomed. Sci. Appl. 740, 227–236.

Birnie-Gauvin, K., Flávio, H., Kristensen, M.L., Walton-Rabideau, S., Cooke, S.J., Willmore, W.G., Koed, A., Aarestrup, K., 2019. Cortisol predicts migration timing and success in both Atlantic salmon and sea trout kelts. Sci. Rep. 9, 2422.

Björnsson, B.T., Stefansson, S.O., McCormick, S.D., 2011. Environmental endocrinology of salmon smoltification. Gen. Comp. Endocrinol. 170, 290–298.

Blackwell, B.R., Ankley, G.T., 2021. Simultaneous determination of a suite of endogenous steroids by LC-APPI-MS: application to the identification of endocrine disruptors in aquatic toxicology. J. Chromatogr. B Analyt. Technol. Biomed. Life Sci. 1163, 122513.

Blanchfield, P.J., Kidd, K.A., Docker, M.F., Palace, V.P., Park, B.J., Postma, L.D., 2015. Recovery of a wild fish population from whole-lake additions of a synthetic estrogen. Environ. Sci. Technol. 49, 3136–3144.

Bock, S.L., Chow, M.I., Forsgren, K.L., Lema, S.C., 2021. Widespread alterations to hypothalamic-pituitary-gonadal (HPG) axis signaling underlie high temperature reproductive inhibition in the eurythermal sheepshead minnow (*Cyprinodon variegatus*). Mol. Cell. Endocrinol. 537, 111447.

Boerrigter, J.G.J., van den Bos, R., van de Vis, H., Spanings, T., Flik, G., 2016. Effects of density, PVC-tubes and feeding time on growth, stress and aggression in African catfish (*Clarias gariepinus*). Aquacult. Res. 47, 2553–2568.

Boggs, A.S.P., Bowden, J.A., Galligan, T.M., Guillette, L.J., Kucklick, J.R., 2016. Development of a multi-class steroid hormone screening method using liquid chromatography/tandem mass spectrometry (LC-MS/MS). Anal. Bioanal. Chem. 408, 4179–4190.

Bond, M.H., Beckman, B.R., Rohrbach, L., Quinn, T.P., 2014. Differential growth in estuarine and freshwater habitats indicated by plasma IGF1 concentrations and otolith chemistry in dolly varden *Salvelinus malma*. J. Fish Biol. 85, 1429–1445.

Booc, F., Thornton, C., Lister, A., MacLatchy, D., Willett, K.L., 2014. Benzo[a]pyrene effects on reproductive endpoints in *Fundulus heteroclitus*. Toxicol. Sci. 140, 73–82.

Bopp, L., Resplandy, L., Orr, J.C., Doney, S.C., Dunne, J.P., Gehlen, M., Halloran, P., Heinze, C., Ilyina, T., Séférian, R., Tjiputra, J., Vichi, M., 2013. Multiple stressors of ocean ecosystems in the 21st century: projections with CMIP5 models. Biogeosciences 10, 6225–6245.

Bortolotti, G.R., Marchant, T.A., Blas, J., German, T., 2008. Corticosterone in feathers is a long-term, integrated measure of avian stress physiology. Funct. Ecol. 22, 494–500.

Braithwaite, V.A., Salvanes, A.G.V., 2005. Environmental variability in the early rearing environment generates behaviourally flexible cod: implications for rehabilitating wild populations. Proc. R. Soc. B Biol. Sci. 272, 1107–1113.

Breitburg, D., Levin, L.A., Oschlies, A., Grégoire, M., Chavez, F.P., Conley, D.J., Garçon, V., Gilbert, D., Gutiérrez, D., Isensee, K., Jacinto, G.S., Limburg, K.E., Montes, I., Naqvi, S.W.A., Pitcher, G.C., Rabalais, N.N., Roman, M.R., Rose, K.A., Seibel, B.A., Telszewski, M., Yasuhara, M., Zhang, J., 2018. Declining oxygen in the global ocean and coastal waters. Science 359, eaam7240.

Breves, J.P., Duffy, T.A., Einarsdottir, I.E., Björnsson, B.T., McCormick, S.D., 2018. *In vivo* effects of 17α-ethinylestradiol, 17β-estradiol and 4-nonylphenol on insulin-like growth-factor binding proteins (igfbps) in Atlantic salmon. Aquat. Toxicol. 203, 28–39.

Brown, C., Day, R.L., 2002. The future of stock enhancements: lessons for hatchery practice from conservation biology. Fish Fish. 3, 79–94.

Brown, C., Davidson, T., Laland, K., 2003. Environmental enrichment and prior experience of live prey improve foraging behaviour in hatchery-reared Atlantic salmon. J. Fish Biol. 63, 187–196.

Bryant, A.S., Greenwood, A.K., Juntti, S.A., Byrne, A.E., Fernald, R.D., 2016. Dopaminergic inhibition of gonadotropin-releasing hormone neurons in the cichlid fish *Astatotilapia burtoni*. J. Exp. Biol. 219, 3861–3865.

Bugel, S.M., White, L.A., Cooper, K.R., 2010. Impaired reproductive health of killifish (*Fundulus heteroclitus*) inhabiting Newark Bay, NJ, a chronically contaminated estuary. Aquat. Toxicol. 96, 182–193.

Burgerhout, E., Lokman, P.M., van den Thillart, G.E.E.J.M., Dirks, R.P., 2019. The time-keeping hormone melatonin: a possible key cue for puberty in freshwater eels (*Anguilla* spp.). Rev. Fish Biol. Fish. 29, 1–21.

Burns, A.C., Sorensen, P.W., Hoye, T.R., 2011. Synthesis and olfactory activity of unnatural, sulfated 5β-bile acid derivatives in the sea lamprey (*Petromyzon marinus*). Steroids 76, 291–300.

Butcher, J.B., Nover, D., Johnson, T.E., Clark, C.M., 2015. Sensitivity of lake thermal and mixing dynamics to climate change. Clim. Change 129, 295–305.

Caballero-Gallardo, K., Olivero-Verbel, J., Freeman, J.L., 2016. Toxicogenomics to evaluate endocrine disrupting effects of environmental chemicals using the zebrafish model. Curr. Genomics 17, 515–527.

Caldarone, E.M., MacLean, S.A., Beckman, B.R., 2016. Evaluation of nucleic acids and plasma IGF1 levels for estimating short-term responses of postsmolt Atlantic salmon (*Salmo salar*) to food availability. Fish. Bull. 114, 288–301.

Campinho, M.A., Saraiva, J., Florindo, C., Power, D.M., 2014. Maternal thyroid hormones are essential for neural development in zebrafish. Mol. Endocrinol. 28, 1136–1149.

Cao, Y., Tveten, A.K., Stene, A., 2017. Establishment of a non-invasive method for stress evaluation in farmed salmon based on direct fecal corticoid metabolites measurement. Fish Shellfish Immunol. 66, 317–324.

Carbajal, A., Reyes-López, F.E., Tallo-Parra, O., Lopez-Bejar, M., Tort, L., 2019a. Comparative assessment of cortisol in plasma, skin mucus and scales as a measure of the hypothalamic-pituitary-interrenal axis activity in fish. Aquaculture 506, 410–416.

Carbajal, A., Soler, P., Tallo-Parra, O., Isasa, M., Echevarria, C., Lopez-Bejar, M., Vinyoles, D., 2019b. Towards non-invasive methods in measuring fish welfare: the measurement of cortisol concentrations in fish skin mucus as a biomarker of habitat quality. Animals 9, 939.

Carnevali, O., Santangeli, S., Forner-Piquer, I., Basili, D., Maradonna, F., 2018. Endocrine-disrupting chemicals in aquatic environment: what are the risks for fish gametes? Fish Physiol. Biochem. 44, 1561–1576.

Carr, J.A., Patiño, R., 2011. The hypothalamus-pituitary-thyroid axis in teleosts and amphibians: endocrine disruption and its consequences to natural populations. Gen. Comp. Endocrinol. 170, 299–312.

Castañeda Cortés, D.C., Padilla, L.F.A., Langlois, V.S., Somoza, G.M., Fernandino, J.I., 2019. The central nervous system acts as a transducer of stress-induced masculinization through corticotropin-releasing hormone B. Development 146, dev172866.

Castro, A.J.G., Baptista, I.E., de Moura, K.R.S., Padilha, F., Tonietto, J., de Souza, A.Z.P., Soares, C.H.L., Silva, F.R.M.B., Van Der Kraak, G., 2018. Exposure to a Brazilian pulp mill effluent impacts the testis and liver in the zebrafish. Comp. Biochem. Physiol., Part C: Toxicol. Pharmacol. 206–207, 41–47.

Chadwick, J.G., Nislow, K.H., McCormick, S.D., 2015. Thermal onset of cellular and endocrine stress responses correspond to ecological limits in brook trout, an iconic cold-water fish. Conserv. Physiol. 3, cov017.

Chamberlin, J.W., Beckman, B.R., Greene, C.M., Rice, C.A., Hall, J.E., 2017. How relative size and abundance structures the relationship between size and individual growth in an ontogenetically piscivorous fish. Ecol. Evol. 7, 6981–6995.

Chang, J.P., Wong, A.O.L., 2009. Growth hormone regulation in fish. In: Bernier, N.J., Van Der Kraak, G., Farrell, A.P., Brauner, C.J. (Eds.), Fish Neuroendocrinology, Fish Physiology. vol. 28. Academic Press, London, pp. 151–195.

Cheek, A.O., Landry, C.A., Steele, S.L., Manning, S., 2009. Diel hypoxia in marsh creeks impairs the reproductive capacity of estuarine fish populations. Mar. Ecol. Prog. Ser. 392, 211–221.

Chen, L., Guo, Y., Hu, C., Lam, P.K.S., Lam, J.C.W., Zhou, B., 2018. Dysbiosis of gut microbiota by chronic coexposure to titanium dioxide nanoparticles and bisphenol A: implications for host health in zebrafish. Environ. Pollut. 234, 307–317.

Cheung, C.H.Y., Chiu, J.M.Y., Wu, R.S.S., 2014. Hypoxia turns genotypic female medaka fish into phenotypic males. Ecotoxicology 23, 1260–1269.

Chiang, G., Mcmaster, M.E., Urrutia, R., Saavedra, M.F., Gavilán, J.F., Tucca, F., Barra, R., Munkittrick, K.R., 2011. Health status of native fish (*Percilia gillissi* and *Trichomycterus areolatus*) downstream of the discharge of effluent from a tertiary-treated elemental chlorine-free pulp mill in Chile. Environ. Toxicol. Chem. 30, 1793–1809.

Chittenden, C.M., Biagi, C.A., Davidsen, J.G., Davidsen, A.G., Kondo, H., McKnight, A., Pedersen, O.P., Raven, P.A., Rikardsen, A.H., Mark Shrimpton, J., Zuehlke, B., Scott McKinley, R., Devlin, R.H., 2010. Genetic versus rearing-environment effects on phenotype: hatchery and natural rearing effects on hatchery- and wild-born coho salmon. PLoS One 5, e12261.

Clark, R.W., Henderson-Arzapalo, A., Sullivan, C.V., 2005. Disparate effects of constant and annually-cycling daylength and water temperature on reproductive maturation of striped bass (*Morone saxatilis*). Aquaculture 249, 497–513.

Clark, T.D., Donaldson, M.R., Drenner, S.M., Hinch, S.G., Patterson, D.A., Hills, J., Ives, V., Carter, J.J., Cooke, S.J., Farrell, A.P., 2011. The efficacy of field techniques for obtaining and storing blood samples from fishes. J. Fish Biol. 79, 1322–1333.

Clements, K.D., Angert, E.R., Montgomery, W.L., Choat, J.H., 2014. Intestinal microbiota in fishes: what's known and what's not. Mol. Ecol. 23, 1891–1898.

Cogliati, K.M., Herron, C.L., Noakes, D.L.G., Schreck, C.B., 2019. Reduced stress response in juvenile chinook salmon reared with structure. Aquaculture 504, 96–101.

Collier, T.K., Anulacion, B.F., Arkoosh, M.R., Dietrich, J.P.J.P., Incardona, J.P., Johnson, L.L., Ylitalo, G.M., Myers, M.S., 2013. Effects on fish of polycyclic aromatic hydrocarbons (PAHS) and naphthenic acid exposures. In: Tierney, K.B., Farrell, A.P., Brauner, C.J. (Eds.), Fish Physiology, Organic Chemical Toxicology of Fishes, vol. 33. Academic Press, London, pp. 195–255.

Corkum, L.D., Belanger, R.M., 2007. Use of chemical communication in the management of freshwater aquatic species that are vectors of human diseases or are invasive. Gen. Comp. Endocrinol. 153, 401–417.

Corkum, L.D., Sapota, M.R., Skora, K.E., 2004. The round goby, *Neogobius melanostomus*, a fish invader on both sides of the Atlantic Ocean. Biol. Invasions 6, 173–181.

Costigan, S.L., Werner, J., Ouellet, J.D., Hill, L.G., Law, R.D., 2012. Expression profiling and gene ontology analysis in fathead minnow (*Pimephales promelas*) liver following exposure to pulp and paper mill effluents. Aquat. Toxicol. 122–123, 44–55.

Culbert, B.M., Ligocki, I.Y., Salena, M.G., Wong, M.Y.L., Hamilton, I.M., Aubin-Horth, N., Bernier, N.J., Balshine, S., 2021. Rank- and sex-specific differences in the neuroendocrine regulation of glucocorticoids in a wild group-living fish. Horm. Behav. 136, 105079.

Cusick, J.A., Wellman, C.L., Demas, G.E., 2021. The call of the wild: using non-model systems to investigate microbiome–behaviour relationships. J. Exp. Biol. 224, jeb224485.

Dabrowski, K., Rinchard, J., Ottobre, J.S., Alcantara, F., Padilla, P., Ciereszko, A., De Jesus, M.J., Kohler, C.C., 2003. Effect of oxygen saturation in water on reproductive performances of pacu *Piaractus brachypomus*. J. World Aquacult. Soc. 34, 441–449.

Dabrowski, R., Ripa, R., Latza, C., Annibal, A., Antebi, A., 2020. Optimization of mass spectrometry settings for steroidomic analysis in young and old killifish. Anal. Bioanal. Chem. 412, 4089–4099.

Davidsen, J.G., Daverdin, M., Sjursen, A.D., Rønning, L., Arnekleiv, J.V., Koksvik, J.I., 2014. Does reduced feeding prior to release improve the marine migration of hatchery brown trout *Salmo trutta* smolts? J. Fish Biol. 85, 1992–2002.

Davis, M.J., Chamberlin, J.W., Gardner, J.R., Connelly, K.A., Gamble, M.M., Beckman, B.R., Beauchamp, D.A., 2020. Variable prey consumption leads to distinct regional differences in chinook salmon growth during the early marine critical period. Mar. Ecol. Prog. Ser. 640, 147–169.

De Mercado, E., Larrán, A.M., Pinedo, J., Tomás-Almenar, C., 2018. Skin mucous: a new approach to assess stress in rainbow trout. Aquaculture 484, 90–97.

de Weerth, C., 2017. Do bacteria shape our development? Crosstalk between intestinal microbiota and HPA axis. Neurosci. Biobehav. Rev. 83, 458–471.

Deal, C.K., Volkoff, H., 2020. The role of the thyroid axis in fish. Front. Endocrinol. (Lausanne) 11, 1–25.

Delbes, G., Blázquez, M., Fernandino, J.I., Grigorova, P., Hales, B.F., Metcalfe, C., Navarro-Martín, L., Parent, L., Robaire, B., Rwigemera, A., Van Der Kraak, G., Wade, M., Marlatt, V., 2022. Effects of endocrine disrupting chemicals on gonad development: mechanistic insights from fish and mammals. Environ. Res. 204, 112040.

Denison, M.S., Nagy, S.R., 2003. Activation of the aryl hydrocarbon receptor by structurally diverse exogenous and endogenous chemicals. Annu. Rev. Pharmacol. Toxicol. 43, 309–334.

Devlin, R.H., Biagi, C.A., Yesaki, T.Y., Smailus, D.E., Byatt, J.C., 2001. Growth of domesticated transgenic fish. Nature 409, 781–782.

Diaz, R.J., Rosenberg, R., 2008. Spreading dead zones and consequences for marine ecosystems. Science 321, 926–929.

Donaldson, E.M., Hunter, G.A., 1983. Induced final maturation, ovulation, and spermiation in cultured fish. Fish Physiol. 9, 351–403.

Douxfils, J., Mandiki, S.N.M., Marotte, G., Wang, N., Silvestre, F., Milla, S., Henrotte, E., Vandecan, M., Rougeot, C., Mélard, C., Kestemont, P., 2011. Does domestication process affect stress response in juvenile Eurasian perch *Perca fluviatilis*? Comp. Biochem. Physiol. A Mol. Integr. Physiol. 159, 92–99.

Duffy, T.A., Iwanowicz, L.R., McCormick, S.D., 2014. Comparative responses to endocrine disrupting compounds in early life stages of Atlantic salmon, *Salmo salar*. Aquat. Toxicol. 152, 1–10.

Dufour, S., Quérat, B., Tostivint, H., Pasqualini, C., Vaudry, H., Rousseau, K., 2020. Origin and evolution of the neuroendocrine control of reproduction in vertebrates, with special focus on genome and gene duplications. Physiol. Rev. 100, 869–943.

Dufour, S., Sebert, M.E., Weltzien, F.A., Rousseau, K., Pasqualini, C., 2010. Neuroendocrine control by dopamine of teleost reproduction. J. Fish Biol. 76, 129–160.

Duguid, W.D.P., Iwanicki, T.W., Journey, M.L., Noel, A.L., Beckman, B.R., Juanes, F., 2018. Assessing indices of growth for field studies of juvenile salmon: an experiment and synthesis. Mar. Coast. Fish. 10, 204–223.

Egerton, S., Culloty, S., Whooley, J., Stanton, C., Ross, R.P., 2018. The gut microbiota of marine fish. Front. Microbiol. 9, 873.

Elisio, M., Chalde, T., Miranda, L.A., 2012. Effects of short periods of warm water fluctuations on reproductive endocrine axis of the pejerrey (*Odontesthes bonariensis*) spawning. Comp. Comp. Biochem. Physiol. A Mol. Integr. Physiol. 163, 47–55.

Ellis, T., James, J.D., Stewart, C., Scott, A.P., 2004. A non-invasive stress assay based upon measurement of free cortisol released into the water by rainbow trout. J. Fish Biol. 65, 1233–1252.

Ellis, T., James, J.D., Scott, A.P., 2005. Branchial release of free cortisol and melatonin by rainbow trout. J. Fish Biol. 67, 535–540.

Ellis, T., James, J.D., Sundh, H., Fridell, F., Sundell, K., Scott, A.P., 2007. Non-invasive measurement of cortisol and melatonin in tanks stocked with seawater Atlantic salmon. Aquaculture 272, 698–706.

Ellis, T., Sanders, M.B., Scott, A.P., 2013. Non-invasive monitoring of steroids in fishes. Wien. Tierarztl. Monatsschr. 100, 255–269.

Ervin, S.M., Li, H., Lim, L., Roberts, L.R., Liang, X., Mani, S., Redinbo, M.R., 2019. Gut microbial β-glucuronidases reactivate estrogens as components of the estrobolome that reactivate estrogens. J. Biol. Chem. 294, 18586–18599.

Ervin, S.M., Simpson, J.B., Gibbs, M.E., Creekmore, B.C., Lim, L., Walton, W.G., Gharaibeh, R.Z., Redinbo, M.R., 2020. Structural insights into endobiotic reactivation by human gut microbiome-encoded sulfatases. Biochemistry 59, 3939–3950.

Fabbri, E., Capuzzo, A., Moon, T.W., 1998. The role of circulating catecholamines in the regulation of fish metabolism: an overview. Comp. Biochem. Physiol. C Pharmacol. Toxicol. Endocrinol. 120, 177–192.

Fabbri, E., Moon, T.W., 2016. Adrenergic signaling in teleost fish liver, a challenging path. Comp. Biochem. Physiol. B Biochem. Mol. Biol. 199, 74–86.

Fairchild, W.L., Swansburg, E.O., Arsenault, J.T., Brown, S.B., 1999. Does an association between pesticide use and subsequent declines in catch of Atlantic salmon (*Salmo salar*) represent a case of endocrine disruption? Environ. Health Perspect. 107, 349–357.

Fanouraki, E., Papandroulakis, N., Ellis, T., Mylonas, C.C., Scott, A.P., Pavlidis, M., 2008. Water cortisol is a reliable indicator of stress in European sea bass, *Dicentrarchus labrax*. Behaviour 145, 1267–1281.

Faria, A.M., Lopes, A.F., Silva, C.S.E., Novais, S.C., Lemos, M.F.L., Gonçalves, E.J., 2018. Reproductive trade-offs in a temperate reef fish under high pCO2 levels. Mar. Environ. Res. 137, 8–15.

Farwell, M., Hughes, G., Smith, J.L., Clelland, E., Loeb, S.J., Semeniuk, C., Scott, A.P., Li, W., Zielinski, B., 2017. Differential female preference for individual components of a reproductive male round goby (*Neogobius melanostomus*) pheromone. J. Great Lakes Res. 43, 379–386.

Faught, E., Aluru, N., Vijayan, M.M., 2016. The molecular stress response. In: Schreck, C.B., Tort, L., Farrell P., A, Brauner, C.J. (Eds.), Fish Physiology: Biology of Stress in Fish, vol. 35. Academic Press, London, pp. 113–166.

Ferretti, F., Worm, B., Britten, G.L., Heithaus, M.R., Lotze, H.K., 2010. Patterns and ecosystem consequences of shark declines in the ocean. Ecol. Lett. 13, 1055–1071.

Ferriss, B.E., Trudel, M., Beckman, B.R., 2014. Regional and inter-annual trends in marine growth of juvenile salmon in coastal pelagic ecosystems of British Columbia, Canada. Mar. Ecol. Prog. Ser. 503, 247–261.

Filby, A.L., Thorpe, K.L., Maack, G., Tyler, C.R., 2007. Gene expression profiles revealing the mechanisms of anti-androgen- and estrogen-induced feminization in fish. Aquat. Toxicol. 81, 219–231.

Fine, J.M., Sorensen, P.W., 2008. Isolation and biological activity of the multi-component sea lamprey migratory pheromone. J. Chem. Ecol. 34, 1259–1267.

Fissette, S.D., Buchinger, T.J., Wagner, C.M., Johnson, N.S., Scott, A.M., Li, W., 2021. Progress towards integrating an understanding of chemical ecology into sea lamprey control. J. Great Lakes Res. 47, S660–S672.

Flagg, T.A., Nash, C.E., 1999. A Conceptual Framework for Conservation Hatchery Strategies for Pacific Salmonids. NOAA Technical Memorandum.

Flagg, T.A., Berejikian, B.A., Colt, J.E., Dickhoff, W.W., Harrell, L.W., Maynard, D.J., Nash, C.E., Strom, M.S., Iwamoto, R.N., Mahnken, C.V.W., 2000. Ecological and Behavioral Impacts of Artificial Production Strategies on the Abundance of Wild Salmon Populations: A Review of Practices in the Pacific Northwest. NOAA Technical Memorandum.

Flores-Valverde, A.M., Horwood, J., Hill, E.M., 2010. Disruption of the steroid metabolome in fish caused by exposure to the environmental estrogen 17α-ethinylestradiol. Environ. Sci. Technol. 44, 3552–3558.

Forsgren, E., Dupont, S., Jutfelt, F., Amundsen, T., 2013. Elevated CO2 affects embryonic development and larval phototaxis in a temperate marine fish. Ecol. Evol. 3, 3637–3646.

Fredricks, K.T., Hubert, T.D., Amberg, J.J., Cupp, A.R., Dawson, V.K., 2021. Chemical controls for an integrated pest management program. North Am. J. Fish. Manag. 41, 289–300.

Friesen, C.N., Aubin-Horth, N., Chapman, L.J., 2012a. The effect of hypoxia on sex hormones in an African cichlid *Pseudocrenilabrus multicolor victoriae*. Comp. Biochem. Physiol. A Mol. Integr. Physiol. 162, 22–30.

Friesen, C.N., Chapman, L.J., Aubin-Horth, N., 2012b. Holding water steroid hormones in the African cichlid fish *Pseudocrenilabrus multicolor victoriae*. Gen. Comp. Endocrinol. 179, 400–405.

Frisk, M.G., Miller, T.J., Fogarty, M.J., 2001. Estimation and analysis of biological parameters in elasmobranch fishes: a comparative life history study. Can. J. Fish. Aquat. Sci. 58, 969–981.

Froehlich, H.E., Gentry, R.R., Halpern, B.S., 2017. Conservation aquaculture: shifting the narrative and paradigm of aquaculture's role in resource management. Biol. Conserv. 215, 162–168.

Frommel, A.Y., Stiebens, V., Clemmesen, C., Havenhand, J., 2010. Effect of ocean acidification on marine fish sperm (Baltic cod: *Gadus morhua*). Biogeosciences 7, 3915–3919.

Fuentes, E.N., Valdés, J.A., Molina, A., Björnsson, B.T., 2013. Regulation of skeletal muscle growth in fish by the growth hormone—insulin-like growth factor system. Gen. Comp. Endocrinol. 192, 136–148.

Fujimoto, Y., Yambe, H., Takahashi, K., Sato, S., 2020. Bile from reproductively mature male largemouth bass micropterus salmoides attracts conspecific females and offers a practical application to control populations. Manag. Biol. Invasions 11, 415–427.

Fuzzen, M.L.M., Bernier, N.J., Van Der Kraak, G., 2011. Stress and reproduction. In: Norris, D.O., Lopez, K.H. (Eds.), Hormones and Reproduction of Vertebrates. Academic Press, Burlington, pp. 103–117.

Fuzzen, M.L.M., Bennett, C.J., Tetreault, G.R., McMaster, M.E., Servos, M.R., 2015. Severe intersex is predictive of poor fertilization success in populations of rainbow darter (*Etheostoma caeruleum*). Aquat. Toxicol. 160, 106–116.

Galligan, T.M., Schwacke, L.H., Houser, D.S., Wells, R.S., Rowles, T., Boggs, A.S.P., 2018. Characterization of circulating steroid hormone profiles in the bottlenose dolphin (*Tursiops truncatus*) by liquid chromatography–tandem mass spectrometry (LC–MS/MS). Gen. Comp. Endocrinol. 263, 80–91.

Galligan, T.M., Boggs, A.S.P., Balmer, B.C., Rowles, T., Smith, C.R., Townsend, F., Wells, R.S., Kellar, N.M., Zolman, E.S., Schwacke, L.H., 2020. Blubber steroid hormone profiles as indicators of physiological state in free-ranging common bottlenose dolphins (*Tursiops truncatus*). Comp. Biochem. Physiol. A Mol. Integr. Physiol. 239, 110583.

Gammon, D.B., Li, W., Scott, A.P., Zielinski, B.S., Corkum, L.D., 2005. Behavioural responses of female *Neogobius melanostomus* to odours of conspecifics. J. Fish Biol. 67, 615–626.

Gamperl, A.K., Vijayan, M.M., Boutilier, R.G., 1994. Experimental control of stress hormone levels in fishes: techniques and applications. Rev. Fish Biol. Fish. 4, 215–255.

Gamperl, A.K., Gillis, T.E., Farrell, A.P., Brauner, C.J., 2017. Fish Physiology. The Cardiovascular System: Morphology, Control and Function, vol. 36A. Academic Press, London.

García-López, Á., Pascual, E., Sarasquete, C., Martínez-Rodríguez, G., 2006. Disruption of gonadal maturation in cultured Senegalese sole *Solea senegalensis* Kaup by continuous light and/or constant temperature regimes. Aquaculture 261, 789–798.

Garner, S.R., Madison, B.N., Bernier, N.J., Neff, B.D., 2011. Behavioural interactions and hormones in naturally and hatchery-spawned chinook salmon. Ethology 117, 37–48.

Geffroy, B., Wedekind, C., 2020. Effects of global warming on sex ratios in fishes. J. Fish Biol. 97, 596–606.

Geven, E.J.W., Nguyen, N.K., Van Den Boogaart, M., Spanings, F.A.T., Flik, G., Klaren, P.H.M., 2007. Comparative thyroidology: thyroid gland location and iodothyronine dynamics in Mozambique tilapia (*Oreochromis mossambicus* Peters) and common carp (*Cyprinus carpio* L.). J. Exp. Biol. 210, 4005–4015.

Ghosal, R., Eichmiller, J.J., Witthuhn, B.A., Sorensen, P.W., 2018. Attracting common carp to a bait site with food reveals strong positive relationships between fish density, feeding activity, environmental DNA, and sex pheromone release that could be used in invasive fish management. Ecol. Evol. 8, 6714–6727.

Gibbons, W.N., Munkittrick, K.R., McMaster, M.E., Taylor, W.D., 1998. Monitoring aquatic environments receiving industrial effluents using small fish species 2: comparison between responses of trout-perch (*Percopsis omiscomaycus*) and white sucker (*Catostomus commersoni*) downstream of a pulp mill. Environ. Toxicol. Chem. 17, 2238–2245.

Goikoetxea, A., Sadoul, B., Blondeau-Bidet, E., Aerts, J., Blanc, M.O., Parrinello, H., Barrachina, C., Pratlong, M., Geffroy, B., 2021. Genetic pathways underpinning hormonal stress responses in fish exposed to short- and long-term warm ocean temperatures. Ecol. Indic. 120, 106937.

Goksøyr, A., 2006. Endocrine disruptors in the marine environment: mechanisms of toxicity and their influence on reproductive processes in fish. J. Toxicol. Environ. Health, Part A 69, 175–184.

Gorissen, M., Flik, G., 2016. The endocrinology of the stress response in fish: an adaptation-physiological view. In: Schreck, C.B., Tort, L., Farrell, A.P., Brauner, C.J. (Eds.), Fish Physiology: Biology of Stress in Fish, vol. 35. Academic Press, London, pp. 75–111.

Gormally, B.M.G., Romero, L.M., 2020. What are you actually measuring? A review of techniques that integrate the stress response on distinct time-scales. Funct. Ecol. 34, 2030–2044.

Graham, M.A., Earley, R.L., Baker, J.A., Foster, S.A., 2018. Evolution of steroid hormones in reproductive females of the threespine stickleback fish. Gen. Comp. Endocrinol. 268, 71–79.

Graham, K.M., Burgess, E.A., Rolland, R.M., 2021. Stress and reproductive events detected in North Atlantic right whale blubber using a simplified hormone extraction protocol. Conserv. Physiol. 9, coaa133.

Gräns, J., Wassmur, B., Celander, M.C., 2010. One-way inhibiting cross-talk between arylhydrocarbon receptor (AhR) and estrogen receptor (ER) signaling in primary cultures of rainbow trout hepatocytes. Aquat. Toxicol. 100, 263–270.

Greytak, S.R., Champlin, D., Callard, G.V., 2005. Isolation and characterization of two cytochrome P450 aromatase forms in killifish (*Fundulus heteroclitus*): differential expression in fish from polluted and unpolluted environments. Aquat. Toxicol. 71, 371–389.

Grosell, M., Farrell, A.P., Brauner, C.J., 2010. In: Fish Physiology: The Multifunctional Gut of Fish, vol. 30. Academic Press, London.

Guardiola, F.A., Cuesta, A., Esteban, M.Á., 2016. Using skin mucus to evaluate stress in gilthead seabream (*Sparus aurata* L.). Fish Shellfish Immunol. 59, 323–330.

Hack, N.L., Cordova, K.L., Glaser, F.L., Journey, M.L., Resner, E.J., Hardy, K.M., Beckman, B.R., Lema, S.C., 2019. Interactions of long-term food ration variation and short-term fasting on insulin-like growth factor-1 (IGF-1) pathways in copper rockfish (*Sebastes caurinus*). Gen. Comp. Endocrinol. 280, 168–184.

Hanke, I., Ampe, B., Kunzmann, A., Gärdes, A., Aerts, J., 2019. Thermal stress response of juvenile milkfish (*Chanos chanos*) quantified by ontogenetic and regenerated scale cortisol. Aquaculture 500, 24–30.

Hara, T.J., 2011. Neurophysiology of olfaction. In: Farrell, A.P. (Ed.), Encyclopedia of Fish Physiology: From Genome to Environment. Elsevier Science & Technology, San Diego, pp. 208–217.

Harbicht, A.B., Fraser, D.J., Ardren, W.R., 2020. Minor shifts towards more natural conditions in captivity improve long-term survival among reintroduced Atlantic salmon. Can. J. Fish. Aquat. Sci. 77, 931–942.

Harris, C.A., Hamilton, P.B., Runnalls, T.J., Vinciotti, V., Henshaw, A., Hodgson, D., Coe, T.S., Jobling, S., Tyler, C.R., Sumpter, J.P., 2011. The consequences of feminization in breeding groups of wild fish. Environ. Health Perspect. 119, 306.

Harstad, D.L., Larsen, D.A., Miller, J., Adams, I., Spangenberg, D.K., Nance, S., Rohrbach, L., Murauskas, J.G., Beckman, B.R., 2018. Winter-rearing temperature affects growth profiles, age of maturation, and smolt-to-adult returns for yearling summer Chinook salmon in the upper Columbia River basin. North Am. J. Fish. Manag. 38, 867–885.

Harter, T.S., Brauner, C.J., 2017. The O_2 and CO_2 transport system in teleosts and the specialized mechanisms that enhance $Hb–O_2$ unloading to tissues. In: Gamperl, A.K., Gillis, T.E., Farrell, A.P., Brauner, C.J. (Eds.), Fish Physiology, The Cardiovascular System: Development, Plasticity and Physiological Responses, vol. 36B. Academic Press, London, pp. 1–106.

Hasegawa, Y., Surugaya, R., Adachi, S., Ijiri, S., 2022. Regulation of 17α-hydroxyprogesterone production during induced oocyte maturation and ovulation in amur sturgeon (*Acipenser schrenckii*). J. Mar. Sci. Eng. 10, 86.

Hayashi, Y., Kobira, H., Yamaguchi, T., Shiraishi, E., Yazawa, T., Hirai, T., Kamei, Y., Kitano, T., 2010. High temperature causes masculinization of genetically female medaka by elevation of cortisol. Mol. Reprod. Dev. 77, 679–686.

Heath, D.D., Bernier, N.J., Heath, J.W., Iwama, G.K., 1993. Genetic, environmental, and interaction effects on growth and stress response of chinook salmon (*Oncorhynchus tshawytscha*) fry. Can. J. Fish. Aquat. Sci. 50, 435–442.

Hecker, M., Hollert, H., 2011. Endocrine disruptor screening: regulatory perspectives and needs. Environ. Sci. Eur. 23, 1–14.

Heimbürge, S., Kanitz, E., Otten, W., 2019. The use of hair cortisol for the assessment of stress in animals. Gen. Comp. Endocrinol. 270, 10–17.

Herranz-Jusdado, J.G., Rozenfeld, C., Morini, M., Pérez, L., Asturiano, J.F., Gallego, V., 2019. Recombinant vs purified mammal gonadotropins as maturation hormonal treatments of European eel males. Aquaculture 501, 527–536.

Hewitt, L.M., Parrott, J.L., McMaster, M.E., 2007. A decade of research on the environmental impacts of pulp and paper mill effluents in Canada: sources and characteristics of bioactive substances. J. Toxicol. J. Toxicol. Environ. Health B Crit. Rev. 9, 341–356.

Hirschenhauser, K., Canário, A.V.M., Ros, A.F.H., Taborsky, M., Oliveira, R.F., 2008. Social context may affect urinary excretion of 11-ketotestosterone in African cichlids. Behaviour 145, 1367–1388.

Höglund, E., Weltzien, F.A., Schjolden, J., Winberg, S., Ursin, H., Døving, K.B., 2005. Avoidance behavior and brain monoamines in fish. Brain Res. 1032, 104–110.

Holzer, G., Laudet, V., 2015. Thyroid hormones: a triple-edged sword for life history transitions. Curr. Biol. 25, R344–R347.

Holzer, G., Besson, M., Lambert, A., François, L., Barth, P., Gillet, B., Hughes, S., Piganeau, G., Leulier, F., Viriot, L., Lecchini, D., Laudet, V., 2017. Fish larval recruitment to reefs is a thyroid hormone-mediated metamorphosis sensitive to the pesticide chlorpyrifos. eLife 6, 1–22.

Honeycutt, J.L., Deck, C.A., Miller, S.C., Severance, M.E., Atkins, E.B., Luckenbach, J.A., Buckel, J.A., Daniels, H.V., Rice, J.A., Borski, R.J., Godwin, J., 2019. Warmer waters masculinize wild populations of a fish with temperature-dependent sex determination. Sci. Rep. 9, 6527.

Hook, S.E., 2010. Promise and progress in environmental genomics: a status report on the applications of gene expression-based microarray studies in ecologically relevant fish species. J. Fish Biol. 77, 1999–2022.

Hook, S.E., Gallagher, E.P., Batleyy, G.E., 2014. The role of biomarkers in the assessment of aquatic ecosystem health. Integr. Environ. Assess. Manag. 10, 327–341.

Houston, A.H., 1990. Blood and circulation. In: Schreck, C.B., Moyle, P.B. (Eds.), Methods for Fish Biology. American Fisheries Society, Bethesda, pp. 273–334.

Howarth, R., Chan, F., Conley, D.J., Garnier, J., Doney, S.C., Marino, R., Billen, G., 2011. Coupled biogeochemical cycles: eutrophication and hypoxia in temperate estuaries and coastal marine ecosystems. Front. Ecol. Environ. 9, 18–26.

Hu, C.K., Southey, B.R., Romanova, E.V., Maruska, K.P., Sweedler, J.V., Fernald, R.D., 2016. Identification of prohormones and pituitary neuropeptides in the African cichlid, *Astatotilapia burtoni*. BMC Genomics 17, 660.

Hughes, T.P., Rodrigues, M.J., Belwood, D.R., Ceccarelli, D., Hoegh-Guldberg, O., McCook, L., Moltschaniwskyj, N., Pratchett, M.S., Steneck, R.S., Wills, B., 2007. Phase shifts, herbivory, and the resilience of coral reefs to climate change. Curr. Biol. 17, 360–365.

Huntingford, F.A., 2004. Implications of domestication and rearing conditions for the behaviour of cultivated fishes. J. Fish Biol. 65, 122–142.

Inaba, K., Dréanno, C., Cosson, J., 2003. Control of flatfish sperm motility by CO_2 and carbonic anhydrase. Cell Motil. Cytoskeleton 55, 174–187.

Ings, J.S., George, N., Peter, M.C.S., Servos, M.R., Vijayan, M.M., 2012. Venlafaxine and atenolol disrupt epinephrine-stimulated glucose production in rainbow trout hepatocytes. Aquat. Toxicol. 106–107, 48–55.

Ivy, C.M., Robertson, C.E., Bernier, N.J., 2017. Acute embryonic anoxia exposure favours the development of a dominant and aggressive phenotype in adult zebrafish. Proc. R. Soc. B Biol. Sci. 284, 20161868.

Jacoby, D.M.P., Casselman, J.M., Crook, V., DeLucia, M.B., Ahn, H., Kaifu, K., Kurwie, T., Sasal, P., Silfvergrip, A.M.C., Smith, K.G., Uchida, K., Walker, A.M., Gollock, M.J., 2015. Synergistic patterns of threat and the challenges facing global anguillid eel conservation. Glob. Ecol. Conserv. 4, 321–333.

Janz, D.M., McMaster, M.E., Munkittrick, K.R., Van Der Kraak, G., 1997. Elevated ovarian follicular apoptosis and heat shock protein-70 expression in white sucker exposed to bleached kraft pulp mill effluent. Toxicol. Appl. Pharmacol. 147, 391–398.

Jarque, S., Piña, B., 2014. Deiodinases and thyroid metabolism disruption in teleost fish. Environ. Res. 135, 361–375.

Jeffrey, J.D., Bernier, N.J., Anderson, W.G., 2022. Endocrinology. In: Midway, S., Hasler, C., Chakrabarty, P. (Eds.), Methods in Fish Biology. American Fisheries Society, Bethesda.

Jenny, J.P., Francus, P., Normandeau, A., Lapointe, F., Perga, M.E., Ojala, A., Schimmelmann, A., Zolitschka, B., 2016. Global spread of hypoxia in freshwater ecosystems during the last three centuries is caused by rising local human pressure. Glob. Chang. Biol. 22, 1481–1489.

Jobling, S., Beresford, N., Nolan, M., Rodgers-Gray, T., Brighty, G.C., Sumpter, J.P., Tyler, C.R., 2002a. Altered sexual maturation and gamete production in wild roach (*Rutilus rutilus*) living in rivers that receive treated sewage effluents. Biol. Reprod. 66, 272–281.

Jobling, S., Coey, S., Whitmore, J.G.G., Kime, D.E.E., Van Look, K.J.W.J.W., McAllister, B.G.G., Beresford, N., Henshaw, A.C.C., Brighty, G., Tyler, C.R.R., Sumpter, J.P.P., 2002b. Wild intersex roach (*Rutilus rutilus*) have reduced fertility. Biol. Reprod. 67, 515–524.

Johnson, N.S., Siefkes, M.J., Wagner, C.M., Dawson, H., Wang, H., Steeves, T., Twohey, M., Li, W., 2013. A synthesized mating pheromone component increases adult sea lamprey (*Petromyzon marinus*) trap capture in management scenarios. Can. J. Fish. Aquat. Sci. 70, 1101–1108.

Johnson, N.S., Tix, J.A., Hlina, B.L., Wagner, C.M., Siefkes, M.J., Wang, H., Li, W., 2015. A sea lamprey (*Petromyzon marinus*) sex pheromone mixture increases trap catch relative to a single synthesized component in specific environments. J. Chem. Ecol. 41, 311–321.

Johnson, N.S., Lewandoski, S.A., Alger, B.J., O'Connor, L., Bravener, G., Hrodey, P., Huerta, B., Barber, J., Li, W., Wagner, C.M., Siefkes, M.J., 2020. Behavioral responses of sea lamprey to varying application rates of a synthesized pheromone in diverse trapping scenarios. J. Chem. Ecol. 46, 233–249.

Journey, M.L., Trudel, M., Young, G., Beckman, B.R., 2018. Evidence for depressed growth of juvenile Pacific salmon (*Oncorhynchus*) in Johnstone and Queen Charlotte Straits, British Columbia. Fish. Oceanogr. 27, 174–183.

Kaabia, Z., Laparre, J., Cesbron, N., Le Bizec, B., Dervilly-Pinel, G., 2018. Comprehensive steroid profiling by liquid chromatography coupled to high resolution mass spectrometry. J. Steroid Biochem. Mol. Biol. 183, 106–115.

Kajimura, S., Duan, C., 2007. Insulin-like growth factor-binding protein-1: an evolutionarily conserved fine tuner of insulin-like growth factor action under catabolic and stressful conditions. J. Fish Biol. 71, 309–325.

Kamei, H., 2020. Oxygen and embryonic growth: the role of insulin-like growth factor signaling. Gen. Comp. Endocrinol. 294, 113473.

Kaneko, N., Taniyama, N., Inatani, Y., Nagano, Y., Fujiwara, M., Torao, M., Miyakoshi, Y., Shimizu, M., 2015. Circulating insulin-like growth factor I in juvenile chum salmon: relationship with growth rate and changes during downstream and coastal migration in northeastern Hokkaido, Japan. Fish Physiol. Biochem. 41, 991–1003.

Kamio, M., Yambe, H., Fusetani, N., 2022. Chemical cues for intraspecific chemical communication and interspecific interactions in aquatic environments: applications for fisheries and aquaculture. Fish. Sci. 88, 203–239.

Kaneko, N., Journey, M.L., Neville, C.M., Trudel, M., Beckman, B.R., Shimizu, M., 2019a. Utilization of an endocrine growth index, insulin-like growth factor binding protein (IGFBP)-1b, for postsmolt coho salmon in the Strait of Georgia, British Columbia, Canada. Fish Physiol. Biochem. 45, 1867–1878.

Kaneko, N., Torao, M., Koshino, Y., Fujiwara, M., Miyakoshi, Y., Shimizu, M., 2019b. Evaluation of growth status using endocrine growth indices, insulin-like growth factor (IGF)-I and IGF-binding protein-1b, in out-migrating juvenile chum salmon. Gen. Comp. Endocrinol. 274, 50–59.

Karels, A., Markkula, E., Oikari, A., 2001. Reproductive, biochemical, physiological, and population responses in perch (*Perca fluviatilis* L.) and roach (*Rutilus rutilus* L.) downstream of two elemental chlorine-free pulp and paper mills. Environ. Toxicol. Chem. 20, 1517–1527.

Keller-Costa, T., Hubbard, P.C., Paetz, C., Nakamura, Y., Da Silva, J.P., Rato, A., Barata, E.N., Schneider, B., Canario, A.V.M., 2014. Identity of a tilapia pheromone released by dominant males that primes females for reproduction. Curr. Biol. 24, 2130–2135.

Kelley, K.M., Schmidt, K.E., Berg, L., Sak, K., Galima, M.M., Gillespie, C., Balogh, L., Hawayek, A., Reyes, J.A., Jamison, M., 2002. Comparative endocrinology of the insulin-like growth factor-binding protein. J. Endocrinol. 175, 3–18.

Kennedy, C.J., Smyth, K.R., 2015. Disruption of the rainbow trout reproductive endocrine axis by the polycyclic aromatic hydrocarbon benzo[a]pyrene. Gen. Comp. Endocrinol. 219, 102–111.

Khansari, A.R., Balasch, J.C., Vallejos-Vidal, E., Parra, D., Reyes-López, F.E., Tort, L., 2018. Comparative immune- and stress-related transcript response induced by air exposure and *Vibrio anguillarum* bacterin in rainbow trout (*Oncorhynchus mykiss*) and gilthead seabream (*Sparus aurata*) mucosal surfaces. Front. Immunol. 9.

Kidd, K.A., Blanchfield, P.J., Mills, K.H., Palace, V.P., Evans, R.E., Lazorchak, J.M., Flick, R.W., 2007. Collapse of a fish population after exposure to a synthetic estrogen. Proc. Natl. Acad. Sci. U. S. A. 104, 8897–8901.

Kidd, C.E., Kidd, M.R., Hofmann, H.A., 2010. Measuring multiple hormones from a single water sample using enzyme immunoassays. Gen. Comp. Endocrinol. 165, 277–285.

Kidd, K.A., Paterson, M.J., Rennie, M.D., Podemski, C.L., Findlay, D.L., Blanchfield, P.J., Liber, K., 2014. Direct and indirect responses of a freshwater food web to a potent synthetic oestrogen. Philos. Trans. R. Soc. B Biol. Sci. 369, 20130578.

Kim, S.G.W., Hong, W.H., Han, S.J., Kwon, J., Ko, H., Lee, S.B., Giri, S.S., Kim, S.G.W., Kim, B.Y., Jang, G., Lee, B.C., Kim, D.W., Park, S.C., 2020. Use of synthetic salmon GnRH and domperidone (Ovaprim®) in sharks: preparation for *ex situ* conservation. Front. Mar. Sci. 7, 998.

Kim, S.S., Kan, H., Hwang, K.S., Yang, J.Y., Son, Y., Shin, D.S., Lee, B.H., Ahn, S.H., Ahn, J.H., Cho, S.H., Bae, M.A., 2021. Neurochemical effects of 4-(2chloro-4-fluorobenzyl)-3-(2-thienyl)-1,2,4-oxadiazol-5(4H)-one in the pentylenetetrazole (ptz)-induced epileptic seizure zebrafish model. Int. J. Mol. Sci. 22, 1–15.

King, H.R., Pankhurst, N.W., Watts, M., 2007. Reproductive sensitivity to elevated water temperatures in female Atlantic salmon is heightened at certain stages of vitellogenesis. J. Fish Biol. 70, 190–205.

Kohl, K.D., Yahn, J., 2016. Effects of environmental temperature on the gut microbial communities of tadpoles. Environ. Microbiol. 18, 1561–1565.

Kottmann, J.S., Jørgensen, M.G.P., Bertolini, F., Loh, A., Tomkiewicz, J., 2020. Differential impacts of carp and salmon pituitary extracts on induced oogenesis, egg quality, molecular ontogeny and embryonic developmental competence in European eel. PLoS One 15, e0235617.

Kroska, A.C., Wolf, N., Planas, J.V., Baker, M.R., Smeltz, T.S., Harris, B.P., 2021. Controlled experiments to explore the use of a multi-tissue approach to characterizing stress in wild-caught Pacific halibut (*Hippoglossus stenolepis*). Conserv. Physiol 9, coab001.

Kumar, V., Govindaswamy, U., 2019. Non-invasive monitoring of steroid hormones in wildlife for conservation and management of endangered species-a review non-invasive reproductive steroid monitoring in Indian wild animals. Indian J. Exp. Biol. 57, 307–314.

La Merrill, M.A., Vandenberg, L.N., Smith, M.T., Goodson, W., Browne, P., Patisaul, H.B., Guyton, K.Z., Kortenkamp, A., Cogliano, V.J., Woodruff, T.J., Rieswijk, L., Sone, H., Korach, K.S., Gore, A.C., Zeise, L., Zoeller, R.T., 2020. Consensus on the key characteristics of endocrine-disrupting chemicals as a basis for hazard identification. Nat. Rev. Endocrinol. 16, 45–57.

Labadie, P., Budzinski, H., 2006a. Alteration of steroid hormone balance in juvenile turbot (*Psetta maxima*) exposed to nonylphenol, bisphenol A, tetrabromodiphenyl ether 47, diallylphthalate, oil, and oil spiked with alkylphenols. Arch. Environ. Contam. Toxicol. 50, 552–561.

Labadie, P., Budzinski, H., 2006b. Alteration of steroid hormone profile in juvenile turbot (*Psetta maxima*) as a consequence of short-term exposure to 17α-ethynylestradiol. Chemosphere 64, 1274–1286.

Laberge, F., Hara, T.J., 2001. Neurobiology of fish olfaction: a review. Brain Res. Rev. 36, 46–59.

Laberge, F., Yin-Liao, I., Bernier, N.J., 2019. Temporal profiles of cortisol accumulation and clearance support scale cortisol content as an indicator of chronic stress in fish. Conserv. Physiol. 7, coz052.

Lai, K.P., Li, J.W., Tse, A.C.K., Chan, T.F., Wu, R.S.S., 2016. Hypoxia alters steroidogenesis in female marine medaka through miRNAs regulation. Aquat. Toxicol. 172, 1–8.

Lai, K.P., Li, J.W., Wang, S.Y., Wan, M.T., Chan, T.F., Lui, W.Y., Au, D.W.T., Wu, R.S.S., Kong, R.Y.C., 2018. Transcriptomic analysis reveals transgenerational effect of hypoxia on the neural control of testicular functions. Aquat. Toxicol. 195, 41–48.

Lai, K.P., Wang, S.Y., Li, J.W., Tong, Y., Chan, T.F., Jin, N., Tse, A., Zhang, J.W., Wan, M.T., Tam, N., Au, D.W.T., Lee, B.Y., Lee, J.S., Wong, A.S.T., Kong, R.Y.C., Wu, R.S.S., 2019. Hypoxia causes transgenerational impairment of ovarian development and hatching success in fish. Environ. Sci. Technol. 53, 3917–3928.

Landry, C.A., Steele, S.L., Manning, S., Cheek, A.O., 2007. Long term hypoxia suppresses reproductive capacity in the estuarine fish, *Fundulus grandis*. Comp. Biochem. Physiol. A Mol. Integr. Physiol. 148, 317–323.

Larsen, D.A., Beckman, B.R., Strom, C.R., Parkins, P.J., Cooper, K.A., Fast, D.E., Dickhoff, W.W., 2006. Growth modulation alters the incidence of early male maturation and physiological development of hatchery-reared spring chinook salmon: a comparison with wild fish. Trans. Am. Fish. Soc. 135, 1017–1032.

Larsson, D.G.J., Hällman, H., Förlin, L., 2000. More male fish embryos near a pulp mill. Environ. Toxicol. Chem. 19, 2911–2917.

Larsson, D.G.J., Adolfsson-Erici, M., Thomas, P., 2006. Characterization of putative ligands for a fish gonadal androgen receptor in a pulp mill effluent. Environ. Toxicol. Chem. 25, 419–427.

Lawrence, M.J., Raby, G.D., Teffer, A.K., Jeffries, K.M., Danylchuk, A.J., Eliason, E.J., Hasler, C.T., Clark, T.D., Cooke, S.J., 2020. Best practices for non-lethal blood sampling of fish via the caudal vasculature. J. Fish Biol. 97, 4–15.

Le Page, Y., Vosges, M., Servili, A., Brion, F., Kah, O., 2011. Neuroendocrine effects of endocrine disruptors in teleost fish. J. Toxicol. Environ. Health B Crit. Rev. 14, 370–386.

Lee, S., Hong, S., Liu, X., Kim, C., Jung, D., Yim, U.H., Shim, W.J., Khim, J.S., Giesy, J.P., Choi, K., 2017. Endocrine disrupting potential of PAHs and their alkylated analogues associated with oil spills. Environ Sci Process Impacts 19, 1117–1125.

Legacki, E.L., Robeck, T.R., Steinman, K.J., Conley, A.J., 2020. Comparative analysis of steroids in cyclic and pregnant killer whales, beluga whales and bottlenose dolphins by liquid chromatography tandem mass spectrometry. Gen. Comp. Endocrinol. 285, 113273.

Leong, H., Ros, A.F.H., Oliveira, R.F., 2009. Effects of putative stressors in public aquaria on locomotor activity, metabolic rate and cortisol levels in the Mozambique tilapia *Oreochromis mossambicus*. J. Fish Biol. 74, 1549–1561.

Lepage, O., Øverli, Ø., Petersson, E., Järvi, T., Winberg, S., 2000. Differential stress coping in wild and domesticated sea trout. Brain Behav. Evol. 56, 259–268.

Lerner, D.T., Björnsson, B.T., McCormick, S.D., 2007. Effects of aqueous exposure to polychlorinated biphenyls (Aroclor 1254) on physiology and behavior of smolt development of Atlantic salmon. Aquat. Toxicol. 81, 329–336.

Lerner, D.T., Sheridan, M.A., McCormick, S.D., 2012. Estrogenic compounds decrease growth hormone receptor abundance and alter osmoregulation in Atlantic salmon. Gen. Comp. Endocrinol. 179, 196–204.

Leusch, F.D.L., MacLatchy, D.L., 2003. In vivo implants of β-sitosterol cause reductions of reactive cholesterol pools in mitochondria isolated from gonads of male goldfish (*Carassius auratus*). Gen. Comp. Endocrinol. 134, 255–263.

Li, K., Scott, A.M., Riedy, J.J., Fissette, S., Middleton, Z.E., Li, W., 2017. Three novel bile alcohols of mature male sea lamprey (*Petromyzon marinus*) act as chemical cues for conspecifics. J. Chem. Ecol. 43, 543–549.

Li, K., Brant, C.O., Huertas, M., Hessler, E.J., Mezei, G., Scott, A.M., Hoye, T.R., Li, W., 2018. Fatty-acid derivative acts as a sea lamprey migratory pheromone. Proc. Natl. Acad. Sci. U. S. A. 115, 8603–8608.

Lim, H., Sorensen, P.W., 2011. Polar metabolites synergize the activity of prostaglandin F2α in a species-specific hormonal sex pheromone released by ovulated common carp. J. Chem. Ecol. 37, 695–704.

Lim, H., Sorensen, P.W., 2012. Common carp implanted with prostaglandin F2α release a sex pheromone complex that attracts conspecific males in both the laboratory and field. J. Chem. Ecol. 38, 127–134.

Lindholm-Lehto, P.C., Knuutinen, J.S., Ahkola, H.S.J., Herve, S.H., 2015. Refractory organic pollutants and toxicity in pulp and paper mill wastewaters. Environ. Sci. Pollut. Res. 22, 6473–6499.

Little, A.G., Seebacher, F., 2014. The evolution of endothermy is explained by thyroid hormone-mediated responses to cold in early vertebrates. J. Exp. Biol. 217, 1642–1648.

Liu, X., Jung, D., Zhou, K., Lee, S., Noh, K., Khim, J.S., Giesy, J.P., Yim, U.H., Shim, W.J., Choi, K., 2018. Characterization of endocrine disruption potentials of coastal sediments of Taean, Korea employing H295R and MVLN assays–Reconnaissance at 5 years after Hebei Spirit oil spill. Mar. Pollut. Bull. 127, 264–272.

Lokman, P.M., Vermeulen, G.J., Lambert, J.G.D., Young, G., 1998. Gonad histology and plasma steroid profiles in wild New Zealand freshwater eels (*Anguilla dieffenbachii* and *A. australis*) before and at the onset of the natural spawning migration. I. Females. Fish Physiol. Biochem. 19, 325–338.

Lokman, P.M., Damsteegt, E.L., Wallace, J., Downes, M., Goodwin, S.L., Facoory, L.J., Wylie, M.J., 2016. Dose-responses of male silver eels, *Anguilla australis*, to human chorionic gonadotropin and 11-ketotestosterone *in vivo*. Aquaculture 463, 97–105.

Lower, N., Moore, A., 2007. The impact of a brominated flame retardant on smoltification and olfactory function in Atlantic salmon (*Salmo salar* L.) smolts. Mar. Freshw. Behav. Physiol. 40, 267–284.

Lu, X., Yu, R.M.K., Murphy, M.B., Lau, K., Wu, R.S.S., 2014. Hypoxia disrupts gene modulation along the brain-pituitary-gonad (BPG)-liver axis. Ecotoxicol. Environ. Saf. 102, 70–78.

Lupica, S.J., Turner, J.W., 2009. Validation of enzyme-linked immunosorbent assay for measurement of faecal cortisol in fish. Aquacult. Res. 40, 437–441.

MacDougall-Shackleton, S.A., Bonier, F., Romero, L.M., Moore, I.T., 2019. Glucocorticoids and "stress" are not synonymous. Integr. Org. Biol. 1, 1–8.

Madison, B.N., Heath, J.W., Heath, D.D., Bernier, N.J., 2015a. Effects of early rearing environment and breeding strategy on social interactions and the hormonal response to stressors in juvenile Chinook salmon. Can. J. Fish. Aquat. Sci. 72, 673–683.

Madison, B.N., Tavakoli, S., Kramer, S., Bernier, N.J., 2015b. Chronic cortisol and the regulation of food intake and the endocrine growth axis in rainbow trout. J. Endocrinol. 226, 103–119.

Madliger, C.L., Love, O.P., 2014. The need for a predictive, context-dependent approach to the application of stress hormones in conservation. Conserv. Biol. 28, 283–287.

Mahanty, A., Purohit, G.K., Mohanty, S., Mohanty, B.P., 2019. Heat stress–induced alterations in the expression of genes associated with gonadal integrity of the teleost *Puntius sophore*. Fish Physiol. Biochem. 45, 1409–1417.

Marcon, M., Mocelin, R., Benvenutti, R., Costa, T., Herrmann, A.P., De Oliveira, D.L., Koakoski, G., Barcellos, L.J.G., Piato, A., 2018. Environmental enrichment modulates the response to chronic stress in zebrafish. J. Exp. Biol. 221, jeb176735.

Martel, P.H., O'Connor, B.I., Kovacs, T.G., Van Den Heuvel, M.R., Parrott, J.L., McMaster, M.E., MacLatchy, D.L., Van Der Kraak, G.J., Hewitt, L.M., 2017. The relationship between organic loading and effects on fish reproduction for pulp mill effluents across Canada. Environ. Sci. Technol. 51, 3499–3507.

Martinović-Weigelt, D., Mehinto, A.C., Ankley, G.T., Denslow, N.D., Barber, L.B., Lee, K.E., King, R.J., Schoenfuss, H.L., Schroeder, A.L., Villeneuve, D.L., 2014. Transcriptomic effects-based monitoring for endocrine active chemicals: assessing relative contribution of treated wastewater to downstream pollution. Environ. Sci. Technol. 48, 2385–2394.

Martyniuk, C.J., Feswick, A., Munkittrick, K.R., Dreier, D.A., Denslow, N.D., 2020. Twenty years of transcriptomics, 17alpha-ethinylestradiol, and fish. Gen. Comp. Endocrinol. 286, 113325.

Matthews, J., Gustafsson, J.-Å., 2006. Estrogen receptor and aryl hydrocarbon receptor signaling pathways. Nucl. Recept. Signal. 4, nrs.04016.

Matthiessen, P., Wheeler, J.R., Weltje, L., 2018. A review of the evidence for endocrine disrupting effects of current-use chemicals on wildlife populations. Crit. Rev. Toxicol. 48, 195–216.

Maule, A.G., Schreck, C.B., Kaattari, S.L., 1987. Changes in the immune system of coho salmon (*Oncorhynchus kisutch*) during the parr-to-smolt transformation and after implantation of cortisol. Can. J. Fish. Aquat. Sci. 44, 161–166.

Mazur, C.F., Iwama, G.K., 1993. Effect of handling and stocking density on hematocrit, plasma cortisol, and survival in wild and hatchery-reared chinook salmon (*Oncorhynchus tshawytscha*). Aquaculture 112, 291–299.

Mazzeo, I., Peñaranda, D.S., Gallego, V., Baloche, S., Nourizadeh-Lillabadi, R., Tveiten, H., Dufour, S., Asturiano, J.F., Weltzien, F.A., Pérez, L., 2014. Temperature modulates the progression of vitellogenesis in the European eel. Aquaculture 434, 38–47.

McConnachie, S.H., Cook, K.V., Patterson, D.A., Gilmour, K.M., Hinch, S.G., Farrell, A.P., Cooke, S.J., 2012. Consequences of acute stress and cortisol manipulation on the physiology, behavior, and reproductive outcome of female Pacific salmon on spawning grounds. Horm. Behav. 62, 67–76.

McCormick, S.D., 2013. Smolt physiology and endocrinology. In: McCormick, S.D., Farrell, A.P., Brauner, C.J. (Eds.), Fish Physiology: Euryhaline Fishes, vol. 32. Academic Press, London, pp. 199–251.

McCormick, M.I., Makey, L., Dufour, V., 2002. Comparative study of metamorphosis in tropical reef fishes. Mar. Biol. 141, 841–853.

McCormick, S.D., Regish, A., O'Dea, M.F., Shrimpton, J.M., 2008. Are we missing a mineralocorticoid in teleost fish? Effects of cortisol, deoxycorticosterone and aldosterone on osmoregulation, gill Na^+,K^+-ATPase activity and isoform mRNA levels in Atlantic salmon. Gen. Comp. Endocrinol. 157, 35–40.

McCormick, S.D., Lerner, D.T., Monette, M.Y., Nieves-Puigdoller, K., Kelly, J.T., Björnsson, B.T., 2009. Taking it with you when you go: how perturbations to the freshwater environment, including temperature, dams, and contaminants, affect marine survival of salmon. Am. Fish. Soc. Symp. 69, 195–214.

McCormick, S.D., Farrell, A.P., Brauner, C.J., 2012. Fish Physiology: Euryhaline Fishes, vol. 32. Academic Press, London.

McCormick, S.D., Romero, M., 2017. Conservation endocrinology. BioScience 67, 429–442.

McMaster, M.E., Van Der Kraak, G.J., Portt, C.B., Munkittrick, K.R., Sibley, P.K., Smith, I.R., Dixon, D.G., 1991. Changes in hepatic mixed-function oxygenase (MFO) activity, plasma steroid levels and age at maturity of a white sucker (*Catostomus commersoni*) population exposed to bleached kraft pulp mill effluent. Aquat. Toxicol. 21, 199–217.

McMaster, M.E., Hewitt, L.M., Parrott, J.L., 2006. A decade of research on the environmental impacts of pulp and paper mill effluents in Canada: field studies and mechanistic research. J. Toxicol. Environ. Health B Crit. Rev. 9, 319–339.

McMaster, M.E., Van Der Kraak, G.J., Munkittrick, K.R., 1995. Exposure to bleached kraft pulp mill effluent reduces the steroid biosynthetic capacity of white sucker ovarian follicles. Comp. Biochem. Physiol. 112, 169–178.

McMenamin, S.K., Parichy, D.M., 2013. Metamorphosis in teleosts. Curr. Top. Dev. Biol. 103, 127–165.

McPhie, R.P., Campana, S.E., 2009. Reproductive characteristics and population decline of four species of skate (*Rajidae*) off the eastern coast of Canada. J. Fish Biol. 75, 223–246.

Meckley, T.D., Wagner, C.M., Gurarie, E., 2014. Coastal movements of migrating sea lamprey (*Petromyzon marinus*) in response to a partial pheromone added to river water: implications for management of invasive populations. Can. J. Fish. Aquat. Sci. 71, 533–544.

Medeiros, L.R., Galbreath, P.F., Knudsen, C.M., Stockton, C.A., Koch, I.J., Bosch, W.J., Narum, S.R., Nagler, J.J., Pierce, A.L., 2018. Plasma 11-ketotestosterone in individual age-1 spring chinook salmon males accurately predicts age-2 maturation status. Trans. Am. Fish. Soc. 147, 1042–1051.

Mendoza Alfaro, R., González, C.A., Ferrara, A.M., 2008. Gar biology and culture: status and prospects. Aquacult. Res. 39, 748–763.

Mennigen, J.A., Stroud, P., Zamora, J.M., Moon, T.W., Trudeau, V.L., 2011. Pharmaceuticals as neuroendocrine disruptors: lessons learned from fish on prozac. J. Toxicol. Environ. Health B Crit. Rev. 14, 387–412.

Mes, D., von Krogh, K., Gorissen, M., Mayer, I., Vindas, M.A., 2018. Neurobiology of wild and hatchery-reared Atlantic salmon: how nurture drives neuroplasticity. Front. Behav. Neurosci. 12, 210.

Mes, D., Van Os, R., Gorissen, M., Ebbesson, L.O.E., Finstad, B., Mayer, I., Vindas, M.A., 2019. Effects of environmental enrichment on forebrain neural plasticity and survival success of stocked Atlantic salmon. J. Exp. Biol. 22, jeb212258.

Meyer, J.S., Novak, M.A., 2012. Minireview: hair cortisol: a novel biomarker of hypothalamic-pituitary- adrenocortical activity. Endocrinology 153, 4120–4127.

Middleton, M.A., Larsen, D.A., Dickey, J.T., Swanson, P., 2019. Evaluation of endocrine and transcriptomic markers of male maturation in winter-run Steelhead Trout (*Oncorhynchus mykiss*). Gen. Comp. Endocrinol. 281, 30–40.

Milazzo, M., Cattano, C., Alonzo, S.H., Foggo, A., Gristina, M., Rodolfo-Metalpa, R., Sinopoli, M., Spatafora, D., Stiver, K.A., Hall-Spencer, J.M., 2016. Ocean acidification affects fish spawning but not paternity at CO_2 seeps. Proc. R. Soc. B Biol. Sci. 283, 20161021.

Milla, S., Pasquet, A., El Mohajer, L., Fontaine, P., 2021. How domestication alters fish phenotypes. Rev. Aquac. 13, 388–405.

Miller, G.M., Watson, S.A., Mccormick, M.I., Munday, P.L., 2013. Increased CO_2 stimulates reproduction in a coral reef fish. Glob. Chang. Biol. 19, 3037–3045.

Miller, G.M., Kroon, F.J., Metcalfe, S., Munday, P.L., 2015. Temperature is the evil twin: effects of increased temperature and ocean acidification on reproduction in a reef fish. Ecol. Appl. 25, 603–620.

Mills, L.J., Chichester, C., 2005. Review of evidence: are endocrine-disrupting chemicals in the aquatic environment impacting fish populations? Sci. Total Environ. 343, 1–34.

Mohammadzadeh, S., Yeganeh, S., Moradian, F., Milla, S., Falahatkar, B., 2021. Spawning induction in Sterlet sturgeon (*Acipenser ruthenus*) with recombinant GnRH: analysis of hormone profiles and spawning indices. Aquaculture 533, 736108.

Molés, G., Hausken, K., Carrillo, M., Zanuy, S., Levavi-Sivan, B., Gómez, A., 2020. Generation and use of recombinant gonadotropins in fish. Gen. Comp. Endocrinol. 299, 113555.

Mommsen, T.P., Vijayan, M.M., Moon, T.W., 1999. Cortisol in teleosts: dynamics, mechanisms of action, and metabolic regulation. Rev. Fish Biol. Fish. 9, 211–268.

Monette, M.Y., Björnsson, B.T., McCormick, S.D., 2008. Effects of short-term acid and aluminum exposure on the parr-smolt transformation in Atlantic salmon (*Salmo salar*): disruption of seawater tolerance and endocrine status. Gen. Comp. Endocrinol. 158, 122–130.

Monfort, S.L., 2002. Non-invasive endocrine measures of reproduction and stress in wild populations. In: Holt, W.V., Pickard, A.R., Rodger, J.C., Wildt, D.E. (Eds.), Reproductive Science and Integrated Conservation. Cambridge University Press, Cambridge, pp. 147–165.

Monteiro, P.R.R., Reis-Henriques, M.A., Coimbra, J., 2000. Plasma steroid levels in female flounder (*Platichthys flesus*) after chronic dietary exposure to single polycyclic aromatic hydrocarbons. Mar. Environ. Res. 49, 453–467.

Morita, K., Saito, T., Miyakoshi, Y., Fukuwaka, M.-A., Nagasawa, T., Kaeriyama, M., 2006. A review of Pacific salmon hatchery programmes on Hokkaido Island, Japan. ICES J. Mar. Sci. 63, 1353–1363.

Mortensen, A.S., Arukwe, A., 2007. Interactions between estrogen- and ah-receptor signalling pathways in primary culture of salmon hepatocytes exposed to nonylphenol and 3,3′,4,4′-tetrachlorobiphenyl (Congener 77). Comp. Hepatol. 6, 1–14.

Mortensen, A.S., Arukwe, A., 2008. Estrogenic effect of dioxin-like aryl hydrocarbon receptor (AhR) agonist (PCB congener 126) in salmon hepatocytes. Mar. Environ. Res. 66, 119–120.

Munday, P.L., Donelson, J.M., Dixson, D.L., Endo, G.G.K., 2009. Effects of ocean acidification on the early life history of a tropical marine fish. Proc. R. Soc. B Biol. Sci. 276, 3275–3283.

Munday, P.L., Gagliano, M., Donelson, J.M., Dixson, D.L., Thorrold, S.R., 2011. Ocean acidification does not affect the early life history development of a tropical marine fish. Mar. Ecol. Prog. Ser. 423, 211–221.

Munkittrick, K.R., Portt, C.B., Van Der Kraak, G.J., Smith, I.R., Rokosh, D.A., 1991. Impact of bleached kraft mill effluent on population characteristics, liver MFO activity, and serum steroid levels of a Lake Superior white sucker (*Catostomus commersoni*) population. Can. J. Fish. Aquat. Sci. 48, 1371–1380.

Murphy, C.A., Stacey, N.E., Corkum, L.D., 2001. Putative steroidal pheromones in the round goby, *Neogobius melanostomus*: olfactory and behavioral responses. J. Chem. Ecol. 27, 443–470.

Myers, R.A., Rosenberg, A.A., Mace, P.M., Barrowman, N., Restrepo, V.R., 1994. In search of thresholds for recruitment overfishing. ICES J. Mar. Sci. 51, 191–205.

Mylonas, C.C., Fostier, A., Zanuy, S., 2010. Broodstock management and hormonal manipulations of fish reproduction. Gen. Comp. Endocrinol. 165, 516–534.

Mylonas, C.C., Duncan, N.J., Asturiano, J.F., 2017. Hormonal manipulations for the enhancement of sperm production in cultured fish and evaluation of sperm quality. Aquaculture 472, 21–44.

Näslund, J., Johnsson, J.I., 2016. Environmental enrichment for fish in captive environments: effects of physical structures and substrates. Fish Fish. 17, 1–30.

Näslund, J., Rosengren, M., Del Villar, D., Gansel, L., Norrgård, J.R., Persson, L., Winkowski, J.J., Kvingedal, E., 2013. Hatchery tank enrichment affects cortisol levels and shelter-seeking in Atlantic salmon (*Salmo salar*). Can. J. Fish. Aquat. Sci. 70, 585–590.

Navarrete, P., Mardones, P., Opazo, R., Espejo, R., Romero, J., 2008. Oxytetracycline treatment reduces bacterial diversity of intestinal microbiota of Atlantic salmon. J. Aquat. Anim. Health 20, 177–183.

Navarro-Martín, L., Viñas, J., Ribas, L., Díaz, N., Gutiérrez, A., Di Croce, L., Piferrer, F., 2011. DNA methylation of the gonadal aromatase (cyp19a) promoter is involved in temperature-dependent sex ratio shifts in the European sea bass. PLoS Genet. 7, e1002447.

Nelson, E.R., Habibi, H.R., 2009. Thyroid receptor subtypes: structure and function in fish. Gen. Comp. Endocrinol. 161, 90–96.

Nesan, D., Vijayan, M.M., 2013. Role of glucocorticoid in developmental programming: evidence from zebrafish. Gen. Comp. Endocrinol. 181, 35–44.

Nesan, D., Kamkar, M., Burrows, J., Scott, I.C., Marsden, M., Vijayan, M.M., 2012. Glucocorticoid receptor signaling is essential for mesoderm formation and muscle development in zebrafish. Endocrinology 153, 1288–1300.

Neuman, H., Debelius, J.W., Knight, R., Koren, O., 2015. Microbial endocrinology: the interplay between the microbiota and the endocrine system. FEMS Microbiol. Rev. 39, 509–521.

Nilsson, S., 1983. Chromaffin tissue. In: Nilsson, S. (Ed.), Autonomic Nerve Function in the Vertebrates. Zoophysiology. vol. 13. Springer, Berlin, pp. 100–111.

Noguera, J.C., Aira, M., Pérez-Losada, M., Domínguez, J., Velando, A., 2018. Glucocorticoids modulate gastrointestinal microbiome in a wild bird. R. Soc. Open Sci. 5, 171743.

Nordgarden, U., Fjelldal, P.G., Hansen, T., Björnsson, B.T., Wargelius, A., 2006. Growth hormone and insulin-like growth factor-I act together and independently when regulating growth in vertebral and muscle tissue of atlantic salmon postsmolts. Gen. Comp. Endocrinol. 149, 253–260.

Nugegoda, D., Kibria, G., 2017. Effects of environmental chemicals on fish thyroid function: implications for fisheries and aquaculture in Australia. Gen. Comp. Endocrinol. 244, 40–53.

O'Connor, C.M., Gilmour, K.M., Arlinghaus, R., Matsumura, S., Suski, C.D., Philipp, D.P., Cooke, S.J., 2011. The consequences of short-term cortisol elevation on individual physiology and growth rate in wild largemouth bass (*Micropterus salmoides*). Can. J. Fish. Aquat. Sci. 68, 693–705.

Ochs, C.L., Laframboise, A.J., Green, W.W., Basilious, A., Johnson, T.B., Zielinski, B.S., 2013. Response to putative round goby (*Neogobius melanostomus*) pheromones by centrarchid and percid fish species in the Laurentian Great Lakes. J. Great Lakes Res. 39, 186–189.

O'Connor, C.M., Gilmour, K.M., Arlinghaus, R., Van Der Kraak, G., Cooke, S.J., 2009. Stress and parental care in a wild teleost fish: insights from exogenous supraphysiological cortisol implants. Physiol. Biochem. Zool. 82, 709–719.

Okuzawa, K., Gen, K., 2013. High water temperature impairs ovarian activity and gene expression in the brain-pituitary-gonadal axis in female red seabream during the spawning season. Gen. Comp. Endocrinol. 194, 24–30.

Olesti, E., Boccard, J., Visconti, G., González-Ruiz, V., Rudaz, S., 2021. From a single steroid to the steroidome: trends and analytical challenges. J. Steroid Biochem. Mol. Biol. 206, 105797.

Oliveira, R.F., Almada, V.C., Canario, A.V.M., 1996. Social modulation of sex steroid concentrations in the urine of male cichlid fish *Oreochromis mossambicus*. Horm. Behav. 30, 2–12.

Oliveira, R.F., Lopes, M., Carneiro, L.A., Canário, A.V.M., 2001. Watching fights raises fish hormone levels. Nature 409, 475.

Olla, B.L., Davis, M.W., Ryer, C.H., 1998. Understanding how the hatchery environment represses or promotes the development of behavioral survival skills. Bull. Mar. Sci. 62, 531–550.

Ondricek, K., Thomas, P., 2018. Effects of hypoxia exposure on apoptosis and expression of membrane steroid receptors, ZIP9, mPRα and GPER in Atlantic croaker ovaries. Comp. Biochem. Physiol. A Mol. Integr. Physiol. 224, 84–92.

Orias, F., Bony, S., Devaux, A., Durrieu, C., Aubrat, M., Hombert, T., Wigh, A., Perrodin, Y., 2015. Tamoxifen ecotoxicity and resulting risks for aquatic ecosystems. Chemosphere 128, 79–84.

Orrego, R., Milestone, C.B., Hewitt, L.M., Guchardi, J., Heid-Furley, T., Slade, A., MacLatchy, D.L., Holdway, D., 2017. Evaluating the potential of effluent extracts from pulp and paper mills in Canada, Brazil, and New Zealand to affect fish reproduction: estrogenic effects in fish. Environ. Toxicol. Chem. 36, 1547–1555.

Ortega-Recalde, O., Goikoetxea, A., Hore, T.A., Todd, E.V., Gemmell, N.J., 2020. The genetics and epigenetics of sex change in fish. Annu. Rev. Anim. Biosci. 8, 47–69.

Ospina-Álvarez, N., Piferrer, F., 2008. Temperature-dependent sex determination in fish revisited: prevalence, a single sex ratio response pattern, and possible effects of climate change. PLoS One 3, e2837.

Øverli, Ø., Winberg, S., Pottinger, T.G., 2005. Behavioral and neuroendocrine correlates of selection for stress responsiveness in rainbow trout—a review. Integr. Comp. Biol. 45, 463–474.

Overturf, M.D., Anderson, J.C., Pandelides, Z., Beyger, L., Holdway, D.A., 2015. Pharmaceuticals and personal care products: a critical review of the impacts on fish reproduction. Crit. Rev. Toxicol. 45, 469–491.

Pait, A.S., Nelson, J.O., 2009. A survey of indicators for reproductive endocrine disruption in *Fundulus heteroclitus* (killifish) at selected sites in the Chesapeake Bay. Mar. Environ. Res. 68, 170–177.

Palace, V.P., Evans, R.E., Wautier, K.G., Mills, K.H., Blanchfield, P.J., Park, B.J., Baron, C.L., Kidd, K.A., 2009. Interspecies differences in biochemical, histopathological, and population responses in four wild fish species exposed to ethynylestradiol added to a whole lake. Can. J. Fish. Aquat. Sci. 66, 1920–1935.

Palme, R., 2019. Non-invasive measurement of glucocorticoids: advances and problems. Physiol. Behav. 199, 229–243.

Pankhurst, N.W., 2011. The endocrinology of stress in fish: an environmental perspective. Gen. Comp. Endocrinol. 170, 265–275.

Pankhurst, N.W., 2016. Reproduction and development. In: Schreck, C.B., Tort, L., Farrell P., A, Brauner, C.J. (Eds.), Fish Physiology: Biology of Stress in Fish, vol. 35. Academic Press, London, pp. 295–331.

Pankhurst, N.W., Munday, P.L., 2011. Effects of climate change on fish reproduction and early life history stages. Mar. Freshw. Res. 62, 1015–1026.

Pankhurst, N.W., King, H.R., Anderson, K., Elizur, A., Pankhurst, P.M., Ruff, N., 2011. Thermal impairment of reproduction is differentially expressed in maiden and repeat spawning Atlantic salmon. Aquaculture 316, 77–87.

Pankhurst, N.W., Van Der Kraak, G., 1997. Effects of stress on growth and reproduction. In: Iwama, G.K., Pickering, A.D., Sumpter, J.P., Schreck, C.B. (Eds.), Fish Stress and Health in Aquaculture. Cambridge University Press, Cambridge, pp. 73–93.

Parrott, J.L., McMaster, M.E., Hewitt, L.M., 2006. A decade of research on the environmental impacts of pulp and paper mill effluents in Canada: development and application of fish bioassays. J. Toxicol. Environ. Health B Crit. Rev. 9, 297–317.

Peñaranda, D.S., Gallego, V., Rozenfeld, C., Herranz-Jusdado, J.G., Pérez, L., Gómez, A., Giménez, I., Asturiano, J.F., 2018. Using specific recombinant gonadotropins to induce spermatogenesis and spermiation in the European eel (*Anguilla anguilla*). Theriogenology 107, 6–20.

Pepin, P., 2016. Reconsidering the impossible—linking environmental drivers to growth, mortality, and recruitment of fish. Can. J. Fish. Aquat. Sci. 73, 205–215.

Pérez, L., Peñaranda, D.S., Dufour, S., Baloche, S., Palstra, A.P., Van Den Thillart, G.E.E.J.M., Asturiano, J.F., 2011. Influence of temperature regime on endocrine parameters and vitellogenesis during experimental maturation of European eel (*Anguilla anguilla*) females. Gen. Comp. Endocrinol. 174, 51–59.

Pérez-Sánchez, J., Terova, G., Simó-Mirabet, P., Rimoldi, S., Folkedal, O., Calduch-Giner, J.A., Olsen, R.E., Sitjà-Bobadilla, A., 2017. Skin mucus of gilthead sea bream (*Sparus aurata* L.). protein mapping and regulation in chronically stressed fish. Front. Physiol. 8, 34.

Perry, S.F., Bernier, N.J., 1999. The acute humoral adrenergic stress response in fish: facts and fiction. Aquaculture 177, 285–295.

Peter, R.E., Lin, H.R., Van Der Kraak, G., 1988. Induced ovulation and spawning of cultured freshwater fish in China: advances in application of GnRH analogues and dopamine antagonists. Aquaculture 74, 1–10.

Peterson, B.C., Waldbieser, G.C., 2009. Effects of fasting on IGF-I, IGF-II, and IGF-binding protein mRNA concentrations in channel catfish (*Ictalurus punctatus*). Domest. Anim. Endocrinol. 37, 74–83.

Petochi, T., Di Marco, P., Priori, A., Finoia, M.G., Mercatali, I., Marino, G., 2011. Coping strategy and stress response of European sea bass *Dicentrarchus labrax* to acute and chronic environmental hypercapnia under hyperoxic conditions. Aquaculture 315, 312–320.

Pfalzgraff, T., Lund, I., Skov, P.V., 2021. Cortisol affects feed utilization, digestion and performance in juvenile rainbow trout (*Oncorhynchus mykiss*). Aquaculture 536, 736472.

Philip, A.M., Vijayan, M.M., 2015. Stress-immune-growth interactions: cortisol modulates suppressors of cytokine signaling and JAK/STAT pathway in rainbow trout liver. PLoS One 10, e0129299.

Picha, M.E., Turano, M.J., Beckman, B.R., Borski, R.J., 2008. Endocrine biomarkers of growth and applications to aquaculture: a minireview of growth hormone, insulin-like growth factor (IGF)-I, and IGF-binding proteins as potential growth indicators in fish. N. Am. J. Aquac. 70, 196–211.

Pickering, A.D., Pottinger, T.G., 1989. Stress responses and disease resistance in salmonid fish: effects of chronic elevation of plasma cortisol. Fish Physiol. Biochem. 7, 253–258.

Pikulkaew, S., Benato, F., Celeghin, A., Zucal, C., Skobo, T., Colombo, L., Dalla Valle, L., 2011. The knockdown of maternal glucocorticoid receptor mRNA alters embryo development in zebrafish. Dev. Dyn. 240, 874–889.

Pottinger, T.G., Carrick, T.R., 1999. Modification of the plasma cortisol response to stress in rainbow trout by selective breeding. Gen. Comp. Endocrinol. 116, 122–132.

Pounder, K.C., Mitchell, J.L., Thomson, J.S., Pottinger, T.G., Buckley, J., Sneddon, L.U., 2016. Does environmental enrichment promote recovery from stress in rainbow trout? Appl. Anim. Behav. Sci. 176, 136–142.

Power, D.M., Llewellyn, L., Faustino, M., Nowell, M.A., Björnsson, B.T., Einarsdottir, I.E., Canario, A.V.M., Sweeney, G.E., 2001. Thyroid hormones in growth and development of fish. Comp. Biochem. Physiol., Part C: Toxicol. Pharmacol. 130, 447–459.

Preus-Olsen, G., Olufsen, M.O., Pedersen, S.A., Letcher, R.J., Arukwe, A., 2014. Effects of elevated dissolved carbon dioxide and perfluorooctane sulfonic acid, given singly and in combination, on steroidogenic and biotransformation pathways of Atlantic cod. Aquat. Toxicol. 155, 222–235.

Rand, P.S., Berejikian, B.A., Bidlack, A., Bottom, D., Gardner, J., Kaeriyama, M., Lincoln, R., Nagata, M., Pearsons, T.N., Schmidt, M., Smoker, W.W., Weitkamp, L.A., Zhivotovsky, L.A., 2012. Ecological interactions between wild and hatchery salmonids and key recommendations for research and management actions in selected regions of the North Pacific. Environ. Biol. Fishes 94, 343–358.

Reid, S.G., Bernier, N.J., Perry, S.F., 1998. The adrenergic stress response in fish: control of catecholamine storage and release. Comp. Biochem. Physiol. C Pharmacol. Toxicol. Endocrinol. 120, 1–27.

Reid, G.M., Hall, H., 2003. Reproduction in fishes in relation to conservation. In: Holt, W.V., Pickard, A.R., Rodger, J.C., Wildt, D.E. (Eds.), Reproductive Science and Integrated Conservation. Cambridge University Press, Cambridge, p. 375. 39.

Reid, N.M., Whitehead, A., 2016. Functional genomics to assess biological responses to marine pollution at physiological and evolutionary timescales: toward a vision of predictive ecotoxicology. Brief. Funct. Genomics 15, 358–364.

Reindl, K.M., Sheridan, M.A., 2012. Peripheral regulation of the growth hormone-insulin-like growth factor system in fish and other vertebrates. Comp. Biochem. Physiol. A Mol. Integr. Physiol. 163, 231–245.

Reinecke, M., Björnsson, B.T., Dickhoff, W.W., McCormick, S.D., Navarro, I., Power, D.M., Gutiérrez, J., 2005. Growth hormone and insulin-like growth factors in fish: where we are and where to go. Gen. Comp. Endocrinol. 142, 20–24.

Rhodes, J.S., Quinn, T.P., 1998. Factors affecting the outcome of territorial contests between hatchery and naturally reared coho salmon parr in the laboratory. J. Fish Biol. 53, 1220–1230.

Riley, S.C., Tatara, C.P., Scheurer, J.A., 2005. Aggression and feeding of hatchery-reared and naturally reared steelhead (*Oncorhynchus mykiss*) fry in a laboratory flume and a comparison with observations in natural streams. Can. J. Fish. Aquat. Sci. 62, 1400–1409.

Robertson, C.E., Wright, P.A., Köblitz, L., Bernier, N.J., 2014. Hypoxia-inducible factor-1 mediates adaptive developmental plasticity of hypoxia tolerance in zebrafish, *Danio rerio*. Proc. R. Soc. B Biol. Sci. 281, 20140637.

Rogers, N.J., Urbina, M.A., Reardon, E.E., McKenzie, D.J., Wilson, R.W., 2016. A new analysis of hypoxia tolerance in fishes using a database of critical oxygen level (Pcrit). Conserv. Physiol. 4, cow012.

Romanova, E.V., Sweedler, J.V., 2015. Peptidomics for the discovery and characterization of neuropeptides and hormones. Trends Pharmacol. Sci. 36, 579–586.

Romero, L.M., Beattie, U.K., 2022. Common myths of glucocorticoid function in ecology and conservation. J. Exp. Zool. A Ecol. Integr. Physiol. 337, 7–14.

Romero, L.M., Fairhurst, G.D., 2016. Measuring corticosterone in feathers: strengths, limitations, and suggestions for the future. Comp. Biochem. Physiol. A Mol. Integr. Physiol. 202, 112–122.

Rosengren, M., Kvingedal, E., Näslund, J., Johnsson, J.I., Sundell, K., 2017. Born to be wild: effects of rearing density and environmental enrichment on stress, welfare, and smolt migration in hatchery-reared Atlantic salmon. Can. J. Fish. Aquat. Sci. 74, 396–405.

Russell, E., Koren, G., Rieder, M., Van Uum, S., 2012. Hair cortisol as a biological marker of chronic stress: current status, future directions and unanswered questions. Psychoneuroendocrinology 37, 589–601.

Sadoul, B., Vijayan, M.M., 2016. Stress and growth. In: Schreck, C.B., Tort, L., Farrell P., A, Brauner, C.J. (Eds.), Fish Physiology: Biology of Stress in Fish, vol. 35. Academic Press, London, pp. 167–205.

Salonius, K., Iwama, G.K., 1993. Effects of early rearing environment on stress response, immune function, and disease resistance in juvenile coho (*Oncorhynchus kisutch*) and chinook salmon (*O. tshawytscha*). Can. J. Fish. Aquat. Sci. 50, 759–766.

Salvanes, A.G.V., Moberg, O., Ebbesson, L.O.E., Nilsen, T.O., Jensen, K.H., Braithwaite, V.A., 2013. Environmental enrichment promotes neural plasticity and cognitive ability in fish. Proc. R. Soc. B Biol. Sci. 280, 20131331.

San Juan, P.A., Hendershot, J.N., Daily, G.C., Fukami, T., 2020. Land-use change has host-specific influences on avian gut microbiomes. ISME J. 14, 318–321.

Sandstrom, O., 1994. Incomplete recovery in a coastal fish community exposed to effluent from a modernized Swedish bleached kraft mill. Can. J. Fish. Aquat. Sci. 51, 2195–2202.

Schade, F.M., Clemmesen, C., Mathias Wegner, K., 2014. Within- and transgenerational effects of ocean acidification on life history of marine three-spined stickleback (*Gasterosteus aculeatus*). Mar. Biol. 161, 1667–1676.

Schiffer, L., Barnard, L., Baranowski, E.S., Gilligan, L.C., Taylor, A.E., Arlt, W., Shackleton, C.H.L., Storbeck, K.H., 2019. Human steroid biosynthesis, metabolism and excretion are differentially reflected by serum and urine steroid metabolomes: a comprehensive review. J. Steroid Biochem. Mol. Biol. 194, 105439.

Schreck, C.B., Scanlon, P.F., 1977. Endocrinology in fisheries and wildlife: biology and management. Fisheries 2, 20–30.

Schultz, D.R., Perez, N., Tan, C.K., Mendez, A.J., Capo, T.R., Snodgrass, D., Prince, E.D., Serafy, J.E., 2005. Concurrent levels of 11-ketotestosterone in fish surface mucus, muscle tissue and blood. J. Appl. Ichthyol. 21, 394–398.

Schultz, D.R., Perez, N., Mendez, A.J., Snodgrass, D., Serafy, J.E., Prince, E.D., Crow, W.A., Capo, T.R., 2007. Tracking gender factors in fish surface mucus: temporal patterns in individual Koi (*Cyprinus carpio*). J. Appl. Ichthyol. 23, 184–188.

Scott, A.P., Ellis, T., 2007. Measurement of fish steroids in water-a review. Gen. Comp. Endocrinol. 153, 392–400.

Scott, A.P., Sorensen, P.W., 1994. Time course of release of pheromonally active gonadal steroids and their conjugates by ovulatory goldfish. Gen. Comp. Endocrinol. 96, 309–323.

Scott, A.P., Hirschenhauser, K., Bender, N., Oliveira, R., Earley, R.L., Sebire, M., Ellis, T., Pavlidis, M., Hubbard, P.C., Huertas, M., Canario, A., 2008. Non-invasive measurement of steroids in fish-holding water: important considerations when applying the procedure to behaviour studies. Behaviour 145, 1307–1328.

Scott, A.P., Sumpter, J.P., Stacey, N., 2010. The role of the maturation-inducing steroid, 17,20β-dihydroxypregn-4-en-3-one, in male fishes: a review. J. Fish Biol. 76, 183–224.

Scott, A.P., Ellis, T., Tveiten, H., 2014. Identification of cortisol metabolites in the bile of Atlantic cod *Gadus morhua* L. Steroids 88, 26–35.

Sebire, M., Katsiadaki, I., Scott, A.P., 2007. Non-invasive measurement of 11-ketotestosterone, cortisol and androstenedione in male three-spined stickleback (*Gasterosteus aculeatus*). Gen. Comp. Endocrinol. 152, 30–38.

Segner, H., 2011. Reproductive and developmental toxicity in fishes. In: Gupta, R.C. (Ed.), Reproductive and Developmental Toxicology. Academic Press, London, pp. 1145–1166.

Servili, A., Canario, A.V.M., Mouchel, O., Muñoz-Cueto, J.A., 2020. Climate change impacts on fish reproduction are mediated at multiple levels of the brain-pituitary-gonad axis. Gen. Comp. Endocrinol. 291, 113439.

Shahjahan, M., Kitahashi, T., Ogawa, S., Parhar, I.S., 2013. Temperature differentially regulates the two kisspeptin systems in the brain of zebrafish. Gen. Comp. Endocrinol. 193, 79–85.

Servos, M.R., Bennie, D.T., Burnison, B.K., Jurkovic, A., McInnis, R., Neheli, T., Schnell, A., Seto, P., Smyth, S.A., Ternes, T.A., 2005. Distribution of estrogens, 17beta-estradiol and estrone, in Canadian municipal wastewater treatment plants. Sci. Total Environ. 336, 155–170.

Shahjahan, M., Kitahashi, T., Ando, H., 2017. Temperature affects sexual maturation through the control of kisspeptin, kisspeptin receptor, GnRH and GTH subunit gene expression in the grass puffer during the spawning season. Gen. Comp. Endocrinol. 243, 138–145.

Shang, E.H.H., Yu, R.M.K., Wu, R.S.S., 2006. Hypoxia affects sex differentiation and development leading to a male-dominated population in zebrafish (Danio rerio). Environ. Sci. Technol. 40, 3118–3122.

Shearer, K., Parkins, P., Gadberry, B., Beckman, B., Swanson, P., 2006. Effects of growth rate/body size and a low lipid diet on the incidence of early sexual maturation in juvenile male spring chinook salmon (Oncorhynchus tshawytscha). Aquaculture 252, 545–556.

Shephard, K.L., 1994. Functions for fish mucus. Rev. Fish Biol. Fish. 4, 401–429.

Shimizu, M., Dickhoff, W.W., 2017. Circulating insulin-like growth factor binding proteins in fish: their identities and physiological regulation. Gen. Comp. Endocrinol. 252, 150–161.

Shimizu, M., Cooper, K.A., Dickhoff, W.W., Beckman, B.R., 2009. Postprandial changes in plasma growth hormone, insulin, insulin-like growth factor (IGF)-I, and IGF-binding proteins in coho salmon fasted for varying periods. Am. J. Physiol. Regul. Integr. Comp. Physiol. 297, 352–361.

Shrimpton, J.M., Bernier, N.J., Iwama, G.K., Randall, D.J., 1994a. Differences in measurements of smolt development between wild and hatchery-reared juvenile coho salmon (Oncorhynchus kisutch) before and after saltwater exposure. Can. J. Fish. Aquat. Sci. 51, 2170–2178.

Shrimpton, J.M., Bernier, N.J., Randall, D.J., 1994b. Changes in cortisol dynamics in wild and hatchery-reared juvenile coho salmon (Oncorhynchus kisutch) during smoltification. Can. J. Fish. Aquat. Sci. 51, 2179–2187.

Shuter, B.J., Post, J.R., 1990. Climate, population viability, and the zoogeography of temperate fishes. Trans. Am. Fish. Soc. 119, 314–336.

Siefkes, M.J., 2017. Use of physiological knowledge to control the invasive sea lamprey (Petromyzon marinus) in the Laurentian Great Lakes. Conserv. Physiol. 5, 1–18.

Simontacchi, C., Poltronieri, C., Carraro, C., Bertotto, D., Xiccato, G., Trocino, A., Radaelli, G., 2008. Alternative stress indicators in sea bass Dicentrarchus labrax, L. J. Fish Biol. 72, 747–752.

Singh, A.K., Chandra, R., 2019. Pollutants released from the pulp paper industry: aquatic toxicity and their health hazards. Aquat. Toxicol. 211, 202–216.

Söffker, M., Tyler, C.R., 2012. Endocrine disrupting chemicals and sexual behaviors in fish a critical review on effects and possible consequences. Crit. Rev. Toxicol. 42, 653–668.

Sogard, S.M., 1997. Size-selective mortality in the juvenile stage of teleost fishes: a review. Bull. Mar. Sci. 60, 1129–1157.

Son, H.H., Yun, W.S., Cho, S.H., 2020. Development and validation of an LC-MS/MS method for profiling 39 urinary steroids (estrogens, androgens, corticoids, and progestins). Biomed. Chromatogr. 34, e4723.

Sopinka, N.M., Donaldson, M.R., O'Connor, C.M., Suski, C.D., Cooke, S.J., 2016. Stress indicators in fish. In: Schreck, C.B., Tort, L., Farrell, A.P., Brauner, C.J. (Eds.), Fish Physiology: Biology of Stress in Fish, vol. 35. Academic Press, London, pp. 405–462.

Sorensen, P.W., Johnson, N.S., 2016. Theory and application of semiochemicals in nuisance fish control. J. Chem. Ecol. 42, 698–715.

Sorensen, P.W., Stacey, N.E., 2004. Brief review of fish pheromones and discussion of their possible uses in the control of non-indigenous teleost fishes. New Zeal. J. Mar. Freshw. Res. 38, 399–417.

Sorensen, P.W., Pinillos, M., Scott, A.P., 2005. Sexually mature male goldfish release large quantities of androstenedione into the water where it functions as a pheromone. Gen. Comp. Endocrinol. 140, 164–175.

Sorensen, P.W., Rue, M.C.P., Leese, J.M., Ghosal, R., Lim, H., 2019. A blend of F prostaglandins functions as an attractive sex pheromone in silver carp. Fishes 4, 27.

Soria, F.N., Strüssmann, C.A., Miranda, L.A., 2008. High water temperatures impair the reproductive ability of the pejerrey fish *Odontesthes bonariensis*: effects on the hypophyseal-gonadal axis. Physiol. Biochem. Zool. 81, 898–905.

Stacey, N.E., Sorensen, P.W., 2002. Hormonal pheromones in fish. In: Pfaff, D.W., Arnold, A.P., Etgen, A.M., Fharbach, A.M., Rubin, R.T. (Eds.), Hormones, Brain and Behavior. Academic Press, London, pp. 375–434.

Stacey, N., Sorensen, P.W., 2011. Hormonal pheromones. In: Farrell, A.P. (Ed.), Encyclopedia of Fish Physiology: From Genome to Environment. Elsevier Science & Technology, San Diego, pp. 1553–1562.

Stacey, N.E., Sorensen, P.W., Van Der Kraak, G.J., Dulka, J.G., 1989. Direct evidence that 17α,20β-dihydroxy-4-pregnen-3-one functions as a goldfish primer pheromone: preovulatory release is closely associated with male endocrine responses. Gen. Comp. Endocrinol. 75, 62–70.

Stothart, M.R., Bobbie, C.B., Schulte-Hostedde, A.I., Boonstra, R., Palme, R., Mykytczuk, N.C.S., Newman, A.E.M., 2016. Stress and the microbiome: linking glucocorticoids to bacterial community dynamics in wild red squirrels. Biol. Lett. 12, 20150875.

Sullam, K.E., Essinger, S.D., Lozupone, C.A., O'Connor, M.P., Rosen, G.L., Knight, R., Kilham, S.S., Russell, J.A., 2012. Environmental and ecological factors that shape the gut bacterial communities of fish: a meta-analysis. Mol. Ecol. 21, 3363–3378.

Sundström, L.F., Löhmus, M., Johnsson, J.I., 2003. Investment in territorial defence depends on rearing environment in brown trout (*Salmo trutta*). Behav. Ecol. Sociobiol. 54, 249–255.

Swanson, P., Campbell, B., Shearer, K., Dickey, J., Beckman, B., Larsen, D., Park, L., Berejikian, B., 2008. Application of reproductive technologies to captive breeding programs for conservation of imperiled stocks of Pacific salmon. Cybium 32, 279–282.

Taranger, G.L., Muncaster, S., Norberg, B., Thorsen, A., Andersson, E., 2015. Environmental impacts on the gonadotropic system in female Atlantic salmon (*Salmo salar*) during vitellogenesis: photothermal effects on pituitary gonadotropins, ovarian gonadotropin receptor expression, plasma sex steroids and oocyte growth. Gen. Comp. Endocrinol. 221, 86–93.

Taylor, M.D., Chick, R.C., Lorenzen, K., Agnalt, A.L., Leber, K.M., Blankenship, H.L., Haegen, G.V., Loneragan, N.R., 2017. Fisheries enhancement and restoration in a changing world. Fish. Res. 186, 407–412.

Tellier, J.M., Kalejs, N.I., Leonhardt, B.S., Cannon, D., Höök, T.O., Collingsworth, P.D., 2022. Widespread prevalence of hypoxia and the classification of hypoxic conditions in the Laurentian Great Lakes. J. Great Lakes Res. 48, 13–23.

Thomas, P., Rahman, M.S., 2009. Chronic hypoxia impairs gamete maturation in atlantic croaker induced by progestins through nongenomic mechanisms resulting in reduced reproductive success. Environ. Sci. Technol. 43, 4175–4180.

Thomas, P., Rahman, M.S., 2012. Extensive reproductive disruption, ovarian masculinization and aromatase suppression in Atlantic croaker in the northern Gulf of Mexico hypoxic zone. Proc. R. Soc. B Biol. Sci. 279, 28–38.

Thomas, P., Rahman, M.S., Khan, I.A., Kummer, J.A., 2007. Widespread endocrine disruption and reproductive impairment in an estuarine fish population exposed to seasonal hypoxia. Proc. R. Soc. B Biol. Sci. 274, 2693–2701.

Thomas, P., Rahman, M.S., Picha, M.E., Tan, W., 2015. Impaired gamete production and viability in Atlantic croaker collected throughout the 20,000 km2 hypoxic region in the northern Gulf of Mexico. Mar. Pollut. Bull. 101, 182–192.

Tintos, A., Gesto, M., Míguez, J.M., Soengas, J.L., 2007. Naphthalene treatment alters liver intermediary metabolism and levels of steroid hormones in plasma of rainbow trout (Oncorhynchus mykiss). Ecotoxicol. Environ. Saf. 66, 139–147.

Tipping, J.M., 1997. Effect of smolt length at release on adult returns of hatchery-reared winter steelhead. Prog. Fish. Cult. 59, 310–311.

Tokarz, J., Möller, G., Hrabě De Angelis, M., Adamski, J., 2015. Steroids in teleost fishes: a functional point of view. Steroids 103, 123–144.

Turner, J.W., Nemeth, R., Rogers, C., 2003. Measurement of fecal glucocorticoids in parrotfishes to assess stress. Gen. Comp. Endocrinol. 133, 341–352.

Tyler, C.R., Filby, A.L., van Aerle, R., Lange, A., Ball, J., Santos, E.M., 2008. Fish toxicogenomics. Adv. Exp. Biol. 2, 75–325.

Ullah, I., Zuberi, A., Rehman, H., Ali, Z., Thörnqvist, P.O., Winberg, S., 2020. Effects of early rearing enrichments on modulation of brain monoamines and hypothalamic–pituitary–interrenal axis (HPI axis) of fish mahseer (Tor putitora). Fish Physiol. Biochem. 46, 75–88.

Uren Webster, T.M., Rodriguez-Barreto, D., Consuegra, S., Garcia de Leaniz, C., 2020. Cortisol-related signatures of stress in the fish microbiome. Front. Microbiol. 11, 1621.

Ussery, E.J., McMaster, M.E., Servos, M.R., Miller, D.H., Munkittrick, K.R., 2021. A 30-year study of impacts, recovery, and development of critical effect sizes for endocrine disruption in white sucker (Catostomus commersonii) exposed to bleached-kraft pulp mill effluent at Jackfish Bay, Ontario, Canada. Front. Endocrinol. (Lausanne) 12, 369.

Vainikka, A., Huusko, R., Hyvärinen, P., Korhonen, P.K., Laaksonen, T., Koskela, J., Vielma, J., Hirvonen, H., Salminen, M., 2012. Food restriction prior to release reduces precocious maturity and improves migration tendency of Atlantic salmon (Salmo salar) smolts. Can. J. Fish. Aquat. Sci. 69, 1981–1993.

Van Camp, K.A., Baggerman, G., Blust, R., Husson, S.J., 2017. Peptidomics of the zebrafish Danio rerio: in search for neuropeptides. J. Proteomics 150, 290–296.

Van Der Kraak, G.J., Munkittrick, K.R., McMaster, M.E., Portt, C.B., Chang, J.P., 1992. Exposure to bleached kraft pulp mill effluent disrupts the pituitary-gonadal axis of white sucker at multiple sites. Toxicol. Appl. Pharmacol. 115, 224–233.

Vargas-Chacoff, L., Regish, A.M., Weinstock, A., McCormick, S.D., 2018. Effects of elevated temperature on osmoregulation and stress responses in Atlantic salmon Salmo salar smolts in fresh water and seawater. J. Fish Biol. 93, 550–559.

Vargas-Chacoff, L., Regish, A.M., Weinstock, A., Björnsson, B.T., McCormick, S.D., 2021. Effects of long-term cortisol treatment on growth and osmoregulation of Atlantic salmon and brook trout. Gen. Comp. Endocrinol. 308, 113769.

Verbeek, P., Iwamoto, T., Murakami, N., 2008. Variable stress-responsiveness in wild type and domesticated fighting fish. Physiol. Behav. 93, 83–88.

Vermeirssen, E.L.M., Scott, A.P., 1996. Excretion of free and conjugated steroids in rainbow trout (*Oncorhynchus mykiss*): evidence for branchial excretion of the maturation-inducing steroid, 17,20β-dihydroxy-4-pregnen-3-one. Gen. Comp. Endocrinol. 101, 180–194.

Vikingstad, E., Andersson, E., Hansen, T.J., Norberg, B., Mayer, I., Stefansson, S.O., Fjelldal, P.G., Taranger, G.L., 2016. Effects of temperature on the final stages of sexual maturation in Atlantic salmon (*Salmo salar* L.). Fish Physiol. Biochem. 42, 895–907.

Vindas, M.A., Madaro, A., Fraser, T.W.K., Höglund, E., Olsen, R.E., Øverli, Ø., Kristiansen, T.S., 2016. Coping with a changing environment: the effects of early life stress. R. Soc. Open Sci. 3, 160382.

Wagner, C.M., Jones, M.L., Twohey, M.B., Sorensen, P.W., 2006. A field test verifies that pheromones can be useful for sea lamprey (*Petromyzon marinus*) control in the Great Lakes. Can. J. Fish. Aquat. Sci. 63, 475–479.

Wang, S., Yuen, S.S., Randall, D.J., Hung, C.Y., Tsui, T.K., Poon, W.L., Lai, J.C., Zhang, Y., Lin, H., 2008. Hypoxia inhibits fish spawning via LH-dependent final oocyte maturation. Comp. Biochem. Physiol. C Toxicol. Pharmacol. 148, 363–369.

Watson, S.E., Hauffe, H.C., Bull, M.J., Atwood, T.C., McKinney, M.A., Pindo, M., Perkins, S.E., 2019. Global change-driven use of onshore habitat impacts polar bear faecal microbiota. ISME J. 13, 2916–2926.

Watts, M., Pankhurst, N.W., King, H.R., 2004. Maintenance of Atlantic salmon (*Salmo salar*) at elevated temperature inhibits cytochrome P450 aromatase activity in isolated ovarian follicles. Gen. Comp. Endocrinol. 135, 381–390.

Waye, A., Trudeau, V.L., 2011. Neuroendocrine disruption: more than hormones are upset. J. Toxicol. Environ. Health B Crit. Rev. 14, 270–291.

Weber, R., Gaus, C., Tysklind, M., Johnston, P., Forter, M., Hollert, H., Heinisch, E., Holoubek, I., Lloyd-Smith, M., Masunaga, S., Moccarelli, P., Santillo, D., Seike, N., Symons, R., Torres, J.P.M., Verta, M., Varbelow, G., Vijgen, J., Watson, A., Costner, P., Woelz, J., Wycisk, P., Zennegg, M., 2008. Dioxin- and POP-contaminated sites— contemporary and future relevance and challenges: overview on background, aims and scope of the series. Environ. Sci. Pollut. Res. 15, 363–393.

Weirup, L., Schulz, C., Seibel, H., Aerts, J., 2021. Scale cortisol is positively correlated to fin injuries in rainbow trout (*Oncorhynchus mykiss*) reared in commercial flow through systems. Aquaculture 543, 736924.

Welch, M.J., Munday, P.L., 2016. Contrasting effects of ocean acidification on reproduction in reef fishes. Coral Reefs 35, 485–493.

Wendelaar Bonga, S.E., 1997. The stress response in fish. Physiol. Rev. 77, 591–625.

White, L.J., Thomson, J.S., Pounder, K.C., Coleman, R.C., Sneddon, L.U., 2017. The impact of social context on behaviour and the recovery from welfare challenges in zebrafish, *Danio rerio*. Anim. Behav. 132, 189–199.

Whitehead, A., Pilcher, W., Champlin, D., Nacci, D., 2012. Common mechanism underlies repeated evolution of extreme pollution tolerance. Proc. R. Soc. B Biol. Sci. 279, 427–433.

Whitehead, A., Clark, B.W., Reid, N.M., Hahn, M.E., Nacci, D., 2017. When evolution is the solution to pollution: key principles, and lessons from rapid repeated adaptation of killifish (*Fundulus heteroclitus*) populations. Evol. Appl. 10, 762–783.

Wilkes, L., Owen, S.F., Readman, G.D., Sloman, K.A., Wilson, R.W., 2012. Does structural enrichment for toxicology studies improve zebrafish welfare? Appl. Anim. Behav. Sci. 139, 143–150.

Winberg, S., Höglund, E., Øverli, Ø., 2016. Variation in the neuroendocrine stress response. In: Schreck, C.B., Tort, L., Farrell, A.P., Brauner, C.J. (Eds.), Fish Physiology: Biology of Stress in Fish, vol. 35. Academic Press, London, pp. 35–74.

Woodward, C.C., Strange, R.J., 1987. Physiological stress responses in wild and hatchery-reared rainbow trout. Trans. Am. Fish. Soc. 116, 574–579.

Wu, R.S.S., Zhou, B.S., Randall, D.J., Woo, N.Y.S., Lam, P.K.S., 2003. Aquatic hypoxia is an endocrine disruptor and impairs fish reproduction. Environ. Sci. Technol. 37, 1137–1141.

Yamaguchi, T., Yoshinaga, N., Yazawa, T., Gen, K., Kitano, T., 2010. Cortisol is involved in temperature-dependent sex determination in the Japanese flounder. Endocrinology 151, 3900–3908.

Yan, Z., Lu, G., He, J., 2012. Reciprocal inhibiting interactive mechanism between the estrogen receptor and aryl hydrocarbon receptor signaling pathways in goldfish (*Carassius auratus*) exposed to 17β-estradiol and benzo[a]pyrene. Comp. Biochem. Physiol., Part C: Toxicol. Pharmacol. 156, 17–23.

Yanong, R.P.E.E., Martinez, C., Watson, A., 2009. Use of Ovaprim in Ornamental Fish Aquaculture. IFAS Extension University of Florida, Gainesville, FL.

Young, G., Björnsson, B.T., Prunet, P., Lin, R.J., Bern, H.A., 1989. Smoltification and seawater adaptation in coho salmon (*Oncorhynchus kisutch*): plasma prolactin, growth hormone, thyroid hormones, and cortisol. Gen. Comp. Endocrinol. 74, 335–345.

Zadmajid, V., Bashiri, S., Sharafi, N., Butts, I.A.E., 2018. Effect of hCG and Ovaprim™ on reproductive characteristics of male Levantine scraper, *Capoeta damascina* (Valenciennes, 1842). Theriogenology 115, 45–56.

Zhou, L.F., Zhao, B.W., Guan, N.N., Wang, W.M., Gao, Z.X., 2017. Plasma metabolomics profiling for fish maturation in blunt snout bream. Metabolomics 13, 40.

Zohar, Y., 2021. Fish reproductive biology—reflecting on five decades of fundamental and translational research. Gen. Comp. Endocrinol. 300, 113544.

Zohar, Y., Mylonas, C.C., 2001. Endocrine manipulations of spawning in cultured fish: from hormones to genes. Aquaculture 197, 99–136.

Zohar, Y., Muñoz-Cueto, J.A., Elizur, A., Kah, O., 2010. Neuroendocrinology of reproduction in teleost fish. Gen. Comp. Endocrinol. 165, 438–455.

Zubair, S.N., Peake, S.J., Hare, J.F., Anderson, W.G., 2012. The effect of temperature and substrate on the development of the cortisol stress response in the lake sturgeon, *Acipenser fulvescens*, Rafinesque (1817). Environ. Biol. Fishes 93, 577–587.

Zuberi, A., Ali, S., Brown, C., 2011. A non-invasive assay for monitoring stress responses: a comparison between wild and captive-reared rainbowfish (*Melanoteania duboulayi*). Aquaculture 321, 267–272.

Chapter 7

Conservation aspects of osmotic, acid-base, and nitrogen homeostasis in fish

Chris M. Wood[a,b,c,*]

[a]*Department of Zoology, University of British Columbia, Vancouver, BC, Canada*
[b]*Department of Biology, McMaster University, Hamilton, ON, Canada*
[c]*Rosenstiel School of Marine and Atmospheric Science, University of Miami, Miami, FL, United States*
[*]*Corresponding author: e-mail: woodcm@zoology.ubc.ca*

Chapter Outline

The basic mechanisms of iono/osmoregulation, acid-base regulation, and homeostasis and excretion of nitrogenous wastes (ammonia and urea) are reviewed for freshwater fish, marine and euryhaline fish, and for special cases (marine hagfish and chondrichthyans). Six different examples are then explored where physiological understanding of environmental impacts on these processes has already informed regulatory and conservation strategies, or should do so in the future. These include: (i) the acid-rain crisis; (ii) survival of fishes in the acidic, ion-poor blackwaters of the Rio Negro; (iii) development of the Biotic Ligand Model for environmental regulation of metals; (iv) survival of fishes in highly alkaline lakes; (v) the commercial hagfish fishery; and (vi) the critical importance of feeding for osmoregulation in chondrichthyans. These lessons argue for a greater awareness of the wide-spread variation in the physical chemistry of natural waters, and the associated physiology of the fish that live there. This knowledge should be incorporated into regulatory strategies for both environmental protection and conservation of resident species at risk.

1   Introduction–General principles

Effective conservation, whether it be due to voluntary measures or legislation, invariably follows rather than precedes scientific investigation. This chapter will highlight examples where scientific understanding of issues of osmotic, acid-base, and/or nitrogen regulation and nitrogenous waste excretion in fish has already led to conservation measures, or should do so in the future. Before discussing these cases, it is important to review the basic principles involved in normal homeostasis of these physiological processes.

For reference throughout this and subsequent sections, Figs. 1 and 2 present summaries of osmolyte levels in the blood plasma (i.e., extracellular fluid) of the four basic types of fish that are featured in this chapter. These are most freshwater fish (Fig. 1A, for which freshwater teleosts are representative), most seawater fish (Fig. 1B, for which marine teleosts are representative), seawater hagfish (myxinids) (Fig. 2A, note that there are no freshwater hagfish), and seawater chondrichthyans (Fig. 2B, for which marine elasmobranchs are representative). These figures also highlight typical osmolyte levels in the environments of these fish, so that the concentration gradients across the gills and skin can be appreciated. In contrast to plasma osmolyte levels which are well documented and relatively consistent among species, data for intracellular ion levels are relatively sparse, but appear to be more variable among both tissues and species, so have not been tabulated. Nevertheless, it is clear that regardless of the intracellular osmolyte composition, under steady-state conditions, the osmolalities of the intracellular and extracellular compartments are virtually identical so that there is no net flux of water between them. This is not the case at the outer and inner surfaces of the fish where dynamic net fluxes of water, ions, and other osmolytes are occurring under steady-state conditions. Four organs at these surfaces (gills, kidney, gut, and skin) are involved in the regulation of these fluxes. The gills, which constitute more than 60% of the outer surface area, are generally considered the most important, and have been the subject of most research, yet the other three also play critical roles in homeostasis.

A FRESHWATER TELEOST

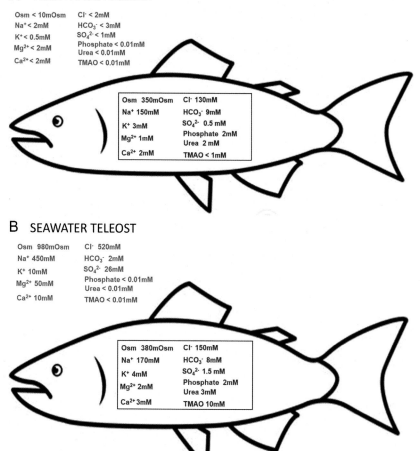

FIG. 1 Typical values for osmolality and the concentrations of major osmolytes in the environment (blue) and the blood plasma/extracellular fluid (red) of (A) most freshwater fish and (B) most seawater fish, for which teleosts are representative in both cases.

1.1 Ionic and osmotic balance in freshwater fish

Freshwater fish live in a hydrating medium yet maintain an internal osmolality (300–375 mOsmol kg^{-1}) many-fold greater than their environment (a few mOsmol kg^{-1}) (Fig. 1A). In the extracellular fluid the major strong ions are Na$^+$ and Cl$^-$ (110–170 mmol L^{-1}), while other essential strong ions (K$^+$, Ca^{2+}, Mg^{2+}, SO$_4^{2-}$, phosphate) are generally less than 5% of these levels (Fig. 1A). The net charge difference between the total cations and total strong anions (Strong Ion Difference, SID) is positive, and is balanced by negative charge on plasma proteins and the dependent variable HCO$_3^-$, the latter

A SEAWATER HAGFISH

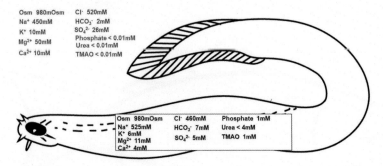

Osm 980mOsm   Cl⁻ 520mM
Na⁺ 450mM   HCO₃⁻ 2mM
K⁺ 10mM   SO₄²⁻ 26mM
Mg²⁺ 50mM   Phosphate < 0.01mM
  Urea < 0.01mM
Ca²⁺ 10mM   TMAO < 0.01mM

Osm 980mOsm   Cl⁻ 460mM   Phosphate 1mM
Na⁺ 525mM   HCO₃⁻ 7mM   Urea < 4mM
K⁺ 6mM
Mg²⁺ 11mM   SO₄²⁻ 5mM   TMAO 1mM
Ca²⁺ 4mM

B SEAWATER ELASMOBRANCH

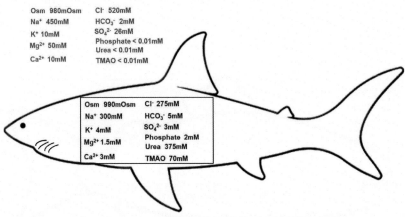

Osm 980mOsm   Cl⁻ 520mM
Na⁺ 450mM   HCO₃⁻ 2mM
K⁺ 10mM   SO₄²⁻ 26mM
Mg²⁺ 50mM   Phosphate < 0.01mM
  Urea < 0.01mM
Ca²⁺ 10mM   TMAO < 0.01mM

Osm 990mOsm   Cl⁻ 275mM
Na⁺ 300mM   HCO₃⁻ 5mM
K⁺ 4mM   SO₄²⁻ 3mM
Mg²⁺ 1.5mM   Phosphate 2mM
  Urea 375mM
Ca²⁺ 3mM   TMAO 70mM

FIG. 2 Typical values for osmolality and the concentrations of major osmolytes in the environment (blue) and the blood plasma/extracellular fluid (red) of (A) seawater hagfish (myxinids) and (B) seawater chondrichthyans, for which elasmobranchs are representative.

playing a critical role in acid-base balance (Fig. 3). As plasma protein generally does not change much, though its net charge is altered by pH, the HCO_3^- concentration is largely determined by the SID (see Section 1.4). In the intracellular fluids, the dominant cations ions are K^+ (100–200 mmol L^{-1}) and Mg^{2+} (35–50 mmol L^{-1}), whereas most of the anionic charge is held by negatively charged organic molecules (proteins, amino acids, nucleic acids, fatty acids, etc.) and by phosphate (35–70 mmol L^{-1}). Intracellular Na^+, Ca^{2+}, Cl^-, and SO_4^{2-} levels are all very low, and intracellular HCO_3^- concentration is generally lower than extracellular HCO_3^-, reflecting lower intracellular pH (pHi) than extracellular pH (pHe).

Hyper-osmoregulation in freshwater is achieved by the active uptake of essential major ions at the gills from the dilute external environment

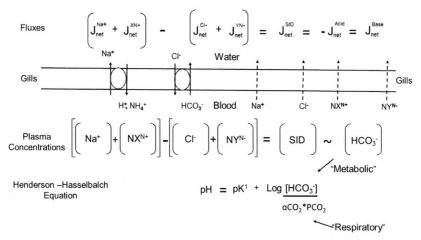

FIG. 3 A conceptual illustration of how acid-base balance is linked to ionoregulation by Strong Ion Difference (SID) theory. Major strong cations such as Na^+ may be actively transported across the gills in direct exchange for acidic equivalents such as H^+ and NH_4^+, and strong anions such as Cl^- may actively transported across the gills in direct exchange for basic equivalents such as HCO_3^-. Other strong cations (X^{N+}) such as Ca^{2+}, Mg^{2+}, and K^+, and strong anions (Y^{N-}) such as SO_4^{2-} may be similarly transported by as yet uncharacterized mechanisms. Additionally, however, all these strong cations and anions may simply diffuse across the gills according to electrochemical gradients. In terms of net acid-base balance, it is the difference between net strong cation fluxes ($J^{Na+}_{net} + J^{XN+}_{net}$) and net strong anion fluxes ($J^{Cl-}_{net} + J^{YN-}_{net}$) with the environment, signs considered, regardless of how these occur, that comprises the net strong ion difference flux (J^{SID}_{net}) and constrains the net flux of acidic equivalents ("metabolic acid") (J^{Acid}_{net}) in the opposite direction at any time. Note that in acid-base terms, the net flux of basic equivalents ("metabolic base") (J^{Base}_{net}) is equivalent to J^{SID}_{net} in the same direction and to the J^{Acid}_{net} in the opposite direction. In the blood plasma, the net difference between the concentrations of strong cations ([Na^+] and [X^{N+}]) and the concentrations of strong anions ([Cl^-] and [Y^{N-}]) is the strong ion difference concentration ([SID]), which constrains and is approximately equal to the bicarbonate concentration ([HCO_3^-]). In turn, via the Henderson-Hasselbalch equation, it can be seen that the [HCO_3^-] is the "metabolic component" directly affecting the blood acid-base status, such that an increase in [HCO_3^-] (i.e., $\sim$ [SID]) raises the pH and vice versa. The other major influence on acid-base status is PCO_2, the "respiratory component", such that an increase in PCO_2 lowers pH, and vice versa. Note that pK' (conditional equilibrium constant) and αCO_2 (CO_2 solubility coefficient) remain essentially unchanged.

(Fig. 1A), but only the branchial pathways for Na^+, Cl^-, and Ca^{2+} have been well characterized. Fig. 4 presents a conceptual model. In most fish, a critical role of the skin is to serve as a low permeability barrier to ion loss and water gain, though in some species extra-branchial ionocytes may be present in the skin and account for a significant fraction of ion uptake, especially in early life stages (reviewed by Glover et al., 2013; Zimmer et al., 2017a). The gills are also the major sites of diffusive loss of ions, greatly exceeding losses through the skin, rectum, and urine. The paracellular pathway (i.e., tight junctions between gill cells) is widely believed to be the major site of diffusive

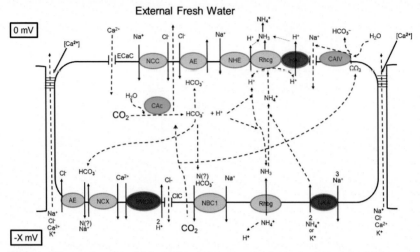

FIG. 4 Conceptual model of branchial ion transport pathways in the ionocytes of freshwater fish, incorporating the Rhesus protein (Rh) pathways for ammonia excretion. Note that while all pathways are diagrammed in a single cell, in reality, the various pathways are usually partitioned among several different ionocyte types within a species, and not all species exhibit all pathways shown. ECaC=epithelial calcium channel; NCC=sodium, chloride co-transporter; AE=anion exchanger; NHE=sodium, hydrogen exchanger; Rhcg=apical Rhesus glycoprotein; HAT=vacuolar type proton ATPase (vH$^+$ATPase); ASIC=acid-sensing ion channel; CAIV=membrane-bound carbonic anhydrase. CAc=cytosolic carbonic anhydrase (sometimes termed CA2); NCX=sodium, calcium exchanger, PMCA=plasma membrane calcium ATPase; Clc=chloride intracellular ion channel; NBC1=sodium bicarbonate co-transporter type one; Rhbg=basolateral Rhesus glycoprotein; NKA=sodium, potassium ATPase. Active transporters are orange/dark, passive transporters are blue/light, carbonic anhydrase is green/intermediate. The possible substitution of NH$_4^+$ for K$^+$ at NKA is shown. Strong ions are shown diffusing out of the freshwater fish through paracellular pathways along their electrochemical gradients, and external waterborne calcium concentration ([Ca^{2+}]) is shown acting on tight junctions to restrict this pathway. The transepithelial potential (TEP) is negative in the ECF relative to the external freshwater as zero mV.

losses of strong ions, but there is no definitive evidence on this point (reviewed by Wood and Eom, 2021). Freshwater fish without access to food are able to stay more or less in ion balance. However, the importance of ion uptake from the food has been generally underappreciated by ionoregulatory physiologists. There is good evidence that in fact the gut may be the dominant route of ion uptake in fish that are actively feeding and growing, and that the diet may be a particularly important source of inorganic electrolytes for fish living in ion-poor waters (reviewed by Wood and Bucking, 2011). Ion loss in the feces is minimal (except for phosphate), as are ion losses in the urine and across the skin.

The kidney serves as a "bilge pump," excreting water by glomerular filtration (coupled with minimal water reabsorption) at the same rate as it enters by osmosis because of the large osmotic gradient across the gills (Fig. 1A). As drinking is negligible, at least in non-stressed fish, the urine flow rate provides a measure of net water uptake rate. Reabsorptive transporters for the strong ions in the nephron tubules plus low tubular water permeability ensure that urinary ion concentrations are only a few percent of plasma levels. However, the reabsorptive pumps may be down-regulated or replaced by active secretory mechanisms in the tubules when dietary loading of selected ions such as Mg^{2+}, K^+, or Ca^{2+} becomes particularly high (reviewed by Takvam et al., 2021).

Detailed reviews of the uptake mechanisms for Na^+, Cl^-, and Ca^{2+} by the mitochondria -rich ionocytes in the freshwater gill have been provided by Perry (1997), Evans et al. (2005), Hwang and Lee (2007), Marshall and Grosell (2005); Evans (2008), Evans (2011), Dymowska et al. (2012), Hiroi and McCormick (2012), Edwards and Marshall (2013), Hwang and Lin (2014), Hsu et al. (2014) and Guh et al. (2015). In brief, while there is great interspecific diversity and multiple types of ionocytes (generally two or more), active Na^+ and Cl^- uptake are largely independent processes in most fish, coupled respectively on a 1:1 basis to metabolic acid (H^+ or NH_4^+) and metabolic base (HCO_3^-) excretion as first hypothesized by August Krogh (1939). This allows them to play key roles in acid-base balance and ammonia excretion (see Sections 1.4 and 1.5). Acid and base excretion processes generally occur in different ionocytes. Na^+ uptake from the water relies on an apical Na^+/H^+ exchanger (NHE) and/or on an as yet unidentified apical Na^+ channel (perhaps the acid-sensing ion channel, ASIC—Dymowska et al., 2015) energized by apical vH^+-ATPase (HAT). As illustrated in Fig. 4, all these appear to be coupled in a metabolon to an apical Rh glycoprotein (Rhcg) that facilitates the diffusion of NH_3 outward (Ito et al., 2013; Wright and Wood, 2009), and to cytoplasmic carbonic anhydrase (CAc) that creates H^+ and HCO_3^-. In the external boundary layer, excreted H^+ reacts with excreted NH_3 so that in effect NH_4^+ is excreted in exchange for Na^+, though it is also possible that direct Na^+/NH_4^+ exchange occurs (Ito et al., 2014). H^+ may also be supplied by hydration of respiratory CO_2 via outward -facing membrane-bound carbonic anhydrase (CAIV) on the apical surface. Cl^- uptake is virtually absent in a few species (e.g., killifish, eels) that rely on dietary Cl^- acquisition, but in most fish, it occurs by an apical Cl^-/HCO_3^- exchanger at slightly lower rates than Na^+ uptake. Additionally, in a few species there is evidence for coupled Na^+ and Cl^- uptake by an apical Na^+,Cl^- co-transporter (NCC), a mechanism that would not affect acid-base balance. Basolateral efflux of Na^+ to the extracellular fluid (ECF) occurs via Na^+, K^+-ATPase (NKA) and also by an electrogenic Na^+ HCO_3^- cotransporter (NBC1), while Cl^- appears to leave by basolateral Cl^- channels. A basolateral anion exchange may export HCO_3^- into the ECF in exchange for Cl^- when there is a need for HCO_3^- retention for acid-base regulation. Ca^{2+} is taken up at about 5–30% of the rate of Na^+ and Cl^- uptake through apical epithelial Ca^{2+} channels (ECaC) and is then exported to the ECF via the basolateral plasma membrane Ca^{2+}-ATPase (PMCA), probably in exchange for H^+, and/or the basolateral Na^+/Ca^{2+} exchanger (NCX). Therefore, the mechanisms of energy input are the NKA (establishing the Na^+ gradient for NHE and NCX and the basolateral membrane potential for NBC1), HAT (establishing the apical membrane potential for channel-mediated Na^+ entry), and PMCA (for basolateral Ca^{2+} export). It remains unclear how Cl^- uptake is energized; high localized HCO_3^- concentration in the subapical region created by intracellular CA and the necessary removal of H^+ by HAT and NHE may contribute.

Unidirectional Na^+ uptake and loss rates are generally higher than Cl^- uptake and loss rates; this reflects the greater passive permeability of the gills to Na^+ than Cl^-, requiring greater Na^+ pumping than Cl^- pumping to counteract diffusive losses. The transepithelial potential (TEP) is largely or entirely a diffusion potential in freshwater fish, and is therefore negative-inside relative to the external water (Fig. 4) (Eddy, 1975; McWilliams and Potts, 1978; Potts, 1984; Wood et al., 2020). Increases in water Ca^{2+} concentration decrease overall branchial permeability, probably by stabilizing tight junctions (Hunn, 1985; McDonald and Rogano, 1986), and importantly have a larger effect in reducing Na^+ permeability than Cl^- permeability, such that the TEP increases (i.e., becomes less negative-inside).

1.2 Ionic and osmotic balance in seawater and euryhaline fish

Seawater fish live in a dehydrating environment, yet most maintain an internal osmolality ($325–400\,mOsmol\,kg^{-1}$) very similar to that of freshwater fish, despite facing seawater-to-ECF osmotic gradients that are about two-fold higher and opposite those facing freshwater fish (Fig. 1A and B). Typical seawater osmolality is $\sim 1000\,mOsmol\,kg^{-1}$, and in addition to high Na^+ and Cl^- levels ($450–550\,mmol\,L^{-1}$), concentrations of essential but other potentially toxic strong ion levels are also very high: Mg^{2+} ($\sim 50\,mmol\,L^{-1}$), SO_4^{2-} ($\sim 26\,mmol\,L^{-1}$), Ca^{2+} ($\sim 10\,mmol\,L^{-1}$), and K^+ ($\sim 10\,mmol\,L^{-1}$ (Fig. 1B). Despite these challenges, most marine fish maintain all plasma (ECF) and ICF ions at levels similar to or only slightly higher than those of freshwater fish (Fig. 1A and B). Exceptions include the ancient Classes Myxini (hagfishes; Fig. 2A) and Chondrichthyes (elasmobranchs and holocephalans; Fig. 2B) which are osmoconformers (see Section 1.3). In all marine fish, the same SID principles as in freshwater fish apply with respect to regulation of ECF and ICF HCO_3^- levels.

A notable feature of most marine fish are the very high unidirectional influx and efflux rates of Na^+ and Cl^- at the gills that can be measured with radiotracers, sometimes >100-fold the rates in freshwater. These matching bidirectional fluxes are reduced almost instantaneously when the fish are transferred to lower salinity. The phenomenon was originally observed by Motais et al. (1966) and explained as an exchange diffusion effect (i.e., rapid carrier-mediated external Na^+ versus internal Na^+ exchange and comparable Cl^-/Cl^- exchange dependent on external ion concentration - essentially futile cycling), an explanation that has been challenged but never disproven (e.g., Marshall, 2013; Wood, 2011; Wood and Marshall, 1994). Regardless, the phenomenon cannot contribute to net ion balance.

While the gills are undoubtedly a site of passive ion loading due to large concentration gradients from seawater to extracellular fluid (Fig. 1B), they are also arguably the most important site enabling hypo-osmoregulation in seawater. This is achieved by the active excretion of Na^+ and Cl^- at the gills, whereas Mg^{2+}, Ca^{2+} and SO_4^{2-} are thought to be mainly excreted by secretory processes in the kidney (Takvam et al., 2021). Mass balance calculations suggest that net K^+, SO_4^{2-}, and Ca^{2+} excretion may also occur at the gills, though there is some uncertainty (Grosell, 2011). The skin serves a similar impermeability function as in freshwater. However, marine fish need to constantly replace the water which is being lost due to the osmotic gradient across the gills (Fig. 1B), as well as a smaller water loss in the low volume urine. This replacement is achieved by the drinking of seawater, and the absorption of ions, principally Na^+ and Cl^-; water follows passively, mainly in the intestine. Most of the K^+ and some Ca^{2+} are also absorbed, whereas Mg^{2+} and SO_4^{2-} are largely excluded. Notably, intestinal HCO_3^- secretion by Cl^-/HCO_3^- exchange results in the precipitation of calcium and magnesium carbonates, thereby removing them from solution and lowering the osmotic pressure in the luminal fluid. This both facilitates water absorption and helps minimize Ca^{2+} and Mg^{2+} absorption from the ingested seawater. Nevertheless, the fish is left with the problem of Na^+, Cl^-, and K^+ loading through the gut as well as the gills, necessitating their active branchial excretion (Grosell, 2011).

Detailed reviews of the transport mechanisms for Na^+, Cl^-, K^+ and Ca^{2+} by the mitochondria-rich ionocytes in the seawater gill have been provided by Wood and Marshall (1994), Marshall and Bryson (1998), Evans et al. (2005), Marshall and Grosell, (2005); Freire and Prodocimo (2007), Evans (2008), Hiroi and McKormick (2012), Edwards and Marshall (2013), and Hwang and Lin (2014). As first demonstrated by Keys and Willmer (1932), the ionocytes ("chloride cells") are the sites of Cl^- excretion, and the model proposed by Silva et al. (1977) has stood the test of time. Fig. 5 presents this model, updated with new information gained since 1977. Ionocytes are bordered by neighboring accessory cells with "leaky" tight junctions. In the ionocytes, basolateral NKA creates a strong electrochemical gradient for Na^+ entry, energizing the basolateral NKCC1 ($Na^+,K^+,2Cl^-$ cotransporter) to bring in one Na^+, one K^+, and two Cl^- ions simultaneously; K^+ and Cl^- are transported in against their gradients so both are above electrochemical equilibrium in the intracellular compartment. Cl^- is therefore excreted to the external seawater via CFTR ("cystic fibrosis transmembrane regulator") channels localized in apical crypts, whereas K^+ leaves by both basolateral and apical K^+ channels. The latter, also located in the apical crypts, are now identified as ROMK ("renal outer medullary K^+") channels and were not part of the original Silva model, though they now explain net K^+ export (Furukawa et al., 2012, 2015). The basolateral extrusion of Na^+ by NKA creates a high

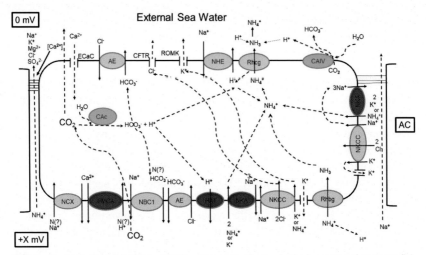

FIG. 5 Conceptual model of branchial ion transport pathways in the ionocytes of seawater fish, incorporating the Rhesus protein (Rh) pathways for ammonia excretion. Note that while all pathways are diagrammed in a single cell, in reality, the various pathways are usually partitioned among several different ionocyte types within a species, and not all species exhibit all pathways shown. Abbreviations as in legend of Fig. 4, with the addition of CFTR = cystic fibrosis transmembrane regulator type chloride channel; ROMK = renal outer medullary type potassium channel; AC = accessory cell; NKCC = sodium, potassium, two chloride cotransporter type one. Active transporters are orange/dark, passive transporters are blue/light, carbonic anhydrase is green/intermediate. The possible substitution of NH_4^+ for K^+ at NKA and NKCC is shown. Strong ions are shown diffusing into the seawater fish through paracellular pathways along their electrochemical gradients, and external waterborne calcium concentration $[Ca^{2+}]$ is shown acting on tight junctions to restrict this pathway. Note the location of NKA, NKCC, and K^+ channels on the specialized paracellular channels with the accessory cell (AC) which terminate in leaky junctions. Na^+ passes out via this route. NKA, NKCC, and K^+ channels also occur on the basolateral surface of the ionocyte. The transepithelial potential (TEP) is positive in the ECF relative to the external seawater as zero mV. In the Magadi tilapia living in saline, highly alkaline water (Section 2.4), by the model of Laurent et al. (1995), HCO_3^- replaces Cl^- at basolateral NKCC and apical CFTR, so is excreted through the transcellular pathway, while Na^+ passes out through the paracellular pathway.

local concentration of Na^+ in the ECF in the paracellular space behind tight junctions to the accessory cells (Fig. 5). This, combined with the inside positive transepithelial potential TEP caused by Cl^- extrusion, is sufficient to drive paracellular efflux of Na^+ in equimolar amounts to transcellular Cl^- efflux. Note that this positive-inside TEP is considered an electrogenic potential, in contrast to the negative-inside TEP of freshwater fish, that is considered to be a diffusion potential (see Section 1.1).

In addition to these mechanisms for the net excretion of Na^+ and Cl^-, which run at very high rates, apical Na^+/H^+ (NH_4^+) and Cl^-/HCO_3^- exchangers similar to those in freshwater fish appear to be present in the ionocytes, running in the opposite direction at much slower rates for the purposes of acid-base regulation (Fig. 5). Indeed, these may be the precursors of the mechanisms which allowed net Na^+ and Cl^- uptake during the evolutionary invasion of freshwater (Evans, 1984). In general, vH^+ ATPase (HAT) appears to be absent from the apical membrane, but is present basolaterally in many marine fish, poised to export H^+ ions to the ECF at times when the apical Cl^-/HCO_3^- exchanger is exporting base to the external water for acid-base regulation. The Rh metabolon system that couples ammonia excretion to Na^+ uptake in freshwater also appears to be present in marine fish, using NHE for H^+ extrusion (Wright and Wood, 2012). In parallel, the same mechanism(s) for Ca^{2+} uptake (apical ECaC and basolateral NCX and PMCA) seem to be present as in freshwater fish, but may be effectively inhibited by the high concentrations of other cations (Na^+ and Mg^{2+}) in seawater (Zimmer et al., 2019); mechanisms for putative net Ca^{2+} excretion at the gills of marine fish remain unknown.

The reader is referred to Volume 32 of the Fish Physiology series (McCormick et al., 2013) for detailed discussions of the iono- and osmoregu-latory mechanisms used by euryhaline fish. In grand overview, euryhaline fish use the basic mechanisms outlined in Section 1.1 when in freshwater (Fig. 4), and the basic mechanisms outlined in the current Section 1.2 when in seawa-ter (Fig. 5). What distinguishes them from their stenohaline relatives is their ability to switch back and forth between hyper- and hypo-regulatory mechan-isms required in the two very different environments (Fig. 1A, B), transitions that can occur either slowly (e.g., salmonids) or rapidly (e.g., most killifishes) based on complex developmental, environmental, and hormonal controls. Clearly the balance point where either hypo- or hyper-regulatory processes must become dominant is the isosmotic salinity (8–11 ppt or about 30% seawa-ter), though mechanistic transitions often start before this threshold is reached.

1.3 Special cases—Ionic and osmotic balance in marine hagfish and chondrichthyans

Two ancient classes, the Myxini (the hagfishes; Fig. 2A) and the Chon-drichthyes (the elasmobranchs and holocephalans; Fig. 2B) are prominent exceptions to the principles laid out in Section 1.2 for hypo-regulating marine fish, though some of the same mechanisms are present. These fish are osmo-conformers, keeping the osmolality of their body fluids equal to (hagfish; Fig. 2A) or very slightly above (chondrichthyans; Fig. 2B) the osmolality of the external seawater.

The jawless Class Myxini branched off the vertebrate root almost 500 mil-lion years ago. There has been little change in their morphology since then, and they have never invaded freshwater. Thus, their ionoregulatory physiol-ogy may be close to that of the ancestors of all other vertebrates, prior to the invasion of freshwater and subsequently land (Clifford et al., 2015; Currie and Edwards, 2010). However, as the phylogenetic position of hagfishes in vertebrate evolution remains controversial, and there is no fossil record for physiology, this statement is speculative. Unlike their distant jawless relatives, the lampreys (Class Hyperoartia), which have an ionoregulatory physiology similar to those of bony fish in both seawater and freshwater, hagfish osmoconform to their seawater environment. Nevertheless, they regulate individual ECF ion levels; Na^+ is kept slightly higher, Cl^- slightly lower, and Mg^{2+}, SO_4^{2-}, Ca^{2+}, and K^+ all much lower than in seawater (Giacomin et al., 2019a; Sardella et al., 2009; Fig. 2A). The ICF composition is also unusual for a vertebrate: intracellular K^+ and phosphate concentrations are similar to those of standard fish, but Na^+ and Cl^- levels are both high (comparable to K^+ concentrations), while other ions, including Mg^{2+} are low, and the inorganic osmolyte trimethylamine oxide (TMAO) accounts for a large portion of the ICF osmolality, together with amino acids and unidentified N-compounds. Curiously TMAO is virtually absent from the ECF (Fig. 2A). Ionocytes are abundant in the gills, and all appear to be of one type. These are rich in NKA, HAT, and CA, and also express Na^+/H^+ (NH_4^+) and Cl^-/HCO_3^- exchangers. Thus, the necessary components for both acid and base excretion appear to be present in a single ionocyte type in this early vertebrate, and are dependent on external Na^+ and Cl^- availability respectively (Evans, 1984). Rh proteins are also present in the gills, though their precise localization is uncertain (Braun and Perry, 2010; Edwards et al., 2015). Hagfish are extremely efficient at correcting acid-base disturbances (see Section 1.4) and most interest has focused on the roles of the ionocytes in this regulation. Virtually nothing is known about how they regulate the concentration of major plasma ions, though hormonal control of the rates of radio-labeled Ca^{2+} and Na^+ efflux through the gills has been demonstrated (Forster and Fenwick, 1994; Tait et al., 2009). The kidney may play a small role in ionoregulation, though limited by a very low urine flow rate (Alt et al., 1981; McInerney, 1974).

The cartilaginous Class Chondrichthyes (elasmobranchs and holocephalans) diverged from the bony fishes more than 400 million years ago. The great majority are restricted to the marine environment; only 10% routinely enter estuaries, and of these only 1% are truly euryhaline, while only 2–3% are restricted to freshwater (Martin, 2005). Iono- and osmoregulation in Chondrichthyans has been reviewed recently by Wright and Wood (2016) and Yancey (2016) and earlier by Shuttleworth (1988), Evans et al. (2004), Hammerschlag (2006), and Anderson et al. (2007). The basic details of osmoregulation in seawater were worked out by Homer Smith almost a century ago (e.g., Smith, 1929, 1931, 1936) and remain true today.

In brief, chondrichthyans are ureotelic, retaining high concentrations of urea and TMAO, so as to raise internal osmolality just above seawater osmolality (Fig. 2B), thereby avoiding osmotic water losses across the gills. This strategy thereby largely obviates the need to drink seawater to obtain free water, and provides enough water (by osmotic entry across the gills) to form a low urine flow, and most importantly avoids the need to transport large amounts of Na^+ and Cl^- at the gut (absorption) and gills (excretion). Nevertheless, very low-level drinking has been detected, and elasmobranchs do absorb ions and water from ingested prey (Anderson et al., 2007; Wood et al., 2007b). The virtual absence of drinking in chondrichthyans differs from the physiology of marine teleosts, and also in the fact that the former regulate somewhat higher Na^+ and Cl^- concentrations (225–350 mmol L^{-1}; Fig. 2B versus Fig. 1B), yet keep other strong ion levels very low, similar to those of "standard" marine fish. Thus urea (30–40%) and TMAO (5–10%) together constitute about 45% of plasma osmolality, whereas Na^+ and Cl^- account for about 50% (Fig. 2B). Because urea production is metabolically expensive (2.5 ATP per urea-N), the energetic cost of the chondrichthyan strategy for osmoregulation in seawater is about the same as the teleost hypo-regulation strategy (Kirschner, 1993). In the ICF, urea is at similar concentration to that in the ECF, whereas the TMAO level may be up to 5-fold higher than in the ECF. Intracellular enzymes in elasmobranchs appear to be adapted to high levels of urea and to require them for optimal kinetic properties (e.g., Yancey and Somero, 1978). At the same time, the presence of TMAO, as a "counteracting osmolyte" helps protect protein function against the destabilizing effects of urea (Yancey, 2016). Free amino acids in the ICF are high relative to other fishes (Forster and Goldstein, 1976). Other intracellular ion concentrations appear to be similar to those in marine teleosts.

As plasma Na^+ and Cl^- levels are only about half those in seawater, there are strong concentration gradients for the latter to enter across the gills (Fig. 2B). The rectal gland, which is an unusual small digitiform organ in the posterior intestine densely packed with ionocytes, is often thought to be the major organ of NaCl excretion. However, it has been shown repeatedly that when this gland is disabled by removal or ligation, the fish is still able to regulate plasma Na^+ and Cl^- concentrations at normal levels, even when urinary excretion is also blocked (reviewed by Wright and Wood, 2016). The normal role of this gland is probably to deal with the sudden load when a salt-laden meal is ingested, quickly excreting a fluid containing $\sim$500 mmol L^{-1} NaCl and little else (Wood et al., 2007c). The gills appear to be completely competent in Na^+ and Cl^- regulation at other times. The branchial epithelium is richly endowed with ionocytes, but accessory cells, which are critically important in the Na^+ and Cl^- excretion mechanism in marine teleosts (see Section 1.2, Fig. 5), appear to be absent There are at least two different types of ionocyte. One exhibits high basolateral Na^+,K^+-ATPase activity and appears to be involved in apical acid-excretion coupled to Na^+ influx from the water, while the other exhibits high basolateral vH^+-ATPase (HAT) activity and appears to be involved in apical base-excretion coupled to Cl^- influx from the water (see Section 1.4). It remains unclear how Chondrichthyan gills actively excrete Na^+ and Cl^-, though it is known that the unidirectional Na^+ and Cl^- efflux rates are many-fold higher than those needed to counteract the "short-circuiting" unidirectional influxes needed for acid base regulation (reviewed by Wright and Wood, 2016). The other critical role of the gills is to minimize the diffusive losses of urea and TMAO; this involves both selective impermeability properties due to an unusual lipid composition, and in the case of urea at least, an active "back-transport" mechanism (Fines et al., 2001; Pärt et al., 1998; Wood et al., 2013a). This appears to short-circuit the leakage of urea across the gills, actively transporting it back into the extracellular fluid. The kidney excretes toxic Mg^{2+} and SO_4^{2-} as well as inorganic phosphate while reabsorbing some filtered urea, TMAO, Na^+, Ca^{2+}, and Cl^-. Renal function plays only a modest role in NaCl excretion as the kidney cannot produce a urine hypertonic to the blood plasma.

The reader is referred to Ballantyne and Fraser (2013) and Wright and Wood (2016) for detailed information on the physiology of euryhalinity in chondrichthyans. In brief, most are partially euryhaline "tolerators," able to invade lower salinity coastal waters and estuaries on feeding forays, but to widely varying degrees. They do this by tolerating osmotic dilution, though they cease being osmoconformers. Internal osmotic pressure is reduced by increasing urea and TMAO excretion, mainly via the kidney, and by turning down the metabolic urea production rate. Plasma Na^+ and Cl^- levels also fall, though to a lesser extent, and this occurs both by dilution and by increased excretion of ions and water through kidney and rectal gland. Internal fluid volumes are allowed to rise. However, osmolality, while reduced, remains above that in the environment. This strategy differs quantitatively from that used by truly euryhaline species that are able to live indefinitely in freshwater, and mechanistically from that seen in the very few stenohaline freshwater elasmobranchs that are restricted to freshwater (see Section 2.6).

1.4 Acid-base regulation

Burggren and Bautista (2019) and Tresguerres et al. (2020) have provided up-to-date overviews of acid-base regulation. In obligate water-breathing fish, the gills are without doubt the major organ of acid-base regulation on a quantitative basis, with the kidney playing a lesser but still critical role, especially in freshwater. Because the CO_2 capacitance of water is approximately $30 \times$ greater than its O_2 capacitance, CO_2 is effectively washed out of the blood by the normal ventilatory water flow required to take up adequate O_2. As a result, normal arterial PCO_2 tensions are very low (a few torr), and water-breathing fish have very little capability for the "respiratory" control of arterial PCO_2 (i.e., by changing ventilation) which is used by higher air-breathing vertebrates for rapid adjustment of internal pH. Interestingly, recent observations on air-breathing fish suggest that alterations in ventilation of the air-breathing organ (ABO) may be driven by internal pH and/or PCO_2 signals, perhaps a precursor of this strategy (Bayley et al., 2019; Tuong et al., 2018, 2019). However, water-breathing fish are often passive victims of increases in environmental PCO_2 which readily diffuses across the gills into the bloodstream, causing "respiratory acidosis." Therefore, they rely on "metabolic" compensation by net excretion of acidic or basic equivalents, mainly across the gills, by mechanisms linked to ion transport through the SID concept (see Section 1.1 and Fig. 3). This is true regardless of whether the disturbance is of "respiratory" origin (i.e., due to PCO_2 variations) or "metabolic" origin (i.e., due to variations in acidic or basic equivalents—e.g., low environmental pH, lactacidosis, or the postprandial alkaline tide).

As described in Sections 1.1–1.3 acidic equivalent excretion (as H^+ or NH_4^+ coupled to Na^+ uptake) and basic equivalent excretion (as HCO_3^- coupled to Cl^- uptake) generally occur in two or more different types of branchial ionocytes, though there is great variation among species in the fine details (Clifford et al. 2015; Dymowska et al., 2012; Evans et al., 2005; Wright and Wood, 2016). The one exception appears to be the hagfishes, which exhibit a single type of ionocyte apparently incorporating both mechanisms. In general, acceleration, slowing, or even reversal of the Cl^-/basic equivalent exchange mechanism appears to be quantitatively more important than reciprocal changes in the Na^+/acidic equivalent mechanism for changing the internal SID (Fig. 3). The net result is that plasma Na^+ concentrations usually remain fairly constant, whereas plasma Cl^- concentrations may change considerably, in reciprocity with changes in plasma HCO_3^- concentrations. The HCO_3^- level is thereby determined by the SID (Fig. 3). While the focus is usually on modulation of Na^+ and Cl^- influx rates, the importance of the differential regulation of Na^+ and Cl^- efflux rates, constraining net acid-base flux by the SID theory should not be overlooked, as emphasized by McDonald et al. (1989). The bottom line is that it is the *net* ion fluxes that matter (Fig. 3). The difference between the *net* branchial flux rates of strong cations (mainly Na^+) and strong anions (mainly Cl^-) constrains and must equal the *net* branchial flux rates of acidic equivalents in the opposite direction, recognizing that in acid-base terms, acidic equivalent efflux equals basic equivalent uptake, and vice versa (Wood, 1989, 1991).

The rates of both Na^+/acidic equivalent and Cl^-/basic equivalent exchange processes clearly depend on the availability of the counterions in the external medium. Indeed, some teleosts living in the very dilute, acidic softwaters of the Amazon show negligible ability to regulate extracellular acid-base status, apparently choosing instead to regulate only intracellular pH in selected tissues (Shartau et al., 2016). Therefore, rates of acid-base compensation are lowest in ion-poor freshwater, higher in ion-rich freshwater, and much higher in seawater that has a relative abundance of Na^+, Cl^-, and HCO_3^- (Brauner et al., 2019; McDonald et al., 1989). Indeed, marine elasmobranchs (e.g., Heisler, 1988 ; Tresguerres et al., 2005 ; Wood et al., 1995) and especially marine hagfishes (e.g., Clifford et al., 2018; McDonald et al., 1991; Parks et al., 2007) are exceptionally proficient at rapid acid-base regulation. In both cases, an additional contributing factor may be the high internal reserves of plasma Na^+ and Cl^- ions. For example, the Pacific hagfish appears to be the current champion of CO_2 tolerance, allowing blood pH to fall by over 1.2 pH units during exposure to a PCO_2 of 50 Torr, then rapidly compensating it by accumulating more than $75 \, mmol \, L^{-1}$ of plasma HCO_3^-, while allowing plasma Cl^- concentration to drop by an equimolar amount (Baker et al., 2015). For most fish, this would be a loss of about 50% of their plasma Cl^- (Fig. 1B), but for the hagfish this is only a 16% loss (Fig. 2A).

The most important function of the kidney in acid-base balance in teleost fish, especially in freshwater fish, is to reabsorb HCO_3^- from the high volume of glomerular filtrate; otherwise, basic equivalents would be quickly lost, and the function of the gills in acid-base regulation would be short-circuited (Wood, 1995). This becomes especially important when high concentrations of plasma HCO_3^- are accumulated due to branchial acidic equivalent excretion (= basic equivalent uptake) during the compensation of respiratory acidosis. In addition, the kidney accounts for between 5% and 33% of total acid-base excretion in most studies on freshwater fish, with the larger portion occurring via the gills. However, the one important exception is the response to low environmental pH (4.0–4.5) in freshwater fish (in waters of high Ca^{2+} concentration only, see Section 2.1), where the ability of the gills to excrete acidic equivalents is blocked by the external acidity, and the kidney is the only route of compensation (McDonald and Wood, 1981; Wood et al., 1999). The renal contribution is much lower in seawater fish where glomerular filtration (GFR) and urine flow rates (UFR) are both very low and the urine is very acidic (Wood, 1995); this low pH is thought to minimize the precipitation of Ca^{2+} and Mg^{2+} salts during long periods of storage in the bladder. The situation is similar in marine chondrichthyans (negligible renal contribution, very acidic urine) though GFR and UFR values are higher than in marine teleosts (reviewed by Wright and Wood, 2016). There is little information available in marine hagfish, but the very low GFR and UFR values suggest a negligible contribution to acid-base regulation (Clifford et al., 2015; Clifford et al., 2018).

In broad plan, the mechanisms of HCO_3^- reabsorption and net acidic equivalent excretion in the fish kidney (Lawrence et al., 2015; Wood, 1995; Wood et al., 1999) are similar to those in higher vertebrates (Weiner and Verlander, 2017). HCO_3^- reabsorption from the glomerular filtrate is achieved by secretion of H^+ in exchange for Na^+ reabsorption via a Na^+/H^+ exchanger and/or a Na^+ channel coupled to a vH^+-ATPase. Intracellular carbonic anhydrase (CA) catalyzes the hydration of CO_2 to form the H^+ ion that is secreted across the apical membrane and the HCO_3^- ion that is returned to the blood across the basolateral membrane, by a Na^+-HCO_3^- co-transporter. As a result, for every one H^+ ion secreted, one HCO_3^- ion and one Na^+ ion are removed from the filtrate, while one HCO_3^- (albeit a different HCO_3^-) and one Na^+ ion are returned to the blood (i.e., no net excretion of acid or base). When secreted H^+ ions exceed filtered HCO_3^- ions, they are buffered in the urine by inorganic phosphate (HPO_4^{2-}, yielding $H_2PO_4^-$) or ammonia (yielding NH_4^+), and are excreted, contributing to net acidic equivalent excretion; as a result, a new basic equivalent (HCO_3^-) is synthesized on a net basis, and returned to the blood (i.e., acidic equivalent excretion = basic equivalent uptake). If H^+ secretion is turned off, filtered HCO_3^- escapes reabsorption and passes out in the urine (i.e., basic equivalent excretion = acid equivalent uptake). Most of the HPO_4^{2-} enters the urine by glomerular filtration, whereas most of the ammonia is produced by the renal tubule cells themselves via deamination and deamidation of glutamine and other amino acids. and enters the urine by secretion.

1.5 Nitrogenous waste excretion

The rate of nitrogenous waste excretion depends on the rate of nitrogenous waste production, as fish are normally more or less in a steady state where the two processes are in approximate balance (reviewed by Altinok and Grizzle, 2004; Bucking, 2017; Ip and Chew, 2010; Wood, 1993, 2001). In turn, both depend largely on the rate at which proteins and amino acids are being used in oxidative metabolism, and to a minor extent on the metabolism of adenylates, nucleotides, purines, pyrimidines, and diverse N-compounds. If 100% of aerobic respiration is fueled by these substrates, then the nitrogen quotient (NQ = N-excretion/O_2 consumption) will be about 0.27, and if only 10%, the NQ will be 0.027, etc. Thus, the NQ and the rates depend on diet, feeding versus fasting, activity levels, and the presence of corticosteroids (e.g., cortisol) that favor proteolysis. In most fish, the great majority of N-waste is produced and excreted across the gills as ammonia, with a small amount (generally less than 20%) as urea-N, the latter produced by uricolysis or arginolysis. In many studies (reviewed by Wood, 2001), total N-excretion to the water often substantially exceeds the sum of ammonia-N excretion plus urea-N excretion, a discrepancy that has never been satisfactorily explained. At least part of the explanation may be sloughing of mucoproteins and leakage of amino acids; in this case, the losses would not be "waste-N," but rather "non-metabolized-N." A great advantage of being ammoniotelic is that the metabolic cost of ammonia excretion is negligible. This contrasts with the ureotelism of many higher vertebrates where most of the ammonia produced is converted into urea by the ornithine urea cycle (OUC) in the liver, a metabolically costly process, prior to excretion through the kidney (see Section 1.3).

Current evidence suggest that the mechanisms of ammonia and urea excretion are at least qualitatively similar among most species, with the gills dominating throughout, regardless of phylogeny or environmental salinity. The Rh proteins facilitate ammonia excretion, also help couple it to Na^+ uptake, and allow active excretion across the gills when gradients are unfavorable for passive diffusion (Sections 1.2–1.4, Figs. 4 and 5). To date, Rh proteins have been found in the gills of all species that have been examined (e.g., Edwards et al., 2015; Nawata et al., 2015; Wright and Wood, 2009), and have also been detected in the skin (e.g., Nawata et al., 2007; Zimmer et al., 2014), kidney (e.g., Nakada et al., 2007a; Lawrence et al., 2015), and gut (e.g., Bucking and Wood, 2012; Bucking et al., 2013) of some species. The normally small role of the kidney in N-excretion increases in freshwater fish at times of acidosis, where increased metabolic production in the renal tubule cells provides ammonia as an important urinary buffer (Section 1.4). However, the kidney appears to be generally less important in marine fish, reflecting low GFR and UFR. Like Rh proteins, UT proteins, which serve as channels to facilitate urea transport, have also been found in the gills of all species that have been examined (e.g., Walsh et al., 2000; Braun and Perry, 2010; McDonald et al., 2006, 2012; Walsh et al., 2001a,b).

However. there are two marked exceptions to these generalities. Firstly, virtually all fish produce and accumulate urea by the OUC early in embryonic life when metabolism is fueled by a "diet" that is exclusively yolk protein (reviewed by Wright and Fyhn, 2001; Zimmer et al., 2017b). Urea-N excretion generally predominates over ammonia excretion at this time. Ammonia accumulation and toxicity are thereby avoided, but this OUC-based capacity is soon lost once the gills develop to provide the major route for ammonia excretion. Secondly, most Chondrichthyans are ureotelic throughout their lives. These animals produce urea in the liver, gut, and skeletal muscle and retain it as a key osmolyte in their body fluids (see Section 1.3; Fig. 2B). Despite branchial, renal, intestinal, and rectal gland mechanisms aimed at urea retention, urea-N accounts for >70% of waste N-excretion, occurring mainly across the gills (reviewed by Wright and Wood, 2016).

2 Conservation issues

2.1 Acid-rain toxicity in North America and Northern Europe – A detective story

Starting in the late 1960's, and continuing through to the present day, ion-poor soft-waters (i.e., low in Ca^{2+} concentration) in many parts of the northern hemisphere have been devastated by "acid rain" resulting from precipitation enriched in H^+, SO_4^{2-} and NO_x from coal-fired power plants, smelters, diverse industries, and vehicular emissions (Schindler, 1988). For overviews of effects on fish populations the reader is referred to extensive compilations edited by Johnson (1982) and Morris et al. (1989). Impacts included recruitment failures of entire year classes, progressive disappearance of fish populations, and highly visible fish kills, especially during spring snowmelt and rainstorm events. The latter represented the periods of greatest runoff of acidity into lakes and streams, and often coincided with the period when the gills started to function in swim-up fry. In general, toxicity started in a few species once pH fell below 6.0, and most species were lost once pH reached 4.0.

Initial laboratory investigations focused on mucification, epithelial cell proliferation, and edema of the gills, resulting in suffocation and hypoxaemia as the mechanism of toxicity (e.g., Daye and Garside, 1976; Packer and Dunson, 1972; Plonka and Neff, 1969), but it soon became clear that this occurred only at severely low pH, and could not explain toxicity in the pH 4.0–6.0 range (Fromm, 1980; Wood and McDonald, 1982). Attention then turned to failure of acid-base regulation, and initial results (e.g., Neville, 1979; McDonald et al., 1980; McDonald and Wood, 1981) were encouraging, showing acidic equivalent uptake at the gills, only partially compensating acidic equivalent excretion at the kidney, and severe metabolic acidosis - large decrements in plasma pH and HCO_3^- with unchanged PCO_2 (see Section 1.4, and Fig. 3). However, these studies were performed in hardwater (high Ca^{2+} concentration), and environmental acidification does not normally occur in hardwater which is invariably well-buffered. As of 1982, only about 25% of all the physiological studies on fish under acute acid stress had been performed in softwater (Wood and McDonald, 1982), despite the fact that low water Ca^{2+} concentration had been clearly identified as the major risk factor for fish death at low pH (Brown, 1982; Haines, 1981; Wright and Snekvik, 1977). When the experiments were subsequently repeated in soft waters of low Ca^{2+} concentration ($\leq 0.1\,mmol\,L^{-1}$) representative of those present in acid-sensitive lakes of North America and Northern Europe, a very different physiological pattern emerged (e.g., McDonald, 1983; McDonald et al., 1980, 1983). Even though toxicity was exacerbated, acid-base disturbance was attenuated or non-existent, while ion losses were greatly increased. Indeed, as soon as Na^+ and Cl^- net losses from the plasma, which occurred almost entirely across the gills, reached about 30%, the fish died. Ion loss induced internal fluid shifts, hemoconcentration, elevation of blood pressure, reduction of blood volume, and eventual cardiovascular collapse as the cause of death (Milligan and Wood, 1982). The nature of the disturbance at low pH turned out to be a direct function of the water Ca^{2+} concentration, and little affected by other water ions.

From a physiological perspective, these data were important in providing a real-world illustration of the SID concept (see Section 1.4 and Fig. 3). Low pH clearly increased the permeability of the gills to strong ions, and higher water Ca^{2+} levels clearly provided protection against this effect. At high Ca^{2+} concentrations (e.g., $>1.0\,mmol\,L^{-1}$) while overall gill permeability was reduced, partially countering the effect of low pH, Cl^- permeability was protected to a greater extent than Na^+ permeability. Thus, net Na^+ losses exceeded net Cl^- losses, constraining a net uptake of acidic equivalents with a resulting metabolic acidosis, signaled by a reduction of the SID and HCO_3^- concentration in the blood plasma. At low water Ca^{2+} concentrations ($\leq 0.1\,mmol\,L^{-1}$) and low pH, the gill became much more permeable to both ions, and net Na^+ and Cl^- losses became much larger and approximately equal. Once net cation losses (mainly Na^+) become equal to net anion losses (mainly Cl^-), there could be no net uptake of acidic equivalents by the SID concept (Fig. 3). Thus, there was no internal acid-base disturbance.

Initial mechanistic focus was on the inhibition of unidirectional Na^+ influx at low pH. This had been observed by several previous workers (e.g., Packer and Dunson, 1972; McWilliams and Potts, 1978) and was attributed to blockade of apical Na^+ uptake versus acidic equivalent efflux transporters in the ionocytes (Fig. 4), but McDonald et al. (1983) were the first to show that Cl^- uptake was also inhibited by low pH, the exact mechanism of which remains unclear even today. More importantly, McDonald et al. (1983) demonstrated that, rather than influx inhibition, it was the differential stimulation of the Na^+ and Cl^- efflux rates, dependent upon water Ca^{2+} concentration, that was quantitatively more important in determining net Na^+ and Cl^- losses at low pH. Later studies at more moderate acidic pH (4.8) in soft water showed some capacity for restoration of both Na^+ and Cl^- influx and efflux rates over long periods of time (2–3 months), but plasma ion concentrations remained depressed, acidic equivalent fluxes remained close to zero and true acclimation did not occur (Audet et al., 1988).

Despite these advances, there was a growing consensus in the 1970s and 1980s that acidity alone was insufficient to explain fish losses in nature—mortality in the field was often much greater than in the lab under the same water Ca^{2+} and pH conditions. Circumstantial evidence pointed to low-pH induced mobilization of aluminum, the third most abundant element in the lithosphere, as the culprit (reviewed by Wilson, 2012). The aquatic geochemistry of aluminum is extremely complex and beyond the scope of this volume. Suffice it to say that aluminum is virtually insoluble at neutral pH, but its solubility increases exponentially below pH 6.0; concentrations of $100–1000 \mu g \ L^{-1}$ have been commonly reported during spring acid pulses. Aluminum hydroxides ($AlOH_2^+$, and $Al(OH)^{2+}$) dominate at pH 5.0–6.0, whereas Al^{3+} becomes the major form at pH $\leq$ 4.0. Toxicity is greatest around 5.5, so it now appears that fish kills occurring in the pH 5.0–6.0 range may have been primarily due to aluminum hydroxide toxicity, with H^+ toxicity being an exacerbating factor.

This was clarified by a decade of research that showed two mechanisms of toxicity, ionoregulatory and respiratory, the severity of each being critically dependent upon both pH and Ca^{2+} levels within the natural ranges for acidified softwaters (e.g., Booth et al., 1988; Neville, 1985; Neville and Campbell, 1988; Playle et al., 1989; Wood et al., 1988; Wood and McDonald, 1987). Fig. 6 illustrates a conceptual model. Ionoregulatory toxicity predominated at lower pH and lower water Ca^{2+} concentrations, where Al^{3+} and cationic aluminum hydroxides exacerbated the disruption of Na^+ and Cl^- balance caused by low pH alone (described above). In addition to direct inhibition of unidirectional Na^+ and Cl^- influxes (Booth et al., 1988), reductions of gill Na^+,K^+-ATPase and carbonic anhydrase activities were seen (Staurnes et al., 1984). Al^{3+} and cationic aluminum hydroxides also further disrupted the tight junctions of the gill epithelium, exacerbating diffusive ion losses caused by low pH alone. Respiratory toxicity predominated at higher pHs, reflecting the precipitation of

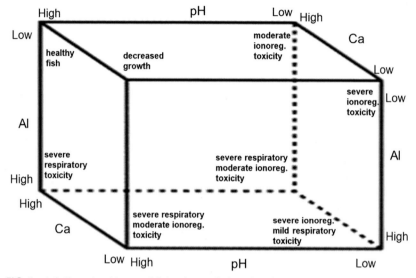

FIG. 6 A 3-dimensional box model showing predictions, based on laboratory data, of qualitative responses of fish to acidified natural waters of varying pH, aluminum (Al), and calcium (Ca) levels, indicating the complex interactions between ionoregulatory and respiratory toxicity. A model similar to this, presented to a US. congressional subcommittee, helped promote the passage of amendments to the US Clean Air Act in 1990, which ultimately resulted in reduced transboundary fluxes of acid emissions.

aluminum hydroxide complexes onto the gill surface and resulting in pathologies (inflammation, edema, epithelial sloughing and mucification) and death by suffocation, rather similar to those seen in the earlier acid-only experiments conducted at unrealistically low pHs (described above). The abrupt rise in water pH in the gill micro-environment due to NH_3 excretion (Fig. 4) was an important trigger for this rapid precipitation when the inspired pH was in the 5.0–6.0 range (Playle and Wood, 1989a, 1989b). Elevations in water Ca^{2+} levels reduced ionoregulatory toxicity but exacerbated respiratory toxicity. Dissolved organic carbon (DOC) furthered complicated the story by reducing the bioavailability of aluminum in a pH-dependent fashion (Gensemer and Playle, 1999; Wilson, 2012).

This mechanistic understanding of the physiology of acid and aluminum toxicity to fish played an important role in conservation. For example, presented as a brief to the US congress, with a summary diagram similar to Fig. 6, it contributed to passage of amendments to the Clean Air Act in 1990, which greatly reduced acidic atmospheric emissions, with similar legislative outcomes in Canada and Europe. It also informed more thoughtful strategies for "liming" acidified lakes; previously, this treatment in itself sometimes had detrimental effects on fish and other organisms by causing aluminum precipitation events (Gensemer and Playle, 1999; Wilson, 2012).

This physiological information also provided an impetus for implementation of water quality criteria and/or revision of outdated criteria, and stricter emission controls in North America, Europe, and China. Acidified lakes and streams are now slowly recovering (e.g., Keller et al., 2019), and regulatory strategies are improving, but again slowly. For example, the US EPA ambient water quality criteria for aluminum dated from 1988, and were renewed unchanged in 2009. Curiously, the criteria covered only the pH range 6.5–9.0, which is largely irrelevant for acidification scenarios. Only in 2018 did the US EPA implement new, more realistic aluminum criteria (US EPA, 2018), taking into account pH, Ca^{2+} and dissolved organic carbon (DOC) levels over the ranges relevant to environmental acidification. These criteria were derived based on a metal bioavailability approach using multiple linear regression modeling, which in turn was based on physiological understanding captured in a Biotic Ligand Model (see Section 2.3) for aluminum (Santore et al., 2018).

2.2 Survival of fishes in the acidic, ion-poor blackwaters of the Rio Negro, a biodiversity hot spot

The so-called "blackwaters" of the Rio Negro and its associated watershed present physicochemical conditions so severe that they would kill most freshwater fish, yet in fact are a hotspot of biodiversity. Endemic fishes represent 8% of all the freshwater teleost species in the world, and these waters also host the only stenohaline freshwater elasmobranchs (stingrays) (Val and Almeida-Val, 1995). The waters are often extremely acidic (pH = 3.0–5.5), mainly due to the presence of high concentrations of DOC (5–35 mg C L^{-1}), more than 50% of which are humic substances (humic and fulvic acids) of terrestrial origin (Kullberg et al., 1993; Leenheer, 1980; Thurman, 1985). These are complex, heterogeneous molecules derived from the breakdown of lignins and other compounds produced by the jungle vegetation. Additional acidification comes from unusually high PCO_2 levels, anywhere from 3 Torr in the main river (Richey et al., 2002) to 60 Torr in floodplains (Furch and Junk, 1997) resulting from respiration of decaying vegetation, and lack of equilibration with the atmosphere (0.3 Torr). The water is also extremely ion-poor, lower in alkalinity and all major ions than even the softwaters of the Northern Hemisphere discussed in Section 2.1. Gonzalez et al. (2006) calculated that on average, Na^+, Cl^-, Mg^{2+}, and Ca^{2+} concentrations of the Rio Negro are only 6%, 22%, 3%, 1% of the mean values in global rivers. On the other hand, natural metal levels (particularly Al, Fe, Mn, and Hg) are unusually high, again reflecting the underlying geochemistry.

Morris et al. (2021) have recently reviewed how endemic species cope with these multiple challenges. Physiological understanding greatly benefitted from the principles learned from the acid rain story (Section 2.1). In short, the high native metal levels are not a problem because they are rendered biologically unavailable by the high DOC concentrations. Interestingly, the particular physicochemical properties of DOC that are effective in metal binding appear to be the same as those which exert direct protective effects on ionoregulation (Al-Reasi et al., 2013), as discussed subsequently. Furthermore, acid-base balance does not appear to be impacted by the water acidity, at least in the absence of elevated PCO_2 (e.g., Wilson et al., 1999; Wood et al., 1998) because water Ca^{2+} levels are extremely low (generally $\leq 20\,\mu mol\,L^{-1}$). As explained in Sections 1.4 and 2.1 via SID theory (Fig. 3), this actually protects against metabolic acidosis by preventing acidic equivalent uptake at the gills. On the other hand, elevated water PCO_2 does equilibrate across the gills, causing respiratory acidosis in the body fluids. There is accumulating evidence that Rio Negro fishes preferentially regulate intracellular pH, prioritizing this over extracellular pH regulation under these circumstances (Brauner et al., 2004; Harter et al., 2014; Shartau et al., 2016, 2019). It is thought that the very low concentrations of ions and lack of alkalinity in the external water may greatly limit the extent to which excretion of acidic equivalents across the gills can occur by processes linked to Na^+ and Cl^- transport. Therefore, this limited capacity for regulation is conserved for intracellular pH regulation, and internal basic equivalents are redistributed to key tissues.

It is now clear that the most important physiological challenge for Rio Negro fishes is that of maintaining ionic homeostasis in the face of extremely low pH and ion levels in the external water, and that this has been solved by important adaptations in the mechanisms of gill ion transport (Morris et al., 2021). In general, gill uptake and diffusive loss pathways are much more resistant to disruption by low water pH than in other species. Additional evolutionary solutions to this problem include regulation of generally lower plasma levels of strong ions (Mangum et al., 1977, 1978), increased retention efficiency of these ions at the kidney (reviewed by Pelster and Wood, 2018), and increased reliance on the digestive tract for ion uptake from food (Pelster et al., 2015) in Rio Negro fishes.

At the gills, two distinct adaptive strategies have been identified so far, as illustrated in Fig. 7, though only 3 of the 17 Orders of fish occurring in the Rio Negro (Beltrão et al., 2019) have yet been studied in any detail. R.J. Gonzalez, A.L. Val and colleagues have used a combination of unidirectional flux measurements, concentration-kinetic analyses, and pharmacology to describe these two strategies (Duarte et al., 2013; Gonzalez et al. 1997, 2002, 2006, 2018, 2020a, 2020b; Gonzalez and Preest, 1999; Gonzalez and Wilson 2001; Preest et al., 2005; Wood et al. 2002a, 2014). The first is characteristic of the Orders Cichliformes (cichlids, third most abundant Order) and Myliobatiformes (formerly Rajiformes, stingrays). Their Na^+ and Cl^- uptake mechanisms seem to be pharmacologically similar to those described in "standard" freshwater fish (see Section 1.1, Fig. 4), and these transport systems have a relatively low affinities (i.e., high Km values), so unidirectional

Unidirectional Na⁺ Flux

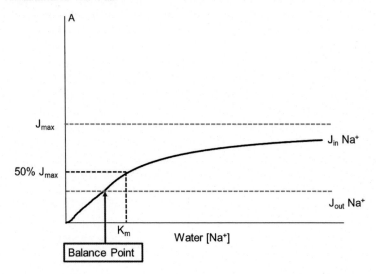

Unidirectional Na Flux

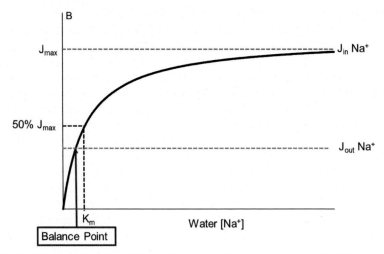

FIG. 7   Gill Na⁺ transport models illustrating the two known strategies used by fish in Amazonian blackwaters to achieve Na⁺ balance in this ion-poor, acidic environment. The solid black lines represent the relationship between unidirectional Na⁺ transport rate (influx $=J_{in}^{Na^+}$) and the external Na⁺ concentration ([Water Na⁺]), as described by the Michaelis-Menten equation: $J_{in}^{Na^+} = \frac{J_{max} \times [\text{Water Na}^+]}{K_m + [\text{Water Na}^+]}$, where J_{max} is the maximum $J_{in}^{Na^+}$ and K_m is the affinity constant, representing the water sodium concentration ([Water Na⁺]) where $J_{in}^{Na^+}$ is 50% of J_{max}. The lower dashed lines represent the unidirectional Na⁺ efflux rates ($J_{out}^{Na^+}$). The intersection point between the influx and efflux lines represents the balance point where $J_{in}^{Na^+} = J_{out}^{Na^+}$. In (A), the strategy used by members of the Orders Cichliformes (cichlids) and Myliobatiformes (stingrays), the high K_m values

influx rates are quite low (Fig. 7A). However, they can be inhibited by low pH (3.5–4.5). Compensation includes low unidirectional efflux rates (Fig. 7), extreme resistance to elevation of these diffusive effluxes by low pH, and probably enhanced uptake from the diet. The latter may be particularly important, as the balance point is shifted to higher external Na^+ concentrations (Fig. 7A). The second strategy is characteristic of the Order Chariciformes (the most abundant Order), where the Na^+ and Cl^- uptake systems appear to be unique and as yet poorly understood, being resistant to most drugs that are effective in other fish (e.g., blockers of the NHE, vH^+-ATPase/ Na^+-channel, and NaCl cotransport systems described in Section 1.1) They exhibit very high affinities (i.e., low Km values) and high capacities (i.e., high Jmax values), so that at moderate pH's, unidirectional uptake rates are high, keeping balance with comparably high rates of unidirectional efflux at low external Na^+ concentrations (Fig. 7B). They also appear to be almost completely resistant to inhibition by very low external pH (e.g., 3.5), whereas efflux rates may show only moderate elevation, so that the balance point is little affected (Fig. 7B).

An additional factor that helps fish to resist the negative effects of low pH, regardless of their gill strategies, is the very same factor that causes the acidity of blackwaters - the presence of the DOC of terrestrial origin (humic substances, "allochthonous DOC"; Kullberg et al., 1993; Wood et al., 2011). While this protective phenomenon was first observed in Rio Negro species (Gonzalez et al., 1998, 2002; Matsuo and Val, 2007; Wood et al., 2003), it is now clear that the protection by allochthonous DOC extends to species and DOCs from other parts of the world (Al-Reasi et al., 2016; Duarte et al., 2016, 2018; Matsuo et al., 2004). In the presence of environmentally realistic concentrations (e.g., $10\,mg\,C\,L^{-1}$) of these DOCs, there is less inhibition of active Na^+, Cl^- uptake rates by low environmental pH, and less stimulation of their diffusive efflux rates, and after acclimation, these protective effects persist even when the DOC is no longer present in the water The mechanisms of protection by these large, heterogeneous amphiphilic molecules remain poorly understood, but the following may all be important (Morris et al., 2021). Low pH titrates the negative charge on these molecules, making it easier for them to interact with the gill membranes (Campbell et al., 1997), which are also negatively charged (Reid and McDonald, 1991).

---

and the low J_{max} values (i.e., low affinity and capacity) result in low $J_{in}^{Na^+}$ rates that are moderately susceptible to inhibition by low pH. However, $J_{out}^{Na^+}$ (efflux rate) is also low and extremely resistant to stimulation by low pH. The balance point is higher than in the second strategy and may be shifted to an even higher [Water Na^+] by low pH. In (B), the strategy used by members of the Order Characiformes, the low K_m values and the high J_{max} values (i.e., high affinity and capacity) result in high $J_{in}^{Na^+}$ rates that are very resistant to inhibition by low pH. The $J_{out}^{Na^+}$ (efflux rate) is higher than in the first strategy and susceptible to some stimulation by low pH. Nevertheless, the balance point is lower than in the first strategy, and not greatly affected by low pH.

In so doing, DOC may act in a Ca^{2+}-like fashion to "tighten" gill membrane permeability (Wood et al., 2003). DOC also makes the TEP across the gill membranes more negative (Fig. 4); this will reverse the depolarizing effect of low pH, reducing the electrochemical gradient opposing Na^+ uptake, thereby favoring Na^+ retention (Galvez et al., 2008; Sadauskas-Henrique et al., 2019). DOC also helps to maintain the coupling of Na^+ uptake to ammonia excretion (Fig. 4) (Duarte et al., 2016, 2018; Wood et al., 2003). This may be related to its ability to act as a buffer in the external micro-environment of the gills so as to minimize the inhibitory effects of low pH on apical Na^+ transporters (Al-Reasi et al., 2016; Wood et al., 2003). Additionally, DOC exposure elevates Na^+, K^+-ATPase and vH^+-ATPase activities in the gills (McGeer et al., 2002; Sadauskas-Henrique et al., 2021); these effects may explain its ability to promote increases in the Jmax values of the concentration-kinetic relationships for Na^+, Cl^-, and Ca^{2+} uptake (Fig. 7) (Al-Reasi et al., 2016; Matsuo et al., 2004; Matsuo and Val, 2007; Wood et al., 2003).

To date, the Rio Negro watershed has not been as badly impacted by anthropogenic threats as other parts of the Amazon region. For example, no major dams have yet been built, but deforestation, oil spills, mining, and sewage discharges are increasing, exotic species have been introduced, and over-fishing for human consumption and collection of live specimens for the ornamental aquarium fish trade continue. The annual water cycle has already changed, with drought events becoming more prevalent, and water temperatures are increasing. Chao (2001), Barletta et al. (2010), Zehev et al. (2015), Beltrão et al. (2019), Evers et al. (2019), Val (2019) and Duarte and Val (2020) provide perspectives on these issues. In only 30 years of detailed study, Rio Negro fishes have provided key lessons to physiology- most importantly illumination of mechanisms of ionoregulatory homeostasis at low pH, and the critical importance of DOC, a water chemistry parameter previously overlooked by both physiologists and toxicologists (Wood et al., 2011). We have much left to learn from this invaluable assemblage of fishes, and the need for their conservation is becoming critical.

2.3 The biotic ligand model (BLM), a regulatory tool for environmental regulation based on physiological understanding of ionoregulatory impacts

The Biotic Ligand Model (BLM; DiToro et al., 2001; Mebane et al., 2020; Niyogi and Wood, 2004; Paquin et al., 2002, 2012; Santore et al., 2001) is a computational tool for predicting the toxicity of metals in the aquatic environment, one that is now being used world-wide for risk assessment and for deriving regulatory criteria. By taking local water chemistry into account, it allows decisions to be made on a site-specific basis. The BLM traces its origin to a workshop of physiologists, toxicologists, geochemists, modelers, and environmental regulators, illustrating the great value of interdisciplinary collaboration (Bergman and Dorward-King, 1997). Like the acid rain story (Section 2.1), the BLM is another excellent example where physiological understanding of toxicant impact on ionoregulatory processes in fish has yielded direct real-world conservation and socioeconomic benefits. On the physiological side, the development of the BLM was greatly aided by the principles learned during the study of acid rain toxicity (Section 2.1) and the importance of DOC in the survival of fish in acidic blackwaters (Section 2.2).

Conceptually, the BLM uses geochemical modeling to predict short-term (e.g., 3-h) binding of metals to key sites ("biotic ligands") on target organisms, and in turn predicts subsequent toxicity (e.g., 96-h LC50) or chronic EC50 values based on resulting physiological impacts (Fig. 8A). Respectively, these represent concentrations lethal to 50% of the population, or concentrations effective in causing sublethal effects in 50% of the population. The concept behind the BLM can be traced back to the idea of Pagenkopf (1983), formulated as the Gill Surface Interaction Model, GSIM), that metal toxicity to fish could be explained by free metal cations (Me^{N+}) binding to biological ligands on the gills, thereby causing toxicity. At about the same time, Morel (1983) formulated the Free Ion Activity Model (FIAM) for algae with very similar assumptions to the GSIM. Both models used geochemical speciation modeling to explain earlier observations that water chemistry parameters such as pH, hardness (i.e., Ca+Mg concentration), salinity, alkalinity, and DOC had major effects on the toxicity of many metals (e.g., Holm-Jensen, 1948; Jones, 1938; Lloyd and Herbert 1962; Zitko et al. 1973). Thanks to the experimental work of Rick Playle and colleagues (Janes and Playle 1995; Playle, 1998; Playle et al., 1993a, 1993b; Richards et al., 2001), we now recognize that these factors can protect against toxicity either by competition with Me^{N+} cations (in the case of naturally occurring cations such as Ca^{2+}, Mg^{2+}, Na^+, and H^+), preventing their binding to key biological ligands on the gills, or by complexation of the Me^{N+} cations (in the case of anions such as Cl^-, SO_4^{2-}, HCO_3^-, DOC) thereby rendering them unavailable for binding (Fig. 8B). By both mechanisms, the amount of metal bound to the gill is reduced, and physiological impact and toxicity are reduced in parallel. Curiously, Pagenkopf (1983) had dismissed "organic material" as lacking importance, but it was incorporated into the FIAM (Morel, 1983). We now recognize that DOC, which is generally anionic and ubiquitous in natural waters, is often the most important water quality parameter protecting against metal toxicity (Al-Reasi et al., 2011; Wood et al., 2011).

Pagenkopf (1983) thought that when Me^{N+} cations bound to critical sites on the gill surface, lethality resulted from respiratory toxicity. However, many studies over the next two decades demonstrated that the proximate cause of lethality was usually interference with the active branchial uptake of either Na^+ (by Cu and Ag) or Ca^{2+} (by Zn, Cd, and Pb) from the water (Fig. 8B),

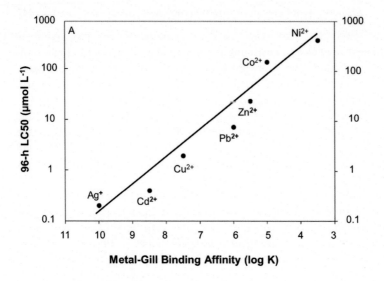

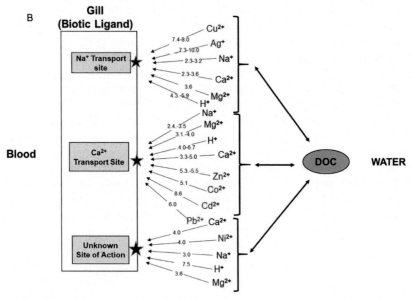

FIG. 8 Illustration of key aspects of the Biotic Ligand Model (BLM) for predicting metal toxicity on a site-specific basis in natural waters. (A). The relationship between the conditional equilibrium constant for gill-metal binding (log K) and acute toxicity (96-h LC50), both determined experimentally, in juvenile rainbow trout for 8 different metals in moderately hard water. The higher the log K value, the greater are both the gill-metal binding affinity and the acute toxicity. (B) Conceptual model illustrating free metal ions (red) binding to physiological sites on the gills and the competitive binding of protective cations (Ca^{2+}, Mg^{2+}, Ca^{2+}, and H^+) at these same sites, all with specific, experimentally determined log K values. Ag and Cu preferentially bind to Na^+ transport sites (inhibiting active Na^+ uptake), while Pb, Co, Cd, and Zn preferentially bind to Ca^{2+} transport sites (inhibiting active Ca^{2+} uptake). The binding site for Ni is unknown. The protective role of dissolved organic carbon (DOC) is also illustrated. DOC binds the free metal ions more strongly than it binds the protective cations, thereby reducing metal bioavailability. *Data from Niyogi, S., Wood, C.M., 2004. The Biotic Ligand Model, a flexible tool for developing site-specific water quality guidelines for metals. Environ. Sci. Technol. 38, 6177-6192.*

at least for metal levels causing toxicity to fish (Niyogi and Wood 2004; Paquin et al. 2002). The basic principle involved is ionic mimicry of nutritive metals by toxic metals (Bury et al., 2003). When Na^+ uptake was inhibited, the fish died from the same sequelae of events (internal fluid shifts leading to cardiovascular collapse) as seen during toxicity due to environmental acidity (Section 2.1). When Ca^{2+} uptake was inhibited, the fish died from hypocalcemia and associated neuromuscular failure (Roch and Maly, 1979). Detailed reviews on the exact mechanisms at the gills have been provided for Cu (Grosell, 2012), Ag (Wood, 2012), Zn (Hogstrand, 2012), Cd (McGeer et al., 2012), and Pb (Mager, 2012). In short, Cu and Ag kill fish by interfering with Na^+ uptake (Lauren and McDonald, 1987b; Wood et al., 1996) by at least three mechanisms: competition with Na^+ for the apical uptake pathway, either at the NHE or Na^+ channel (Bury and Wood, 1999; Grosell and Wood., 2002), inhibition of intracellular carbonic anhydrase that provides the H^+ counter-ion for apical Na^+ uptake (Morgan et al., 1997; Zimmer et al., 2012), and inhibition of the basolateral Na^+,K^+-ATPase (Lauren and McDonald, 1987a; Li et al. 1996; Morgan et al., 2004) (cf. Fig. 4). Zn, Cd and Pb inhibit Ca^{2+} uptake via both competition at the apical epithelial Ca^{2+} channel (ECaC) and blockade of the basolateral plasma membrane Ca^{2+}-ATPase (PMCA) and/or the basolateral Na^+/Ca^{2+} exchanger (NCX) (Hogstrand et al., 1996; Rogers and Wood, 2004; Verbost et al., 1987, 1988; Rogers et al., 2003) (cf. Fig. 4). The toxic mechanism and binding site for Ni remains unclear (Brix et al. 2017).

As originally formulated in the GSIM of Pakenkopf (1983) and the FIAM of Morel (1983), steady-state conditions are assumed. BLMs therefore assign conditional equilibrium constants (log K values) to all the reactions of the metals in solution, including those at the biotic ligands on the gills (Fig. 8B), as well as the interactions between protective cations and anions with one another in order to predict how much metal is going to bind to the gills in any site-specific water (Fig. 8B). The higher the log K value, the stronger the binding. Therefore, highly toxic metals such as Ag have much higher log K values at the gills than moderately toxic metals such as Zn (Niyogi and Wood, 2004) (Fig. 8A and B). Similarly, the higher the log K value for the competing natural cation at the biotic ligand, the greater is its ability to protect against metal toxicity. Within a metal, log K values for the biotic ligands are generally kept constant among different species, reflecting their original physiological determination, and an assumed commonality in the affinity of the proteins involved. They do however vary between models for acute versus chronic toxicity. Traditionally, the relationships between short-term gill metal burden and resulting toxicity were determined experimentally (e.g., MacRae et al., 1999; Meyer et al., 1999; McGeer et al., 2000), allowing estimation of the LA50, the gill metal burden associated with 50% mortality on an acute or chronic basis. However, this has proven impractical for many target organisms, especially small ones such as daphnia which are the most sensitive to metals because ionoregulatory impacts are greatest (Grosell et al., 2002). Therefore, modern BLMs often skip this step, instead fitting LA50 values, now often called sensitivity parameters, by iteration to observed toxicity data (Mebane et al., 2020). The greater the sensitivity of the organism, the lower the LA50 value. This in silico approach has allowed the development of BLMs for a wide range of organisms (i.e., not just fish), and media (e.g., seawater BLMs for marine organisms, pore water BLMs for sediment organisms, soil BLMs for terrestrial plants), and for chronic toxicity where toxic mechanisms are unknown. It has also facilitated the development of BLMs for metals such as Co, Ni, Mn, and U where the mechanism(s) of toxicity are unclear, and metals such as Al (Santore et al., 2018) where compound mechanisms are involved (see Section 2.1).

The computational complexity of the BLM, together with the requirement for many water chemistry measurements, has been a barrier to its implementation in certain jurisdictions, and this has been remedied by the development of Multiple Linear Regression Models (MLRs; Brix et al., 2020) and simplified BLM versions, look-up tables and translation tools (Adams et al., 2020; Mebane et al., 2020) which retain the BLM concept of metal bioavailability as determined by water chemistry. Regardless, the BLM owes its origin to a mechanistic understanding of the effects of specific metals on fish gill ionoregulatory function. Just as with the acid rain story (Section 2.2), the existence of this mechanistic understanding has played a key role in its acceptance by regulatory authorities (Mebane et al., 2020; Niyogi and Wood, 2004; Paquin et al., 2002). This approach is now aiding environmental protection and conservation world-wide. Its greatest impact so far has been its adoption in the European Union and the United Kingdom for a variety of risk assessment and regulatory processes aimed at protection against chronic toxicity for a range of metals (Cu-VRAR., 2008; Peters et al., 2020; Rüdel et al., 2015). In the US, the BLM is used for the same purposes for both acute and chronic Cu toxicity (Gensamer et al., 2016), in Canada for chronic Cu toxicity (ECCC, 2021), and its applications are increasing in China (Feng et al., 2012) and Australia/New Zealand (Peters et al., 2018).

2.4 Survival of fishes at high pH

Alkaline lakes (pH 9.0–10.5) are usually quite saline, and are widespread throughout the world. Brauner et al. (2013) have provided a water chemistry summary of some of those most studied by fish physiologists. There is also an ongoing trend for progressive alkalization of formerly circumneutral natural waters, often accompanied by salinization. This is caused by dehydration associated with climate change, especially in areas where the surrounding watershed and geology are rich in alkalinity (Schindler, 1997), and by anthropogenic changes in land use such as urbanization and farming practices (Kaushal et al., 2017). Many lakes that have been alkaline for historical periods harbor endemic species that have unique adaptations (reviewed by Brauner et al., 2013, Danulat, 1995). These species are providing ongoing lessons in physiology. This in itself is a strong argument for their conservation, for they provide guidance on the traits and treatments that will be useful in stocking and selection programs for recreational and commercial fisheries in alkalinizing lakes (e.g., Bigelow et al., 2010; Northrup, 2017; Thompson et al., 2015).

Most of the key physiological studies on the problems of fish at high pH were done in non-saline freshwater, on rainbow trout, a species that is only moderately tolerant, surviving up to pH 9.0–9.5 (e.g., McGeer and Eddy, 1998; Wilkie and Wood, 1991, 1994; Wilson et al., 1998; Wright and Wood, 1985; Yesaki and Iwama, 1992). These have identified three major physiological problems: disturbance of ionoregulation, alkalinization of the body fluids, and inhibition of branchial ammonia excretion (reviewed by Wilkie and Wood, 1996).

Of these, the ammonia excretion problem appears to be the most challenging, and has been solved in part by an increased reliance on urea-N excretion—i.e., greater relative ureotely (Table 1). The studies cited above occurred before the discovery of the Rh glycoproteins in fish gills (Nakada et al., 2007b; Nawata et al., 2007) and their importance in ammonia excretion and Na^+ uptake (Wright and Wood, 2009; Wright and Wood, 2012). However, we now interpret the inhibition phenomenon in light of the Rh proteins. In brief, as explained in Sections 1.1 and 1.5, ammonia excretion is coupled to Na^+ uptake via a metabolon involving an apical Rh protein (Rhcg) that facilitates the diffusion of NH_3 outward (Fig. 4). H^+ provision in the external boundary layer water, via transporters (NHE, HAT) coupled to Na^+ uptake, traps excreted NH_3 as NH_4^+, thereby sustaining the PNH_3 diffusion gradient, so that in effect NH_4^+ is excreted in exchange for Na^+. Once bulk water pH reaches >9.0, the pH in the external boundary layer rises closer to or eventually above the pK (9.1–9.5) of the $NH_4^+ \leftrightarrow NH_3$ interconversion (Playle and Wood, 1989a). As a result, a greater and greater fraction of excreted ammonia stays as NH_3, thereby raising boundary layer PNH_3, and inhibiting diffusive NH_3 flux through the Rh proteins. Net ammonia excretion may drop close to zero, or even reverse if the inspired water contains significant amounts of ammonia. However metabolic ammonia production continues, so internal ammonia levels rise and alkalinize the body fluids. Internal PNH_3 levels in particular increase because the rise in blood pH is exacerbated by respiratory alkalosis (see below). If the fish does not die of ammonia toxicity, the PNH_3 gradients are reset to a level which allows ammonia excretion to resume at normal rates. This is aided by upregulation of the Rh metabolon (Sashaw et al., 2010; Northrup, 2017), thereby also allowing restoration of Na^+ uptake, though evidence for the latter is conflicting (McGeer and Eddy, 1998; Wilkie and Wood, 1994; Yesaki and Iwama, 1992). A frequent observation (summarized by Wilkie and Wood, 1996) is that during the period of

TABLE 1 N-metabolism in a range of fish species in relation to the alkalinity and pH levels of their environments

Water Body	Alkalinity (mmol L^{-1})	pH	Species	%Ureotelism[a]	Plasma Ammonia-N (µmol/L^{-1})	Plasma Urea-N (mmol/L^{-1})	Ornithine-Urea Cycle (OUC)	References#
Freshwater	0.1	6.0	Potamotrygon sp.	9%	306	1.22	No	1
Freshwater	2.3	8.1	Oncorhynchus mykiss	13%	100	5.00	No	2
Pyramid Lake, USA	23	9.4	Oncorhynchus clarkii henshawi	34%	300	8.15	No	3
Qinghai Lake, China	29	9.3	Gymnocypris przewalski	<5%	2430	5.87	No	4.
Lake Van, Turkey	153	9.8	Alburnus tarichi	37%	1250	43	No	5
Lake Magadi, Kenya	300	10.0	Alcalicus grahami	100%	860	8.11	Yes	6
Seawater	2.3	8.0	Oncorhynchus mykiss	22%	180	2.15	No	7
Seawater	2.3	8.0	Squalus acanthias suckleyi	95%	80	640	yes	8

#References 1. Wood et al. (2002a); 2. Wilkie and Wood (1991); 3. Wright et al. (1993); 4. Wang et al. (2003), Wood et al. (2007d); 5. Danulat and Selcuk (1992); Danulat and Kempe (1992); 6. Wood et al. (1989), Wood et al. (1994); 7. Wood and Nawata (2011); 8. Wood et al. (1995).
[a]Calculated as 100% x Urea-N Excretion Rate/[Ammonia-N+Urea-N Excretion Rates].

inhibition of ammonia excretion (up to several days), there is a transient elevation in urea-N excretion which ameliorates a small but significant fraction of the N-load. This occurs in adult fish lacking expression of a functional OUC, and circumstantial evidence suggests that an upregulation of uricolysis (Section 1.5) is responsible.

Acid-base disturbance occurs at high pH, because the alkaline external water acts as a "vacuum" for CO_2; excreted CO_2 is immediately converted to HCO_3^- or CO_3^{2-}, so blood PCO_2 is washed out to very low levels, resulting in respiratory alkalosis (Johansen et al., 1975). In many alkaline waters, HCO_3^- and CO_3^{2-} concentrations are also very high, favoring an entry of basic equivalents across the gills and further raising blood pH by metabolic alkalosis. In the short-term, an enhanced production of lactic acid (i.e., metabolic acidic equivalents) by glycolysis in the white muscle offers partial compensation (Wilkie and Wood, 1991), but this is a costly process, and during prolonged exposure, lactacidosis attenuates (Wilkie et al., 1996; Wilson et al., 1998). Compensatory fluxes of acidic and basic equivalents across the gills, linked to ion fluxes, eventually stabilize or reduce the elevated blood pH (Wilkie et al., 1996; McGeer and Eddy, 1998).

Ionoregulatory disturbances at high water pH are significant, but probably insufficient to be the direct cause of mortality, in contrast to the lethal ionoregulatory failures seen in response to low water pH (Sections 2.1 and 2.2) and metals (Section 2.3). The usual initial response to high pH exposure is a modest drop (e.g., 5–15%) in plasma Na^+ and Cl^- concentrations attributable to an inhibition of active Na^+ and Cl^- uptake at the gills, with little effect on unidirectional efflux rates. The inhibition of Cl^- uptake likely reflects competition by external basic equivalents at the apical Cl^-/HCO_3^- anion exchanger (Fig. 4); the reason for decreased Na^+ uptake is less clear. In addition to the inhibition of ammonia excretion discussed above, internal alkalosis in the ionocytes may decrease H^+ availability for export by NHE or HAT against Na^+ uptake (Fig. 4). Thereafter Cl^- uptake recovers more quickly, in concert with a marked increase in the total fractional surface area of exposure of the branchial ionocytes, suggesting a greater activation of Cl^- versus basic equivalent exchange than Na^+ versus acidic equivalent exchange in order to relieve metabolic alkalosis (McGeer and Eddy, 1998; Laurent et al., 2000; Wilkie et al., 1999; Wilkie and Wood, 1994). Nevertheless, partly restored Na^+ uptake and decreased unidirectional Na^+ efflux eventually restore net Na^+ balance.

Pyramid Lake (Nevada, USA) is a highly alkaline remnant (pH = 9.4, salinity = 4.5 ppt, alkalinity = 23 mmol L^{-1}) of Pleistocene Lake Lahontan (Minckley et al., 1986). The Lahontan cutthroat trout (*Oncorhynchus clarkii henshawi*), a subspecies classified as "threatened" in the USA since 1995, has been reduced to 1–11% of its historical range in the arid inland west of the country, but is now the subject of an active restoration program (Al-Chockhachy et al., 2020). Historically, the largest population was in Pyramid Lake, and the fish spawned in the inflowing freshwater Truckee River

(Galat et al., 1981). This was stopped by anthropogenic water diversion early in the last century, resulting in extirpation. However, the subspecies is now raised in freshwater hatcheries, for stocking for recreational fishing in both Pyramid Lake and nearby alkaline lakes, and limited natural spawning has been re-established in the Truckee River since 2014 (Al-Chockhachy et al., 2020).

Physiological investigations have shown that in comparison to the rainbow trout, this salmonid exhibits only modest changes in its physiology, rather than wholesale re-organization, so as to cope with chronically high water pH throughout its adult life in the alkaline lake. Ammonia excretion is difficult because of the high pH, and the buffering provided by the high alkalinity undoubtedly limits boundary layer acidification. Lahontan cutthroat trout produce less nitrogenous waste, and excrete a higher proportion of it as urea-N (34%; Table 1, which is produced by uricolysis (Wilkie et al., 1993; Wright et al., 1993). Interestingly, three other species (all Cypriniformes) endemic to the same waters also have low rates of N-waste excretion with a high proportion ($\sim$ 30%) in the form of urea-N, again originating from uricolysis (McGeer et al., 1994). Lahontan cutthroat trout also exhibit higher rates of ammonia excretion via the urine, and have higher plasma pH and higher blood ammonia and urea-N levels (Table 1) than standard freshwater salmonids (McGeer et al., 1994; Wilkie et al., 1993, 1994; Wright et al., 1993). The elevated plasma pH and total ammonia concentrations increase the blood-to-water PNH_3 gradient, thus facilitating the diffusive efflux of NH_3 via the Rh metabolon in spite of the alkaline conditions (Wright et al., 1993). The hardness cations ($Ca^{2+} = 0.2\,mmol\,L^{-1}$, $Mg^{2+} = 7.3\,mmol\,L^{-1}$) in Pyramid Lake appear to be important in promoting ammonia excretion in this fish, because their selective reduction (to $Ca^{2+} = 0.1\,mmol\,L^{-1}$, $Mg^{2+} = 0.8\,mmol\,L^{-1}$) greatly reduced ammonia excretion (Iwama et al., 1997). Interestingly this finding is in accord with one previous study on rainbow trout where elevated water Ca^{2+} concentration alone (from $0.04\,mmol\,L^{-1}$ to $8.0\,mmol\,L^{-1}$) supported ammonia excretion at high pH (Yesaki and Iwama, 1992), but contrasted with another where smaller increases (up to $3.75\,mmol\,L^{-1}$) had no effect (Wilson et al., 1998). The potential mechanism remains unclear. Additionally, Lahontan cutthroat trout appear to have an enhanced ability to control blood ions and acid-base status at high pH (Wilkie et al., 1994). This is probably associated with a marked hyperplasia of ionocytes in the gills, with much greater exposed fractional surface of these cells (Galat et al., 1985; Wilkie et al., 1994).

Qinghai Lake on the Tibetan plateau is a high-altitude alkaline lake (pH = 9.3, salinity = 13 ppt, alkalinity = 29 mmol L^{-1}), the largest inland water body in China. The lake underwent progressive shrinking, salinization and alkalinization in the second half of the 20th century due to climate change, dam-building on inflowing rivers, and water diversion for agriculture, Przewalski's naked carp (*Gymnocypris przewalskii*), which are planktivores, return annually to their natal rivers for spawning (O'Bryan et al.,

2010), but forage for most of the year in the lake. They were formerly the most abundant fish in the lake but stocks declined precipitously from the 1960's onwards due to commercial overfishing and blockage of access to spawning rivers, and possibly also due to the changing chemistry of the water (Walker et al., 1996). The species is now red-listed as "endangered" in China.

In contrast to the Cyprinformes of Pyramid Lake (McGeer et al., 1994), this cyprinid excretes negligble urea-N as an adult (Wang et al., 2003; Wood et al., 2007d) (Table 1) but does so as a juvenile (Yi et al., 2017). Ammonia excretion of adults drops by up to 70% when exposed to lake water, and this persists throughout lake residence, indicating a long-lasting reduction in N-waste production. In this respect the naked carp is very similar to the Lahontan cutthroat trout (Wright et al., 1993; Wilkie et al., 1994). However, plasma total ammonia stabilizes at much higher levels (2000–2500 $\mu mol\,L^{-1}$) than in the Lahontan cutthroat trout ($\sim$300 $\mu mol\,L^{-1}$) (Table 1), despite an ability of juveniles to upregulate Rh proteins in the gills during high pH exposure (Yi et al., 2017). This indicates an exceptional tolerance of internal ammonia in adult naked carp, perhaps associated with high levels of the ammonia-scavenging enzymes glutamate dehydrogenase and glutamine synthetase in liver and brain (Wang et al., 2003). These may provide a detoxification mechanism at critical sites. At least part of the reduction in N-waste production can be attributed to a lower metabolic rate (40% reduction in MO_2) in lake water, probably because the fish allow plasma osmolality, and concentrations of Na^+, Cl^-, and K^+ to virtually equilibrate with the environment, so that osmoregulatory costs are greatly reduced (Wood et al., 2007d). This conclusion is reinforced by marked decreases in Na^+, K^+- ATPase activities in gill (by 70%) and kidney (by 30%) and 95% declines in UFR. The species appears to have little problem with acid-base regulation when in the lake. Gill intracellular carbonic anhydrase activity is reduced, limiting the fall in PCO_2 (Yao et al., 2016) and the moderately elevated blood pH (respiratory alkalosis) appears to improve tolerance of hypoxia by increasing the affinity of the blood for O_2 (Li et al., 2020). Nevertheless, Mg^{2+} concentrations in the lake (36 $mmol\,L^{-1}$) are currently approaching seawater levels (cf. Fig. 1B), though Ca^{2+} concentration is only 0.2 $mmol\,L^{-1}$, and large persistent increases in plasma Mg^{2+} levels to about 12 $mmol\,L^{-1}$ (about 6-fold greater than seen in most marine fish, Fig. 1B) may pose a threat to the ongoing health of the species. Interestingly Mg^{2+} is the only major ion for which urinary excretion increases after exposure to lake water, despite the massive reduction in UFR (Wood et al., 2007d). Furthermore, cyprinids are generally intolerant of salinity (DeBoeck et al., 2000), yet plasma Na^+ and Cl^- concentrations now exceed 200mM and osmolality is above 440 $mOsmol\,kg^{-1}$, all slightly higher than the present chemistry of the lake water (Wood et al., 2007d). Indeed, toxicity tests performed in 1992, when salinity was probably lower, indicted that the 5-day LC50 for lake water was 18.4 ppt (Walker et al., 1996), only about 5–6 ppt above the present salinity.

Happily, the situation has now improved dramatically, due in part to this physiological research influencing government policy. Commercial fishing has been banned, farming practices have been altered to reduce water demand, dams have been removed and fish ladders installed on two of the major rivers, and two hatcheries have been built (Chen et al., 2012; Wu and Shi, 2014). These regulatory measures have coincided with a change in the climatic regime, such that Lake Qinghai water levels have increased since 2004, and progressive alkalization and salinization have ceased (Cui and Li, 2016; Fang et al., 2019). The fish population has exhibited a marked recovery in the past 15 years (Wang, 2016; Qinghai Ministry of Agriculture, 2020).

Lake Van in Turkey is the largest alkaline lake in the world, and is the sole home of another endemic cyprinid, the tarek ("pearl mullet," *Alburnus tarichi*, formerly *Chalcalburnus tarichi*) The tarek faces even more severe ionic and pH challenges than the naked carp of Qinghai Lake. Like the latter, this planktivore migrates annually into its natal freshwater rivers to spawn, but lives the rest of the year in the alkaline lake water (salinity $= 17$–22 ppt, pH $= 9.8$, alkalinity $= 153$ mmol L^{-1}), where concentrations of Mg^{2+} (4 mmol L^{-1}) and Ca^{2+} (0.1 mmol L^{-1}) are at moderate levels (Arabacı et al., 2001). In fish adapted to lake water, plasma ion concentrations approach those of the external environment, suggesting that like the naked carp, the tarek is saving ionoregulatory costs while in the lake (Danulat, 1995; Arabacı et al., 2001; Oğuz, 2013). Indeed plasma osmolality (470–550 mOsmol kg^{-1}) is the highest ever reported in a teleost fish (Oğuz, 2015b; cf. Fig. 1A and B). This interpretation is reinforced by a reduction in branchial ionocytes (Oğuz, 2013) and mucocytes (Oğuz, 2015a), and lower Na^+, K^+ ATPase abundance in the kidneys of lakewater-adapted fish (Oğuz, 2015b). Blood pH appears to be high (8.2 at 18 °C in fish sampled by caudal puncture; Danulat and Kempe, 1992), and a large apparent SID in the plasma suggests that plasma HCO_3^- levels are also high (Arabacı et al., 2001). Danulat and Kempe (1992) have provided the only investigation of N-metabolism in this species. As in fish native to Pyramid Lake, urea-N excretion in the tarek accounts for a relatively high fraction of N-waste excretion (37%; Table 1), produced by uricolysis or arginolysis. The OUC appears to be absent. Plasma and tissue urea-N concentrations are exceptionally high (36–50 mmol urea-N L^{-1}; Table 1, suggesting that they use urea as an inert osmolyte. Plasma total ammonia concentrations are also high (1000–2000 μmol L^{-1}) and high glutamine synthetase activity in the brain may help protect against ammonia toxicity. The population has been decimated by overfishing and sand extraction from its spawning rivers, and has been red-listed by the IUCN since 1996, though some conservation efforts have been initiated (Bozaoğlu et al., 2019; Sari, 2008).

Lake Magadi in the Rift Valley of Kenya is so far the most alkaline water body on earth in which fish are known to thrive. Lake Magadi is one of the remnants of much larger Paleolake Orolongo, which started progressively drying and shrinking at the end of Pleisocene (Butzer et al., 1972). Most of the lake is covered with a floating crust of "trona" (precipitated sodium carbonate salts). Relatively small, genetically and morphologically distinct subpopulations of what is presently a single species, the Magadi tilapia (*Alcolapia grahami*, formerly *Oreochromis alcalicus grahami*, *Sarotheradon alcalicus grahami*, and *Tilapia grahami*) are found in isolated hot-springs at the edges of the trona that vary immensely in their physical chemistry (pH = 9.1–10.2, salinity = 16–51 ppt, alkalinity = 184–1625 mmol L^{-1}) (Coe, 1966; Kavembe et al., 2014; Maina et al., 2019; Reite et al., 1974; Wilson et al., 2004; Wood et al., 2002c, 2016). However, both Ca^{2+} and Mg^{2+} concentrations are very low (<1 mmol L^{-1}). Molecular and systematic studies suggest that despite isolation for only a few 100 years, some of these subpopulations are well on their way to forming separate species for which individual conservation strategies are needed (Kavembe et al., 2014, 2016a; Seegers et al., 1999; Seegers and Tichy, 1999; Tichy and Seegers, 1999; Wilson et al., 2000, 2004).

Indeed, the various hot-springs of Lake Magadi may well be the harshest aquatic environments on earth harboring fish life. Day-time temperatures exceeding 44 °C, night-time temperatures dropping to 22 °C, night-time anoxia, day-time hyperoxia, and immensely high day-time concentrations of reactive oxygen species (ROS) have been reported in various areas where the fish are known to live (Coe, 1966; Johannsson et al., 2014; Narahara et al., 1996; Wood et al., 2016). A host of documented morphological and physiological adaptations have been reviewed by Pörtner et al. (2010) and Kavembe et al. (2016b). These include facultative air-breathing with a physostomous swim bladder (Franklin et al., 1995; Johannsson et al., 2014; Maina et al., 1995; Narahara et al., 1996), a high-affinity blood O_2 dissociation curve which is insensitive to pH fluctuations in the physiological range (Lykkeboe et al., 1975; Narahara et al., 1996), a unique gastro-intestinal anatomy that allows imbibed alkaline lake-water to bypass an acidic stomach (Bergman et al., 2003), an exceptionally high gill O_2 diffusing capacity (Maina et al., 1996), high mitochondrial respiration rates (Johnston et al., 1994), the highest temperature tolerance of any fish (Wood et al., 2016), and the highest mass-specific metabolic rates of any fish (Franklin et al., 1995; Narahara et al., 1996; Wood et al., 2016). However, here we will focus on their evolutionary solutions to challenges in N-waste excretion, acid-base balance, and ionoregulation.

In light of the very high alkalinity of most hot-spring waters, which will completely stabilize the pH in the gill water thereby preventing boundary layer acidification, the species has abandoned ammonia excretion. Instead, even as adults, they excrete 100% of their waste-N as urea-N (Table 1) at immensely high rates (Randall et al., 1989; Wood et al., 1989), reflecting their very high metabolic rates (Franklin et al., 1995; Narahara et al., 1996; Wood et al., 2016), and voracious consumption of blue-green algae (N-fixing cyanobacteria) which comprise 90% of their diet (Coe, 1966; Bergman et al., 2003). While the uricolytic pathway is present, the vast majority of this urea-N is produced by the OUC, which is expressed not only in the liver

(Randall et al., 1989; Walsh et al., 1993) but also throughout the muscle (Lindley et al., 1999) and gills (Kavembe et al., 2015). This is an excellent physiological example of neoteny, retention into adult life of an embryonic characteristic present during early development in all teleosts (the OUC, see Section 1.5). Essentially, the whole body of the Magadi tilapia is a urea-production factory (Lindley et al., 1999), undoubtedly necessary because of their high N-ingestion rate (Coe, 1966). Based on the NQ (see Section 1.5), protein-like substrates support 67–100% of the aerobic metabolic rate (De Boeck et al., 2019; Wood et al., 1994). Although urea-N synthesis is costly (2.5 ATP per urea-N; see Section 1.3), and N-excretion rates are up to 20-fold greater in the Magadi tilapia than in standard teleosts, Wood et al. (2002b) calculated that the cost of this strategy accounted for less than 6% of the very high routine MO_2. The urea-N is excreted continuously, with about 80% passing out through the gills (Wood et al., 1994) via a facilitated diffusion urea transporter (UT) similar to that in the mammalian kidney (Walsh et al., 2000; 2001b). Even though ammonia is always present in the blood at elevated levels (700–1600 $\mu mol L^{-1}$; Table 1), no ammonia excretion occurs even when the Magadi tilapia is provided with the opportunity by exposure to neutralized Magadi water or gradual acclimation to circumneutral freshwater; 100% urea-N excretion continues, indicating obligate ureotelism (Wood et al., 1989, 1994, 2002b). High levels of glutamine synthetase and glutamate dehydrogenase may act as a "trap", funneling ammonia into glutamine and then into the OUC (Walsh et al., 1993; Wilson et al., 2004; Wood et al., 1989).

The rate of urea-N production appears to be set by the supply of nitrogen in the form of ammonia and/or glutamine (Lindley et al., 1999; Wood et al., 1994). Thus, when environmental ammonia levels are elevated, urea-N production and excretion rates increase quickly (Wood et al., 1989, 2013b), providing a detoxification mechanism such that exceptionally high ammonia levels can be tolerated; indeed, the Magadi tilapia appears to hold the teleost record for NH_3 tolerance (Walsh et al., 1993). This is presumably adaptive, as these fish thrive in some sites rich in flamingo guano, where both pH and total ammonia concentrations are high (Wilson et al., 2004). Interestingly, although this species appears to never excrete ammonia, Rh proteins are expressed in the gills, and their mRNA expressions are up-regulated during exposure to high environmental ammonia (HEA; Wood et al., 2013b), just as in standard teleosts (see Sections 1.1 and 1.5). Furthermore, the basolateral Na^+, K^+ ATPase in the gills is preferentially activated by NH_4^+, which is more potent than K^+ in the physiological range of elevated plasma NH_4^+ levels seen during high HEA. Wood et al. (2013b) have presented a speculative model whereby the combination of upregulated Rh proteins and "Na^+, NH_4^+ ATPase" activity may actively excrete at least some of the ammonia that is entering across the gills, thereby lessening the ammonia burden, though not necessarily achieving net ammonia excretion against prevailing gradients during HEA.

As first pointed out by Johansen et al. (1975), the Magadi tilapia lives not only in exceptionally high alkalinity, but also in a "CO_2 vacuum" created by the very basic water pH, and therefore is likely subject to both "respiratory" and "metabolic" acidosis (see Section 1.4). The small size (typically $<10\,g$) and high metabolic activity of the fish have prevented reliable determinations of blood acid-base status. To date, measurements have been made only on anesthetized fish sampled by caudal puncture. However, the general consensus is that true "resting" blood pH at $30–36\,°C$ is likely above 8.4, and plasma HCO_3^- concentration above $15\,mmol\,L^{-1}$, an extreme state of alkalosis (0.5–1.0 pH unit higher) relative to standard teleosts at comparable temperature (Johansen et al., 1975; Wood et al., 1989, 1994). Intracellular pH appears to be similarly elevated (Johnston et al., 1983; Wood et al., 1994). Atkinson (1992) argued that high levels of ureagenesis in this species protected the organism from being "flooded by HCO_3^- from the ambient water", in support of his controversial idea that HCO_3^- consumption for urea production is the primary role of the OUC in vertebrates. This possibility cannot be eliminated, but Wood et al. (1994) reported experimental evidence against the idea. Later studies demonstrated that the gills are unusually impermeant to HCO_3^- entry from the environment (Wood et al., 2012), that HCO_3^- uptake through the gut is in fact the major route of metabolic base loading (Bergman et al., 2003), and that lactic acid production is activated during HEA exposure, probably as a mechanism to prevent blood pH from rising too high (Wood et al., 2013b). As described below, active excretion of HCO_3^- occurs through the gills against very strong opposing electrochemical gradients (Wood et al., 2012). As a result, the cost of acid-base regulation accounts for about 50% of the fish's routine metabolic rate (Wood et al., 2002b).

Ionoregulation has been characterized in detail in only one subpopulation of Magadi tilapia, the residents of Fish Springs Lagoon (GPS coordinates $=1°$ $53'30.2''$S, $36°18'09.9''$E, pH ~9.9, salinity $\sim24\,ppt$, alkalinity $\sim380\,mmol\,L^{-1}$, osmolality $\sim580\,mOm\,kg^{-1}$). Living in an environment ($Na^+\sim355$, Cl^- $\sim110\,mmol\,L^{-1}$) with a salinity $\sim65\%$ seawater, but where much of the Cl^- normally present in seawater is replaced by titratable alkalinity (HCO_3^- and CO_3^{2-}; cf. Fig. 1B), poses unusual ionoregulatory challenges for this fish. Nevertheless, they exhibit plasma osmolyte values ($Na^+=182$, $Cl^-=171\,mmol\,L^{-1}$, $K^+=5$, $Ca^{2+}=2.3$, $Mg^{2+}=1.0\,mmol\,L^{-1}$, osmolality $=375\,mOsmol\,kg^{-1}$) very typical of standard teleosts hypo-regulating in seawater (see Section 1.2; Fig. 1B). These data are from Wood et al. (2013b) but generally similar values for this subpopulation have been reported by Leatherland et al. (1974), Eddy and Maloiy (1984), Maloiy et al. (1978), Skadhauge et al. (1980), Eddy et al. (1981), Wright et al. (1990), Wood et al. (2002b), and Wood et al.(2002c). Thus, unlike the naked carp of Lake Qinghai and the tarek of Lake Van, Magadi tilapia of Fish Springs Lagoon do not profit metabolically by osmotic and ionic equilibration with their environment. Interestingly, despite the fact that urea-N production is such an important part of the Magadi tilapia's adaptation to this hostile environment, it makes only very modest use of it as an osmolyte. Plasma and tissue urea concentrations account for only about $10 mOsmol kg^{-1}$ (2–3% of internal osmolality) in fish acclimated to Fish Springs water, but it increases several-fold (up to 8% of internal osmolality) in fish subjected to long-term hyper-osmotic challenge (Wilson et al., 2004; Wood et al., 2002c).

In order to hypo-regulate, Magadi tilapia drink the hot-spring water at rates $(8–24 \mu L g^{-1} h^{-1})$ greater than seen in most seawater teleosts (Maloiy et al., 1978; Wood et al., 2002c). Like urea-N excretion, drinking appears to be obligate, because it continues unchanged even when the fish are acclimated to freshwater. The imbibed hot-spring water is absorbed, together with most of its ions, largely by an apparent $Na^+ + HCO_3^-$ co-transport mechanism in the intestine (Bergman et al., 2003). This appears to be the major route of metabolic base loading into the fish. Yet at the same time, the fish are in the unusual situation of regulating plasma Na^+ and HCO_3^- below and plasma Cl^- concentrations above environmental levels. The main need is therefore to actively excrete Na^+ and HCO_3^- at the gills. As in standard seawater teleosts that actively excrete Na^+ and Cl^- (see Section 1.2 and Fig. 5), the gill TEP is inside-positive and of electrogenic origin. Analyses of electrochemical gradients demonstrate that there is a very large gradient driving HCO_3^- inwards, about 8-fold greater than that driving Na^+ inwards, whereas Cl^- distribution is close to equilibrium (slight outward gradient) across the gills (Eddy et al., 1981; Wood et al., 2012). Unidirectional flux rates of Na^+ and Cl^- measured with radiotracers are unusually low relative to other euryhaline teleosts at comparable salinity (Eddy et al., 1981; Eddy and Maloiy, 1984; Maetz and DeRenzis, 1978; Wright et al., 1990). These data, together with electrochemical modeling, indicate that gill permeability is low, and particularly low for HCO_3^-, clearly adaptive to counter the strong inwardly directed gradient for HCO_3^- entry (Wood et al., 2012). Nevertheless, most of the high metabolic costs must be in the active excretion of HCO_3^- against this gradient.

The mitochondria-rich ionocytes responsible for this regulation are abundant on both the gills and opercular epithelia, and have a morphology virtually identical to those of standard seawater teleosts (see Section 1.2) (Laurent et al., 1995; Maina, 1990, 1991). They are located below apical pits, bordered by overlying pavement cells, and flanked by typical accessory cells with "leaky" tight junctions. Laurent et al. (1995) proposed a hypothetical scheme based on the classic "chloride cell model" of Silva et al. (1977; (Section 1.2 and Fig. 5) in which electrogenic HCO_3^- excretion replaces electrogenic Cl^- excretion, while HCO_3^- substitutes for Cl^- at key sites in the model (basolateral NKCC, apical CFTR). As a result, HCO_3^- is exported through the transcellular pathway and Na^+ is exported through the paracellular pathway. Electrophysiological experiments (Wood et al., 2012) and recent immunohistochemical studies (J.M. Wilson, pers. comm.) have confirmed the basic elements of the Laurent model, with the addition of an apical NCC facilitating Cl^- uptake from the water, driven by the inward Na^+ gradient. This is a remarkable example of evolutionary plasticity in a transport system.

The multitude of unique adaptations outlined above, together with the rapid ongoing speciation of the various subpopulations, have made the Magadi tilapia a model for evolutionary physiology in the face of environmental extremity (Kavembe et al., 2014, 2015, 2016a; Pinho and Faria, 2016; Pörtner et al., 2010; Wilson et al., 2004). Thanks to petitions based on their physiological uniqueness, *Alcolapia grahami* was red-listed as a threatened species by the IUCN in 2006. However, siltation of anthropogenic origin, filling in the small hot-spring pools and covering their major cyanobacterial food source (*Arthrospira* sp.), is now a major threat to the fish (De Boeck et al., 2019; Maina et al., 2019). Aggressive conservation efforts based on habitat preservation are now required.

2.5 Osmoregulatory consequences of the commercial fishery for hagfish

The commercial fishery for hagfish is rapidly expanding, mainly to satisfy the South-East Asian market for live hagfish meat and "fish leather"; 9 of the known 76 species worldwide are now on the IUCN red list (Ellis et al., 2015). Hagfish are generally caught in baited traps at great depth, then brought to the surface where the water is often much warmer, of lower salinity, and of much higher O_2 content. The hagfish are then size-sorted on deck, either in air or in this surface water, so that only animals within certain size limits are retained. Many are subsequently transported in this surface water for live sale, with considerable mortality. Little is known about the survival of the animals that are released. However, we do know that the Pacific hagfish *Eptatretus stoutii* may suffer significant osmoregulatory challenges under these circumstances. Even short-term exposure (3–6h) of these osmoconformers (Section 1.3; Fig. 2A) to a 20% drop in salinity will significantly lower plasma osmolality, Na^+, Cl^-, and Mg^{2+} concentrations (Hastey, 2011), with more severe decreases during longer term 48-h exposures (Giacomin et al., 2019a; Sardella et al., 2009). These salinity effects are exacerbated by simultaneous elevations in temperature; indeed, exposures to 30% and 45% decreases in salinity at 25°C caused 14% and 100% mortality respectively (Hastey, 2011). Rather than being protective, hyperoxia (often used in holding) appears to increase the stress on the animals (Giacomin et al., 2019b). A more detailed understanding of these osmoregulatory challenges will help minimize unnecessary mortalities in both released and retained hagfish, thereby promoting conservation efforts.

2.6 Osmoregulatory threats to elasmobranchs; the critical importance of feeding

As of 2014, of the 1040–1050 extant species of Chondrichthyes, approximately 25% were in imminent danger of extinction and less than 40% were "safe" (Dulvy et al., 2014). The global abundance of oceanic sharks and rays has declined by 71% since 1970, and three-quarters of these species are now in danger of extinction (Pacoureau et al., 2021). These studies have identified rampant over-fishing as the major causative factor in the current precipitous decline in chondrichthyan populations. While this conclusion is unquestionable, a possible contributary factor that has been largely overlooked is the intimate relationship between feeding and osmoregulation in this Class. In terms of water balance, the osmoregulatory benefits for a marine chondrichthyan (internal osmolality $\sim$990 mOsmol kg^{-1}; Fig. 2B) to eat a marine teleost (internal osmolality $\sim$380 mOsmol kg^{-1}; Fig. 1B) are obvious. However, there is another important aspect of feeding. The osmoregulatory strategy of marine chondrichvthans is built around the use of "waste" nitrogen, specifically in the form of urea-N (Table 1), and to a lesser extent TMAO-N, as inert osmolytes to reduce the costs of ion and water homeostasis (Smith, 1936; see Section 1.3). Most of these animals are predators, sourcing N from their protein-rich diet for both osmoregulation and growth, thereby differing from teleosts and other fish that need the nitrogen only for protein growth. As the world's prey-fish populations collapse through over-exploitation, habitat destruction, and climate change, the availability of protein-N for these dual purposes becomes scarcer. There is growing physiological evidence that chondrichthyans may be N-limited in nature (reviewed by Wright and Wood, 2016).

In brief, this idea can be traced to the work of Haywood (1973) who demonstrated that plasma urea concentrations progressively declined and osmoregulation was impaired when pajama sharks (*Poroderma africanum*) were fasted, but both recovered quickly upon refeeding. The idea was reinforced by later studies showing that lesser-spotted dogfish (*Scyliorhinus canicula*) on prolonged low protein diets could not increase plasma urea concentrations in response to osmotic challenge (Armour et al., 1993) and that Pacific spiny dogfish (*Squalus acanthias suckleyi*) subjected to prolonged starvation (56 days) could not reduce urea loss rates to the environment below a relatively high baseline level amounting to about 3% of the body urea-N pool per day (Kajimura et al., 2008). Interestingly, in both of these studies, and in contrast to the pajama shark, plasma urea and TMAO levels were maintained constant. In the starved dogfish, plasma urea and TMAO concentrations were sustained by wasting of the muscle and liver, which are rich in these compounds, as well as in protein-N. The problem appears to be that urea losses cannot be reduced below the high baseline level, despite well-documented urea retention mechanisms at gills (Boylan, 1967; Fines et al., 2001; Hill et al., 2004; Pärt et al., 1998; Wood et al., 2013a), gut (Anderson et al., 2010, 2012, 2015; Liew et al., 2013; Weinrauch et al., 2020; Wood et al., 2019), kidney (Boylan, 1972; Forster et al., 1972; Hays et al., 1977; Schmidt-Nielsen et al., 1972; Smith, 1931), and rectal gland (Burger and Hess, 1960; Zeidel et al., 2005). The same appears to be true of TMAO losses, though here the evidence is fragmentary (Kajimura et al., 2008; Wright and Wood, 2016; Yancey, 2016). For both, the gills appear to be the major sites of irreducible loss rates (Wood et al., 1995). This is not surprising given the massive concentration gradients (ECF→ seawater) for both molecules (Δ urea $=\sim$ 375 mmol L^{-1}; Δ TMAO $=\sim$ 70 mmol L^{-1}; Fig. 2B) across an epithelium that must be kept permeable for respiratory gas exchange. Interestingly, the absolute permeability of the gills (at least in the efflux direction) is even lower to ammonia (the third respiratory gas) than to urea (Wood et al., 1995).

Nevertheless, the metabolism of these animals appears to be set up to deal effectively with new nitrogen that comes on board from the meal. In contrast to teleosts (Bucking, 2017; Wood, 2001), N-losses to the environment (above baseline) during the processing of a meal are negligible in elasmobranchs (Anderson et al., 2010; Wood et al., 2005, 2007a), and there is a rapid activation of urea synthesis, via the OUC so that plasma urea concentration rises (Kajimura et al., 2006). This OUC activation may be signaled, at least in part, by the post-prandial alkaline tide (Wood et al., 2005, 2008). In Chondrichthyes, the first step of the OUC is the trapping of ammonia into glutamine, by glutamine synthetase (Anderson, 2001), and the affinity of glutamine synthetase for its substrate appears to be exceptionally high, which may help explain the efficiency of N-trapping (Shankar and Anderson, 1985). In the Pacific spiny dogfish, there is also emerging evidence of an alternative nitrogen source for urea synthesis - direct uptake of ammonia across the gills from the environmental seawater (Nawata et al., 2015). The influx transport system may be active and rectifying, because it operates against apparent outwardly directed gradients, and efflux permeability is low. This "scavenging for nitrogen" obeys Michaelis-Menten kinetics, appears to be dependent upon both Rh proteins and glutamine synthetase, and has high affinity and capacity, such that net N accretion occurs at environmentally realistic ammonia concentrations (100–400 μmol L^{-1}; Fig. 9) (Wood and Giacomin, 2016).

A particular concern associated with the nitrogen-dependent osmoregulatory system of marine chondrichthyans is its sensitivity to waterborne metals such as Ag, Cu, and Pb (De Boeck et al., 2007; DeBoeck et al., 2001; Eyckmans et al., 2013; Grosell et al., 2003). Sensitivity to Ag is particularly marked (about 10-fold greater than in marine teleosts; DeBoeck et al., 2001), and for all three metals, the urea and TMAO retention properties of the gills appear to be the target, such that loss rates across the gills increase and plasma concentrations fall. The mechanistic basis is as unclear, and it is not yet known whether ammonia permeability and ammonia uptake are also affected; these are important topics for future research.

Obligate freshwater stingrays (Order Myliobatiformes), which comprise $\sim$35 species within the Family Potomotrigonidae of South America, are at the other end of the physiological spectrum in the Chondrichthyes

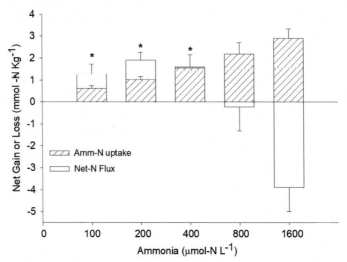

FIG. 9 Scavenging nitrogen from seawater by the uptake of environmental ammonia in the Pacific spiny dogfish shark (*Squalus acanthias suckleyi*). Sharks were exposed to ammonia-N concentrations of 100, 200, 400, 800, or 1600 μmol-N L^{-1} in the external seawater for 10 h. The hatched bars show the total uptake of ammonia-N over that period. The open bars show the total net gain or loss of N (ammonia-N+urea-N) over that period relative to non-exposed control animals. Means ±1 SEM (*N* = 10). Asterisks indicate a significant net positive N-balance. *Data from Wood, C.M., Giacomin, M., 2016. Feeding through your gills and turning a toxicant into a resource: how the dogfish shark scavenges ammonia from its environment. J. Exp. Biol. 219, 3218-3226.*

(Ballantyne and Fraser, 2013; Wright and Wood, 2016). The OUC does not appear to operate in adult life (Goldstein and Forster, 1971). Plasma ion, osmolality, and urea concentrations (Table 1) are kept at levels typical of other freshwater fish (Fig. 1A), and TMAO is not detectable (Thorson, 1967; Treberg et al., 2006). When challenged with elevated external osmolality or HEA, they exhibit negligible ability to elevate plasma urea or TMAO levels (Gerst and Thorson, 1977; Ip et al., 2003; Ip and Chew, 2010; Tam et al., 2003; Thorson, 1970). Unidirectional flux experiments with ^{22}Na$^+$ and ^{36}Cl$^-$ radioisotopes have shown active branchial uptake mechanisms for both ions (Carrier and Evans, 1973) that are pharmacologically similar to those of freshwater fish (Wood et al., 2002a), as outlined in Section 1.1. and Fig. 4. Many of these stingrays are endemic to the acidic, ion-poor blackwaters of the Amazon basin. As described in Section 2.2, their adaptive strategy includes relatively low affinity uptake systems for Na$^+$ and Cl$^-$ at the gills, coupled with extreme resistance to elevation of their diffusive effluxes by low pH (Fig. 7A). These fish have furnished some of the first evidence for the supportive effects of humic-rich, allochthonous DOC on ionoregulation at low pH (Section 2.2). The presence of DOC in natural blackwater at low pH elevates Na$^+$ and Cl$^-$ influx rates and decreases diffusive efflux rates (Wood et al., 2003). Nevertheless, the balance point (cf. Fig. 7A) remains well above the native Na^+ and Cl^- concentrations in the blackwaters where these stingrays live. Therefore, these animals appear to be living on the edge with respect to ionic homeostasis. As with marine elasmobranchs, feeding again appears to play an important role in osmoregulation, but in this case in supplying essential ions rather than essential nitrogen. Conservation information is fragmentary, but there is evidence of population declines due to habitat degradation, overfishing for food, collection for the export aquarium trade, and targeted removal for the protection of tourists (Rosa et al., 2010).

3 Future directions and concluding remarks

Two recent reviews (Pinheiro et al., 2021; Zimmer et al., 2021) have documented the immense variability that exists in the physical chemistries of so-called "fresh" waters throughout the world, and the effects that these have in dictating both the species that can live there, and their responses to waterborne toxicants. Pinheiro et al. (2021) assessed nine different parameters (pH, temperature, PO_2, PCO_2, divalent cations (Ca^{2+}, Mg^{2+}), inorganic anions, carbonate alkalinity, salinity, and DOC) that vary in concentration by $10+$fold (temperature, oxygen) to $10,000–100,000$-fold (H^+, Ca^{2+}, Na^+) in natural freshwaters, and that are known to greatly affect pollutant toxicity and physiological homeostasis. In the present chapter, six of these (pH, PCO_2, divalent cations, alkalinity, salinity, and DOC levels) have been explicitly discussed as parameters having major effects on iono/osmoregulation, acid-base balance, and/or N-waste excretion. There is also abundant evidence summarized elsewhere that the other three, temperature (Bucking, 2017; Burggren and Bautista, 2019), PO_2 (Gilmour and Perry, 2018; Wood and Eom, 2021), and inorganic anions such as Cl^- and SO_4^{2-} (Griffith, 2017) also can have major influences on these same physiological processes. Zimmer et al. (2021) focused on variation in the pH of natural waters, and found that ionoregulatory capacity significantly correlated with realized chemical niche for individual species, as defined by species distribution surveys. Both reviews argued for incorporating this variation in natural chemical environment, and the associated physiology of the resident species, into both environmental regulatory strategies and conservation strategies for fish presently living at the limits of their realized niches. Indeed, this has already been done with bioavailability modeling for metals (Mebane et al., 2020). These views are heartily endorsed by the author.

Acknowledgments

Supported by a Natural Sciences and Engineering Research Council of Canada (NSERC) Discovery grant to CMW (RGPIN-2017-03843). I thank Adalberto Val, Yuxiang Wang, Mike Wilkie, Bob Gensamer, Bill Adams, Carrie Claytor, Chris Schlekat, Michele Nawata, Alex Zimmer, Steve Perry, Martin Grosell, Jonathan Wilson, and Sunita Nadella for their advice and assistance, and two constructive reviewers whose input improved the chapter.

References

Adams, W., Blust, R., Dwyer, R., Mount, D., Nordheim, E., Rodriguez, P.H., Spry, D., 2020. Bioavailability assessment of metals in freshwater environments: a historical review. Environ. Toxicol. Chem. 39, 48–59.

Al-Chokhachy, R., Heki, L., Loux, T., Peka, R., 2020. Return of a giant: coordinated conservation leads to the first wild reproduction of Lahontan cutthroat trout in the Truckee River in nearly a Century. Fisheries 45 (2), 63–73.

Al-Reasi, H.A., Wood, C.M., Smith, D.S., 2011. Physicochemical and spectroscopic properties of natural organic matter (NOM) from various sources and implications for ameliorative effects on metal toxicity to aquatic biota. Aquat. Toxicol. 103, 179–190.

Al-Reasi, H.A., Wood, C.M., Smith, D.S., 2013. Characterization of freshwater natural dissolved organic matter (DOM): mechanistic explanations for protective effects against metal toxicity and direct effects on organisms. Environ. Int. 59, 201–207.

Al-Reasi, H.A., Smith, D.S., Wood, C.M., 2016. The influence of dissolved organic matter (DOM) on sodium regulation and nitrogenous waste excretion in the zebrafish (*Danio rerio*). J. Exp. Biol. 219, 2289–2299.

Alt, J.M., Stolte, H., Eisenbach, G.M., Walvig, F., 1981. Renal electrolyte and fluid excretion in the Atlantic hagfish *Myxine glutinosa*. J. Exp. Biol. 91, 323–330.

Altinok, I., Grizzle, J.M., 2004. Excretion of ammonia and urea by phylogenetically diverse fish species in low salinities. Aquaculture 238, 499–507.

Anderson, P.M., 2001. Urea and glutamine synthesis: environmental influences on nitrogen excretion. In: Wright, P.M., Anderson, P.M. (Eds.), Nitrogen Excretion. Fish Physiology. vol. 20. Academic Press, Orlando, pp. 239–277.

Anderson, W.G., Taylor, J.R., Good, J.P., Hazon, N., Grosell, M., 2007. Body fluid volume regulation in elasmobranch fish. Comp. Biochem. Physiol. 148A, 3–13.

Anderson, W.G., Dasiewicz, P.J., Liban, S., Ryan, C., Taylor, J.R., Grosell, M., Weihrauch, D., 2010. Gastro-intestinal handling of water and solutes in 3 species of elasmobranch fish, the white-spotted bamboo shark, *Chiloscyllium plagiosum*, little skate, *Leucoraja erinacea*, and the clear nose skate, *Raja eglanteria*. Comp. Biochem. Physiol. A 155, 493–502.

Anderson, W.G., Nawata, C.M., Wood, C.M., Piercy-Normone, M.D., Weinrauch, D., 2012. Body fluid osmolytes and urea and ammonia flux in the colon of two chondrichthyan fishes, the ratfish, *Hydrolagus colliei*, and spiny dogfish, *Squalus acanthias*. Comp. Biochem. Physiol. A 161, 27–35.

Anderson, W.G., McCabe, C., Brandt, C., Wood, C.M., 2015. Examining urea flux across the intestine of the spiny dogfish, *Squalus acanthias*. Comp. Biochem. Physiol. A 181, 71–78.

Armour, K.J., O'Toole, L.B., Hazon, N., 1993. The effect of dietary protein restriction on the secretory dynamics of 1α-hydroxycorticosterone and urea in the dogfish, *Scyliorhinus canicula*: a possible role for 1α-hydroxycorticosterone in sodium retention. J. Endocrinol. 138, 275–282.

Atkinson, D.E., 1992. Functional roles of urea synthesis in vertebrates. Physiol. Zool. 65, 243–267.

Arabacı, M., Cağırgan, H., Sarı, M., Şekeroğlu, R., 2001. Serum ionic content of endemic *Chalcalburnus tarichi* during spawning, prespawning and postspawning terms, living in highly alkaline waters of lake Van (pH 9.8), Turkey. Turk. J. Fish. Aquat. Sci. 1, 53–57.

Audet, C., Munger, R.S., Wood, C.M., 1988. Long-term sublethal acid exposure in rainbow trout (*Salmo gairdneri*) in soft water: effects on ion exchanges and blood chemistry. Can. J. Fish. Aquat. Sci. 45, 1387–1398.

Baker, D.W., Sardella, B., Rummer, J.L., Sackville, M., Brauner, C.J., 2015. Hagfish: champions of CO_2 tolerance question the origins of vertebrate gill function. Sci. Rep. 5, 11182.

Ballantyne, J.S., Fraser, D.I., 2013. Euryhaline elasmobranchs. In: McCormick, S.D., Farrell, A.P., Brauner, C.J. (Eds.), Euryhaline Fishes. Fish Physiology. vol. 32. Academic Press, San Diego, pp. 125–197.

Barletta, M., Jaureguizar, A.J., Baigun, C., Fontoura, N.F., Agostinho, A.A., Almeida-Val, V.M.F., Val, A.L., Torres, R.A., Jimenes-Segura, L.F., Giarrizzo, T., Fabré, N.N., Batista, V.-S., Lasso, C., Taphorn, D.C., Costa, M.F., Chaves, P.T., Vieira, J.P., Corrêa, M.F.M., 2010. Fish and aquatic habitat conservation in South America: a continental overview with emphasis on neotropical systems. J. Fish Biol. 76, 2118–2176.

Bayley, M., Damsgaard, C., Thomsen, M., Malte, H., Wang, T., 2019. Learning to air-breathe: the first steps. Physiology 34, 14–29.

Beltrão, H., Zuanon, J., Ferreira, E., 2019. Checklist of the ichthyofauna of the Rio Negro basin in the Brazilian Amazon. ZooKeys 881, 53–89.

Bergman, H.L., Dorward-King, E.J. (Eds.), 1997. Reassessment of Metals Criteria for Aquatic Life Protection: Priorities for Research and Implementation. SETAC Pellston Workshop on Reassessment of Metals Criteria for Aquatic Life Protection. SETAC Press, Pensacola, p. 114.

Bergman, A.N., Laurent, P., Otiang'a-Owiti, G., Bergman, H.L., Walsh, P.J., Wilson, P., Wood, C.M., 2003. Physiological adaptations of the gut in the Lake Magadi tilapia, *Alcolapia grahami*, and alkaline- and saline-adapted fish. Comp. Biochem. Physiol. A 136, 701–715.

Bigelow, J.P., Rauw, W.M., Gomez-Raya, L., 2010. Acclimation improves short-term survival of hatchery Lahontan cutthroat trout in water from saline, alkaline Walker Lake, Nevada. J. Fish Wildl. Manag. 1, 86–92.

Booth, C.E., McDonald, D.G., Simons, B.P., Wood, C.M., 1988. The effects of aluminum and low pH on net ion fluxes and ion balance in the brook trout, *Salvelinus fontinalis*. Can. J. Fish. Aquat. Sci. 45, 1563–1574.

Boylan, J.W., 1967. Gill permeability in *Squalus acanthias*. In: Gilbert, P.W., Mathewson, R.F., Rall, D.P. (Eds.), Sharks, Skates, and Rays. John Hopkins Press, Baltimore, pp. 197–206.

Boylan, J.W., 1972. A model for passive urea reabsorption in the elasmobranch kidney. Comp. Biochem. Physiol. A 42, 27–30.

Bozaoğlu, A.S., Akkus, M., Ysil, A., 2019. Pearl Mullet (*Alburnus tarichi* (Guldenstaedtii, 1814)). Fishing with trammel nets in Lake Van. Commagene J. Biol. 3, 27–31.

Braun, M.H., Perry, S.F., 2010. Ammonia and urea excretion in the Pacific hagfish *Eptatretus stoutii*: evidence for the involvement of Rh and UT proteins. Comp. Biochem. Physiol. A 157, 405–415.

Brauner, C.J., Wang, T., Wang, Y., Richards, J.G., Gonzalez, R.J., Bernier, N.J., Xi, W., Patrick, M., Val, A.L., 2004. Limited extracellular but complete intracellular acid-base regulation during short-term environmental hypercapnia in the armoured catfish, *Liposarcus pardalis*. J. Exp. Biol. 207, 3381–3390.

Brauner, C.J., Gonzalez, R.J., Wilson, J.M., 2013. Extreme environments: hypersaline, alkaline, and ion-poor waters. In: McCormick, S.D., Farrell, A.P., Brauner, C.J. (Eds.), Euryhaline Fishes, Fish Physiology. vol. 32. Academic Press, San Diego, pp. 435–476.

Brauner, C.J., Shartau, R.B., Damsgaard, C., Esbaugh, A.J., Wilson, R.W., Grosell, M., 2019. Acid-base physiology and CO_2 homeostasis: Regulation and compensation in response to elevated environmental CO_2. In: Grosell, M., Munday, P.L., Farrell, A.P., Brauner, C.J. (Eds.), Fish Physiology. vol. 37. Academic Press, San Diego, pp. 69–132.

Brix, K.V., Schlekat, C.E., Garman, E.R., 2017. The mechanisms of nickel toxicity in aquatic environments: an adverse outcome pathway analysis. Environ. Toxicol. Chem. 36, 1128–1137.

Brix, K.V., DeForest, D.K., Tear, L., Peijnenburg, W., Peters, A., Middleton, E.T., Erickson, R., 2020. Development of empirical bioavailability models for metals. Environ. Toxicol. Chem. 39, 85–100.

Brown, D.J.A., 1982. The effect of pH and calcium on fish and fisheries. Water Air Soil Pollut. 18, 343–351.

Bucking, C., 2017. A broader look at ammonia production, excretion, and transport in fish: a review of impacts of feeding and the environment. J. Comp. Physiol. B. 187, 1–18.

Bucking, C., Wood, C.M., 2012. Digestion of a single meal affects gene expression of ion and ammonia transporters and glutamine synthetase activity in the gastrointestinal tract of freshwater rainbow trout. J. Comp. Physiol. B. 182, 341–350.

Bucking, C., Edwards, S.L., Tickle, P., Smith, C.P., McDonald, M.D., Walsh, P.J., 2013. Immunohistochemical localization of urea and ammonia transporters in two confamilial fish species, the ureotelic gulf toadfish (*Opsanus beta*) and the ammoniotelic plainfin midshipman (*Porichthys notatus*). Cell Tissue Res. 352, 623–637.

Burger, J.W., Hess, W.N., 1960. Function of the rectal gland in the spiny dogfish. Science 131, 670–671.

Burggren, W., Bautista, N., 2019. Development of acid-base regulation in vertebrates. Comp. Biochem. Physiol. A 236, 110518.

Bury, N.R., Wood, C.M., 1999. Mechanism of branchial apical silver uptake by rainbow trout is via the proton-coupled Na^+ channel. Am. J. Physiol. 277, R1385–R1391.

Bury, N.R., Walker, P.A., Glover, C.N., 2003. Review: nutritive metal uptake in teleost fish. J. Exp. Biol. 206, 11–23.

Butzer, K.W., Isaac, G.L., Richardson, J.L., Washbourn-Kamau, C., 1972. Radiocarbon dating of East African lake levels. Science 175, 1069–1076.

Campbell, P.G.C., Twiss, M.R., Wilkinson, K.J., 1997. Accumulation of natural organic matter on the surfaces of living cells: implications for the interaction of toxic solutes with aquatic biota. Can. J. Fish. Aquat. Sci. 54, 2543–2554.

Carrier, J.C., Evans, D.H., 1973. Ion and water turnover in the fresh-water elasmobranch *Potamotrygon* sp. Comp. Biochem. Physiol. A 45, 667–670.

Chao, N.L., 2001. The fishery, diversity, and conservation of ornamental fishes of the Rio Negro Basin, Brazil: a review of Project Piaba (1989-99). In: Chao, N.L., Petry, P., Prang, G., Sonnenschein, L., Thusty, M. (Eds.), Conservation and Management of Ornamental Fish Resources of the Rio Negro Basin, Amazonia, Brazil—Project Piaba. Editora da Universidade do Amazonas, Manaus, pp. 161–204.

Chen, D., Li, S., Wang, K., 2012. Enhancement and conservation of inland fisheries resources in China. Environ. Biol. Fish 93, 531–545.

Clifford, A.M., Goss, G.G., Roa, J.N., Tresguerres, M., 2015. Acid/base and ionic regulation in hagfish. In: Edwards, S.L., Goss, G.G. (Eds.), Hagfish Biology. CRC Press, Boca Raton, pp. 277–298.

Clifford, A.M., Weinrauch, A.M., Goss, G.G., 2018. Dropping the base: recovery from extreme hypercarbia in the CO_2-tolerant Pacific hagfish (*Eptatretus stoutii*). J. Comp. Physiol. B. 188, 421–435.

Coe, M.J., 1966. The biology of *Tilapia grahami* Boulenger in Lake Magadi, Kenya. Acta Trop. 23, 146–177.

Cui, B.L., Li, X.Y., 2016. The impact of climate changes on water level of Qinghai Lake in China over the past 50 years. Hydrol. Res. 47, 532–542.

Currie, S., Edwards, S.L., 2010. The curious case of the chemical composition of hagfish tissues—50 years on. Comp. Biochem. Physiol. A 157, 111–115.

Cu-VRAR, 2008. Voluntary Risk Assessment Report (VRAR) of Copper, Copper Sulphate Pentahydrate, Copper (I) Oxide, Copper (II) Oxide, Dicopper Chloride Trihydroxide. http://echa.europa.eu/nl/copper-voluntary-risk-assessment-reports.

Danulat, E., 1995. Biochemical-physiological adaptations of teleosts to highly alkaline, saline lakes. In: Mommsen, T.P., Hochachka, P.W. (Eds.), Biochemistry and Molecular Biology of Fishes. vol. 5. Elsevier, Amsterdam, pp. 229–249.

Danulat, E., Kempe, S., 1992. Nitrogenous waste excretion and accumulation of urea and ammonia in *Chalcalburnus tarichi* (Cyprinidae), endemic to the extremely alkaline Lake Van (Eastern Turkey). Fish Physiol. Biochem. 9, 377–386.

Danulat, E., Selcuk, B., 1992. Life history and environmental conditions of the anadromous *Chalcalburnus tarichi* (Cyprinidae) in the highly alkaline Lake Van, Eastern Anatolia, Turkey. Arch. Hydrobiol. 126, 105–125.

Daye, P.G., Garside, E.T., 1976. Histopathologic changes in surficial tissues of brook trout, *Salvelinus fontinalis* (Mitchill), exposed to acute and chronic levels of pH. Can. J. Zool. 54, 2140–2155.

De Boeck, G., Vlaeminck, A., Van der Linden, A., Blust, R., 2000. The energy metabolism of common carp (*Cyprinis carpio*) when exposed to salt stress: an increase in energy expenditure or effects of starvation? Physiol. Biochem. Zool. 73, 102–111.

De Boeck, G., Hattink, J., Franklin, N.M., Bucking, C.P., Wood, S., Walsh, P.J., Wood, C.M., 2007. Metal toxicity in the spiny dogfish (*Squalus acanthias*): role of urea loss. Aquat. Toxicol. 84, 133–142.

De Boeck, G., Wood, C.M., Brix, K.V., Sinha, A., Matey, V., Johannsson, O.E., Bianchini, A., Bianchini, L.F., Maina, J.N., Kavembe, G.D., Papah, M.B., Kisipan, M.L., Ojoo, R.O., 2019. Fasting in the ureotelic Lake Magadi tilapia, *Alcolapia grahami*, does not reduce its high metabolic demand. Conserv. Physiol. https://doi.org/10.1093/conphys/coz060.

DeBoeck, G., Grosell, M., Wood, C.M., 2001. Sensitivity of the spiny dogfish (*Squalus acanthias*) to waterborne silver exposure. Aquat. Toxicol. 54, 261–275.

Di Toro, D.M., Allen, H.E., Bergman, H.L., Meyer, J.S., Paquin, P.R., Santore, R.C., 2001. Biotic ligand model of the acute toxicity of metals. 1 Technical basis. Environ. Toxicol. Chem. 20, 2383–2396.

Duarte, R.M., Val, A.L., 2020. Water-related problem with special reference to global climate change in Brazil. In: Singh, P., Milshina, Y., Tian, K., Gusain, D., Bassin, J.P. (Eds.), Water Conservation and Wastewater Treatment in BRICS Nations: Technologies, Challenges, Strategies, and Policies. Elsevier, Amsterdam, pp. 3–21.

Duarte, R.M., Ferreira, M.S., Wood, C.M., Val, A.L., 2013. Effect of low pH exposure on Na^+ regulation in two cichlid fish species of the Amazon. Comp. Biochem. Physiol. A 166, 441–448.

Duarte, R.M., Smith, D.S., Val, A.L., Wood, C.M., 2016. Dissolved organic carbon from the upper Rio Negro protects zebrafish (*Danio rerio*) against ionoregulatory disturbances caused by low pH exposure. Sci. Rep. 6, 1–10. https://doi.org/10.1038/srep20377.

Duarte, R.M., Wood, C.M., Val, A.L., Smith, D.S., 2018. Physiological protective action of dissolved organic carbon on ion regulation and nitrogenous waste excretion of zebrafish (*Danio rerio*) exposed to low pH in ion-poor water. J. Comp. Physiol. B. 188, 793–807.

Dulvy, N.K., Fowler, S.L., Musick, J.A., Cavanagh, R.D., Kyne, P.M., Harrison, L.R., Carlson, J.-K., Davidson, L.N., Fordham, S.V., Francis, M.P., Pollock, C.M., 2014. Extinction risk and conservation of the world's sharks and rays. elife 3, e00590.

Dymowska, A.K., Hwang, P.P., Goss, G.G., 2012. Structure and function of ionocytes in the freshwater fish gill. Respir. Physiol. Neurobiol. 184, 282–292.

Dymowska, A.K., Boyle, D., Schultz, A.G., Goss, G.G., 2015. The role of acid-sensing ion channels in epithelial Na⁺ uptake in adult zebrafish (*Danio rerio*). J. Exp. Biol. 218, 1244–1251.

ECCC (Environment and Climate Change Canada), 2021. Federal Water Quality Guideline for Copper: Biotic Ligand Model (BLM) Tool and User Manual. National Guidelines and Standards Office, Environment and Climate Change Canada, Gatineau, QC, Canada.

Eddy, F.B., 1975. The effect of calcium on gill potentials and on sodium and chloride fluxes in the goldfish, *Carassius auratus*. J. Comp. Physiol. 96, 131–142.

Eddy, F.B., Maloiy, G.M.O., 1984. Ionic content of body fluids and sodium efflux in *Oreochromis alcalicus grahami*, a fish living at temperatures above 30°C and in conditions of extreme alkalinity. Comp. Biochem. Physiol. A 78, 359–361.

Eddy, F.B., Bamford, O.S., Maloiy, G.M.O., 1981. Na⁺ and Cl⁻ effluxes and ionic regulation in *Tilapia grahami*, a fish living in conditions of extreme alkalinity. J. Exp. Biol. 91, 349–353.

Edwards, S.L., Marshall, W.S., 2013. Principles and patterns of osmoregulation and euryhalinity in fishes. In: McCormick, S.D., Farrell, A.P., Brauner, C.J. (Eds.), Euryhaline Fishes. Fish Physiology. vol. 32. Academic Press, San Diego, pp. 1–44.

Edwards, S.L., Arnold, J., Blair, S.D., Pray, M., Bradley, R., Erikson, O., Walsh, P.J., 2015. Ammonia excretion in the Atlantic hagfish (*Myxine glutinosa*) and responses of an Rhc glycoprotein. Am. J. Physiol. 308, R769–R778.

Ellis, J.E., Rowe, S., Lotze, H.K., 2015. Expansion of hagfish fisheries in Atlantic Canada and worldwide. Fish. Res. 161, 24–33.

Evans, D.H., 1984. Gill Na⁺/H⁺ and Cl⁻/HCO₃ exchange systems evolved before the vertebrates entered fresh water. J. Exp. Biol. 113, 465–469.

Evans, D.H., 2008. Teleost fish osmoregulation: what have we learned since August Krogh, Homer Smith, and Ancel Keys? Am. J. Physiol. 295, R704–R713.

Evans, D.H., 2011. Freshwater fish gill ion transport: August Krogh to morpholinos and microprobes. Acta Physiol. 202, 349–359.

Evans, D.H., Piermarini, P.M., Choe, K.P., 2004. Homeostasis: osmoregulation, pH regulation, and nitrogen excretion. In: Carrier, J.C., Musick, J.A., Heithaus, M.R. (Eds.), Biology of Sharks and their Relatives. CRC Press, Boca Raton, pp. 247–268.

Evans, D.H., Piermarini, P.M., Choe, K.P., 2005. The multifunctional fish gill: dominant site of gas exchange, osmoregulation, acid-base regulation, and excretion of nitrogenous waste. Physiol. Rev. 85, 97–177.

Evers, H.G., Pinnegar, J.K., Taylor, M.I., 2019. Where are they all from? – sources and sustainability in the ornamental freshwater fish trade. J. Fish Biol. 94, 909–916.

Eyckmans, M., Lardon, I., Wood, C.M., De Boeck, G., 2013. Physiological effects of waterborne lead exposure in spiny dogfish (*Squalus acanthias*). Aquat. Toxicol. 126, 373–381.

Fang, J., Li, G., Rubinato, M., Ma, G., Zhou, J., Jia, G., Yu, X., Wang, H., 2019. Analysis of long-term water level variations in Qinghai Lake in China. Water 11, 2136.

Feng, C., Wu, F., Zheng, B., Meng, W., Paquin, P.R., Wu, K.B., 2012. Biotic Ligand Models for Metals; a practical application in the revision of water quality standards in China. Environ. Sci. Technol. 46, 10877–10878.

Fines, G.A., Ballantyne, J.S., Wright, P.A., 2001. Active urea transport and an unusual basolateral membrane composition in the gills of a marine elasmobranch. Am. J. Physiol. 280, R16–R24.

Forster, M.E., Fenwick, J.C., 1994. Stimulation of calcium efflux from the hagfish, *Eptatretus cirrhatus*, gill pouch by an extract of corpuscles of Stannius from an eel (*Anguilla dieffenbachii*): Teleostei. Gen. Comp. Endocrinol. 94, 92–103.

Forster, R.P., Goldstein, L., 1976. Intracellular osmoregulatory role of amino acids and urea in marine elasmobranchs. Am. J. Physiol. 230, 925–931.

Forster, R.P., Goldstein, L., Rosen, J.K., 1972. Intrarenal control of urea reabsorption by renal tubules of the marine elasmobranch, *Squalus acanthias*. Comp. Biochem. Physiol. A 42, 3–12.

Franklin, C.E., Crockford, T., Johnston, I.A., Kamunde, C., 1995. Scaling of oxygen consumption in Lake Magadi tilapia, *Oreochromis alcalicus grahami*: a fish living at 37°C. J. Fish Biol. 46, 829–834.

Freire, C.A., Prodocimo, V., 2007. Special challenges to teleost fish osmoregulation in environmentally extreme or unstable habitats. In: Baldisserotto, B., Mancera, J.M., Kapoor, B.G. (Eds.), Fish Osmoregulation. Science Publishers, Enfield, pp. 249–276.

Fromm, P.O., 1980. A review of some physiological and toxicological responses of freshwater fish to acid stress. Environ. Biol. Fish 5, 79–93.

Furch, K., Junk, W.J., 1997. Physicochemical conditions in the floodplains. In: Junk, W.J. (Ed.), The Central Amazon Floodplain: Ecology of a Pulsing System. Springer–Verlag, Berlin, pp. 69–108.

Furukawa, F., Watanabe, S., Kimura, S., Kaneko, T., 2012. Potassium excretion through ROMK potassium channel expressed in gill mitochondrion-rich cells of Mozambique tilapia. Am. J. Physiol. 302, R568–R576.

Furukawa, F., Watanabe, S., Seale, A.P., Breves, J.P., Lerner, D.T., Grau, E.G., Kaneko, T., 2015. *In vivo* and *in vitro* effects of high-K$^+$ stress on branchial expression of ROMKa in seawater-acclimated Mozambique tilapia. Comp. Biochem. Physiol. A 187, 111–118.

Galat, D.L., Lider, E.L., Vigg, S., Robertson, S.R., 1981. Limnology of a large, deep, North American terminal lake, Pyramid Lake, Nevada, USA. Hydrobiologia 82, 281–317.

Galat, D.L., Post, G., Keefe, T.J., Bouck, G.R., 1985. Histological changes in the gill, kidney and liver of Lahontan cutthroat trout, *Salmo clarki henshawi*, living in lakes of different salinity-alkalinity. J. Fish Biol. 27, 533–552.

Galvez, F., Donini, A., Playle, R.C., Smith, D.S., O'Donnell, M.J., Wood, C.M., 2008. A matter of potential concern: natural organic matter alters the electrical properties of fish gills. Environ. Sci. Technol. 42, 9385–9390.

Gensemer, R.W., Playle, R.C., 1999. The bioavailability and toxicity of aluminum in aquatic environments. Crit. Rev. Environ. Sci. Technol. 29, 315–450.

Gensemer, R.W., Gondek, J., Canton, S.P., Kovach, A., Claytor, C.A., 2016. Regulatory implementation of the copper biotic ligand model for aquatic life protection: what have we learned and how are we doing? Proc. Water Environ. Fed. 2016, 1877–1894.

Gerst, J.W., Thorson, T.B., 1977. Effects of saline acclimation on plasma electrolytes, urea excretion, and hepatic urea biosynthesis in a freshwater stingray, *Potamotrygon* sp. Garman, 1877. Comp. Biochem. Physiol. A 56, 87–93.

Giacomin, M., Dal Pont, G., Eom, J., Schulte, P.M., Wood, C.M., 2019a. The effects of salinity and hypoxia exposure on oxygen consumption, ventilation, diffusive water exchange and ionoregulation in the Pacific hagfish *(Eptatretus stoutii)*. Comp. Biochem. Physiol. A 232, 47–59.

Giacomin, M., Eom, J., Schulte, P.M., Wood, C.M., 2019b. Acute temperature effects on metabolic rate, ventilation, diffusive water exchange, osmoregulation, and acid-base status in the Pacific hagfish *(Eptatretus stoutii)*. J. Comp. Physiol. B. 189, 17–35.

Gilmour, K.M., Perry, S.F., 2018. Conflict and compromise: using reversible remodeling to manage competing physiological demands at the fish gill. Physiology 33, 412–422.

Glover, C., Bucking, C., Wood, C.M., 2013. The skin of fish as a transport epithelium: a review. J. Comp. Physiol. B. 183, 877–891.

Goldstein, L., Forster, R.P., 1971. Urea biosynthesis and excretion in fresh-water and marine elasmobranchs. Comp. Biochem. Physiol. B 39, 415–421.

Gonzalez, R.J., Preest, M., 1999. Mechanisms for exceptional tolerance of ion-poor, acidic waters in the neon tetra (*Paracheirodon innesi*). Physiol. Biochem. Zool. 72, 156–163.

Gonzalez, R.J., Wilson, R.W., 2001. Patterns of ion regulation in acidophilic fish native to the ion-poor, acidic Rio Negro. J. Fish Biol. 58, 1680–1690.

Gonzalez, R.J., Dalton, V.M., Patrick, M.L., 1997. Ion regulation in ion-poor acidic water by the blackskirt tetra (*Gymnocorymbus ternetzi*), a fish native to the Amazon River. Physiol. Zool. 70, 428–435.

Gonzalez, R.J., Wood, C.M., Wilson, R.W., Patrick, M.L., Bergman, H.L., Narahara, A., Val, A.L., 1998. Effects of water pH and Ca^{2+} concentration on ion balance in fish of the Rio Negro, Amazon. Physiol. Zool. 71, 15–22.

Gonzalez, R.J., Wood, C.M., Patrick, M.L., Val, A.L., 2002. Diverse strategies for ion regulation in fish collected from the ion-poor, acidic Rio Negro. Physiol. Biochem. Zool. 75, 37–47.

Gonzalez, R.J., Wilson, R.W., Wood, C.M., 2006. Ionoregulation in tropical fish from ion- poor, acidic blackwaters. In: Val, A.L., Almeida-Val, V.M., Randall, D.J. (Eds.), The Physiology of Tropical Fish, Fish Physiology. vol. 22. Academic Press, San Diego, pp. 397–437.

Gonzalez, R.J., Cradeur, A., Guinnip, M., Mitchell, A., Reduta, V., 2018. South American characids share very similar ionoregulatory characteristics. Comp. Biochem. Physiol. A 226, 17–21.

Gonzalez, R.J., Hsu, R., Mahaffey, L., Rebagliatti, D., Shami, R., 2020a. Examination of ionoregulatory characteristics of South American cichilids. Comp. Biochem. Physiol. A 253, 110854.

Gonzalez, R.J., Patrick, M.L., Duarte, R.M., Casciato, A., Thackeray, J., Day, N., Val, A.L., 2020b. Exposure to pH 3.5 water has no effect on the gills of the Amazonian tambaqui (*Colossoma macropomum*). J. Comp. Physiol. B. 191, 493–502.

Griffith, M.B., 2017. Toxicological perspective on the osmoregulation and ionoregulation physiology of major ions by freshwater animals: teleost fish, Crustacea, aquatic insects, and Mollusca. Environ. Toxicol. Chem. 36, 576–600.

Grosell, M., 2011. The role of the gastrointestinal tract in salt and water balance. In: Grosell, M., Farrell, A.P., Brauner, C.J. (Eds.), The Multifunctional Gut of Fish, Fish Physiology. vol. 30. Academic Press, London, pp. 135–164.

Grosell, M., 2012. Copper. In: Wood, C.M., Farrell, A.P., Brauner, C.J. (Eds.), Homeostasis and Toxicology of Essential Metals, Fish Physiology. vol. 31A. Academic Press, London, pp. 53–133.

Grosell, M., Wood, C.M., 2002. Copper uptake across rainbow trout gills: mechanisms of apical entry. J. Exp. Biol. 205, 1179–1188.

Grosell, M., Nielsen, C., Bianchini, A., 2002. Sodium turnover rates in freshwater animals determine sensitivity to acute copper and silver exposure. Comp. Biochem. Physiol. C 133, 287–303.

Grosell, M., Wood, C.M., Walsh, P.J., 2003. Copper homeostasis and toxicity in the elasmobranch *Raja erinacea* and the teleost *Myoxocephalus octodecemspinosus* during exposure to elevated water-borne copper. Comp. Biochem. Physiol. C 135, 179–190.

Guh, Y.J., Lin, C.H., Hwang, P.P., 2015. Osmoregulation in zebrafish: ion transport mechanisms and functional regulation. EXCLI J. 14, 627–659.

Haines, T.A., 1981. Acidic precipitation and its consequences for aquatic ecosystems: a review. Trans. Am. Fish. Soc. 110, 669–707.

Hammerschlag, N., 2006. Osmoregulation in elasmobranchs: a review for fish biologists, behaviourists, and ecologists. Mar. Freshw. Behav. Physiol. 39, 209–228.

Harter, T.S., Shartau, R.B., Baker, D.W., Jackson, D.C., Val, A.L., Brauner, C.J., 2014. Preferential intracellular pH regulation represents a general pattern of pH homeostasis during acid–base disturbances in the armoured catfish, *Pterygoplichthys pardalis*. J. Comp. Physiol. B. 184, 709–718.

Hastey, J.P., 2011. Effects of Acute Salinity and Temperature Change on Pacific Hagfish, *Eptatretus Stoutii*; Implications for Bycatch Post Release Survival (M.Sc. Thesis). Dept. of Zoology. University of British Columbia, Vancouver, Canada.

Hays, R.M., Levine, S.D., Myers, J.D., Heinemann, H.O., Kaplan, M.A., Franki, N., Berliner, H., 1977. Urea transport in the dogfish kidney. J. Exp. Zool. 199, 309–316.

Haywood, G.P., 1973. Hypo-osmotic regulation coupled with reduced metabolic urea in the dogfish *Poroderma africanum*: an analysis of serum osmolarity, chloride, and urea. Mar. Biol. 23, 121–127.

Heisler, N., 1988. Acid–base regulation. In: Shuttleworth, T.J. (Ed.), Physiology of Elasmobranch Fishes. Springer, New York, pp. 213–252.

Hill, W.G., Mathai, J.C., Gensure, R.H., Zeidel, J.D., Apodaca, G., Saenz, J.P., Kinne-Saffran, E., Kinne, R., Zeidel, M.L., 2004. Permeabilities of teleost and elasmobranch gill apical membranes: evidence that lipid bilayers alone do not account for barrier function. Am. J. Physiol. 287, C235–C242.

Hiroi, J., McCormick, S.D., 2012. New insights into gill ionocyte and ion transporter function in euryhaline and diadromous fish. Respir. Physiol. Neurobiol. 184, 257–268.

Hogstrand, C., 2012. Zinc. In: Wood, C.M., Farrell, A.P., Brauner, C.J. (Eds.), Homeostasis and Toxicology of Essential Metals, Fish Physiology. vol. 31A. Academic Press, London, pp. 136–200.

Hogstrand, C., Verbost, P.M., Bonga, S.E., Wood, C.M., 1996. Mechanisms of zinc uptake in gills of freshwater rainbow trout: interplay with calcium transport. Am. J. Physiol. 270, R1141–R1147.

Holm-Jensen, I., 1948. Osmotic regulation in *Daphnia magna* under physiological conditions and in the presence of heavy metals. Det Kongelige Danske Videnskabernes Selskab Biologiske Meddelelser 11, 69.

Hsu, H.H., Lin, L.Y., Tseng, Y.C., Horng, J.L., Hwang, P.P., 2014. A new model for fish ion regulation: identification of ionocytes in freshwater-and seawater-acclimated medaka (*Oryzias latipes*). Cell Tissue Res. 357, 225–243.

Hunn, J.B., 1985. Role of calcium in gill function in freshwater fishes. Comp. Biochem. Physiol. A 82, 543–547.

Hwang, P.P., Lee, T.H., 2007. New insights into fish ion regulation and mitochondrion-rich cells. Comp. Biochem. Physiol. A 148, 479–497.

Hwang, P.P., Lin, L.Y., 2014. Gill ion transport, acid-base regulation and nitrogen excretion. In: Evans, D.H., Claiborne, J.B., Currie, S. (Eds.), The Physiology of Fishes, fourth ed. CRC Press, Boca Raton, pp. 205–233.

Ip, Y.K., Chew, S.F., 2010. Ammonia production, excretion, toxicity, and defense in fish: a review. Front. Physiol. 1, 1–20.

Ip, Y.K., Tam, W.L., Wong, W.P., Loong, A.M., Hiong, K.C., Ballantyne, J.S., Chew, S.F., 2003. A comparison of the effects of environmental ammonia exposure on the Asian freshwater stingray *Himanutra signifer* and the Amazonian freshwater stingray *Potamotrygon motoro*. J. Exp. Biol. 206, 3625–3633.

Ito, Y., Kobayashi, S., Nakamura, N., Miyagi, H., Esaki, M., Hoshijima, K., Hirose, S., 2013. Close association of carbonic anhydrase (CA2a and CA15a), Na^+/H^+ Exchanger (Nhe3b), and ammonia transporter Rhcg1 in zebrafish ionocytes responsible for Na^+ uptake. Front. Physiol. 4, 59. https://doi.org/10.3389/fphys.2013.00059.

Ito, Y., Kato, A., Hirata, T., Hirose, S., Romero, M.F., 2014. Na^+/H^+ and Na^+/NH_4^+ activities of zebrafish NHE3b expressed in *Xenopus* oocytes. Am. J. Physiol. 306, R315–R327.

Iwama, G.K., McGeer, J.C., Wright, P.A., Wilkie, M.P., Wood, C.M., 1997. Divalent cations enhance ammonia excretion in Lahontan cutthroat trout in highly alkaline waters. J. Fish Biol., 1061–1073.

Janes, N., Playle, R.C., 1995. Modeling silver binding to gills of rainbow trout (*Oncorhynchus mykiss*). Environ. Toxicol. Chem. 14, 1847–1858.

Johannsson, O.E., Bergman, H.L., Wood, C.M., Laurent, P., Kavembe, D.G., Bianchini, A., Maina, J.N., Chevalier, C., Bianchini, L.F., Papah, M.B., Ojoo, R.O., 2014. Air breathing in the Lake Magadi tilapia *Alcolapia grahami*, under normoxic and hyperoxic conditions, and the association with sunlight and ROS. J. Fish Biol. 84, 844–863.

Johansen, K., Maloiy, G.M.O., Lykkeboe, G., 1975. A fish in extreme alkalinity. Respir. Physiol. 24, 156–162.

Johnson, R.E. (Ed.), 1982. Acid Rain/Fisheries: Proceedings of an International Symposium on Acidic Precipitation and Fishery Impacts in Northeastern North America, Cornell University, Ithaca, New York, August 2-5, 1981. American Fisheries Society (USA), Bethesda, p. 357.

Johnston, I.A., Eddy, F.B., Maloiy, G.M.O., 1983. The effects of temperature on muscle pH, adenylate and phosphogen concentrations in *Oreochromis alcalicus grahami*, a fish adapted to an alkaline hot-spring. J. Fish Biol. 23, 717–724.

Johnston, I.A., Guderley, H., Franklin, C.E., Crockford, T., Kamunde, C., 1994. Are mitochondria subject to evolutionary temperature adaptation? J. Exp. Biol. 195, 293–306.

Jones, J.R.E., 1938. The relative toxicity of salts of lead, zinc and copper to the stickleback (*Gasterosteus aculeatus* L.) and the effect of calcium on the toxicity of lead and zinc salts. J. Exp. Biol. 15, 394–407.

Kajimura, M., Walsh, P.J., Mommsen, T.P., Wood, C.M., 2006. The dogfish shark (*Squalus acanthias*) increases both hepatic and extrahepatic ornithine urea cycle enzyme activities for nitrogen conservation after feeding. Physiol. Biochem. Zool. 79, 602–613.

Kajimura, M., Walsh, P.J., Wood, C.M., 2008. The spiny dogfish *Squalus acanthias* L. maintains osmolyte balance during long-term starvation. J. Fish Biol. 72, 656–670.

Kaushal, S.S., Duan, S., Doody, T.R., Haq, S., Smith, R.M., Johnson, T.A.N., Newcomb, K.D., Gorman, J., Bowman, N., Mayer, P.M., Wood, K.L., 2017. Human-accelerated weathering increases salinization, major ions, and alkalinization in fresh water across land use. Appl. Geochem. 83, 121–135.

Kavembe, G.D., Machado-Schiaffino, G., Meyer, A., 2014. Pronounced genetic differentiation of small, isolated and fragmented tilapia populations inhabiting the Magadi Soda Lake in Kenya. Hydrobiologia 739, 55–71.

Kavembe, G.D., Franchini, P., Irisarri, I., Machado-Schiaffino, G., Meyer, A., 2015. Genomics of adaptation to multiple concurrent stresses: insights from comparative transcriptomics of a Cichlid fish from one of earth's most extreme environments, the hypersaline soda Lake Magadi in Kenya, East Africa. J. Mol. Evol. 81, 90–109.

Kavembe, G.D., Kautt, A.F., Machado-Schiaffino, G., Meyer, A., 2016a. Eco-morphological differentiation in Lake Magadi tilapia, an extremophile cichlid fish living in hot, alkaline and hypersaline lakes in East Africa. Mol. Ecol. 25, 1610–1625.

Kavembe, G.D., Meyer, A., Wood, C.M., 2016b. Fish populations in East African saline lakes. In: Schagerl, M. (Ed.), Soda Lakes of East Africa. Springer, Cham, New York, pp. 227–257.

Keller, W., Heneberry, J., Edwards, B.A., 2019. Recovery of acidified Sudbury, Ontario, Canada, lakes: a multi-decade synthesis and update. Environ. Rev. 27, 1–16.

Keys, A., Willmer, E.N., 1932. "Chloride secreting cells" in the gills of fishes, with special reference to the common eel. J. Physiol. 76, 368–378.

Kirschner, L., 1993. The energetics of osmotic regulation in ureotelic and hypoosmotic fishes. J. Exp. Zool. 267, 19–26.

Krogh, A., 1939. Osmotic Regulation in Aquatic Animals. Cambridge University Press, Cambridge, p. 242.

Kullberg, A., Bishop, K.H., Hargeby, A., Jansson, M., Petersen, R.C., 1993. The ecological significance of dissolved organic carbon in acidified waters. Ambio 22, 331–337.

Lauren, D.J., McDonald, D.G., 1987a. Acclimation to copper by rainbow trout, *Salmo gairdneri* – biochemistry. Can. J. Fish. Aquat. Sci. 44, 105–111.

Lauren, D.J., McDonald, D.G., 1987b. Acclimation to copper by rainbow trout, *Salmo gairdneri* – physiology. Can. J. Fish. Aquat. Sci. 44, 99–104.

Laurent, P.L., Maina, J.N., Bergman, H.L., Narahara, A., Walsh, P.J., Wood, C.M., 1995. Gill structure of a fish from an alkaline lake. Effect of exposure to pH 7. Can. J. Zool. 73, 1170–1181.

Laurent, P., Wilkie, M.P., Chevalier, C., Wood, C.M., 2000. The effects of highly alkaline water (pH = 9.5) on the morphology and morphometry of chloride cells and pavement cells in the gills of freshwater rainbow trout: relationship to ionic transport and ammonia excretion. Can. J. Zool. 78, 307–319.

Lawrence, M.J., Wright, P.A., Wood, C.M., 2015. Physiological and molecular responses of the goldfish (*Carassius auratus*) kidney to metabolic acidosis, and potential mechanisms of renal ammonia transport. J. Exp. Biol. 218, 2124–2135.

Leatherland, J.F., Hyder, M., Ensor, D.M., 1974. Regulation of plasma Na^+ and K^+ concentrations in five species of *Tilapia* fishes. Comp. Biochem. Physiol. A 48, 699–710.

Leenheer, J.A., 1980. Origin and nature of humic substances in the waters of the Amazon River basin. Acta Amazon. 10, 513–526.

Li, J., Lock, R.A.C., Klaren, P.H.M., Swarts, H.G.P., Stekhoven, F.M.A.H., Bonga, S.E.W., Flik, G., 1996. Kinetics of Cu^{2+} inhibition of Na^+/K^+-ATPase. Toxicol. Lett. 87, 31–38.

Li, H., Lai, Q., Yao, Z., Liu, Y., Gao, P., Zhou, K., Sun, Z., 2020. Ammonia excretion and blood gas variation in naked carp (*Gymnocypris przewalskii*) exposed to acute hypoxia and high alkalinity. Fish Physiol. Biochem. 46, 1981–1990.

Liew, H.T., DeBoeck, G., Wood, C.M., 2013. An *in vitro* study of urea, water, ion, and CO_2/HCO_3^- transport in the gastrointestinal tract of the dogfish shark (*Squalus acanthias*): the influence of feeding. J. Exp. Biol. 216, 2063–2072.

Lindley, T.E., Scheiderer, C.L., Walsh, P.J., Wood, C.M., Bergman, H.L., Bergman, A.L., Laurent, P., Wilson, P., Anderson, P.M., 1999. Muscle as the primary site of urea cycle enzyme activity in an alkaline lake-adapted tilapia, *Oreochromis alcalicus grahami*. J. Biol. Chem. 274, 29858–29861.

Lloyd, R., Herbert, D.W.M., 1962. The effect of the environment on the toxicity of poisons to fish. J. Inst. Public Health Eng. 61, 132–145.

Lykkeboe, G., Johansen, K., Maloiy, G.M.O., 1975. Functional properties of hemoglobins in the teleost *Tilapia grahami*. J. Comp. Physiol. 104, 1–11.

MacRae, R.K., Smith, D.D., Swoboda-Colberg, N., Meyer, J.S., Bergman, H.L., 1999. Copper binding affinity of rainbow trout (*Oncorhynchus mykiss*) and brook trout (*Salvelinus fontinalis*) gills: implications for assessing bioavailable metal. Environ. Toxicol. Chem. 18, 1180–1189.

Maetz, J., DeRenzis, J., 1978 Aspects of the adaptation to high external alkalinity: comparison of *Tilapia grahami* and *T. mossambica*. In: Schmidt-Nielsen, K., Bolis, L., Maddrell, S.H.P. (Eds.), Comparative Physiology: Water, Ions and Fluid Mechanics. Cambridge University Press, Cambridge, pp. 213–228.

Mager, E.M., 2012. Lead. In: Wood, C.M., Farrell, A.P., Brauner, C.J. (Eds.), Homeostasis and Toxicology of Essential Metals, Fish Physiology. vol. 31A. Academic Press, London, pp. 185–236.

Maina, J.N., 1990. A study of the morphology of the gills of an extreme alkalinity and hyperosmotic adapted teleost *Oreochromis alcalicus grahami* (Boulenger) with particular emphasis on the ultrastructure of the chloride cells and their modifications with water dilution. A SEM and TEM study. Anat. Embryol. 181, 83–98.

Maina, J.N., 1991. A morphometric analysis of chloride cells in the gills of the teleosts *Oreochromis alcalicus* and *Oreochromis niloticus* and a description of presumptive urea-excreting cells in *O. alcalicus*. J. Anat. 175, 131–145.

Maina, J.N., Wood, C.M., Narahara, A.B., Bergman, H.L., Laurent, P., Walsh, P.J., 1995. Morphology of the swim bladder of a cichlid teleost: *Oreochromis alcalicus grahami* (Trewavas, 1983), a fish adapted to a hyperosmotic, alkaline and hypoxic environment: a brief outline of the structure and function of the swim bladder. In: Munshi, J.S., Dutta, H.M. (Eds.), Horizons of New Research in Fish Morphology in the 20th Century. Oxford and IBH Publishing Co., New Delhi, pp. 179–192.

Maina, J.N., Kisia, S.M., Wood, C.M., Narahara, A.B., Bergman, H.L., Laurent, P., Walsh, P.J., 1996. A comparative allometric study of the morphometry of the gills of an alkalinity adapted cichilid fish, *Oreochromis alcalicus grahami*. Int. J. Salt Lake Res. 5, 131–156.

Maina, J.N., Kavembe, G.D., Papah, M.B., Mashiteng, R., Wood, C.M., Bianchini, A., Bianchini, L.F., Bergman, H.L., Johannsson, O.E., Laurent, P., Chevalier, C., Ojoo, R.O., 2019. Sizes, condition factors and sex ratios of the scattered populations of the small cichlid fish, *Alcolapia grahami*, that inhabits the lagoons and sites of Lake Magadi (Kenya), one of the most extreme aquatic habitats on Earth. Environ. Biol. Fish 102, 1265–1280.

Maloiy, G.M.O., Lykkeboe, G., Johansen, K., Bamford, O.S., 1978. Osmoregulation in *Tilapia grahami*: a fish in extreme alkalinity. In: Schmidt-Nielsen, K., Bolis, L., Maddrell, S.H.P. (Eds.), Comparative Physiology: Water, Ions and Fluid Mechanics. Cambridge University Press, Cambridge, pp. 229–238.

Mangum, C.P., Haswell, M.S., Johansen, K., 1977. Low salt and high pH in the blood of Amazon fishes. J. Exp. Zool. 200, 163–168.

Mangum, C.P., Haswell, M.S., Johansen, K., Towle, D.W., 1978. Inorganic ions and pH in the body fluids of Amazon animals. Can. J. Zool. 56, 907–916.

Marshall, W.S., 2013. Osmoregulation in estuarine and intertidal fishes. In: McCormick, S.D., Farrell, A.P., Brauner, C.J. (Eds.), Euryhaline Fishes. Fish Physiology. vol. 32. Academic Press, San Diego, pp. 396–434.

Marshall, W.S., Bryson, S.E., 1998. Transport mechanisms of seawater teleost chloride cells: an inclusive model of a multifunctional cell. Comp. Biochem. Physiol. A 119, 97–106.

Marshall, W.S., Grosell, M., 2005. Ion transport, osmoregulation and acid-base balance. In: Evans, D., Claiborne, J.B. (Eds.), Physiology of Fishes, third ed. CRC Press, Boca Raton, pp. 177–230.

Martin, R.A., 2005. Conservation of freshwater and euryhaline elasmobranchs. J. Mar. Biol. Assoc. U. K. 85, 1049–1073.

Matsuo, A.Y.O., Val, A.L., 2007. Acclimation to humic substances prevents whole body sodium loss and stimulates branchial calcium uptake capacity in cardinal tetras *Paracheirodon axelrodi* (Schultz) subjected to extremely low pH. J. Fish Biol. 70, 989–1000.

Matsuo, A.Y.O., Playle, R.C., Val, A.L., Wood, C.M., 2004. Physiological action of dissolved organic matter in rainbow trout in the presence and absence of copper: sodium uptake kinetics and unidirectional flux rates in hard and softwater. Aquat. Toxicol. 70, 63–81.

McCormick, S.D., Farrell, A.P., Brauner, C.J. (Eds.), 2013. Euryhaline Fishes, Fish Physiology. vol. 32. Academic Press, New York, p. 559.

McDonald, D.G., 1983. The interaction of environmental calcium and low pH on the physiology of the rainbow trout, *Salmo gairdneri*: I. Branchial and renal net ion and H^+ fluxes. J. Exp. Biol. 102, 123–140.

McDonald, D.G., Rogano, M.S., 1986. Ion regulation by the rainbow trout, *Salmo gairdneri*, in ion-poor water. Physiol. Zool. 59, 318–331.

McDonald, D.G., Wood, C.M., 1981. Branchial and renal acid and ion fluxes in the rainbow trout, *Salmo gairdneri*, at low environmental pH. J. Exp. Biol. 93, 101–118.

McDonald, D.G., Hobe, H., Wood, C.M., 1980. The influence of calcium on the physiological responses of the rainbow trout, *Salmo gairdneri*, to low environmental pH. J. Exp. Biol. 88, 109–131.

McDonald, D.G., Walker, R.L., Wilkes, P.R.H., 1983. The interaction of environmental calcium and low pH on the physiology of the rainbow trout, *Salmo gairdneri*: II. Branchial ionoregulatory mechanisms. J. Exp. Biol. 102, 141–155.

McDonald, D.G., Tang, Y., Boutilier, R.G., 1989. Acid and ion transfer across the gills of fish: mechanisms and regulation. Can. J. Zool. 67, 3046–3054.

McDonald, D.G., Cavdek, V., Calvert, L., Milligan, C.L., 1991. Acid–base regulation in the Atlantic hagfish *Myxine glutinosa*. J. Exp. Biol. 161, 201–215.

McDonald, M.D., Smith, C.P., Walsh, P.J., 2006. The physiology and evolution of urea transport in fishes. J. Membr. Biol. 212, 93–107.

McDonald, M.D., Gilmour, K.M., Walsh, P.J., 2012. New insights into the mechanisms controlling urea excretion in fish gills. Respir. Physiol. Neurobiol. 184, 241–248.

McGeer, J.C., Eddy, F.B., 1998. Ionic regulation and nitrogenous excretion in rainbow trout exposed to buffered and unbuffered freshwater of pH 10.5. Physiol. Zool. 71, 179–190.

McGeer, J.C., Wright, P.A., Wood, C.M., Wilkie, M.P., Mazur, C.F., Iwama, G.K., 1994. Nitrogen excretion in four species of fish from an alkaline/saline lake. Trans. Am. Fish. Soc. 123, 824–829.

McGeer, J.C., Playle, R.C., Wood, C.M., Galvez, F., 2000. A physiologically based biotic ligand model for predicting the acute toxicity of waterborne silver to rainbow trout in fresh waters. Environ. Sci. Technol. 34, 4199–4207.

McGeer, J.C., Szebedinszky, C., McDonald, D.G., Wood, C.M., 2002. The role of dissolved organic carbon in moderating the bioavailability and toxicity of Cu to rainbow trout during chronic waterborne exposure. Comp. Biochem. Physiol. C 133, 147–160.

McGeer, J.C., Niyogi, S., Smith, D.S., 2012. Cadmium. In: Wood, C.M., Farrell, A.P., Brauner, C.J. (Eds.), Homeostasis and Toxicology of Essential Metals, Fish Physiology. vol. 31A. Academic Press, London, pp. 125–184.

McInerney, J.E., 1974. Renal sodium reabsorption in the hagfish, *Eptatretus stoutii*. Comp. Biochem. Physiol. A 49, 273–280.

McWilliams, P.G., Potts, W.T.W., 1978. The effects of pH and calcium concentrations on gill potentials in the brown trout, *Salmo trutta*. J. Comp. Physiol. 126, 277–286.

Mebane, C.A., Chowdhury, M.J., De Schamphelaere, K.A.C., Lofts, S., Paquin, P.R., Santore, R.C., Wood, C.M., 2020. Metal bioavailability models: current status, lessons learned, considerations for regulatory use, and the path forward. Environ. Toxicol. Chem. 39, 60–84.

Meyer, J.S., Santore, R.C., Bobbitt, J.P., Debrey, L.D., Boese, C.J., Paquin, P.R., Allen, H.E., Bergman, H.L., Di Toro, D.M., 1999. Binding of nickel and copper to fish gills predicts toxicity when water hardness varies, but free-ion activity does not. Environ. Sci. Technol. 33, 913–916.

Milligan, C.L., Wood, C.M., 1982. Disturbances in haematology, fluid volume distribution and circulatory function associated with low environmental pH in the rainbow trout, *Salmo gairdneri*. J. Exp. Biol. 99, 397–415.

Minckley, W.L., Hendrickson, D.A., Bond, C.E., 1986. Geography of western North American freshwater fishes: description and relationships to intracontinental tectonism. In: Hocutt, C.H., Wiley, E.O. (Eds.), The Zoogeography of North American Freshwater Fishes. Wiley, Toronto, pp. 519–613.

Morel, F.M.M., 1983. Principles of Aquatic Chemistry. Wiley Interscience, New York, p. 446.

Morgan, I.J., Henry, R.P., Wood, C.M., 1997. The mechanism of acute silver nitrate toxicity in freshwater rainbow trout (*Oncorhynchus mykiss*) is inhibition of gill Na^+ and Cl^- transport. Aquat. Toxicol. 38, 145–163.

Morgan, T.P., Grosell, M., Gilmour, K.M., Playle, R.C., Wood, C.M., 2004. Time course analysis of the mechanism by which silver inhibits active Na^+ and Cl^- uptake in gills of rainbow trout. Am. J. Physiol. 287, R234–R242.

Morris, R., Taylor, E.W., Brown, D.J.A., Brown, J.A. (Eds.), 1989. Acid Toxicity and Aquatic Animals, Society of Experimental Biology Seminar Series No. 34. Cambridge University Press, Cambridge, p. 282.

Morris, C., Val, A.L., Brauner, C.J., Wood, C.M., 2021. The physiology of fish in acidic waters rich in dissolved organic carbon (DOC), with specific reference to the Amazon basin: ionoregulation, acid-base regulation, ammonia excretion, and metal toxicity. J. Exp. Zool. A. https://doi.org/10.1002/jez.2468.

Motais, R., Garcia-Romeu, F., Maetz, J., 1966. Exchange diffusion effect and euryhalinity in teleosts. J. Gen. Physiol. 50, 391–442.

Nakada, T., Hoshijima, K., Esaki, M., Nagayoshi, S., Kawakami, K., Hirose, S., 2007a. Localization of ammonia transporter Rhcg1 in mitochondrion-rich cells of yolk sac, gill, and kidney of zebrafish and its ionic strength-dependent expression. Am. J. Physiol. 293, R1743–R1753.

Nakada, T., Westhoff, C.M., Kato, A., Hirose, S., 2007b. Ammonia secretion from fish gill depends on a set of Rh glycoproteins. FASEB J. 21, 1067–1074.

Narahara, A., Bergman, H.L., Laurent, P., Maina, J.N., Walsh, P.J., Wood, C.M., 1996. Respiratory physiology of the Lake Magadi tilapia (*Oreochromis alcalilcus grahami*), a fish adapted to a hot, alkaline, and frequently hypoxic environment. Physiol. Zool. 69, 1114–1136.

Nawata, C.M., Hung, C.C.Y., Tsui, T.K.N., Wilson, J.M., Wright, P.A., Wood, C.M., 2007. Ammonia excretion in rainbow trout (*Oncorhynchus mykiss*): evidence for Rh glycoprotein and H^+-ATPase involvement. Physiol. Genomics 31, 463–474.

Nawata, C.M., Walsh, P.J., Wood, C.M., 2015. Physiological and molecular responses of the spiny dogfish shark (*Squalus acanthias*) to high environmental ammonia: scavenging for nitrogen. J. Exp. Biol. 218, 238–248.

Neville, C.M., 1979. Influence of mild hypercapnia on the effects of environmental acidification on rainbow trout (*Salmo gairdneri*). J. Exp. Biol. 83, 345–349.

Neville, C.M., 1985. Physiological response of juvenile rainbow trout, *Salmo gairdneri*, to acid and aluminum—prediction of field responses from laboratory data. Can. J. Fish. Aquat. Sci. 42, 2004–2019.

Neville, C.M., Campbell, P.G., 1988. Possible mechanisms of aluminum toxicity in a dilute, acidic environment to fingerlings and older life stages of salmonids. Water Air Soil Pollut. 42, 311–327.

Niyogi, S., Wood, C.M., 2004. The Biotic Ligand Model, a flexible tool for developing site-specific water quality guidelines for metals. Environ. Sci. Technol. 38, 6177–6192.

Northrup, S.L., 2017. Contributions of Genetic Variations and Phenotypic Plasticity to Variation in High pH Tolerance in Rainbow Trout (Ph.D. Thesis). University of British Columbia.

O'Bryan, D.M., Xie, Z., Wang, Y., Du, J.-Z., Brauner, C.J., Richards, J.G., Wood, C.M., Chen, X., Murray, B.W., 2010. Phylogeography and conservation genetics of Lake Qinghai scaleless carp (*Gymnocypris przewalskii*). J. Fish Biol. 77, 2072–2092.

Oğuz, A.R., 2013. Environmental regulation of mitochondria-rich cells in *Chalcalburnus tarichi* (Pallas, 1811) during reproductive migration. J. Membr. Biol. 246, 183–188.

Oğuz, A.R., 2015a. Histological changes in the gill epithelium of endemic Lake Van Fish (*Chalcalburnus tarichi*) during migration from alkaline water to freshwater. NW J. Zool. 11, 51–57.

Oğuz, A.R., 2015b. A histological study of the kidney structure of Van fish (*Alburnus tarichi*) acclimated to highly alkaline water and freshwater. Mar. Freshw. Behav. Physiol. 48, 135–144.

Packer, R.K., Dunson, W.A., 1972. Anoxia and sodium loss associated with the death of brook trout at low pH. Comp. Biochem. Physiol. A 41, 17–26.

Pacoureau, N., Rigby, C.L., Kyne, P.M., Sherley, R.B., Winker, H., Carlson, J.K., Fordham, S.V., Barreto, R., Fernando, D., Francis, M.P., Jabado, R.W., Herman, K.B., Liu, K.-M., Marshall, A.D., Pollom, R.A., Romanov, E.V., Simpfendorfer, C.A., Yin, J.S., Kindsvater, H.K., Dulvy, N.K., 2021. Half a century of global decline in oceanic sharks and rays. Nature 589, 567–571.

Pagenkopf, G.K., 1983. Gill surface interaction model for trace-metal toxicity to fishes: role of complexation, pH, and water hardness. Environ. Sci. Technol. 17, 342–347.

Paquin, P.R., Gorsuch, J.W., Apte, S., Batley, G.E., Bowles, K.C., Campbell, P.G.C., Delos, C.G., Di Toro, D.M., Dwyer, R.L., Galvez, F., Gensemer, R.W., Goss, G.G., Hogstrand, C., Janssen, C.R., McGeer, J.C., Naddy, R.B., Playle, R.C., Santore, R.C., Schneider, U., Stubblefield, W.A., Wood, C.M., Wu, K.B., 2002. The biotic ligand model: a historical overview. Comp. Biochem. Physiol. C 133, 3–35.

Paquin, P.R., Redman, A., Ryan, A., Antore, R., 2012. Modeling the physiology and toxicology of metals. In: Wood, C.M., Farrell, A.P., Brauner, C.J. (Eds.), Homeostasis and Toxicology of Essential Metals, Fish Physiology. vol. 31A. Academic Press, London, pp. 429–484.

Parks, S.K., Tresguerres, M., Goss, G.G., 2007. Blood and gill responses to HCl infusions in the Pacific hagfish (*Eptatretus stoutii*). Can. J. Zool. 85, 855–862.

Pärt, P., Wright, P.A., Wood, C.M., 1998. Urea and water permeability in dogfish gills (*Squalus acanthias*). Comp. Biochem. Physiol. A 119, 117–123.

Pelster, B., Wood, C.M., 2018. Ionoregulatory and oxidative stress issues associated with the evolution of air-breathing. Acta Histochem. 120, 667–679.

Pelster, B., Wood, C.M., Speers-Roesch, B., Driedzic, W.R., Almeida-Val, V., Val, A.L., 2015. Gut transport characteristics in herbivorous and carnivorous serrasalmid fish from ion poor Rio Negro water. J. Comp. Physiol. B. 185, 225–241.

Perry, S.F., 1997. The chloride cell: structure and function in the gills of freshwater fishes. Annu. Rev. Physiol. 59, 325–347.

Peters, A., Merrington, G., Schlekat, C., De Schamphelaere, K., Stauber, J., Batley, G., Harford, A., Van Dam, R., Pease, C., Mooney, T., Warne, M., 2018. Validation of the nickel biotic ligand model for locally relevant species in Australian freshwaters. Environ. Toxicol. Chem. 37, 2566–2574.

Peters, A., Nys, C., Merrington, G., Verdonck, F., Baken, S., Cooper, C.A., Van Assche, F., Schlekat, C., Garman, E., 2020. Demonstrating the reliability of bio-met for determining compliance with environmental quality standards for metals in Europe. Environ. Toxicol. Chem. 39, 2361–2377.

Pinheiro, J.P.S., Windsor, F.M., Wilson, R.W., Tyler, C.R., 2021. Global variation in freshwater physico-chemistry and its influence on chemical toxicity in aquatic wildlife. Biol. Rev. 96, 1528–1546. https://doi.org/10.1111/brv.12711.

Pinho, C., Faria, R., 2016. Magadi tilapia ecological specialization: filling the early gap in the speciation continuum. Mol. Ecol. 25, 1420–1422.

Playle, R.C., 1998. Modelling metal interactions at fish gills. Sci. Total Environ. 219, 147–163.

Playle, R.C., Wood, C.M., 1989a. Water chemistry changes in the gill microenvironment of rainbow trout: experimental observations and theory. J. Comp. Physiol. B. 159, 527–537.

Playle, R.C., Wood, C.M., 1989b. Water pH and aluminum chemistry in the gill micro-environment of rainbow trout during acid and aluminum exposures. J. Comp. Physiol. B. 159, 539–550.

Playle, R.C., Goss, G.G., Wood, C.M., 1989. Physiological disturbances in rainbow trout (*Salmo gairdneri*) during acid and aluminum exposures in soft water of two calcium concentrations. Can. J. Zool. 67, 314–324.

Playle, R.C., Dixon, D.G., Burnison, B.K., 1993a. Copper and cadmium binding to fish gills: estimates of metal–gill stability constants and modelling of metal accumulation. Can. J. Fish. Aquat. Sci. 50, 2678–2687.

Playle, R.C., Dixon, D.G., Burnison, B.K., 1993b. Copper and cadmium binding to fish gills: modification by dissolved organic carbon and synthetic ligands. Can. J. Fish. Aquat. Sci. 50, 2667–2677.

Plonka, A.C., Neff, W.H., 1969. Mucopolysaccharide histochemistry of gill epithelial secretions in brook trout (*Salvelinus fontinalis*) exposed to acidic conditions. Pa. Acad. Sci. 43, 53–56.

Pörtner, H.O., Schulte, P.M., Wood, C.M., Schiemer, F., 2010. Niche dimensions in fishes: an integrative view. Illustrating the role of physiology in understanding ecological realities. Physiol. Biochem. Zool. 83, 808–826.

Potts, W.T.W., 1984. Transepithelial potentials in fish gills. In: Hoar, W.S., Randall, D.J. (Eds.), Gills, Fish Physiology. vol. 10B. Academic Press, San Diego, pp. 105–128.

Preest, M.R., Gonzalez, R.J., Wilson, R.W., 2005. A pharmacological examination of Na^+ and Cl^- transport in two species of freshwater fish. Physiol. Biochem. Zool. 78, 259–272.

Qinghai Ministry of Agriculture, 2020. Qinghai fish recover. J. Qinghai Agric. Anim. Hus. 4, 35 (in Chinese).

Randall, D.J., Wood, C.M., Perry, S.F., Bergman, H., Maloiy, G.M.O., Mommsen, T.P., Wright, P.A., 1989. Urea excretion as a strategy for survival in a fish living in a very alkaline environment. Nature (London) 337, 165–166.

Reid, S.D., McDonald, D.G., 1991. Metal binding activity of the gills of rainbow trout (*Oncorhynchus mykiss*). Can. J. Fish. Aquat. Sci. 48, 1061–1068.

Reite, O.B., Maloiy, G.M.O., Aasehaug, B., 1974. pH, salinity, and temperature tolerance of Lake Magadi *Tilapia*. Nature (London) 274, 315–316.

Richards, J.G., Curtis, P.J., Burnison, B.K., Playle, R.C., 2001. Effects of natural organic matter source on reducing metal toxicity to rainbow trout (*Oncorhynchus mykiss*) and on metal binding to their gills. Environ. Toxicol. Chem. 20, 1159–1166.

Richey, J.E., Melack, J.M., Aufdenkampe, A.K., Ballester, V.M., Hess, L.L., 2002. Outgassing from Amazonian rivers and wetlands as a large tropical source of atmospheric CO_2. Nature 416, 617–620.

Roch, M., Maly, E.J., 1979. Relationship of cadmium-induced hypocalcemia with mortality in rainbow trout (*Salmo gairdneri*) and the influence of temperature on toxicity. J. Fish. Res. Board Can. 36, 1297–1303.

Rogers, J.T., Wood, C.M., 2004. Characterization of branchial lead–calcium interaction in the freshwater rainbow trout *Oncorhynchus mykiss*. J. Exp. Biol. 207, 813–825.

Rogers, J.T., Richards, J.G., Wood, C.M., 2003. Ionoregulatory disruption as the acute toxic mechanism for lead in the rainbow trout (*Oncorhynchus mykiss*). Aquat. Toxicol. 64, 215–234.

Rosa, R.S., Charvet-Almeida, P., Quijada, C.C.D., 2010. Biology of the South American potamotrygonid stingrays. In: Carrier, J.C., Musick, J.A., Heithaus, M.R. (Eds.), Sharks and their Relatives II: Biodiversity, Adaptive Physiology, and Conservation. CRC Press, Boca Raton, pp. 241–281.

Rüdel, H., Muñiz, C.D., Garelick, H., Kandile, N.G., Miller, B.W., Munoz, L.P., Peijnenburg, W.J., Purchase, D., Shevah, Y., Van Sprang, P., Vijver, M., 2015. Consideration of the bioavailability of metal/metalloid species in freshwaters: experiences regarding the implementation of biotic ligand model-based approaches in risk assessment frameworks. Environ. Sci. Pollut. Res. 22, 7405–7421.

Sadauskas-Henrique, H., Wood, C.M., Souza-Bastos, L.R., Duarte, R., Smith, D., Val, A.L., 2019. Does dissolved organic carbon from Amazon black water (Brazil) help a native species, the tambaqui *Colossoma macropomum* to maintain ionic homeostasis in acidic water? J. Fish Biol. 94, 595–605.

Sadauskas-Henrique, H., Smith, D.S., Val, A.L., Wood, C.M., 2021. Physicochemical properties of the dissolved organic carbon can lead to different physiological responses of zebrafish (*Danio rerio*) under neutral and acidic conditions. J. Exp. Zool. A 335, 864–878.

Santore, R.C., Paquin, P.R., Di Toro, D.M., Allen, H.E., Meyer, J.S., 2001. Biotic ligand model of the acute toxicity of metals. 2. Application to acute copper toxicity in freshwater fish and *Daphnia*. Environ. Toxicol. Chem. 20, 2397–2402.

Santore, R.C., Ryan, A.C., Kroglund, F., Teien, H.C., Rodriguez, P.H., Stubblefield, W.A., Cardwell, A.S., Adams, W.J., Nordheim, E., 2018. Development and application of a biotic ligand model for predicting the chronic toxicity of dissolved and precipitated aluminum to aquatic organisms. Environ. Toxicol. Chem. 37, 70–79.

Sardella, B.A., Baker, D.W., Brauner, C.J., 2009. The effects of variable water salinity and ionic composition on the plasma status of the Pacific hagfish (*Eptatretus stoutii*). J. Comp. Physiol. B. 179, 721–728.

Sari, M., 2008. Threatened fishes of the world: *Chalcalburnus tarichi* (Pallas 1811) (Cyprinidae) living in the highly alkaline Lake Van, Turkey. Environ. Biol. Fish 81, 21–23.

Sashaw, J., Nawata, M., Thompson, S., Wood, C.M., Wright, P.A., 2010. Rhesus glycoprotein and urea transporter genes in rainbow trout embryos are upregulated in response to alkaline water (pH 9.7) but not elevated water ammonia. Aquat. Toxicol. 196, 308–313.

Schindler, D.W., 1988. Effects of acid rain on freshwater ecosystems. Science 239, 149–157.

Schindler, D.W., 1997. Widespread effects of climatic warming on freshwater ecosystems in North America. Hydrol. Process. 11, 1043–1067.

Schmidt-Nielsen, B., Truniger, B., Rabinowitz, L., 1972. Sodium-linked urea transport by the renal tubule of the spiny dogfish *Squalus acanthias*. Comp. Biochem. Physiol. A 42, 13–25.

Seegers, L., Tichy, H., 1999. The *Oreochromis alcalicus* flock (Teleostei: Cichlidae) from lakes Natron and Magadi, Tanzania and Kenya, with descriptions of two new species. Ichthyol. Explor. Freshw. 10, 97–146.

Seegers, L., Sonnenberg, R., Yamamoto, R., 1999. Molecular analysis of the Alcolapia flock from lakes Natron and Magadi, Tanzania and Kenya (Teleostei: Cichlidae), and implications for their systematics and evolution. Ichthyol. Explor. Freshw. 10, 175–199.

Shankar, R.A., Anderson, P.M., 1985. Purification and properties of glutamine synthetase from liver of *Squalus acanthias*. Arch. Biochem. Biophys. 239, 248–259.

Shartau, R.B., Baker, D.W., Crossley, D.A., Brauner, C.J., 2016. Preferential intracellular pH regulation: hypotheses and perspectives. J. Exp. Biol. 219, 2235–2244.

Shartau, R.B., Damsgaard, C., Brauner, C.J., 2019. Limits and patterns of acid-base regulation during elevated environmental CO_2 in fish. Comp. Biochem. Physiol. A 236, 110524.

Shuttleworth, T.J., 1988. Salt and water balance – extrarenal mechanisms. In: Shuttleworth, T.J. (Ed.), Physiology of Elasmobranch Fishes. Springer-Verlag, Berlin, pp. 171–199.

Silva, P., Solomon, R., Spokes, K., Epstein, F.H., 1977. Ouabain inhibition of gill Na-K-ATPase: relationship to active chloride transport. J. Exp. Zool. 199, 419–426.

Skadhauge, E., Lechene, C.P., Maloiy, G.M.O., 1980. *Tilapia grahami*: role of intestine in osmoregulation under conditions of extreme alkalinity. In: Lahlou, B. (Ed.), Epithelial Transport in the Lower Vertebrates. Cambridge University Press, Cambridge, pp. 133–142.

Smith, H.W., 1929. The composition of body fluids of elasmobranchs. J. Biol. Chem. 81, 407–419.

Smith, H.W., 1931. The absorption and excretion of water and salts by the elasmobranch fishes II. Marine elasmobranchs. Am. J. Physiol. 98, 296–310.

Smith, H.W., 1936. The retention and physiological role of urea in the elasmobranchii. Biol. Rev. 11, 49–82.

Staurnes, M., Sigholt, T., Reite, O.B., 1984. Reduced carbonic anhydrase and Na-K ATPase activity in gills of salmonids exposed to aluminium-containing acid water. Experientia 40, 226–227.

Tait, L.W., Simpson, C.W.C., Takei, Y., Forster, M.E., 2009. Hagfish natriuretic peptide changes urine flow rates and vascular tensions in a hagfish. Comp. Biochem. Physiol. C 150, 45–49.

Takvam, M., Wood, C.M., Kryvi, H., Nilsen, T.O., 2021. Ion transporters and osmoregulation in the kidney of teleost fishes as a function of salinity. Front. Physiol. 12, 664588.

Tam, W.L., Wong, W.P., Loong, A.M., Hiong, K.C., Chew, S.F., Ballantyne, J.S., Ip, Y.K., 2003. The osmotic response of the Asian freshwater stingray (*Himantura signifer*) to increased salinity: a comparison with marine (*Taeniura lymma*) and Amazonian freshwater (*Potamotrygon motoro*) stingrays. J. Exp. Biol. 206, 2931–2940.

Thompson, W.A., Rodela, T.M., Richards, J.G., 2015. The effects of strain and ploidy on the physiological responses of rainbow trout (*Oncorhynchus mykiss*) to pH 9.5 exposure. Comp. Biochem. Physiol. B 183, 22–29.

Thorson, T.B., 1967. Osmoregulation in fresh-water elasmobranchs. In: Gilbert, P.W., Mathewson, R.F., Rall, D.P. (Eds.), Sharks, Skates and Rays. Johns Hopkins Press, Baltimore, pp. 265–270.

Thorson, T.B., 1970. Freshwater stingrays, *Potamotrygon* spp.: failure to concentrate urea when exposed to saline medium. Life Sci. 9, 893–900.

Thurman, E.M., 1985. Geochemistry of Natural Waters. Martinus Nijhof/Dr. W. Junk Publishers, Dordrecht, p. 497.

Tichy, H., Seegers, L., 1999. The *Oreochromis alcalicus* flock (Teleostei: Cichlidae) from lakes Natron and Magadi, Tanzania and Kenya: a model for the evolution of "new" species flocks in historical times? Ichthyol. Explor. Freshw. 10, 147–174.

Treberg, J.R., Speers-Roesch, B., Piermarini, P.M., Ip, Y.K., Ballantyne, J.S., Driedzic, W.R., 2006. The accumulation of methylamine counteracting solutes in elasmobranchs with differing levels of urea: a comparison of marine and freshwater species. J. Exp. Biol. 209, 860–870.

Tresguerres, M., Katoh, F., Fenton, H., Jasinska, E., Goss, G.G., 2005. Regulation of branchial V-H⁺-ATPase, Na⁺/K⁺-ATPase and NHE2 in response to acid and base infusions in the Pacific spiny dogfish (*Squalus acanthias*). J. Exp. Biol. 208, 345–354.

Tresguerres, M., Clifford, A.M., Harter, T.S., Roa, J.N., Thies, A.B., Yee, D.P., Brauner, C.J., 2020. Evolutionary links between intra-and extracellular acid–base regulation in fish and other aquatic animals. J. Exp. Zool. A. https://doi.org/10.1002/jez.2367.

Tuong, D.D., Borowiec, B., Clifford, A.M., Filogonio, R., Somo, D., Huong, D.T.T., Phuong, N.T., Wang, T., Bayley, M., Milsom, W.K., 2018. Ventilatory responses of the clown knifefish, *Chitala ornata*, to hypercarbia and hypercapnia. J. Comp. Physiol. B. 188, 581–589.

Tuong, D.D., Phuong, N.T., Bayley, M., Milsom, W.K., 2019. Ventilatory responses of the clown knifefish, *Chitala ornata*, to arterial hypercapnia remain after gill denervation. J. Comp. Physiol. B. 189, 673–683.

US EPA, 2018. Final Aquatic Life Ambient Water Quality Criteria for Aluminum 2018. United States Environmental Protection Agency Office of Water.4304T. (EPA-822-R-18-001, Washington).

Val, A.L., 2019. Fishes of the Amazon: diversity and beyond. An. Acad. Bras. Cienc. 91. https://doi.org/10.1590/0001-3765201920190260.

Val, A.L., Almeida-Val, V.M.F., 1995. Fishes of the Amazon and Their Environment. Physiological and Biochemical Features, Zoophysiology. vol. 32 Springer, Heidelberg, p. 224.

Verbost, P.M., Flik, G., Lock, R.A., Wendelaar Bonga, S.E., 1987. Cadmium inhibition of Ca²⁺ uptake in rainbow trout gills. Am. J. Physiol. 253, 216–221.

Verbost, P.M., Flik, G., Lock, R.A.C., Wendelaar Bonga, S.E., 1988. Cadmium inhibits plasma-membrane calcium-transport. J. Membr. Biol. 102, 97–104.

Walker, K.F., Dunn, I.G., Edwards, D., Petr, T., Yang, H.Z., 1996. A fishery in a changing lake environment: The naked carp *Gymnocypris przewalskii* (Kessler) (Cyprinidae: Schizothoracinae) in Qinghai Hu, China. Int. J. Salt Lake Res. 4, 169–222.

Walsh, P.J., Bergman, H.L., Narahara, A., Wood, C.M., Wright, P.A., Randall, D.J., Maina, J.N., Laurent, P., 1993. Effects of ammonia on survival, swimming, and activities of enzymes of nitrogen metabolism in the Lake Magadi tilapia *Oreochromis alcalicus grahami*. J. Exp. Biol. 180, 323–327.

Walsh, P.J., Heitz, M.J., Campbell, C.E., Cooper, G.J., Medina, M., Wang, Y.S., Goss, G.G., Vincek, V., Wood, C.M., Smith, C.P., 2000. Molecular characterization of a urea transporter in the gill of the gulf toadfish (*Opsanus beta*). J. Exp. Biol. 203, 2357–2364.

Walsh, P.J., Wang, Y., Campbell, C.E., De Boeck, G., Wood, C.M., 2001a. Patterns of nitrogenous waste excretion and gill urea transporter mRNA expression in several species of marine fish. Mar. Biol. 139, 839–844.

Walsh, P.J., Grosell, M., Goss, G.G., Bergman, H.L., Bergman, A.N., Wilson, P., Laurent, P., Alper, S.L., Smith, C.P., Kamunde, C., Wood, C.M., 2001b. Physiological and molecular characterization of urea transport by the gills of the Lake Magadi tilapia (*Alcolapia grahami*). J. Exp. Biol. 204, 509–520.

Wang, M., 2016. Qinghai was sealed for 12 years and it brought up fish in the lake. Qinghai Sci. Technol. 3, 90–91 (In Chinese).

Wang, Y.S., Gonzalez, R.J., Patrick, M.L., Grosell, M., Zhang, C., Feng, Q., Du, J.Z., Walsh, P.J., Wood, C.M., 2003. Unusual physiology of scaleless carp, *Gymnocypris przewalskii*, in Lake Qinghai: a high-altitude saline lake. Comp. Biochem. Physiol. A 134, 409–421.

Weiner, I.D., Verlander, J.W., 2017. Ammonia transporters and their role in acid-base balance. Physiol. Rev. 97, 465–494.

Weinrauch, A.M., Folkerts, E.J., Blewett, T.A., Bucking, C., Anderson, W.G., 2020. Impacts of low salinity exposure and antibiotic application on gut transport activity in the Pacific spiny dogfish, *Squalus acanthias suckleyi*. J. Comp. Physiol. D. 190, 535–545.

Wilkie, M.P., Wood, C.M., 1991. Nitrogenous waste excretion, acid-base regulation, and ionoregulation in rainbow trout (*Onchorhynchus mykiss*) exposed to extremely alkaline water. Physiol. Zool. 64, 1069–1086.

Wilkie, M.P., Wood, C.M., 1994. The effects of extremely alkaline water (pH 9.5) on rainbow trout gill function and morphology. J. Fish Biol. 45, 87–98.

Wilkie, M.P., Wood, C.M., 1996. The adaptations of fish to extremely alkaline environments. Comp. Biochem. Physiol. B 113, 665–673.

Wilkie, M.P., Wright, P.A., Iwama, G.K., Wood, C.M., 1993. The physiological responses of the Lahontan cutthroat trout (*Oncorhynchus clarki henshawi*), a resident of highly alkaline Pyramid Lake (pH 9.4), to challenge at pH 10. J. Exp. Biol. 175, 173–194.

Wilkie, M.P., Wright, P.A., Iwama, G.K., Wood, C.M., 1994. The physiological adaptations of the Lahontan cutthroat trout (*Oncorhynchus clarki henshawi*) following transfer from well water to the highly alkaline waters of Pyramid Lake, Nevada (pH 9.4). Physiol. Zool. 67, 355–380.

Wilkie, M.P., Simmons, H.E., Wood, C.M., 1996. Physiological adaptations of rainbow trout to chronically elevated water pH (pH = 9.5). J. Exp. Zool. 274, 1–14.

Wilkie, M.P., Laurent, P., Wood, C.M., 1999. The physiological basis for altered Na^+ and Cl^- movements across the gills of rainbow trout (*Oncorhynchus mykiss*) in alkaline (pH = 9.5) water. Physiol. Biochem. Zool. 72, 360–368.

Wilson, R.W., 2012. Aluminum. In: Wood, C.M., Farrell, A.P., Brauner, C.J. (Eds.), Homeostasis and Toxicology of Essential Metals, Fish Physiology. vol. 31A. Academic Press, London, pp. 67–123.

Wilson, J.M., Iwata, K., Iwama, G.K., Randall, D.J., 1998. Inhibition of ammonia excretion and production in rainbow trout during severe alkaline exposure. Comp. Biochem. Physiol. B 121, 99–109.

Wilson, R.W., Wood, C.M., Gonzalez, R.J., Patrick, M.L., Bergman, H.L., Narahara, A., Val, A.L., 1999. Ion and acid-base balance in three species of Amazonian fish during gradual acidification of extremely soft water. Physiol. Biochem. Zool. 72, 277–285.

Wilson, P.J., Wood, C.M., Maina, J.N., White, B.N., 2000. Genetic structure of Lake Magadi Tilapia (*Oreochromis alcalicus grahami*) populations. J. Fish Biol. 56, 590–603.

Wilson, P.J., Wood, C.M., Walsh, P.J., Bergman, A.L., Bergman, H.L., Laurent, P., White, B.N., 2004. Discordance between genetic structure and morphological, ecological, and physiological adaptation in Lake Magadi tilapia. Physiol. Biochem. Zool. 77, 537–555.

Wood, C.M., 1989. The physiological problems of fish in acid waters. In: Morris, R., Brown, D.J.A., Taylor, E.W., Brown, J.A. (Eds.), Acid Toxicity and Aquatic Animals, Society for Experimental Biology Seminar Series, No. 34. Cambridge University Press, Cambridge, pp. 125–148.

Wood, C.M., 1991. Branchial ion and acid-base transfer in freshwater teleost fish: environmental hyperoxia as a probe. Physiol. Zool. 64, 68–102.

Wood, C.M., 1993. Ammonia and urea metabolism and excretion. In: Evans, D. (Ed.), The Physiology of Fishes. CRC Press, Boca Raton, pp. 379–425.

Wood, C.M., 1995. Excretion. In: Groot, C., Margolis, L., Clarke, W.C. (Eds.), Physiological Ecology of the Pacific Salmon. UBC Press, Vancouver, pp. 381–438.

Wood, C.M., 2001. The influence of feeding, exercise, and temperature on nitrogen metabolism and excretion. In: Anderson, P.A., Wright, P.A. (Eds.), Nitrogen Excretion. Fish Physiology. vol. 20. Academic Press, Orlando, pp. 201–238.

Wood, C.M., 2011. Rapid regulation of Na$^+$ and Cl$^-$ flux rates in killifish after acute salinity challenge. J. Exp. Mar. Biol. Ecol. 409, 62–69.

Wood, C.M., 2012. Silver. In: Wood, C.M., Farrell, A.P., Brauner, C.J. (Eds.), Homeostasis and Toxicology of Essential Metals, Fish Physiology. vol. 31A. Academic Press, London, pp. 1–65.

Wood, C.M., Bucking, C., 2011. The role of feeding in salt and water balance. In: Grosell, M., Farrell, A.P., Brauner, C.J. (Eds.), The Multifunctional Gut of Fish: Fish Physiology. vol. 30. Academic Press, San Diego, pp. 165–212.

Wood, C.M., Eom, J., 2021. The osmorespiratory compromise in the fish gill. Comp. Biochem. Physiol. A 254, 110895.

Wood, C.M., Giacomin, M., 2016. Feeding through your gills and turning a toxicant into a resource: how the dogfish shark scavenges ammonia from its environment. J. Exp. Biol. 219, 3218–3226.

Wood, C.M., Marshall, W.S., 1994. Ion balance, acid-base regulation, and chloride cell function in the common killifish, *Fundulus heteroclitus* - a euryhaline estuarine teleost. Estuaries 17, 34–52.

Wood, C.M., McDonald, D.G., 1982. Physiological mechanisms of acid toxicity in fish. In: Johnson, R.E. (Ed.), Acid Rain/Fisheries, Proceedings of an International Symposium on Acidic Precipitation and Fishery Impacts in North-Eastern North America. American Fisheries Society, Bethesda, pp. 197–226.

Wood, C.M., McDonald, D.G., 1987. The physiology of acid/aluminium stress in trout. Ann. Soc. R. Zool. Belg. 117, 399–410.

Wood, C.M., Nawata, C.M., 2011. Nose-to-nose comparison of physiological and molecular responses of trout to high environmental ammonia in seawater vs. freshwater. J. Exp. Biol. 214, 3557–3569.

Wood, C.M., Playle, R.C., Simons, B.P., Goss, G.G., McDonald, D.G., 1988. Blood gases, acid–base status, ions, and hematology in adult brook trout (*Salvelinus fontinalis*) under acid/aluminum exposure. Can. J. Fish. Aquat. Sci. 45, 1575–1586.

Wood, C.M., Perry, S.F., Wright, P.A., Bergman, H.L., Randall, D.J., 1989. Ammonia and urea dynamics in the Lake Magadi tilapia, a ureotelic teleost fish adapted to an extremely alkaline environment. Respir. Physiol. 77, 1–20.

Wood, C.M., Bergman, H.L., Laurent, P., Maina, J.N., Narahara, A., Walsh, P.J., 1994. Urea production, acid-base regulation and their interactions in the Lake Magadi tilapia, a unique teleost adapted to a highly alkaline environment. J. Exp. Biol. 189, 13–36.

Wood, C.M., Pärt, P., Wright, P.A., 1995. Ammonia and urea metabolism in relation to gill function and acid-base balance in a marine elasmobranch, the spiny dogfish (*Squalus acanthias*). J. Exp. Biol. 198, 1545–1558.

Wood, C.M., Hogstrand, C., Galvez, F., Munger, R.S., 1996. The physiology of waterborne silver toxicity in freshwater rainbow trout (*Oncorhynchus mykiss*) 1. The effects of ionic Ag$^+$. Aquat. Toxicol. 35, 93–109.

Wood, C.M., Wilson, R.W., Gonzalez, R.J., Patrick, M.L., Bergman, H.L., Narahara, A., Val, A.L., 1998. Responses of an Amazonian teleost, the tambaqui (*Colossoma macropomum*), to low pH in extremely soft water. Physiol. Zool. 71, 658–670.

Wood, C.M., Milligan, C.L., Walsh, P.J., 1999. Renal responses of trout to chronic respiratory and metabolic acidoses and metabolic alkalosis. Am. J. Physiol. 277, R482–R492.

Wood, C.M., Matsuo, A.Y.O., Gonzalez, R.J., Wilson, R.W., Patrick, M.L., Val, A.L., 2002a. Mechanisms of ion transport in *Potamotrygon*, a stenohaline freshwater elasmobranch native to the ion-poor blackwaters of the Rio Negro. J. Exp. Biol. 205, 3039–3054.

Wood, C.M., Wilson, P.W., Bergman, H.L., Bergman, A.N., Laurent, P., Otiang'a-Owiti, G., Walsh, P.J., 2002b. Obligatory urea production and the cost of living in the Magadi tilapia revealed by acclimation to reduced salinity and alkalinity. Physiol. Biochem. Zool. 75, 111–122.

Wood, C.M., Wilson, P.W., Bergman, H.L., Bergman, A.N., Laurent, P., Otiang'a-Owiti, G., Walsh, P.J., 2002c. Ionoregulatory strategies and the role of urea in the Magadi tilapia (*Alcolapia grahami*). Can. J. Zool. 80, 503–515.

Wood, C.M., Matsuo, A.Y.O., Wilson, R.W., Gonzalez, R.J., Patrick, M.L., Playle, R.C., Val, A.L., 2003. Protection by natural blackwater against disturbances in ion fluxes caused by low pH exposure in freshwater stingrays endemic to the Rio Negro. Physiol. Biochem. Zool. 76, 12–27.

Wood, C.M., Kajimura, M., Mommsen, T.P., Walsh, P.J., 2005. Alkaline tide and nitrogen conservation after feeding in an elasmobranch (*Squalus acanthias*). J. Exp. Biol. 208, 2693–2705.

Wood, C.M., Bucking, C.P., Fitzpatrick, J., Nadella, S.R., 2007a. The alkaline tide goes out and the nitrogen stays in after feeding in the dogfish shark, *Squalus acanthias*. Respir. Physiol. Neurobiol. 159, 163–170.

Wood, C.M., Kajimura, M., Bucking, C., Walsh, P.J., 2007b. Osmoregulation, ionregulation and acid-base regulation by the gastrointestinal tract after feeding in the elasmobranch (*Squalus acanthias*). J. Exp. Biol. 210, 1335–1349.

Wood, C.M., Munger, R.S., Thompson, J., Shuttleworth, T.J., 2007c. Control of rectal gland secretion by blood acid–base status in the intact dogfish shark (*Squalus acanthias*). Respir. Physiol. Neurobiol. 156, 220–228.

Wood, C.M., Du, J., Rogers, J., Brauner, C.J., Richards, J.G., Semple, J.W., Murray, B.W., Chen, X.-Q., Wang, Y., 2007d. Przewalski's naked carp (*Gymnocyypris przewalski*): an endangered species taking a metabolic holiday in Lake Qinghai, China. Physiol. Biochem. Zool. 80, 59–77.

Wood, C.M., Kajimura, M., Mommsen, T.P., Walsh, P.J., 2008. Is the alkaline tide a signal to activate metabolic or ionoregulatory enzymes in the dogfish shark (*Squalus acanthias*)? Physiol. Biochem. Zool. 81, 278–287.

Wood, C.M., Al-Reasi, H.A., Smith, D.S., 2011. The two faces of DOC. Aquat. Toxicol. 105, 3–8.

Wood, C.M., Bergman, H.L., Bianchini, A., Laurent, P., Maina, J., Johannsson, O.E., Bianchini, L.F., Chevalier, C., Kavembe, G.D., Papah, M.B., Ojoo, R.O., 2012. Transepithelial potential in the Magadi tilapia, a fish in extreme alkalinity. J. Comp. Physiol. B. 182, 247–258.

Wood, C.M., Liew, H.J., DeBoeck, G., Walsh, P.J., 2013a. A perfusion study of the handling of urea and urea analogues by the gills of the dogfish shark (*Squalus acanthias*). Peer J. 1 (e33). https://doi.org/10.7717/peerj.33.

Wood, C.M., Nawata, C.M., Wilson, J.M., Laurent, P., Chevalier, C., Bergman, H.L., Bianchini, A., Maina, J.N., Johannsson, O.E., Bianchini, L.F., Kavembe, G.D., Papah, M.B., Ojoo, R.O., 2013b. Rh proteins and NH_4^+-activated Na^+ATPase in the Magadi Tilapia (*Alcolapia grahami*), a 100% ureotelic teleost fish. J. Exp. Biol. 216, 2998–3007.

Wood, C.M., Robertson, L.M., Johannsson, O.E., Val, A.L., 2014. Mechanisms of Na^+ uptake, ammonia excretion, and their potential linkage in native Rio Negro tetras (*Paracheirodon axelrodi*, *Hemigrammus rhodostomus*, and *Moenkhausia diktyota*). J. Comp. Physiol. B. 184, 877–890.

Wood, C.M., Brix, K.V., De Boeck, G., Bergman, H.L., Bianchini, A., Bianchini, L.F., Maina, J.N., Johannsson, O.E., Kavembe, G.D., Papah, M.B., Letura, K.M., Ojoo, R.O., 2016. Mammalian metabolic rates in the hottest fish on earth. Sci. Rep. 6, 26990.

Wood, C.M., Liew, H.J., De Boeck, G., Hoogenboom, J.L., Anderson, W.G., 2019. Nitrogen handling in the elasmobranch gut: a role for microbial urease. J. Exp. Biol. 222. https://doi. org/10.1242/jeb.194787, jeb194787.

Wood, C.M., McDonald, M.D., Grosell, M., Mount, D.R., Adams, W.J., Po, B.H., Brix, K.V., 2020. The potential for salt toxicity: can the trans-epithelial potential (TEP) across the gills serve as a metric for major ion toxicity in fish? Aquat. Toxicol. 226, 105568.

Wright, P.A., Fyhn, H.J., 2001. Ontogeny of nitrogen metabolism and excretion. In: Wright, P.A., Anderson, P.M. (Eds.), Nitrogen Excretion., Fish Physiology. vol. 20. Academic Press, New York, pp. 149–200.

Wright, R.F., Snekvik, E., 1977. Chemistry and fish populations in 700 lakes in southernmost Norway. Verh. Int. Ver. Theor. Angew. Limnol. 20, 765–775.

Wright, P.A., Wood, C.M., 1985. An analysis of branchial ammonia excretion in the freshwater rainbow trout: effects of environmental pH change and sodium uptake blockade. J. Exp. Biol. 114, 329–353.

Wright, P.A., Wood, C.M., 2009. A new paradigm for ammonia excretion in aquatic animals: role of rhesus (Rh) glycoproteins. J. Exp. Biol. 212, 2303–2312.

Wright, P.A., Wood, C.M., 2012. Seven things fish know about ammonia and we don't. Respir. Physiol. Neurobiol. 184, 231–240.

Wright, P.A., Wood, C.M., 2016. Regulation of ions, acid–base, and nitrogenous wastes in elasmobranchs. In: Shadwick, R.E., Farrell, A.P., Brauner, C.J. (Eds.), Physiology of Elasmobranch Fishes: Internal Processes, Fish Physiology. vol. 34B. Academic Press, New York, pp. 279–345.

Wright, P.A., Perry, S.F., Randall, D.J., Wood, C.M., Bergman, H., 1990. The effects of reducing water pH and total CO_2 on a teleost fish adapted to an extremely alkaline environment. J. Exp. Biol. 151, 361–369.

Wright, P.A., Iwama, G.K., Wood, C.M., 1993. Ammonia and urea excretion in Lahontan cutthroat trout (*Oncorhynchus clarki henshawi*) adapted to highly alkaline Pyramid Lake (pH 9.4). J. Exp. Biol. 175, 153–172.

Wu, X., Shi, J., 2014. Construction and management of fish passage on Shaliu River adjacent to Qinghai Lake based on ecological restoration. Trans. Chin. Soc. Agric. Eng. 30, 130–136.

Yancey, P.H., 2016. Organic osmolytes in elasmobranchs. In: Shadwick, R.E., Farrell, A.P., Brauner, C.J. (Eds.), Physiology of Elasmobranch Fishes: Internal Processes, Fish Physiology. vol. 34B. Academic Press, New York, pp. 221–277.

Yancey, P.H., Somero, G.N., 1978. Urea-requiring lactate dehydrogenases of marine elasmobranch fishes. J. Comp. Physiol. 125, 135–141.

Yao, Z., Guo, W., Lai, Q., Shi, J., Zhou, K., Qi, H., Lin, T., Li, Z., Wang, H., 2016. *Gymnocypris przewalskii* decreases cytosolic carbonic anhydrase expression to compensate for respiratory alkalosis and osmoregulation in the saline-alkaline Lake Qinghai. J. Comp. Physiol. B. 186, 83–95.

Yesaki, T.Y., Iwama, G.K., 1992. Survival, acid-base regulation, ion regulation, and ammonia excretion in rainbow trout in highly alkaline hard water. Physiol. Zool. 65, 763–787.

Yi, X., Lai, Q., Shi, J., Gao, P., Zhou, K., Qi, H., Wang, H., Yao, Z., 2017. Nitrogenous waste excretion and gene expression of nitrogen transporter in *Gymnocypris przewalskii* in high alkaline environment. J. Fish. Sci. China 24, 681–689.

Zehev, B.S., Almeida, V., Asher, B., Raimundo, R., 2015. Ornamental fishery in Rio Negro (Amazon region), Brazil: combining social, economic and fishery analyses. Fish. Aquac. J. 6, 1000143.

Zeidel, J.D., Mathai, J.C., Campbell, J.D., Ruiz, W.G., Apodaca, G.L., Riordan, J., Zeidel, M.L., 2005. Selective permeability barrier to urea in shark rectal gland. Am. J. Physiol. 289, F83–F89.

Zimmer, A., Baracolli, I.F., Wood, C.M., Bianchini, A., 2012. Waterborne copper exposure inhibits ammonia excretion and branchial carbonic anhydrase activity in euryhaline guppies acclimated to both fresh water and sea water. Aquat. Toxicol. 122-123, 172–180.

Zimmer, A.M., Brauner, C.J., Wood, C.M., 2014. Ammonia transport across the skin of adult rainbow trout (*Oncorhynchus mykiss*) exposed to high environmental ammonia (HEA). J. Comp. Physiol. B. 184, 77–90.

Zimmer, A.M., Wilson, J.M., Wright, P.A., Hiroi, J., Wood, C.M., 2017a. Different mechanisms of Na^+ uptake and ammonia excretion by the gill and yolk sac epithelium of early life stage rainbow trout. J. Exp. Biol. 220, 775–786.

Zimmer, A.M., Wright, P.A., Wood, C.M., 2017b. Ammonia and urea handling by early life stages of fishes. J. Exp. Biol. 220, 3843–3855.

Zimmer, A.M., Brix, K.V., Wood, C.M., 2019. Mechanisms of Ca^{2+} uptake in freshwater and seawater-acclimated killifish, *Fundulus heteroclitus*, and their response to acute salinity transfer. J. Comp. Physiol. B. 189, 47–60.

Zimmer, A., Goss, G.G., Glover, C.N., 2021. Chemical niches and ionoregulatory traits; Applying ionoregulatory physiology to the conservation management of freshwater fishes. Conservation. Physiology 9, coab066.

Zitko, P., Carson, W.V., Carson, W.G., 1973. Prediction of incipient lethal levels of copper to juvenile Atlantic salmon in the presence of humic acid by cupric electrode. Bull. Environ. Contam. Toxicol. 10, 265–271.

Chapter 8

Applied aspects of gene function for the conservation of fishes

Ken M. Jeffries[a,*], Jennifer D. Jeffrey[a,†], and Erika B. Holland[b]
[a]Department of Biological Sciences, University of Manitoba, Winnipeg, MB, Canada
[b]Department of Biological Sciences, California State University of Long Beach, Long Beach, CA, United States
[*]Corresponding author: e-mail: ken.jeffries@umanitoba.ca

Chapter Outline

†Current address: Department of Biology, University of Winnipeg, 515 Portage Ave., Winnipeg, MB R3B 2E9, Canada; 1-204-786-9904.

Fish Physiology, Vol. 39A. https://doi.org/10.1016/bs.fp.2022.04.008

Fishes respond to different abiotic and biotic stressors through changes in gene expression as a part of an integrated organismal response. Gene expression can be described as the way that the gene displays as a phenotype, which includes aspects of the transcription, translation, and the protein function of genes. The gene expression response to environmental stressors in fishes is a complex process that occurs relatively rapidly (i.e., within minutes to hours after acute exposure events); however, the response can be influenced by numerous factors that operate across different temporal scales, from evolutionary, to transgenerational, to within the lifespan and the recent acclimation history of the individual. All these factors need to be considered when interpreting gene expression patterns in wild fishes. There are also distinct challenges for quantifying gene expression in fishes that include the large number of species, polyploidy in some species, and incomplete functional annotation of the genes. In this chapter, we describe what regulates gene expression, how it is quantified, and what emerging gene editing tools may be potentially useful in the conservation of wild fishes. As technologies continue to improve, assessing the gene-level response to environmental stressors in wild fishes and fishes of conservation concern will continue to be a powerful tool in conservation physiology.

1 Gene expression and the integrated organismal response

Aquatic ecosystems are impacted by a variety of human-related activities, which can affect fishes and lead to population declines over time. Therefore, understanding how different species and populations respond to changing environmental conditions throughout their distributions is fundamental to the conservation of fishes. An environmental stressor can be defined as any variable that can disturb the physiological and biochemical systems in an organism leading to reduced performance or fitness (Schulte, 2014). These stressors consist of abiotic (e.g., temperature, salinity, pH, dissolved oxygen), biotic (e.g., pathogens, invasive species, predators, conspecifics) and anthropogenic (e.g., pollution, habitat degradation, fisheries) factors, which can act on their own or as part of a complex interaction. The type of response that is observed depends on the magnitude and the duration of the exposure to the stressful conditions. Relatively low levels of exposure to some stressors can have a positive or stimulatory effect on a fish (i.e., eustress), whereas chronic exposure or a relatively higher magnitude exposure to the stressor may have adverse effects (i.e., distress; Schreck and Tort, 2016). The physiological response observed after exposure to a stressor is typically reflective of an attempt to cope with or overcome the stressor in the short-term or to establish homeostasis (i.e., acclimate) in the longer-term. Differences in physiological responses observed in wild fishes are generally influenced by local environmental factors and genomic variation (i.e., differences in the genetic sequences between individuals) that may be associated with adaptations to environmental conditions (Crozier and Hutchings, 2014; Whitehead et al., 2011). The processes of acclimation and adaptation can lead to differences in the physiological responses to environmental stressors between species and populations. Physiological differences in the response to environmental stressors between organisms can determine which species or populations can persist in a changing environment (Somero, 2010).

Fishes respond to different stressors through changes in gene expression as a part of an integrated organismal response. Proteins already present in cells at the time of exposure to a stressor are typically required to rapidly respond (i.e., within minutes) to acute stressors, however, chronic exposure (i.e., weeks to months) to stressors or recovery from an acute exposure to stressors can require alterations in the expression of certain genes to help maintain homeostasis or reduce longer-term fitness impacts. The gene expression response to environmental stressors in fishes is a complex process that generally occurs relatively rapidly (i.e., often within tens of minutes to hours after exposure to the stressor), however, the response can be influenced by numerous factors that operate across different temporal scales, from evolutionary, to transgenerational, to within the lifespan and the recent acclimation history of the individual. Because of the many different factors that regulate gene expression (some of which will be discussed in this chapter), there can be substantial variation in gene expression that contributes to phenotypic variation within and between populations. Gene expression can be described as the way that the gene displays as a phenotype, which includes aspects of the transcription, translation, and the protein function of genes (Buccitelli and Selbach, 2020; Fig. 1). The abundance of mRNA transcripts, the intermediate step between the transcription of genes and the translation into proteins, is often measured as an estimate of gene expression in comparative fish physiology. Quantifying mRNA responses (i.e., the transcriptomic response) provides valuable insight into the cellular response to a wide variety of acute and chronic stressors in fishes in captivity and in the wild (Connon et al., 2018). Transcriptomic responses tend to respond within hours and they better reflect transcription factor responses, epigenetic regulation of transcription (e.g., DNA methylation), and RNA processing than protein-level changes (i.e., the proteomic response) when assessing gene expression (Buccitelli and Selbach, 2020). Proteomic methods provide an estimate of protein abundance at a given time, which mRNAs are translated into protein, and the post-translational modifications of proteins (Martyniuk and Denslow, 2009). Gene function occurs at the level of the proteome through protein interactions and protein function. The temporal delay between the transcriptome and proteome responses must be considered when interpreting gene expression responses in fishes.

The gene expression patterns of fishes observed in response to environmental stressors are influenced by the genotype of the individual (over evolutionary time scales) and the environmental conditions the individual has

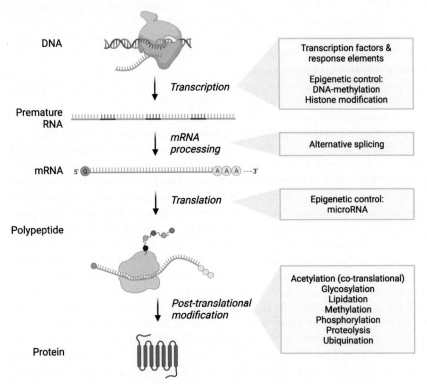

DNA

Transcription

Transcription factors & response elements

Epigenetic control:
DNA-methylation
Histone modification

Premature RNA

mRNA processing

Alternative splicing

mRNA   5' G ⋯⋯⋯⋯⋯⋯ A A A --- 3'

Translation

Epigenetic control:
microRNA

Polypeptide

Post-translational modification

Acetylation (co-translational)
Glycosylation
Lipidation
Methylation
Phosphorylation
Proteolysis
Ubiquination

Protein

FIG. 1 A simplified representation of the different components of gene expression from transcription to protein function and some of the mechanisms that can regulate gene expression at multiple stages. In eukaryotes, transcription by RNA polymerase II results in the synthesis of premature RNA that includes both exons (the coding regions of a gene; light blue) and introns (dark blue). Premature RNA is processed into mature mRNA through the addition of a 5' cap and a 3' poly-A tail, and the removal of introns and combination of exons through splicing. Polypeptide chains that consist of amino acids are translated from the mRNA template in ribosomes. Polypeptide chains are then modified post-translationally to produce mature three-dimensionally folded proteins.

experienced (over physiological time scales). While this chapter is not intended to be a comprehensive review of the evolutionary processes that can alter the genomic architecture of a species, we will address several processes that are the result of evolution and that directly impact gene expression patterns in fishes. The genome of a population or species contributes to the types of genes that can be expressed in response to various types of stressors. This response is influenced by aspects of the genome such as different isoforms that are expressed in different environmental conditions, or the diversity of genes associated with gene superfamilies. For example, gene diversity can be important from a detoxification perspective as some species have different suites of detoxification genes (e.g., cytochrome P450 genes; Arellano-Aguilar et al., 2009; Burnett et al., 2007). A greater gene diversity can also be facilitated by genome duplication events in the evolutionary history of fishes, which has contributed to the complexity of fish genomes (Braasch and Postlethwait, 2012; discussed further in Section 5 of this chapter). Furthermore, genome-level processes control how responsive certain genes are to environmental stressors through phenotypic flexibility (i.e., reversible plasticity; Schulte and Healy, 2022). For example, the genome of a species can contribute to transcriptome flexibility that facilitates enhanced tolerance to environmental stressors and may be an adaptive trait of successful invasive species (Komoroske et al., 2021; Wellband and Heath, 2017). Additionally, the genome determines the acclimatization capacity of a species or population. The ability to acclimatize to environmental conditions can influence what genes respond, when they begin to respond, and the magnitude of the change in expression levels in response to a given stressor (e.g., Komoroske et al., 2015; Logan and Somero, 2011). Acclimatization history can be important for gene families that have inducible and constitutive isoforms, such as some heat shock proteins (Hsps; Deane and Woo, 2011). While acute exposures to elevated temperatures can lead to increased expression of inducible isoforms, chronic exposures can increase the expression of constitutive isoforms. Consequently, the acute temperature that leads to increased expression of inducible Hsp genes is dependent on the acclimatization history of the individual (Logan and Somero, 2011). As a result, acclimatization history is important for understanding the gene expression response to environmental stressors in fishes.

In this chapter, we discuss common ways that the expression of certain genes is regulated and some of the methods used in quantifying gene expression. Throughout the text, when possible, we provide examples of the usefulness of these methods for quantifying gene expression for the conservation of fishes. However, it is important to recognize the effects of genomic divergence between species and populations, and the acclimatization history of the individuals, that influence the gene expression patterns observed in fish physiological studies. We then discuss methods for manipulating gene expression and end by discussing potential future applications of genetic tools for the conservation of fishes. As is common in fisheries research with the incorporation of new technologies, the approaches are often first applied in a laboratory setting using model species before eventually being applied to wild fishes or fishes of conservation concern. Note that this is not an exhaustive review of the literature, as there is an extensive body of literature for every section in this chapter; however, we attempt to describe what regulates gene expression, how it is quantified, and what emerging tools may be potentially useful in the conservation of wild fishes.

2 Genomic factors regulating gene expression

2.1 Genomic divergence and sequence variation

Genomic divergence between species and populations can result from muta
tions that occur as single nucleotide polymorphisms (SNPs). These SNPs are
the most common type of genetic difference between individuals and populations (e.g., Salem et al., 2012) and there are numerous examples of SNPs
being used to assess adaptive divergence in fishes (reviewed in Bernatchez,
2016). Interestingly, many of the alleles associated with local adaptation
rarely become fixed in fishes, which therefore maintains genomic variation
in natural populations (Bernatchez, 2016). The majority of SNPs occur in
non-coding regions of the DNA or do not alter the amino acid sequence of
a protein (i.e., synonymous SNPs). For example, a genome-wide association
study (GWAS) on rainbow trout (*Oncorhynchus mykiss*) estimated that
91.9% of SNPs discovered were in non-coding regions or were synonymous
(Salem et al., 2018). However, when SNPs occur within the coding region
of the genome or interfere with transcription or translation of a gene, they
can directly affect gene expression. For example, SNPs within a coding
sequence that change the amino acid sequence (i.e., non-synonymous SNPs)
can directly alter protein function; whereas SNPs that occur in promoter
regions of the gene or in the 5′ or 3′ untranslated region (UTR) of a transcript
that lead to altered secondary structure in the RNA, can affect the transcription or translation of a gene. Even synonymous SNPs within a coding
sequence can lead to secondary structure of the mRNA that affects translation
(Shabalina et al., 2006). However, it is important to recognize that many
genes are critical for survival and, therefore, there can be significant levels
of stabilizing selection that occur on SNPs within the coding region of the
genome. Consequently, genes associated with responses to environmental
stressors are less likely to undergo significant sequence changes (e.g.,
Thorstensen et al., 2021). Further, the incidences of large-effect SNPs are relatively rare and many of the traits in fishes are polygenic (i.e., influenced by
more than one gene) such as weight and length in Atlantic salmon (*Salmo
salar*; Tsai et al., 2015), and temperature and hypoxia tolerance in Atlantic
killifish (*Fundulus heteroclitus*; Healy et al., 2018).

2.2 Variation through alternative splicing

The magnitude of change in expression of genes of known function in
response to an environmental stressor is important to quantify; however, it
is important to assess the types of genes that are differentially expressed as
well. One of the processes that can contribute to differences in the types of
transcripts that are involved in a response to a stressor is post-transcriptional
modification of RNA via alternative splicing (Fig. 2). Splicing factors interact
with premature RNA transcripts to determine the specific exons that are

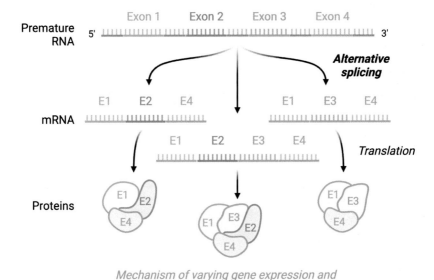

FIG. 2 Modulation of transcript variation through alternative splicing. Premature RNA is produced by transcription of a target gene and includes protein coding regions, exons, as well as introns (in gray). Introns are removed by splicing factors, and exons can be spliced together in alternative ways to produce different mRNAs and ultimately proteins. This example shows how a premature RNA containing four exons could be alternatively spliced to form three different proteins.

combined as part of the final mature mRNAs. Differential exon usage associated with alternative splicing is therefore a possible mechanism of varying gene expression patterns associated with an environmental stressor between populations and species, as well as a potential way of developing novel phenotypic variation (Bush et al., 2017). The role of alternative splicing on the transcriptomic response to cold acclimation has been studied in Atlantic killifish, threespine stickleback (*Gasterosteus aculeatus*), and zebrafish (*Danio rerio*) (Healy and Schulte, 2019). Depending on the species, 426–866 mRNA transcripts were linked to alternative splicing and 284–355 alternatively spliced genes were also differentially expressed due to cold acclimation (Healy and Schulte, 2019). Furthermore, numerous studies that have examined the transcriptomic response to high water temperature have shown that specific splicing factors are differentially expressed when fish are exposed to acute or chronic temperature changes (e.g., Jeffries et al., 2014b; Komoroske et al., 2015). These differential expression patterns highlight the important role of alternative splicing in the gene expression response to high temperatures. Indeed, alternative splicing has been linked to the transcriptomic response to elevated temperatures in a hybrid between channel catfish (*Ictalurus punctatus*) and blue catfish (*I. furcatus*), a relevant hybrid for aquaculture (Tan et al., 2019). Likewise, alternative splicing has been suggested as a potential mechanism contributing to phenotypic divergence in salinity tolerance in Atlantic cod (*Gadus morhua*; Kijewska et al., 2018) and Sacramento splittail (*Pogonichthys macrolepidotus*; Thorstensen et al., 2021). As common with many aspects of gene expression, the relative role of alternative splicing in determining the effects of acclimation may be tissue specific (Li et al., 2020). Further, beyond a physiological response to environmental stressors, alternative splicing has been discussed as a mechanism contributing to morphological differences between species of cichlid (Singh et al., 2017) and the development of different ecotypes of Arctic charr (*Salvelinus alpinus*; Jacobs and Elmer, 2021). Collectively, these studies all suggest the importance of alternative splicing as a mechanism contributing to fish physiology and ecology. With our ability to examine differential exon usage using RNA-sequencing (RNA-seq) data, research in this area will likely uncover more examples of the role of alternative splicing in adaptive traits in fishes.

2.3 Epigenetic regulation

Epigenetics refers to processes that affect gene expression and can lead to phenotypic variation but do not require genetic change. While epigenetic mechanisms play a critical role in developmental processes (see Schulte and Healy, 2022), environmental conditions can also alter epigenetic processes in wild animals and leave what are referred to as epigenetic marks (Angers et al., 2010). Epigenetic control of gene expression is generally considered through three different mechanisms: microRNA (miRNA) control, DNA-methylation, and histone modification (Best et al., 2018). These processes vary significantly in the time scale of how they influence gene expression patterns. For example, miRNA are non-coding RNAs that can influence the immediate response to an environmental stressor (e.g., Ikert et al., 2021), whereas some changes in methylation patterns that occur during early development can influence the gene expression response at later life-stages (e.g., Anastasiadi et al., 2021). Additionally, DNA-methylation and histone modifications can influence patterns across generations (see Schulte and Healy, 2022) and at evolutionary timescales. Indeed, some methylation patterns have been shown to be conserved across generations (e.g., DeCourten et al., 2020; Hu et al., 2021). The processes of DNA-methylation and histone modifications can interfere with the transcription of genes. This is in contrast to miRNAs that can bind to transcribed RNA and prevent translation into a protein. Given the role of epigenetic regulation of gene expression, epigenetics likely contributes to phenotypic variation as well as affects the adaptive response to environmental stressors in natural populations (Rey et al., 2020).

Arguably, the most studied form of epigenetic control of gene expression in fishes is through quantifying the methylation of the genome (e.g., Rey et al., 2020). The process of DNA-methylation involves adding a methyl functional group to cytosines at CpG sites in the genome through DNA methyltransferases. These enzymes are commonly broken up into two groups, enzymes that add a methyl group to the genome (*DNA methyltransferase 3*; Dnmt3) and enzymes that are responsible for re-methylating the genome (*DNA methyltransferase 1*; Dnmt1). Because of differences in the ploidy level of different families of fishes, there can be several types of the different DNA methyltransferase enzymes in the groups (Best et al., 2018). If methylation interferes with gene transcription (e.g., methylation in the promoter region of the gene; Laing et al., 2018), it can lead to a silencing or reduced expression of the gene. However, gene silencing through methylation is reversible by demethylation, which contributes to that gene being expressed later in life. In zebrafish, there are differential methylation patterns that are consistent with a gradual demethylation during early development, possibly as a mechanism to regulate genes during key periods of development (Mhanni and McGowan, 2004). Interestingly, maternal influences on methylation patterns may decrease during development from egg to fry in Chinook salmon (*O. tshawytscha*) (Venney et al., 2020).

Methylation of DNA has the potential to contribute to phenotypic variation in gene expression patterns even in the absence of genomic differences between individuals. This phenotypic variation through DNA-methylation has been shown in a clonal hybrid dace (*Chrosomus eos-neogaeus*) in natural populations (Massicotte et al., 2011). Further, because methylation is a key component of "genomic imprinting" during early development, early rearing conditions that contribute to differential methylation patterns can potentially contribute to developmental plasticity (i.e., which often leads to non-reversible changes; see Schulte and Healy, 2022) and phenotypic variation observed in fishes (Jonsson and Jonsson, 2019; Fig. 3). Therefore, epigenetic marks from DNA-methylation may be useful indicators of historical and current environmental conditions within a conservation perspective (Rey et al., 2020). Indeed, different salinities during early rearing conditions can alter methylation patterns in threespine stickleback with individuals reared at 21 PSU (practical salinity units) having reduced methylation (i.e., hypomethylation) relative to individuals at 2 PSU (Metzger and Schulte, 2018). Interestingly, these authors also found differential methylation patterns between sexes with increased methylation on the female sex chromosome that may contribute to developmental differences (Metzger and Schulte, 2018), consistent with sex-specific methylation patterns in zebrafish (Laing et al., 2018). Temperature changes have been shown to lead to increased methylation (i.e., hypermethylation) in threespine stickleback, suggesting that differential methylation of the genome is a general response to temperature changes (Metzger and Schulte, 2017). Even small temperature

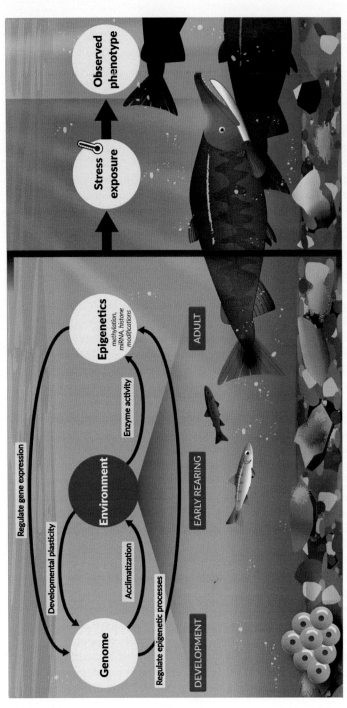

FIG. 3 Conceptual diagram of several processes that operate on different time scales that can regulate the gene expression patterns observed in fishes. For example, the evolutionary history of the individual/population/species causes genomic changes that lead to differences in the individual's ability to acclimatize to environmental conditions. During development and early life stages, environmental conditions can lead to irreversible phenotypic changes associated with developmental plasticity as well as the differential methylation patterns and histone modifications that can affect gene expression in later life stages. Lastly, recently experienced environmental conditions can alter methylation patterns and miRNA expression, and contribute to the physiological changes associated with acclimatization, that will also determine which genes are expressed in response to environmental stressors. All these factors need to be considered when interpreting gene expression patterns in wild fishes.

increases can alter the expression of DNA methyltransferase genes, as well as alter methylation patterns in larval European sea bass (*Dicentrarchus labrax*), a pattern not observed in juvenile fish (Anastasiadi et al., 2017). Exposure to endocrine disrupting compounds during early rearing in inland silverside (*Menidia beryllina*) showed different methylation patterns that persisted across generations, highlighting the lasting effects of the exposure to stressors during early development (DeCourten et al., 2020). Additionally, DNA-methylation differences between populations have been observed and were suggested to contribute to local adaptation in Chinook salmon (Venney et al., 2021). Given the importance of early rearing conditions on development in fishes, alterations in methylation patterns may be an important driver of developmental plasticity and phenotypic variation in response to climate change and anthropogenic influences on aquatic ecosystems.

Histone modifications can influence chromatin structure, which can effectively make genes inaccessible for transcription. These alterations of histones are primarily through post-translational modifications; however, there can also be an incorporation of different types of histones, which can both alter chromatin structure and function (Beal et al., 2018; Best et al., 2018). Similar to DNA-methylation, the number of genes that contribute to histone modifications depends on the ploidy level of the species (Best et al., 2018). The role of histone modification as a mechanism of epigenetic regulation of gene expression is arguably the least studied with respect to wild fishes, possibly due to challenges with studying chromatin structure in non-model species (Beal et al., 2018). However, histone variants can be differentially expressed seasonally in common carp (*Cyprinus carpio*), which may then regulate the expression of genes involved in thermal acclimatization (Araya et al., 2010; Pinto et al., 2005). Furthermore, histone modifications are linked with the expression of immune response genes in larval zebrafish (Galindo-Villegas et al., 2012). Therefore, similar to DNA-methylation, different patterns in histone modification are dependent on the type of environmental stressor, which then contributes to alterations in gene expression response patterns.

Like other epigenetic processes, several environmental factors can elicit differential miRNA responses that subsequently influence gene expression patterns. Complicating miRNA function, miRNAs are capable of binding to multiple RNA transcripts and therefore are not always specific to a particular gene (e.g., Ikert et al., 2021). Further, miRNAs are quite evolutionarily conserved and polyploid species have potential paralogs of different miRNAs (e.g., salmonids; Berthelot et al., 2014). Regardless, miRNAs provide an important regulatory mechanism for gene expression responses to environmental conditions. For example, the influence of miRNA expression on the gene expression response have been demonstrated after hypoxia and high-water temperature exposure in genetically modified tilapia (*Oreochromis niloticus*; Bao et al., 2018; Qiang et al., 2020), heat stress in rainbow trout (*O. mykiss*; Zhou et al., 2019) and to metal exposure in common carp

(Liu et al., 2020). Additionally, miRNAs have been shown to respond to handling stress in rainbow trout (Ikert et al., 2021), suggesting that miRNAs may be involved in the gene expression patterns associated with recovery from acute stressors. These studies demonstrate that determining the causal link between miRNA expression and gene expression will be an important epigenetic mechanism for understanding how fishes respond to changing environmental conditions in the wild.

2.4 Receptor-mediated gene expression

Arguably the first sensors of environmental change are the cellular receptors that are already expressed, where the receptor is defined in the broadest sense as any protein that interacts with a ligand or senses intra- or extracellular changes to drive the signaling transduction necessary for a cellular response. There are numerous signaling cascades and types of cellular responses, but many lead to changes in gene transcription (Latchman, 1997). Signal transduction pathways will activate or repress regulatory transcription factors that bind to select DNA sequences, to control transcription of select genes (Fig. 4). Transcription factors are proteins that will interact with response elements, which are non-coding DNA sequences in proximity (i.e., *cis*) or distant (i.e., *trans*) to the gene of interest. Response elements may be enhancer or silencer elements that activate or repress transcription, respectively (Stallcup and Poulard, 2020). A particular response element may be present in multiple genes such that a single transcription factor can alter the expression of multiple target genes. Individual transcription factors may also increase transcription of one gene while decreasing the transcription of another gene due to the presence of other response elements. Overall, transcription factor binding to response elements will recruit other co-regulatory proteins to activate or repress transcription (Stallcup and Poulard, 2020). Together this creates the regulatory networks needed to control gene expression in different tissues, different stages of development, and in response to environmental stressors.

A well characterized example of receptor-mediated transcriptional regulation includes hormone receptors that sense changes in circulating hormone levels or the presence of hormone-mimicking compounds. For example, there has been considerable work regarding how estrogens control gene expression in vertebrate species. While more complex than originally considered, estrogen-mediated cellular responses are in large part driven by estrogen receptor (ER) transcriptional activity in the nucleus (Yaşar et al., 2017). Here, ERs bound to an estrogen (e.g., 17β-estradiol), or xenoestrogen, create an estrogen-ER complex that acts as a transcription factor. The estrogen-ER complex interacts with the ER response element consensus sequence (5′-GGTCAnnnTGACC-3′; which varies slightly by species (Mushirobira et al., 2018)) present in ER-regulated genes (Fig. 4A). Once bound to the DNA, the estrogen-ER complex recruits co-regulatory proteins such as histone

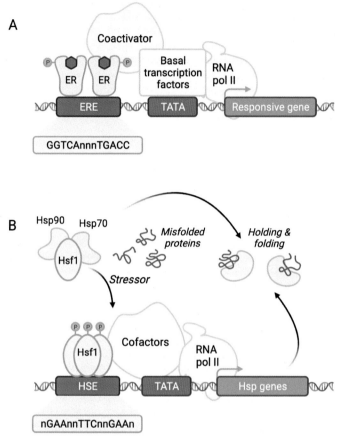

FIG. 4 Receptor-mediated regulation of gene expression. (A) *Estrogen Receptor*: When bound to a ligand, like endogenous 17β-estradiol, ERs form a dimer that binds to the estrogen receptor response element (ERE) in the promoter region upstream of target genes acting as a transcription factor. With the help of coactivators, basal transcription factors, and RNA polymerase II (RNA pol II), binding of ER to the ERE initiates the transcription of responsive genes. (B) *Heat Shock Proteins*: Exposure to a stressor can result in an increase in protein misfolding/unfolding that requires heat shock proteins (Hsp) for protein refolding. The increase in misfolded proteins releases Hsp90 and Hsp70 from Hsf1 allowing Hsf1 to form a trimer that can then bind to heat shock element (HSE) sequences as a transcription factor. The binding of Hsf1 to HSE, along with cofactors and RNA pol II, increases transcription of Hsf1-responsive genes, further enhancing the heat shock response to a stressor.

modifying enzymes and coactivators that are needed for the activation of RNA polymerase II and the transcriptional activity. The binding of ER to ER response elements and the subsequent induction of gene expression has been confirmed in fishes. This includes ER response elements in the promoter region of vitellogenin (*Vtg*), which is an egg precursor important for oocyte development (Mushirobira et al., 2018). Additionally, ER response elements have been found in the promoter region of the aromatase gene, *Cyp19a1b* (Gorelick and Halpern, 2011), which is important for the formation of sex steroids (Tokarz et al., 2015). When fish are exposed to an endogenous estrogen or estrogen mimic, they display increased mRNA expression of *vtg* and *cyp19a1b* related to altered hormone levels, which can lead to the presence of ovotestes and altered secondary sex characteristics in male fishes (Hao et al., 2013; Scholz and Mayer, 2008; Sumpter and Jobling, 1995; Tokarz et al., 2015).

Fish display both short-term inducible (i.e., over mins to hours) and long-term changes (i.e., over weeks to months) in gene expression in response to changing water temperatures. An adaptive heat shock response can help fishes respond to changing temperatures by inducing the expression of protein chaperones, including Hsps, that help maintain protein integrity and prevent protein aggregation during periods of cellular stress (Li et al., 2017). Heat shock factor 1 (Hsf1; Fig. 4B) is a transcription factor that binds to heat shock elements (HSEs) in the promoter region of Hsf1 responsive genes (Vihervaara and Sistonen, 2014). Here, under non-stress conditions the Hsf1 monomer is bound to protein chaperones that include Hsp70 or Hsp90 (Kijima et al., 2018; Li et al., 2017), preventing transcriptional activity. Stress-inducing events, such as altered temperatures, will cause chaperones to be removed from Hsf1 allowing Hsf1 trimerization, which binds to the HSE and induces transcription (Joutsen and Sistonen, 2019). Both *Hsp70* and *Hsp90* genes have HSE sequences in their promoter regions and are regulated by Hsf1 trimerization (Yang et al., 2020, 2021). In fact, Hsf1 is now known to be involved in a cells' response to various stressors and is involved in numerous aspects of general physiology (Li et al., 2017). In fish, *hsp70* mRNA overexpression is an important biomarker for exposure to elevated temperatures but has also been documented in fish exposed to varying salinities, pathogens, and environmental pollutants along with developmental processes (Deane and Woo, 2011).

Salinity is a major abiotic factor that defines suitable habitats for fishes. Changes in salinity of different water bodies is predicted to increase due to climate change and also from increased extreme climatic events (Kültz, 2015). Fishes have several physiological strategies to adjust to changing salinity, however, the degree of salinity tolerance varies greatly by species and population. As such, molecular osmosensors respond to changes in environmental salinity levels and this will initiate the signal transduction needed to maintain favorable osmotic conditions (Kültz, 2015). While the detailed mechanisms are still poorly understood, several transcription factors that participate in osmosensing have been identified. The protein known as the osmotic stress transcription factor 1 (Ostf1) is induced by salinity stress (Fiol et al., 2006) and is a prime candidate for binding to the response element known as the osmotic/salinity sensor enhancer 1 (OSRE1; Wang and Kültz,

2017). The OSRE1 enhancer element promotes mRNA transcription of inositol monophosphatase 1 (*impa1.1*) and myo-inositol phosphate synthase (*mips*), which are osmoprotective enzymes (Wang and Kültz, 2017). The OSRE1 enhancer element also increases the mRNA expression of glutamine synthetase, which synthesizes glutamine that can act as an osmolyte thereby helping to maintain osmotic homeostasis (Kim and Kültz, 2020). Collectively, the examples presented here highlight the importance of receptor-mediated gene expression on responses to major environmental stressors that impact fishes in the wild.

3 Methods of quantifying gene expression

3.1 mRNA transcript abundance

Examining gene expression at the mRNA level can provide valuable information for model and non-model species. Because there are numerous studies that have examined the mRNA response to environmentally-relevant stressors, we attempted to focus on studies of wild fishes or fishes of conservation concern. Applications of assessing mRNA transcript abundance in wild fishes range from targeted quantitative reverse transcription PCR (qPCR) approaches to assess biomarkers of a particular response to assessing whole-transcriptome responses to stressors (Connon et al., 2018). Microarray or RNA-seq approaches can be used to examine the whole-transcriptome profiles of fish exposed to different conditions. Where sequence information is sparse for a target species, whole-transcriptome sequencing can provide a wealth of information. With improvements in RNA-seq approaches, increased depth of sequencing has increased the feasibility of generating *de novo* reference transcriptomes for non-model species. Additionally, non-lethal biopsies such as gill filaments (Chapman et al., 2020; Cooke et al., 2005; Jeffrey et al., 2020; Jeffries et al., 2014a; McCormick, 1993), caudal fin clips (Veldhoen et al., 2013, 2014), and epidermal mucus (Greer et al., 2019) can provide sufficient RNA for qPCR, or even RNA-seq approaches, allowing sensitive or species of conservation concern to be sampled without the need for lethal sampling. Whole-transcriptome approaches are especially useful when specific gene targets are unknown, such as in the case of complex environments where an array of potential environmental factors may be present. For instance, the limited range expansion of invasive silver carp (*Hypophthalmichthys molitrix*) in the Illinois River near Chicago, Illinois, USA, was linked to poor water quality using transcriptomics as fish near the leading edge of their distribution exhibited genotoxic and detoxification responses in their livers because of contaminant inputs compared to fish sampled further downstream (Jeffrey et al., 2019). Similarly, RNA-seq was used to examine the status of yellow perch (*Perca flavescens*) at multiple sites along the St. Lawrence River, Quebec, Canada, to better understand the cumulative effects of complex anthropogenic stressors on the health of the system (Defo et al., 2018). Comparative transcriptomics was also used to examine baseline patterns of transcript expression variation among populations of *Nannoperca australis* across a gradient of hydroclimatic variability in the Murray-Darling Basin in southeastern Australia (Brauer et al., 2017). These whole transcriptome assessments can be used for developing targeted qPCR assays or for the development of biomarkers. The sensitivity of these biomarkers can subsequently be explored in response to stressors of interest using qPCR (e.g., Miller et al., 2014). For example, Akbarzadeh et al. (2020) used RNA-seq to develop qPCR assays for biomarkers sensitive to hypoxia across several salmonid species. Similarly, Jesus et al. (2017) investigated the impacts of a climate change scenario on 14 genes linked to warming or acidification in two fish, *Squalius carolitertii* and *S. torgalensis*, endemic to the northern Atlantic (i.e., Mondego River, Portugal) and southern Mediterranean regions (i.e., Mira River, Portugal) of the Iberian Peninsula, respectively. The 14 genes were chosen because of differential whole-transcriptome responses from a previous study (Jesus et al., 2016). Transcriptomics at the whole-transcriptome or targeted gene level has the potential to act as a valuable tool for examining mechanisms underlying organismal responses to environmental change.

Targeted approaches using qPCR are often used to examine the responses of wild fishes to various environmental stressors. Examining several genes across biological processes of interest and targeting genes that are less sensitive to acute capture and handling stressors are ideal (e.g., Jeffrey et al., 2020; Jesus et al., 2017). Using microfluidic technologies, high-throughput qPCR provides an intermediary to qPCR and whole-transcriptome approaches, allowing for 10s to 100s of genes to be examined simultaneously over more individuals than typically used in a whole-transcriptome study. Gill biopsies taken from tagged sockeye salmon (*O. nerka*) leaving their nursery lake were used to relate migration survival to immune response patterns and pathogen presence using qPCR to examine 69 genes of interest (Jeffries et al., 2014a). Similarly, a qPCR array was developed for pikeperch (*Sander lucioperca*) to examine the impact of elevated temperature on 38 potential biomarkers (Swirplies et al., 2019). Measuring 10s of genes using qPCR may be more cost effective and informative to managers than whole-transcriptome approaches, particularly when the stressors or expected responses are better understood. For instance, several genes associated with specific metabolic pathways (e.g., oxidative stress, stress response and detoxification, energy metabolism, retinoids, reproductive hormones) were examined in pikeperch sampled at six locations along the St. Lawrence River to examine the cumulative impacts of human activities on the system (Bruneau et al., 2016). The effects of an immune challenge and hatchery-rearing techniques on lake sturgeon (*Acipenser fulvescens*) were also examined using targeted qPCR (Bugg et al., 2020; Earhart et al., 2020). A qPCR approach was also used to explore multiple sublethal thresholds of estuarine fish to thermal stressors by evaluating 15 target genes related to stress, cell proliferation, and osmoregulation (Jeffries et al., 2018). Together, these targeted qPCR approaches provide insight on gene expression responses of fishes to stressors in controlled environments that can be potentially applied to studying wild fishes.

Studies have examined the role of the transcriptome in mediating phenotypic responses of wild-caught fishes to ecologically-relevant stressors, such as those associated with climate change. Sandoval-Castillo et al. (2020) exposed Australian rainbow fish (*Melanotaenia duboulayi*, *M. fluviatillis*, *M. spendida tatei*) from temperate, subtropical, and desert locations to contemporary and projected summer temperatures. Using transcriptomics, the authors identified several genes that showed signals of adaptive plasticity due to ecotype-specific directional selection, as well as genes that responded to heat stress (Sandoval-Castillo et al., 2020). Similarly, studies on delta smelt (*Hypomesus transpacificus*), longfin smelt (*Spirinchus thaleichthys*), and inland silverside have examined the species-specific responses to thermal stress of native and invasive species residing in the San Francisco Estuary, California, USA (Jeffries et al., 2016; Komoroske et al., 2015, 2021). Population-specific effects on thermal physiology have also been examined using targeted qPCR in lake sturgeon (Bugg et al., 2020). Additionally, the transcriptomic profiles of genetically differentiated northern and southern populations of tambaqui (*Colossoma macropomum*) have been examined in relation to thermal stress and adaptation (Fé-Gonçalves et al., 2020a,b). Comparative transcriptomics has further been used as a valuable tool in examining the status and responses of Pacific salmon to environmental stressors (Akbarzadeh et al., 2018, 2020; Chapman et al., 2020; Jeffries et al., 2014b). Together, these studies emphasize the usefulness of transcriptomic tools in examining the impacts of climatic factors on fishes of conservation concern.

Laboratory-based experiments using transcriptomics to evaluate responses to ecologically-relevant stressors can further support conservation science. Numerous conservation-based studies have collected fish from the wild and held them in animal holding facilities to conduct both acute and chronic exposure studies (e.g., Bugg et al., 2020; Jesus et al., 2017; Swirplies et al., 2019). Laboratory-based studies have the benefit of providing a more controlled setting, allowing for stressor-specific effects of more complex environments to be teased apart. For instance, Brinkmann et al. (2016) characterized the transcriptional responses of the roach (*Rutilus rutilus*) to dioxins and dioxin-like contaminants. The effects of a broad array of acclimation temperatures on not only transcript abundance, but also blood parameters and whole-animal metabolic responses, allowed for sublethal thresholds to be assessed across ecologically-relevant temperatures in brook trout (*Salvelinus fontinalis*) (Mackey et al., 2021). These laboratory-based studies can inform field-based studies by providing methods to examine the mechanisms underlying the responses of fishes to environmental stressors in the wild.

In addition to the evaluation of transcript levels through RNA-seq and qPCR, *in situ* hybridization also quantifies mRNA levels. Like qPCR, *in situ* hybridization requires sequence information for the genes of interest but requires the additional use of technique-specific sample preservation and preparation (Jeffrey et al., 2022). More qualitative and location-specific detection of mRNA are possible using *in situ* hybridization, which may be particularly useful in situations where the location of mRNA producing cells may change due to an environmental stressor. For instance, glucocorticoid receptor 1 and 2 and mineralocorticoid receptor transcripts were localized in gill, liver, intestine, and brain, as well as corticotropin-releasing factor and its receptor in the brain of tilapia to examine their role in saltwater acclimation (Aruna et al., 2012a,b). *In situ* hybridization was also used to examine the potential developmental plasticity of myogenic regulatory factors with temperature in Atlantic salmon embryos (Macqueen et al., 2007). Additional studies have used *in situ* hybridization to examine acid-base and osmoregulatory aspects of the gill and embryonic skin. For example, the cytoplasmic carbonic anhydrase isoform was localized in gill pavement and mitochondrial-rich cells, supporting the role of carbonic anhydrase in acid-base regulation (Georgalis et al., 2006). Similarly, using *in situ* hybridization, different types of ionocytes were identified in the embryonic skin of medaka during saltwater acclimation (Hsu et al., 2014) and subtypes of Na^+/K^+-ATPase were localized to different ionocytes in the zebrafish gill (Liao et al., 2009). Although not used as much as other transcriptomic techniques in conservation physiology, *in situ* hybridization techniques have potential for examining targeted gene expression in tissues such as the gill under extreme environmental conditions (e.g., salinity, temperature, toxicants).

3.2 Protein abundance and enzyme activity

Measuring proteins in non-model species is difficult due to the lack of genomic and molecular tools available. Protein levels are often measured using targeted antibodies through Western blotting, immunohistochemistry, or with enzyme-linked immunosorbent assays (ELISA). In contrast, RNA-based methodologies can be relatively easily modified for different species, but it is important to recognize that mRNA transcript abundance does not always correlate with increased protein abundance. For example, there can be an increase in *hsp70* mRNA levels at moderate levels of heat shock in zebrafish with no corresponding increase in Hsp70 protein levels (Mottola et al., 2020). The production of gene- and species-specific antibodies can be extensive; however, in the case where fish-specific antibodies are available, the measurement of target proteins can be informative for functional gene expression. For instance, plasma Vtg was quantified using a grouper-specific Vtg ELISA as part of a larger study to examine the baseline health indices of wild-caught goliath grouper (*Epinephelus itajara*; Malinowski et al., 2020).

Because monoclonal or polyclonal antibodies are not available for many fish endemic to South America, Bahamonde et al. (2019) developed a colorimetric assay to measure Vtg-like phosphoproteins using mucus sampling techniques as an alternative to the standard ELISA approach for measuring Vtg levels. Plasma biochemical analytes can also be measured using dry chemistry analyzers that examine several proteins simultaneously, such as alkaline phosphatase, aspartate aminotransferase, creatine phosphokinase, lactate dehydrogenase, lipase, as well as several other biochemical analytes (e.g., calcium, cholesterol, triglycerides). These types of broad plasma analyses allow for non-lethal assessments of fish health (e.g., Malinowski et al., 2020; Roh et al., 2020). Together, targeted protein measurements can provide useful information regarding functional gene expression, where molecular resources (i.e., antibodies, ELISA kits) are available.

Proteomic approaches examine whole-proteome patterns of gene expression. The proteome is highly structured, where protein function is time and location specific and may be dependent on the association with other proteins or biomolecules (Aebersold and Mann, 2016; Parker and Pratt, 2020). High-throughput proteomic approaches are based on mass spectrometry that allows for several aspects of thousands of proteins to be examined, such as abundance, tissue distribution, post-translational modification, and protein–protein interactions (Parker and Pratt, 2020). As such, proteomic-based approaches can provide unique insight into the responses of fish to environmentally-relevant stressors (e.g., Moreira et al., 2021). The advancement of proteomic resources for non-model species has lagged behind transcriptomics, with the annotation of proteomic libraries being a limiting factor. However, there have recently been some promising studies for the use of proteomics to examine and identify useful biomarkers for measuring fish stress. Proteomics was used to identify putative novel protein biomarkers in the epidermal mucus of Atlantic sturgeon (*A. oxyrinchus oxyrhinchus*) in response to by-catch and surgery stressors (Murphy et al., 2020). This study used one-dimensional sodium dodecyl sulfate polyacrylamide gel electrophoresis (SDS-PAGE) to identify protein bands of interest and the protein profiles of these bands were examined using mass spectrometry (Murphy et al., 2020). A similar approach, using two-dimensional gel electrophoresis (2-DE) and tandem mass spectrometry, was used to examine the muscle proteome of gilt-head sea bream (*Sparus aurata*) in response to thermal stress, as a biological model for some key fisheries species (Madeira et al., 2017). Population-specific differences in the liver proteome response to elevated temperature and hypoxia were also examined in European flounder (*Platichthys flesus*) using 2-DE based mass spectrometry (Pédron et al., 2017). The liver proteome response of Atlantic salmon to chronically elevated temperature was examined using shotgun proteomics, a combination of liquid chromatography and mass spectrometry, which provided valuable information regarding their ability to cope with temperatures near their upper thermal tolerance (Nuez-Ortín et al., 2018).

With the increased genomic tools available for threespine stickleback, a previously constructed and validated data-independent acquisition assay library was used in combination with liquid chromatography and mass spectrometry to examine the effects of salinity on the gill proteome in mesocosm experiments (Li and Kültz, 2020). Liquid chromatography and tandem mass spectrometry are used in combination with labeling of digested peptides with an iTRAQ reagent, allowing for unbiased and untargeted biomarker discovery. Wen et al. (2019b) used an iTRAQ-based proteomic approach to examine the response of *Takifugu fasciatus* to cold-stress. With the increasing number of molecular tools becoming available for fishes, high-throughput proteomics approaches have the potential to provide valuable information on gene expression in the context of conservation and management of fishes.

Differences in physical and chemical parameters of proteins can be linked to the capacity for organisms to adapt to challenging environments. The theoretical structure of protein parameters can be predicted using tools such as ProtParam and RaptorX. For instance, Jesus et al. (2017) found that *S. torgalensis*, that inhabits a warmer southern environment compared to *S. carolitertii* in the Iberian Peninsula, exhibited a higher thermostability for Hsp90 and guanylate binding protein 1 as predicted by the structural and functional characteristics of the protein, that may allow them to cope better in warmer environments. Theoretical models of protein structure and function, particularly if paired with gene expression analyses, potentially provide valuable mechanistic information regarding organisms' capacity to cope with climatic changes.

Enzyme activity is another layer in which functional gene expression can be measured. Antioxidant capacity is often measured by examining the activity of enzymes such as glutathione peroxidase (Gpx), superoxide dismutase (Sod), glutathione-S-transferase, glutathione reductase, and catalase (Cat) using commercially available kits or laboratory-derived methods (Campos et al., 2019; Chen et al., 2021; Giuliani et al., 2021; Islam et al., 2021; Jia et al., 2020; Malinowski et al., 2020). Metabolic enzymes can also be examined at the activity level with lactate dehydrogenase, malate dehydrogenase, and citrate synthase being common targets (e.g., Campos et al., 2019). Additionally, immune function can be examined by measuring serum or plasma lysozyme activity (Islam et al., 2021; Malinowski et al., 2020). Thus, measurements of enzyme activity have the potential to add another dimension to measurements of gene expression in the context of fish responses to ecologically-relevant stressors.

3.3 Integrating gene expression assessments across levels of biological organization

Although measurements at the mRNA level can provide useful information regarding regulation of gene expression, these data become even stronger when coupled with protein or enzyme activity measurements, or data across other levels of biological organization. Studies that combine gene expression data with plasma or serum variables such as cortisol, glucose, or lactate, and make connections to whole-animal physiology (e.g., metabolic rate, behavior) allow for a better overall picture of the integrated organismal response. For instance, the physiological response of juvenile turbot (*Scophthalmus maximus*) to heat stress was thoroughly evaluated with hepatic transcript levels of Hsps and apoptosis-related genes, as well as glycogen, antioxidant enzyme activities (Sod, Cat, Gpx), and malondialdehyde content, serum cortisol and glucose, and respiratory frequency (Jia et al., 2020). Giuliani et al. (2021) examined if climate change-related warming and pH reduction would impact the sensitivity of Antarctic rockcod (*Trematomus bernacchii*) to cadmium exposure. The authors integrated the analysis of cadmium accumulation in the liver and gills with the assessment of the antioxidant defense and cellular responses at the mRNA level (metallothionein, *nrf2*, *keap1*, antioxidant genes) as well as metallothionein protein levels, and the catalytic activities of antioxidants, total antioxidant scavenging capacity, onset of lipid peroxidation and genotoxic damage (Giuliani et al., 2021). Similarly, the whole-transcriptome responses of pikeperch were paired with gill histological analyses, qPCR of target genes (*hsps*, antioxidant genes), antioxidant enzyme activity, and malondialdehyde content over a broader time course of exposure to examine the integrated physiological response of heat stress in this cool-water fish (Chen et al., 2021). Multi-omics approaches are also being used more frequently to examine the impacts of environmental stressors on transcriptomic, proteomic, as well as metabolomic responses (e.g., Roh et al., 2020; Wen et al., 2019a). With a broad array of biological responses being examined, researchers can examine patterns in the data representative of whole-organismal responses. Being able to connect gene expression changes to physiological and whole-animal responses provides a more comprehensive understanding of organismal responses to environmental change, better informing conservation and management biology.

4 Methods for manipulating gene expression

With the increased understanding of functional biology and genetics across disciplines, fish conservation biologists, physiologists and aquaculture researchers have begun investigating and implementing the use of environmental variables or genetic tools as ways to enhance desired phenotypic characteristics directly. Early examples in aquaculture used exogenous hormones including growth hormone, insulin-like growth factor, and thyroid hormones to enhance growth in commercially important fish species (Devlin et al., 2004). This was impractical due to challenges with dosage regimes, overall cost, and the potential introduction of hormones into the environment. As such, work shifted to breeding and tools from biotechnology that could be applied to increase or decrease the expression of beneficial or deleterious genes in order to enhance or hinder success of species (Chen et al., 2015;

Devlin et al., 2004; Gratacap et al., 2019; Phelps et al., 2020). These ideas have been around for decades, but for hatcheries in particular, these concepts remain controversial due to potential impacts on wild populations, where ethical challenges will need to be addressed before they have wide-spread use in a conservation context.

4.1 Artificial selection

Management practices designed to aid in the availability or conservation of recreational, economical, or ecologically important fish species have long been recognized as potential selective pressures to fish populations. Implemented fisheries management guidelines, captive breeding and translocation programs may lead to intentional or unintentional pressures driving phenotypic change. These often occur over numerous generations of fishing practices or captive breeding protocols where selective practices have led to a loss of genetic diversity, altered sex ratios, altered time to maturity or reduced survival (Devlin et al., 2004; Holsman et al., 2012). Here, we briefly discuss selective pressure and gene expression in hatcheries. Undoubtedly, there is selective pressure present due to fisheries management practices (Heino et al., 2015), but there is relatively limited research measuring changes in gene expression related to changing physiology under variable fishing management strategies (e.g., size-selective fishing, effects of marine protected areas). This will be an interesting, yet complex, area for future research.

Hatcheries can aid in the conservation of fish species that have experienced population declines (Berejikian and Van Doornik, 2018). Selective breeding practices and domestication selection are common in hatcheries and are known to lead to variable traits between wild and hatchery-reared individuals (Glover et al., 2017; Kim et al., 2019). Such practices have led to increased size in hatchery-reared coho salmon (*O. kisutch*), rainbow trout, and Atlantic salmon and increased size was correlated with altered mRNA expression of growth hormone (*gh*) or insulin growth factor 1 (Devlin et al., 2009; Fleming et al., 2002; Tymchuk et al., 2009). For positive phenotype selection, there is great interest in marker assisted selective breeding, which uses genetic tools to detect quantitative trait loci (QTL) that contributes to a trait of interest in particular individuals (see Schulte and Healy, 2022). Successful QTL application in selective breeding includes resistance of species to specific diseases or viruses including resistance to pancreatic necrosis and salmonid alphavirus in Atlantic salmon, and resistance of Japanese flounder (*Paralichthys olivaceus*) to lymphocystis disease (reviewed in Houston, 2017). However, the application of QTL is relatively limited because beneficial traits are often polygenic (Houston, 2017).

There are also numerous challenges to selective breeding and related practices. It is now known that the positive selection characteristic acquired through selective breeding potentially leads to negative traits that include increased disease susceptibility (Bui et al., 2018; Kim et al., 2019). Here, domesticated fish with increased *gh* mRNA expression have altered responses to infections (Bui et al., 2018; Kim et al., 2019) as Gh has an immunomodulatory role (Harris and Bird, 2000; Yada, 2007). Additionally, Atlantic salmon reared in hatcheries display altered stress responses compared to individuals from wild populations (Mes et al., 2018). Additional studies show that after just one generation in captivity Nile tilapia display signs of differential methylation and differential expression of genes involved in immune function (Konstantinidis et al., 2020). Similar findings were observed in steelhead trout (*O. mykiss*) where immune function genes were differentially expressed between offspring from wild fish and first-generation hatchery fish (Christie et al., 2016). These studies potentially support changes in the immune function and infection susceptibility for fish originating from hatchery versus wild populations.

4.2 Genetic tools for altering gene expression or modifying phenotypes

While well designed selective breeding can be beneficial to a population, it is time-consuming, costly, and may not be effective, which limits its application in conservation (Phelps et al., 2020). With increased numbers of tools in biotechnology, there is interest in assisted gene flow to introduce beneficial or deleterious traits in fish. Altering RNA concentrations offers a unique, often non-heritable, mechanism of controlling gene expression. Here, morpholino and small interfering RNA (siRNA) molecules have received significant interest in functional biology (Zhu and Ge, 2018), disease regulation (Smith and Zain, 2019), and conservation (Thresher et al., 2018). These exogenous RNA-based molecules display complementary base pairing with endogenous RNA and once bound, change protein expression by blocking RNA translation (morpholinos) or inducing RNA degradation (siRNAs; Fig. 5). Morpholinos are strings of 25 phosphorodiamidate modified nucleotides that bind to premature RNA and block the subsequent binding of splicing or translational machinery (Bill et al., 2009; Kole et al., 2016; Fig. 5A). In contrast, siRNAs are 20–25 nucleotides long and double-stranded DNA, that once unwound, binds to the mRNA sequence of interest through complementary base pairing, and an endonuclease leads to mRNA cleavage (Kole et al., 2016; Smith and Zain, 2019; Fig. 5B). These approaches are particularly promising because antisense oligomers can be designed for virtually all RNA sequences. There are, however, major challenges with the application of these antisense oligomers including the pharmacokinetics or delivery, where cellular or organismal uptake can be a challenge. The half-life in the environment is likely in the order of hours, and if ingested by a fish, it is unclear whether effects would be localized to the gut or whole organism (Smith and Zain, 2019; Thresher et al., 2018). Many of these challenges are being investigated for application

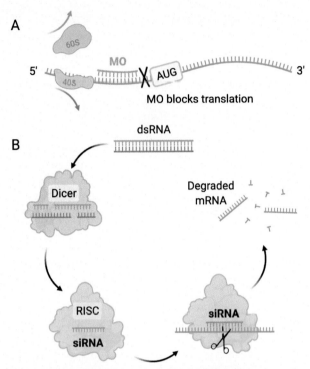

FIG. 5 Morpholino (A) and small interfering RNA (siRNA; B) knockdown of gene expression. Morpholino oligomers (MO) block the translation of proteins by blocking the movement of the initiation complex (shown here) or by affecting accurate splicing of premature RNA. Double-stranded siRNA is fragmented by the endonuclease dicer and incorporated into the RNA-induced silencing complex (RISC). The sense siRNA strand is degraded and the RISC and antisense siRNA bind to complementary mRNA sequences where components of the RISC complex catalyze mRNA cleavage leading to subsequent protein downregulation.

in different fields. For example, siRNA causing lethality, sterilization or altered sex determination has been suggested as a feasible option for controlling invasive sea lamprey (*Petromyzon marinus*) populations in the Laurentian Great Lakes of North America either through inclusion of synthetic siRNA molecules into an ammocoete food source (e.g., yeast) or by genetically engineering algae, another food source, to produce sea lamprey targeting siRNA (Thresher et al., 2018). However, additional concerns with these techniques arise because the impact of knockdown approaches on gene expression is temporary (Zhu and Ge, 2018) and thus newer gene editing techniques are replacing some of the applications of RNA-targeting molecules.

Targeted nuclease activity has arisen as a particularly promising genome editing technique able to create highly specific double-stranded breaks or single-stranded breaks in DNA. Endogenous repair mechanisms can then create insertions or deletions through non-homologous end joining (NHEJ) or can incorporate desirable nucleotide mutations through homology directed repair (HDR), leading to loss or gain of function, respectively (Zhu and Ge, 2018). Once developed for a specific gene sequence, targeted nuclease constructs are introduced into fertilized fish eggs through microinjection or electroporation along with a homologous template of desired DNA for HDR (Chen et al., 2015).

The first sequence-specific nucleases developed were zinc finger nucleases (ZFN), which are composed of three to six zinc finger DNA-binding motifs attached to the FokI nuclease. Synthetic zinc finger arrays are designed to link several zinc finger domains in order to target 9–18 base pair sequences, where a right and left ZFN will attach to the target DNA sequence and dimerize allowing *Fok*I to cleave DNA leading to NHEJ or HDR (Gaj et al., 2013; Li and Wang, 2017; Zhu and Ge, 2018). The application of ZFNs can be limited by the number of cleavage sites that can be targeted, the inefficiency of making the target nuclease arrays, and the potential toxicity due to off-target binding. Transcription activator-like effector nucleases (TALENs), developed in 2011, offer greater flexibility than ZFNs by using the FokI nuclease attached to TAL effector proteins that recognize a single nucleotide (Gaj et al., 2013; Li and Wang, 2017; Zhu and Ge, 2018). Then TAL effector protein arrays can be designed by linking numerous TAL effector proteins to recognize contiguous DNA sequences. Similar to ZFNs, a right and a left TALEN attached to target sequences will dimerize causing DNA double-stranded breaks (Gaj et al., 2013; Li and Wang, 2017; Zhu and Ge, 2018). Because TALEN has limited off-target effects, it is a superior tool compared to ZFNs; however, TALEN generation may be cost prohibitive and delivery methods can be inefficient and time-consuming (Fig. 6).

Clustered regularly interspaced short palindromic repeat (CRISPR; Fig. 6) became available in 2013 providing an inexpensive and convenient genome editing technique with increased efficiency, especially compared to ZFN (Zhu and Ge, 2018). Unlike ZFN and TALEN, which require advanced protein engineering to create targeted nucleases, CRISPR systems use RNA-based molecules and bacteria endonucleases to drive genome editing. CRISPR systems are RNA-guided immune systems that allow bacteria to recognize and cleave foreign genetic nucleic acid sequences using CRISPR associated (Cas) endonucleases (Knott and Doudna, 2018). The first Cas endonuclease to be utilized in genome editing was the type II Cas protein, Cas9, which recognizes short sequences of DNA called spacers. Synthetic chimeric small guide RNAs (sgRNAs) can be developed based on a sequence of interest. The sgRNA will guide the Cas9 nuclease to the appropriate DNA region allowing for cleavage of virtually any DNA sequence of interest. One limitation is that Cas9 DNA binding requires that the sgRNA have an NGG upstream of the target sequence, which is called a protospacer adjacent motif. While Cas9 still remains as the most common endonuclease applied in

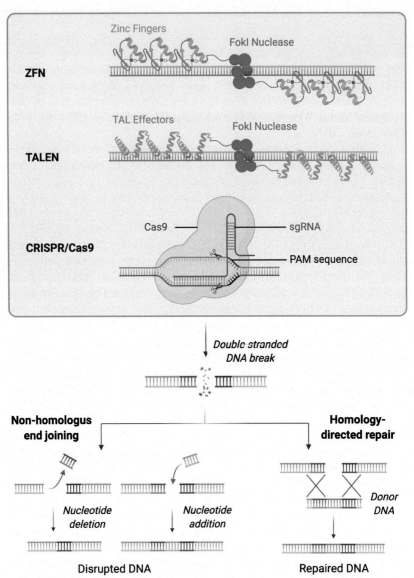

FIG. 6 Mechanisms of genome editing. Zinc finger nucleases (ZFN), composed of three zinc finger DNA-binding motifs attached to a FokI nuclease, attach to a target DNA sequence. Transcription activator-like effector nucleases (TALEN) are composed of a FokI nuclease attached to TAL effector proteins that are designed to recognize contiguous DNA sequences. Both ZFNs and TALENs bind to a target DNA sequence, dimerize, resulting in DNA double-stranded breaks. Clustered regularly interspaced short palindromic repeat (CRISPR) use CRISPR associated endonucleases (e.g., Cas9) to cleave nucleic acid sequences. The synthetic chimeric small guide RNA (sgRNA) recognizes a specific sequence of interest that must be followed by a protospacer adjacent motif (PAM). The sgRNA recruits the Cas9 nuclease to the sequence of interest and Cas9 generates a double-stranded DNA break. Double-stranded breaks can be repaired through non-homologous end joining (NHEJ) or homology-directed repair (HDR). The NHEJ approach can lead to gene knockout and HDR can result in insertion and deletion mutations, where researchers can use donor DNA sequences to incorporate mutations or sequence changes of interest.

CRISPR studies, Cas12a and Cas13a are receiving increased attention for applications beyond programmable genome editing (Knott and Doudna, 2018) to include controlled transcription, RNA editing, and epigenetic control (Knott and Doudna, 2018; Nuñez et al., 2021).

The application of ZFN, TALEN and CRISPR/Cas has primarily been used to understand the cellular and organismal roles of proteins in model fish, such as the zebrafish or Japanese medaka (*Oryzias latipes*). Work in model fishes has been essential for establishing the application in non-mammalian species and understanding specific fish phenotypes. For example, many studies with ZFN, TALEN or CRISPR/Cas have outlined the physiological role of proteins along the hypothalamic-pituitary-gonadal axis contributing to our understanding of fish sex determination and reproduction (Zhu and Ge, 2018). Work in larger-bodied fishes has focused on proof-of-concept applications and beginning uses for desired phenotypes. Genome editing tools have received a lot of attention in aquaculture, where editing has now been utilized to produce sterile individuals or individuals with enhanced reproduction, growth or immunity, and disease resistance (Gratacap et al., 2019). For example, researchers utilized ZFN in channel catfish to knockout the luteinizing hormone (Lh), where the Lh protein plays a major role in the process of sperm production and ovulation in fish (Qin et al., 2016). Knockout of Lh led to sterile individuals that did not spawn even when injected with inducing agents (Qin et al., 2016); however, the study had a small sample size and showed minimal ZFN efficiency, such that it is unclear if this approach is amenable to a large hatchery facility. Outside of aquaculture, studies in cyclostome species have utilized targeted genome editing to understand early vertebrate evolution and development outlining the role of specific proteins in neural crest differentiation and migration, neuronal gene transcription and neuronal cell differentiation. These proof-of-concept and functional studies have increased the possibility of utilizing CRISPR/Cas as an editing tool in sea lamprey, especially as it applies to invasive sea lamprey populations in the Laurentian Great Lakes (York and McCauley, 2020; York et al., 2021).

The studies in the literature to date have mainly used gene editing to complete gene knockout through NHEJ. Template-derived insertion of a desired nucleotide sequence with HDR creates numerous challenges, which partly arise from a lack of understanding of the single or polygenic loci driving specific traits and how that may vary across species (Phelps et al., 2020). As such, extensive research on the genetic and epigenetic determinants of adaptive traits is needed to fully take advantage of the newer highly precise genome-editing techniques such as CRISPR/Cas for single nucleotide insertion for gain of function mutations. Here, NHEJ is the dominant mechanism for DNA repair, where HDR shows extremely low efficiency (Ryu et al., 2019) and would need to be enhanced to be a viable application in desired fields. As such, HDR efficiency has been an important area of focus in mammalian and fish models species (Prykhozhij and Berman, 2018; Ryu et al., 2019). Advancements have included changing the donor DNA length or type, using small molecules to hinder NHEJ repair pathways or enhance HDR pathways, and using base editing that is not affected by HDR efficiency (Ryu et al., 2019). Recent work in Atlantic salmon has used changes in donor DNA to increase knock-in efficiency showing that asymmetrical donor oligonucleotides with short homology arms are particularly effective at increasing the percentage of single nucleotide replacement (Straume et al., 2021).

5   Limitations and challenges for examining gene expression in fishes

5.1   Genomic variation through ploidy levels

A complicating factor when comparing genomes across fishes is differences in ploidy level. Teleosts experienced a genome duplication, which was the third in their evolutionary history, that then reverted back into a more stable diploid state (Braasch and Postlethwait, 2012). Several groups of fishes experienced additional lineage-specific genome duplications, such as salmonids, carps, and sturgeons (Braasch and Postlethwait, 2012; David et al., 2003; Moghadam et al., 2005). Salmonids are well-known to be polyploids, however, may be in the process of reverting into a diploid state through the process of rediploidization (Lien et al., 2016). Sturgeons represent an even more complicated situation as many species are stable tetraploids that are functionally diploids (Vasil'ev, 2009). Several species of Acipenserid sturgeons then experienced genome duplication events making them octoploids that are functional tetraploids (Vasil'ev, 2009). Furthermore, several sturgeons can experience variable ploidy levels in captivity (e.g., 8n, 10n and 12n white sturgeon; Drauch Schreier et al., 2011). When considering polyploid species, there are duplicate copies of genes that typically obtain mutations that lead to a loss of function for duplicate genes creating pseudogenes (Braasch and Postlethwait, 2012). However, mutations or other evolutionary mechanisms can occur such as subfunctionalization that leads to tissue-specific isoforms or neofunctionalization that leads to new forms of genes that can be differentially expressed in different environmental conditions (Braasch and Postlethwait, 2012; Lien et al., 2016). These processes contribute to gene diversity in fishes. As an example of subfunctionalization, different isoforms of *Na/K-ATPase alpha subunit 1a* exhibit tissue-specific expression in salmonids (i.e., rainbow trout; Richards et al., 2003) and Senegalese sole (*Solea senegalensis*; Armesto et al., 2014). Furthermore, greater gene diversity can contribute to specialized functions of a gene, as there is evidence of specific isoforms of *Na/K-ATPase alpha subunit 1a* and *1b*, that alter expression levels during transitions between fresh and salt water (Richards et al., 2003). Therefore, it is likely that aspects of polyploidy contribute to adaptation to different environmental conditions in fishes. Another example of subfunctionalization is the presence of six ryanodine receptors (Ryr) in teleosts with an a and b paralog for the Ryr1, Ryr2 and Ryr3 isoforms found in tetrapods and basal ray-finned fish species (Holland et al., 2017). The differential function of each teleost Ryr is unknown; however, they each display tissue-specific expression supporting subfunctionalization. For example, Ryr2a and Ryr3b are highly expressed in the brain, while Ryr2b and Ryr3a are highly expressed in the heart and skeletal muscle, respectively (Holland et al., 2017).

5.2 Challenges with annotation

Many "omics" approaches can characterize the abundances of all the RNA transcripts and proteins expressed in a tissue or population of cells (i.e., transcriptomics and proteomics, respectively). However, interpreting these patterns can be challenging from a functional standpoint, as much of what is known about cellular functions of genes and gene expression processes has come from mammalian or model invertebrate research. The information for cellular functions is then extrapolated to fish genomes and used to understand the physiological responses to environmental stressors. This can be even more challenging when considering highly specialized tissues like the gill where there is no comparable tissue in mammals. Further, the genomic complexity of fishes has led to fish-specific genes or fish-specific functions of certain genes that may not always be clear from a physiological perspective. Although the power of omics studies to examine genome-wide processes in non-model species is attractive from a logistical standpoint, there are limitations in the interpretation of these data. Therefore, a fundamental limitation of transcriptomic and proteomic studies is the functional annotation of the genes of interest. As functional annotations of genes for fish are generally based on distantly related species, and the conserved function of genes among divergent taxa are largely unknown, these data should be interpreted with caution (Primmer et al., 2013). However, a broad assessment of the functional categories using differentially regulated genes can still provide valuable information in an ecological context and provide a means to develop hypotheses to further examine environmental regulation of gene expression patterns (Pavey et al., 2012). Large-scale collaborative efforts, such as the Functional Annotation of All Salmonid Genomes consortium (FAASG; Macqueen et al., 2017) and the Fundulus Genomics Consortium (FGC; Burnett et al., 2007), to characterize some of the physiological functions of genes is required for researchers to take advantage of emerging tools for quantifying gene expression in fishes. As gene annotations and our understanding of the gene functions improve, the utility of gene expression profiling in wild fishes and fishes of conservation concern will continue to increase.

5.3 Challenges with the implementation of gene editing tools

From a conservation perspective, there are few examples demonstrating the applicability of gene editing and RNA interference in non-model ecologically-relevant fishes. This may be due to limited genetic tools, sizes less amenable for captive breeding needed for current application practices, complicated life histories (e.g., migratory, longevity), and less funding compared to more economically important species (Phelps et al., 2020). There has been wide proof of concept application in larger aquaculture species (e.g., Atlantic salmon), but they are not directly being used in a conservation setting due to delivery, efficiency, or ethical issues discussed elsewhere in the chapter (e.g., see Section 6). For example, there is a major challenge regarding the delivery of gene editing or RNA interference materials to embryos of species of conservation concern. Current practices mainly utilize microinjection of RNA interfering or nuclease gene editing solutions in early stage (i.e., 1–2 cell stage) embryos, which is time consuming and restrictive to the numbers of individuals that can feasibly be injected at one time. Additionally, microinjection techniques are less useful for fish embryos with tougher chorions potentially limiting its application to select species. Electroporation of 1–2 cell stage embryos is also an effective way to introduce editing solutions, which may result in lower embryo survival compared to microinjection but increase the number of individuals that can be processed at one time (Chen et al., 2015). However, to our knowledge the application of either of these techniques to a fish for use in a conservation context, rather than a proof-of-concept application, has not been completed to date. Additional proposed methods of introduction include injection into the somatic tissue or the inclusion in food such that the changes in phenotype would not be incorporated in the germline (Thresher et al., 2018). Again, whether these ideas provide real-world applicability is unknown. In summary, gene-editing tools provide great promise in fish conservation, but research toward the realistic application needs to be expanded.

6 Future directions

6.1 The potential for gene editing for the conservation of fishes

Compared to other early gene-editing methods, CRISPR/Cas represents a particularly cost effective and efficient editing technique, where significant interest has developed over its applicability in conservation. Once realized, the potential uses of CRISPR/Cas systems could contribute to a wide array of questions including (1) targeted sequencing to assess genetic diversity; (2) genetic barcoding, rather than traditional tagging, by inducing neutral mutations in non-coding regions; (3) sensitive species detection for use in environmental DNA (eDNA) studies; and (4) "gene drives" of deleterious or positive phenotypes to aid in invasive species eradication or enhance adaptive capabilities in threatened or endangered species (Phelps et al., 2020). Of these possibilities, gene drives have received the most attention, but similar to other areas where CRISPR technology is being applied, application raises many ethical concerns. A gene drive, or a selfish gene, is inherited in a greater frequency than that predicted by traditional Mendelian genetics (i.e., 50% chance of being inherited) and can cause rapid expansion of a deleterious or beneficial genotype, and thus phenotype, through a population. Gene drives that use CRISPR homing incorporate an endonuclease guided by a sgRNA to a site of interest in the DNA, once cleaved the CRISPR homing construct copies itself into the cleaved site through HDR (Champer et al., 2018). The CRISPR homing construct can be designed to carry "cargo," or a desired trait to aid in conservation (Rode et al., 2019), where cargo alleles could be for the suppression, eradication, or rescue of a population. This strategy has been presented as a theoretical CRISPR application in conservation and has been developed in other fields, including, for example, the suppression of malaria carrying *Anopheles gambiae* (Hammond et al., 2021).

The practice of population rescue in conservation biology is well established where intervention has aimed to help a struggling population rebound through transplant practices or selective breeding (Novak et al., 2018). Many challenges created in traditional rescue practices (e.g., genetic drift) could potentially be avoided with the use of gene editing tools like CRISPR; however, significant roadblocks exist regarding the actual application to populations outside of a laboratory setting. For example, the long-standing concerns over the release of genetically modified organisms into a natural habitat remain, including potential contamination of non-target populations or related species, unforeseen impacts on the fitness of the genetically modified species, and legal and cultural considerations (Phelps et al., 2020). Additionally, there is currently a lack of information regarding the basic biology and ecology for species of concern, and outside of these challenges in research, the risk assessment review process and final implementation would be time consuming and costly. It is likely that each gene editing application will need an independent evaluation to establish a cost benefit analysis with a review board to identify unforeseen risks prior to implementation. Part of the cost benefit analysis would need to consider the importance of preserving the genetic integrity of the species such that each application would need to be ranked on the severity of population decline. Predictably, the level of intervention necessary or appropriate for a threatened species vs a near extinct species would differ. Furthermore, when considering gene editing in model species, germline editing is often utilized, but is considered unethical for human intervention and similar concerns potentially exist in conservation. Thus, researchers and conservation managers may need to consider the usefulness of somatic gene editing, which would produce physiological benefits for the life of the edited individuals but is not heritable. This somatic gene editing approach has received far less attention in a conservation setting but might allow the modified individuals to overcome selective pressures, potentially helping population status, and would likely receive less resistance from the research and management communities (Phelps et al., 2020). As such, there is a strong need for a regulatory framework that considers gene editing in wildlife species (Novak et al., 2018). There has been some development in agriculture, where a modified species that does not have introduced DNA may fall under the same regulatory considerations of selective breeding practices (Novak et al., 2018). These types of rulings are limited; however, and to move forward, an interagency and international approach will be essential, especially for migratory species like anadromous salmon or large marine sport fish (Novak et al., 2018, 2020). These approaches will need to be thoroughly validated in the laboratory before they will be appropriate for use in wild fishes; however, as gene editing technology continues to improve, there will likely be more discussion regarding how these approaches can be effectively used in the conservation physiology of fishes.

6.2 Non-lethal sampling as a key strategy for conservation research

As gene expression approaches have become more commonly used to examine the physiological status of fishes, the ability to study these cellular processes using non-lethally sampled tissue will be important for limiting the impacts on wild populations. Collecting tissue samples from wild fishes using a non-lethal sampling protocol provides the opportunity to relate "snap-shots" of gene expression patterns with fitness-relevant ecological endpoints, such as reproduction, movement patterns, and survival. This approach allows researchers to interpret the ecological significance of gene expression patterns. Through non-lethal sampling, researchers can also sample the same individuals multiple times in a chronic study to investigate the changes in expression patterns over time. Telemetry is a widely used approach in fisheries to understand movement patterns, behavior, and survival in the wild, and this approach has been used to monitor fish in the wild post-sampling. Telemetry has been used in combination with gene expression profiling to study physiological and transcriptomic signatures associated with migration failure (Drenner et al., 2018; Miller et al., 2011), responses to abiotic factors during migration (Evans et al., 2011; Miller et al., 2009) and pathogen infection (Bass et al., 2019; Jeffries et al., 2014b; Teffer et al., 2018) in adult and juvenile Pacific salmon. Another interesting potential avenue for non-lethal sampling is the development of epigenetic clocks that uses epigenetic marks to accurately age wild fishes. The most accurate method for aging fishes is to lethally sample otoliths, however, this is not feasible or appropriate in some circumstances. Comparing age-related DNA-methylation patterns with captive fish of known age, researchers have shown a strong relationship between the methylation patterns and the age of the individual in European sea bass and zebrafish (Anastasiadi and Piferrer, 2020; Mayne et al., 2020). A non-lethal method to estimate ages in wild fish populations will provide a valuable resource to monitor age structure without removing individuals from the populations (e.g., Mayne et al., 2021; Weber et al., 2021).

7 Conclusions

Gene expression profiling in fishes will contribute to a greater understanding of the processes leading to population declines or distribution expansions in the future. As previously discussed, there are numerous challenges associated with studying the effects of environmental stressors on fishes. For example, fishes are the most speciose group of vertebrates (approximately 35,000 described species; https://www.fishbase.de/, accessed in 2021) and have colonized almost every type of aquatic habitat. While this makes fishes a fascinating group for studying physiological adaptations to environmental conditions, it also makes it difficult to find general patterns to describe responses to environmental stressors as there are numerous factors that contribute to the gene expression response in fishes. Additionally, many species of conservation concern are not conducive to standard physiological experimental work because of sensitivities to handling, confinement and experimental conditions (e.g., endangered delta smelt and threatened longfin smelt; Jeffries et al., 2016). In these circumstances, gene expression profiling may be a particularly powerful approach to characterize physiological tolerances in sensitive species or species where there are a limited number of individuals that can be studied.

Future conservation physiology research will benefit from combining multiple omics approaches (e.g., transcriptomics, genomics, proteomics, epigenomics) for comprehensively studying gene expression in wild fish populations. These types of multi-omics studies will allow researchers to assess the mechanisms associated with the regulation of gene expression patterns in fishes. Furthermore, multi-omics research will help provide a functional link between gene sequence variation and gene expression patterns in wild fishes. Lastly, with whole genome assembly becoming more common for non-model species, the combination of different omics technology such as transcriptomics, genomics and epigenomics will provide valuable insight into the factors that regulate phenotypic variation and the physiological responses to environmental conditions. Such multi-omics approaches will dramatically improve our understanding of the significance of gene expression changes in wild fishes.

Acknowledgments

The authors wish to thank Chloé Schmidt from Pineapples and Whales for producing Fig. 3. All other figures were produced using resources from BioRender.com. We also thank Kyra Shewchuk for help with the references, Matthew Thorstensen for help with proof-reading the chapter, and the two reviewers of an earlier version of this chapter.

References

Aebersold, R., Mann, M., 2016. Mass-spectrometric exploration of proteome structure and function. Nature 537, 347–355.

Akbarzadeh, A., Günther, O.P., Houde, A.L., Li, S., Ming, T.J., Jeffries, K.M., Hinch, S.G., Miller, K.M., 2018. Developing specific molecular biomarkers for thermal stress in salmonids. BMC Genomics 19, 749.

Akbarzadeh, A., Houde, A.L.S., Sutherland, B.J.G., Günther, O.P., Miller, K.M., 2020. Identification of hypoxia-specific biomarkers in salmonids using RNA-sequencing and validation using high-throughput qPCR. G3 (g3.401487.2020).

Anastasiadi, D., Piferrer, F., 2020. A clockwork fish: age prediction using DNA methylation-based biomarkers in the European seabass. Mol. Ecol. Resour. 20, 387–397.

Anastasiadi, D., Díaz, N., Piferrer, F., 2017. Small ocean temperature increases elicit stage-dependent changes in DNA methylation and gene expression in a fish, the European sea bass. Sci. Rep-UK. 7, 1–12.

Anastasiadi, D., Shao, C., Chen, S., Piferrer, F., 2021. Footprints of global change in marine life: inferring past environment based on DNA methylation and gene expression marks. Mol. Ecol. 30, 747–760.

Angers, B., Castonguay, E., Massicotte, R., 2010. Environmentally induced phenotypes and DNA methylation: how to deal with unpredictable conditions until the next generation and after. Mol. Ecol. 19, 1283–1295.

Araya, I., Nardocci, G., Morales, J.P., Vera, M.I., Molina, A., Alvarez, M., 2010. MacroH2A subtypes contribute antagonistically to the transcriptional regulation of the ribosomal cistron during seasonal acclimatization of the carp fish. Epigenet. Chromatin. 3, 1–8.

Arellano-Aguilar, O., Montoya, R.M., Garcia, C.M., 2009. Endogenous functions and expression of cytochrome P450 enzymes in teleost fish: a review. Rev. Fish. Sci. 17, 541–556.

Armesto, P., Campinho, M.A., Rodríguez-Rúa, A., Cousin, X., Power, D.M., Manchado, M., Infante, C., 2014. Molecular characterization and transcriptional regulation of the Na$^+$/K$^+$ ATPase α subunit isoforms during development and salinity challenge in a teleost fish, the Senegalese sole (*Solea senegalensis*). Comp. Biochem. Physiol. B 175, 23–38.

Aruna, A., Nagarajan, G., Chang, C.-F., 2012a. Differential expression patterns and localization of glucocorticoid and mineralocorticoid receptor transcripts in the osmoregulatory organs of tilapia during salinity stress. Gen. Comp. Endocrinol. 179, 465–476.

Aruna, A., Nagarajan, G., Chang, C.-F., 2012b. Involvement of corticotrophin-releasing hormone and corticosteroid receptors in the brain–pituitary–gill of tilapia during the course of seawater acclimation. J. Neuroendocrinol. 24, 818–830.

Bahamonde, P., Berrocal, C., Barra, R., Mcmaster, M.E., Munkittrick, K.R., Chiang, G., 2019. Mucus phosphoproteins as an indirect measure of endocrine disruption in native small-bodied freshwater fish, exposed to wastewater treatment plant and pulp and paper mill effluents. Gayana (Concepc). 83, 10–20.

Bao, J.W., Qiang, J., Tao, Y.F., Li, H.X., He, J., Xu, P., Chen, D.J., 2018. Responses of blood biochemistry, fatty acid composition and expression of microRNAs to heat stress in genetically improved farmed tilapia (*Oreochromis niloticus*). J. Therm. Biol. 73, 91–97.

Bass, A.L., Hinch, S.G., Teffer, A.K., Patterson, D.A., Miller, K.M., 2019. Fisheries capture and infectious agents are associated with travel rate and survival of Chinook salmon during spawning migration. Fish. Res. 209, 156–166.

Beal, A., Rodriguez-Casariego, J., Rivera-Casas, C., Suarez-Ulloa, V., Eirin-Lopez, J.M., 2018. Environmental epigenomics and its applications in marine organisms. In: Oleksiak, M.F., Rajora, O.P. (Eds.), Population Genomics: Marine Organisms. Springer, Cham, Switzerland, pp. 325–359.

Berejikian, B.A., Van Doornik, D.M., 2018. Increased natural reproduction and genetic diversity one generation after cessation of a steelhead trout (*Oncorhynchus mykiss*) conservation hatchery program. PLoS One 13, e0190799.

Bernatchez, L., 2016. On the maintenance of genetic variation and adaptation to environmental change: considerations from population genomics in fishes. J. Fish Biol. 89, 2519–2556.

Berthelot, C., Brunet, F., Chalopin, D., Juanchich, A., Bernard, M., Noël, B., Bento, P., Da Silva, C., Labadie, K., Alberti, A., Aury, J.M., 2014. The rainbow trout genome provides novel insights into evolution after whole-genome duplication in vertebrates. Nat. Commun. 5, 1–10.

Best, C., Ikert, H., Kostyniuk, D.J., Craig, P.M., Navarro-Martin, L., Marandel, L., Mennigen, J.A., 2018. Epigenetics in teleost fish: from molecular mechanisms to physiological phenotypes. Comp. Biochem. Physiol. B. 224, 210–244.

Bill, B.R., Petzold, A.M., Clark, K.J., Schimmenti, L.A., Ekker, S.C., 2009. A primer for morpholino use in zebrafish. Zebrafish 6, 69–77.

Braasch, I., Postlethwait, J.H., 2012. Polyploidy in fish and the teleost genome duplication. In: Soltis, P.S., Soltis, D.E. (Eds.), Polyploidy and Genome Evolution. Springer, Berlin, Heidelberg, pp. 341–383.

Brauer, C.J., Unmack, P.J., Beheregaray, L.B., 2017. Comparative ecological transcriptomics and the contribution of gene expression to the evolutionary potential of a threatened fish. Mol. Ecol. 26, 6841–6856.

Brinkmann, M., Koglin, S., Eisner, B., Wiseman, S., Hecker, M., Eichbaum, K., Thalmann, B., Buchinger, S., Reifferscheid, G., Hollert, H., 2016. Characterisation of transcriptional responses to dioxins and dioxin-like contaminants in roach (*Rutilus rutilus*) using whole transcriptome analysis. Sci. Total Environ. 541, 412–423.

Bruneau, A., Landry, C., Giraudo, M., Douville, M., Brodeur, P., Boily, M., Gagnon, P., Houde, M., 2016. Integrated spatial health assessment of yellow perch (*Perca flavescens*) populations from the St. Lawrence River (QC, Canada), part B: cellular and transcriptomic effects. Environ. Sci. Pollut. Res. 23, 18211–18221.

Buccitelli, C., Selbach, M., 2020. mRNAs, proteins and the emerging principles of gene expression control. Nat. Rev. Genet. 21, 630–644.

Bugg, W.S., Yoon, G.R., Schoen, A.N., Laluk, A., Brandt, C., Anderson, W.G., Jeffries, K.M., 2020. Effects of acclimation temperature on the thermal physiology in two geographically distinct populations of lake sturgeon (*Acipenser fulvescens*). Conserv. Physiol. 8, coaa087.

Bui, S., Dalvin, S., Dempster, T., Skulstad, O.F., Edvardsen, R.B., Wargelius, A., Oppedal, F., 2018. Susceptibility, behaviour, and retention of the parasitic salmon louse (*Lepeophtheirus salmonis*) differ with Atlantic salmon population origin. J. Fish Dis. 41, 431–442.

Burnett, K.G., Bain, L.J., Baldwin, W.S., Callard, G.V., Cohen, S., Di Giulio, R.T., Evans, D.H., Gómez-Chiarri, M., Hahn, M.E., Hoover, C.A., Karchner, S.I., 2007. Fundulus as the premier teleost model in environmental biology: opportunities for new insights using genomics. Comp. Biochem. Phys. D. 2, 257–286.

Bush, S.J., Chen, L., Tovar-Corona, J.M., Urrutia, A.O., 2017. Alternative splicing and the evolution of phenotypic novelty. Philos. T. Roy. Soc. B. 372, 20150474.

Campos, D.F., Braz-Mota, S., Val, A.L., Almeida-Val, V.M.F., 2019. Predicting thermal sensitivity of three Amazon fishes exposed to climate change scenarios. Ecol. Indic. 101, 533–540.

Champer, J., Liu, J., Oh, S.Y., Reeves, R., Luthra, A., Oakes, N., et al., 2018. Reducing resistance allele formation in CRISPR gene drive. Proc. Natl. Acad. Sci. U. S. A. 115 (21), 5522–5527.

Chapman, J.M., Teffer, A.K., Bass, A.L., Hinch, S.G., Patterson, D.A., Miller, K.M., Cooke, S.J., 2020. Handling, infectious agents and physiological condition influence survival and post-release behaviour in migratory adult coho salmon after experimental displacement. Conserv. Physiol. 8, coaa033.

Chen, T.T., Lin, C.-M., Chen, M.J., Lo, J.H., Chiou, P.P., Gong, H.Y., Wu, J.L., Chen, M.H.-C., Yarish, C., 2015. Transgenic Technology in Marine Organisms. In: Kim, S.-K. (Ed.), Springer Handbook of Marine Biotechnology. Springer, Berlin, Heidelberg, pp. 387–412.

Chen, Y., Liu, E., Li, C., Pan, C., Zhao, X., Wang, Y., Ling, Q., 2021. Effects of heat stress on histopathology, antioxidant enzymes, and transcriptomic profiles in gills of pikeperch *Sander lucioperca*. Aquaculture 534, 736277.

Christie, M.R., Marine, M.L., Fox, S.E., French, R.A., Blouin, M.S., 2016. A single generation of domestication heritably alters the expression of hundreds of genes. Nat. Commun. 7, 10676.

Connon, R.E., Jeffries, K.M., Komoroske, L.M., Todgham, A.E., Fangue, N.A., 2018. The utility of transcriptomics in fish conservation. J. Exp. Biol. 221, jeb148833.

Cooke, S.J., Crossin, G.T., Patterson, D.A., English, K.K., Hinch, S.G., Young, J.L., Alexander, R.F., Healey, M.C., Van Der Kraak, G., Farrell, A.P., 2005. Coupling non-invasive physiological assessments with telemetry to understand inter-individual variation in behaviour and survivorship of sockeye salmon: development and validation of a technique. J. Fish Biol. 67, 1342–1358.

Crozier, L.G., Hutchings, J.A., 2014. Plastic and evolutionary responses to climate change in fish. Evo. Appl. 7, 68–87.

David, L., Blum, S., Feldman, M.W., Lavi, U., Hillel, J., 2003. Recent duplication of the common carp (*Cyprinus carpio* L.) genome as revealed by analyses of microsatellite loci. Mol. Biol. Evol. 20, 1425–1434.

Deane, E.E., Woo, N.Y., 2011. Advances and perspectives on the regulation and expression of piscine heat shock proteins. Rev. Fish Biol. Fisher. 21, 153–185.

DeCourten, B.M., Forbes, J.P., Roark, H.K., Burns, N.P., Major, K.M., White, J.W., Li, J., Mehinto, A.C., Connon, R.E., Brander, S.M., 2020. Multigenerational and transgenerational effects of environmentally relevant concentrations of endocrine disruptors in an estuarine fish model. Environ. Sci. Technol. 54, 13849–13860.

Defo, M.A., Douville, M., Giraudo, M., Brodeur, P., Boily, M., Houde, M., 2018. RNA-sequencing to assess the health of wild yellow perch (*Perca flavescens*) populations from the St. Lawrence River. Canada. Environ. Pollut. 243, 1657–1668.

Devlin, R.H., Biagi, C.A., Yesaki, T.Y., 2004. Growth, viability and genetic characteristics of GH transgenic coho salmon strains. Aquaculture 236, 607–632.

Devlin, R.H., Sakhrani, D., Tymchuk, W.E., Rise, M.L., Goh, B., 2009. Domestication and growth hormone transgenesis cause similar changes in gene expression in coho salmon (*Oncorhynchus kisutch*). Proc. Natl. Acad. Sci. 106, 3047–3052.

Drauch Schreier, A., Gille, D., Mahardja, B., May, B., 2011. Neutral markers confirm the octoploid origin and reveal spontaneous autopolyploidy in white sturgeon, *Acipenser transmontanus*. J. Appl. Ichthyol. 27, 24–33.

Drenner, S.M., Hinch, S.G., Furey, N.B., Clark, T.D., Li, S., Ming, T., Jeffries, K.M., Patterson, D.A., Cooke, S.J., Robichaud, D., Welch, D.W., 2018. Transcriptome patterns and blood physiology associated with homing success of sockeye salmon during their final stage of marine migration. Can. J. Fish. Aquat. Sci. 75, 1511–1524.

Earhart, M.L., Bugg, W.S., Wiwchar, C.E., Kroeker, J.R.L., Jeffries, K.M., Anderson, W.G., 2020. Shaken, rattled and rolled: the effects of hatchery-rearing techniques on endogenous cortisol production, stress-related gene expression, growth and survival in larval Lake sturgeon, *Acipenser fulvescens*. Aquaculture 522, 735116.

Evans, T.G., Hammill, E.D.D., Kaukinen, K., Schulze, A.D., Patterson, D.A., English, K.K., Curtis, J.M., Miller, K.M., 2011. Transcriptomics of environmental acclimatization and survival in wild adult Pacific sockeye salmon (*Oncorhynchus nerka*) during spawning migration. Mol. Ecol. 20, 4472–4489.

Fé-Gonçalves, L.M., Araújo, J.D.A., Santos, C.H.D.A.D., de Almeida-Val, V.M.F., 2020a. Transcriptomic evidences of local thermal adaptation for the native fish *Colossoma macropomum* (Cuvier, 1818). Genet. Mol. Biol. 43, e20190377.

Fé-Gonçalves, L.M., Alves Araújo, J.D., dos Anjos dos Santos, C.H., Val, A.L., de Almeida-Val, V.M.F., 2020b. How will farmed populations of freshwater fish deal with the extreme climate scenario in 2100? Transcriptional responses of *Colossoma macropomum* from two Brazilian climate regions. J. Therm. Biol. 89, 102487.

Fiol, D.F., Chan, S.Y., Kültz, D., 2006. Regulation of osmotic stress transcription factor 1 (Ostf1) in tilapia (*Oreochromis mossambicus*) gill epithelium during salinity stress. J. Exp. Biol. 209, 3257–3265.

Fleming, I.A., Agustsson, T., Finstad, B., Johnsson, J.I., Björnsson, B.T., 2002. Effects of domestication on growth physiology and endocrinology of Atlantic salmon (*Salmo salar*). Can. J. Fish. Aquat. Sci. 59, 1323–1330.

Gaj, T., Gersbach, C.A., Barbas, C.F., 2013. ZFN, TALEN, and CRISPR/Cas-based methods for genome engineering. Trends Biotechnol. 31, 397–405.

Galindo-Villegas, J., García-Moreno, D., De Oliveira, S., Meseguer, J., Mulero, V., 2012. Regulation of immunity and disease resistance by commensal microbes and chromatin modifications during zebrafish development. Proc. Natl. Acad. Sci. 109, E2605–E2614.

Georgalis, T., Perry, S.F., Gilmour, K.M., 2006. The role of branchial carbonic anhydrase in acid-base regulation in rainbow trout (*Oncorhynchus mykiss*). J. Exp. Biol. 209, 518–530.

Giuliani, M.E., Nardi, A., Di Carlo, M., Benedetti, M., Regoli, F., 2021. Transcriptional and catalytic responsiveness of the Antarctic fish *Trematomus bernacchii* antioxidant system toward multiple stressors. Antioxidants. 10, 410.

Glover, K.A., Solberg, M.F., McGinnity, P., Hindar, K., Verspoor, E., Coulson, M.W., Hansen, M.M., Araki, H., Skaala, Ø., Svåsand, T., 2017. Half a century of genetic interaction between farmed and wild Atlantic salmon: status of knowledge and unanswered questions. Fish Fish. 18, 890–927.

Gorelick, D.A., Halpern, M.E., 2011. Visualization of estrogen receptor transcriptional activation in zebrafish. Endocrinology 152, 2690–2703.

Gratacap, R.L., Wargelius, A., Edvardsen, R.B., Houston, R.D., 2019. Potential of genome editing to improve aquaculture breeding and production. Trends Genet. 35, 672–684.

Greer, J.B., Andrzejczyk, N.E., Mager, E.M., Stieglitz, J.D., Benetti, D., Grosell, M., Schlenk, D., 2019. Whole-transcriptome sequencing of epidermal mucus as a novel method for oil exposure assessment in juvenile mahi-mahi (*Coryphaena hippurus*). Environ. Sci. Technol. Lett. 6, 538–544.

Hammond, A., Karlsson, X., Morianou, I., Kyrou, K., Beaghton, A., Gribble, M., Kranjc, N., Galizi, R., Burt, A., Crisanti, A., Nolan, T., 2021. Regulating the expression of gene drives is key to increasing their invasive potential and the mitigation of resistance. PLoS Genet. 17, e1009321.

Hao, R., Bondesson, M., Singh, A.V., Riu, A., McCollum, C.W., Knudsen, T.B., Gorelick, D.A., Gustafsson, J.Å., 2013. Identification of estrogen target genes during zebrafish embryonic development through transcriptomic analysis. PLoS One 8, e79020.

Harris, J., Bird, D.J., 2000. Modulation of the fish immune system by hormones. Vet. Immunol. Immunopathol. 77, 163–176.

Healy, T.M., Schulte, P.M., 2019. Patterns of alternative splicing in response to cold acclimation in fish. J. Exp. Biol. 222, jeb193516.

Healy, T.M., Brennan, R.S., Whitehead, A., Schulte, P.M., 2018. Tolerance traits related to climate change resilience are independent and polygenic. Glob. Chang. Biol. 24, 5348–5360.

Heino, M., Díaz Pauli, B., Dieckmann, U., 2015. Fisheries-induced evolution. Annu. Rev. Ecol. Evol. Syst. 46, 461–480.

Holland, E.B., Goldstone, J.V., Pessah, I.N., Whitehead, A., Reid, N.M., Karchner, S.I., Hahn, M.E., Nacci, D.E., Clark, B.W., Stegeman, J.J., 2017. Ryanodine receptor and FK506 binding protein 1 in the Atlantic killifish (*Fundulus heteroclitus*): a phylogenetic and population-based comparison. Aquat. Toxicol. 192, 105–115.

Holsman, K.K., Scheuerell, M.D., Buhle, E., Emmett, R., 2012. Interacting effects of transloca-tion, artificial propagation, and environmental conditions on the marine survival of Chinook salmon from the Columbia River, Washington, U.S.a. Conserv. Biol. 26, 912–922.

Houston, R.D., 2017. Future directions in breeding for disease resistance in aquaculture species. Rev. Bras. Zootec. 46, 545–551.

Hsu, H.-H., Lin, L.-Y., Tseng, Y.-C., Horng, J.-L., Hwang, P.-P., 2014. A new model for fish ion regulation: identification of ionocytes in freshwater- and seawater-acclimated medaka (*Ory-zias latipes*). Cell Tissue Res. 357, 225–243.

Hu, J., Wuitchik, S.J., Barry, T.N., Jamniczky, H.A., Rogers, S.M., Barrett, R.D., 2021. Heritabil-ity of DNA methylation in threespine stickleback (*Gasterosteus aculeatus*). Genetics 217, 1–15.

Ikert, H., Lynch, M.D., Doxey, A.C., Giesy, J.P., Servos, M.R., Katzenback, B.A., Craig, P.M., 2021. High throughput sequencing of microRNA in rainbow trout plasma, mucus, and sur-rounding water following acute stress. Front. Physiol. 11, 1821.

Islam, M.J., Slater, M.J., Thiele, R., Kunzmann, A., 2021. Influence of extreme ambient cold stress on growth, hematological, antioxidants, and immune responses in European seabass, *Dicentrarchus labrax* acclimatized at different salinities. Ecol. Indic. 122, 107280.

Jacobs, A., Elmer, K.R., 2021. Alternative splicing and gene expression play contrasting roles in the parallel phenotypic evolution of a salmonid fish. Mol. Ecol. 30, 4955–4969.

Jeffrey, J.D., Jeffries, K.M., Suski, C.D., 2019. Physiological status of silver carp (*Hypophthal-michthys molitrix*) in the Illinois River: an assessment of fish at the leading edge of the inva-sion front. Comp. Biochem. Physiol. D. 32, 100614.

Jeffrey, J.D., Carlson, H., Wrubleski, D., Enders, E.C., Treberg, J.R., Jeffries, K.M., 2020. Apply-ing a gene-suite approach to examine the physiological status of wild-caught walleye (*Sander vitreus*). Conserv. Physiol. 8 (coaa099).

Jeffrey, J.D., Bernier, N.J., Anderson, W.G., 2022. Endocrinology. In: Midway, S.R., Hasler, C.T., Chakrabarty, P. (Eds.), Methods for Fish Biology, Second Edition. American Fisheries Society.

Jeffries, K.M., Hinch, S.G., Gale, M.K., Clark, T.D., Lotto, A.G., Casselman, M.T., Li, S., Rechisky, E.L., Porter, A.D., Welch, D.W., Miller, K.M., 2014a. Immune response genes and pathogen presence predict migration survival in wild salmon smolts. Mol. Ecol. 23, 5803–5815.

Jeffries, K.M., Hinch, S.G., Sierocinski, T., Pavlidis, P., Miller, K.M., 2014b. Transcriptomic responses to high water temperature in two species of Pacific salmon. Evol. Appl. 7, 286–300.

Jeffries, K.M., Connon, R.E., Davis, B.E., Komoroske, L.M., Britton, M.T., Sommer, T., Todgham, A.E., Fangue, N.A., 2016. Effects of high temperatures on threatened estuarine fishes during periods of extreme drought. J. Exp. Biol. 219, 1705–1716.

Jeffries, K.M., Fangue, N.A., Connon, R.E., 2018. Multiple sub-lethal thresholds for cellular responses to thermal stressors in an estuarine fish. Comp. Biochem. Physiol. A. 225, 33–45.

Jesus, T.F., Grosso, A.R., Almeida-Val, V.M.F., Coelho, M.M., 2016. Transcriptome profiling of two Iberian freshwater fish exposed to thermal stress. J. Therm. Biol. 55, 54–61.

Jesus, T.F., Moreno, J.M., Repolho, T., Athanasiadis, A., Rosa, R., Almeida-Val, V.M.F., Coelho, M.M., 2017. Protein analysis and gene expression indicate differential vulnerability of Iberian fish species under a climate change scenario. PLoS One 12, e0181325.

Jia, Y., Chen, X., Wang, Z., Meng, Z., Huang, B., Guan, C., 2020. Physiological response of juvenile turbot (*Scophthalmus maximus* L) during hyperthermal stress. Aquaculture 529, 735645.

Jonsson, B., Jonsson, N., 2019. Phenotypic plasticity and epigenetics of fish: embryo temperature affects later-developing lift-history traits. Aquat. Biol. 28, 21–32.

Joutsen, J., Sistonen, L., 2019. Tailoring of proteostasis networks with heat shock factors. Cold Spring Harb. Perspect. Biol. 11 (4), a034066.

Kijewska, A., Malachowicz, M., Wenne, R., 2018. Alternatively spliced variants in Atlantic cod (*Gadus morhua*) support response to variable salinity environment. Sci. Rep.-UK. 8, 1–11.

Kijima, T., Prince, T.L., Tigue, M.L., Yim, K.H., Schwartz, H., Beebe, K., Lee, S., Budzynski, M.A., Williams, H., Trepel, J.B., Sistonen, L., Calderwood, L., Neckers, L., 2018. HSP90 inhibitors disrupt a transient HSP90-HSF1 interaction and identify a noncanonical model of HSP90-mediated HSF1 regulation. Sci. Rep-UK. 8, 6976.

Kim, C., Kültz, D., 2020. An osmolality/salinity-responsive enhancer 1 (OSRE1) in intron 1 promotes salinity induction of tilapia glutamine synthetase. Sci. Rep.-UK. 10, 12103.

Kim, J.H., Macqueen, D.J., Winton, J.R., Hansen, J.D., Park, H., Devlin, R.H., 2019. Effect of growth rate on transcriptomic responses to immune stimulation in wild-type, domesticated, and GH-transgenic coho salmon. BMC Genomics 20, 1024.

Knott, G.J., Doudna, J.A., 2018. CRISPR-Cas guides the future of genetic engineering. Science 361, 866–869.

Kole, R., Krainer, A.R., Altman, S., 2016. RNA therapeutics: beyond RNA interference and antisense oligonucleotides. Nat. Rev. Drug Discov. 11, 125–140.

Komoroske, L.M., Connon, R.E., Jeffries, K.M., Fangue, N.A., 2015. Linking transcriptional responses to organismal tolerance reveals mechanisms of thermal sensitivity in a mesothermal endangered fish. Mol. Ecol. 24, 4960–4981.

Komoroske, L.M., Jeffries, K.M., Whitehead, A., Roach, J.L., Britton, M., Connon, R.E., Verhille, C., Brander, S.M., Fangue, N.A., 2021. Transcriptional flexibility during thermal challenge corresponds with expanded thermal tolerance in an invasive compared to native fish. Evol. Appl. 14, 931–949.

Konstantinidis, I., Sætrom, P., Mjelle, R., Nedoluzhko, A.V., Robledo, D., Fernandes, J.M., 2020. Major gene expression changes and epigenetic remodelling in Nile tilapia muscle after just one generation of domestication. Epigenetics 15, 1052–1067.

Kültz, D., 2015. Physiological mechanisms used by fish to cope with salinity stress. J. Exp. Biol. 218, 1907–1914.

Laing, L.V., Viana, J., Dempster, E.L., Webster, T.U., van Aerle, R., Mill, J., Santos, E.M., 2018. Sex-specific transcription and DNA methylation profiles of reproductive and epigenetic associated genes in the gonads and livers of breeding zebrafish. Comp. Biochem. Phys. A. 222, 16–25.

Latchman, D.S., 1997. Transcription factors: an overview. Int. J. Biochem. Cell Biol. 29, 1305–1312.

Li, J., Kültz, D., 2020. Proteomics of osmoregulatory responses in threespine stickleback gills. Integr. Comp. Biol. 60, 304–317.

Li, M., Wang, D., 2017. Gene editing nuclease and its application in tilapia. Sci. Bull. 62, 165–173.

Li, J., Labbadia, J., Morimoto, R.I., 2017. Rethinking HSF1 in stress, development, and organismal health. Trends Cell Biol. 27 (12), 895–905.

Li, B.J., Zhu, Z.X., Qin, H., Meng, Z.N., Lin, H.R., Xia, J.H., 2020. Genome-wide characterization of alternative splicing events and their responses to cold stress in tilapia. Front. Genet. 11, 244.

Liao, B.-K., Chen, R.-D., Hwang, P.-P., 2009. Expression regulation of Na + -K + -ATPase α1-subunit subtypes in zebrafish gill ionocytes. Am J Physiol. Regul. Integr. Comp. Physiol. 296, R1897–R1906.

Lien, S., Koop, B.F., Sandve, S.R., Miller, J.R., Kent, M.P., Nome, T., Hvidsten, T.R., Leong, J.S., Minkley, D.R., Zimin, A., Grammes, F., Grove, H., Gjuvsland, A., Walenz, B., Hermansen, R.A., von Schalburg, K., Rondeau, E.B., Di Genova, A., Samy, J.K.A., Vik, J.O., Vigeland, M.D., Caler, L., Grimholt, U., Jentoft, S., Våge, D.I., de Jong, P., Moen, T., Baranski, M., Palti, Y., Smith, D.R., Yorke, J.A., Nederbragt, A.J., Tooming-Klunderud, A., Jakobsen, K.S., Jiang, X., Fan, D., Hu, Y., Liberles, D.A., Vidal, R., Iturra, P., Jones, S.J.M., Jonassen, I., Maass, A., Omholt, S.W., Davidson, W.S., 2016. The Atlantic salmon genome provides insights into rediploidization. Nature 533, 200–205.

Liu, Q., Yang, J., Gong, Y., Cai, J., Zheng, Y., Zhang, Y., Yu, D., Zhang, Z., 2020. MicroRNA profiling identifies biomarkers in head kidneys of common carp exposed to cadmium. Chemosphere 247, 125901.

Logan, C.A., Somero, G.N., 2011. Effects of thermal acclimation on transcriptional responses to acute heat stress in the eurythermal fish *Gillichthys mirabilis* (Cooper). Am. J. Physiol.-Reg. I. 300, R1373–R1383.

Mackey, T.E., Hasler, C.T., Durhack, T., Jeffrey, J.D., Macnaughton, C.J., Ta, K., Enders, E.C., Jeffries, K.M., 2021. Molecular and physiological responses predict acclimation limits in juvenile brook trout (*Salvelinus fontinalis*). J. Exp. Biol. 224. p.jeb241885.

Macqueen, D., Robb, D., Johnston, I.A., 2007. Temperature influences the coordinated expression of myogenic regulatory factors during embryonic myogenesis in Atlantic salmon (*Salmo salar*). J. Exp. Biol. 210, 2781–2794.

Macqueen, D.J., Primmer, C.R., Houston, R.D., Nowak, B.F., Bernatchez, L., Bergseth, S., Davidson, W.S., Gallardo-Escárate, C., Goldammer, T., Guiguen, Y., Iturra, P., Kijas, J.W., Koop, B.F., Lien, S., Maass, A., Martin, S.A.M., McGinnity, P., Montecino, M., Naish, K.A., Nichols, K.M., Ólafsson, K., Omholt, S.W., Palti, Y., Plastow, G.S., Rexroad 3rd, C.E., Rise, M.L., Ritchie, R.J., Sandve, S.R., Schulte, P.M., Tello, A., Vidal, R., Vik, J.O., Wargelius, A., Yáñez, J.M., The FAASG Consortium, 2017. Functional annotation of all salmonid genomes (FAASG): an international initiative supporting future salmonid research, conservation and aquaculture. BMC Genomics 18, 484.

Madeira, D., Araújo, J.E., Vitorino, R., Costa, P.M., Capelo, J.L., Vinagre, C., Diniz, M.S., 2017. Molecular plasticity under ocean warming: proteomics and fitness data provides clues for a better understanding of the thermal tolerance in fish. Front. Physiol. 8, 825.

Malinowski, C.R., Perrault, J.R., Coleman, F.C., Koenig, C.C., Stilwell, J.M., Cray, C., Stacy, N.I., 2020. The iconic Atlantic goliath grouper (*Epinephelus itajara*): a comprehensive assessment of health indices in the southeastern United States population. Front. Vet. Sci. 7, 635.

Martyniuk, C.J., Denslow, N.D., 2009. Towards functional genomics in fish using quantitative proteomics. Gen. Comp. Endocrinol. 164, 135–141.

Massicotte, R., Whitelaw, E., Angers, B., 2011. DNA methylation: a source of random variation in natural populations. Epigenetics 6, 421–427.

Mayne, B., Korbie, D., Kenchington, L., Ezzy, B., Berry, O., Jarman, S., 2020. A DNA methylation age predictor for zebrafish. Aging 12, 24817.

Mayne, B., Espinoza, T., Roberts, D., Butler, G.L., Brooks, S., Korbie, D., Jarman, S., 2021. Nonlethal age estimation of three threatened fish species using DNA methylation: Australian lungfish, Murray cod and Mary River cod. Mol. Ecol. Resour. 21, 2324–2332.

McCormick, S.B., 1993. Methods for nonlethal gill biopsy and measurement of Na+, K+ -ATPase activity. Can. J. Fish. Aquat. Sci. 50, 656–658.

Mes, D., Von Krogh, K., Gorissen, M., Mayer, I., Vindas, M.A., 2018. Neurobiology of wild and hatchery-reared Atlantic salmon: how nurture drives neuroplasticity. Front. Behav. Neurosci. 12, 210.

Metzger, D.C., Schulte, P.M., 2017. Persistent and plastic effects of temperature on DNA methylation across the genome of threespine stickleback (*Gasterosteus aculeatus*). P. Roy. Soc. B- Biol. Sci. 284 (20171667).

Metzger, D.C., Schulte, P.M., 2018. The DNA methylation landscape of stickleback reveals patterns of sex chromosome evolution and effects of environmental salinity. Genome Biol. Evol. 10, 775–785.

Mhanni, A.A., McGowan, R.A., 2004. Global changes in genomic methylation levels during early development of the zebrafish embryo. Dev. Genes Evol. 214, 412–417.

Miller, K.M., Schulze, A.D., Ginther, N., Li, S., Patterson, D.A., Farrell, A.P., Hinch, S.G., 2009. Salmon spawning migration: metabolic shifts and environmental triggers. Comp. Biochem. Physiol. D. 4, 75–89.

Miller, K.M., Li, S., Kaukinen, K.H., Ginther, N., Hammill, E., Curtis, J.M., Patterson, D.A., Sierocinski, T., Donnison, L., Pavlidis, P., Hinch, S.G., Hruska, K.A., Cooke, S.J., English, K.K., Farrell, A.P., 2011. Genomic signatures predict migration and spawning failure in wild Canadian salmon. Science 331, 214–217.

Miller, K.M., Teffer, A., Tucker, S., Li, S., Schulze, A.D., Trudel, M., Juanes, F., Tabata, A., Kaukinen, K.H., Ginther, N.G., Ming, T.J., 2014. Infectious disease, shifting climates, and opportunistic predators: cumulative factors potentially impacting wild salmon declines. Evol. Appl. 7, 812–855.

Moghadam, H.K., Ferguson, M.M., Danzmann, R.G., 2005. Evidence for hox gene duplication in rainbow trout (*Oncorhynchus mykiss*): a tetraploid model species. J. Mol. Evol. 61, 804–818.

Moreira, M., Schrama, D., Farinha, A.P., Cerqueira, M., Raposo de Magalhães, C., Carrilho, R., Rodrigues, P., 2021. Fish pathology research and diagnosis in aquaculture of farmed fish; a proteomics perspective. Animals 11, 125.

Mottola, G., Nikinmaa, M., Anttila, K., 2020. Hsp70s transcription-translation relationship depends on the heat shock temperature in zebrafish. Comp. Biochem. Physiol. A. 240, 110629.

Murphy, A.E., Stokesbury, M.J.W., Easy, R.H., 2020. Exploring epidermal mucus protease activity as an indicator of stress in Atlantic sturgeon (*Acipenser oxyrinchus oxyrhinchus*). J. Fish Biol. 97, 1354–1362.

Mushirobira, Y., Nishimiya, O., Nagata, J., Todo, T., Hara, A., Reading, B.J., Hiramatsu, N., 2018. Molecular cloning of vitellogenin gene promoters and in vitro and in vivo transcription profiles following estradiol-17β administration in the cutthroat trout. Gen. Comp. Endocrinol. 267, 157–166.

Novak, B.J., Maloney, T., Phelan, R., 2018. Advancing a new toolkit for conservation: from science to policy. CRISPR J. 1, 11–15.

Novak, B.J., Fraser, D., Maloney, T.H., 2020. Transforming Ocean conservation: applying the genetic rescue toolkit. Gene 11, 209.

Nuez-Ortín, W.G., Carter, C.G., Nichols, P.D., Cooke, I.R., Wilson, R., 2018. Liver proteome response of pre-harvest Atlantic salmon following exposure to elevated temperature. BMC Genomics 19, 133.

Nuñez, J.K., Chen, J., Pommier, G.C., Cogan, J.Z., Replogle, J.M., Adriaens, C., Ramadoss, G.N., Shi, Q., Hung, K.L., Samelson, A.J., Pogson, A.N., 2021. Genome-wide programmable transcriptional memory by CRISPR-based epigenome editing. Cell 184, 2503–2519.

Parker, C.G., Pratt, M.R., 2020. Click chemistry in proteomic investigations. Cell 180, 605–632.

Pavey, S.A., Bernatchez, L., Aubin-Horth, N., Landry, C.R., 2012. What is needed for next-generation ecological and evolutionary genomics? Trends Ecol. Evolution 27, 673–678.

Pédron, N., Artigaud, S., Infante, J.-L.Z., Le Bayon, N., Charrier, G., Pichereau, V., Laroche, J., 2017. Proteomic responses of European flounder to temperature and hypoxia as interacting stressors: differential sensitivities of populations. Sci. Total Environ. 586, 890–899.

Phelps, M.P., Seeb, L.W., Seeb, J.E., 2020. Transforming ecology and conservation biology through genome editing. Conserv. Biol. 34, 54–65.

Pinto, R., Ivaldi, C., Reyes, M., Doyen, C., Mietton, F., Mongelard, F., Alvarez, M., Molina, A., Dimitrov, S., Krauskopf, M., Vera, M.I., 2005. Seasonal environmental changes regulate the expression of the histone variant macroH2A in an eurythermal fish. FEBS Lett. 579, 5553–5558.

Primmer, C.R., Papakostas, S., Leder, E.H., Davis, M.J., Ragan, M.A., 2013. Annotated genes and nonannotated genomes: cross-species use of gene ontology in ecology and evolution research. Mol. Ecol. 22, 3216–3241.

Prykhozhij, S.V., Berman, J.N., 2018. Zebrafish knock-ins swim into the mainstream. Dis. Model. Mech. 11 (dmm037515).

Qiang, J., Zhu, X.W., He, J., Tao, Y.F., Bao, J.W., Zhu, J.H., Xu, P., 2020. miR-34a regulates the activity of HIF-1a and P53 signaling pathways by promoting GLUT1 in genetically improved farmed tilapia (GIFT, *Oreochromis niloticus*) under hypoxia stress. Front. Physiol. 11, 670.

Qin, Z., Li, Y., Su, B., Cheng, Q., Ye, Z., Perera, D.A., Fobes, M., Shang, M., Dunham, R.A., 2016. Editing of the luteinizing hormone gene to sterilize channel catfish, *Ictalurus punctatus*, using a modified zinc finger nuclease technology with electroporation. Marine Biotechnol. 18, 255–263.

Rey, O., Eizaguirre, C., Angers, B., Baltazar-Soares, M., Sagonas, K., Prunier, J.G., Blanchet, S., 2020. Linking epigenetics and biological conservation: towards a conservation epigenetics perspective. Funct. Ecol. 34, 414–427.

Richards, J.G., Semple, J.W., Bystriansky, J.S., Schulte, P.M., 2003. Na+/K+-ATPase α-isoform switching in gills of rainbow trout (*Oncorhynchus mykiss*) during salinity transfer. J. Exp. Biol. 206, 4475–4486.

Rode, N., Estoup, A., Bourguet, D., Courtier-Orgogozo, V., Débarre, F., 2019. Population management using gene drive: molecular design, models of spread dynamics and assessment of ecological risks. Conserv. Genet. 20, 671–690.

Roh, H., Kim, A., Kim, N., Lee, Y., Kim, D.-H., 2020. Multi-omics analysis provides novel insight into immuno-physiological pathways and development of thermal resistance in rainbow trout exposed to acute thermal stress. Int. J. Mol. Sci. 21, 9198.

Ryu, S.-M., Hur, J.W., Kim, K., 2019. Evolution of CRISPR towards accurate and efficient mammal genome engineering. BMB Rep. 52, 475–481.

Salem, M., Vallejo, R.L., Leeds, T.D., Palti, Y., Liu, S., Sabbagh, A., Rexroad III, C.E., Yao, J., 2012. RNA-seq identifies SNP markers for growth traits in rainbow trout. PLoS One 7, e36264.

Salem, M., Al-Tobasei, R., Ali, A., Lourenco, D., Gao, G., Palti, Y., Kenney, B., Leeds, T.D., 2018. Genome-wide association analysis with a 50K transcribed gene SNP-chip identifies QTL affecting muscle yield in rainbow trout. Front. Genet. 9, 387.

Sandoval-Castillo, J., Gates, K., Brauer, C.J., Smith, S., Bernatchez, L., Beheregaray, L.B., 2020. Adaptation of plasticity to projected maximum temperatures and across climatically defined bioregions. Proc. Natl. Acad. Sci. 117, 17112–17121.

Scholz, S., Mayer, I., 2008. Molecular biomarkers of endocrine disruption in small model fish. Mol. Cell. Endocrinol. 293, 57–70.

Schreck, C.B., Tort, L., 2016. The concept of stress in fish. In: Schreck, C.B., Tort, L., Farrell, A.P., Brauner, C.J. (Eds.), Fish Physiology (Vol. 35): Biology of Stress in Fish. Academic Press, pp. 1–34.

Schulte, P.M., 2014. What is environmental stress? Insights from fish living in a variable environment. J. Exp. Biol. 217, 23–34.

Schulte, P.M., Healy, T.M., 2022. Physiological diversity and its importance for fish conservation and management in the Anthropocene. In: Cooke, S.J., et al. (Eds.), Fish Physiology: Conservation Physiology for the Anthropocene—A Systems Approach Part A, Vol. 39A. Academic Press, pp. 435–477.

Shabalina, S.A., Ogurtsov, A.Y., Spiridonov, N.A., 2006. A periodic pattern of mRNA secondary structure created by the genetic code. Nucleic Acids Res. 34, 2428–2437.

Singh, P., Börger, C., More, H., Sturmbauer, C., 2017. The role of alternative splicing and differential gene expression in cichlid adaptive radiation. Genome Biol. Evol. 9, 2764–2781.

Smith, C.I.E., Zain, R., 2019. Therapeutic oligonucleotides: state of the art. Annu. Rev. Pharmacol. Toxicol. 59, 605–630.

Somero, G.N., 2010. The physiology of climate change: how potentials for acclimatization and genetic adaptation will determine 'winners' and 'losers'. J. Exp. Biol. 213, 912–920.

Stallcup, M.R., Poulard, C., 2020. Gene-specific actions of transcriptional coregulators facilitate physiological plasticity: evidence for a physiological coregulator code. Trends Biochem. Sci. 45, 497–510.

Straume, A.H., Kjærner-Semb, E., Skaftnesmo, K.O., Güralp, H., Lillico, S., Wargelius, A., Edvardsen, R.B., 2021. A refinement to gene editing in Atlantic salmon using asymmetrical oligonucleotide donors. bioRxiv. 2021.02.08.430296.

Sumpter, J.P., Jobling, S., 1995. Vitellogenesis as a biomarker for estrogenic contamination of the aquatic environment. Environ. Health Perspect. 103, 173–178.

Swirplies, F., Wuertz, S., Baßmann, B., Orban, A., Schäfer, N., Brunner, R.M., Hadlich, F., Goldammer, T., Rebl, A., 2019. Identification of molecular stress indicators in pikeperch *Sander lucioperca* correlating with rising water temperatures. Aquaculture 501, 260–271.

Tan, S., Wang, W., Tian, C., Niu, D., Zhou, T., Jin, Y., Yang, Y., Gao, D., Dunham, R., Liu, Z., 2019. Heat stress induced alternative splicing in catfish as determined by transcriptome analysis. Comp. Biochem. Phys. D. 29, 166–172.

Teffer, A.K., Bass, A.L., Miller, K.M., Patterson, D.A., Juanes, F., Hinch, S.G., 2018. Infections, fisheries capture, temperature, and host responses: multistressor influences on survival and behaviour of adult Chinook salmon. Can. J. Fish. Aquat. Sci. 75, 2069–2083.

Thorstensen, M.J., Baerwald, M.R., Jeffries, K.M., 2021. RNA sequencing describes both population structure and plasticity-selection dynamics in a non-model fish. BMC Genomics 22, 1–12.

Thresher, R.E., Jones, M., Drake, D.A.R., 2018. Evaluating active genetic options for the control of sea lamprey (*Petromyzon marinus*) in the Laurentian Great Lakes. Can. J. Fish. Aquat. Sci. 76, 1186–1202.

Tokarz, J., Möller, G., Hrabě de Angelis, M., Adamski, J., 2015. Steroids in teleost fishes: a functional point of view. Steroids 103, 123–144.

Tsai, H.Y., Hamilton, A., Tinch, A.E., Guy, D.R., Gharbi, K., Stear, M.J., Matika, O., Bishop, S.C., Houston, R.D., 2015. Genome wide association and genomic prediction for growth traits in juvenile farmed Atlantic salmon using a high density SNP array. BMC Genomics 16, 1–9.

Tymchuk, W.E., Beckman, B., Devlin, R.H., 2009. Altered expression of growth hormone/insulin-like growth factor I axis hormones in domesticated fish. Endocrinology 150, 1809–1816.

Vasil'ev, V.P., 2009. Mechanisms of polyploid evolution in fish: polyploidy in sturgeons. In: Carmona, R., Domezain, A., García-Gallego, M., Hernando, J.A., Rodríguez, F., Ruiz-Rejón, M. (Eds.), Biology, Conservation and Sustainable Development of Sturgeons. Springer, Dordrecht, pp. 97–117.

Veldhoen, N., Stevenson, M.R., Skirrow, R.C., Rieberger, K.J., van Aggelen, G., Meays, C.L., Helbing, C.C., 2013. Minimally invasive transcriptome profiling in salmon: detection of biological response in rainbow trout caudal fin following exposure to environmental chemical contaminants. Aquat. Toxicol. 142–143, 239–247.

Veldhoen, N., Beckerton, J.E., Mackenzie-Grieve, J., Stevenson, M.R., Truelson, R.L., Helbing, C.C., 2014. Development of a non-lethal method for evaluating transcriptomic endpoints in Arctic grayling (*Thymallus arcticus*). Ecotox. Environ. Safe. 105, 43–50.

Venney, C.J., Love, O.P., Drown, E.J., Heath, D.D., 2020. DNA methylation profiles suggest intergenerational transfer of maternal effects. Mol. Biol. Evol. 37, 540–548.

Venney, C.J., Sutherland, B.J., Beacham, T.D., Heath, D.D., 2021. Population differences in Chinook salmon (*Oncorhynchus tshawytscha*) DNA methylation: genetic drift and environmental factors. Ecol. Evol. 11, 6846–6861.

Vihervaara, A., Sistonen, L., 2014. HSF1 at a glance. J. Cell Sci. 127 (2), 261–266.

Wang, X., Kültz, D., 2017. Osmolality/salinity-responsive enhancers (OSREs) control induction of osmoprotective genes in euryhaline fish. Proc. Natl. Acad. Sci. 114, E2729–E2738.

Weber, D.N., Fields, A.T., Patterson III, W.F., Barnett, B.K., Hollenbeck, C.M., Portnoy, D.S., 2021. Novel epigenetic age estimation in wild-caught Gulf of Mexico reef fishes. Can. J. Fish. Aquat. Sci. 79, 1–5.

Wellband, K.W., Heath, D.D., 2017. Plasticity in gene transcription explains the differential performance of two invasive fish species. Evol. Appl. 10, 563–576.

Wen, X., Hu, Y., Zhang, X., Wei, X., Wang, T., Yin, S., 2019a. Integrated application of multi-omics provides insights into cold stress responses in pufferfish *Takifugu fasciatus*. BMC Genomics 20, 563.

Wen, X., Zhang, X., Hu, Y., Xu, J., Wang, T., Yin, S., 2019b. iTRAQ-based quantitative proteomic analysis of *Takifugu fasciatus* liver in response to low-temperature stress. J. Proteomics 201, 27–36.

Whitehead, A., Roach, J.L., Zhang, S., Galvez, F., 2011. Genomic mechanisms of evolved physiological plasticity in killifish distributed along an environmental salinity gradient. Proc. Natl. Acad. Sci. 108, 6193–6198.

Yada, T., 2007. Growth hormone and fish immune system. Gen. Comp. Endocrinol. 152, 353–358.

Yang, X., Gao, Y., Zhao, M., Wang, X., Zhou, H., Zhang, A., 2020. Cloning and identification of grass carp transcription factor HSF1 and its characterization involving the production of fish HSP70. Fish Physiol. Biochem. 46 (6), 1933–1945.

Yang, S., Zhao, T., Ma, A., Huang, Z., Yang, J., Yuan, C., et al., 2021. Heat stress-induced HSP90 expression is dependent on ERK and HSF1 activation in turbot (Scophthalmus maximus) kidney cells. Cell Stress Chaperon 26 (1), 173–185.

Yaşar, P., Ayaz, G., User, S.D., Güpür, G., Muyan, M., 2017. Molecular mechanism of estrogen–estrogen receptor signaling. Reprod. Med. Biol. 16, 4–20.

York, J.R., McCauley, D.W., 2020. Functional genetic analysis in a jawless vertebrate, the sea lamprey: insights into the developmental evolution of early vertebrates. J. Exp. Biol. 223, jeb.206433.

York, J.R., Thresher, R.E., McCauley, D.W., 2021. Applying functional genomics to the study of lamprey development and sea lamprey population control. J. Great Lakes Res. 47, S639–S649.

Zhou, C.Q., Zhou, P., Ren, Y.L., Cao, L.H., Wang, J.L., 2019. Physiological response and miRNA-mRNA interaction analysis in the head kidney of rainbow trout exposed to acute heat stress. J. Therm. Biol. 83, 134–141.

Zhu, B., Ge, W., 2018. Genome editing in fishes and their applications. Gen. Comp. Endocrinol. 257, 3–12.

Chapter 9

Physiological diversity and its importance for fish conservation and management in the Anthropocene

Patricia M. Schulte[a,*] and Timothy M. Healy[b]

[a]*Department of Zoology and Biodiversity Research Centre, The University of British Columbia, Vancouver, BC, Canada*
[b]*University of California-San Diego, Scripps Institution of Oceanography, La Jolla, CA, United States*
[*]*Corresponding author: e-mail: pschulte@zoology.ubc.ca*

Chapter Outline

Variation in physiological traits can influence the relative resilience of individuals, populations and species of fish to environmental change. Thus, understanding the extent, origins, and consequences of physiological diversity is a key challenge for fish conservation. Physiological diversity within a species can arise as a result of ontogeny, growth, and differences between sexes, and can also be generated by various types of phenotypic plasticity including acclimation, developmental plasticity and transgenerational plasticity. Genetic differences among individuals, populations, and species are also crucial drivers of physiological diversity, and interact with all of the other processes to result in the physiological diversity of living systems. Describing and

Fish Physiology, Vol. 39A. https://doi.org/10.1016/bs.fp.2022.04.009
435

preserving physiological diversity not only can help target management actions, but also has intrinsic value because high physiological diversity increases ecosystem resilience and influences the capacity for evolutionary adaptation of species to changing environments. However, quantifying the extent of physiological diversity among fishes and assessing its impacts on resilience to environmental change is a daunting task. Despite this, there have already been success stories for fish conservation to date, and taking physiological diversity into account will continue to be critically important for the management and conservation of fish in the Anthropocene.

1 Introduction

Variation is a fundamental property of living systems, and is apparent across levels of organization from individuals to ecosystems and from subspecies to kingdoms. In this chapter we discuss the importance of this variation for the conservation of fish in the Anthropocene, an era in which human activities at both local and global scales are rapidly reshaping aquatic ecosystems (Albert et al., 2021; Cahill et al., 2013; Luypaert et al., 2020; Su et al., 2021; Urban, 2015). We show how taking this variation into account will be vital for the development of strategies to protect fish biodiversity.

In their insightful book "*Physiological Diversity and Its Ecological Implications,*" Spicer and Gaston (1999) discuss the concept of variation in physiological traits, defining physiological diversity as the extent of variation in physiological traits within or among individuals, populations or species (Spicer and Gaston, 1999). Traits that influence organismal performance and their interaction with the environment are of great importance in conservation biology because these traits determine where species can live, how species interact with each other, and how ecosystems function (Cadotte et al., 2011). These characteristics are referred to as "functional traits" by ecologists, and they are often subdivided into categories such as life history traits, morphological traits, physiological traits, and behavioral traits (Gallagher et al., 2020; Lecocq et al., 2019; Martini et al., 2021; Violle et al., 2007). We consider physiological traits to be the subset of functional traits that allow organisms to adjust to their environments and regulate critical functions at the tissue, system, cellular and molecular levels (Westneat, 2001).

Traits may be measured at levels of biological organization from the molecule to the individual, as variation in traits at the molecular, biochemical and cellular levels inevitably integrates upwards to affect the performance of individuals (Fig. 1). Similarly, physiological diversity can be assessed at a variety of hierarchical levels with the extent of variation in physiological traits among individuals contributing to the physiological diversity of a population, the extent of variation in physiological traits among populations contributing to the physiological diversity of a species, and the extent of variation in physiological traits among species contributing to the physiological diversity of a community. In addition, because an individual's traits can vary over time,

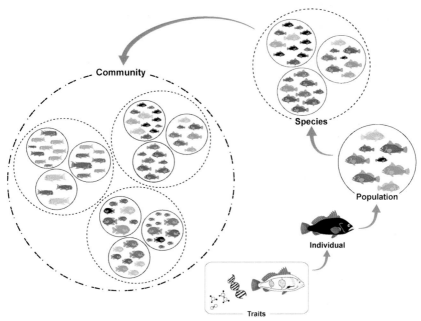

FIG. 1 Physiological diversity across levels of organization. Physiological diversity is the result of variation within and among individuals in physiological traits, which can be assessed from the molecular to the organismal level. The extent of variation in physiological traits among individuals (represented here with fish of different shades or sizes) contributes to the physiological diversity of a population. Variation within and among populations contributes to the physiological diversity of species. Variation within and among species contributes to the physiological diversity of a community. *Figure created by Sylvia Heredia.*

physiological diversity can also be generated by trait variation within an individual across its lifespan. Thus, we can think of the study of physiological diversity as occurring in two dimensions, one representing the level of biological organization of physiological traits from molecule to individual and the other representing the level of comparison of the extent of variation in those traits, from individual to ecosystem.

2 The causes of physiological diversity

Physiological diversity can arise through a variety of processes (Burggren, 2014; Spicer and Gaston, 1999) acting over a range of timescales. These processes include acute responses to environmental change, the effects of developmental programs across ontogeny and as a result of growth, various types of plasticity such as acclimation and epigenetic effects, and processes acting at the scale of generations to millennia such as evolutionary adaptation (Des Roches et al., 2018; Merilä and Hendry, 2014). Processes acting at different timescales can also interact. For example, the presence of plasticity due to acclimation and epigenetic effects can affect rates of evolutionary adaptation, in some cases enhancing the rates of adaptive evolution, and in other cases blunting evolutionary responses (Ashe et al., 2021; Fox et al., 2019). Similarly, the capacity for plasticity itself can evolve (Chevin and Hoffmann, 2017).

2.1 Ontogeny, growth, and sex

The early life stages of fish tend to be more sensitive to environmental change than other life stages (Dahlke et al., 2020; Finn and Kapoor, 2008; Rombough, 1997), which has important implications for predicting the responses of individuals, populations, and species to environmental change (Fig. 2). Embryonic stages of fishes appear to have much higher mass-specific metabolic rate and much lower aerobic scope than adults (Barrionuevo and Burggren, 1999). During early development, and particularly at later developmental stages, there is an increase in oxygen demand that is not fully compensated by increases in

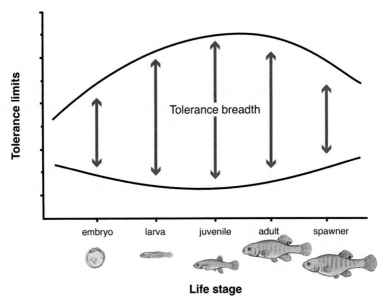

FIG. 2 The effects of life-stage on physiological diversity. The limits of tolerance of an abiotic stressor such as temperature can vary across life stages, with embryos and spawners often having the narrowest range of tolerated temperatures. Solid lines represent the maximum and minimum tolerated temperatures (e.g. CTMax and CTMin), and double-headed arrows represent tolerance breadth. *Based on data from Dahlke, F.T., Wohlrab, S., Butzin, M., Pörtner, H.-O., 2020. Thermal bottlenecks in the life cycle define climate vulnerability of fish. Science 369 (80), 65–70. https:// doi.org/10.1126/science.aaz3658; Rombough, P.J., 1997. The effects of temperature on embryonic and larval development. In: Wood, C.M., McDonald, D.G. (Eds.), Global Warming: Implications for Freshwater and Marine Fish. Cambridge University Press, Cambridge, pp. 177–223. https:// doi.org/10.3389/conf.fmars.2019.08.00073. Figure created by Rashpal Dhillon.*

oxygen supply, and this constraint has been linked to decreases in thermal tolerance in developing Chinook salmon (*Oncorhynchus tshawytscha*) (Martin et al., 2020). Similarly, the larval stages of fish have been shown to have a much lower aerobic scope (i.e., difference between maximal and resting rates of oxygen consumption) than do adults (Killen et al., 2007; Post and Lee, 1996), possibly due to the high energetic demands of growth (i.e., relatively high resting metabolic rates) at these early stages. The rate of development of fish embryos is also much more temperature sensitive than is the rate of growth at later developmental stages (Rombough, 1997). The high metabolic rate, low aerobic scope, low thermal tolerance, and high thermal sensitivity of embryos and larvae could represent a physiological constraint by limiting the energy available for important behaviors and physiological functions. Thus, taking into account this physiological diversity across life stages within individuals may be of key importance for forecasting the likely responses of populations or species to climate change.

Body size also has profound effects on a variety of physiological traits (Peters, 1983), and thus has the potential to contribute in important ways to the physiological diversity of individuals, populations, and species. The effect of body size on metabolic rate provides a particularly clear example of the importance of body size for physiological diversity (Beamish, 1964a; White et al., 2006). Metabolic rate does not increase isometrically with body size and as a result, metabolic rate per unit body mass is higher in smaller individuals than larger individuals, and smaller individuals require relatively more energy input than do larger ones to maintain their biological functions (Glazier, 2015). Interestingly, the value of this scaling relationship may vary depending on environmental temperature and the lifestyle of the species, which suggests that there are important interactions between body size, life history and the environment (Killen et al., 2010; Ohlberger et al., 2012). This observation is consistent with the fact that body size also has important impacts on how fish respond to environmental stressors. For example, in coho salmon (*Oncorhynchus kisutch*) body size influences both the response to a thermal challenge and the rate of recovery from exhaustive exercise (Clark et al., 2012), with smaller individuals being more resilient. However, the effects of body size on physiological traits are not always consistent across species. For example, the maximum tolerated temperature measured using an acute test (CTMax; critical thermal maximum) declines with body mass in some fish species, but not in others (McKenzie et al., 2020). The effects of body size and life stage on phenotypic traits are of particular importance because laboratory assessments are often conducted on fish at a particular body size or life stage (possibly determined by experimental convenience), and it can be difficult to generalize from these measurements to the impacts on performance across life stages.

Many animals display profound differences between sexes in life history, behavior and physiology, although this aspect of physiological diversity has received surprisingly little attention in fish (Hanson et al., 2008). The clearest evidence for sex-related differences is in the responses to environmental contaminants (Hanson et al., 2008). However, differences between sexes have also been detected in acclimation to hypoxia (Rees et al., 2001), standard metabolic rate (Beamish, 1964b), cardiovascular performance (Altimiras et al., 1996; Battiprolu et al., 2007; Rodnick et al., 2014; Sandblom et al., 2009), and responses to food deprivation (Harmon et al., 2011). Most of these studies have been conducted during breeding seasons, but differences in behavior and the cortisol-mediated stress response exist between sexes even in juvenile rainbow trout, *Oncorhynchus mykiss* (Øverli et al., 2006). The environment also has strong impacts on biological sex in fishes, with many species having environmental sex determination (Godwin et al., 2003; Ospina-Álvarez and Piferrer, 2008), and even in those species with genetic sex determination the environment can alter sex ratios (Ospina-Álvarez and Piferrer, 2008; Robertson et al., 2014; Shang et al., 2006).

Differences between sexes in physiological traits can have profound conservation importance. For example, in sockeye salmon (*Oncorhynchus nerka*), mortality during the spawning migration is much greater in females than in males (Jeffries et al., 2012; Martins et al., 2012), and these differences in survival are associated with differences in a variety of physiological traits including the ability to recover from stress (Eliason et al., 2020). Higher female mortality during the spawning migration is also observed in coho salmon (*Oncorhynchus kisutch*), and in this species the sexes differ in swimming performance, aerobic metabolism, and the ability to recover from exhaustive exercise (Kraskura et al., 2020).

2.2 Phenotypic plasticity

Phenotypic plasticity is defined as the ability of individual genotypes to produce different phenotypes when exposed to different environmental conditions (DeWitt et al., 1998; Whitman, 2009; Whitman and Agrawal, 2009). This plasticity can be an important contributor to physiological diversity among organisms in different microhabitats or across populations. Plasticity acts at a variety of timescales from acute physiological responses to effects that persist across generations (Schulte, 2011). The vast majority of studies of plasticity in fish have focussed on reversible plasticity as a result of acclimation to changing environmental conditions over the course of weeks or months, and generally assume that this plasticity is beneficial, whereas plasticity at other timescales has been less extensively studied. It is also important to distinguish between changes in phenotype due to regulated responses of the organism and phenotypic changes due to biophysical effects on chemical reactions within the organism (Havird et al., 2020). For example, temperature has direct effects on reaction rates and these effects integrate upwards to affect whole-animal properties such as metabolic rate. These acute phenotypic changes generally reflect the constraints of biophysical laws, rather than active regulatory processes induced by the organism, and are sometimes termed "passive plasticity" to distinguish them from the active plasticity that is observed through acclimation and epigenetic effects. Active plasticity, on the other hand, requires regulatory mechanisms induced by the organism, and is shaped by adaptive evolution. Distinguishing passive and active plasticity can be challenging because they may act together to shape trait values. Careful attention to experimental design is required to determine which type of plasticity is responsible for any observed difference in phenotype (Havird et al., 2020).

2.2.1 Reversible plasticity

There are various types of reversible plasticity that act at short timescales and contribute to variation in physiological traits within and among individuals. For example, acute responses such as changes in heart rate can be considered to be a form of rapidly acting plasticity. Although extremely rapid plastic effects have the potential to increase physiological diversity, they may also act to increase measurement "noise" that is not informative in a conservation context. For example, physiological trait values are highly dependent both on the measurement conditions (e.g., the rate of ramping in a thermal challenge trial (Moyano et al., 2017), or the design of a swimming challenge test (Farrell, 2008)) as well as the recent environmental experience of the fish being tested (Fangue et al., 2006; Griffiths and Alderdice, 1972). Thus, strict standards for measurements and holding conditions are essential to minimize the influence technical variation on estimates of physiological diversity. Similarly, the health of the fish being studied is an under-appreciated potential source of variation in studies in conservation physiology.

Physiological acclimation is an example of reversible plasticity acting at intermediate timescales. The literature on physiological acclimation in fish is vast, and we will not attempt to review it here. However, thermal acclimation provides a very clear example of the importance of adaptive plasticity in physiological traits and the role of temperature in promoting physiological diversity. The effects of thermal acclimation can be visualized using a Fry thermal tolerance polygon (Fig. 3) (Brett, 1956; Cossins and Bowler, 1987; Fry, 1971; Schulte, 2011), which clearly illustrates that physiological diversity across all acclimation temperatures is greater than the diversity that would be observed at any individual temperature. Furthermore, there is substantial variation in the capacity for acclimation among fish species (Morley et al., 2019; Seebacher et al., 2015), and even among populations or individuals within a species (Seebacher et al., 2012) for a variety of performance traits. As a result, variation in the capacity for acclimation itself contributes to physiological diversity. On the other hand, acclimation also has the potential to reduce observed physiological diversity. The phenomenon of "countergradient variation" provides an example of this effect in practice (Conover and

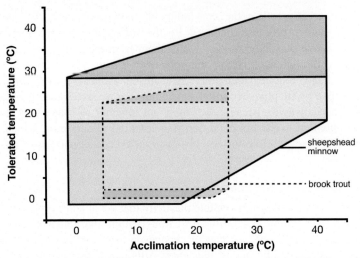

FIG. 3 Representative thermal tolerance polygons for two species of fish. Solid lines outline the thermal tolerance polygon for sheepshead minnow (*Cyprinodon variegatus*), and dotted lines outline the thermal tolerance polygon for brook trout (*Salvelinus fontinalis*). The central (light/yellow) zone represents the acclimation-independent, or intrinsic, tolerance, while the shaded (blue/pink) areas represent the additional physiological diversity that is acquired via acclimation. The eurythermal sheepshead minnow has a much broader range of tolerance and a greater extent of plasticity (a larger area of acquired tolerance) than does the more stenothermal brook trout. *Data from Beitinger, T.L., Bennett, W.A., 2000. Quantification of the role of acclimation temperature in temperature tolerance of fishes. Environ. Biol. Fishes 58, 277–288. Figure created by Rashpal Dhillon.*

Schultz, 1995). In multiple species of fish along the Atlantic coast of North America, fish from northern populations have higher metabolic rates, growth rates, and developmental rates than do fish from southern populations, when compared under common temperature conditions in the laboratory (Chung et al., 2018; Conover and Present, 1990). However, when tested at their normal (environmental) operating temperatures, these differences are removed. In essence, the genetic differentiation in these traits is opposite to the environmental influence on the traits, which minimizes the observed phenotypic differentiation between the populations in nature (Conover and Schultz, 1995).

Many physiological mechanisms have the potential to contribute to effects of acclimation, which involve a complex stream of regulators and their downstream targets; however, one common observation is that acclimation occurs coincident with substantial changes in gene expression (Logan and Somero, 2010; see Chapter 8, Volume 39A: Jeffries et al., 2022), including changes in the expression of different splice variants of mRNA through the process of alternative splicing (Healy and Schulte, 2019; see Chapter 8, Volume 39A: Jeffries et al., 2022). This sensitivity of the transcriptome to environmental conditions means changes in gene expression are potential indicators of the physiological health and performance of organisms, which is of particular utility for conservation of fishes. A clear example of the use of gene expression signatures in fish conservation comes from sockeye salmon (*Oncorhynchus nerka*). In this species, gene expression signatures across a panel of genes were associated with a 50% increase in mortality before reaching the spawning grounds in one stock (Miller et al., 2011). A variety of gene expression marker panels have been developed for use in conservation monitoring of wild populations of Pacific salmon (Patterson et al., 2016), including indicators of viral disease (Miller et al., 2017), thermal stress (Akbarzadeh et al., 2018), and smoltification status (Houde et al., 2019), and for many other fish species as well (see Chapter 8, Volume 39A: Jeffries et al., 2022).

2.2.2 Developmental plasticity

Developmental plasticity occurs when the environment during early development causes an alteration in a trait that persists across a long period, and often for the remainder of the animal's lifetime (Georgakopoulou et al., 2007; Scott and Johnston, 2012). However, early developmental effects can either be irreversible, partly reversible, or fully reversible at various timescales, although whether these more reversible forms of plasticity should be considered developmental plasticity is a matter of debate (Burggren, 2020). Regardless, differences in developmental conditions undoubtedly lead to variation among individuals later in life that in turn contributes to diversity in physiological traits in populations and communities. A variety of mechanisms can result in developmental plasticity, but they are often grouped under the general heading of epigenetic mechanisms and include processes such as DNA methylation, histone modification, and the actions of non-coding RNAs (Best et al., 2018; Metzger and Schulte, 2016a; see Chapter 8, Volume 39A: Jeffries et al., 2022).

Evidence is accumulating of significant developmental plasticity in physiological traits in response to a variety of stressors (Vagner et al., 2019) including hypoxia (Cadiz et al., 2018; Chapman et al., 2008; Cossins and Bowler, 1987; Mendez-Sanchez and Burggren, 2019; Robertson et al., 2014; Vanderplancke et al., 2015; Wood et al., 2017; Wu, 2009), ocean acidification (Rodriguez-Dominguez et al., 2018), and behavioral stress (Chouinard-Thuly et al., 2018). Below we highlight some of the work on developmental plasticity in response to temperature, as the most consistent and long-lasting effects of developmental exposure have been observed in response to this environmental variable (Vagner et al., 2019).

Temperature during early development affects a wide range of physiological traits in adult fish (Jonsson and Jonsson, 2014) including thermal tolerance (Schaefer and Ryan, 2006), swimming performance (Scott and Johnston, 2012; Seebacher et al., 2014; Sfakianakis et al., 2010), energy metabolism (Schnurr et al., 2014), growth (Finstad and Jonsson, 2012) and the capacity for thermal acclimation of metabolic enzymes (Schnurr et al., 2014; Scott and Johnston, 2012; Seebacher et al., 2014). Early developmental exposure to altered temperatures has been shown to alter DNA methylation and gene expression in adult stickleback (Metzger and Schulte, 2017), and similar effects have been detected in a tropical reef fish, *Acanthochromis polyacanthus* (Ryu et al., 2020). Taken together, these studies suggest that changes in DNA methylation are likely mediators of the persistent effects of altered temperatures during development. This hypothesis is supported by the observation in zebrafish that knockout of an enzyme that catalyzes the addition of methyl groups to DNA (DNA methyltransferase 3a; DNMT3a), alters the extent of developmental plasticity in response to temperature in swimming performance and muscle metabolic phenotype (Loughland et al., 2021).

2.2.3 Transgenerational plasticity

Transgenerational plasticity occurs when the environment experienced by the parent affects the phenotype of the offspring (Burton and Metcalfe, 2014; Donelson et al., 2018; Metzger and Schulte, 2016a; Salinas et al., 2013), although this word is sometimes reserved for plasticity that persists across multiple generations, while the word intergenerational plasticity is used to distinguish plasticity that occurs from parent to offspring (Perez and Lehner, 2019). Several potential mechanisms can result in intergenerational or transgenerational effects on physiological traits in offspring, including transmission of nutrients, hormones, or cellular structures (e.g., membranes, mitochondria) via the egg, as well as a variety of epigenetic mechanisms (Best et al., 2018; Metzger and Schulte, 2016a). In fishes, there is strong evidence for intergenerational plasticity as a result of both maternal (Green, 2008; Hellmann et al., 2020a; Metzger and Schulte, 2016b; Shama et al., 2014) and paternal effects (Crean and Bonduriansky, 2014; Hellmann et al., 2020b; Immler, 2018), Interestingly, this intergenerational plasticity can act differently between the sexes. For example, in threespine stickleback maternal stress exposure results in almost entirely opposite changes in gene expression in the brains of male and female offspring (Hellmann et al., 2020a; Metzger and Schulte, 2016b) and maternal and paternal stress induce different effects on brain gene expression in offspring (Hellmann et al., 2020a). There is also strong evidence for multigenerational plasticity in fishes (Beemelmanns and Roth, 2017; Burton and Metcalfe, 2014; Donelson et al., 2018; Hellmann et al., 2020b; Le Roy and Seebacher, 2018; Salinas and Munch, 2012; Shama et al., 2016; Shama and Wegner, 2014), although the patterns of multigenerational plasticity can be complex. For example, multigenerational transmission occurs from F1 males to F2 females and from F1 females to F2 males in stickleback (Hellmann et al., 2020b).

The intergenerational and transgenerational effects of temperature provide a particularly clear example of the relevance of physiological diversity to conservation in the Anthropocene, as these effects have the potential to allow fishes to adjust to a warming climate. For example, in sheepshead minnow, *Cyprinodon variegatus*, juvenile fish grow faster at higher temperatures if their parents were exposed to increased temperatures for at least 30 days (Salinas and Munch, 2012). Similarly, juvenile threespine stickleback reach a larger size when reared in the same thermal environment as their mothers (Shama et al., 2014). This intergenerational plasticity is related to differences in mitochondrial respiration rate in the heart tissue of the offspring, (Shama et al., 2016). This effect of maternal environment is particularly important because elevated summer water temperatures have detrimental effects on growth in the absence of prior maternal exposure to high temperature, which strongly suggests that this intergenerational plasticity could provide a buffer against climate warming. Transgenerational effects of temperature on growth persist across at least one additional generation in stickleback (i.e. there are detectable effects of grandparental thermal experience), although these effects are small compared to the effects of parental thermal exposure (Shama and Wegner, 2014). Additionally, in a tropical reef fish, *Acanthochromis polyacanthus*, both developmental and intergenerational plasticity in response to temperature affect metabolic rate and aerobic scope (Donelson et al., 2012, 2018), and these effects are associated with changes in the expression of a variety of metabolic genes (Veilleux et al., 2015).

2.3 Genetic variation

Much of the physiological diversity within and among populations and species is assumed to be the result of genetic variation across different hierarchical scales, but studies of the heritability of physiological traits suggest that genetic influences on most physiological traits are moderate (Sella and Barton, 2019). Nevertheless, there have been substantial successes in identifying the genes associated with variation in physiological traits in crops and livestock (Georges et al., 2019; Henry, 2020), including in fish in aquaculture (Laghari et al., 2014; You et al., 2020). In contrast, relatively limited progress has been made in connecting genetic variation with variation in physiological traits in natural populations (Bernatchez, 2016; Bernatchez et al., 2017; McMahon et al., 2014; Pearse, 2016; Shafer et al., 2015).

Physiological traits tend to have a continuous distribution over a range of numerical values and are, therefore, termed quantitative traits. Quantitative traits are typically determined by many genes and gene interactions, each with a relatively small effect. This polygenic nature of quantitative traits makes identifying the causal genetic variants challenging (Mackay, 2001; MacKay et al., 2009). This situation is likely to change rapidly over the next decade, as the decrease in the cost of whole-genome sequencing is enabling an increasing number of studies in this area (Amish et al., 2019; Brennan et al., 2018; Chen and Narum, 2021; Healy et al., 2018).

A recent metanalysis suggests that next-generation sequencing approaches have been used to address conservation questions in over 250 species of fish (Bernos et al., 2020). Although many of these studies involve the development of methods, many have used either genome scans (Vasemägi and Primmer, 2005) or landscape genomics to attempt to identify genetic variants that are likely to have arisen via natural selection. Studies in marine fishes have identified genomic regions associated with adaptation to abiotic environmental variables such as temperature and salinity (Dallaire et al., 2021; Dalongeville et al., 2018; Guo et al., 2016; Hoey and Pinsky, 2018; Li et al., 2019; Limborg et al., 2012; Pujolar et al., 2014; Therkildsen et al., 2013; Zhang et al., 2019; Zhao et al., 2018). In freshwater fishes, temperature and precipitation patterns have been established as major drivers of environmental adaptation, and multiple loci that may be under selection in response to these environmental variables have been identified (Bourret et al., 2013a,b; Brauer et al., 2016, 2018; Dallaire et al., 2021; Grummer et al., 2019; Harrisson et al., 2017; Hecht et al., 2015; Rougeux et al., 2019). In addition, studies in taxa that have invaded freshwater habitats have identified loci associated with freshwater adaptation (Defaveri and Merila, 2013; Dennenmoser et al., 2017; Flanagan et al., 2021; Guo et al., 2016; Raeymaekers et al., 2017). Taken together, these studies confirm that adaptation to environmental variables is often polygenic, and involves different genomic regions in different species. This last point is clearly demonstrated in a study of freshwater adaptation in threespine and ninespine stickleback (*Pungitius pungitius*) in which the loci associated with adaptation to freshwater are not the same across species despite their shared habitats (Raeymaekers et al., 2017).

In the context of understanding the genetic basis of physiological diversity, however, the primary limitation of genome scans and landscape genomic studies is that they are unable to make a direct connection between genetic differentiation and the functional traits that are the target of selection. To make these connections, several potential approaches are available, including quantitative trait locus (QTL) mapping and genome-wide association studies (GWAS), which use either experimental crosses or studies of outbred and natural populations, respectively (Darvasi, 1998; Vasemägi and Primmer, 2005). Studies that integrate multiple approaches including QTL mapping, GWAS, and landscape genomics have particular potential for identifying the genetic basis of variation in physiological traits that are relevant to conservation biology (Berg and Coop, 2014; Brennan et al., 2018; Vasemägi and Primmer, 2005). However, these types of combined studies are still the exception. Below we highlight a few examples in which the differing strengths of these approaches have been leveraged to make direct connections between genetic variation and variation in physiological traits that are relevant to adaptation to environmental change.

The vast majority of genomic studies in the context of fish conservation have been conducted in salmonids (Bernos et al., 2020), and these provide excellent examples of the utility of identifying the genetic basis of variation in functional traits in a conservation context. Many species of salmonids exhibit variation in life history traits such as the propensity to migrate, and this trait has long been known to at least in part have a genetic basis (Kendall et al., 2015; Quinn and Myers, 2004). One such migratory decision is the movement of juvenile salmonids from freshwater to saltwater, which requires that they first undergo the physiological process of smoltification (Stefansson et al., 2008). There is substantial variation in the propensity to smoltify in rainbow trout (*Oncorhynchus mykiss*) and both resident and anadromous (migratory steelhead) forms often co-occur within rivers. This creates a substantial challenge for conservation because anadromous steelhead are listed under the U.S. Endangered Species Act, whereas resident populations are not. Additionally, these two types of rainbow trout cannot be easily distinguished morphologically prior to migration, and they often display limited genetic differentiation at neutral molecular markers (McPhee et al., 2007). The first study to identify loci associated with various aspects of the smoltification phenotype used a QTL approach that involved laboratory breeding of two domesticated clonal lines of steelhead and rainbow trout (Nichols et al., 2008). A subsequent QTL study using wild migratory and resident rainbow trout identified 19 QTL each of which explained 4–13.63% of the variation in smoltification-related traits (Hecht et al., 2012). However, only five of these loci were shared across studies (Hecht et al., 2012; Nichols et al., 2008). These results illustrate some of the challenges with interpreting the results of QTL studies for physiological traits. First, these traits tend to be associated with multiple markers (i.e., polygenic); second, the markers associated with the traits can differ depending on the genetic background of the fish used in the crosses; and finally, QTL studies tend to identify fairly broad regions of chromosomes rather than associations with specific candidate genes. Subsequent studies using GWAS in wild populations of rainbow trout identified several hundred single nucleotide polymorphisms (SNPs) associated with smolt status, which indicated specific physiological traits such as sodium potassium ATPase activity (Hecht et al., 2013). This work provides multiple markers that can be used to distinguish resident and anadromous rainbow trout despite the low differentiation between populations across much of the genome, demonstrating the utility of physiological genomic approaches for the conservation of fishes.

The timing of the return migration to freshwater to breed also varies significantly within species in salmonids, and variation at the GREB1L locus, which encodes an estrogen receptor cofactor, has been associated with run timing in both rainbow trout (Hess et al., 2016; Narum et al., 2018) and chinook salmon (*Oncorhynchus tshawytscha*), (Prince et al., 2017). Although variation at this locus explains the bulk of variation in this trait, variation at multiple other SNPs is also important, emphasizing that even physiological traits that are influenced by genes of major effect have an underlying polygenic basis. In contrast, the GREB1L locus does not appear to be associated with migration timing in Atlantic salmon (*Salmo salar*) (Barson et al., 2015; Cauwelier et al., 2018; Johnston et al., 2014).

Combinations of methods have also been used to identify loci involved in variation in thermal tolerance in rainbow trout (Chen et al., 2018a,b; Chen and Narum, 2021). In general, these studies have detected polygenic associations, but a study using whole-genome resequencing (Chen and Narum, 2021) identified a very strong association with a position on chromosome 4 that contains a gene encoding a ceramide kinase (CERK), which is involved in sphingolipid metabolism. This locus was not previously identified using reduced-representation genotyping approaches, which illustrates the power of whole-genome sequencing to identify the loci involved in generating physiological diversity.

In Atlantic killifish (*Fundulus heteroclitus*), various combinations of GWAS, genome scans, and transcriptomics have been used to identify loci associated with variation in salinity tolerance (Brennan et al., 2018), thermal tolerance (Healy et al., 2018), hypoxia tolerance (Healy et al., 2018), routine metabolic rate (Healy et al., 2019) and pollution tolerance (Reid et al., 2016). Across all these traits, it is clear that their underlying genetic basis is polygenic, and that combinations of alleles across multiple markers can explain a substantial proportion of the observed phenotypic variation (often >50%). In the case of pollution tolerance, several populations that had independently adapted to highly polluted sites were compared (Reid et al., 2016), and although multiple loci associated with adaptation to pollution were identified at each site, only 0.5% of loci exhibited signals of selection across all polluted sites. The shared genomic regions had the strongest signals of selection, suggesting their importance in adaptation to polluted sites, and these regions contain genes coding for the aryl hydrocarbon receptor (AHR), a ligand-activated transcription factor that plays an important role in regulating immune responses, cell growth, and the cell cycle. In addition to binding to a variety of endogenous ligands, the AHR is best known for its ability to bind to and be activated by toxic chemicals such as polynuclear aromatic hydrocarbons (PAHs), and halogenated aromatic hydrocarbons such as polychlorinated biphenyls (PCBs), 2,3,7,8-tetrachlorodibenzo-p-dioxin (TCDD, "dioxin") and other dioxin-like compounds (DLCs), resulting in the formation of even more toxic by-products (Aluru et al., 2015). The importance of these loci in determining pollution sensitivity and resistance is supported by transcriptomic data with tolerant populations exhibiting reduced inducibility of genes regulated by the AHR. In the closely related Gulf killifish (*Fundulus grandis*), AHR genes that have reduced ability to bind or respond to dioxins and PCBs are also implicated in rapid adaptation to polluted sites, but in this case these alleles appear to have originated via introgression from *F. heteroclitus* (Oziolor et al., 2019). Whether adaptation as a result of introgression is likely to be a common mechanism is unclear, but there is no doubt that introgression within the *Fundulus* genus demonstrates the potential benefits of preserving physiological diversity, as variation in traits contributing to pollution tolerance among species was vital for the local persistence of the less-tolerant species following a change in environment.

The examples outlined here highlight that variation in physiological traits is often polygenic, and that our current methods for detecting this genetic basis are only able to detect the genetic variants with the strongest effects. The low sensitivity of current methods results in a potential risk in conservation biology of focussing on preserving the adaptive genetic variation that can be measured, while failing to protect important variation that we cannot yet identify, but which may represent the majority of loci that could be important for responses to environmental change (Fernandez-Fournier et al., 2021; Kardos et al., 2021; Pearse, 2016). In this context, a recent study by Oliveira et al. (2021) is potentially instructive. These authors showed that inter-individual variation in thermal tolerance in rainbow darters (*Etheostoma caeruleum*) is positively correlated with average heterozygosity across the genome (Oliveira et al., 2021), which suggests that genetic diversity itself may be an important mediator of physiological diversity.

3 The importance of physiological diversity

Physiological diversity has important implications for conservation biology both for intrinsic and utilitarian reasons. The primary intrinsic importance of physiological diversity is that it can influence community and ecosystem processes (Bolnick et al., 2011; Cadotte et al., 2011; Des Roches et al., 2018; McGill et al., 2006; Spicer and Gaston, 1999; Violle et al., 2007). There has been a long history of studying the impacts of diversity at the level of the species for community and ecosystem processes, but it is becoming clear that diversity at the population and individual level is also critically important (Bolnick et al., 2011; Des Roches et al., 2018, 2021; Mimura et al., 2017; Ward et al., 2016). Similarly, physiological diversity is key in determining the capacity of populations or species to adapt to a changing environment (Chown, 2012; Chown et al., 2010; Merilä and Hendry, 2014). Thus, characterizing and understanding the causes of physiological diversity within and among populations and species also has important practical value for identifying appropriate targets for conservation action.

3.1 Physiological diversity increases ecosystem resilience

High diversity acts as a buffer against the effects of environmental change and increases ecosystem performance across changing environments, as different species or populations within the assemblage respond to the change in different ways (Hooper et al., 2005; Loreau et al., 2021). These general ideas have been termed the "insurance hypothesis" and the "portfolio effect" because the underlying ecological theories have parallels with concepts from economics

(Figge, 2004; Loreau et al., 2021; Yachi and Loreau, 1999). The insurance hypothesis (Yachi and Loreau, 1999) postulates that if an ecosystem contains more diversity, it will be more likely to have multiple redundant members (either species, populations, or individuals) and a greater number of members that respond to environmental change in different ways. This high functional diversity will tend to enhance an ecosystem's ability to buffer perturbations. The concept of the portfolio effect reaches the same conclusion using a different mathematical approach by analogy to financial portfolios in which diversification across assets can stabilize returns and reduce risk over time (Schindler et al., 2010, 2015).

The majority of studies of the effects of diversity on ecosystem stability have emphasized diversity in terms of the number of species and their relative abundance, rather than focussing on the functional characteristics of the members of the assemblage (Fedor and Zvaríková, 2018; Hillebrand et al., 2018). However, there is an increasing consensus among ecologists that it is not only the number of species or their abundance that influences ecosystem function, but also the diversity of the functional traits among the organisms in that ecosystem (Aubin et al., 2016; Cadotte et al., 2011; Legras et al., 2018; Martini et al., 2021; McGill et al., 2006). This return to trait-based approaches in ecology highlights the importance of physiological perspectives for addressing key issues in conservation science (Madliger et al., 2021).

Studies on the impacts of functional diversity among species are best developed for plant functional diversity, but there has been increasing attention to functional diversity within fish assemblages (Villéger et al., 2017). Functional trait diversity among fish species can affect ecosystem biodiversity in contexts including human-caused habitat degradation (Henderson et al., 2020; Teichert et al., 2018; Teresa and Casatti, 2017), habitat restoration (Manfrin et al., 2019), dam construction (Arantes et al., 2019; Oliveira et al., 2018; Zhang et al., 2020), and invasive species (Colin et al., 2018; Milardi et al., 2019; Toussaint et al., 2018). There have also been attempts to predict the likely effects of climate change on the functional diversity of fish species (Buisson et al., 2013), and to categorize global scale patterns of the functional diversity of fishes (Stuart-Smith et al., 2013; Toussaint et al., 2016).

Although the majority of studies of the ecosystem impacts of functional diversity have focussed on diversity among species, variation among populations within a species can often have as strong or stronger effects on ecosystem function than physiological diversity among species (Bolnick et al., 2011; Des Roches et al., 2018, 2021). A clear example of the ecosystem-level effects of variation among populations within a species comes from alewives (*Alosa pseudoharengus*) in which differences in migratory strategies among populations drive changes in community structure among lakes (Huss et al., 2014). Lakes that harbor migratory vs non-migratory alewives differ in zooplankton community composition (Post et al., 2008), and these differences have cascading effects down the food web onto phytoplankton communities (Howeth et al., 2013) that also result in variation among lakes in the physiological traits of another fish species, the bluegill (*Lepomis macrochirus*) (Huss et al., 2014). Similar effects of intraspecific variation in fish functional traits among populations on ecosystem properties have been demonstrated for Trinidadian guppies (*Poecilia reticulata*) (Bassar et al., 2010), and threespine stickleback (*Gasterosteus aculeatus*) (Bassar et al., 2010; Des Roches et al., 2013; Harmon et al., 2009; Rudman and Schluter, 2016). Interestingly, the ecosystem effects of among-population variation in functional traits may also depend on environmental context, as the ecosystem effects of intraspecific variation in mosquitofish (*Gambusia affinis*) depend on the temperature at which they are tested (Fryxell and Palkovacs, 2017).

Functional diversity among individuals within a population also has important effects on ecosystem processes, although this level of variation is often ignored in studies of community diversity (Bolnick et al., 2011; Des Roches et al., 2018, 2021; Raffard et al., 2019). These effects have been clearly demonstrated in plants (e.g. Crutsinger et al., 2006), but similar effects have also been shown when genetic and phenotypic diversity is manipulated in fish (Raffard et al., 2021). Using *a priori* knowledge of the genetic and phenotypic diversity among populations of European minnows (*Phoxinus phoxinus*), Raffard et al. (2021) constructed artificial populations with varying levels of genetic diversity and variation in body size, and tested the effects of these differences on ecosystem function using mesocosm experiments. They detected effects of genetic and phenotypic diversity on zooplankton and benthic invertebrate diversity and abundance with cascading effects on community decomposition rate and the amount of algae (Raffard et al., 2021). These authors suggest that the genetic diversity of the population is a metric of cryptic (unmeasured) phenotypic differences among individuals in key traits such as behavior or metabolic rate. Other studies have identified the importance of diversity in fish body size on community and ecosystem processes (Lindmark et al., 2018), which may have effects on population, community and ecosystem responses to climate change (Lindmark et al., 2018, 2019; Ohlberger, 2013). These data convincingly demonstrate that the genetic and phenotypic diversity of fish species within an ecosystem has substantial effects on ecosystem processes.

Most studies of the ecosystem impact of functional diversity in fishes have focussed on easily measured morphological, trophic, or life history traits, in part because of the difficulty of measuring physiological traits and the lack of easily accessible and comparable data for physiological traits across species (Villéger et al., 2017). However, some studies have included tolerance traits in their assessments, demonstrating the importance of physiological diversity for ecosystem processes (de Oliveira et al., 2019; Milardi et al., 2019). In any case, many of the morphological and life history traits used in studies assessing the effects of functional trait variation on ecosystem processes are underlain by variation in physiological traits such as metabolic rate, swimming performance, aerobic scope, or tolerance to environmental stressors, but the link between diversity in these physiological traits and ecosystem processes is largely inferred indirectly.

The limited incorporation of data on physiological diversity into studies on the effects of functional diversity in fish on ecosystem processes is in contrast to the long history of incorporating important physiological traits such as drought tolerance and photosynthetic capacity into studies of functional diversity in plants (Cornelissen et al., 2003; Tilman et al., 1997). Studies of functional diversity in plants are facilitated by the existence of a curated Plant Trait Database (https:/www.try-db.org/TryWeb/Home.php) that provides a central repository for global-scale data on plant functional traits. In contrast, in fishes, there are limited central repositories for functional traits. Some functional trait data can be obtained from FishBase (https:/www.fishbase.de/home.htm), but much of the functional and physiological data in FishBase lacks vital information about the environmental context or methodological details of the measurements. This is an important issue because functional traits, and particularly physiological traits, exhibit substantial phenotypic plasticity and as a result the trait value depends on the conditions under which the measurements are made and even the methodology used to determine the trait value (Huey et al., 2012). There have been some attempts to develop trait-based databases for fishes focused on particular geographical areas (Côte et al., 2019; Frimpong and Angermeier, 2009; Mims et al., 2010; Schmidt-Kloiber and Hering, 2015) or for particular physiological traits (e.g. STOREFISH; STrategies Of REproduction in FISH) (Teletchea and Teletchea, 2020; Teletchea et al., 2007), and there are a number of broad taxonomic databases of specific physiological traits such as thermal tolerance that include significant representation of fishes (Bennett et al., 2018), but much work remains to develop consistent standards to allow interoperability of these databases. There have been increasing calls to develop standards for functional trait databases across the tree of life (Gallagher et al., 2020) and for fishes (Villéger et al., 2017), and recently there has been an attempt to develop a comprehensive database of fish functional traits in the TOFF (Traits OF Fish) database (Lecocq et al., 2019). The current instance of this database (as of April 2021) contains data for 241 traits for 248 species of fishes (https:/toff-project.univ-lorraine.fr/), but it currently includes only freshwater species, and given that there are more than 30,000 species of fish, the extent of the challenge becomes obvious. However, this database is an excellent step in the right direction, as it contains relevant environmental and methodological data. Yet, there is still progress to be made in ensuring that taxon-specific databases such as TOFF fully adhere to the principles of the Open Trait Network, which is attempting to standardize metadata structure and interoperability across different organismal databases (Gallagher et al., 2020).

The construction of a P_{crit} database (Rogers et al., 2016) highlights the challenges involved in making a comprehensive database of physiological traits (Rogers et al., 2016). P_{crit}, or the critical oxygen tension, is defined at the oxygen tension at which a fish is no longer able to maintain its metabolic rate. This database provides 331 measurements of P_{crit} for 151 species of fish (Rogers et al., 2016) and includes essential parameters that are necessary to interpret P_{crit} trait data, such as the acclimation and holding conditions of the fish, the life stage, size and sex of the fish in the sample, the type of P_{crit} measurement (e.g. P_{crit} for maximum metabolic rate, routine metabolic rate, or standard metabolic rate) and the method of P_{crit} estimation from the raw data, including the statistical approach used (Rogers et al., 2016), all of which have the potential to influence the trait value for P_{crit}. The plasticity of physiological traits, and their dependence on the specific conditions under which the measurements are made makes including them in ecological meta-analyses problematic. Many studies that have included physiological traits such as thermal tolerance have neglected these key details (Pinsky et al., 2019; Sunday et al., 2012, 2019).

3.2 Physiological diversity influences adaptation to environmental change

Although adaptive evolution is often considered to be a slow process that would be irrelevant in the context of rapid anthropogenic environmental change, there is growing realization that evolution can occur on timescales that are relevant to human-caused environmental change (Hendry and Kinnison, 1999; Merilä and Hendry, 2014; Merilä and Hoffmann, 2016). Therefore, understanding the factors that may promote or constrain adaptive evolution is of great importance to predicting the responses of fish in the Anthropocene. One important factor that may influence the rate of evolutionary adaptation is the existence of pre-existing ("standing") genetic variation (Barrett and Schluter, 2008). Adaptation from standing genetic variation is thought to lead to much faster evolution than waiting for *de novo* mutations, and thus adaptation from standing genetic variation may be the most relevant to species conservation in the Anthropocene. Consequently, there have been calls to monitor and conserve intraspecific variation, or the standing genetic variation within populations, to preserve this potential for evolutionary adaptation (Laikre, 2010; Mimura et al., 2017). This argument has also been made for fish populations and aquatic ecosystems (Kenchington et al., 2003; Smith and Chesser, 1981), as physiological diversity and the associated underlying genetic diversity within a population or among populations represents the critical raw material needed for rapid adaptation (Bolnick et al., 2011).

Despite widespread research on physiological traits, we know relatively little about the genetic basis of variation in the vast majority of these traits or their capacity for evolutionary adaptation. Furthermore, it is unclear whether most species have sufficient standing genetic variation to support the extent and rate of adaptation that may be necessary for species persistence in the face of global climate change (Hoffmann and Sgró, 2011). In a recent study of the potential for adaptive evolution in thermal tolerance Morgan et al (2020) used artificial selection in the laboratory to determine whether it was possible to alter CTMax in zebrafish (*Danio rerio*). In this experiment, evolution of increases in upper thermal tolerance was slow (Morgan et al., 2020), despite the fact that they started with a large outbred population of wild zebrafish and imposed strong selection. This suggests that this population of zebrafish harbors limited standing genetic variation associated with upper thermal tolerance, and that evolutionary adaptation of thermal tolerance is unlikely to allow zebrafish to cope with anthropogenic warming through increasing acute thermal limits. Whether this represents a general limitation across multiple fish species is less clear. For example, a recent macrophysiological investigation of intraspecific variation in CTMax found that freshwater tropical fish species have lower inter-individual variation in tolerance than temperate species in the northern hemisphere (Nati et al., 2020). This suggests that tropical species such as zebrafish may have a reduced capacity to respond to global climate change through evolutionary adaptation, but that ample variation may exist within temperate species. There are relatively few physiological traits for which we have sufficient information on the extent of intraspecific variation at either the phenotypic or genotypic level to allow predictions of the likelihood of adaptation. This represents a signficant knowledge gap that will be important for studies in conservation physiology going forward (Cooke et al., 2014; McKenzie et al., 2016, 2020; Ward et al., 2016).

3.3 Understanding physiological diversity can shape fish conservation and management

There are many success stories from conservation physiology that illustrate the importance of using a physiological perspective in conservation management decisions or for the development of tools for conservation biology (Cooke et al., 2012, 2013, 2014; McKenzie et al., 2016; Wikelski and Cooke, 2006). Less appreciated is the importance of taking physiological diversity into account in a conservation decision-making framework. Because species, populations, and individuals vary in their ability cope with environmental change, characterizing and understanding the extent of physiological diversity in these processes is key to predicting organismal responses to a changing environment and developing appropriate conservation and management actions. In this section we highlight selected case studies where understanding physiological diversity has been useful for conservation and management of fishes.

3.3.1 Management of Pacific salmon

Nowhere is the importance of assessing physiological diversity for conservation management in the Anthropocene more apparent than in the Pacific salmon (Cooke et al., 2012). Pacific salmon have a tendency to home to their natal streams, which results in reproductive isolation and gradual divergence among populations. The recognition of this differentiation has meant that Pacific salmon have been managed as a series of distinct stocks for many decades (Begg et al., 1999; Larkin, 1981). These genetic differences can be easily detected using molecular markers, which facilitates this management approach (Shaklee et al., 1999). As a result, management of the Pacific salmon fishery has a long history of consideration of genetic diversity, which is considered to be a proxy for unmeasured physiological diversity that might represent important adaptive differences among stocks. More recently, studies by conservation physiologists have provided important information that can feed into this decision-making process, including information on differences in thermal tolerance during the upriver migration and other metrics of performance among stocks (Eliason et al., 2011, 2020; Hinch et al., 2012; Steinhausen et al., 2008; Twardek et al., 2021), which have been incorporated into management decision-making (Patterson et al., 2016). Similarly, the strong effects of sex on differential mortality that have been detected during up-river migration are only beginning to be incorporated into management actions such as the regulation of flow from hydroelectric dams (Eliason et al., 2020; Hinch et al., 2021; Jeffries et al., 2012; Kraskura et al., 2020; Martins et al., 2012; Patterson et al., 2016; see Chapter 9, Volume 39B: Hinch et al., 2022), although this work has not yet been fully incorporated into decision-making frameworks. Studies using approaches from conservation and physiological genomics are rapidly identifying genetic markers of important physiological and life history traits in salmon (Hess et al., 2016; Prince et al., 2017), which provide the ability to distinguish stocks in a way that was not possible with neutral markers. These examples provide a clear case for the utility of considering physiological diversity for conservation decision making, and will facilitate the management of these socially, culturally, and economically important fish species. In turn, conserving this physiological diversity has important consequences by increasing ecosystem resilience and the stability of the fishery through processes such as the portfolio effect (Schindler et al., 2010).

3.3.2 Predicting responses to climate change

Assessing and predicting the likely vulnerability of populations, species or ecosystems to anthropogenic environmental change is a key challenge for the coming decades, and taking into account physiological diversity is a crucial component of these assessments (Huey et al., 2012; Madliger et al., 2021; Williams et al., 2008). For example, population and conservation ecologists interested in forecasting how particular species will respond to climate change are beginning to incorporate mechanistic links between the functional traits of organisms and their environments into the species distributions models that are used to forecast the likely effects of future climate change on species distribution and abundance (Buckley et al., 2010; Elith and Leathwick, 2009; Kearney and Porter, 2009; Melo-Merino et al., 2020). These individual species-level assessments can then be combined to make forecasts of the climate-change vulnerability of larger taxonomic groups (Chin et al., 2010), communities and ecosystems (Gaichas et al., 2014; Hare et al., 2016; Johnson and Welch, 2010), or economic sectors (Doubleday et al., 2013; Mathis et al., 2015; Pecl et al., 2014). The majority of these assessments are made using information about physiological diversity at the species level, but some studies incorporate information on physiological diversity among populations into species-level models (Bennett et al., 2019; Moran et al., 2016) and community-level risk assessments (Payne et al., 2020). In fishes, assessments of physiological diversity may be particularly important for the management of stocking programs, as different strains within a species may have different resilience to climate change. For example, strains used in the rainbow trout stocking program in British Columbia differ substantially in mortality following extreme heat stress events (Strowbridge et al., 2021); however, there are only modest differences among these strains in CTMax and hypoxia tolerance, indicating that different measures of resistance to stressors may lead to different conclusions about the relative resilience to climate change among strains. However, the observed differences in mortality among strains of rainbow trout demonstrate that there is physiological diversity that should be considered when making stocking decisions. Similarly, regional lineages of the endangered redside dace (*Clinostomus elongatus*) differ in CTMax suggesting that fish from the central lineage may be more suitable for re-introduction into warm urbanized habitats (Turko et al., 2021). However, neither of these studies have gone beyond recommendations to conservation action. Thus, there remains substantial uncertainty as to whether these types of laboratory studies will translate into improved success of stocking and re-introduction in nature.

3.3.3 Hatchery effects

Hatchery production is used for conservation or stock enhancement for hundreds of species of fish (Brown and Day, 2002), but it has long been known that hatchery fish are not as successful as wild fish, and differ from their wild relatives in a range of physiological traits (Araki et al., 2008; Sopinka et al., 2016; Twardek et al., 2021). Some of these differences are the result of unintentional domestication selection during hatchery production (Araki et al., 2008). Domestication in salmonids is known to have profound effects on morphology, physiology, and life history traits (Milla et al., 2021), and these genetic effects of domestication can be observed as early as after a single generation in the hatchery (Christie et al., 2012, 2016). Such rapid responses to selection must result from selection on standing genetic variation, which demonstrates the importance of taking into account genetic and physiological diversity in conservation management programs.

In addition to these genetic effects of hatchery rearing, there are also significant epigenetic effects on hatchery-reared fish, with hatchery effects on DNA methylation having been observed in species including Atlantic salmon (Rodriguez Barreto et al., 2019) coho salmon (Le Luyer et al., 2017), chinook salmon (Venney et al., 2021), and steelhead trout (Gavery et al., 2018, 2019). It is of particular interest that some of these epigenetic marks are present in the gametes of hatchery fish when they return to spawn, suggesting the possibility of transgenerational effects of hatchery practice, and the possibility of introgression of these epigenetic marks and any associated phenotypes into wild populations (Leitwein et al., 2021; Rodriguez Barreto et al., 2019). These observations provide a potent caution against ignoring physiological diversity and the processes that shape it, as they can have an important influence on attempts to conserve fish populations in the Anthropocene.

4 Conclusions and perspectives

When faced with environmental change that is beyond their capacity to cope, there are only a few fundamental ways in which fish can respond (Catullo et al., 2019; Williams et al., 2008). They may be able to use behavioral responses to find more favorable conditions. However, this can result in local extinction of the species, which can have devastating effects on fisheries-dependent communities (Cheung, 2018). In addition, for many species of fish, and particularly for those in freshwater habitats, migration is not an option because of habitat fragmentation—a problem that has been exacerbated by dam construction in many ecosystems (Fuller et al., 2015). Some species may be able to adjust their tolerance limits through phenotypic plasticity, and in these cases phenotypic plasticity has been suggested to act a buffer against environmental change. However, phenotypic plasticity has limits and once these limits are reached the only available option that could allow a species to cope with environmental change and avoid extinction is adaptation via natural selection. Species, populations, and individuals vary in their extent of tolerance, their capacity for phenotypic plasticity, and the extent of standing genetic variation that is available as a substrate for adaptive evolution. Therefore, characterizing the extent of variation in these processes is key to predicting organismal responses to a changing environment and developing appropriate conservation and management actions. This task is made challenging by the fact that phenotypic plasticity, genetic variation, and the effects of ontogeny, growth, and sex all interact to influence physiological diversity, but distinguishing these effects is important because the necessary management actions differ depending on the source of the observed variation in physiological traits. Despite these complexities, there are success stories that demonstrate the importance of taking physiological diversity into account for the management of fish in a changing environment. However, there is still much work to be done before the full potential of this approach can be realized for the conservation of fish in the Anthropocene.

References

Akbarzadeh, A., Günther, O.P., Houde, A.L., Li, S., Ming, T.J., Jeffries, K.M., Hinch, S.G., Miller, K.M., 2018. Developing specific molecular biomarkers for thermal stress in salmonids. BMC Genomics 19. https://doi.org/10.1186/s12864-018-5108-9

Albert, J.S., Destouni, G., Duke-Sylvester, S.M., Magurran, A.E., Oberdorff, T., Reis, R.E., Winemiller, K.O., Ripple, W.J., 2021. Scientists' warning to humanity on the freshwater biodiversity crisis. Ambio 50, 85–94. https://doi.org/10.1007/s13280-020-01318-8.

Altimiras, J., Johnstone, A.D.F., Lucas, M.C., Priede, I.G., 1996. Sex differences in the heart rate variability spectrum of free-swimming atlantic salmon (*Salmo salar* L.) during the spawning season. Physiol. Zool. 69, 770–784. https://doi.org/10.1086/physzool.69.4.30164229.

Aluru, N., Karchner, S.I., Franks, D.G., Nacci, D., Champlin, D., Hahn, M.E., 2015. Targeted mutagenesis of aryl hydrocarbon receptor 2a and 2b genes in Atlantic killifish (*Fundulus heteroclitus*). Aquat. Toxicol. 158, 192–201. https://doi.org/10.1016/j.aquatox.2014.11.016.

Amish, S.J., Ali, O., Peacock, M., Miller, M., Robinson, M., Smith, S., Luikart, G., Neville, H., 2019. Assessing thermal adaptation using family-based association and FST outlier tests in a threatened trout species. Mol. Ecol. 28, 2573–2593. https://doi.org/10.1111/mec.15100.

Araki, H., Berejikian, B.A., Ford, M.J., Blouin, M.S., 2008. Fitness of hatchery-reared salmonids in the wild. Evol. Appl. 1, 342–355. https://doi.org/10.1111/j.1752-4571.2008.00026.x.

Arantes, C.C., Fitzgerald, D.B., Hoeinghaus, D.J., Winemiller, K.O., 2019. Impacts of hydroelectric dams on fishes and fisheries in tropical rivers through the lens of functional traits. Curr. Opin. Environ. Sustain. 37, 28–40. https://doi.org/10.1016/j.cosust.2019.04.009.

Ashe, A., Colot, V., Oldroyd, B.P., 2021. How does epigenetics influence the course of evolution? Philos. Trans. R. Soc. B Biol. Sci. 376. https://doi.org/10.1098/rstb.2020.0111.

Aubin, I., Munson, A.D., Cardou, F., Burton, P.J., Isabel, N., Pedlar, J.H., Paquette, A., Taylor, A.R., Delagrange, S., Kebli, H., Messier, C., Shipley, B., Valladares, F., Kattge, J., Boisvert-Marsh, L., McKenney, D., 2016. Traits to stay, traits to move: a review of functional traits to assess sensitivity and adaptive capacity of temperate and boreal trees to climate change. Environ. Rev. 24, 164–186. https://doi.org/10.1139/er-2015-0072.

Barrett, R.D.H., Schluter, D., 2008. Adaptation from standing genetic variation. Trends Ecol. Evol. 23, 38–44. https://doi.org/10.1016/j.tree.2007.09.008.

Barrionuevo, W.R., Burggren, W.W., 1999. O_2 consumption and heart rate in developing zebrafish (*Danio rerio*): Influence of temperature and ambient O_2. Am. J. Physiol. Regul. Integr. Comp. Physiol. 276, 505–513. https://doi.org/10.1152/ajpregu.1999.276.2.r505.

Barson, N.J., Aykanat, T., Hindar, K., Baranski, M., Bolstad, G.H., Fiske, P., Jacq, C., Jensen, A.J., Johnston, S.E., Karlsson, S., Kent, M., Moen, T., Niemelä, E., Nome, T., Næsje, T.F., Orell, P., Romakkaniemi, A., Sægrov, H., Urdal, K., Erkinaro, J., Lien, S., Primmer, C.R., 2015. Sex-dependent dominance at a single locus maintains variation in age at maturity in salmon. Nature 528, 405–408. https://doi.org/10.1038/nature16062.

Bassar, R.D., Marshall, M.C., López-Sepulcre, A., Zandonà, E., Auer, S.K., Travis, J., Pringle, C.M., Flecker, A.S., Thomas, S.A., Fraser, D.F., Reznick, D.N., 2010. Local adaptation in Trinidadian guppies alters ecosystem processes. Proc. Natl. Acad. Sci. U. S. A. 107, 3616–3621. https://doi.org/10.1073/pnas.0908023107.

Battiprolu, P.K., Harmon, K.J., Rodnick, K.J., 2007. Sex differences in energy metabolism and performance of teleost cardiac tissue. Am. J. Physiol. Regul. Integr. Comp. Physiol. 292, 827–836. https://doi.org/10.1152/ajpregu.00379.2006.

Beamish, F.W.H., 1964a. Respiration of fishes with special emphasis on standard oxygen consumption. II. Influence of weight and temperature on respiration of several species. Can. J. Zool. 42, 177–188.

Beamish, F.W.H., 1964b. Seasonal changes in the of oxygen consumption of fishes. Can. J. Zool. 42, 189–194.

Beemelmanns, A., Roth, O., 2017. Grandparental immune priming in the pipefish *Syngnathus typhle*. BMC Evol. Biol. 17, 1–15. https://doi.org/10.1186/s12862-017-0885-3.

Begg, G.A., Friedland, K.D., Pearce, J.B., 1999. Stock identification and its role in stock assessment and fisheries management: an overview. Fish. Res. 43, 1–8. https://doi.org/10.1016/S0165-7836(99)00062-4.

Bennett, J.M., Calosi, P., Clusella-Trullas, S., Martínez, B., Sunday, J., Algar, A.C., Araújo, M.B., Hawkins, B.A., Keith, S., Kühn, I., Rahbek, C., Rodríguez, L., Singer, A., Villalobos, F., Ángel Olalla-Tárraga, M., Morales-Castilla, I., 2018. GlobTherm, a global database on thermal tolerances for aquatic and terrestrial organisms. Sci. Data. https://doi.org/10.1038/sdata.2018.22.

Bennett, S., Duarte, C.M., Marbà, N., Wernberg, T., 2019. Integrating within-species variation in thermal physiology into climate change ecology. Philos. Trans. R. Soc. B Biol. Sci. 374. https://doi.org/10.1098/rstb.2018.0550.

Berg, J.J., Coop, G., 2014. A population genetic signal of polygenic adaptation. PLoS Genet. 10, e1004412. https://doi.org/10.1371/journal.pgen.1004412.

Bernatchez, L., 2016. On the maintenance of genetic variation and adaptation to environmental change: considerations from population genomics in fishes. J. Fish Biol. 89, 2519–2556. https://doi.org/10.1111/jfb.13145.

Bernatchez, L., Wellenreuther, M., Araneda, C., Ashton, D.T., Barth, J.M.I., Beacham, T.D., Maes, G.E., Martinsohn, J.T., Miller, K.M., Naish, K.A., Ovenden, J.R., Primmer, C.R., Young Suk, H., Therkildsen, N.O., Withler, R.E., 2017. Harnessing the power of genomics to secure the future of seafood. Trends Ecol. Evol. 32, 665–680. https://doi.org/10.1016/j.tree.2017.06.010.

Bernos, T.A., Jeffries, K.M., Mandrak, N.E., 2020. Linking genomics and fish conservation decision making: a review. Rev. Fish Biol. Fish. 30, 587–604. https://doi.org/10.1007/s11160-020-09618-8.

Best, C., Ikert, H., Kostyniuk, D.J., Craig, P.M., Navarro-Martin, L., Marandel, L., Mennigen, J.A., 2018. Epigenetics in teleost fish: From molecular mechanisms to physiological phenotypes. Comp. Biochem. Physiol. Part B Biochem. Mol. Biol. 224, 210–244. https://doi.org/10.1016/j.cbpb.2018.01.006.

Bolnick, D.I., Amarasekare, P., Araújo, M.S., Bürger, R., Levine, J.M., Novak, M., Rudolf, V.H.W., Schreiber, S.J., Urban, M.C., Vasseur, D.A., 2011. Why intraspecific trait variation matters in community ecology. Trends Ecol. Evol. 26, 183–192. https://doi.org/10.1016/j.tree.2011.01.009.

Bourret, V., Dionne, M., Kent, M.P., Lien, S., Bernatchez, L., 2013a. Landscape genomics in Atlantic salmon (*Salmo salar*): searching for gene-environment interactions driving local adaptation. Evolution (N.Y.) 67, 3469–3487. https://doi.org/10.1111/evo.12139.

Bourret, V., Kent, M.P., Primmer, C.R., Vasemägi, A., Karlsson, S., Hindar, K., McGinnity, P., Verspoor, E., Bernatchez, L., Lien, S., 2013b. SNP-array reveals genome-wide patterns of geographical and potential adaptive divergence across the natural range of Atlantic salmon (*Salmo salar*). Mol. Ecol. 22, 532–551. https://doi.org/10.1111/mec.12003.

Brauer, C.J., Hammer, M.P., Beheregaray, L.B., 2016. Riverscape genomics of a threatened fish across a hydroclimatically heterogeneous river basin. Mol. Ecol. 25, 5093–5113. https://doi.org/10.1111/mec.13830.

Brauer, C.J., Unmack, P.J., Smith, S., Bernatchez, L., Beheregaray, L.B., 2018. On the roles of landscape heterogeneity and environmental variation in determining population genomic structure in a dendritic system. Mol. Ecol. 27, 3484–3497. https://doi.org/10.1111/mec.14808.

Brennan, R.S., Healy, T.M., Bryant, H.J., Van La, M., Schulte, P.M., Whitehead, A., 2018. Integrative population and physiological genomics reveals mechanisms of adaptation in killifish. Mol. Biol. Evol. 35, 2639–2653. https://doi.org/10.1093/molbev/msy154.

Brett, J.R., 1956. Some principles in the thermal requirements of fishes. Q. Rev. Biol. 31, 75–87. https://doi.org/10.1086/401257,

Brown, C., Day, R.L., 2002. The future of stock enhancements: Lessons for hatchery practice from conservation biology. Fish Fish. 3, 79–94. https://doi.org/10.1046/j.1467-2979.2002.00077.x.

Buckley, L.B., Urban, M.C., Angilletta, M.J., Crozier, L.G., Rissler, L.J., Sears, M.W., 2010. Can mechanism inform species' distribution models? Ecol. Lett. 13, 1041–1054. https://doi.org/10.1111/j.1461-0248.2010.01479.x.

Buisson, L., Grenouillet, G., Villéger, S., Canal, J., Laffaille, P., 2013. Toward a loss of functional diversity in stream fish assemblages under climate change. Glob. Chang. Biol. 19, 387–400. https://doi.org/10.1111/gcb.12056.

Burggren, W.W., 2014. Epigenetics as a source of variation in comparative animal physiology—or—Lamarck is lookin' pretty good these days. J. Exp. Biol. 217, 682–689. https://doi.org/10.1242/jeb.086132.

Burggren, W.W., 2020. Phenotypic switching resulting from developmental plasticity: fixed or reversible? Front. Physiol. 10, 1–13. https://doi.org/10.3389/fphys.2019.01634.

Burton, T., Metcalfe, N.B., 2014. Can environmental conditions experienced in early life influence future generations? Proc. R. Soc. B Biol. Sci. 281. https://doi.org/10.1098/rspb.2014.0311.

Cadiz, L., Ernande, B., Quazuguel, P., Servili, A., Zambonino-Infante, J.L., Mazurais, D., 2018. Moderate hypoxia but not warming conditions at larval stage induces adverse carry-over effects on hypoxia tolerance of European sea bass (*Dicentrarchus labrax*) juveniles. Mar. Environ. Res. 138, 28–35. https://doi.org/10.1016/j.marenvres.2018.03.011.

Cadotte, M.W., Carscadden, K., Mirotchnick, N., 2011. Beyond species: functional diversity and the maintenance of ecological processes and services. J. Appl. Ecol. 48, 1079–1087. https://doi.org/10.1111/j.1365-2664.2011.02048.x.

Cahill, A.E., Aiello-Lammens, M.E., Fisher-Reid, M.C., Hua, X., Karanewsky, C.J., Ryu, H.Y., Sbeglia, G.C., Spagnolo, F., Waldron, J.B., Warsi, O., Wiens, J.J., 2013. How does climate change cause extinction? Proc. R. Soc. Biol. Sci. 280, 1–10.

Catullo, R.A., Llewelyn, J., Phillips, B.L., Moritz, C.C., 2019. The potential for rapid evolution under anthropogenic climate change. Curr. Biol. 29, R996–R1007. https://doi.org/10.1016/j.cub.2019.08.028.

Cauwelier, E., Gilbey, J., Sampayo, J., Stradmeyer, L., Middlemas, S.J., 2018. Identification of a single genomic region associated with seasonal river return timing in adult Scottish atlantic salmon (*Salmo salar*), using a genome-wide association study. Can. J. Fish. Aquat. Sci. 75, 1427–1435. https://doi.org/10.1139/cjfas-2017-0293.

Chapman, L., Albert, J., Galis, F., 2008. Developmental plasticity, genetic differentiation, and hypoxia-induced trade-offs in an African cichlid fish. Open Evol. J. 2, 75–88. https://doi.org/10.2174/1874404400802010075.

Chen, Z., Narum, S.R., 2021. Whole genome resequencing reveals genomic regions associated with thermal adaptation in redband trout. Mol. Ecol. 30, 162–174. https://doi.org/10.1111/mec.15717.

Chen, Z., Farrell, A.P., Matala, A., Hoffman, N., Narum, S.R., 2018a. Physiological and genomic signatures of evolutionary thermal adaptation in redband trout from extreme climates. Evol. Appl. 11, 1686–1699. https://doi.org/10.1111/eva.12672.

Chen, Z., Farrell, A.P., Matala, A., Narum, S.R., 2018b. Mechanisms of thermal adaptation and evolutionary potential of conspecific populations to changing environments. Mol. Ecol. 27, 659–674. https://doi.org/10.1111/mec.14475.

Cheung, W.W.L., 2018. The future of fishes and fisheries in the changing oceans. J. Fish Biol. 92, 790–803. https://doi.org/10.1111/jfb.13558.

Chevin, L.M., Hoffmann, A.A., 2017. Evolution of phenotypic plasticity in extreme environments. Philos. Trans. R. Soc. B Biol. Sci. 372. https://doi.org/10.1098/rstb.2016.0138.

Chin, A., Kyne, P.M., Walker, T.I., McAuley, R.B., 2010. An integrated risk assessment for climate change: analysing the vulnerability of sharks and rays on Australia's Great Barrier Reef. Glob. Chang. Biol. 16, 1936–1953. https://doi.org/10.1111/j.1365-2486.2009.02128.x.

Chouinard-Thuly, L., Reddon, A.R., Leris, I., Earley, R.L., Reader, S.M., 2018. Developmental plasticity of the stress response in female but not in male guppies. R. Soc. Open Sci. 5. https://doi.org/10.1098/rsos.172268.

Chown, S.L., 2012. Trait-based approaches to conservation physiology: forecasting environmental change risks from the bottom up. Philos. Trans. R. Soc. B Biol. Sci. 367, 1615–1627. https://doi.org/10.1098/rstb.2011.0422.

Chown, S., Hoffmann, A., Kristensen, T., Angilletta, M., Stenseth, N., Pertoldi, C., 2010. Adapting to climate change: a perspective from evolutionary physiology. Clim. Res. 43, 3–15. https://doi.org/10.3354/cr00879.

Christie, M.R., Marine, M.L., French, R.A., Blouin, M.S., 2012. Genetic adaptation to captivity can occur in a single generation. Proc. Natl. Acad. Sci. U. S. A. 109, 238–242. https://doi.org/10.1073/pnas.1111073109.

Christie, M.R., Marine, M.L., Fox, S.E., French, R.A., Blouin, M.S., 2016. A single generation of domestication heritably alters the expression of hundreds of genes. Nat. Commun. 7, 1–6. https://doi.org/10.1038/ncomms10676.

Chung, D.J., Healy, T.M., McKenzie, J.L., Chicco, A.J., Sparagna, G.C., Schulte, P.M., 2018. Mitochondria, temperature, and the pace of life. Integr. Comp. Biol. 58, 578–590. https://doi.org/10.1093/icb/icy013.

Clark, T.D., Donaldson, M.R., Pieperhoff, S., Drenner, S.M., Lotto, A., Cooke, S.J., Hinch, S.G., Patterson, D.A., Farrell, A.P., 2012. Physiological benefits of being small in a changing world: responses of Coho salmon (*Oncorhynchus kisutch*) to an acute thermal challenge and a simulated capture event. PLoS One 7, e39079. https://doi.org/10.1371/journal.pone.0039079.

Colin, N., Villéger, S., Wilkes, M., de Sostoa, A., Maceda-Veiga, A., 2018. Functional diversity measures revealed impacts of non-native species and habitat degradation on species-poor freshwater fish assemblages. Sci. Total Environ. 625, 861–871. https://doi.org/10.1016/j.scitotenv.2017.12.316.

Conover, D.O., Present, T.M.C., 1990. Countergradient variation in growth rate: compensation for length of the growing season among Atlantic Silversides from different latitudes. Oecologia 83, 316–324.

Conover, D.O., Schultz, E.T., 1995. Phenotypic similarity and the evolutionary significance of countergradient variation. Trends Ecol. Evol. 10, 248–252. https://doi.org/10.1016/S0169-5347(00)89081-3.

Cooke, S.J., Hinch, S.G., Donaldson, M.R., Clark, T.D., Eliason, E.J., Crossin, G.T., Raby, G.D., Jeffries, K.M., Lapointe, M., Miller, K., Patterson, D.A., Farrell, A.P., 2012. Conservation physiology in practice: How physiological knowledge has improved our ability to sustainably manage Pacific salmon during up-river migration. Philos. Trans. R. Soc. B Biol. Sci. 367, 1757–1769. https://doi.org/10.1098/rstb.2012.0022.

Cooke, S.J., Sack, L., Franklin, C.E., Farrell, A.P., Beardall, J., Wikelski, M., Chown, S.L., 2013. What is conservation physiology? Perspectives on an increasingly integrated and essential science. Conserv. Physiol. 1, 1–23. https://doi.org/10.1093/conphys/cot001.

Cooke, S.J., Killen, S.S., Metcalfe, J.D., McKenzie, D.J., Mouillot, D., Jørgensen, C., Peck, M.A., 2014. Conservation physiology across scales: Insights from the marine realm. Conserv. Physiol. 2, 1–15. https://doi.org/10.1093/conphys/cou024.

Cornelissen, J.H.C., Lavorel, S., Garnier, E., Díaz, S., Buchmann, N., Gurvich, D.E., Reich, P.B., Ter Steege, H., Morgan, H.D., Van Der Heijden, M.G.A., Pausas, J.G., Poorter, H., 2003. A handbook of protocols for standardised and easy measurement of plant functional traits worldwide. Aust. J. Bot. 51, 335–380. https://doi.org/10.1071/BT02124.

Cossins, A.R., Bowler, K., 1987. Temperature Biology of Animals. Chapman and Hall, London. https://doi.org/10.1017/CBO9781107415324.004.

Côte, J., Kuczynski, L., Grenouillet, G., 2019. Spatial patterns and determinants of trait dispersion in freshwater fish assemblages across Europe. Glob. Ecol. Biogeogr. 28, 826–838. https://doi.org/10.1111/geb.12896.

Crean, A.J., Bonduriansky, R., 2014. What is a paternal effect? Trends Ecol. Evol. 29, 554–559. https://doi.org/10.1016/j.tree.2014.07.009.

Crutsinger, G.M., Collins, M.D., Fordyce, J.A., Gompert, Z., Nice, C.C., Sanders, N.J., 2006. Plant genotypic diversity predicts community structure and governs an ecosystem process. Science 313, 966–968.

Dahlke, F.T., Wohlrab, S., Butzin, M., Pörtner, H.-O., 2020. Thermal bottlenecks in the life cycle define climate vulnerability of fish. Science 369 (80), 65–70. https://doi.org/10.1126/science.aaz3658.

Dallaire, X., Normandeau, É., Mainguy, J., Tremblay, J., Bernatchez, L., Moore, J., 2021. Genomic data support management of anadromous Arctic Char fisheries in Nunavik by highlighting neutral and putatively adaptive genetic variation. Evol. Appl. 1–18. https://doi.org/10.1111/eva.13248.

Dalongeville, A., Benestan, L., Mouillot, D., Lobreaux, S., Manel, S., 2018. Combining six genome scan methods to detect candidate genes to salinity in the Mediterranean striped red mullet (*Mullus surmuletus*). BMC Genomics 19, 1–13. https://doi.org/10.1186/s12864-018-4579-z.

Darvasi, A., 1998. Experimental strategies for the genetic dissection of complex traits in animal models. Nat. Genet. 18, 19–24. https://doi.org/10.1038/ng0198-19.

de Oliveira, A.G., Bailly, D., Cassemiro, F.A.S., do Couto, E.V., Bond, N., Gilligan, D., Rangel, T.F., Agostinho, A.A., Kennard, M.J., 2019. Coupling environment and physiology to predict effects of climate change on the taxonomic and functional diversity of fish assemblages in the Murray-Darling Basin, Australia. PLoS One 14, 1–21. https://doi.org/10.1371/journal.pone.0225128.

Defaveri, J., Merila, J., 2013. Evidence for adaptive phenotypic differentiation in Baltic Sea sticklebacks. J. Evol. Biol. 26, 1700–1715. https://doi.org/10.1111/jeb.12168.

Dennenmoser, S., Vamosi, S.M., Nolte, A.W., Rogers, S.M., 2017. Adaptive genomic divergence under high gene flow between freshwater and brackish-water ecotypes of prickly sculpin (*Cottus asper*) revealed by Pool-Seq. Mol. Ecol. 26, 25–42. https://doi.org/10.1111/mec.13805.

Des Roches, S., Shurin, J.B., Schluter, D., Harmon, L.J., 2013. Ecological and evolutionary effects of stickleback on community structure. PLoS One 8. https://doi.org/10.1371/journal.pone.0059644.

Des Roches, S., Post, D.M., Turley, N.E., Bailey, J.K., Hendry, A.P., Kinnison, M.T., Schweitzer, J.A., Palkovacs, E.P., 2018. The ecological importance of intraspecific variation. Nat. Ecol. Evol. 2, 57–64. https://doi.org/10.1038/s41559-017-0402-5.

Des Roches, S., Pendleton, L.H., Shapiro, B., Palkovacs, E.P., 2021. Conserving intraspecific variation for nature's contributions to people. Nat. Ecol. Evol. 5, 574–582. https://doi.org/10.1038/s41559-021-01403-5.

DeWitt, T.J., Sih, A., Wilson, D.S., 1998. Costs and limits of phenotypic plasticity. Trends Ecol. Evol. 13, 77–81. https://doi.org/10.1016/S0169-5347(97)01274-3.

Donelson, J.M., Munday, P.L., McCormick, M.I., Pitcher, C.R., 2012. Rapid transgenerational acclimation of a tropical reef fish to climate change. Nat. Clim. Chang. 2, 30–32. https://doi.org/10.1038/nclimate1323.

Donelson, J.M., Salinas, S., Munday, P.L., Shama, L.N.S., 2018. Transgenerational plasticity and climate change experiments: where do we go from here? Glob. Chang. Biol. 24, 13–34. https://doi.org/10.1111/gcb.13903.

Doubleday, Z.A., Clarke, S.M., Li, X., Pecl, G.T., Ward, T.M., Battaglene, S., Frusher, S., Gibbs, P.J., Hobday, A.J., Hutchinson, N., Jennings, S.M., Stoklosa, R., 2013. Assessing the risk of climate change to aquaculture: a case study from south-east Australia. Aquac. Environ. Interact. 3, 163–175. https://doi.org/10.3354/aei00058.

Eliason, E.J., Clark, T.D., Hague, M.J., Hanson, L.M., Gallagher, Z.S., Jeffries, K.M., Gale, M.K., Patterson, D.A., Hinch, S.G., Farrell, A.P., 2011. Differences in thermal tolerance among sockeye salmon populations. Science 332 (80), 109–112. https://doi.org/10.1126/science.1199158.

Eliason, E.J., Dick, M., Patterson, D.A., Robinson, K.A., Lotto, J., Hinch, S.G., Cooke, S.J., 2020. Sex-specific differences in physiological recovery and short-term behaviour following fisheries capture in adult sockeye salmon (*Oncorhynchus nerka*). Can. J. Fish. Aquat. Sci. 77, 1749–1757. https://doi.org/10.1139/cjfas-2019-0258.

Elith, J., Leathwick, J.R., 2009. Species distribution models: Ecological explanation and prediction across space and time. Annu. Rev. Ecol. Evol. Syst. 40, 677–697. https://doi.org/10.1146/annurev.ecolsys.110308.120159.

Fangue, N.A., Hofmeister, M., Schulte, P.M., 2006. Intraspecific variation in thermal tolerance and heat shock protein gene expression in common killifish, Fundulus heteroclitus. J. Exp. Biol. 209, 2859–2872. https://doi.org/10.1242/jeb.02260.

Farrell, A.P., 2008. Comparisons of swimming performance in rainbow trout using constant acceleration and critical swimming speed tests. J. Fish Biol. 72, 693–710. https://doi.org/10.1111/j.1095-8649.2007.01759.x.

Fedor, P., Zvaríková, M., 2018. Biodiversity indices. Encycl. Ecol. 1, 337–346. https://doi.org/10.1016/B978-0-12-409548-9.10558-5.

Fernandez-Fournier, P., Lewthwaite, J.M.M., Mooers, A., 2021. Do we need to identify adaptive genetic variation when prioritizing populations for conservation? Conserv. Genet. 22, 205–216. https://doi.org/10.1007/s10592-020-01327-w.

Figge, F., 2004. Bio-folio: applying portfolio theory to biodiversity. Biodivers. Conserv. 13, 827–849.

Finn, R., Kapoor, B.G. (Eds.), 2008. Fish Larval Physiology. Science Publishers, Enfield, NH.

Finstad, A., Jonsson, B., 2012. Effect of incubation temperature on growth performance in Atlantic salmon. Mar. Ecol. Prog. Ser. 454, 75–82. https://doi.org/10.3354/meps09643.

Flanagan, S.P., Rose, E., Jones, A.G., 2021. The population genomics of repeated freshwater colonizations by Gulf pipefish. Mol. Ecol. 30, 1672–1687. https://doi.org/10.1111/mec.15841.

Fox, R.J., Donelson, J.M., Schunter, C., Ravasi, T., Gaitán-Espitia, J.D., 2019. Beyond buying time: the role of plasticity in phenotypic adaptation to rapid environmental change. Philos. Trans. R. Soc. B Biol. Sci. 374. https://doi.org/10.1098/rstb.2018.0174.

Frimpong, E.A., Angermeier, P.L., 2009. Fish traits: a database of ecological and life-history traits of freshwater fishes of the United States. Fisheries 34, 487–495. https://doi.org/10.1577/1548-8446-34.10.487.

Fry, F.E.J., 1971. The effects of environmental factors on the physiology of fish. Fish Physiol. 6, 1–98. https://doi.org/10.1016/S1546-5098(08)60146-6.

Fryxell, D.C., Palkovacs, E.P., 2017. Warming strengthens the ecological role of intraspecific variation in a predator. Copeia 105, 523–532. https://doi.org/10.1643/ce-16-527.

Fuller, M.R., Doyle, M.W., Strayer, D.L., 2015. Causes and consequences of habitat fragmentation in river networks. Ann. N. Y. Acad. Sci. 1355, 31–51. https://doi.org/10.1111/nyas.12853.

Gaichas, S.K., Link, J.S., Hare, J.A., 2014. A risk-based approach to evaluating northeast US fish community vulnerability to climate change. ICES J. Mar. Sci. 71, 2323–2341. https://doi.org/10.1093/icesjms/fsu048. Contribution.

Gallagher, R.V., Falster, D.S., Maitner, B.S., Salguero-Gómez, R., Vandvik, V., Pearse, W.D., Schneider, F.D., Kattge, J., Poelen, J.H., Madin, J.S., Ankenbrand, M.J., Penone, C., Feng, X., Adams, V.M., Alroy, J., Andrew, S.C., Balk, M.A., Bland, L.M., Boyle, B.L., Bravo-Avila, C.H., Brennan, I., Carthey, A.J.R., Catullo, R., Cavazos, B.R., Conde, D.A., Chown, S.L., Fadrique, B., Gibb, H., Halbritter, A.H., Hammock, J., Hogan, J.A., Holewa, H., Hope, M., Iversen, C.M., Jochum, M., Kearney, M., Keller, A., Mabee, P., Manning, P., McCormack, L., Michaletz, S.T., Park, D.S., Perez, T.M., Pineda-Munoz, S., Ray, C.A., Rossetto, M., Sauquet, H., Sparrow, B., Spasojevic, M.J., Telford, R.J., Tobias, J.A., Violle, C., Walls, R., Weiss, K.C.B., Westoby, M., Wright, I.J., Enquist, B.J., 2020. Open science principles for accelerating trait-based science across the tree of life. Nat. Ecol. Evol. 4, 294–303. https://doi.org/10.1038/s41559-020-1109-6.

Gavery, M.R., Nichols, K.M., Goetz, G.W., Middleton, M.A., Swanson, P., 2018. Characterization of genetic and epigenetic variation in sperm and red blood cells from adult hatchery and natural-origin steelhead, *Oncorhynchus mykiss*. G3 Genes, Genomes, Genet. 8, 3723–3736. https://doi.org/10.1534/g3.118.200458.

Gavery, M.R., Nichols, K.M., Berejikian, B.A., Tatara, C.P., Goetz, G.W., Dickey, J.T., Van Doornik, D.M., Swanson, P., 2019. Temporal dynamics of DNA methylation patterns in response to rearing juvenile steelhead (*Oncorhynchus mykiss*) in a hatchery versus simulated stream environment. Genes (Basel) 10. https://doi.org/10.3390/genes10050356.

Georgakopoulou, E., Sfakianakis, D.G., Kouttouki, S., Divanach, P., Kentouri, M., Koumoundouros, G., 2007. The influence of temperature during early life on phenotypic expression at later ontogenetic stages in sea bass. J. Fish Biol. 70, 278–291. https://doi.org/10.1111/j.1095-8649.2007.01305.x.

Georges, M., Charlier, C., Hayes, B., 2019. Harnessing genomic information for livestock improvement. Nat. Rev. Genet. 20, 135–156. https://doi.org/10.1038/s41576-018-0082-2.

Glazier, D.S., 2015. Is metabolic rate a universal "pacemaker" for biological processes? Biol. Rev. 90, 377–407. https://doi.org/10.1111/brv.12115.

Godwin, J., Luckenbach, J.A., Borski, R.J., 2003. Ecology meets endocrinology: environmental sex determination in fishes. Evol. Dev. 5, 40–49. https://doi.org/10.1046/j.1525-142X.2003.03007.x.

Green, B.S., 2008. Maternal effects in fish populations. In: Sims, D. (Ed.), Advances in Marine Biology. Academic Press, pp. 1–105. https://doi.org/10.1016/S0065-2881(08)00001-1.

Griffiths, J., Alderdice, D., 1972. Effects of acclimation and acute temperature experience on the swimming speed of juvenile coho salmon. J. Fish. Board Canada 29, 251–264.

Grummer, J.A., Beheregaray, L.B., Bernatchez, L., Hand, B.K., Luikart, G., Narum, S.R., Taylor, E.B., 2019. Aquatic landscape genomics and environmental effects on genetic variation. Trends Ecol. Evol. 34, 641–654. https://doi.org/10.1016/j.tree.2019.02.013.

Guo, B., Li, Z., Merilä, J., 2016. Population genomic evidence for adaptive differentiation in the Baltic Sea herring. Mol. Ecol. 25, 2833–2852. https://doi.org/10.1111/mec.13657.

Hanson, K.C., Gravel, M.A., Graham, A., Shoji, A., Cooke, S.J., 2008. Sexual variation in fisheries research and management: when does sex matter? Rev. Fish. Sci. 16, 421–436. https://doi.org/10.1080/10641260802013866.

Hare, J.A., Morrison, W.E., Nelson, M.W., Stachura, M.M., Teeters, E.J., Griffis, R.B., Alexander, M.A., Scott, J.D., Alade, L., Bell, R.J., Chute, A.S., Curti, K.L., Curtis, T.H., Kircheis, D., Kocik, J.F., Lucey, S.M., McCandless, C.T., Milke, L.M., Richardson, D.E., Robillard, E., Walsh, H.J., McManus, M.C., Marancik, K.E., Griswold, C.A., 2016. A vulnerability assessment of fish and invertebrates to climate change on the northeast U.S. continental shelf. PLoS One 11, 1–30. https://doi.org/10.1371/journal.pone.0146756.

Harmon, L.J., Matthews, B., Des Roches, S., Chase, J.M., Shurin, J.B., Schluter, D., 2009. Evolutionary diversification in stickleback affects ecosystem functioning. Nature 458, 1167–1170. https://doi.org/10.1038/nature07974.

Harmon, K.J., Bolinger, M.T., Rodnick, K.J., 2011. Carbohydrate energy reserves and effects of food deprivation in male and female rainbow trout. Comp. Biochem. Physiol. A Mol. Integr. Physiol. 158, 423–431. https://doi.org/10.1016/j.cbpa.2010.11.017.

Harrisson, K.A., Amish, S.J., Pavlova, A., Narum, S.R., Telonis-Scott, M., Rourke, M.L., Lyon, J., Tonkin, Z., Gilligan, D.M., Ingram, B.A., Lintermans, M., Gan, H.M., Austin, C.M., Luikart, G., Sunnucks, P., 2017. Signatures of polygenic adaptation associated with climate across the range of a threatened fish species with high genetic connectivity. Mol. Ecol. 26, 6253–6269. https://doi.org/10.1111/mec.14368.

Havird, J.C., Neuwald, J.L., Shah, A.A., Mauro, A., Marshall, C.A., Ghalambor, C.K., 2020. Distinguishing between active plasticity due to thermal acclimation and passive plasticity due to Q10 effects: why methodology matters. Funct. Ecol. 34, 1015–1028. https://doi.org/10.1111/1365-2435.13534.

Healy, T.M., Schulte, P.M., 2019. Patterns of alternative splicing in response to cold acclimation in fish. J. Exp. Biol. 222. https://doi.org/10.1242/jeb.193516.

Healy, T.M., Brennan, R.S., Whitehead, A., Schulte, P.M., 2018. Tolerance traits related to climate change resilience are independent and polygenic. Glob. Chang. Biol. 24, 5348–5360. https://doi.org/10.1111/gcb.14386.

Healy, T.M., Brennan, R.S., Whitehead, A., Schulte, P.M., 2019. Mitochondria, sex and variation in routine metabolic rate. Mol. Ecol. 28, 4608–4619. https://doi.org/10.1111/mec.15244.

Hecht, B.C., Thrower, F.P., Hale, M.C., Miller, M.R., Nichols, K.M., 2012. Genetic architecture of migration-related traits in rainbow and steelhead trout, *Oncorhynchus mykiss*. G3 Genes, Genomes, Genet. 2, 1113–1127. https://doi.org/10.1534/g3.112.003137.

Hecht, B.C., Campbell, N.R., Holecek, D.E., Narum, S.R., 2013. Genome-wide association reveals genetic basis for the propensity to migrate in wild populations of rainbow and steelhead trout. Mol. Ecol. 22, 3061–3076. https://doi.org/10.1111/mec.12082.

Hecht, B.C., Matala, A.P., Hess, J.E., Narum, S.R., 2015. Environmental adaptation in Chinook salmon (*Oncorhynchus tshawytscha*) throughout their North American range. Mol. Ecol. 24, 5573–5595. https://doi.org/10.1111/mec.13409.

Hellmann, J.K., Bukhari, S.A., Deno, J., Bell, A.M., 2020a. Sex-specific plasticity across generations I: maternal and paternal effects on sons and daughters. J. Anim. Ecol. 89, 2788–2799. https://doi.org/10.1111/1365-2656.13364.

Hellmann, J.K., Carlson, E.R., Bell, A.M., 2020b. Sex-specific plasticity across generations II: Grandpaternal effects are lineage specific and sex specific. J. Anim. Ecol. 89, 2800–2812. https://doi.org/10.1111/1365-2656.13365.

Henderson, C.J., Gilby, B.L., Schlacher, T.A., Connolly, R.M., Sheaves, M., Maxwell, P.S., Flint, N., Borland, H.P., Martin, T.S.H., Gorissen, B., Olds, A.D., 2020. Landscape transformation alters functional diversity in coastal seascapes. Ecography (Cop.). 43, 138–148. https://doi.org/10.1111/ecog.04504.

Hendry, A.P., Kinnison, M.T., 1999. The pace of modern life: measuring rates of contemporary microevolution. Evolution (N. Y). 53, 1637–1653. https://doi.org/10.1111/j.1558-5646.1999.tb04550.x.

Henry, R.J., 2020. Innovations in plant genetics adapting agriculture to climate change. Curr. Opin. Plant Biol. 56, 168–173. https://doi.org/10.1016/j.pbi.2019.11.004.

Hess, J.E., Zendt, J.S., Matala, A.R., Narum, S.R., 2016. Genetic basis of adult migration timing in anadromous steelhead discovered through multivariate association testing. Proc. R. Soc. B Biol. Sci., 283. https://doi.org/10.1098/rspb.2015.3064.

Hillebrand, H., Blasius, B., Borer, E.T., Chase, J.M., Downing, J.A., Eriksson, B.K., Filstrup, C.T., Harpole, W.S., Hodapp, D., Larsen, S., Lewandowska, A.M., Seabloom, E.W., Van de Waal, D.B., Ryabov, A.B., 2018. Biodiversity change is uncoupled from species richness trends: consequences for conservation and monitoring. J. Appl. Ecol. 55, 169–184. https://doi.org/10.1111/1365-2664.12959.

Hinch, S.G., Cooke, S.J., Farrell, A.P., Miller, K.M., Lapointe, M., Patterson, D.A., 2012. Dead fish swimming: a review of research on the early migration and high premature mortality in adult Fraser River sockeye salmon *Oncorhynchus nerka*. J. Fish Biol. 81, 576–599. https://doi.org/10.1111/j.1095-8649.2012.03360.x.

Hinch, S.G., Bett, N.N., Eliason, E.J., Farrell, A.P., Cooke, S.J., Patterson, D.A., 2021. Exceptionally high mortality of adult female salmon: a large-scale pattern and a conservation concern. Can. J. Fish. Aquat. Sci. 78, 639–654. https://doi.org/10.1139/cjfas-2020-0385.

Hinch, S.G., Bett, N.N., Farrell, A.P., 2022. A conservation physiological perspective on dam passage by fishes. Fish Physiol. 39B (In press).

Hoey, J.A., Pinsky, M.L., 2018. Genomic signatures of environmental selection despite near-panmixia in summer flounder. Evol. Appl. 11, 1732–1747. https://doi.org/10.1111/eva.12676.

Hoffmann, A.A., Sgró, C.M., 2011. Climate change and evolutionary adaptation. Nature. https://doi.org/10.1038/nature09670.

Hooper, D.U., Chapin, F.S.S., Ewel, J.J., Hector, A., Inchausti, P., Lavorel, S., Lawton, J.H., Lodge, D.M., Loreau, M., Naeem, S., Schmid, B., Setälä, H., Symstad, A.J., Vandermeer, J., Wardle, D.A., Setala, H., Symstad, A.J., Vandermeer, J., Wardle, D.A., 2005. Effects of biodiversity on ecosystem functioning: a consensus of current knowledge. Ecol. Monogr. 75, 3–35. https://doi.org/10.1890/04-0922.

Houde, A.L.S., Günther, O.P., Strohm, J., Ming, T.J., Li, S., Kaukinen, K.H., Patterson, D.A., Farrell, A.P., Hinch, S.G., Miller, K.M., 2019. Discovery and validation of candidate smoltification gene expression biomarkers across multiple species and ecotypes of Pacific salmonids. Conserv. Physiol. 7, 1–21. https://doi.org/10.1093/conphys/coz051.

Howeth, J.G., Weis, J.J., Brodersen, J., Hatton, E.C., Post, D.M., 2013. Intraspecific phenotypic variation in a fish predator affects multitrophic lake metacommunity structure. Ecol. Evol. 3, 5031–5044. https://doi.org/10.1002/ece3.878.

Huey, R.B., Kearney, M.R., Krockenberger, A., Holtum, J.A.M., Jess, M., Williams, S.E., 2012. Predicting organismal vulnerability to climate warming: roles of behaviour, physiology and adaptation. Philos. Trans. R. Soc. B Biol. Sci. 367, 1665–1679. https://doi.org/10.1098/rstb.2012.0005.

Huss, M., Howeth, J.G., Osterman, J.I., Post, D.M., 2014. Intraspecific phenotypic variation among alewife populations drives parallel phenotypic shifts in bluegill. Proc. R. Soc. B Biol. Sci. 281. https://doi.org/10.1098/rspb.2014.0275.

Immler, S., 2018. The sperm factor: paternal impact beyond genes. Heredity (Edinb). 121, 239–247. https://doi.org/10.1038/s41437-018-0111-0.

Jeffries, K.M., Hinch, S.G., Martins, E.G., Clark, T.D., Lotto, A.G., Patterson, D.A., Cooke, S.J., Farrell, A.P., Miller, K.M., 2012. Sex and proximity to reproductive maturity influence the survival, final maturation, and blood physiology of pacific salmon when exposed to high temperature during a simulated migration. Physiol. Biochem. Zool. 85, 62–73. https://doi.org/10.1086/663770.

Jeffries, K.M., Jeffrey, J.D., Holland, E.B., 2022. Applied aspects of gene function for the conservation of fishes. Fish Physiol. 39A, 389–433.

Johnson, J.E., Welch, D.J., 2010. Marine fisheries management in a changing climate: a review of vulnerability and future options. Rev. Fish. Sci. 18, 106–124. https://doi.org/10.1080/10641260903434557.

Johnston, S.E., Orell, P., Pritchard, V.L., Kent, M.P., Lien, S., Niemelä, E., Erkinaro, J., Primmer, C.R., 2014. Genome-wide SNP analysis reveals a genetic basis for sea-age variation in a wild population of Atlantic salmon (*Salmo salar*). Mol. Ecol. 23, 3452–3468. https://doi.org/10.1111/mec.12832.

Jonsson, B., Jonsson, N., 2014. Early environment influences later performance in fishes. J. Fish Biol. 85, 151–188. https://doi.org/10.1111/jfb.12432.

Kardos, M., Armstrong, E., Fitzpatrick, S.W., Hauser, S., Hedrick, P., Miller, J., Tallmon, D., Funk, W.C., 2021. The crucial role of genome-wide genetic variation in conservation. bioRxiv. 2021.07.05.451163.

Kearney, M., Porter, W., 2009. Mechanistic niche modelling: combining physiological and spatial data to predict species' ranges. Ecol. Lett. 12, 334–350. https://doi.org/10.1111/j.1461-0248.2008.01277.x.

Kenchington, E., Heino, M., Nielsen, E.E., 2003. Managing marine genetic diversity: time for action? ICES J. Mar. Sci. 60, 1172–1176. https://doi.org/10.1016/S1054-3139(03)00136-X.

Kendall, N.W., McMillan, J.R., Sloat, M.R., Buehrens, T.W., Quinn, T.P., Pess, G.R., Kuzishchin, K.V., McClure, M.M., Zabel, R.W., 2015. Anadromy and residency in steelhead and rainbow trout (*Oncorhynchus mykiss*): a review of the processes and patterns. Can. J. Fish. Aquat. Sci. 342, 319–342. https://doi.org/10.1139/cjfas-2014-0192.

Killen, S.S., Costa, I., Brown, J.A., Gamperl, A.K., 2007. Little left in the tank: metabolic scaling in marine teleosts and its implications for aerobic scope. Proc. Biol. Sci. 274, 431–438. https://doi.org/10.1098/rspb.2006.3741.

Killen, S.S., Atkinson, D., Glazier, D.S., 2010. The intraspecific scaling of metabolic rate with body mass in fishes depends on lifestyle and temperature. Ecol. Lett. 13, 184–193. https://doi.org/10.1111/j.1461-0248.2009.01415.x.

Kraskura, K., Hardison, E.A., Little, A.G., Dressler, T., Prystay, T.S., Hendriks, B., Farrell, A.P., Cooke, S.J., Patterson, D.A., Hinch, S.G., Eliason, E.J., 2020. Sex-specific differences in swimming, aerobic metabolism and recovery from exercise in adult coho salmon (*Oncorhynchus kisutch*) across ecologically relevant temperatures. Conserv. Physiol. 9, 1–22. https://doi.org/10.1093/conphys/coab016.

Laghari, M.Y., Lashari, P., Zhang, Y., Sun, X., 2014. Identification of quantitative trait loci (QTLs) in aquaculture species. Rev. Fish. Sci. Aquac. 22, 221–238. https://doi.org/10.1080/23308249.2014.931172.

Laikre, L., 2010. Genetic diversity is overlooked in international conservation policy implementation. Conserv. Genet. 11, 349–354, https://doi.org/10.1007/s10592-009-0037-4.

Larkin, P.A., 1981. A perspective on population genetics and salmon management. Can. J. Fish. Aquat. Sci. 38, 1469–1475. https://doi.org/10.1139/f81-198.

Le Luyer, J., Laporte, M., Beacham, T.D., Kaukinen, K.H., Withler, R.E., Leong, J.S., Rondeau, E.B., Koop, B.F., Bernatchez, L., 2017. Parallel epigenetic modifications induced by hatchery rearing in a Pacific salmon. Proc. Natl. Acad. Sci. U. S. A. 114, 12964–12969. https://doi.org/10.1073/pnas.1711229114.

Le Roy, A., Seebacher, F., 2018. Transgenerational effects and acclimation affect dispersal in guppies. Funct. Ecol. 32, 1819–1831. https://doi.org/10.1111/1365-2435.13105.

Lecocq, T., Benard, A., Pasquet, A., Nahon, S., Ducret, A., Dupont-Marin, K., Lang, I., Thomas, M., 2019. TOFF, a database of traits of fish to promote advances in fish aquaculture. Sci. Data 6, 1–5. https://doi.org/10.1038/s41597-019-0307-z.

Legras, G., Loiseau, N., Gaertner, J.C., 2018. Functional richness: overview of indices and underlying concepts. Acta Oecol. 87, 34–44. https://doi.org/10.1016/j.actao.2018.02.007.

Leitwein, M., Laporte, M., Le Luyer, J., Mohns, K., Normandeau, E., Withler, R., Bernatchez, L., 2021. Epigenomic modifications induced by hatchery rearing persist in germ line cells of adult salmon after their oceanic migration. Evol. Appl. 1–12. https://doi.org/10.1111/eva.13235.

Li, Y.L., Xue, D.X., Zhang, B.D., Liu, J.X., 2019. Population genomic signatures of genetic structure and environmental selection in the catadromous roughskin sculpin *Trachidermus fasciatus*. Genome Biol. Evol. 11, 1751–1764. https://doi.org/10.1093/gbe/evz118.

Limborg, M.T., Helyar, S.J., De Bruyn, M., Taylor, M.I., Nielsen, E.E., Ogden, R., Carvalho, G.R., Bekkevold, D., 2012. Environmental selection on transcriptome-derived SNPs in a high gene flow marine fish, the Atlantic herring (*Clupea harengus*). Mol. Ecol. 21, 3686–3703. https://doi.org/10.1111/j.1365-294X.2012.05639.x.

Lindmark, M., Huss, M., Ohlberger, J., Gårdmark, A., 2018. Temperature-dependent body size effects determine population responses to climate warming. Ecol. Lett. 21, 181–189. https://doi.org/10.1111/ele.12880.

Lindmark, M., Ohlberger, J., Huss, M., Gårdmark, A., 2019. Size-based ecological interactions drive food web responses to climate warming. Ecol. Lett. 22, 778–786. https://doi.org/10.1111/ele.13235.

Logan, C.A., Somero, G.N., 2010. Transcriptional responses to thermal acclimation in the eurythermal fish *Gillichthys mirabilis* (Cooper 1864). Am. J. Physiol. Regul. Integr. Comp. Physiol. 299, R843–R852. https://doi.org/10.1152/ajpregu.00306.2010.

Loreau, M., Barbier, M., Filotas, E., Gravel, D., Isbell, F., Miller, S.J., Montoya, J.M., Wang, S., Aussenac, R., Germain, R., Thompson, P.L., Gonzalez, A., Dee, L.E., 2021. Biodiversity as insurance: from concept to measurement and application. Biol. Rev. https://doi.org/10.1111/brv.12756.

Loughland, I., Little, A., Seebacher, F., 2021. DNA methyltransferase 3a mediates developmental thermal plasticity. BMC Biol. 19, 1–11. https://doi.org/10.1186/s12915-020-00942-w.

Luypaert, T., Hagan, J.G., McCarthy, M.L., Poti, M., 2020. Status of marine biodiversity in the anthropocene. In: Jungblut, S., Liebich, V., Bode-Dalby, M. (Eds.), YOUMARES 9—The Oceans: Our Research, Our Future: Proceedings of the 2018 Conference for YOUng MArine RESearcher in Oldenburg, Germany. Springer International Publishing, Cham, pp. 57–82. https://doi.org/10.1007/978-3-030-20389-4_4.

Mackay, T.F.C., 2001. The genetic architecture of quantitative traits. North 35, 303–339.

MacKay, T.F.C., Stone, E.A., Ayroles, J.F., 2009. The genetics of quantitative traits: challenges and prospects. Nat. Rev. Genet. 10, 565–577. https://doi.org/10.1038/nrg2612.

Madliger, C.L., Franklin, C.E., Chown, S.L., Fuller, A., Hultine, K.R., Costantini, D., Hopkins, W.A., Peck, M.A., Rummer, J.L., Sack, L., Willis, C.K.R., Cooke, S.J., 2021. The second warning to humanity: contributions and solutions from conservation physiology. Conserv. Physiol. 9, 1–18. https://doi.org/10.1093/conphys/coab038.

Manfrin, A., Teurlincx, S., Lorenz, A.W., Haase, P., Marttila, M., Syrjänen, J.T., Thomas, G., Stoll, S., 2019. Effect of river restoration on life-history strategies in fish communities. Sci. Total Environ. 663, 486–495. https://doi.org/10.1016/j.scitotenv.2019.01.330.

Martin, B.T., Dudley, P.N., Kashef, N.S., Stafford, D.M., Reeder, W.J., Tonina, D., Del Rio, A.M., Scott Foott, J., Danner, E.M., 2020. The biophysical basis of thermal tolerance in fish eggs: thermal tolerance in fish eggs. Proc. R. Soc. B Biol. Sci. 287. https://doi.org/10.1098/rspb.2020.1550rspb20201550.

Martini, S., Larras, F., Boyé, A., Faure, E., Aberle, N., Archambault, P., Bacouillard, L., Beisner, B.E., Bittner, L., Castella, E., Danger, M., Gauthier, O., Karp-Boss, L., Lombard, F., Maps, F., Stemmann, L., Thiébaut, E., Usseglio-Polatera, P., Vogt, M., Laviale, M., Ayata, S.D., 2021. Functional trait-based approaches as a common framework for aquatic ecologists. Limnol. Oceanogr. 66, 965–994. https://doi.org/10.1002/lno.11655.

Martins, E.G., Hinch, S.G., Patterson, D.A., Hague, M.J., Cooke, S.J., Miller, K.M., Robichaud, D., English, K.K., Farrell, A.P., 2012. High river temperature reduces survival of sockeye salmon (*Oncorhynchus nerka*) approaching spawning grounds and exacerbates female mortality. Can. J. Fish. Aquat. Sci. 69, 330–342. https://doi.org/10.1139/F2011-154.

Mathis, J.T., Cooley, S.R., Lucey, N., Colt, S., Ekstrom, J., Hurst, T., Hauri, C., Evans, W., Cross, J.N., Feely, R.A., 2015. Ocean acidification risk assessment for Alaska's fishery sector. Prog. Oceanogr. 136, 71–91. https://doi.org/10.1016/j.pocean.2014.07.001.

McGill, B.J., Enquist, B.J., Weiher, E., Westoby, M., 2006. Rebuilding community ecology from functional traits. Trends Ecol. Evol. 21, 178–185. https://doi.org/10.1016/j.tree.2006.02.002.

McKenzie, D.J., Axelsson, M., Chabot, D., Claireaux, G., Cooke, S.J., Corner, R.A., de Boeck, G., Domenici, P., Guerreiro, P.M., Hamer, B., Jørgensen, C., Killen, S.S., Lefevre, S., Marras, S., Michaelidis, B., Nilsson, G.E., Peck, M.A., Perez-Ruzafa, A., Rijnsdorp, A.D., Shiels, H.A., Steffensen, J.F., Svendsen, J.C., Svendsen, M.B.S., Teal, L.R., van der Meer, J., Wang, T., Wilson, J.M., Wilson, R.W., Metcalfe, J.D., 2016. Conservation physiology of marine fishes: state of the art and prospects for policy. Conserv. Physiol. 4, 1–20. https://doi.org/10.1093/conphys/cow046.

McKenzie, D.J., Zhang, Y., Eliason, E.J., Schulte, P.M., Claireaux, G., Blasco, F.R., Nati, J.J.H., Farrell, A.P., 2020. Intraspecific variation in tolerance of warming in fishes. J. Fish Biol., 1–20. https://doi.org/10.1111/jfb.14620.

McMahon, B.J., Teeling, E.C., Hoglund, J., 2014. How and why should we implement genomics into conservation? Evol. Appl. 7, 999–1007. https://doi.org/10.1111/eva.12193.

McPhee, M.V., Utter, F., Stanford, J.A., Kuzishchin, K.V., Savvaitova, K.A., Pavlov, D.S., Allendorf, F.W., 2007. Population structure and partial anadromy in *Oncorhynchus mykiss* from Kamchatka: relevance for conservation strategies around the Pacific Rim. Ecol. Freshw. Fish 16, 539–547. https://doi.org/10.1111/j.1600-0633.2007.00248.x.

Melo-Merino, S.M., Reyes-Bonilla, H., Lira-Noriega, A., 2020. Ecological niche models and species distribution models in marine environments: a literature review and spatial analysis of evidence. Ecol. Modell. 415, 108837. https://doi.org/10.1016/j.ecolmodel.2019.108837.

Mendez-Sanchez, J.F., Burggren, W.W., 2019. Hypoxia-induced developmental plasticity of larval growth, fill and labyrinth organ morphometrics in two anabantoid fish: the facultative air breather Siamese fighting fish (*Betta splendens*) and the obligate air-breather the blue gourami. J. Morphol. 280, 193–204. https://doi.org/10.1002/jmor.20931.

Merilä, J., Hendry, A.P., 2014. Climate change, adaptation, and phenotypic plasticity: the problem and the evidence. Evol. Appl. 7, 1–14. https://doi.org/10.1111/eva.12137.

Merilä, J., Hoffmann, A.A., 2016, Evolutionary impacts of climate change. Oxford Res. Encycl. Environ. Sci., 1–15. https://doi.org/10.1093/acrefore/9780199389414.013.136.

Metzger, D.C.H., Schulte, P.M., 2016a. Epigenomics in marine fishes. Mar. Genomics 30, 43–54. https://doi.org/10.1016/j.margen.2016.01.004.

Metzger, D.C.H., Schulte, P.M., 2016b. Maternal stress has divergent effects on gene expression patterns in the brains of male and female threespine stickleback. Proc. R. Soc. B Biol. Sci. 283, 20161734. https://doi.org/10.1098/rspb.2016.1734.

Metzger, D.C.H., Schulte, P.M., 2017. Persistent and plastic effects of temperature on DNA methylation across the genome of threespine stickleback (*Gasterosteus aculeatus*). Proc. R. Soc. B Biol. Sci. 284, 20171667. https://doi.org/10.1098/rspb.2017.1667.

Milardi, M., Gavioli, A., Soininen, J., Castaldelli, G., 2019. Exotic species invasions undermine regional functional diversity of freshwater fish. Sci. Rep. 9, 17921. https://doi.org/10.1038/s41598-019-54210-1.

Milla, S., Pasquet, A., El Mohajer, L., Fontaine, P., 2021. How domestication alters fish phenotypes. Rev. Aquac. 13, 388–405. https://doi.org/10.1111/raq.12480.

Miller, K.M., Li, S., Kaukinen, K.H., Ginther, N., Hammill, E., Curtis, J.M.R., Patterson, D.A., Sierocinski, T., Donnison, L., Pavlidis, P., Hinch, S.G., Hruska, K.A., Cooke, S.J., English, K.K., Farrell, A.P., 2011. Genomic signatures predict migration and spawning failure in wild Canadian salmon. Science 331 (80), 214–217.

Miller, K.M., Günther, O.P., Li, S., Kaukinen, K.H., Ming, T.J., 2017. Molecular indices of viral disease development in wild migrating salmon. Conserv. Physiol. 5. https://doi.org/10.1093/conphys/cox036.

Mims, M.C., Olden, J.D., Shattuck, Z.R., Poff, N.L., 2010. Life history trait diversity of native freshwater fishes in North America. Ecol. Freshw. Fish 19, 390–400. https://doi.org/10.1111/j.1600-0633.2010.00422.x.

Mimura, M., Yahara, T., Faith, D.P., Vázquez-Domínguez, E., Colautti, R.I., Araki, H., Javadi, F., Núñez-Farfán, J., Mori, A.S., Zhou, S., Hollingsworth, P.M., Neaves, L.E., Fukano, Y., Smith, G.F., Sato, Y.I., Tachida, H., Hendry, A.P., 2017. Understanding and monitoring the consequences of human impacts on intraspecific variation. Evol. Appl. 10, 121–139. https://doi.org/10.1111/eva.12436.

Moran, E.V., Hartig, F., Bell, D.M., 2016. Intraspecific trait variation across scales: implications for understanding global change responses. Glob. Chang. Biol. 22, 137–150. https://doi.org/10.1111/gcb.13000.

Morgan, R., Finnøen, M.H., Jensen, H., Pélabon, C., Jutfelt, F., 2020. Low potential for evolutionary rescue from climate change in a tropical fish. Proc. Natl. Acad. Sci. U. S. A. 117, 33365–33372. https://doi.org/10.1073/pnas.2011419117.

Morley, S.A., Peck, L.S., Sunday, J.M., Heiser, S., Bates, A.E., 2019. Physiological acclimation and persistence of ectothermic species under extreme heat events. Glob. Ecol. Biogeogr. 28, 1018–1037. https://doi.org/10.1111/geb.12911.

Moyano, M., Candebat, C., Ruhbaum, Y., Álvarez-Fernández, S., Claireaux, G., Zambonino-Infante, J.L., Peck, M.A., 2017. Effects of warming rate, acclimation temperature and ontogeny on the critical thermal maximum of temperate marine fish larvae. PLoS One 12, 1–23. https://doi.org/10.1371/journal.pone.0179928.

Narum, S.R., Di Genova, A., Micheletti, S.J., Maass, A., 2018. Genomic variation underlying complex life-history traits revealed by genome sequencing in Chinook salmon. Proc. R. Soc. B Biol. Sci., 285. https://doi.org/10.1098/rspb.2018.0935.

Nati, J.J.H., Svendsen, M.B., Marras, S., Killen, S.S., Steffensen, J.F., McKenzie, D.J., Domenici, P., 2020. Intraspecific variation in thermal tolerance differs between tropical and temperate fishes. bioRxiv 1–9. https://doi.org/10.1101/2020.12.07.414318. this.

Nichols, K.M., Edo, A.F., Wheeler, P.A., Thorgaard, G.H., 2008. The genetic basis of smoltification-related traits in Oncorhynchus mykiss. Genetics 179, 1559–1575. https://doi.org/10.1534/genetics.107.084251.

Ohlberger, J., 2013. Climate warming and ectotherm body size - from individual physiology to community ecology. Funct. Ecol. 27, 991–1001. https://doi.org/10.1111/1365-2435.12098.

Ohlberger, J., Mehner, T., Staaks, G., Hölker, F., 2012. Intraspecific temperature dependence of the scaling of metabolic rate with body mass in fishes and its ecological implications. Oikos 121, 245–251. https://doi.org/10.1111/j.1600-0706.2011.19882.x.

Oliveira, A.G., Baumgartner, M.T., Gomes, L.C., Dias, R.M., Agostinho, A.A., 2018. Long-term effects of flow regulation by dams simplify fish functional diversity. Freshw. Biol. 63, 293–305. https://doi.org/10.1111/fwb.13064.

Oliveira, D.R., Reid, B.N., Fitzpatrick, S.W., 2021. Genome-wide diversity and habitat underlie fine-scale phenotypic differentiation in the rainbow darter (*Etheostoma caeruleum*). Evol. Appl. 14, 498–512. https://doi.org/10.1111/eva.13135.

Ospina-Álvarez, N., Piferrer, F., 2008. Temperature-dependent sex determination in fish revisited: prevalence, a single sex ratio response pattern, and possible effects of climate change. PLoS One 3, 2–4. https://doi.org/10.1371/journal.pone.0002837.

Øverli, Ø., Sørensen, C., Nilsson, G.E., 2006. Behavioral indicators of stress-coping style in rainbow trout: Do males and females react differently to novelty? Physiol. Behav. 87, 506–512. https://doi.org/10.1016/j.physbeh.2005.11.012.

Oziolor, E.M., Reid, N.M., Yair, S., Lee, K.M., Guberman Verplog, S., Bruns, P.C., Shaw, J.R., Whitehead, A., Matson, C.W., 2019. Adaptive introgression enables evolutionary rescue from extreme environmental pollution. Science 364 (80), 455–457. https://doi.org/10.1126/science.aav4155.

Patterson, D.A., Cooke, S.J., Hinch, S.G., Robinson, K.A., Young, N., Farrell, A.P., Miller, K.M., 2016. A perspective on physiological studies supporting the provision of scientific advice for the management of Fraser River sockeye salmon (*Oncorhynchus nerka*). Conserv. Physiol. 4, 1–15. https://doi.org/10.1093/conphys/cow026.

Payne, M.R., Kudahl, M., Engelhard, G.H., Peck, M.A., Pinnegar, J.K., 2020. Climate-risk to European fisheries and coastal communities. bioRxiv. 2020.08.03.234401 https://doi.org/10.1101/2020.08.03.234401.

Pearse, D.E., 2016. Saving the spandrels?Adaptive genomic variation in conservation and fisheries management. J. Fish Biol. 89, 2697–2716. https://doi.org/10.1111/jfb.13168.

Pecl, G.T., Ward, T.M., Doubleday, Z.A., Clarke, S., Day, J., Dixon, C., Frusher, S., Gibbs, P., Hobday, A.J., Hutchinson, N., Jennings, S., Jones, K., Li, X., Spooner, D., Stoklosa, R., 2014. Rapid assessment of fisheries species sensitivity to climate change. Clim. Change 127, 505–520. https://doi.org/10.1007/s10584-014-1284-z.

Perez, M.F., Lehner, B., 2019. Intergenerational and transgenerational epigenetic inheritance in animals. Nat. Cell Biol. 21, 143–151. https://doi.org/10.1038/s41556-018-0242-9.

Peters, R.H., 1983. The Ecological Implications of Body Size. Cambridge University Press, Cambridge.

Pinsky, M.L., Eikeset, A.M., McCauley, D.J., Payne, J.L., Sunday, J.M., 2019. Greater vulnerability to warming of marine versus terrestrial ectotherms. Nature 569, 108–111. https://doi.org/10.1038/s41586-019-1132-4.

Post, J.R., Lee, J.A., 1996. Metabolic ontogeny of teleost fishes. Can. J. Fish. Aquat. Sci. 53, 910–923.

Post, D.M., Palkovacs, E.P., Schielke, E.G., Dodson, S.I., 2008. Intraspecific variation in a predator affects community structure and cascading trophic interactions. Ecology 89, 2019–2032. https://doi.org/10.1890/07-1216.1.

Prince, D.J., O'Rourke, S.M., Thompson, T.Q., Ali, O.A., Lyman, H.S., Saglam, I.K., Hotaling, T.-J., Spidle, A.P., Miller, M.R., 2017. The evolutionary basis of premature migration in Pacific salmon highlights the utility of genomics for informing conservation. Sci. Adv. 3. https://doi.org/10.1126/sciadv.1603198.

Pujolar, J.M., Jacobsen, M.W., Als, T.D., Frydenberg, J., Munch, K., Jönsson, B., Jian, J.B., Cheng, L., Maes, G.E., Bernatchez, L., Hansen, M.M., 2014. Genome-wide single-generation signatures of local selection in the panmictic European eel. Mol. Ecol. 23, 2514–2528. https://doi.org/10.1111/mec.12753.

Quinn, T.P., Myers, K.W., 2004. Anadromy and the marine migrations of Pacific salmon and trout: Rounsefell revisited. Rev. Fish Biol. Fish. 14, 421–442. https://doi.org/10.1007/s11160-005-0802-5.

Raeymaekers, J.A.M., Chaturvedi, A., Hablützel, P.I., Verdonck, I., Hellemans, B., Maes, G.E., De Meester, L., Volckaert, F.A.M., 2017. Adaptive and non-adaptive divergence in a common landscape. Nat. Commun. 8, 1–8. https://doi.org/10.1038/s41467-017-00256-6.

Raffard, A., Santoul, F., Cucherousset, J., Blanchet, S., 2019. The community and ecosystem consequences of intraspecific diversity: a meta-analysis. Biol. Rev. 94, 648–661. https://doi.org/10.1111/brv.12472.

Raffard, A., Cucherousset, J., Montoya, J.M., Richard, M., Acoca-Pidolle, S., Poésy, C., Garreau, A., Santoul, F., Blanchet, S., 2021. Intraspecific diversity loss in a predator species alters prey community structure and ecosystem functions. PLoS Biol. 19, 1–17. https://doi.org/10.1371/journal.pbio.3001145.

Rees, B.B., Sudradjat, F.A., Love, J.W., 2001. Acclimation to hypoxia increases survival time of zebrafish, *Danio rerio*, during lethal hypoxia. J. Exp. Zool. 289, 266–272. https://doi.org/10.1002/1097-010X(20010401/30)289:4<266::AID-JEZ7>3.0.CO;2-5.

Reid, N.M., Proestou, D.A., Clark, B.W., Warren, W.C., Colbourne, J.K., Shaw, J.R., Karchner, S.I., Hahn, M.E., Nacci, D., Oleksiak, M.F., Crawford, D.L., Whitehead, A., 2016. The genomic landscape of rapid repeated evolutionary adaptation to toxic pollution in wild fish. Science 354 (80), 1305–1308. https://doi.org/10.1126/science.aah4993.

Robertson, C.E., Wright, P.A., Köblitz, L., Bernier, N.J., 2014. Hypoxia-inducible factor-1 mediates adaptive developmental plasticity of hypoxia tolerance in zebrafish, *Danio rerio*. Proc. R. Soc. B Biol. Sci. 281. https://doi.org/10.1098/rspb.2014.0637.

Rodnick, K.J., Gamperl, A.K., Nash, G.W., Syme, D.A., 2014. Temperature and sex dependent effects on cardiac mitochondrial metabolism in Atlantic cod (Gadus morhua L.). J. Therm. Biol. 44, 110–118. https://doi.org/10.1016/j.jtherbio.2014.02.012.

Rodriguez Barreto, D., Garcia De Leaniz, C., Verspoor, E., Sobolewska, H., Coulson, M., Consuegra, S., Mulligan, C., 2019. DNA methylation changes in the sperm of captive-reared fish: a route to epigenetic introgression in wild populations. Mol. Biol. Evol. 36, 2205–2211. https://doi.org/10.1093/molbev/msz135.

Rodriguez-Dominguez, A., Connell, S.D., Baziret, C., Nagelkerken, I., 2018. Irreversible behavioural impairment of fish starts early: embryonic exposure to ocean acidification. Mar. Pollut. Bull. 133, 562–567. https://doi.org/10.1016/j.marpolbul.2018.06.004.

Rogers, N.J., Urbina, M.A., Reardon, E.E., McKenzie, D.J., Wilson, R.W., 2016. A new analysis of hypoxia tolerance in fishes using a database of critical oxygen level (Pcrit). Conserv. Physiol. 4, 1–19. https://doi.org/10.1093/conphys/cow012.

Rombough, P.J., 1997. The effects of temperature on embryonic and larval development. In: Wood, C.M., McDonald, D.G. (Eds.), Global Warming: Implications for Freshwater and Marine Fish. Cambridge University Press, Cambridge, pp. 177–223. https://doi.org/10.3389/conf.fmars.2019.08.00073.

Rougeux, C., Gagnaire, P.A., Praebel, K., Seehausen, O., Bernatchez, L., 2019. Polygenic selection drives the evolution of convergent transcriptomic landscapes across continents within a Nearctic sister species complex. Mol. Ecol. 28, 4388–4403. https://doi.org/10.1111/mec.15226.

Rudman, S.M., Schluter, D., 2016. Ecological impacts of reverse speciation in threespine stickleback. Curr. Biol. 26, 490–495. https://doi.org/10.1016/j.cub.2016.01.004.

Ryu, T., Veilleux, H.D., Munday, P.L., Jung, I., Donelson, J.M., Ravasi, T., 2020. An epigenetic signature for within-generational plasticity of a reef fish to ocean warming. Front. Mar. Sci. 7, 1–15. https://doi.org/10.3389/fmars.2020.00284.

Salinas, S., Munch, S.B., 2012. Thermal legacies: transgenerational effects of temperature on growth in a vertebrate. Ecol. Lett. 15, 159–163. https://doi.org/10.1111/j.1461-0248.2011.01721.x.

Salinas, S., Brown, S.C., Mangel, M., Munch, S.B., 2013. Non-genetic inheritance and changing environments. Non-Genetic Inherit. 1. https://doi.org/10.2478/ngi-2013-0005.

Sandblom, E., Clark, T.D., Hinch, S.G., Farrell, A.P., 2009. Sex-specific differences in cardiac control and hematology of sockeye salmon (*Oncorhynchus nerka*) approaching their spawning grounds. Am. J. Physiol. Regul. Integr. Comp. Physiol. 297, 1136–1143. https://doi.org/10.1152/ajpregu.00363.2009.

Schaefer, J., Ryan, A., 2006. Developmental plasticity in the thermal tolerance of zebrafish *Danio rerio*. J. Fish Biol. 69, 722–734. https://doi.org/10.1111/j.1095-8649.2006.01145.x.

Schindler, D.E., Hilborn, R., Chasco, B., Boatright, C.P., Quinn, T.P., Rogers, L.A., Webster, M.-S., 2010. Population diversity and the portfolio effect in an exploited species. Nature 465, 609–612. https://doi.org/10.1038/nature09060.

Schindler, D.E., Armstrong, J.B., Reed, T.E., 2015. The portfolio concept in ecology and evolution. Front. Ecol. Environ. 13, 257–263. https://doi.org/10.1890/140275.

Schmidt-Kloiber, A., Hering, D., 2015. www.freshwaterecology.info—an online tool that unifies, standardises and codifies more than 20,000 European freshwater organisms and their ecological preferences. Ecol. Indic. 53, 271–282. https://doi.org/10.1016/j.ecolind.2015.02.007.

Schnurr, M.E., Yin, Y., Scott, G.R., 2014. Temperature during embryonic development has persistent effects on metabolic enzymes in the muscle of zebrafish. J. Exp. Biol. 217, 1370–1380. https://doi.org/10.1242/jeb.094037.

Schulte, P.M., 2011. Effects of temperature: an Introduction. In: Farrell, A.P., Cech, J.J., Stevens, E.D., Richards, J.G. (Eds.), Encyclopedia of Fish Physiology: From Genome to Environment. Elsevier Inc, pp. 1688–1694. https://doi.org/10.1016/B978-0-1237-4553-8.00159-3.

Scott, G.R., Johnston, I.A., 2012. Temperature during embryonic development has persistent effects on thermal acclimation capacity in zebrafish. Proc. Natl. Acad. Sci. U. S. A. 109, 14247–14252. https://doi.org/10.1073/pnas.1205012109.

Seebacher, F., Holmes, S., Roosen, N.J., Nouvian, M., Wilson, R.S., Ward, A.J.W., 2012. Capacity for thermal acclimation differs between populations and phylogenetic lineages within a species. Funct. Ecol. 26, 1418–1428. https://doi.org/10.1111/j.1365-2435.2012.02052.x.

Seebacher, F., Beaman, J., Little, A.G., 2014. Regulation of thermal acclimation varies between generations of the short-lived mosquitofish that developed in different environmental conditions. Funct. Ecol. 28, 137–148. https://doi.org/10.1111/1365-2435.12156.

Seebacher, F., White, C.R., Franklin, C.E., 2015. Physiological plasticity increases resilience of ectothermic animals to climate change. Nat Clim. Chang. 5, 61 66. https://doi.org/10.1038/nclimate2457.

Sella, G., Barton, N.H., 2019. Thinking about the evolution of complex traits in the era of genome-wide association studies. Annu. Rev. Genomics Hum. Genet. 20, 461–493. https://doi.org/10.1146/annurev-genom-083115-022316.

Sfakianakis, D.G., Leris, I., Kentouri, M., 2010. Effect of developmental temperature on swimming performance of zebrafish (*Danio rerio*) juveniles. Environ. Biol. Fishes 90, 421–427. https://doi.org/10.1007/s10641-010-9751-5.

Shafer, A.B.A., Wolf, J.B.W., Alves, P.C., Bergström, L., Bruford, M.W., Brännström, I., Colling, G., Dalén, L., De Meester, L., Ekblom, R., Fawcett, K.D., Fior, S., Hajibabaei, M., Hill, J.A., Hoezel, A.R., Höglund, J., Jensen, E.L., Krause, J., Kristensen, T.N., Krützen, M., McKay, J.K., Norman, A.J., Ogden, R., Österling, E.M., Ouborg, N.J., Piccolo, J., Popović, D., Primmer, C.R., Reed, F.A., Roumet, M., Salmona, J., Schenekar, T., Schwartz, M.K., Segelbacher, G., Senn, H., Thaulow, J., Valtonen, M., Veale, A., Vergeer, P., Vijay, N., Vilà, C., Weissensteiner, M., Wennerström, L., Wheat, C.W., Zieliński, P., 2015. Genomics and the challenging translation into conservation practice. Trends Ecol. Evol. 30, 78–87. https://doi.org/10.1016/j.tree.2014.11.009.

Shaklee, J.B., Beacham, T.D., Seeb, L., White, B.A., 1999. Managing fisheries using genetic data: Case studies from four species of Pacific salmon. Fish. Res. 43, 45–78. https://doi.org/10.1016/S0165-7836(99)00066-1.

Shama, L.N.S., Wegner, K.M., 2014. Grandparental effects in marine sticklebacks: transgenerational plasticity across multiple generations. J. Evol. Biol. 27, 2297–2307. https://doi.org/10.1111/jeb.12490.

Shama, L.N.S., Strobel, A., Mark, F.C., Wegner, K.M., 2014. Transgenerational plasticity in marine sticklebacks: maternal effects mediate impacts of a warming ocean. Funct. Ecol. 28, 1482–1493. https://doi.org/10.1111/1365-2435.12280.

Shama, L.N.S., Mark, F.C., Strobel, A., Lokmer, A., John, U., Mathias Wegner, K., 2016. Transgenerational effects persist down the maternal line in marine sticklebacks: gene expression matches physiology in a warming ocean. Evol. Appl. 9, 1096–1111. https://doi.org/10.1111/eva.12370.

Shang, E.H.H., Yu, R.M.K., Wu, R.S.S., 2006. Hypoxia affects sex differentiation and development leading to a male-dominated population in zebrafish (*Danio rerio*). Environ. Sci. Technol. 40, 3118–3122. https://doi.org/10.1021/es0522579.

Smith, M.H., Chesser, R.K., 1981. Rationale for conserving genetic variation in fish: preservation of genetic resources in relation to wild fish stocks. Ecol. Bull. 34, 13–20.

Sopinka, N.M., Middleton, C.T., Patterson, D.A., Hinch, S.G., 2016. Does maternal captivity of wild, migratory sockeye salmon influence offspring performance? Hydrobiologia 779, 1–10. https://doi.org/10.1007/s10750-016-2763-1.

Spicer, J.I., Gaston, K.J., 1999. Physiological Diversity and Its Ecological Implications. Blackwell Publishing Ltd, Oxford.

Stefansson, S.O., Bjornsson, B.T., Ebbesson, L.O.E., McCormick, S.D., 2008. Smoltification. In: Fish Larval Physiology. CRC Press, pp. 639–681.

Steinhausen, M.F., Sandblom, E., Eliason, E.J., Verhille, C., Farrell, A.P., 2008. The effect of acute temperature increases on the cardiorespiratory performance of resting and swimming sockeye salmon (*Oncorhynchus nerka*). J. Exp. Biol. 211, 3915–3926. https://doi.org/10.1242/jeb.019281.

Strowbridge, N., Northrup, S.L., Earhart, M.L., Blanchard, T.S., Schulte, P.M., 2021. Acute measures of upper thermal and hypoxia tolerance are not reliable predictors of mortality following environmental challenges in rainbow trout (*Oncorhynchus mykiss*). Conserv. Physiol. 9, 1–16. https://doi.org/10.1093/conphys/coab095.

Stuart-Smith, R.D., Bates, A.E., Lefcheck, J.S., Duffy, J.E., Baker, S.C., Thomson, R.J., Stuart-Smith, J.F., Hill, N.A., Kininmonth, S.J., Airoldi, L., Becerro, M.A., Campbell, S.J., Dawson, T.P., Navarrete, S.A., Soler, G.A., Strain, E.M.A., Willis, T.J., Edgar, G.J., 2013. Integrating abundance and functional traits reveals new global hotspots of fish diversity. Nature 501, 539–542. https://doi.org/10.1038/nature12529.

Su, G., Logez, M., Xu, J., Tao, S., Villéger, S., Brosse, S., 2021. Human impacts on global freshwater fish biodiversity. Science 371 (80), 835–838. https://doi.org/10.1126/science.abd3369.

Sunday, J.M., Bates, A.E., Dulvy, N.K., 2012. Thermal tolerance and the global redistribution of animals. Nat. Clim. Chang. 2, 686–690. https://doi.org/10.1038/nclimate1539.

Sunday, J., Bennett, J.M., Calosi, P., Clusella-Trullas, S., Gravel, S., Hargreaves, A.L., Leiva, F.P., Verberk, W.C.E.P., Olalla-Tárraga, M.Á., Morales-Castilla, I., 2019. Thermal tolerance patterns across latitude and elevation. Philos. Trans. R. Soc. B Biol. Sci. 374. https://doi.org/10.1098/rstb.2019.0036.

Teichert, N., Lepage, M., Lobry, J., 2018. Beyond classic ecological assessment: the use of functional indices to indicate fish assemblages sensitivity to human disturbance in estuaries. Sci. Total Environ. 639, 465–475. https://doi.org/10.1016/j.scitotenv.2018.05.179.

Teletchea, S., Teletchea, F., 2020. STOREFISH 2.0: a database on the reproductive strategies of teleost fishes. Database 2020, 1–17. https://doi.org/10.1093/database/baaa095.

Teletchea, F., Fostier, A., Le Bail, P.Y., Jalabert, B., Gardeur, J.N., Fontaine, P., 2007. STOREFISH: a new database dedicated to the reproduction of temperate freshwater teleost fishes. Cybium 31, 227–235.

Teresa, F.B., Casatti, L., 2017. Trait-based metrics as bioindicators: responses of stream fish assemblages to a gradient of environmental degradation. Ecol. Indic. 75, 249–258. https://doi.org/10.1016/j.ecolind.2016.12.041.

Therkildsen, N.O., Hemmer-Hansen, J., Als, T.D., Swain, D.P., Morgan, M.J., Trippel, E.A., Palumbi, S.R., Meldrup, D., Nielsen, E.E., 2013. Microevolution in time and space: SNP analysis of historical DNA reveals dynamic signatures of selection in Atlantic cod. Mol. Ecol. 22, 2424–2440. https://doi.org/10.1111/mec.12260.

Tilman, D., Knops, J., Wedin, D., Reich, P., Ritchie, M., Siemann, E., 1997. The influence of functional diversity and composition on ecosystem processes. Science 277 (80), 1300–1302. https://doi.org/10.1126/science.277.5330.1300.

Toussaint, A., Charpin, N., Brosse, S., Villéger, S., 2016. Global functional diversity of freshwater fish is concentrated in the neotropics while functional vulnerability is widespread. Sci. Rep. 6, 1–9. https://doi.org/10.1038/srep22125.

Toussaint, A., Charpin, N., Beauchard, O., Grenouillet, G., Oberdorff, T., Tedesco, P.A., Brosse, S., Villéger, S., 2018. Non-native species led to marked shifts in functional diversity of the world freshwater fish faunas. Ecol. Lett. 21, 1649–1659. https://doi.org/10.1111/ele.13141.

Turko, A.J., Leclair, A.T.A., Mandrak, N.E., Drake, D.A.R., Scott, G.R., Pitcher, T.E., 2021. Choosing source populations for conservation reintroductions: lessons from variation in thermal tolerance among populations of the imperilled redside dace. Can. J. Fish. Aquat. Sci. 9, 1–9. https://doi.org/10.1139/cjfas-2020-0377.

Twardek, W.M., Ekström, A., Eliason, E.J., Lennox, R.J., Tuononen, E., Abrams, A.E.I., Jeanson, A.L., Cooke, S.J., 2021. Field assessments of heart rate dynamics during spawning migration of wild and hatchery-reared chinook salmon. Philos. Trans. R Soc. B 376 (1830).

Urban, M.C., 2015. Accelerating extinction risk from climate change. Science 348 (80), 571–573. https://doi.org/10.1126/science.aaa4984

Vagner, M., Zambonino-Infante, J.L., Mazurais, D., 2019. Fish facing global change: are early stages the lifeline? Mar. Environ. Res. 147, 159–178. https://doi.org/10.1016/j.marenvres. 2019.04.005.

Vanderplancke, G., Claireaux, G., Quazuguel, P., Madec, L., Ferraresso, S., Sévère, A., Zambonino-Infante, J.L., Mazurais, D., 2015. Hypoxic episode during the larval period has long-term effects on European sea bass juveniles (*Dicentrarchus labrax*). Mar. Biol. 162, 367–376. https://doi.org/10.1007/s00227-014-2601-9.

Vasemägi, A., Primmer, C.R., 2005. Challenges for identifying functionally important genetic variation: the promise of combining complementary research strategies. Mol. Ecol. 14, 3623–3642. https://doi.org/10.1111/j.1365-294X.2005.02690.x.

Veilleux, H.D., Ryu, T., Donelson, J.M., Van Herwerden, L., Seridi, L., Ghosheh, Y., Berumen, M.L., Leggat, W., Ravasi, T., Munday, P.L., 2015. Molecular processes of transgenerational acclimation to a warming ocean. Nat. Clim. Chang. 5, 1074–1078. https://doi.org/10.1038/nclimate2724.

Venney, C.J., Sutherland, B.J.G., Beacham, T.D., Heath, D.D., 2021. Population differences in Chinook salmon (Oncorhynchus tshawytscha) DNA methylation: genetic drift and environmental factors. Ecol. Evol. 11, 6846–6861. https://doi.org/10.1002/ece3.7531.

Villéger, S., Brosse, S., Mouchet, M., Mouillot, D., Vanni, M.J., 2017. Functional ecology of fish: current approaches and future challenges. Aquat. Sci. 79, 783–801. https://doi.org/10.1007/s00027-017-0546-z.

Violle, C., Navas, M.L., Vile, D., Kazakou, E., Fortunel, C., Hummel, I., Garnier, E., 2007. Let the concept of trait be functional! Oikos 116, 882–892. https://doi.org/10.1111/j.0030-1299. 2007.15559.x.

Ward, T.D., Algera, D.A., Gallagher, A.J., Hawkins, E., Horodysky, A., Jørgensen, C., Killen, S.S., McKenzie, D.J., Metcalfe, J.D., Peck, M.A., Vu, M., Cooke, S.J., 2016. Understanding the individual to implement the ecosystem approach to fisheries management. Conserv. Physiol. 4, 1–18. https://doi.org/10.1093/conphys/cow005.

Westneat, M.W., 2001. Vertebrate functional morphology and physiology. Encycl. Life Sci., 1–4. https://doi.org/10.1038/npg.els.0001815.

White, C.R., Phillips, N.F., Seymour, R.S., 2006. The scaling and temperature dependence of vertebrate metabolism. Biol. Lett. 2, 125–127. https://doi.org/10.1098/rsbl.2005.0378.

Whitman, D.W., 2009. Acclimation. In: Whitman, D.W., Ananthshrishnan, T.N. (Eds.), Phenotypic Plasticity of Insects: Mechanisms and Consequences. Science Publishers, Enfield, NH, pp. 675–739.

Whitman, D., Agrawal, A., 2009. What is phenotypic plasticity and why is it important? In: Whitman, D.W., Ananthshrishnan, T.N. (Eds.), Phenotypic Plasticity of Insects: Mechanisms and Consequences. Science Publishers, Enfield, NH, pp. 1–63.

Wikelski, M., Cooke, S.J., 2006. Conservation physiology. Trends Ecol. Evol. 21, 38–46. https://doi.org/10.1016/j.tree.2005.10.018.

Williams, S.E., Shoo, L.P., Isaac, J.L., Hoffmann, A.A., Langham, G., 2008. Towards an integrated framework for assessing the vulnerability of species to climate change. PLoS Biol. 6. https://doi.org/10.1371/journal.pbio.0060325.

Wood, A.T., Clark, T.D., Andrewartha, S.J., Elliott, N.G., Frappell, P.B., 2017. Developmental hypoxia has negligible effects on long-term hypoxia tolerance and aerobic metabolism of Atlantic salmon (*Salmo salar*). Physiol. Biochem. Zool. 90, 494–501. https://doi.org/10. 1086/692250.

Wu, R.S.S., 2009. Effects of hypoxia on fish reproduction and development. In: Hypoxia, Fish Physiology. Elsevier Inc, pp. 79–141. https://doi.org/10.1016/S1546-5098(08)00003-4.

Yachi, S., Loreau, M., 1999. Biodiversity and ecosystem productivity in a fluctuating environment: the insurance hypothesis. Proc. Natl. Acad. Sci. U. S. A. 96, 1463–1468. https://doi. org/10.1073/pnas.96.4.1463.

You, X., Shan, X., Shi, Q., 2020. Research advances in the genomics and applications for molecular breeding of aquaculture animals. Aquaculture 526, 735357. https://doi.org/10.1016/j. aquaculture.2020.735357.

Zhang, B.D., Xue, D.X., Li, Y.L., Liu, J.X., 2019. RAD genotyping reveals fine-scale population structure and provides evidence for adaptive divergence in a commercially important fish from the northwestern Pacific Ocean. PeerJ 2019. https://doi.org/10.7717/peerj.7242.

Zhang, C., Fujiwara, M., Pawluk, M., Liu, H., Cao, W., Gao, X., 2020. Changes in taxonomic and functional diversity of fish communities after catastrophic habitat alteration caused by construction of Three Gorges Dam. Ecol. Evol. 10, 5829–5839. https://doi.org/10.1002/ ece3.6320.

Zhao, Y., Peng, W., Guo, H., Chen, B., Zhou, Z., Xu, J., Zhang, D., Xu, P., 2018. Population genomics reveals genetic divergence and adaptive differentiation of Chinese sea bass (*Lateolabrax maculatus*). Mar. Biotechnol. 20, 45–59. https://doi.org/10.1007/s10126-017-9786-0.

Other volumes in the Fish Physiology series

Index

Note: Page numbers followed by "*f*" indicate figures and "*t*" indicate tables.

Printed in the United States
by Baker & Taylor Publisher Services